Chalcogenide glasses

Related titles:

Carbon nanotubes and graphene for photonic applications
(ISBN 978-0-85709-417-9)

Handbook of organic materials for optical and optoelectronic devices
(ISBN 978-0-85709-265-6)

Laser growth and processing of photonic devices
(ISBN 978-1-84569-936-9)

The team responsible for publishing this book:

Commissioning Editor: Laura Pugh
Publications Coordinator: Adam Davies
Project Editor: Anneka Hess
Editorial and Production Manager: Mary Campbell
Production Editor: Mandy Kingsmill
Cover Designer: Terry Callanan

Woodhead Publishing Series in Electronic and Optical Materials:
Number 44

Chalcogenide glasses

Preparation, properties and applications

Edited by

Jean-Luc Adam and Xianghua Zhang

WOODHEAD PUBLISHING

Oxford Cambridge Philadelphia New Delhi

Published by Woodhead Publishing Limited,
80 High Street, Sawston, Cambridge CB22 3HJ, UK
www.woodheadpublishing.com
www.woodheadpublishingonline.com

Woodhead Publishing, 1518 Walnut Street, Suite 1100, Philadelphia, PA 19102-3406, USA

Woodhead Publishing India Private Limited, 303, Vardaan House, 7/28 Ansari Road, Daryaganj, New Delhi – 110002, India
www.woodheadpublishingindia.com

First published 2014, Woodhead Publishing Limited

British Library Cataloguing in Publication Data
A catalogue record for this book is available from the British Library.

Library of Congress Control Number: 2013948573

ISBN 978-0-85709-345-5 (print)
ISBN 978-0-85709-356-1 (online)
ISSN 2050-1501 Woodhead Publishing Series in Electronic and Optical Materials (print)
ISSN 2050-151X Woodhead Publishing Series in Electronic and Optical Materials (online)

The publisher's policy is to use permanent paper from mills that operate a sustainable forestry policy, and which has been manufactured from pulp which is processed using acid-free and elemental chlorine-free practices. Furthermore, the publisher ensures that the text paper and cover board used have met acceptable environmental accreditation standards.

Typeset by Replika Press Pvt. Ltd, India
Printed by TJ International Ltd, Padstow, Cornwall, UK

Contents

Contributor contact details

(* = main contact)

Editors

J.-L. Adam and X. H. Zhang
University of Rennes
Institut des Sciences Chimiques de Rennes – UMR6226
263 Avenue du Général Leclerc
Campus de Beaulieu – Bâtiment 10B
Rennes Cedex, 35042, France

E-mail: jean-luc.adam@univ-rennes1.fr; xzhang@univ-rennes1.fr

Chapter 1

V. S. Shiryaev* and M. F. Churbanov
Institute of Chemistry of High-Purity Substances
Russian Academy of Sciences
49 Tropinin St.
Nizhny Novgorod, 603950, Russia

E-mail: shiryaev@ihps.nnov.ru

Chapter 2

B. Bureau
Equipe Verres et Céramiques
UMR CNRS Sciences Chimiques de Rennes
University of Rennes
Rennes Cedex, 35042, France

E-mail: bruno.bureau@univ-rennes1.fr

Chapter 3

P. Lucas
Department of Materials Science and Engineering
University of Arizona
4715 E. Fort Lowell Road
Tucson, AZ 85712, USA

E-mail: pierre@u.arizona.edu

Chapter 4

J. D. Musgraves*
IRradiance Glass Inc.
3564 Avalon Park Blvd E
Suite 1 226
Orlando, FL 32828, USA

S. Danto and K. Richardson
College of Optics and Photonic
University of Central Florida
4000 Central Florida Blvd
Orlando, FL 32816, USA

E-mail: musgraves@irradianceglass.com

Chapter 5

J. Sanghera and D. Gibson*
Code 5620 – Optical Materials and Devices Branch
Naval Research Laboratory
4555 Overlook Ave SW
Washington, DC 20375, USA

E-mail: jas.sanghera@nrl.navy.mil; dan.gibson@nrl.navy.mil

Chapter 6

K. Tanaka
Department of Applied Physics
Graduate School of Engineering
Hokkaido University
Kita-ku
060-8628, Sapporo, Japan

E-mail: keiji@eng.hokudai.ac.jp

Chapter 7

A. Pradel* and M. Ribes
Institut Charles Gerhardt CNRS
University of Montpellier
CC 1503
Place Eugène Bataillon
Montpellier cedex 5, 34095, France

E-mail: annie.pradel@univ-montp2.fr

Chapter 8

O. Shpotyuk*
Lviv Scientific Research Institute of Materials of SRC 'Carat'
Ukraine

and

Institute of Physics
Jan Dlugosz University of Częstochowa
Poland

E-mail: shpotyuk@novas.lviv.ua

R. Golovchak
Department of Physics and Astronomy
Austin Peay State University
USA

A. Kozdras
Opole University of Technology
Poland

Chapter 9

J. Orava
Department of Materials Science and Metallurgy
University of Cambridge
27 Charles Babbage Road
Cambridge CB3 OFS, UK

and

WPI-Advanced Institute for Materials Research (WPI-AIMR)
Tohoku University
2-1-1 Katahira
Aoba-ku
Sendai, 980-8577, Japan

E-mail: jo316@cam.ac.uk

T. Kohoutek
Optical Functional Materials Laboratory
Toyota Technological Institute
2-12-1 Hisakata
Tempaku
Nagoya
Aichi, 468-8511, Japan

and

Department of General and Inorganic Chemistry
Faculty of Chemical Technology
University of Pardubice
Legion's Sq. 565
Pardubice, 532 10, Czech Republic

E-mail: tomas.kohoutek@upce.cz

T. Wagner*
Department of General and Inorganic Chemistry
Faculty of Chemical Technology
University of Pardubice
Legion's Sq. 565
Pardubice, 532 10, Czech Republic

E-mail: tomas.wagner@upce.cz

Chapter 10

L. Calvez
Equipe Verres et Céramiques
UMR CNRS 6226 Institut des Sciences Chimiques de Rennes
University of Rennes
Rennes Cedex, 35042, France

E-mail: laurent.calvez@univ-rennes1.fr

Chapter 11

J. Heo*
Center for Information Materials
Department of Materials Science and Engineering
Pohang University of Science and Technology
San 31
Hyoja-dong
Pohang
Gyeongbuk, 790-784, Republic of Korea

E-mail: jheo@postech.ac.kr

W. J. Chung
Division of Advanced Materials Engineering
Kongju National University
275 Budae-dong, Cheonan
Chungnam, 331-717, Republic of Korea

E-mail: wjin@kongju.ac.kr

Chapter 12

C. Boussard-Plédel
Equipe Verres et Céramiques
UMR CNRS 6226 Institut des Sciences Chimiques de Rennes
University of Rennes
Rennes Cedex, 35042, France

E-mail: catherine.boussard@univ-rennes1.fr

Chapter 13

J. Troles*
Equipe Verres et Céramiques,
UMR-CNRS 6226 Institut des Sciences Chimiques de Rennes,
University of Rennes
Rennes Cedex, 35042, France
E-mail: johann.troles@univ-rennes1.fr

L. Brilland
PERFOS
R&D Plateform of Photonics Bretagne
11 rue Louis de Broglie
Lannion, 22300, France

E-mail: lbrilland@perfos.com

Chapter 14

R. Pant* and B. J. Eggleton,
CUDOS
University of Sydney
Australia

E-mail: rpant@physics.usyd.edu.au

Chapter 15

D. W. Hewak
Optoelectronics Research Centre
University of Southampton
University Road
Southampton SO16 3PN, UK

E-mail: dh@orc.soton.ac.uk

Chapter 16

V. Nazabal*
Glass and Ceramics Team
UMR CNRS 6226 Chemical Sciences Institute of Rennes
University of Rennes
Rennes Cedex, 35042, France

E-mail: virginie.nazabal@univ-rennes1.fr

I. Kityk
Electrical Engineering Department
Częstochowa University of Technology
Armii Krajowej 17
Częstochowa, Poland

E-mail: ikityk@el.pcz.czest.pl

Chapter 17

H. Jain*
Department of Materials Science and Engineering
Lehigh University
Bethlehem, PA 18015, USA
E-mail: h.jain@lehigh.edu

A. Kovalskiy
Department of Physics and Astronomy
Austin Peay State University
Clarksville, TN 37044, USA
E-mail: kovalskyya@apsu.edu

M. Vlcek
Department of General and Inorganic Chemistry
Faculty of Chemical Technology
University of Pardubice
Pardubice, 53210, Czech Republic

E-mail: miroslav.vleek@upce.cz

Chapter 18

B. Hyot
CEA, LETI
Minatec Campus
Grenoble Cedex 9, 38054, France

E-mail: berangere.hyot@cea.fr

Chapter 19

M. Tatsumisago* and A. Hayashi
Department of Applied Chemistry
Faculty of Engineering
Osaka Prefecture University
1-1 Gakuen-cho
Naka-ku
Sakai
Osaka, 599-8531, Japan

E-mail: tatsu@chem.osakafu-u.ac.jp

Woodhead Publishing Series in Electronic and Optical Materials

1 **Circuit analysis**
J. E. Whitehouse

2 **Signal processing in electronic communications: For engineers and mathematicians**
M. J. Chapman, D. P. Goodall and N. C. Steele

3 **Pattern recognition and image processing**
D. Luo

4 **Digital filters and signal processing in electronic engineering: Theory, applications, architecture, code**
S. M. Bozic and R. J. Chance

5 **Cable engineering for local area networks**
B. J. Elliott

6 **Designing a structured cabling system to ISO 11801: Cross-referenced to European CENELEC and American Standards**
Second edition
B. J. Elliott

7 **Microscopy techniques for materials science**
A. Clarke and C. Eberhardt

8 **Materials for energy conversion devices**
Edited by C. C. Sorrell, J. Nowotny and S. Sugihara

9 **Digital image processing: Mathematical and computational methods**
Second edition
J. M. Blackledge

10 **Nanolithography and patterning techniques in microelectronics**
Edited by D. Bucknall

11 **Digital signal processing: Mathematical and computational methods, software development and applications Second edition**
J. M. Blackledge

12 **Handbook of advanced dielectric, piezoelectric and ferroelectric materials: Synthesis, properties and applications**
Edited by Z.-G. Ye

13 **Materials for fuel cells**
Edited by M. Gasik

14 **Solid-state hydrogen storage: Materials and chemistry**
Edited by G. Walker

15 **Laser cooling of solids**
S. V. Petrushkin and V. V. Samartsev

16 **Polymer electrolytes: Fundamentals and applications**
Edited by C. A. C. Sequeira and D. A. F. Santos

17 **Advanced piezoelectric materials: Science and technology**
Edited by K. Uchino

18 **Optical switches: Materials and design**
Edited by S. J. Chua and B. Li

19 **Advanced adhesives in electronics: Materials, properties and applications**
Edited by M. O. Alam and C. Bailey

20 **Thin film growth: Physics, materials science and applications**
Edited by Z. Cao

21 **Electromigration in thin films and electronic devices: Materials and reliability**
Edited by C.-U. Kim

22 ***In situ* characterization of thin film growth**
Edited by G. Koster and G. Rijnders

23 **Silicon-germanium (SiGe) nanostructures: Production, properties and applications in electronics**
Edited by Y. Shiraki and N. Usami

24 **High-temperature superconductors**
Edited by X. G. Qiu

25 **Introduction to the physics of nanoelectronics**
S. G. Tan and M. B. A. Jalil

26 **Printed films: Materials science and applications in sensors, electronics and photonics**
Edited by M. Prudenziati and J. Hormadaly

27 **Laser growth and processing of photonic devices**
Edited by N. A. Vainos

28 **Quantum optics with semiconductor nanostructures**
Edited by F. Jahnke

29 **Ultrasonic transducers: Materials and design for sensors, actuators and medical applications**
Edited by K. Nakamura

30 **Waste electrical and electronic equipment (WEEE) handbook**
Edited by V. Goodship and A. Stevels

31 **Applications of ATILA FEM software to smart materials: Case studies in designing devices**
Edited by K. Uchino and J.-C. Debus

32 **MEMS for automotive and aerospace applications**
Edited by M. Kraft and N. M. White

33 **Semiconductor lasers: Fundamentals and applications**
Edited by A. Baranov and E. Tournie

34 **Handbook of terahertz technology for imaging, sensing and communications**
Edited by D. Saeedkia

35 **Handbook of solid-state lasers: Materials, systems and applications**
Edited by B. Denker and E. Shklovsky

36 **Organic light-emitting diodes (OLEDs): Materials, devices and applications**
Edited by A. Buckley

37 **Lasers for medical applications: Diagnostics, therapy and surgery**
Edited by H. Jelínková

38 **Semiconductor gas sensors**
Edited by R. Jaaniso and O. K. Tan

39 **Handbook of organic materials for optical and (opto)electronic devices: Properties and applications**
Edited by O. Ostroverkhova

40 **Metallic films for electronic, optical and magnetic applications: Structure, processing and properties**
Edited by K. Barmak and K. Coffey

41 **Handbook of laser welding technologies**
Edited by S. Katayama

42 **Nanolithography: The art of fabricating nanoelectronic and nanophotonic devices and systems**
Edited by M. Feldman

43 **Laser spectroscopy for sensing: Fundamentals, techniques and applications**
Edited by M. Baudelet

44 **Chalcogenide glasses: Preparation, properties and applications**
Edited by J.-L. Adam and X. Zhang

45 **Handbook of MEMS for wireless and mobile applications**
Edited by D. Uttamchandani

46 **Subsea optics and imaging**
Edited by J. Watson and O. Zielinski

47 **Carbon nanotubes and graphene for photonic applications**
Edited by S. Yamashita, Y. Saito and J. H. Choi

48 **Optical biomimetics: Materials and applications**
Edited by M. Large

49 **Optical thin films and coatings**
Edited by A. Piegari and F. Flory

50 **Computer design of diffractive optics**
Edited by V. A. Soifer

51 **Smart sensors and MEMS: Intelligent devices and microsystems for industrial applications**
Edited by S. Nihtianov and A. Luque

52 **Fundamentals of femtosecond optics**
S. A. Kozlov and V. V. Samartsev

53 **Nanostructured semiconductor oxides for the next generation of electronics and functional devices: Properties and applications**
S. Zhuiykov

54 **Nitride semiconductor light-emitting diodes (LEDs): Materials, performance and applications**
Edited by J. J. Huang, H. C. Kuo and S. C. Shen

55 **Sensor technologies for civil infrastructures Volume 1: Sensing hardware and data collection for performance assessment**
Edited by M. Wang, J. Lynch and H. Sohn

56 **Sensor technologies for civil infrastructures Volume 2: Applications in structural health monitoring**
Edited by M. Wang, J. Lynch and H. Sohn

57 **Graphene: Properties, preparation, characterisation and devices**
Edited by V. Skákalová and A. B. Kaiser

58 **Handbook of silicon-on-insulator (SOI) technology**
Edited by O. Kononchuk and B.-Y. Nguyen

59 **Biological identification: DNA amplification and sequencing, optical sensing, lab-on-chip and portable systems**
Edited by P. Schaudies

60 **High performance silicon imaging: Fundamentals and applications of CMOS and CCD sensors**
Edited by D. Durini

61 **Nanosensors for chemical and biological applications: Sensing with nanotubes, nanowires and nanoparticles**
Edited by K. C. Honeychurch

62 **Composite magnetoelectrics: Materials, structures, and applications**
G. Srinivasan, S. Priya, and N. Sun

63 **Quantum information processing with diamond: Principles and applications**
Edited by S. Prawer and I. Aharonovich

64 **Advances in nonvolatile memory and storage technology**
Edited by Y. Nishi

65 **Laser surface engineering: Processes and applications**
Edited by J. Lawrence, C. Dowding, D. Waugh, J. Griffiths

Part I

Preparation and properties of chalcogenide glasses

1
Preparation of high-purity chalcogenide glasses

V. S. SHIRYAEV and M. F. CHURBANOV, Institute of Chemistry of High-Purity Substances of the Russian Academy of Sciences, Russia

DOI: 10.1533/9780857093561.1.3

Abstract: The methods of preparation of chalcogenide glasses for infrared fiber optics with low content of limiting impurities have been developed. The methods are based on solidification of glass-forming melts prepared by melting the initial high-purity elements in evacuated silica ampoule and purified additionally with chemical and distillation procedures. The effect of impurities (oxygen, hydrogen, carbon, sulfur, inclusions) on optical transmission of glasses was determined. The optical, thermal and crystallization properties of glasses were studied. Chalcogenide glass-based, low loss, multimode and single-mode optical fibers are produced with technical and operation characteristics sufficient for a number of cutting-edge applications. Factors affecting the optical absorption of glasses and optical fibers are analyzed.

Key words: chalcogenide glass, limiting impurities, chemical and distillation purification, optical transmission, low loss optical fibers.

1.1 Introduction

Chalcogenide glasses are effective optical photonic materials. Chalcogenide glasses formed by individual chalcogenides of the III–V group elements of the Periodic Table or by their compounds have been known as optical materials for more than 50 years [1]. The first report on their application in fiber optics was published in 1965 [2, 3]. Active investigations of chalcogenide glasses as the materials for fiber optics in the middle infrared (IR) range have been carried out for the last 20–25 years [4–7]. A number of glasses with chalcogenides of III–V group elements were tested for production of optical fibers. The most important results were obtained for glasses based on chalcogenides of arsenic and germanium. These glasses have interesting properties, e.g. a wide transparency range, low optical losses, stability to atmospheric moisture, high non-linearity of optical properties, etc. There are a great number of technical problems in optics and optoelectronics which can be efficiently solved using chalcogenide optical fibers with low optical losses. The main efforts of researchers are directed to the increase of chemical and phase purity of glasses, to the development of optical fibers in view of special features of chalcogenide glasses, to investigation of the properties of glasses and optical fibers. This review presents the up-to-date status of developments of high-purity glasses from arsenic and germanium chalcogenides, as well as of optical fibers based on them.

1.2 Preparation of vitreous chalcogenides

Bulk samples of vitreous chalcogenides of arsenic and germanium of optical quality are produced by solidification of glass-forming melt. The initial charge of chalcogenide glass with a given composition is melted in evacuated sealed quartz ampoules. The melt solidification is carried out at time–temperature modes precluding crystallization, liquation and fluctuation of macro-composition. Being individual for different glass systems, these conditions are additionally governed by charge mass, glass and melt stability to crystallization and liquation. For example, the values for the critical cooling rate in vitreous As_2S_3 and As_2Se_3 differ by approximately three orders and are equal to 2.4×10^{-6} and 9×10^{-3} K/s, respectively [8].

The initial charge for the synthesis of glass-forming chalcogenides can be produced by different techniques [9, 10], i.e., from chemical elements (the traditional method); by decomposition of volatile inorganic compounds (hydrides, chlorides or iodides); via arsenic monosulfide for arsenic-containing components; or by deposition of chalcogenides from a solution.

Sulfide and selenide layers of arsenic and germanium can be produced by plasma chemical decomposition of hydrides and chlorides of these elements [5]. The advantage of this method is the lower rate of impurity income into the glass due to the lower temperature and shorter synthesis time as well as to the possibility to combine in one process the glass synthesis and production of preform for drawing optical fiber with a given index profile. A description is given in Refs [5] and [9] for production of glasses with GeS_2, $GeSe_2$, $GeSe_3$, $Ge_{27}As_{18}Se_{55}$, $Ge_{30}As_{15}Se_{55}$, and $Ge_{33}As_{12}Se_{45}$ composition by thermal decomposition of GeH_4, AsH_3, H_2Se, and H_2S mixtures. The produced glasses contained less than 0.1–0.3 ppm of transition metals and silicon impurities. The production of glasses of the Ge-Se system via chlorides by chemical vapor decomposition (CVD) is described in Refs [11] and [12]. Gaseous chlorides of glass elements together with H_2 passed in argon flow through the furnace heated up to 800°C. Then the ampoule with deposited particles of Ge and Se is sealed off in a vacuum and synthesized. However, hydrogen and oxygen impurities are present in the sample transmission spectra. High content of hydrogen (several atomic percent) in the produced glasses is a serious problem for manufacturing optical-grade glass. In the process of thermal treatment of glass, including fiber fabrication, hard-to-remove hydrogen bubbles are formed. However, this method can be a promising one for the production of thin films from arsenic, antimony and germanium chalcogenides [13]. For plasma chemical production of As_2Se_3 glass from the hydrides of As and Se, the dependence of the presence of arsenic and selenium in vitreous products upon the gas mixture composition was found [5].

Sulfur, selenium, tellurium, and arsenic chalcogenides have a comparable and sufficiently high volatility, and the binding energy of As–As, chalcogen–chalcogen, and chalcogen–arsenic are comparable. For these reasons, it has not yet been possible to realize a sufficiently efficient vapor-phase chemical deposition of arsenic chalcogenides similar to vapor-phase chemical deposition of high-purity silicon dioxide.

1.3 Properties of chalcogenide glasses determining their application as optical materials

1.3.1 Composition and structure

Chalcogenide glasses of As-S, As-Se, As-S-Se, Ge-S, Ge-Se, Ge-As-S, As-Se-Te, Ge-As-Se, Ge-Se-Te, Ge-As-Se-Te, Ga-La-S, Ge-(Ga)-Sb-S, and Ge-Ga-Te systems were tested as materials for fiber optics. The ranges of glass formation and structures of glasses are investigated and described in detail in a number of papers [6, 8, 14–17]. Most attention was given to glasses based on chalcogenides of arsenic and germanium, i.e., As_2S_3, As_2Se_3, GeS_2, and $GeSe_2$. The structure of vitreous As_2S_3 and As_2Se_3 is described as a layered two-dimensional network mainly consisting of structural units $[AsS_{3/2}]$ and $[AsSe_{3/2}]$ in the shape of a trigonal pyramid. Inside the layers, the atoms (structural units) are connected with strong covalent bonds while there are only van der Waals bonds between the layers. The structure of vitreous GeS_2 and $GeSe_2$ is described as a spatial three-dimensional network consisting of tetragonal structural units $[GeS_{4/2}]$ and $[GeSe_{4/2}]$.

Structural skeletons of glasses of more complex systems (Ge-As-S, Ge-As-Se, As-Se-Te, Ge-As-Se-Te) can contain different groups of short-range ordering (e.g., $SeSe_{2/2}$, $AsAs_{3/3}$, $GeGe_{4/4}$, $AsSe_{3/2}$, $GeSe_{4/2}$, $As_2Se_{4/2}$, $GeSe_{2/2}$), as well as the mixed structural units ($[AsSe_{3-x}Te_x]$, $[GeSe_{4-x}Te_x]$). Glasses of Ge-Sb-S system consist of $[GeS_{4/2}]$ and $[SbS_{3/2}]$ structural units.

A set of requirements for glasses suitable for mid-IR fiber optics includes a wide spectral transmission range, stability towards the effect of moisture and to crystallization, high mechanical strength, non-toxicity of glasses, and a broad range of operating temperatures. There is no chalcogenide glass which has a complete set of these characteristics at a maximum level. The glasses of As-S, As-Se, and As-Se-Te systems, as well as of Ge-As-S and Ge-As-Se systems with low germanium content, have a relatively low softening temperature (<185°C). The glasses of Ge-S, Ge-Se, As-Se systems as well as tellurium-containing glasses can crystallize in the fiber drawing process.

Arsenic-containing glasses are considered to be toxic due to the presence of arsenic. Glasses with sulfur have a limited transparency, in the 0.6–7 μm wavelength range. Thus, optical fibers made from one-two chalcogenide glasses can be suitable for each target application. Optical fibers from selenide and selenium-telluride glasses are suitable for transmission in the whole range of 2–12 μm wavelengths. Optical fibers from glasses of Ge-S, Ge-Se, Ge-As-S, Ge-As-Se, Ge-As-Se-Te systems with high content of germanium (15–30 at.%) are potentially suitable for operation at 200–300°C. Ga-La-S, Ge-Ga-La-S chalcogenide glasses containing rare-earth elements as macrocomponents, or the dopants (Ga, I) which increase their solubility, are most suitable for manufacturing fiber-optic lasers and amplifiers.

To use the glasses in fiber optics the following optical and physico-chemical properties should be known: the transparency range, the minimum optical losses, the refractive index and its spectral dependence, the non-linear refractive index, the tendency to crystallization, the characteristic temperatures (the glass transition temperature T_g, the temperature for the onset of crystallization T_c, the temperature of the maximum of crystallization peak T_p, the liquidus temperature T_l), the density, the coefficient of thermal expansion, heat capacity, heat conductivity, the Young's modulus, microhardness, and viscosity.

1.3.2 Thermal and crystallization properties

The key properties of glasses are their thermal properties, knowledge of which makes it possible to select glass compositions with low tendency to crystallization, optimum time–temperature modes for the production process of the glass samples and optical fibers, the range of operating temperatures, and the threshold magnitude for transmitted power of laser radiation.

There is a broad group of chalcogenide glasses with relatively close vitrification and crystallization temperatures, which are thus inclined to crystallization. Glasses with low tendency to crystallize are promising for fiber optics. However, glasses with increased crystallization ability also find their application in engineering, for example as optical switches and storage devices.

The tendency of glasses to crystallize is determined by a number of factors, including the difference between the temperatures of crystallization T_c and glass transition T_g. Another known criterion is the critical cooling rate (CCR) of glass-forming melt. The higher the CCR, the more probable is crystallization at melt solidification. Table 1.1 gives thermal characteristics and CCR values for some chalcogenide glasses. The glass transition temperature for glasses with two-dimensional network (As_2S_3, As_2Se_3) is 170–185°C, and that of glasses with three-dimensional network (GeS_2, $GeSe_2$) is above 300°C.

As_2S_3, $As_2S_{1.5}Se_{1.5}$ and $As_{30}Se_{50}Te_{20}$ glasses have low tendency to

Table 1.1 Thermal characteristics and CCR of some chalcogenide glasses [8]

Glass	T_g(°C)	T_c(°C)	T_l(°C)	$T_c - T_g$(°C)	CCR (K/s)
Se	30.2	–	217		≥ 0.33
As_2S_3	185		310		2.4×10^{-6}
As_2Se_3	170	298	370	128	9×10^{-3}
GeS_2	495	550	820	55	17
$GeSe_2$	380	580	740	200	≅ 1
$As_2S_{1.5}Se_{1.5}$	175	–	–	–	–
$As_2Se_{1.5}Te_{1.5}$	150	231	300	81	$\cong 5 \times 10^{-3}$

crystallize. These glasses do not crystallize during differential scanning calorimetry (DSC) analysis, while As_2Se_3 and $As_2SE_{1.5}Te_{1.5}$ glasses manifest the peaks of crystallization and melting on DSC thermograms. Vitreous As_2Se_3 crystallizes at 2 K/min heating rate with the formation of crystals of the same composition. The difference between the temperatures of crystallization and glass transition in As_2Se_3 glass is 128°C. According to the experimental data [18], its critical cooling rate is 0.009 K/s, and according to the calculated data [18] it is equal to 0.0034 K/s. Using differential thermal analysis (DTA), the critical heating rate (CHR) is determined for As-Se glass system as the minimum heating rate, at which the part of crystalline phase in the sample does not exceed 5% of the total sample volume [19]. CHR is 0.04 K/s for As_2Se_3 composition. The experimental dependencies of crystal growth rate and nucleation rate were determined for As_2Se_3 glass [20]. The maximum crystal growth is 7×10^{-5} cm/s at 344°C. The maximum nucleation rate W_{max} is equal to 130 $cm^{-3}s^{-1}$ and corresponds to 210°C. The nucleation rate of the crystalline phase, the linear rate of crystal growth, as well as the temperature dependence of these parameters are the most informative characteristics of the 'glass-crystal' transition.

Binary As_2Te_3 glasses have a lower tendency to form glass compared with As_2Se_3 glass, which, in turn, shows a lower glass-forming ability than As_2S_3. This is related to the increase in delocalization of arsenic-chalcogen covalent bonds in the S-Se-Te series. Since the tendency to crystallize decreases with the increase in number of components in the alloy, the As-Se-Te glasses of triple system are more inclined to glass formation compared with As-Se and As-Te glasses. This is proved by the experimental dependence of the CCR upon composition [18]. The CCR minimum in the As_2Se_3-As_2Te_3 system is shifted towards As_2Se_3. It was impossible to crystallize the melts containing 70 and 80 mol.% of As_2Se_3 at the applied minimum cooling rates. In $As_2Se_{1.5}Te_{1.5}$ glass, the CCR is about 5–10^{-3} K/s.

Glasses of the As-Se-Te system are attractive as high-transparency materials in the long-wave spectral range (3–16 μm). That is why their crystallization in the process of heating was the subject of detailed investigations. Crystallization of high-purity $As_{40}Se_{60-x}Te_x$ ($x = 0$–40) glasses was investigated using DSC [21].

Characteristic temperatures of these glasses with dependence on composition at a heating rate of 5 K/min are given in Table 1.2, and DSC thermograms of glasses at a heating rate of 10 K/min are given in Fig. 1.1. Glasses with x = 30, 35, and 40 exhibit an exothermic peak of crystallization at all heating rates. $As_{40}Se_{40}Te_{20}$ does not show any crystallization peak. DSC thermograms of all investigated glasses are characterized by one peak of crystallization only at T_c = 245–320°C. Investigation of crystallization kinetics for a number of alloys in the As-Se-Te system produced from commercial substances,

Table 1.2 Temperature points and criteria of glass stability at heating rate of 5 K/min for $As_{40}Se_{60x}Te_x$ glasses

Glass composition	T_g(°C)	T_c(°C)	T_p(°C)	T_s(°C)	T_l(°C)	ΔT(C)	K_{gl}	S, K
$As_{40}Se_{60}$	178	322	356	376	387	145	2.23	10.9
$As_{40}Se_{55}Te_5$	166	306	339	355	368	110	2.26	10.5
$As_{40}Se_{50}Te_{10}$	158	–	–	–	–	–	–	–
$As_{40}Se_{45}Te_{15}$	147	–	–	–	–	–	–	–
$As_{40}Se_{40}Te_{20}$	144	–	–	–	–	–	–	–
$As_{40}Se_{35}Te_{25}$	140	290	309	330	348	150	2.58	6.9
$As_{40}Se_{30}Te_{30}$	130	245	271	297	313	115	1.69	7.4
$As_{40}Se_{25}Te_{35}$	131	244	271	304	327	113	1.36	7.4
$As_{40}Se_{20}Te_{40}$	132	246	277	309	336	114	1.28	8.5

$\Delta T = T_c - T_g$; $K_{gl} = (T_c - T_g)/(T_l - T_c)$; $S = (T_p - T_c)(T_c - T_g)/T_g$.

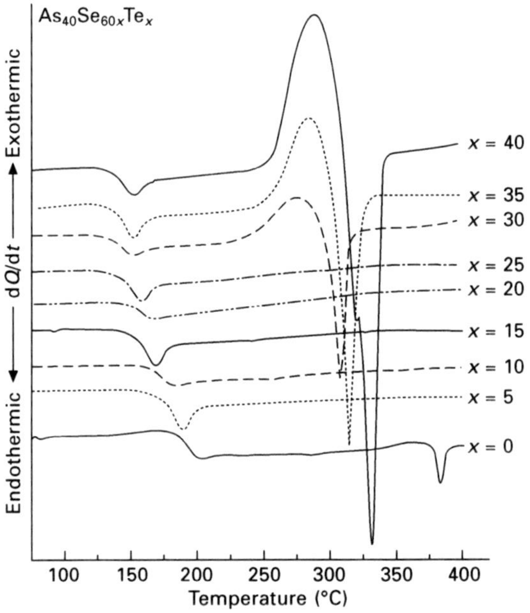

1.1 DSC thermograms for $As_{40}Se_{60x}Te_x$ glasses at a heating rate of 10 K/min.

leads to thermograms with two crystallization peaks [22]. The second, more intensive peak of crystallization observed at lower temperatures (e.g., T_{c1} = 188°C for $As_{40}Se_{20}Te_{40}$ [22]) could be due to heterogeneous crystallization in glass from impurity inclusions.

Adding Ge atoms into As-Se-Te glass improves mechanical strength and expands the operating temperature range [14]. For laser radiation power delivery, glasses having T_g > 150°C and an expansion coefficient less than 17×10^{-6} C^{-1} are necessary. Unfortunately, the glass transition temperature of As-Se-Te glasses is not sufficiently high. This limits the fields of application of fibers based on this glass. The addition of Ge to As-Se-Te glass can improve the thermal-mechanical properties, by decreasing the expansion coefficient and increasing the glass transition temperature.

As $As_{40}Se_{40}Te_{20}$ glass is the most stable against crystallization in the $As_{40}Se_{60-x}Te_x$ series, it was used as the basic glass to substitute a part of As for Ge [23]. Glasses of $Ge_xAs_{40-x}Se_{40}Te_{20}$ (x = 0–40) composition have T_g values in the range from 140 to 320°C, depending upon germanium content and heating rate. Glasses with $x \leq 35$ show no exothermal peaks of crystallization, whatever the heating rate, indicating a high glass-forming ability. Glass with $Ge_{40}Se_{40}Te_{20}$ composition presents one-stage glass transition and double-stage crystallization process during phase change.

In the composition zone of the Ge-As-Se-Te system, with high tellurium content, there is a small range of glass compositions with low tendency to crystallization [24]. The glasses, containing 16–24 at.% Ge and 42–48 at.% Te, have glass transition temperatures above 200°C and do not crystallize during annealing for 20 hours at temperatures exceeding T_g by 50–100°C.

Glasses of the Ge-S system are more inclined to crystallization as compared with glasses of As-S, As-Se, and As-S-Se systems [14]. Ge_xS_{1-x} glasses are stable against crystallization at x values of 0.11–0.33 and 0.39–0.43. As an example, for GeS_2 glass the T_c–T_g difference is 55°C, and CCR in the range of glass transition temperatures is 17 K/s. The melt of $GeS_{2.06}$ composition with a mass of 50–100 g can be transformed into a vitreous state by cooling in air (CCR ≈ ~3 K/s) and 2 g of $GeS_{1.94}$ melt can be transformed just by annealing in water (CCR ≈ 40 K/s). The melt of a composition close to GeS/GeS_2 eutectics has a CCR equal to 8 K/s.

Glasses of the Ge-Se system are more stable against crystallization than their sulfide analogues. Ge_xSe_{1-x} melts ($0 < x < 0.33$ and $0.388 < x < 0.417$) with a mass of 20–30 g are vitrified by a cooling process at ≈2 K/s rate.

For $GeS_{3.3}$ glass composition, the nucleation rate (W) and the linear rate of crystal growth (V) are experimentally defined in the 300–500°C temperature range (Fig. 1.2) [25]. During the experiment, the numbers and size of crystals are determined by laser ultramicroscopy using the same of glass sample as a function of time at constant temperature. The nucleation rate has its maximum of $8–10^2$ $cm^{-3}\cdot s^{-1}$ at a temperature close to 425°C. The nucleation

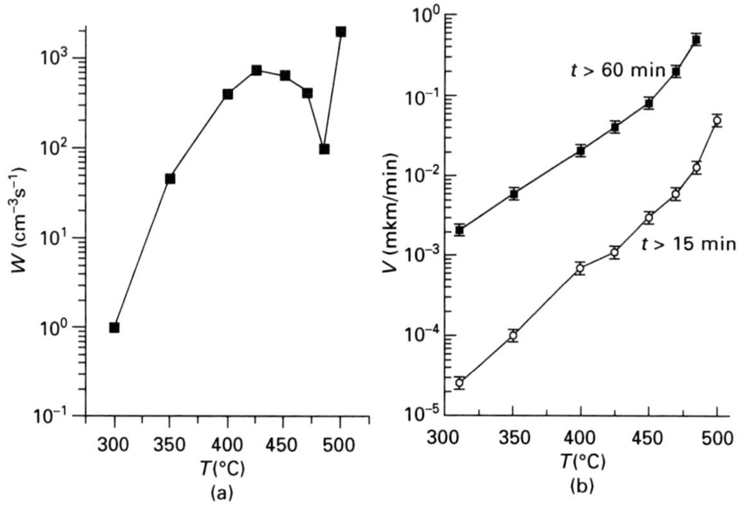

1.2 Temperature dependence of nucleation rate (a) and linear rate of the crystal growth (b) for $GeS_{3.3}$ glass sample.

rate at 500°C, i.e., at the point corresponding to crystallization temperature according to DTA data, is 2×10^3 $cm^{-3} \cdot s^{-1}$, which is substantially higher than the value corresponding to the maximum of the *W(T)* curve. This can be due to the occurrence of a large number of subcritical nuclei as a result of high crystal growth rate at crystallization temperature. Non-linearity and treatment-dependency of the *V(T)* curve may be related to sub-liquidus phase micro-separation before crystallization.

Glasses of the Ge-Sb-S system are less inclined to crystallization than Ge-S glasses. The boundaries of glass-formation range in the Ge-Sb-S system have been investigated [17, 26, 27]. It is found that glasses with compositions close to $Ge_{25}Sb_{10}S_{65}$ are mostly stable to crystallization. DSC thermograms for this glass do not show any crystallization peaks. Stability towards crystallization is maintained with replacement of germanium with gallium up to 5 at.%. With deviation by about 5 at.% with respect to Ge or Sb contents in $Ge_{25}Sb_{10}S_{65}$ composition, glasses are inclined to micro-liquation and to crystallization with formation of α-GeS_2 and/or Sb_2S_3 crystals phases.

The crystallization kinetic of Ga-La-S glass has been investigated [28]. In this system, the $(Ga_2S_3)_{0.7}(La_2S_3)_{0.3}$ composition shows the highest glass-forming ability. Figure 1.3 gives the temperature dependences of nucleation rate *W* and linear rate of crystal growth *V* for Pr-doped $(Ga_2S_3)_{0.7}(La_2S_3)_{0.3}$ glass in comparison with undoped glass.

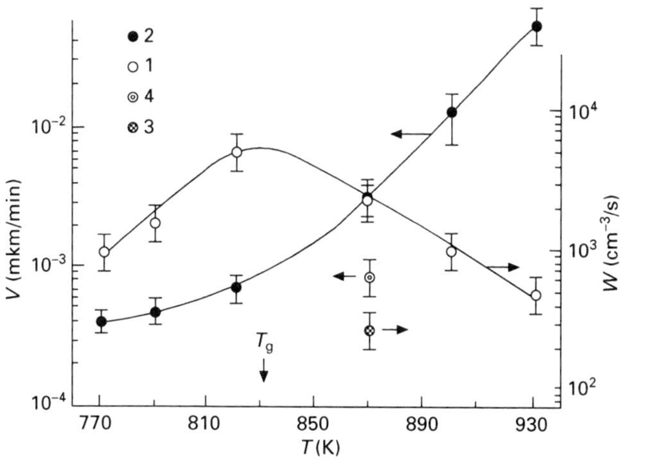

1.3 Temperature dependences of nucleation rate W (1) and linear rate of crystal growth V (2) for Pr-doped $(Ga_2S_3)_{0.7}(La_2S_3)_{0.3}$ glass; 3, nucleation rate for undoped glass at 870 K; 4, linear rate of crystal growth for undoped glass at 870 K.

Table 1.3 Properties of some chalcogenide glasses (T_g, glass transition temperature; T_c, onset of crystallization temperature; n, refractive index; n_2, non-linear refractive index)

Glass	Transmission range*, (μm)	T_g(°C)	$T_c - T_g$(°C)	n (λ, μm)	$n_2 \times 10^{-18}$ (m²/W)
As_2S_3	0.62–11.53	185	–	2.42 (3)	4–6
As_2Se_3	0.85–17.5	178	147	2.83 (3)	14–30
$As_{40}S_{30}Se_{30}$	0.75–12.5	180	–	2.61 (3)	14.5
$As_{40}Se_{40}Te_{20}$	1.23–18.52	140	–	2.9 (5)	>20
$GeSe_4$	0.75	163	–	2.48 (1.55)	13
$Ge_{25}Sb_{10}S_{65}$	0.65–11.0	315	>200	2.25 (1.55)	2–5
$Ge_{30}As_{10}Se_{30}Te_{30}$	1.2–17.0	260	225	2.8 (10.6)	>20
$Ge_{21}Te_{76}Se_3$	2–20	160	123		
$Ge_{15}Ga_{10}Te_{75}$	2–25	172	≥ 113	3.415	

* At 1 cm^{-1} absorption coefficient.

1.3.3 Optical and physico-chemical properties

Optical and physico-chemical properties of glasses, based on arsenic chalcogenides, were thoroughly investigated. Table 1.3 gives the main properties of glasses with As_2S_3, As_2Se_3, $As_2S_{1.5}Se_{1.5}$, $As_2Se_{1.5}Te_{1.5}$, and $Ge_{25}Sb_{10}S_{65}$ compositions. Short-wave and long-wave absorption edges are shifted to the left with substitution of Se for S and to the right with substitution of Se for Te, which is related to the change in equivalent mass of glass components affecting the multi-phonon absorption edge, electron absorption and refractive index.

Figures 1.4 and 1.5 give the experimental dependencies of short-wave and long-wave absorption edges of $x$$As_2S_3$-(100-$x$)$As_2Se_3$ glasses. It can be seen that a conversion of impurity hydrogen in As-S-Se glass into S-H groups makes the 2200–1800 cm^{-1} spectral region noticeably more transparent.

Figure 1.6 shows the experimental spectral dependence of absorption coefficient for As_2S_3, As_2Se_3, $As_{40}S_{30}Se_{30}$, $As_{30}Se_{50}Te_{20}$, and $Ge_{15}Ga_{10}Te_{75}$ glasses and the main absorption bands. The multi-phonon absorption band is shifted to long-wave range at an absorption level of 1 cm^{-1} by 2.76 μm, with substitution of 30 at.% S for Se in As_2S_3 glass and by 2.38 μm with substitution of 20% Se for Te in $As_{30}Se_{70}$ glass.

Figure 1.7 shows the transparency spectra of three glasses: $As_{30}Se_{50}Te_{20}$, $Ge_{21}Te_{76}Se_3$, and $Ge_{15}Te_{75}Ga_{10}$.There are no intense absorption bands in these spectra. In glass $As_{30}Se_{50}Te_{20}$, the long wavelength edge is limited by absorption due to two-phonon valent vibration of As-Se bonds at wavelength of 21 μm, and absorption due to two-phonon deformative vibration of As-Se-As bonds at 14.5 μm affects the transparency near 10.6 μm. Transmission spectra for $Te_{76}Ge_{21}Se_3$ and $Te_{75}Ge_{15}Ga_{10}$ glasses possess wider transmission range than $As_{30}Se_{50}Te_{20}$ glass. The transparency range of $Te_{75}Ge_{15}Ga_{10}$ extends up to 27 μm.

Chalcogenide glasses are characterized by a high refractive index value, which varies from 2.2 for germanium sulfide to 3.0 for arsenic selenium tellurides. The optical nonlinearity of chalcogenide glasses is 2–3 orders

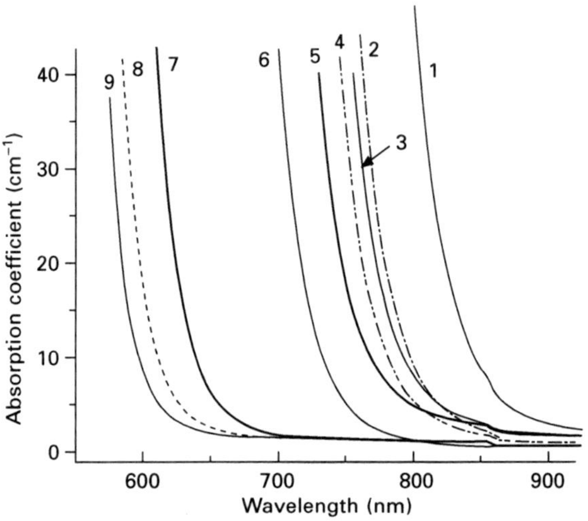

1.4 Short-wave transmission edges in $x$$As_2S_3$-(1-$x$)$As_2Se_3$ glasses. x =: 1, 0; 2, 5; 3, 10; 4, 15; 5, 25; 6, 40; 7, 85; 8, 95; 9, 100.

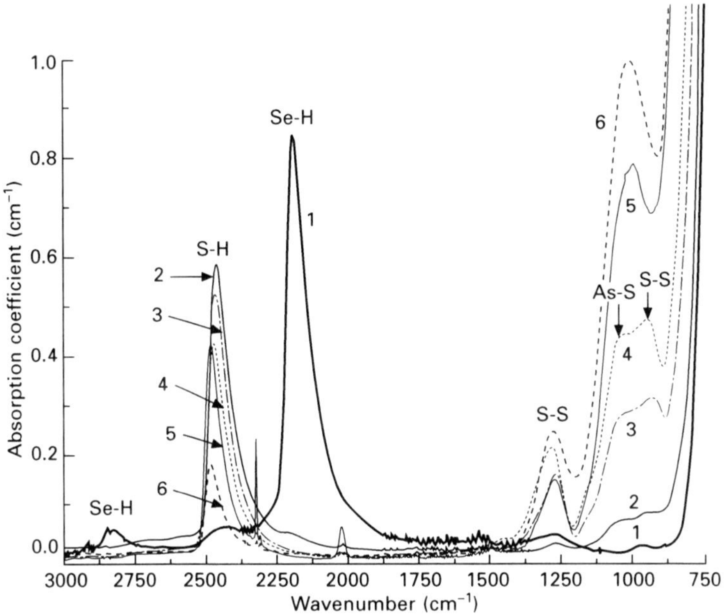

1.5 Spectral dependence of long-wavelength edge and absorption coefficient of SH and SeH groups in xAs_2S_3-$(1-x)As_2Se_3$ glasses. x =: 1, 0; 2, 10; 3, 30; 4, 50; 5, 80; 6, 100.

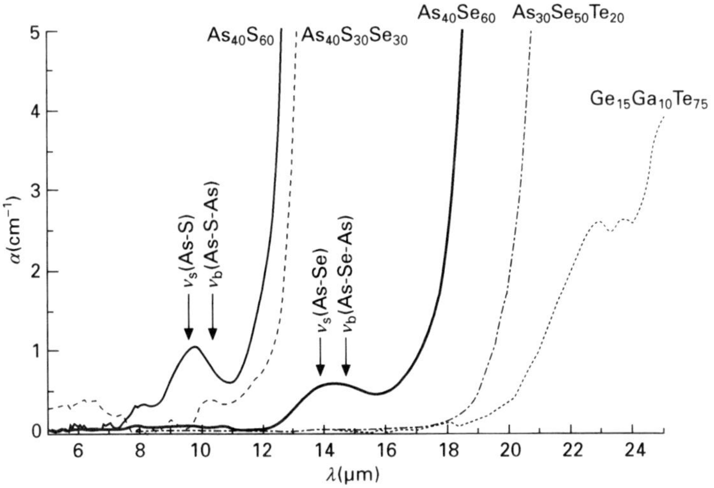

1.6 Spectral dependence of absorption composition in glasses: As_2S_3, As_2Se_3, $As_4S_3Se_3$, $As_{30}Se_{50}Te_{20}$, and $Ge_{15}Ga_{10}Te_{75}$.

higher than in silica glass. The Young's modulus of chalcogenide glasses is from 17 to 18.5 GPa which is several times less than that in silica glass (72.1 GPa).

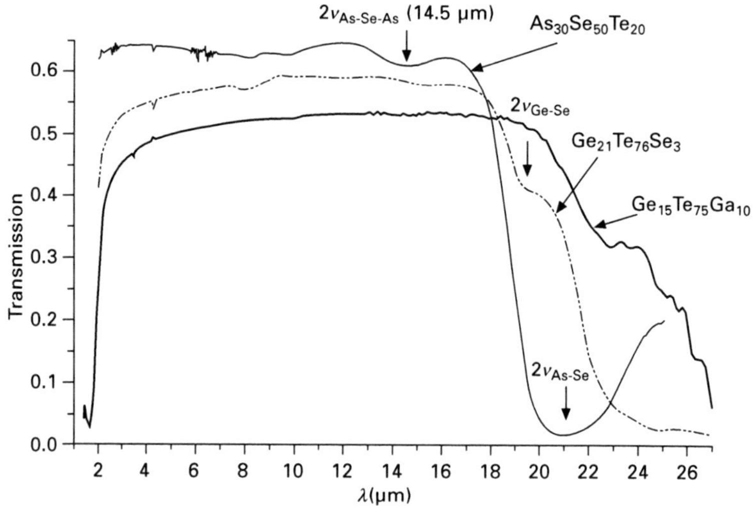

1.7 Transparency spectra of $As_{30}Se_{50}Te_{20}$, $Te_{76}Ge_{21}Se_3$, and $Te_{75}Ge_{15}Ga_{10}$ glasses (thickness is 2 mm).

The physico-chemical properties of glasses are affected by the history of sample production. In optical fibers drawn at different rates from the glasses produced by quick annealing of the melt and annealed at a temperature less than T_g, structural changes are observed leading to modifications of density, refractive index, Young's modulus, thermal expansion and other physical properties. In this process, the glass structure is relaxed to equilibrium structure at a lower temperature [29, 30].

1.4 Preparation of high-purity chalcogenide glasses

1.4.1 Transparency of chalcogenide glasses as an impurity-sensitive property

According to *a priori* estimation [31], which accounts only for electron and multi-phonon absorptions as well as for Rayleigh scattering, the minimum optical losses in As_2S_3 and As_2Se_3 are equal to (6–7) × 10^{-2} dB/km at 4–6 μm. Consideration of the 'weak absorption tail' increases the theoretical minimum of optical losses to 1–5 dB/km.

The measured optical losses in chalcogenide glasses are substantially higher than the intrinsic losses, mainly because of the presence of impurities in glasses. There are three main sources of impurities in chalcogenide glasses. The first source is the initial components used for synthesis of the glass. Arsenic, germanium and chalcogens were developed as pure materials for semiconductor applications with attention paid mainly to electrically active impurities, i.e., metal impurities. That is why commercial samples of As, Ge, S, Se, and Te contain 0.1–0.01 ppm wt. of metal impurities and a higher content (1–100 ppm wt.) of hydrogen, oxygen and silicon.

The second source of impurities is the container utilized for the synthesis of glass-forming melt. Silica glass is the main container material used for the synthesis of chalcogenide glasses. The presence of hydrogen and silicon dioxide impurities is connected with their presence in the walls of the quartz ampoule. The content of silicon in germanium-containing chalcogenide glasses, produced by direct synthesis, can attain hundredth parts of a percent [9]. Impurities with a high diffusion coefficient value, especially hydrogen, enter the chalcogenide melt. It was found [9] that hydrogen, from 100-ppm hydrogen-containing silica glass, starts to enter chalcogenide glass melt at 650°C. The temperature dependence of hydrogen entrance rate υ (g·cm^{-2}s^{-1}) is given by the following expression

$$\ln \upsilon = -11.2 - (15300/T) \quad (873 \text{ K} \leq T \leq 1023 \text{ K}).$$

Thus, the measured rate of hydrogen entrance into selenium melt is 4×10^{-12} g/(cm^{2}s) at 750°C.

Chemical interaction of chalcogenides and impurities with silica glass at high temperature leads to formation of thin layers of reaction products on its surface. Due to difference in thermal and mechanical properties between the silica substrate and layers, the latter can separate from silica glass and enter the melt in the form of heterophase inclusions in the form of α- and β-cristobalite [5]. In the case of tellurium, the oxygen-containing tellurium compounds are detected in interaction products by mass spectrometry. Size and concentration of inclusions are experimentally determined by laser ultramicroscopy [32]. This method makes it possible to determine particles with a size of 50–200 nm in a quantity of 10^2–10^8 cm^{-3} in glasses, including those non-transparent in the visible range. Contamination with heterophase particles due to interaction of the melt with the walls of a silica container increases with the increase in temperature, especially in the case of glasses containing Ge, Sb or Te. Figure 1.8 presents microphotos of a laser beam passing through samples of As_2S_3 glass made at different temperatures as well as the corresponding histograms for size distribution of particles. Heterophase inclusions lead to optical losses in chalcogenide glass due to absorption and scattering.

The third source of impurities is the environment and residual vacuum gases during the synthesis and treatment of glass. Arsenic and chalcogens are inclined to oxidation at elevated temperature. As follows from thermodynamic calculations and from mass spectrometric investigation of oxide vapors, the equilibrium partial pressure of oxygen during dissociation of the oxides of arsenic, selenium, and tellurium at ~1000K is at the level of 10^{-4}–10^{-6} Pa (10^{-6}–10^{-8} torr). This value will also be less at lower temperatures. For this

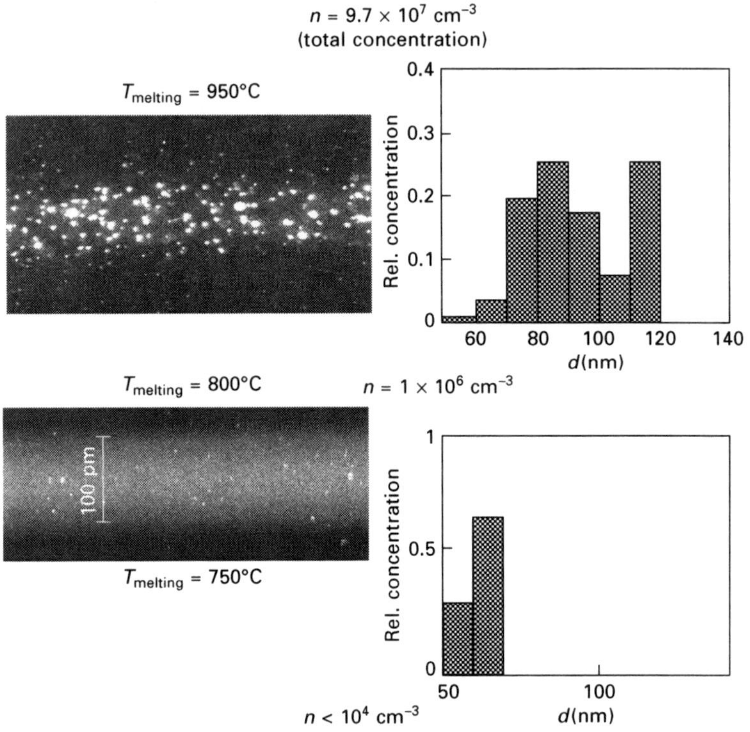

1.8 Microphotographs and histograms of distribution of sizes of inclusions in As_2S_3 glass samples prepared at different temperatures.

reason, the oxidation of chalcogens and arsenic in the open vacuum system will be precluded at vacuum less than 10^{-4}–10^{-6} Pa.

The impurities embedded into the glass network or the dissolved impurities are observed in the loss spectra of optical fibers as selective absorption bands, the position of which is determined by the nature of the impurity. Table 1.4 gives the position of impurity maxima of absorption bands in the transmission spectra of chalcogenide glasses. They have a weak dependence upon glass macrocomposition. It is seen that in the transparency range of chalcogenide glasses, there is a large number of impurity absorption bands. The main limiting impurities are oxygen, carbon, hydrogen, and silicon connected with each other and with components of the glass matrix. Some known values for extinction coefficients in chalcogenide glasses are given in Table 1.5.

As an example, Fig. 1.9 gives the spectral dependence of extinction coefficient of SeH-groups, As_2O_3, and sulfur in As_2Se_3 glass [33–35]. It follows from Tables 1.4, 1.5, and Fig. 1.9 that the effect of hydrogen impurity upon glass transparency is mostly present in the 2.5–6.5 μm spectral range

Table 1.4 Position of the maxima of absorption impurity bands in chalcogenide glasses

Component or functional group leading to absorption	Position of the maximum of absorption band (μm)
OH^-	2.92
S-H	4.01; 3.65; 3.11; 2.05
Se-H	7.8; 4.57; 4.12; 3.53; 2.32
Ge-H	4.95
As-H	5.02
P-H	4.35
H_2O	6.31; 2.86; 2.79
Ge-O	12.8; 7.9
P-O	8.3
CO_2	4.33; 4.31; 15.0
COS	4.95
CSe_2	7.8
CS_2	6.68; 4.65
Arsenic oxides (different forms)	15.4; 12.7; 9.5; 8.9; 7.9; 7.5
Se-O	10.67; 11.06
Si-O	9.1–9.6
Non-identified bands	4.65; 5.17; 5.56; 6.0

Table 1.5 Values for extinction coefficients ε for impurities in chalcogenide glasses [33–35]

Impurity compound or functional group	Glass	Maximum of absorption band (μm)	Extinction coefficient (dB/km/ppm)*	Calculated content of impurity leading to optical losses equal to intrinsic losses (ppb)*
SH	As_2S_3	4.0	2500 ppm at	0.3
SeH	As_2S_3	4.5	1000 ppm at	0.1
CO_2	As_2S_3	4.33	1.5×10^4	0.05
COS	As_2S_3	4.95	10^5	0.008
CS_2	As_2S_3	6.68	4.8×10^5	0.2
As_2O_3	As_2Se_3	12.65	4.3×10^4	100
		9.5	1030	400
Se-O	As_2Se_3	10.6	380	2000
S	As_2Se_3	10.6	0.52 ppm at	10^6
	$As_2Se_{1.5}Te_{1.5}$	14.5	32 ppm at	10^4

* Intrinsic losses of glass calculated accounting for the 'tail of weak absorption'.

and of oxygen is in the longer wavelength range (7–14 μm). Quantitatively, the magnitude of excess oxygen absorption is approximately the same as that of optical losses connected with the presence of hydrogen impurity in the

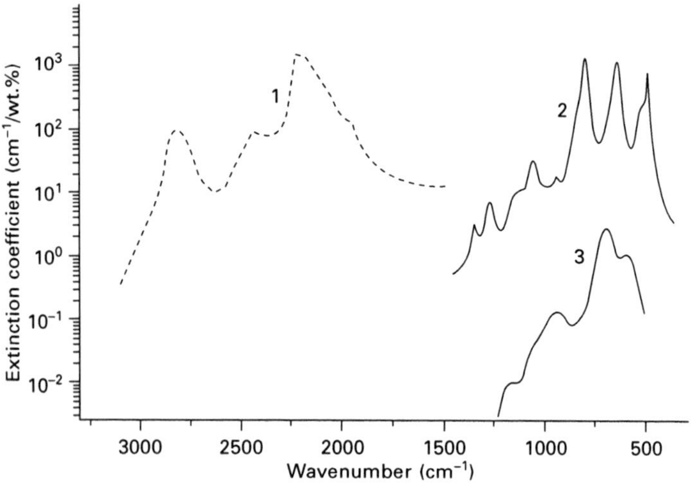

1.9 Spectral dependence of extinction coefficient in impurities of hydrogen (1), oxygen (2), and sulfur (3) in As_2Se_3 glass.

form of Se-H bonds. The absorption of sulfur and oxygen impurity refers to the same spectral range; however, sulfur has a substantially lower effect on glass transmission.

On the basis of extinction coefficients, requirements were set for the admissible content of impurities in chalcogenide glasses for fiber optics. To obtain optical losses as low as the level of intrinsic losses, the content of oxygen, carbon, and silicon impurities should not exceed 0.1–10 ppb (Table 1.5).

Nano- and micro-crystals of macro-components, nano- and micro-drops of the second glass phase, as well as extensive corrugated inhomogeneities (striaes) manifest themselves as inhomogeneities in chalcogenide glasses [36, 37]. Their formation is due to inclination of chalcogenide glasses and their melts to crystallization and micro-segregation. Cooling of the melt into glass, annealing and fabrication of glass into optical fiber and other optical products in non-optimum time–temperature modes are accompanied by a formation of nano- and micro-crystals, which scatter the radiation transmitted through the optical fiber. Presence of the striaes is due to non-optimum thermal fields and time–temperature modes during the melt solidification, especially for macro-components with a higher volatility.

Glasses containing several individual or conditional glass-formers are inclined to micro-segregation. Micro-segregation of $As_2S_{1.5}Se_{1.5}$ glass fiber was observed during drawing from a double crucible [38]. Figure 1.10 illustrates the formation and development of optical inhomogeneities in the core of $As_{40}S_{30}Se_{30}$ glass fiber during drawing, originating from phase segregation

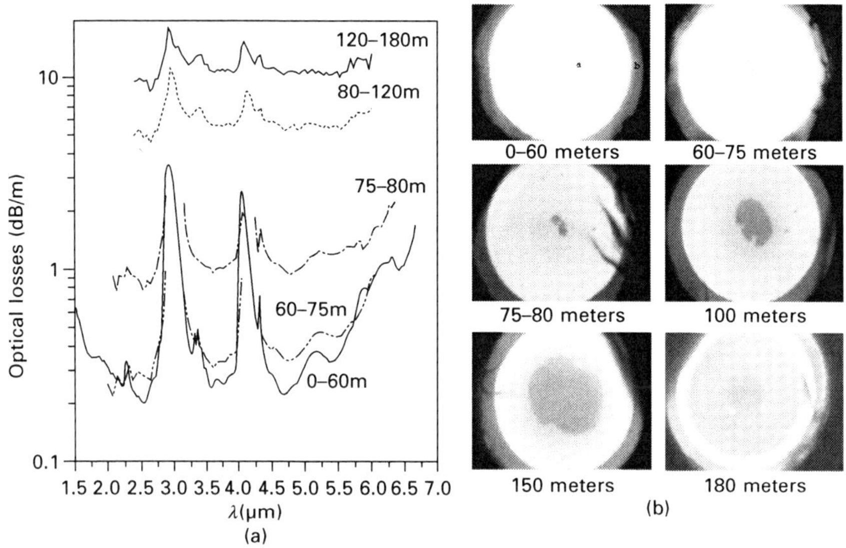

1.10 (a) Evolution of the total optical loss spectrum along the length of fiber; (b) cross-sectional micrographs of different parts of $As_{35}S_{35}Se_{30}$ glass fiber; the distance from the first-drawn portion of the fiber is <60, 60–75, 75–80, 100, 150, and 180 m.

of the molten glass. Differential thermal analysis of glass after fiber drawing indicated the appearance of a second glass phase. In the process of drawing, the amount and size of new phases increases, which leads to the increase in optical losses mainly due to scattering in the optical fiber produced.

The presence of metal–metal and chalcogen–chalcogen homo-bonds is a specific feature of the atomic molecular structure of chalcogenide glasses [39]. It is due to relatively close energies of homo- and hetero-bonds. Homo-bonds are responsible for selective absorption bands in the transparency range. For example, in As_2S_3 glass the absorption bands with maxima at 5.1 and 5.5 μm, which are attributed to combined vibrations with the participation of S–S bonds, become as intense as impurity absorption bands with total losses of 10–100 dB/km in optical fiber.

Impurities in the form of heterophase inclusions in chalcogenide glass lead to optical losses due to absorption and scattering, as well as a decrease in mechanical and laser damage resistance of optical fibers. The variation of these physical parameters is determined by concentration and size of scattering particles. Ideally, glasses designed for application in power optics should not contain any particles in the volume passed through by the intense radiant flux.

1.4.2 The features of preparation of chalcogenide glasses

Production of glasses with increased transparency

A general scheme for production of chalcogenide glasses for fiber optics comprises the preparation of initial high-purity substances and charge on their basis, melting the charge with conservation of the attained purity of substances, and cooling the melt into glass. During charge melting, the optimum temperature and time of melting are selected, as well as the melt mixing conditions (providing homogeneity of macro-composition) and minimization of the contamination effect of the reactor walls (with respect to the mass of the sample produced).

The following features must be considered for the preparation of high-purity chalcogenide glasses suitable for fiber optics:

- predominantly covalent bond in chalcogens, arsenic and in glass-forming compounds, as well as proximity of energy in their homo- and heteroatomic bonds;
- high sensitivity of optical properties to the presence of gas-forming impurities;
- specific ensemble of limiting impurities (C, H, O) and ability of such impurities to change their chemical form as a results of reaction with each other and with glass macrocomponents;
- tendency of glasses to crystallization and liquation which limits the selection of non-crystallizing compositions and of optimum time–temperature modes for the production of glasses and optical fibers;
- dissociative nature of glass-forming compounds during melt evaporation leading to a change in macrocomposition of the melt after distillation purification.

It follows from the above-mentioned considerations that a compulsory condition for obtaining high-purity chalcogenide glasses is the low content of impurities and heterophase inclusions. At present, there are two efficient techniques for production of chalcogenide glass melts with low content of impurities, leading to optically homogeneous, high-purity glasses after solidification. The first is the synthesis of glass-forming compounds or of their mixture by vacuum melting of the charge produced by the above-mentioned methods. This variant can be used for the production of all chalcogenide glasses.

The present problems are connected with a high content (up to several tens of ppm) of such impurities as oxygen, carbon, and hydrogen in commercial samples of high-purity chalcogens and arsenic. Selenium, tellurium, and arsenic can oxidize in normal environments, even at room temperature. That is why the initial high-purity substances are loaded into containers under oil-free vacuum.

Thus, high-purity As, S (and Se) elements were loaded into the synthesis reactor by evaporation from intermediate ampoules in an evacuated all-sealed system, in an oil-free high vacuum (10^{-7} torr). The scheme of this system is shown in Fig. 1.11(a). The melt was homogenized in the rocking muffle furnace at 750°C for 7 hours. Solidification of the melt was carried out at the rate of 1 K/min.

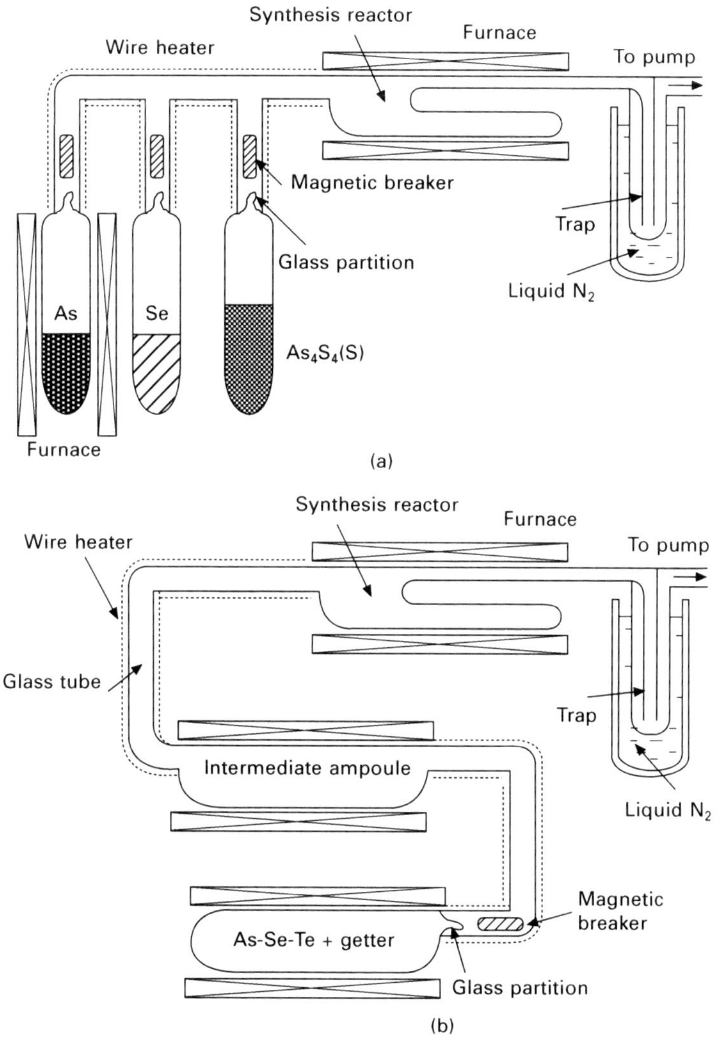

1.11 Set-ups for preparing high-purity chalcogenide glasses: (a) from high-purity initial elements using intermediate ampoules; (b) using chemical getters.

For arsenic and selenium, reducing the presence of submicron particles, mainly consisting of carbon, by the method of vacuum sublimation or distillation is ineffective [40]. Arsenic monosulfide, As_4S_4, was used as the arsenic-containing component for production of glasses with As/S ratio equal to 1/1 or less. This compound is more suitable for ultrapurification, especially for decreasing the number of submicron particles, due to low viscosity of the melts [41]. Two- and three-component glasses were produced by melting purified arsenic monosulfide with the required amount of chalcogens, e.g.:

$$As_4S_4 + S_2 \rightarrow 2\ As_2S_3$$

$$As_4S_4 + Se_2 \rightarrow 2\ As_2S_2Se$$

The second technique (Fig. 1.11(b)) includes the purification of glass-forming chalcogenides of technical quality by chemical distillation methods, the melting of the purified fractions up to homogeneous state and the melt solidification. Vacuum distillation in the open and closed system decreases the content of impurities of light-volatile substances and submicron particles by 3–10 times [42]. The particles behave like as heavy volatile impurities, and their content decreases in the final fraction of distillate. The efficiency of ultrapurification is highly sensitive to the evaporation rate and melt viscosity. Depending on the type of distillate fraction taken for a subsequent treatment, it is possible to produce the glasses with different content of various impurities. The glass made from the initial and final fractions of distillate is enriched with the compounds of hydrogen, oxygen, and heterogeneous particles, respectively. To remove some impurities (carbon, oxygen, hydrogen), the preliminary chemical treatment of arsenic and germanium chalcogenides can be used. For example, for binding the oxygen-containing impurities, it is possible to use aluminum and magnesium, while for binding the hydrogen-containing ones it is possible to use chlorine and chlorides of glass-forming elements.

In the melting process, in the charge with micro amounts of Al or Mg, oxygen is bound with a formation of low-volatile oxides Al_2O_3 or MgO, and in the case of micro amounts of $TeCl_4$, hydrogen is bound to HCl. After melting at a temperature of 700–800°C, the melt is subjected to a vacuum distillation. To decrease the content of carbon, similar procedures can be used with addition of small amounts of arsenic and selenium oxides. Experimental results on the production of high-purity chalcogenide glasses by chemical distillation indicate a high efficiency of this method [43–46]. This process presents some complications due to a difference in macrocomposition of the distillate and the initial chalcogenide melt. Arsenic chalcogenides are evaporated with dissociation, which leads to a separation of vapor components during condensation [47]. Figure 1.12 shows the average content of distillate as a function of the distilled part during vacuum distillation of As_2S_3 in the

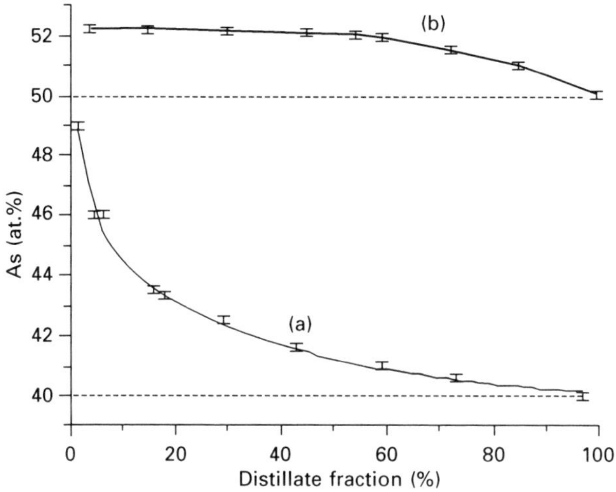

1.12 Average distillate composition as a function of a part of distilled substance during vacuum distillation of As_2S_3 (a) and As_4S_4 (b).

closed system. It is seen that As_2S_3 distillate is enriched with arsenic in the process of vacuum distillation.

Influence of conditions of glass preparation on the impurities content in samples has been studied [43]. Three purity grades of Te-As-Se glasses were obtained: by using only simple melting of high-purity elements without oxygen getter, or with addition of a small quantity of Al (from 100–300 ppm wt), or with addition of 700 ppm wt Al with subsequent triple distillation. The preparation of Te-As-Se glasses with low content of oxygen, carbon and water impurities was achieved by a combination of chemical and distillation methods. The residue after distillation did not exceed 5 wt.%. When high purity elements were melted without oxygen getter, there are some oxide, hydroxide and water impurity bands on the absorption spectrum. An addition of a small quantity of Al (100–300 ppm wt) was also not enough to remove all oxide bands. Only the glass melted with 700 ppm wt. of Al with subsequent triple distillation gave spectra without these absorption bands. The disadvantage of this method is the increase of SeH band intensity. The main reason for the presence of hydrogen could be the appearance of free hydrogen after the chemical reaction of Al with impurities.

In the process of production of chalcogenide glasses with reproducible optical properties, it is important to provide the identity of impurities contained in initial substances and the conditions of glass production. It has been shown that the increase of synthesis temperature of the $Ge_{25}Sb_{10}S_{65}$ glass from 800°C up to 950°C resulted in the increase of the content of H, Na, Al, Si, K, Ca and transition metal impurities in the prepared glasses [37]. Glasses subjected to chemical-distillation purification were characterized by a low content of gas-forming impurities and an increased content of Al, Si, and Cl. The glasses contained inclusions of heterophase impurities, mainly consisting of SiO_2, and their concentration and size depended on the conditions of glass preparation. It was found that $Ge_{25}Sb_{10}S_{65}$ glass samples prepared at a temperature of 800°C contained microscopic inclusions 5–50 μm in size of a concentration of 10^4–10^6 cm^{-3}; the samples prepared at temperature of 950°C contained inclusions 80–500 nm in size at a concentration of 10^8–10^9 cm^{-3}. The additional vacuum distillation of glass samples in the closed system made it possible to decrease the content of submicron impurity inclusions to 10^2 cm^{-3}.

Impurities in the form of heterophase inclusions in chalcogenide glasses can be the centers of nucleation and initiate heterogeneous crystallization. Using laser ultra-microscopy, changes in the size distribution of micro-inclusions in $Ge_{25}Sb_{10}S_{65}$ glass sample were investigated in the process of annealing at 350°C [37]. The isothermal treatment of the glass for 3–5 hours caused the increase in average diameter of inclusions, while their numerical concentration increased insignificantly. This may be evidence of the beginning of glass crystallization with SiO_2 inclusions as centers of heterogeneous crystallization.

Chalco-iodide glasses with high content of iodine can be prepared by low-temperature melting of the charge. The development of a new method for preparing glasses with low concentration of iodine at low temperatures is described in Ref. [48]. Ge–Sb–S–I glasses with iodine content less than 15 at.% were prepared by interaction of germanium iodide (IV) and antimony iodide (III) with the sulfur melt at temperatures not more than 650°C. A scheme of the preparation set-up for glass-forming melt, made from quartz glass, is shown in Fig. 1.13. Chemical interaction of charge components was carried out in reactor 7 under heating. Iodine vapors, together with vapors of volatile iodides, escaped through mass-transfer section (9) where separation took place. Iodine, as a low-boiling component, concentrated in the upper part of the mass-transfer section and was collected in receiver 10. Vapors of high-boiling iodides condensed in the bottom part of the separating section and drained down into the melt. It provided the selective removal of iodine from the glass-forming melt with simultaneous retention of other macro-components in the melt. Iodine was periodically extracted in the 300–650°C temperature range. Additional homogenization of the melt was carried out in the rocking furnace at 650°C for 3 hours. The prepared glasses were in the form of cylinders with diameter of 10 mm and length of 30–60 mm. Such samples are characterized by low content of impurities of transition metals (≤0.1 ppm wt.), silicon (1–2 ppm wt.) and hydrogen (0.3 ppm mol) in the form of SH groups.

For creation of middle-infrared tunable lasers, composite materials based on As_2S_3 glass containing nanocrystals of ZnS(ZnSe) are a promising perspective.

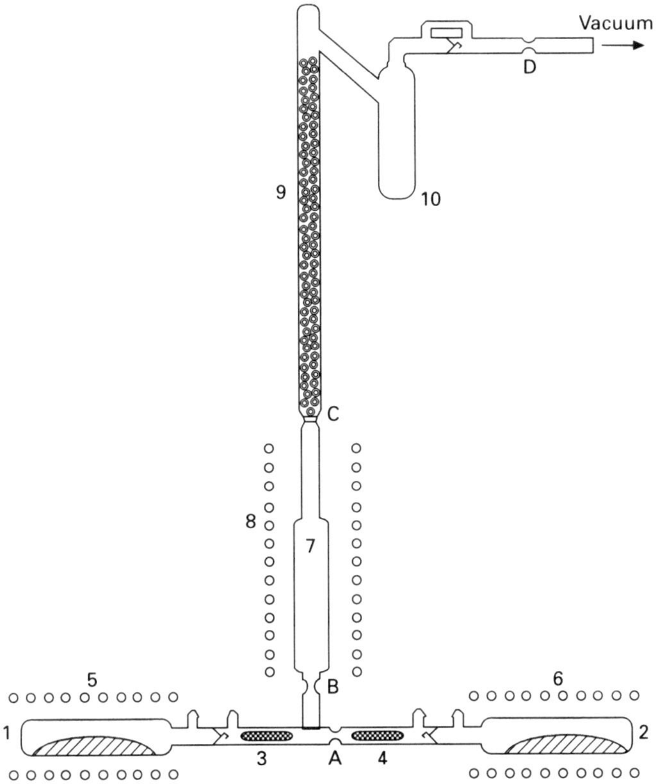

1.13 Set-up for preparation of Ge-Sb-S-I glass: 1, 2, loading ampoules; 3, 4, magnetic hammer-breaker; 5, 6, 8, resistance heating furnaces; 7, reactor; 9, mass-transfer section; 10, iodine receiver.

The method of solid phase synthesis of ZnS (ZnSe):Cr^{2+}, which allows the material to be received with high concentration of Cr^{2+} (up to 6.4 wt.%) was developed [49]. This material was ground to a highly refined powder with average size of particles of 1 μm, and it was then introduced in high-purity As_2S_3. The mixture was heated, held at fixed temperature and cooled at different rates. The synthesis conditions of the composite material are the following: temperature is 550–750°C, exposure time in rotating furnace is 1–24 h, annealing time is 1–18 h. The dependences of optical characteristics of composite materials versus the concentration of initial crystalline A_2B_6 compounds, varying from 0.08 to 4 wt.%, were investigated. For the first time, the mono-index fibers based on new composite material As_2S_3-ZnS (ZnSe): Cr^{2+} exhibited luminescence in the 1.8–3 μm wavelength range.

Table 1.6 gives the content of impurities in the best of the chalcogenide samples. The content of metals and silicon, determined by laser mass spectrometry, does not exceed 0.1–0.5 ppm wt. The content of carbon,

Table 1.6 Content of impurities (ppm wt.) in the chalcogenide glasses produced

Impurity	Content of impurity (ppm) in					
	As_2S_3	$As_2S_{1.5}Se_{1.5}$	As_2Se_3	$As_{40}Se_{40}Te_{20}$	$Ge_{25}Sb_{10}S_{65}$	$Ge_{21}Te_{76}Se_3$
Hydrogen	0.05	0.2	0.1	0.02	0.13	0.1
Oxygen	0.001	–	0.5	0.6	–	< 0.02
Carbon	≤ 0.2	<0.02	0.5	0.6	≤5	≤ 0.14
Silicon	0.1	<0.4	0.1	0.05	0.6	4
Al, Fe, Ni, Cr, Mg, Mn, Ni	10^{-3}–0.1	< 0.2–2	10^{-3}–0.1	10^{-3}–0.1	<0.1–1	<0.1
Inclusions of 0.06–0.1 μm size	≤ 2×10^4 cm^{-3}	5×10^6 cm^{-3}	2×10^6 cm^{-3}	2×10^6 cm^{-3}	1×10^2 cm^{-3}	

determined by the method of reaction gas chromatography and by activation method, is 0.2–0.6 ppm wt. The content of hydrogen and oxygen, determined by spark time-of-flight tandem laser mass spectrometry and IR spectroscopy in the bulk samples and optical fibers, is 0.02–0.2 ppm wt. The lowest hydrogen content (≤0.02 ppm) is observed in the glass melted with addition of $AlCl_3$ or mixture of ($TeCl_4$+Al). The content of inclusions with a diameter of 0.1 μm, determined by laser ultramicroscopy, is less than 1×10^4 cm^{-3} in As_2S_3 and $Ge_{25}Sb_{10}S_{65}$ glasses. The content of submicron inclusions in As-Se and As-Se-Te glasses is $1 \times 10^5 - 5 \times 10^6$ cm^{-3}.

1.5 Preparation and characterization of chalcogenide optical fibers

Chalcogenide optical fibers are manufactured by drawing from the melt (the method of single and double crucible) and from single-layer or double-layer rod (preform). The required difference in refractive index between the core and cladding glasses is provided by addition of cationic or anionic agent to the base glass. Substitution of a part of arsenic for germanium or a part of sulfur for selenium in As_2S_3 glass decreases or increases the refractive index, respectively.

Figure 1.14 shows the set-up for optical fiber drawing with the double crucible technique. The core and cladding glasses, heated up to the plastic state, are supplied from crucibles into a concentric double drawing nozzle. At the output of the drawing nozzle, a conical part is formed, the 'drawing cone', from which a double-layered optical fiber is drawn by a pulling mechanism. Then it passes through the system for application of protective coating and, finally, is stowed on the reel (or fiber receiver). The drawing rate and diameter of optical fibers are determined by the rotation rate of the pulling mechanism and by the pressure of inert gas over the melt. The core

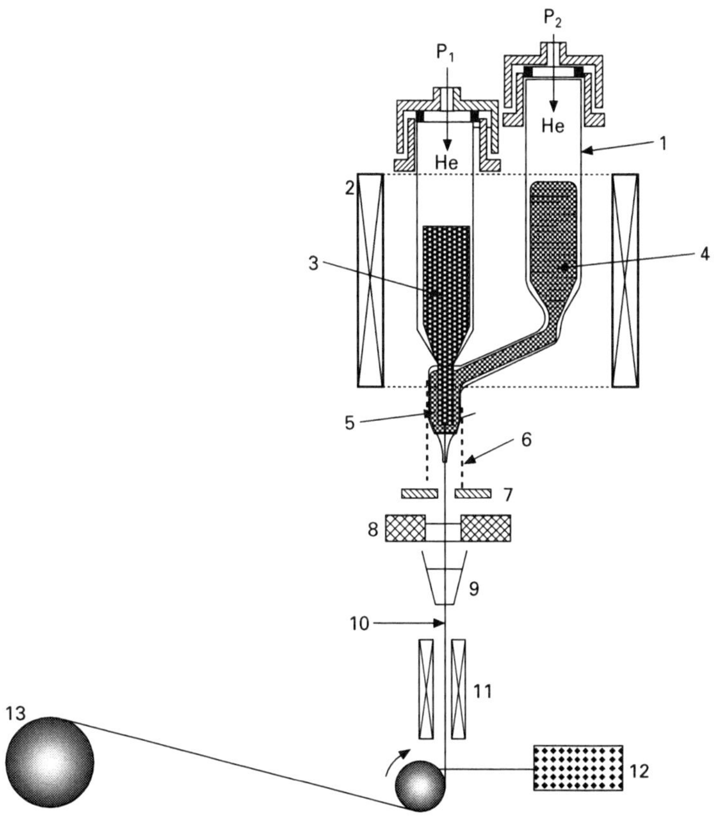

1.14 Set-up for drawing optical fibers by the double crucible method: 1, double crucible; 2, thermostat; 3, core glass; 4, cladding glass; 5, drawing nozzle; 6, furnace for heating the drawing nozzle; 7, diaphragm; 8, meter of the diameter; 9, drawing nozzle for application of polymer coating; 10, optical fiber; 11, furnace for polymer drying; 12, tension meter; 13, take-up reel.

and cladding diameter ratio is determined by the ratio in diameters of circular and annular channels of the drawing nozzle and by the volumetric flow rate of the core and cladding melts. This method was applied for drawing optical fibers As-S, Ge-Se, Ge-Se-Sb, As-S-Se, and As-Se-Te glasses.

In the process of optical fiber drawing from the preform, the upper end of the preform is fixed in the movable holder of the set-up and the lower one is placed inside the tubular resistance heater. While heating up the lower part of the preform to the softening temperature of glass, its thickness decreases to fiber thickness under the effect of gravity or drawing force with a formation of a drawing cone. As in the double-crucible method, the fiber with applied protective coating is wound on the take-up drum. A purified inert gas is supplied into the drawing zone to preclude oxidation of the heated chalcogenide glass. The drawing rate of optical fibers in both methods is from 1 to 10 m/min., the diameter of the produced optical fibers is from 0.1 to 0.8 mm with a core diameter from 3 to 600 μm. The single-index (without the glass cladding) and multimode (with the glass cladding) optical fibers from As-S, As-Se, As-S-Se, As-Se-Te, Ge-As-S, Ge-As-Se, and Ge-Sb-S glasses were produced by drawing from 'rod-in-tube' or monolithic preforms. Several methods were developed to manufacture monolithic and 'rod-in-tube' preforms [50–52]. While using these methods, special attention is paid to achieve optically smooth and chemically pure internal and external surfaces of the core rod and cladding tube, as well as to maintain glass purity.

In the drawing process, a primary protective coating from polymer or melted metal is applied to optical fibers. F-42 fluoroplastic is often used as a primary polymeric coating and is applied from methylethylketone solution. Polyvinylchloride and photocuring polyacrylates are often used as a secondary coating.

Each of these methods of optical fiber production has its own merits and limitations. Drawing from the crucible provides a high-quality surface and, as a result, the higher mechanical strength of the optical fibers. It is easier to control the ratio of core-cladding diameters in 'rod-in-tube' method. Drawing from the crucible is preferable when manufacturing optical fibers from glasses with low tendency to crystallization.

When developing the double-crucible method, it was found that in the optimum temperature range for producing optical fibers, the behavior of the melts of As-S, As-Se glass systems is viscous-plastic but not Newtonian. To describe the flow of these melts in the channels of circular and annular section, it was necessary to determine such characteristics as plastic viscosity and ultimate shear stress [53, 54].

The main characteristics of fibers are optical losses, their spectral dependence and their mechanical strength. Figures 1.15–1.18 give the spectra of the total optical losses in optical fibers based on arsenic and germanium chalcogenides. It is seen from the figures that the transparency range of optical fibers shifts towards the long-wave region with heavier chalcogens.

The optical losses for single-index optical fibers can be considered as an indirect characteristic of purity and microhomogeneity of glass used for their production and are presented in Table 1.7.

Multimode optical fibers with glass reflective and protective polymer coatings, produced by the double crucible method, had a continuous length of a hundred meters with core and cladding diameters in the range of 50–600 μm and 150–800 μm, respectively. Table 1.8 gives the minimum optical losses and the losses at some laser wavelengths. As can be seen from the table and the spectra of total optical losses, the lowest content of impurities and the lowest losses are in arsenic sulfide and arsenic selenide glasses.

With the double-crucible technique, minimum optical losses of 12 dB/km

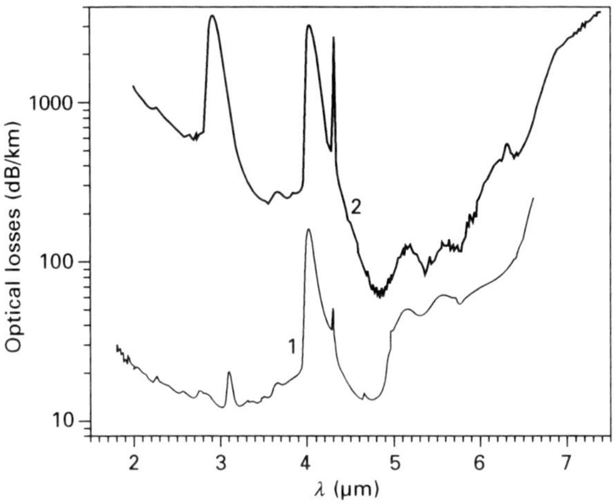

1.15 Spectra of the total optical losses in multi-mode optical fiber from the glasses of As-S (curve 1) and As-S-Se (curve 2) systems.

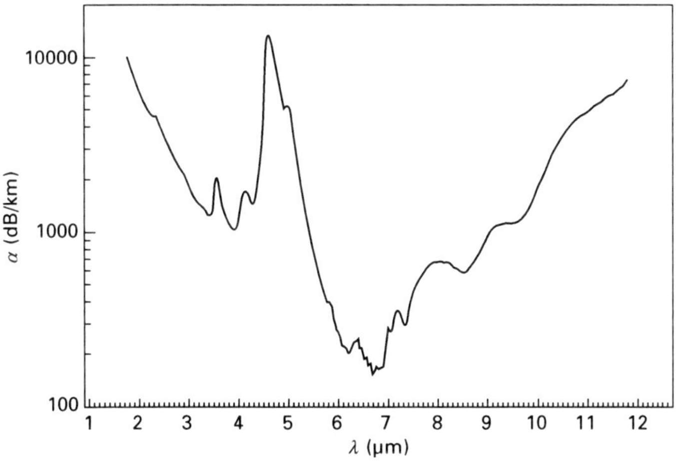

1.16 Spectrum of the total optical losses in double-layered optical $Ge_2As_{38}Se_{40}Te_{20}/Ge_2As_{36}Se_{44}Te_{18}$ glass fiber.

were obtained at 3.0 μm and 14 dB/km at 4.8 μm in a high-purity As_2S_3, double-layered, optical fiber (Fig. 1.15) [36]. The production of optical fibers from $As_{38}S_{25}Se_{37}/As_{38}S_{27}Se_{35}$ glasses with minimum optical losses of 60 dB/km at 4.8 μm has been reported. [58] Optical fibers were drawn by the double-crucible method, had core/cladding diameters of 300–400 and 200–400 μm, with continuous length of more than 150 meters and a double polymer coating. In optical fibers of $As_{39}Se_{61}/As_{36}Se_{64}$ and $Ge_1As_{39}Se_{60}/Ge_2As_{38}Se_{60}$ compositions, produced by the double-crucible technique, minimum optical

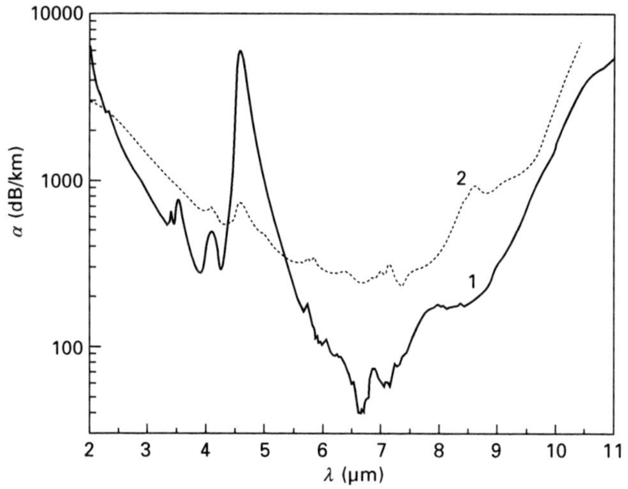

1.17 Spectra of the total optical losses in optical fibers based on selenium-telluride glasses and produced by different methods of ultrapurification: 1, $As_{30}Se_{50}Te_{20}$, distillation of a glass-forming compound purified by chemical method with addition of 700 ppm of Al; 2, $Ge_{1.3}As_{37.1}Se_{41.3}Te_{20.3}/Ge_{1.2}As_{40.4}Se_{41.2}Te_{17.2}$, distillation of a glass-forming compound purified by chemical method with addition of $TeCl_4$+Al.

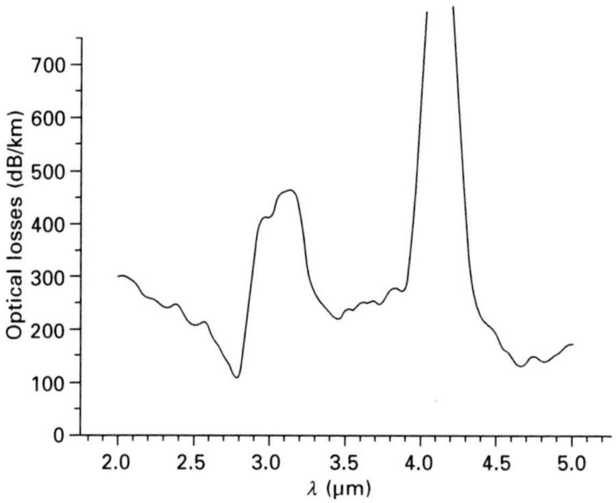

1.18 Spectra of optical losses of single-mode optical fibers from the glass of As-S system (core diameter is 3 μm).

losses were 0.2–0.3 dB/m at 2–6 μm and about 5 dB/m at 10.6 μm [59]. In optical fiber from $Ge_2As_{38}Se_{40}Te_{20}/Ge_2As_{36}Se_{44}Te_{18}$ glasses, also produced by the double-crucible method, the minimum optical losses were 150 dB/

Table 1.7 Optical losses for single-index optical fibers

Glass	$As_{35}Se_{65}$	$As_{30}Se_{50}Te_{20}$	$Ga_5Ge_{20}Sb_{10}S_{65}$	$GeSe_4$	$Ge_{30}As_{10}Se_{30}Te_{30}$
Losses (dB/km)	76	40	100	100	110
Wavelength (μm)	4.0	6.7	3.4	6.6	6.6
Reference	[5]	[55]	[56]	[57]	[44]

Table 1.8 Optical transmission of the best samples of chalcogenide glasses

Glass	Minimum optical losses		Optical losses at laser wavelengths, dB/km				
	β (dB/km)	λ (μm)	2.08 (μm)	2.9 (μm)	3.7 (μm)	5.6 (μm)	10.6 (μm)
As_2S_3	12	3.0	23	13	17	75	–
$As_2S_{1.5}Se_{1.5}$	60	4.8	1000	3800	400	200	–
As_2Se_3	78	4.2	160	160	80	85	650
$As_{40}Se_{30}Te_{50}$	40	6.6	4000	950	380	160	3500
$GeSe_4$	100	5.9	300	500	320	180	–

km at 6.6 μm (Fig. 1.16) [43]. Minimum optical loss in multimode double-layered optical fiber from $Ge_{30}As_{10}Se_{30}Te_{30}$ glass was 700 dB/km at 6.56 μm [44].

A single-mode optical fiber from As_2S_3 glass had a core diameter of 3–20 μm, a cladding diameter of 125 μm, a numerical aperture of 0.17, and minimum losses of 100 dB/km at 2.1–2.2 μm [60] (Fig. 1.18). In a single-mode optical fiber from $As_{39}Se_{42.4}Te_{18.6}/As_{39}Se_{42.7}Te_{18.3}$ glasses with a core diameter of 20 μm and total fiber diameter of 450 μm, the minimum losses were 330 dB/km at 7.5 μm [61].

Investigation of time–stability of chalcogenide glass optical fibers indicated that the optical and strength characteristics of such fibers with glass cladding and well-protected side surface are sufficiently stable in time from the viewpoint of their actual application [62]. Thus, As_2S_3 optical fibers with metal or double-layered polymer protective coating have been stable in terms of mechanical strength and optical losses for 3 years of storage at normal conditions, within margins of error. Degradation of optical and mechanical properties is characteristic of optical fibers with no protective coating, which results in a degradation of their side surface.

1.6 Conclusion

Production of high-purity chalcogenide glasses and investigation of their properties is an actively developed division of optical materials science. A number of the key properties of chalcogenide vitreous materials, first of all, transparency, mechanical strength, and laser damage resistance, are the impurity-sensitive properties. The limiting impurities in glasses are the compounds of oxygen, carbon, hydrogen, silicon, transition metals, and nanosized heterophase inclusions. The admissible content of impurity elements is 1–100 ppb wt., of heterophase inclusions is 10^3–10^4 cm^{-3}. Up to the present moment the purest glasses are produced using arsenic chalcogenides. In the best glass samples, the content of limiting impurities is 10–100 ppb wt., and that of the heterophase inclusions of 100 nm size is less than 10^3 cm^{-3}. These values are higher in routine samples. Chalcogenide glass single-mode and multi-mode optical fibers are produced with technical and operating characteristics sufficient for a number of actual applications. The minimum optical losses of 12–14 dB/km at 3–5 μm are attained in the optical fiber made from arsenic-sulfide glass. The level of losses in standard optical fibers is 50–300 dB/km in 2–9 μm spectral range. Increase in the degree of purity of glasses, especially germanium- and tellurium-containing glasses, via the advancement of present synthesis methods and the development of new methods, can be considered as the priority direction for future research.

1.7 References

1. Frerics R.J. (1953) New optical glasses with good transparency in the infrared, *J. Opt. Soc. Am.* 43, 1153–1154.
2. Standler R.R. and Henderson R.E. (1963) Infrared fiber optics technique. *Infrared Physics* 3, 223–227.
3. Kapany N.S. and Simms R.S. (1965) Recent developments in infrared fiber optics. *Infrared Physics* 5, 69–80.
4. Katsuayama T. and Matsumura H. (1988) *Infrared Optical Fibers*. Adam Hilger, Bristol.
5. Devyatykh G.G., Dianov E.M., Plotnichenko V.G., Skripachev I.V. and Churbanov M.F. (1991) Fiber waveguides based on high-purity chalcogenide glasses. *High-Purity Substances* 5(1), 1–27.
6. Nishii J. and Yamashita T. (1998) Chalcogenide glass-based fibers. In *Infrared Fiber Optics* (J.S. Sanghera and I.D. Aggarwal, eds CRC), Press, Boca Raton, FL, pp. 143–183.
7. Churbanov M.F. and Plotnichenko V.G. (2004) Optical fibers from high-purity arsenic chalcogenide glasses. In *Semiconducting Chalcogenide Glass. III. Application of chalcogenide fibers. Glasses Semiconductor and Semi-metals*, vol. 80 (R. Fairman and B. Ushkov, eds), Elsevier Academic Press, Oxford, pp. 209–230.
8. Feltz, A. (1993) *Amorphous Inorganic Materials and Glasses*. VCH Verlag, Weinheim.
9. Churbanov M.F. (1998) Purification of chalcogenide glasses. In *Properties, Processing and Applications of Glass and Rare Earth Doped Glasses for Optical Fibers* (D. Hewak, ed.) EMIS Date-reviews Series No. 22, London, pp. 340–343.
10. Skripachev I.V., Devyatykh G.G., Churbanov M.F., Boiko V.A. and Bagrov A.M. (1987) High-purity chalcogenide glasses for fiber optics. *Vysokochistye Veshchestva* 1, 120–129.
11. Katsuyama T., Ishida K., Satoh S. and Matsumura H. (1984) Low-loss Ge-Se chalcogenide glass optical fibres. *Appl. Phys. Lett.* 45, 925–927.

12. Le Sergent C. (1987) Chalcogenide glass optical fibers – an overview. *Proc. SPIE* 799, 18–24.
13. Hu J., Tarasov V., Carlie N., Petit L., Agarwal A., Richardson K. and Kimerling L. (2008) Exploration of waveguide fabrication from thermally evaporated Ge-Sb-S glass films. *Optical Materials* 30(10), 1560–1566.
14. Inagawa I., Iizuka R., Yamagishi T. and Yokota R. (1987) Optical and thermal properties of chalcogenide Ge-As-Se-Te glasses for IR fibers. *J. Non-Crystal. Solids* 95–96, 801–808.
15. Kokorina V.F. (1996). *Glasses for Infrared Optics*, CRC Press, Boca Raton.
16. Minaev V.S. (1991) *Stekloobraznye poluprovodnikovye splavy (Vitreous Semiconductor Alloys)* Metallurgiya, Moscow.
17. Bletskan D.I. (2006) Glass formation in binary and ternary chalcogenide systems. *Chalcogenide Letters*. 3(11), 81–119.
18. Mikhailov M.D. and Tver'yanovich A.S. (1986) Kriticheskie skorosti okhlazhdeniya nekotorykh khalkogenidnykh stekloobrazuyushchikh rasplavov (Critical Cooling Rates of Chalcogenide Glass-forming Melts). *Fizika i khimiya stekla* (Russ.) 12(3), 274–284.
19. Lapin Yu.K., Mikhailov M.D., Ananichev V.A., Baidakov L.A. and Tetereva V.A. (1991) Kristallizatsiya stekol v sistemakh As-Se i Tl-As-Se pri nagrevanii (Crystallization of As-Se and Tl-As-Se Glasses during Heating). *Fizika i khimiya stekla* (Russ.) 17(1), 3–7.
20. Churbanov M.F., Shiryaev V.S. (1994) Kristallizatsiya khalkogenidnykh stekol (Crystallization of Chalcogenide Glasses). *Vysokochistye Veshchestva* (Russ.) 4, 21–33.
21. Shiryaev V.S., Adam J.-L., Zhang X.H. and Churbanov M.F. (2005) Study of characteristic temperatures and non-isothermal crystallization kinetics in As-Se-Te glass system. *Solid State Sciences* 7, 209–215.
22. Wahab L.A. and Fayek S.A. (1996) Study of non-isothermal kinetics and thermal characterization of As-Se-Te system. *Solid State Commun*. 100(5) 345–350.
23. Shiryaev V.S., Adam J.-L. and Zhang X.H. (2004) Calorimetric study of characteristic temperatures and crystallization behaviour in Ge-As-Se-Te glass system. *J. Physics and Chemistry of Solids* 65(10), 1737–1744.
24. Tikhomirov V.K., Furniss D., Seddon A.B., Savage J.A., Mason P.D., Orchard D.A. and Lewis K.L. (2004) Glass formation in the Te-enriched part of the quaternary Ge-As-Se-Te system and its implication for mid-infrared optical fibres. *Infrared Physics & Technology* 45, 115–123.
25. Devyatykh G.G., Churbanov M.F., Shiryaev V.S. and Shipunov V.A. and Kuznetsov V.V. (1993) Kinetics of crystallization of GeS3.3 Glass. *High-Purity Substances* 7(6), 638–643.
26. Baro M.D., Clavaguera N., Surinach S., Barta C., Rysava N. and Triska A. (1991) DSC study of some Ge-Sb-S glasses. *J. Mat. Sci*. 26, 3680–3684.
27. Takebe H., Hirakawa T., Ichiki T. and Morinaga K. (2003) Thermal stability and structure of Ge-Sb-S glasses. *J. Cer. Soc. Jap*. 111(8), 572–575.
28. Dianov E.M., Shiryaev V.S., Izyneev A.A., Plotnichenko G.E. and Churbanov M.F. (1997) Crystallization kinetics of praseodymium-doped $(Ga_2S_3)_{0.7}(La_2S_3)_{0.3}$ glass. *Inorganic Materials* 33(9), 975–977.
29. Hari P., Taylor P.C., King W.A., LaCourse W.C. (1998) Metastable, drawing-induced crystallization in As_2Se_3 fibers. *J. Non-Crystal. Solids* 227–230, 789–793.
30. Griffiths J.E., Espinosa G.P., Remeika J.P. and Phillips J.C. (1981) Reversible

reconstruction and crystallization of $GeSe_2$ glass. *Solid State Commun.* 40(12), 1077–1080.
31. Lines M.E. (1984) Scattering losses in optical fiber materials. II. Numerical estimates. *J. Appl. Phys.* 55, 4058–4063.
32. Ketkova L.A., Kurilin A.V. and Churbanov M.F. (2008) Characterization of heterophase inclusions in glasses opaque in the visible range. *Inorganic Materials* 44, 2008, 773–777.
33. Churbanov M.F. (1995) High-purity chalcogenide glasses as materials for fiber optics. *J. Non-Cryst. Solids* 184, 25–29.
34. Shiryaev V.S., Smetanin S.V., Ovchinnikov D.K., Churbanov M.F., Krukova E.B. and Plotnichenko V.G. (2005) Effect of oxygen and carbon impurities on the optical transparency of As_2Se_3 glass. *Inorganic Materials* 41(3), 308–314.
35. Churbanov M.F., Shiryaev V.S., Smetanin S.V, Dianov E.M., Plotnichenko V.G., Hua Q., Li G., Shao H. (1999) Effect of sulfur on the optical transmission of As_2Se_3 and $As_2Se_{1.5}Te_{1.5}$ glasses in the range 500–1100 cm^{-1}. *Inorganic Materials* 35(12), 1229–1234.
36. Churbanov M.F., Snopatin G.E., Shiryaev V.S., Plotnichenko V.G. and Dianov E.M. (2011) Recent advances in preparation of high-purity chalcogenide glasses for fiber optics. *J. Non-Cryst. Solids* 357, 2352–2357.
37. Shiryaev V.S., Ketkova L.A., Churbanov M.F., Potapov A.M., Troles J., Houizot P. and Adam J.-L. (2009) Heterophase inclusions and dissolved impurities in $Ge_{25}Sb_{10}S_{65}$ glasses. *J. Non-Cryst. Solids* 355, 2640–2646.
38. Churbanov M.F., Shiryaev V.S., Pushkin A.A., Gerasimenko V.V., Suchkov A.I., Polyakov V.S., Koltashev V.V. and Plotnichenko V.G. (2007) Origin of microinhomogeneities in As-S-Se glass fibers fabricated by the double-crucible method. *Inorganic Materials* 43(4), 436–440.
39. Kozek F., Chlebny J., Cimpl Z. and Masek J. (1983) Structural and physical properties of the A_sS_x system, *Phil. Mag. B.* 47, (6), 627–639.
40. Devyatykh G.G., Churbanov M.F., Shiryaev V.S., Snopatin G.E. and Gerasimenko V.V. (1998) Impurity inclusions in extra-pure arsenic and chalcogens. *Inorganic Materials* 34(9), 902–906.
41. Devyatykh G.G., Churbanov M.F., Scripachev I.V., Snopatin G.E., Dianov E.M. and Plotnichenko V.G. (1999) Recent development in As-S glass fibers. *J. Non-Cryst. Solids* 256–257, 318–322.
42. Churbanov M.F. (1992) Recent advances in preparation of high-purity chalcogenide glasses. *J. Non-Cryst. Solids* 140, 324–326.
43. Shiryaev V.S., Churbanov M.F., Dianov E.M., Plotnichenko V.G., Adam J.-L. and Lucas J. (2005) Recent progress in preparation of chalcogenide As-Se-Te glasses with low impurity content. *J. Optoelectronics and Advanced Materials* 7(4) 1773–1780.
44. Sanghera J.S., Nguyen V.Q., Pureza P.C., Kung F.H., Miklos R. and Aggarwal I.D. (1994) Fabrication of low-loss IR-transmitting $Ge_{30}As_{10}Se_{30}Te_{30}$ glass fibers. *J. Lightwave Technol* 12(5), 737–741.
45. Nguyen V.Q., Sanghera J.A., Pureza P., Kung F.H. and Aggarwal I.D. (2002) Fabrication of arsenic selenide optical fiber with low hydrogen impurities. *J. Am. Ceram. Soc.* 85, 2849–2851.
46. Shibata S., Manabe T. and Horiguichi M. (1981) Preparation of Ge-S glass fibers with reduced OH, SH content. *Jap. J. Appl. Phys.* 20(1) 13–16.
47. Churbanov M.F. (2001) High-purity glasses based on arsenic chalcogenides. *J. Optoelectronics and Advanced Materials* 3, 341–349.

48. Velmuzhov A.P., Sibirkin A.A., Shiryaev V.S., Churbanov M.F., Suchkov A.I., Potapov A.M., Shaposhnikov R.M., Plotnichenko V.G. and Koltashev V.V. (2011) Production of Ge–Sb–S–I glass system via volatile iodides. *J. Optoelectronics and Advanced Materials* 13(8), 936–939.
49. Mironov R.A., Karaksina E.V., Zabezhailov A.O., Shaposhnikov R.M., Churbanov M.F. and Dianov E.M. (2010) Mid-IR luminescence of Cr2+ : II–VI crystals in chalcogenide glass fibres. *Quantum Electronics* 40(9), 828–829.
50. Smectala F., Le Foulgoc K., Le Neindre L., Blanchetiere C, Zhang X.H. and Lucas J. (1999) TeX-glass infrared optical fibers delivering medium power from a CO_2 laser. *Optical Materials* 13, 271–276.
51. Sanghera J., Aggarwal I., Busse L., Pureza P., Nguyen V., Miklos R., Kung F. and Mossadegh R. (1995) Development of low loss IR transmitting chalogenide glass fibers. *SPIE* 2396 (5)71–77.
52. Le Coq D., Boussard-Pledel C., Fonteneau G., Pain T., Bureau B. and Adam J.-L. (2003) A new approach of preform fabrication for chalcogenide fibers. *J. Non-Crystal. Solids* 326–327, 451–454.
53. Churbanov M.F., Shaposhnikov R.M., Snopatin G.E., Shabarov V.V. and Plotnichenko V.G. (2005) Flow of viscoplastic arsenic selenide melt in circular-cylindrical channels. *Inorganic Materials* 41(11), 1301–1306.
54. Churbanov M.F., Shaposhnikov R.M., Shabarov V.V., Snopatin G.E. and Plotnichenko V.G. (2006) Flow of a viscoplastic arsenic selenide melt in annular channels. *Inorganic Materials* 42(2), 215–219.
55. Shiryaev V.S., Adam J.-L., Zhang X.H., Boussard-Pledel C., Lucas J. and Churbanov M.F. (2004) Infrared fibers based on Te-As-Se glass system with low optical losses. *J. Non-Cryst. Solids* 336, 113–119.
56. Troles J., Niu Y., Duverger-Arfuso C., Smectala F., Brilland L., Nazabal V., Moizan V., Desevedavy F. and Houizot P. (2008) Synthesis and characterization of chalcogenide glasses from the system Ga-Ge-Sb-S and preparation of a single-mode fiber at 1.55 μm. *Mat. Res. Bull.* 43, 976–982.
57. Troles J., Shiryaev V., Churbanov M., Houizot P., Brilland L., Desevedavy F., Charpentier F., Pain T., Snopatin G. and Adam J.-L. (2009) Preparation of low losses $GeSe_4$ fibers. *Optical Materials* 32, 212–215.
58. Churbanov M.F., Shiryaev V.S., Skripachev I.V., Snopatin G.E., Pimenov V.G., Smetanin S.V., Shaposhnikov R.M., Fadin I.E., Pyrkov Yu.N. and Plotnichenko V.G. (2002) High-purity $As_2S_{1.5}Se_{1.5}$ glass optical fibers. *Inorganic Materials* 38(2), 193–197.
59. Dianov E.M., Plotnichenko V.G., Pyrkov Yu.N., Smol'nikov I.V., Koleskin S.B., Devyatykh G.G., Churbanov M.F., Snopatin G.E., Skripachev I.V. and Shaposhnikov R.M. (2003) Single-Mode As–S Glass Fibers. *Inorganic Materials* 39(7), 627–630.
60. Dianov E.M., Plotnichenko V.G., Pyrkov Yu.N., Smol'nikov I.V., Koleskin S.B., Devyatykh G.G., Churbanov M.F., Snopatin G.E., Skripachev I.V. and Shaposhnikov R.M. (2003) Single-mode As–S glass fibers. *Inorganic Materials* 39(7), 627–630.
61. Shiryaev V.S., Boussard-Pledel C., Houizot P., Jouan T., Adam J.-L. and Lucas J. (2006) Single-mode infrared fibers based on Te-As-Se glass system. *Mat. Sci. Eng. B* 127(2–3), 138–143.
62. Churbanov M.F., Shiryaev V.S., Gerasimenko V.V., Pushkin A.A., Skripachev I.V., Snopatin G.E. and Plotnichenko V.G. (2002) Stability of the optical and mechanical properties of chalcogenide fibers. *Inorganic Materials* 38(10), 1063–1068.

2
Structure of chalcogenide glasses characterized by nuclear magnetic resonance (NMR) spectroscopy

B. BUREAU, University of Rennes, France

DOI: 10.1533/9780857093561.1.36

Abstract: Solid state NMR is a matchless tool to probe amorphous material. It provides direct information on the local and the medium range order in complex networks. Unfortunately, chalcogenide glasses do not contain nuclei which are easily probed by NMR due to their low concentration, low sensitivity, large quadrupole moments and/or long relaxation times. Nevertheless, throughout the last decade, increasingly sophisticated NMR sequences have been implemented enabling a better understanding of the structure of these exciting materials. In this chapter, an overview is presented of the NMR results obtained on chalcogenide glasses. Most of them were obtained from ^{77}Se spectroscopy, but results based on ^{125}Te, ^{75}As, ^{73}Ge and ^{71}Ga are also reported.

Key words: solid state NMR, chalcogenide, glass, structure, ^{77}Se.

2.1 Introduction

Solid state nuclear magnetic resonance (NMR) is an efficient and matchless tool to investigate the structure of non-crystalline compounds. It has been widely used during the past 30 years to study the local and medium range order in glass structures. Indeed, with the arrival in the 1980s, of a new generation of modern Fourier transform spectrometers, equipped with magic angle spinning (MAS) facilities, NMR provides spectra with high resolution even in disordered solid state materials. Generally speaking, solid state NMR is a structural tool which probes locally and selectively the neighborhood of a selected element by soliciting its nucleus. Exhaustive descriptions of the resonance phenomena are given in many books [1–7]. Many nuclei possess a nuclear magnetic spin, sensitive enough to the magnetic resonance phenomena. Basically, their abilities to respond to the NMR solicitation are governed by their natural abundance and their sensitivity compared to ^{1}H.

From the mid-1980s, most NMR structural works have dealt with oxide glasses due to their high technological and industrial importance in our everyday lives. Indeed, most of the glasses produced in the world are silicate glasses, which have been used for about 4000 years. Silicon is the most abundant element in the earth's crust and, historically, ^{29}Si has probably received the greatest amount of attention due to its ubiquity in solid state materials: silicates, zeolites, and glasses [3]. Thus, from the mid-1980s, many studies have been carried out that established precise relationships between the structural data and the NMR results. Moreover, silicate and phosphate glasses contain several nuclei playing a strategic structural role in the network and easy to probe by NMR: ^{29}Si, ^{31}P, ^{27}Al, ^{11}B, for instance. So, essential results were obtained on the dimensionality of the structural network, the degree of reticulation of the polyhedra, the role of the elements (glass former or modifier) in these materials. A complete glass structural description then became achievable based only on solid state NMR. For a more exhaustive review of these glasses, readers are referred to Ref. [7], p. 230.

More recently, during the 1990s, some authors showed an interest in more exotic glasses. For examples, because of their potential application in optics, fluoride glasses were studied by ^{19}F NMR spectroscopy. Thus, some interesting results were obtained on the fluorozirconnate glasses [8, 9] and also on transition metal fluoride glasses using ^{19}F [10–12], ^{69,71}Ga [13, 14] and ^{207}Pb [15] as local probes. To get relevant information on the network, correlations have had to be established between the chemical shift of the lines and the structural neighborhood of the selected element. To achieve this goal, a database of related crystalline compound spectra had to be recorded.

The same strategy has been followed more recently on chalcogenide glasses. Indeed, chalcogenide glasses are also developed for their optical properties, in particular in the mid-infrared range. As such, their structural study by NMR is in continuity with oxide and fluoride ones. Chronologically, the first papers devoted to NMR in chalcogenide glass were published in the early 2000s. Indeed, the very first paper on glass containing selenium and probing the ^{77}Se was presented during the 1990s. These pioneer studies concerned the P-S and P-Se systems which are very exotic glasses, but are of no real technological interest. Nevertheless, these papers showed the feasibility of work based on ^{77}Se nucleus in solid state disordered materials [16, 17]. The 'standard' chalcogenide glasses are built up with at least one chalcogen, meaning sulfur (S) selenium (Se) and/or tellurium (Te), associated with other elements close to them in the periodic table (As, Ge, Sb, Ga, I,). Compared to the previous families of glasses, chalcogenides are very covalent systems. Thus, their structures are commonly described as a network of covalent bonds obeying the 8-N rule. Nevertheless, no experimental result had clearly described the actual organization of their network.

In order to develop their applications, to improve their properties and to better understand their physical behavior, a good description of their structure is needed. However, while silicate and phosphate glasses have been widely studied, chalcogenides and, in particular, selenides have received little attention due to the low NMR sensitivity of ^{77}Se which is the most common and easiest nucleus to probe in these systems. Nevertheless, high-quality spectra can be obtained using extended acquisition time up to several days. Finally, ^{77}Se NMR spectroscopy appeared to be a well-appropriated tool to probe the structure of chalcogenide glasses. Thus, various glass systems have already been investigated by ^{77}Se such as As_xSe_{1-x} [18–23], Ge_xSe_{1-x} [24–29] or Te_xSe_{1-x} [30], giving crucial information on the connectivity between Se atoms. In addition, some pioneers and preliminary studies have opened the way to the study of more unusual and tricky nuclei: ^{75}As [31–33], ^{125}Te [30, 34] and ^{73}Ge [35].

In this chapter, we present a review of these recent NMR results, which are very helpful to get a real view of the vitreous network of these exciting materials. The chapter is divided into two main parts. The first is devoted to ^{77}Se NMR spectroscopy, and the second to other more exotic nuclei.

2.2 The ^{77}Se nuclear magnetic resonance (NMR) spectroscopy in chalcogenide glasses

2.2.1 The As–Se binary system

Se-rich region

The works published on this system are probably the most sophisticated and the best example of what solid state NMR could bring to our view of the chalcogenide glass structure. Indeed, the very first ^{77}Se NMR experiments on this system were carried out about 10 year ago by. Rosenhahn *et al.* at high temperature on molten samples [18]. The study showed a tendency to chemical ordering for the network in which the $AsSe_3$ pyramids are randomly distributed.

The first spectra on solid state samples were obtained a couple of years later [19]. ^{77}Se 15 kHz magic angle spinning NMR spectra for five vitreous materials were compared. Three different line positions appear for this glass family at about 850 ppm, 550 ppm and 380 ppm on the chemical shift scale at room temperature. Figure 2.1 shows the ^{77}Se NMR spectra of the Se, $AsSe_9$, $AsSe_3$, and As_2Se_3 samples in the Se-rich region ($0 < x < 0.4$). A comparison with the parent crystalline phases has been proposed. The pure Se glass shows a single peak at 850 ppm associated with the Se-**Se**-Se (**a**) environment while the stoichiometric As_2Se_3 shows a single peak centered at 380 ppm associated with the As-**Se**-As (**c**) environment. Instead, the intermediate composition $AsSe_3$ shows a wide peak centered at 550 ppm which indicates a predominant distribution of Se-**Se**-As (**b**) environment, but also suggests the simultaneous existence of the two other environments. NMR is basically a quantitative structural technique. Then, the reconstruction of the spectra provides the relative intensities of the line which proves the validity of the intuitive model, the chain crossing model (CCM), that predicts a homogeneous distribution of the $AsSe_{3/2}$ pyramids. Indeed, the selenium

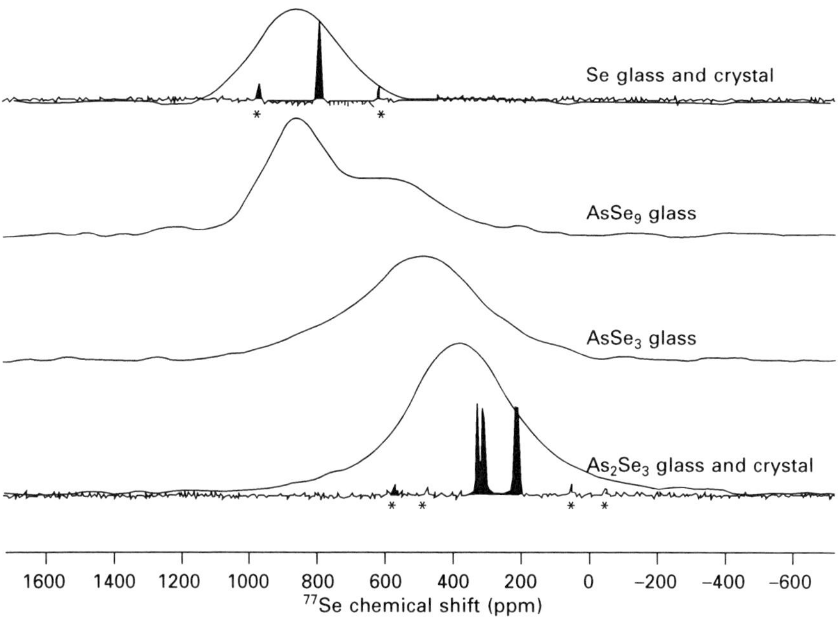

2.1 ^{77}Se NMR spectra of Se-rich glass composition: Se, $AsSe_9$, $AsSe_3$, and As_2Se_3 shown with the parent crystalline phases.

chains cross-linking arsenic atoms become shorter and shorter until they eventually disappear from the totally reticulated As_2Se_3 glass.

In order to go further and refine the model, a new campaign of ^{77}Se NMR measurements has been carried out [21]. Generally speaking, it appeared that the low S/N ratio combined with the weak resolution is a limiting factor. One of the most promising ways to probe the glassy network relies on the use of through-bond correlation experiments based on J-coupling, such as the J-Resolved or refocused INADEQUATE experiments. However, these multi-dimensional experiments suffer from the lower efficiency for low abundant spins, making their application to selenide glasses rather difficult. To overcome these difficulties, the ^{77}Se spectra have been acquired with CPMG detection (train of rotor-synchronized p pulses during the detection of ^{77}Se magnetization). A comparison between the Hahn echo spectrum and the CPMG spectra is presented in Fig. 2.2, and a clear sensitivity enhancement is observed. However, the spectra cannot be rigorously superimposed, as large CSA and offset effects may prevent the perfect refocusing of the ^{77}Se magnetization during the CPMG pulse train. Hence, CPMG spectra have only been used to confirm the presence of the expected NMR lines. To quantify the proportions of **a**, **b** and **c** moieties, the Hahn echo spectra have been recorded. A whole echo has been acquired and Fourier transformed in order to increase the S/N ratio and to obtain a pure absorption line shape.

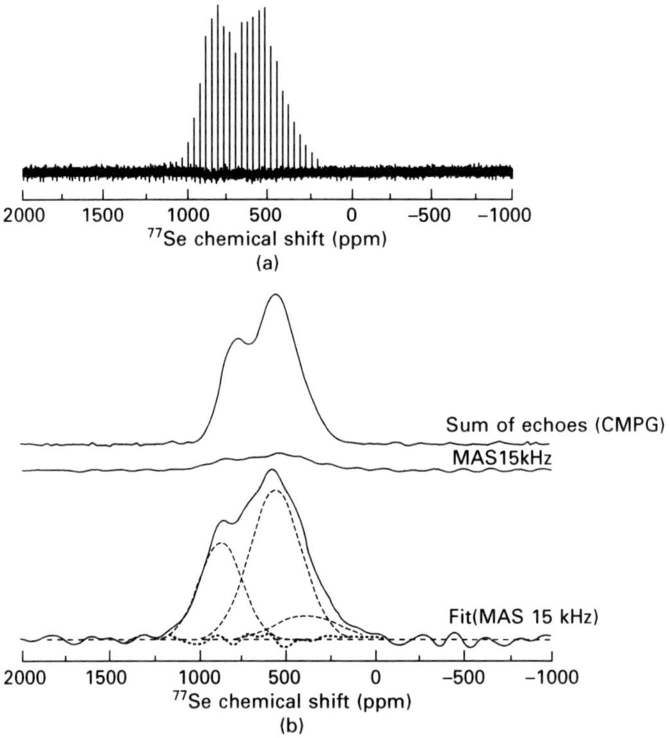

2.2 (a) Spikelet spectrum of the $AsSe_{4.5}$ glass obtained with the CPMG acquisition and a 22 kHz MAS rate at 9.4T. (b) The corresponding MAS–NMR spectra obtained either with a spin echo experiment (MAS 15 kHz and 7 T) or with the sum of the CPMG echoes displayed with a similar noise level. The spin echo spectrum has been fitted with three Gaussian lines (from [21]).

Due to the low sensitivity, the numbers of scans ranged between 2,048 and 5,120 depending on the glass composition and the Se concentration (from 17 to 42 h experimental time).

As an example, the 2D spectra obtained for the As_xSe_{1-x} glasses are presented in Fig. 2.3. The direct dimension is equivalent to the dimension of the 1D spectra previously discussed. The As_2Se_3 spectrum exhibits a single line located at ~377 ppm assigned to the **c** units. In the indirect dimension, the line does not show any splitting in full agreement with the absence of Se–Se bonds and thus Se–Se ^{1}J-coupling. This result confirms the hypothesis that this glass is built up with $AsSe_3$ pyramids directly linked by one atom of selenium. In the other CPMG spectra, the **a** and **b** lines are broadened in the indirect dimension by what appears to be an effect of the ^{1}J (^{77}Se–^{77}Se) coupling.

For $AsSe_3$, the 2D CPMG spectrum exhibits at least two contributions.

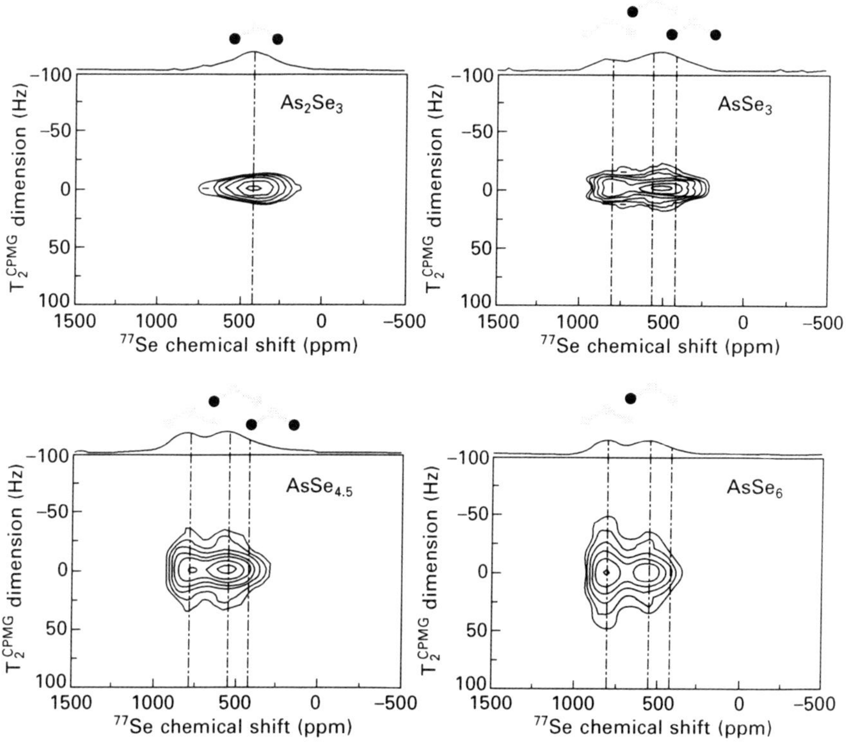

2.3 Two-dimension CPMG spectra of As_2Se_3, $AsSe_3$, $AsSe_{4.5}$, and $AsSe_6$ glasses (from [21]).

This result was not expected because in the CCM, the network is supposed to be constructed with $AsSe_3$ pyramids connected to each other by Se–Se bridges, thereby leading to an NMR spectrum with a single **b** line. Clearly, there is also a contribution of **a** entities together with **c** bridging selenium in the $AsSe_3$ network. The Hahn echo spectrum confirms the co-existence of the three contributions, indicating that the chain length is not uniform and that the presence of each **c** has been compensated by one **c** to ensure the consistency of the results with the formulation of the glass, and corresponding to the following equilibrium:

$$2\text{AsSeSeAs} \leftrightarrow \text{AsSeAs} + \text{AsSeSeSeAs}$$

Indeed, such a model considers AsSeSeAs as the most probable chain (as in the CCM); however, some disorder exists, and two chains of length N can be transformed into two chains of lengths N–1 and N+1.

The $AsSe_{4.5}$ and $AsSe_6$ glass CPMG spectrum clearly show that the three lines are also detected; nevertheless, their integrated intensities measured from the Hahn echo reconstruction are not in full agreement with the above reasoning. Indeed, it has been shown that T_1 decreases with increasing As content. Therefore, in Se-rich glasses, the recovery delay should be longer to allow the Se-**Se**-Se spins to recover their magnetization, leading to proportions lower than expected. Moreover, one can also infer that a supplementary mechanism may be at work, involving As coordination higher than three when a large excess of Se is present.

Finally the 2D CPMG results confirmed the previous assignments as **a**, **b**, and **c**, and the quantitative Hahn echo enabled us to refine the structural description of the $AsSe_3$ glassy network and show that the agreement with the CCM is not perfect, in particular for the $AsSe_3$ composition. Indeed the best description of the Se-rich composition is probably not far from the CCM but assuming also some distribution of the chain length between two As. Moreover, this work also showed that T1 is a key parameter that has to be large enough to get some quantitative information from ^{77}Se solid state NMR experiments.

The As-rich region

Two recent papers have been published on the general physical properties of the As_xSe_{1-x} from pure Se to As_3Se_2 [22, 23]. This last glass is the richest composition in As. The structural part of the study is based mainly on ^{77}Se NMR results. The As-rich samples in Fig. 2.4 show a majority of As-**Se**-As environment along with a surprisingly large fraction of Se-**Se**-As and a notable contribution at 850 ppm. This last contribution could have been assigned to Se-**Se**-Se but that would not be consistent with low Se content of these compositions. Indeed, the NMR spectra of crystalline As_4Se_4 and As_4Se_3 show a strong signal in the same domain which suggests the presence of these molecular components in the glass. The As_4Se_4 cages contain four distinct Se with specific crystallographic features, each of them giving rise to a specific NMR line. Similarly, each As_4Se_3 cages contains three Se, two of them being crystallographically equivalent. These two lines are denoted by arrows in Fig. 2.4; the other lines denoted by stars correspond to the spinning side bands due to the magic angle spinning and will average out in the glass. These results emphasize the ability of ^{77}Se NMR to accurately resolve structure information in solid state materials. The presence of these As-rich molecules is also consistent with the relative intensities of the NMR peaks. Indeed, the presence of large fractions of Se-**Se**-As fragments requires the presence of As-rich clusters in order to compensate for the Se-poor stoichiometry.

Finally the NMR interpretation indicates that the structure follows the CCM in the range of $0 < x < 0.40$ (x in As_xSe_{1-x}), but subsequently loses its network reticulation and generates As-rich cage molecules in the range of $0.40 < x < 0.6$. The reconstructed NMR spectra of the AsSe and As_3Se_2

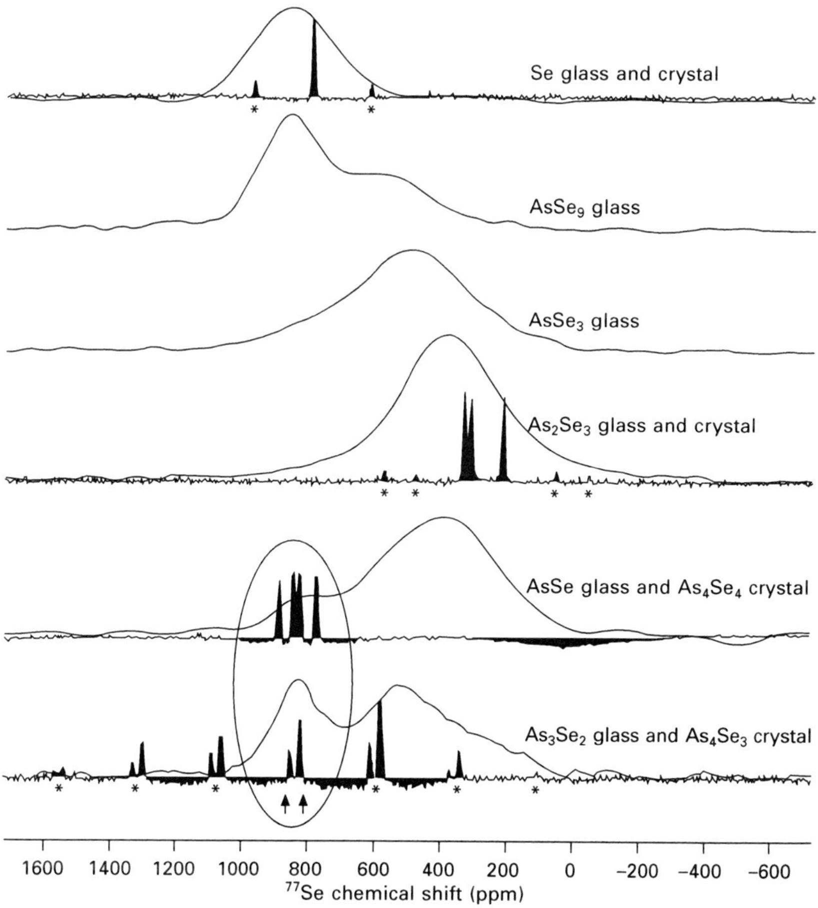

2.4 ^{77}Se NMR spectra of different crystalline and glassy phases of the As_xSe_{1-x} system, including the As-rich compositions. The As_4Se_4 and As_4Se_3 crystalline phases give rise to lines at a surprising position around 800 ppm. It appears that this range of chemical shift corresponds also to molecular cages built with As and Se (adapted from [22]).

glasses indicated that 20% and 40%, respectively, of Se atoms are involved in such molecules in the glass structure.

These NMR results are fully consistent with Raman spectroscopy which are also shown in Refs [22 and 23]. They also clearly exhibited the presence of very sharp features indicative of As_4Se_4 and As_4Se_3 molecular units. Moreover, Raman spectra also showed a sharp and intense peak growing continuously with increasing arsenic content, which is characteristic of amorphous elemental arsenic.

Finally, these results indicates a dramatic structural change at $x = 0.4$, beyond As_2Se_3 and confirmed previous work [36]. Indeed, the glass network evolves from a 1D chain structure to a 2D pyramidal network in the range $0 < x < 0.4$ more or less following the CCM, and then transitions to a lower dimension structure composed of a pyramidal backbone containing an increasing number of 0D molecular inclusions in the range $0.4 < x < 0.6$. This structural description is in full agreement with the evolution of all the physical and mechanical properties which exhibit a single sharp extremum at $x = 0.4$.

2.2.2 The Ge–Se binary system

Se-rich compositions were studied by ^{77}Se about ten years ago [24]. Figure 2.5 displays the recorded spectra. Despite the lack of resolution on these very first acquisitions, only two main broad lines are visible in static mode respectively at 400 and 800 ppm. In this paper, it appeared also that the resolution did not increase significantly in magic angle spinning mode (spinning speed 15 kHz, magnetic field 7 T). In particular, compared to As_xSe_{1-x}, no line is visible at an intermediate value of chemical shift around 600 ppm. Both lines were attributed to Se-**Se**-Se neighborhood (at 800 ppm) on one hand, and to Se bonded to one or two Ge (line at 400 ppm) on the other. Indeed, due to the broadness of the experimental NMR peaks, it is conceivable that an additional peak could be hidden between the two main lines observed in the room temperature spectra. The relative intensities of both lines were evaluated from the reconstruction of the spectra. They showed that, except for $GeSe_2$, the relative intensity of the Se-**Se**-Se lines is high, in agreement with a model based on the existence of domains rich in Se chains on the one hand, and rich in interconnected $GeSe_4$ tetraedra on the other. For example, in $GeSe_4$, the integrated intensity of the Se-**Se**-Se line at 800 ppm is not far from 50%, meaning that the glass composition can be rewritten as $(GeSe_2)Se_2$, $GeSe_2$ corresponding to domains of $GeSe_4$ tetrahedra connected by corner or edge. This view was confirmed later [25], where high temperature NMR was performed on the Ge_xSe_{1-x} samples above their respective T_g. Increased structural mobility above T_g should result in line narrowing and help provide a greater resolution for the identification of experimental lines. The series of spectra reveal a significant narrowing of the Se-**Se**-Se line with increasing temperature. On the other hand, the line at 400 ppm does not exhibit a significant change in width. This behavior is consistent with the increased mobility of Se chain fragments while the over-constrained domains remain rigid. It is worth noting that on these spectra, no additional significant peaks appear at an intermediate position attributable to selenium bonded to one Ge and one Se.

More recently, high-resolution ^{77}Se MAS NMR spectroscopy has been conducted at 11.7 T and high spinning speed [26]. Based on the $GeSe_2$

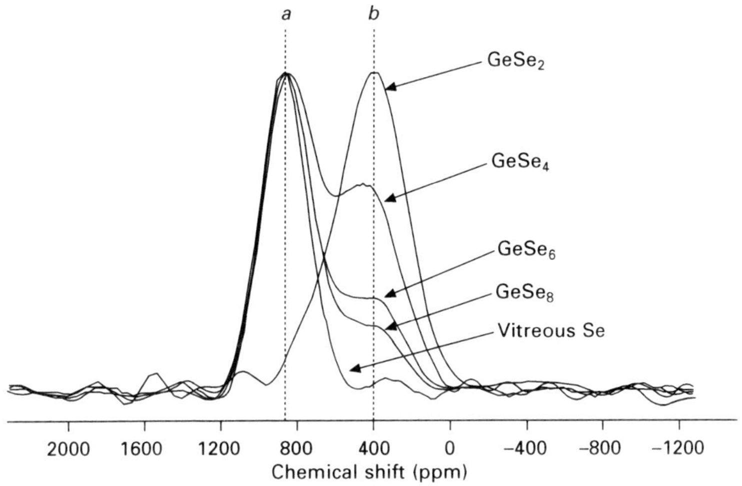

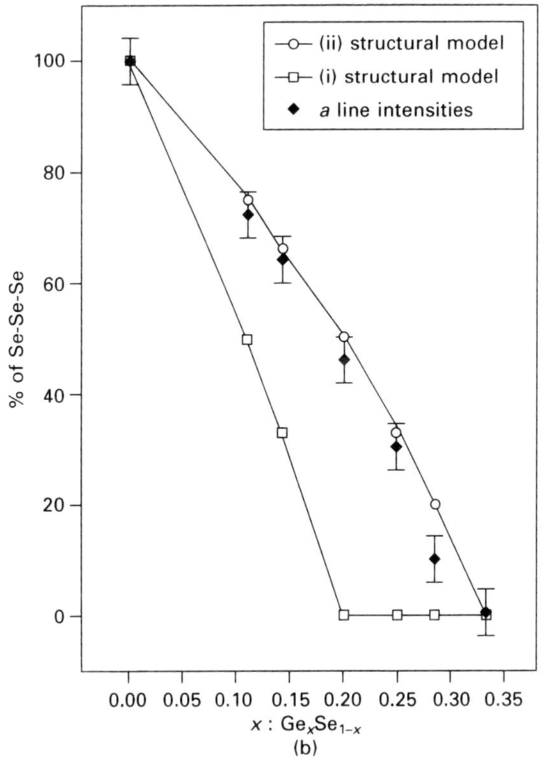

2.5 (a) ^{77}Se spectra of Ge_xSe_{1-x} glasses showing two broad lines. The *a* line around 800 ppm is assigned to Se-Se-Se sequences. (b) Experimental compositional variation of the relative fraction of this Se-Se-Se site compared to the predictions of two models. (i), resp. (ii), structural model corresponds to the chain-crossing model, resp. clustered model. These results are extracted from [24].

crystalline phase study, the authors observed four distinct Se environments. They attributed them respectively to Se-**Se**-Se and Ge-**Se**-Se linkages as well as Ge-**Se**-Ge sites where the Se atoms are shared by two $GeSe_4$ tetrahedra in either corner-sharing or edge-sharing configuration. Thanks to the use of magic angle spinning and longer relaxation time, they succeed in measuring more accurately the integrated intensities of the Se neighborhoods. In particular, the relative weight of Se-**Se**-Se would rather be around 35% instead of 50% leading to a randomly interconnected network of tetrahedra and chains of Se atoms. This description is also more or less corroborated by *ab initio* calculation proposed by different authors on these glasses [28, 29]. In all these cases, the above structural models are very far from the ideal description of the vitreous network, consisting in tetrahedra linked to each other by Se–Se bridges for $GeSe_4$, for instance.

Very recently, as has been achieved previously for As_xSe_{1-x} [21], some CPMG acquisitions have been carried out under static mode (no spinning) [27]. For the first time, the Se-deficient binary Ge_xSe_{1-x} glasses have been investigated. The resolution remains weak, but the authors propose the following assignment for the three observed lines: Ge-Ge-**Se**-Ge-Ge at 136 ppm, Ge-**Se**-Ge (corner shared) at 380 ppm, Se-**Se**-Ge and Ge-**Se**-Ge (edge shared) at 600 ppm. This assignation partially contradicts previous work [26] where edge-shared type Se sites were suggested to appear around 150 ppm. In addition, the Se-excess glasses were also revisited. Compared to [24], the reconstruction is refined by the introduction of the above-mentioned contribution at 600 ppm. It leads to a network description which is in between the chemically ideally organized model, the CCM, and a clustered model. Figure 2.6 displays these new CPMG spectra together with the relative intensities of the Se-**Se**-Se lines which are positioned in between both models.

At this stage, and as brought out from the above studies, supplementary experiments are needed to definitively elucidate the attribution of the NMR lines. Thus, some two-dimensional acquisitions with the same strategy used for studying As_xSe_{1-x} glasses are being implemented. These works are ongoing, taking care also of the relaxation times which play a great role in the final results.

2.2.3 The Te–Se binary system

For final applications, this glassy system is clearly not as important as both the previously described systems. Nevertheless, in view of future works on tellurium-based glasses, it is interesting to probe this basic system, given the way the Se and Te atoms are spread over the network. Indeed, both chalcogens are expected to be two-fold coordinated and are assumed to build a network constituted of chains within distributed Se and Te atoms.

The ^{77}Se 15 kHz MAS NMR spectra for six vitreous materials exhibit three types of line positions at about 880 ppm, 710 ppm, and 510 ppm [30]. Following the strategy assumed for the above binary systems, these lines were attributed respectively to Se-**Se**-Se, Se-**Se**-Te and Te-**Se**-Te. Moreover, from the spectral reconstruction, it appeared that the evolutions of the integrated intensities of the three lines are in agreement with the above assignments: a higher Te content in the glass composition induces a higher intensity of the line at 710 ppm, and then at 510 ppm. Then, different structural models were considered: clustering models (Se and Te do not mix with each other within the chains), homogeneous distribution (privileging heteropolar bonds)

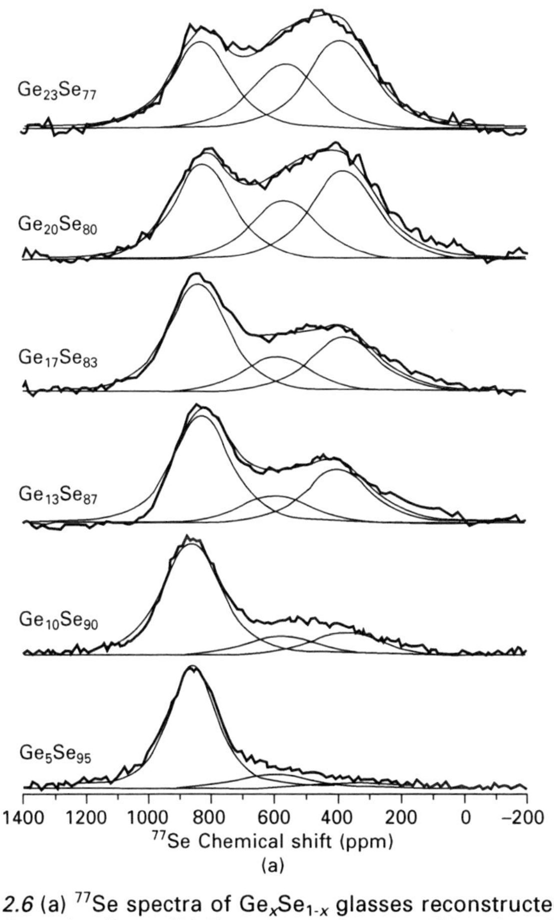

2.6 (a) ^{77}Se spectra of Ge_xSe_{1-x} glasses reconstructed with three Gaussian lines. (b) Experimental compositional variation of the relative fraction of this Se-Se-Se site compared to the predictions of two models. (i), resp. (ii), structural model corresponds to the chain-crossing model, resp. clustered model. These results are extracted from [27].

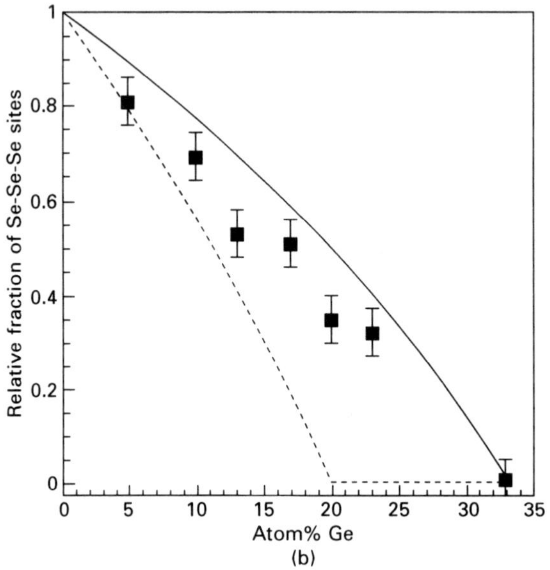

2.6 Continued

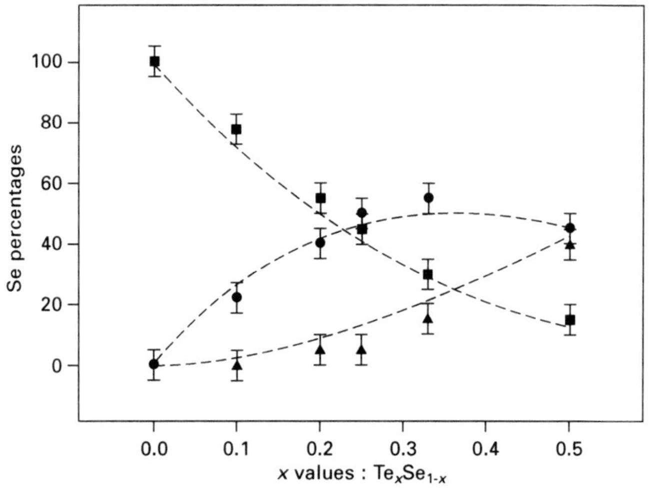

2.7 Compositional variation of the relative fraction of Se-Se-Se, Se-Se-Te and Te-Se-Te sites as obtained from the reconstruction of the ^{77}Se spectra and corresponding prediction of the best model, i.e., a random distribution privileging heteropolar bonds (see [30]).

and a model based on a pure random distribution of Se and Te. Finally, a more sophisticated model had to be assumed, which is a kind of random distribution wherein heteropolar bonds have to be privileged. As shown in Fig. 2.7, the integrated intensities of the lines fit well with the intensities expected in the framework of this model.

2.3 Other nuclei: ^{125}Te, ^{75}As, ^{73}Ge, ^{71}Ga

2.3.1 The ^{125}Te NMR in chalcogenide glasses

The ^{125}Te abundance is 7% and its sensitivity about 2.2×10^{-3}, not very different compared to ^{77}Se. Moreover, it is characterized by a long spin-lattice relaxation time T_1. Overall, the main issue is associated with the large electronic cloud of the tellurium which spreads out the chemical shift over a very wide range. These properties make the acquisition of high resolution ^{125}Te solid state NMR spectra challenging. Thus, Fig. 2.8 displays the ^{125}Te spectrum of the crystalline phase of pure tellurium, showing large chemical shift anisotropy, equal to about 1,200 ppm [30]. ^{125}Te spectra were recorded in the above-mentioned Te-Se binary systems [30]. The lines were very broad and unresolved and did not permit drawing any conclusion regarding the structure of the glasses.

More recently, three interesting papers have been published including ^{125}Te NMR [34, 37, 38] used as tool to investigate telluride glasses which are phase change materials for rewritable data storage applications: Ge-Sb-Te and Ge-As-Sb-Te systems, for example. Once again, the NMR lines are very broad; nevertheless, through the study of various crystalline phases, in [34] the authors succeeded in proposing a useful scale of chemical shifts for this nucleus. Figure 2.9 shows different ^{125}Te spectra collected in crystalline phases. It was shown that the chemical shift decreases with increasing coordination and a distinct range of chemical shifts is established versus the coordination number.

2.3.2 The ^{75}As NMR in chalcogenide glasses

^{75}As is a really tricky nucleus to probe. Indeed, it possesses a very large quadrupolar constant, making ^{75}As NMR experiments very challenging. Thus, very few articles have been published on this nucleus [31–33]. On the other hand, ^{75}As can also be probed by nuclear quadrupole resonance (NQR). For instance, Taylor *et. al*. proposed a study on As_2Se_3 and As_2S_3 crystalline and vitreous phases by coupling NMR and NQR [31]. NMR experiments needed a very high field spectrometer (around 27T). Due to the quadrupolar effect, the conclusions were that the symmetry of the arsenic sites are nearly axial. In the glass, the neighborhoods of the arsenic atoms keep this symmetry, and the lines are just slightly smoothed by the structural disorder as shown in Fig. 2.10.

The same strategy has been implemented more recently to probe the arsenic sites of more complex glass belonging to the Ge-As-Se system [33]. In $Ge_2As_2Se_7$, the arsenic sites exhibit an axial symmetry, whereas in $Ge_2As_2Se_5$ the distribution of the electric field gradient becomes highly asymmetric. The authors conclude that for the first composition, the arsenic

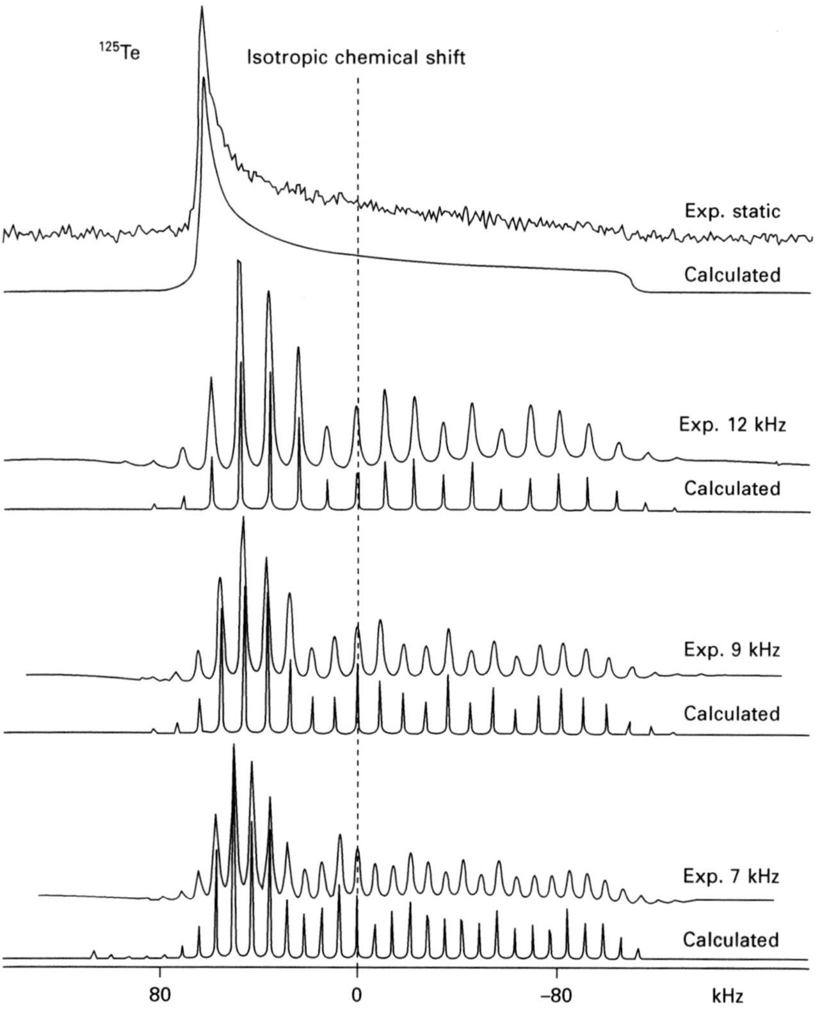

2.8 ^{125}Te spectra in pure tellurium crystalline phase with different magic angle spinning speed [30]. For this nucleus, the large electronic cloud also gives rise to large chemical shift anisotropies and widespread isotropic chemical shifts.

atoms form pyramidal structural units, while with the second composition, homopolar As–As bonds explain the symmetry breaking.

Thus, despite the lack of resolution and the difficulty in carrying out such experiments, ^{75}As NMR can provide a fruitful (additional) point of view on the structural description.

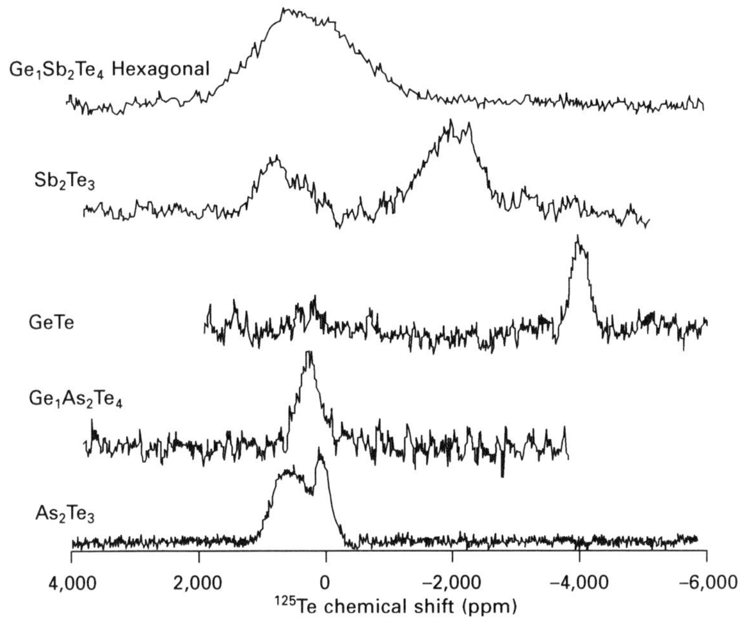

2.9 ^{125}Te spectra in different crystalline phases providing a relevant calibration of the chemical shift with the tellurium neighborhood [34].

2.3.3 The ^{73}Ge NMR in chalcogenide glasses

As for arsenic, germanium plays a crucial structural role at the center of the polyhedra in chalcogenide glasses. In particular, it would be essential to probe ^{73}Ge in order to check its coordination numbers. Nevertheless, ^{73}Ge NMR experiments are experimentally challenging because it is a low gamma nucleus and it possesses a large quadrupole moment. To our knowledge, only one paper has been published on ^{73}Ge in chalcogenide glasses [35]. This work was carried out on the Ge_xSe_{1-x} system using a high field spectrometer (19.6T) and QCPMG sequences. The main conclusion deals with the observation of a broad line superimposed on a sharper one as shown in Fig. 2.11. This last contribution has been attributed to the $GeSe_4$ tetrahedra which are supposed to be highly symmetric, whereas the broad line is assigned to $GeSe_{4-n}Ge_n$ tetrahedra in which homopolar Ge–Ge bonds break the tetrahedral symmetry around Ge, and, thus, generate much larger quadrupolar interactions.

2.3.4 The ^{71}Ga NMR in chalcogenide glasses

^{71}Ga also possesses a quadrupole moment but a little smaller than for ^{73}Ge and ^{75}As nuclei. In a recent paper, ^{71}Ga solid-state NMR spectra recorded

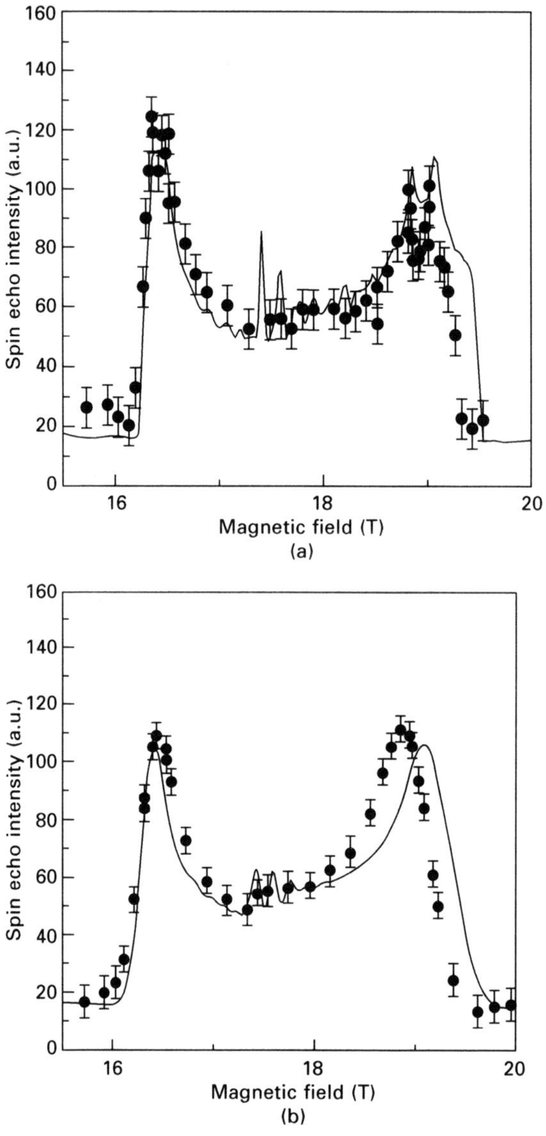

2.10 ^{75}As spectra collected in the crystalline (a) and vitreous (b) phases of As_2Se_3 [31]. Due to its large quadrupolar constant, ^{75}As is highly sensitive to the symmetry of the site occupied by As. This comparison clearly showed that, in the amorphous phase, As keeps its axial neighborhood due the $AsSe_3$ pyramids.

on chalco-halide glasses and glass ceramics were presented [39]. Few works in the literature deal with ^{71}Ga solid state NMR in amorphous materials and only one, to the best of our knowledge, has been reported on chalcogenide

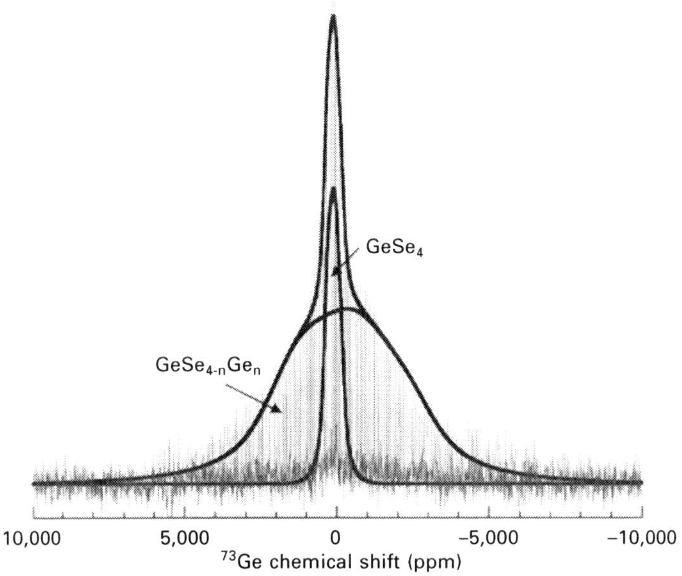

2.11 ^{73}Ge spectrum collected in $GeSe_4$ [35].

glasses [40]. The NMR lines of quadrupolar nuclei (^{27}Al, ^{23}Na, ^{71}Ga) in disordered solids often exhibit an asymmetric shape tailing towards the high field side. Independent Gaussian distributions of quadrupolar parameters are often assumed to reconstruct the line shape in the static or in the magic angle spinning mode. This method is convenient to obtain the relative intensity of each contribution, but its physical basis is not justified. On the other hand, the Czjzek model accounts better for the experimental spectra than the Gaussian approximation, while being based on clear assumptions often discussed in the literature [41–43].

The Czjzek distribution is especially appropriate for an efficient reconstruction of quadrupolar nucleus NMR spectra in glasses whose chemical bonds are dominated by ionic interactions, typically oxide and halide glasses. Indeed, ionic bonding is generally associated with higher coordination numbers around the quadrupolar probe with a somewhat more isotropic local arrangement than in covalent materials. In the case of chalcogenide glasses, gallium is four-fold coordinated and involved in directional covalent chemical bonds. The first neighbors of Ga impose a preferential local symmetry which explains why the Czjzek-type distribution cannot account for the observed line shape. Very recently, an extension of the Czjzek model was proposed by Le Caër *et al.* to retrieve relevant information about the local environment of quadrupolar nuclei in covalent frameworks [44]. This simple extended model was investigated, intending to mimic a well-defined local environment (typically tetrahedra) perturbed by disorder of more remote atoms whose effect is rendered by a Czjzek noise with an adjustable weight. That extension was shown to be helpful to describe some ^{71}Ga NMR spectra of chalcogenide glasses as shown in Fig. 2.12, where the reconstructed spectrum is presented together with the quadrupolar parameter distribution provided by the extended model.

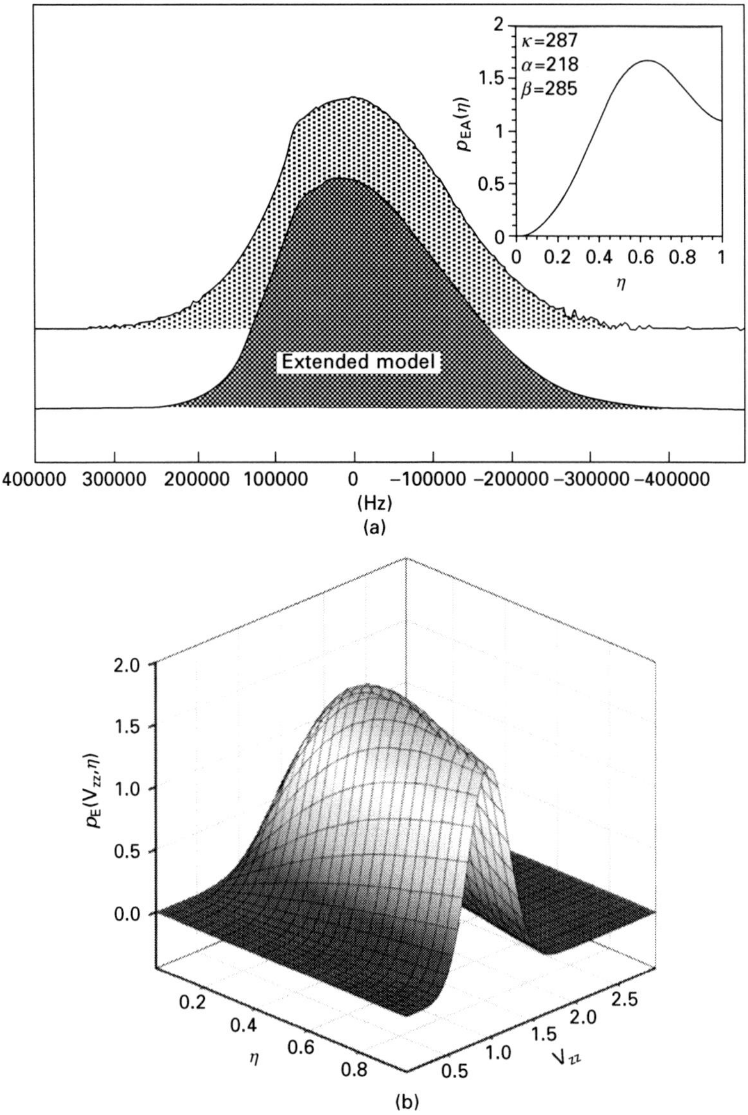

2.12 Broad asymmetric ^{71}Ga lines collected in a GeS_2/Ga_2S_3 glass [39]. The reconstruction has been implemented thanks to an extended Czjzek model [44] providing the V_{ZZ} and η quadrupolar parameter distribution depicted in (b).

2.4 Conclusion

Chalcogenide glasses contain nuclei that are difficult to probe by NMR. Thus, ^{77}Se, ^{125}Te, ^{75}As, ^{73}Ge, ^{71}Ga are characterized by their low abundances, their low sensitivities, their large quadrupole moments and/or their long relaxation time. Nevertheless, for about ten years, solid state NMR has been successfully used to investigate basic chalcogenide glasses, providing a new point of view on the structure of this material. In the recent past, they were described on the basis of chemical assumption rather than from clear experimental data. Then, during the last decade, some disputes arose regarding the description of this glassy network, especially Ge_xSe_{1-x}. Very recently, more sophisticated NMR sequences have been implemented, CPMG and J-resolved, for example, enabling some of these controversies to be resolved. At this time, new ^{77}Se two-dimensional acquisitions are being achieved, taking care also of the relaxation times, which play a large role in the final results. Clearly, we are living in pioneering times. In the future, it is probable that ^{77}Se NMR will become increasingly popular and indispensable in describing more complex chalcogenide glass composition, as has been the case for oxide glasses. Moreover, other nuclei, especially ^{33}S, will also be probed in the future thanks to the new generation of higher field spectrometers.

2.5 References

1. A. Abragham, *Principles of Nuclear Magnetism*, Oxford University Press Oxford (1983).
2. M. Mehring, *High Resolution NMR of Solids*, Springer-Verlag, Berlin (1983).
3. G. Engelhardt and D. Michel, *High Resolution Solid State NMR of silicates and Zeolithes*, Wiley, Chichester (1987).
4. R.R. Ernst, G. Bodenhausen and A. Wokaun, *Principle of Nuclear Magnetic Resonance in One and Two Dimensions*, Oxford University Press, Oxford (1988).
5. C.P. Slichter, *Principle of Magnetism Resonance*, Springer-Verlag, Berlin (1990).
6. K. Schmidt-Rohr and H.W. Speiss, *Multidimensional Solid State NMR and Polymers*, Academic Press, San Diego (1996).
7. K.J.D. MacKenzie and M.E. Smith, *Multinuclear Solid State NMR of Inorganic Materials*, Pergamon, Oxford (2002).
8. S. Sen and R.E. Youngman, *Phys. Rev. B*, 66, 134209 (2002) 1.
9. R.E. Youngman and S. Sen, *Solid State NMR*, 27 (2005) 77.
10. B. Bureau, G. Silly, J.Y. Buzaré and J. Emery, *J. Phys.: Condens Matter*, 9 (1997) 6719–6736.
11. B. Bureau, G. Silly, J.Y. Buzaré and C. Jacoboni, *J. Non-cryst. Solids*, 15 (1999) 129–138.
12. B. Bureau, G. Silly, J.Y. Buzaré and J. Emery, *Chem. Phys.*, 249 (1999) 89–104.
13. B. Bureau, G. Silly, J.Y. Buzaré, C. Legein and D. Massiot, *Solid State NMR*, 15 (1999) 129–138.

14. B. Bureau, G. Silly, J.Y. Buzaré, B. Boulard and C. Legein, *J. Phys.: Condens. Matter*, 12 (2000) 5775–5788.
15. B. Bureau, G. Silly and J.Y. Buzaré, *Solid State NMR*, 15 (1999) 79–89.
16. R. Maxwell, D. Lathrop and H. Eckert, *J. Non-Cryst. Solids*, 180 (1995) 244.
17. P.F. Mutolo, M. Witschas, G. Regelsky, J. Schmedt auf der Guenne and H. Eckert, *J. Non-Cryst. Solids*, 256–257 (1999) 63.
18. C. Rosenhahn, S. E. Hayes, B. Rosenhahn and H. Eckert, *J. Non-Cryst. Solids*, 284 (2001) 1.
19. B. Bureau, J. Troles, M. LeFloch, F. Smektala, G. Silly and J. Lucas, *Solid State Sciences*, 5 (2003) 219.
20. R. Golovchak, O. Shpotyuk, A. Kozdras, B. Bureau, M. Vlcek, A. Ganjoo and H. Jain, *Phil. Mag*., 87 (2007) 4323.
21. M. Deschamps, C. Roiland, B. Bureau, G. Yang, L. Le Pollès and D. Massiot, *Solid State NMR*, 40, 2 (2011) 72–77.
22. G. Yang, B. Bureau, T. Rouxel, Y. Gueguen, O. Gulbiten, C. Roiland, E. Soignard, J.L. Yarger, J. Troles, J.C. Sangleboeuf and P. Lucas, *Phys. Rev. B*, 82 (2010) 195206.
23. G. Yang, O. Gulbiten, Y. Gueguen, B. Bureau, J.C. Sangleboeuf, C. Roiland, E.A. King and P. Lucas, *Phys. Rev. B*, 85 (2012) 144107.
24. B. Bureau, J. Troles, M. Le Floch, F. Smektala and J. Lucas, J. *Non-Cryst. Solids*, 326–327 (2003) 58.
25. P. Lucas, E. King, O. Gulbiten, J.L. Yarger, E. Soignard and B. Bureau, *Phys. Rev. B*, 80 (2009) 1.
26. L. Gjersing, S. Sen and B. G. Aitken, *J. Phys. Chem. C*, 114 (2010) 8601.
27. T. G. Edwards, S. Sen and E.L. Gjersing, *J. Non-Cryst. Solids*, 358, 3, 1 (2012) 609–614.
28. M. Kibalchenko, J.R. Yates, C. Massobrio and A. Pasquarello, *Phys. Rev. B*, 82 (2010) 020202.
29. M. Kibalchenko, J.R. Yates, C. Massobrio and A. Pasquarello, *J. Phys. Chem. C*, 115 (2011) 7755.
30. B. Bureau, C. Boussard, M. LeFloch, J. Troles, F. Smektala and J. Lucas, *J. Phys. Chem B*, 109 (2005) 6130.
31. P. Taylor, P. Hari, A. Kleinhammes, P.L. Kuhns, W.G. Moulton and N.S. Sullivan, *J. Non-Cryst. Solids*, 227–230 (1998) 770–774.
32. E. Ahn, G. Williams and P. Taylor, *Phys. Rev. B*, 74 (2006) 1.
33. E. Mammadov, D. Bobela, A. Reyes, S. Mehdiyeva and P.C. Taylor, *Solid State Com*., 151 (2011) 1459–1462.
34. T.G. Edwards, E.L. Gjersing, S. Sen, S.C. Currie and B.G. Aitken, *J. Non-Cryst. Solids*, 357 (2011) 3036–3041.
35. S. Sen and Z. Gan, *J. Non-Cryst. Solids*, 356 (2010) 1519–1521.
36. D.G. Georgiev, P. Boolchand and M. Micoulaut, *Phys. Rev. B*, 62 (2000) 9228.
37. D.C. Bobela and P.C. Taylor, *J. Non-Cryst. Solids*, 354 (2008) 2665–2668.
38. D.C. Bobela and P.C. Taylor, *Jap. J Appl. Phys*., 47 (2008) 8162.
39. Y. Ledemi, B. Bureau, L. Calvez, M. Le Floch, M. Roze, C. G. Lin, X. H. Zhang, M. Allix, G. Matzen, and Y. Messaddeq, *J. Phys. Chem. B*, 113 (2009) 14574–14580.
40. R.E. Youngman and B.G. Aitken, *J. Non-Cryst. Solids*, 345 (2004) 50–55.
41. G. Czjzek, J. Fink, F. Gotz, H. Schmidt, J. M. D. Coey, J.P. Rebouillat and A. Lienard, *Phys. Rev. B*, 23 (1981) 2513–2530.

42. J.B.D. de Lacaillerie, C. Fretigny and D. Massiot, *J. Magnetic Reson.*, 192 (2008) 244–251.
43. G. Le Caer and R. A. Brand, *J. Phys.: Condens. Matter*, 10 (1998) 10715–10774.
44. G. Le Caër, B. Bureau and D. Massiot, *J. Phys.: Condens. Matter*, 22 (2010) 065402.

3
Mean coordination and topological constraints in chalcogenide network glasses

P. LUCAS, University of Arizona, USA

DOI: 10.1533/9780857093561.1.58

Abstract: This chapter discusses the general ideas of mean coordination and topological constraints that have been successfully applied to chalcogenide glasses in the last few decades. The chapter first reviews the basic principles of mean coordination and the early discovery of the rigidity percolation theory. Experimental confirmations are presented along with the limitations of the model stemming from the ideal network assumption. Modification of the model are reviewed and critically discussed. The chapter then presents recent modeling advances accounting for the temperature dependence of topological constraints.

Key words: mean coordination, topological constraints, rigidity percolation, average coordination number, network glass.

3.1 Introduction

Relatively few inorganic materials have the ability to vitrify upon moderate cooling rates and only three families of inorganic glass-forming solids are generally recognized, namely: oxides, fluorides, and chalcogenides. While oxides and fluorides are usually transparent insulators, chalcogenides belong to the family of amorphous semiconductors and are the focus of many technological applications ranging from microelectronics to photonics (Wuttig and Yamada, 2007; Eggleton *et al.*, 2011).

Chalcogenide glasses are traditionally composed of divalent elements S, Se and Te combined with Ge, As, Sb, Si, P or other neighboring atoms on the periodic table. Two characteristics about these component elements should be readily emphasized. First, they have high atomic masses which generate low energy phonons within the amorphous network and confer wide optical transparency to the glass extending far into the infrared (Lucas, 1999). This property is a defining characteristic of chalcogenide glasses and has been the source of much research for infrared optics applications (Sanghera *et al.*, 2000). Second, these component elements all have a relatively similar electronegativity close to $X = 2$ and consequently form well-defined directional covalent bonds following the 8-N octet rule (Mott and Davis, 1979). The glass structure can then be conveniently described as a network of covalent bonds which lends itself well to modeling and simulations. Indeed, chalcogenide glasses have been the topic of many modeling investigations aimed at establishing some correlations between macroscopic physical properties and the microscopic connectivity of the amorphous network. The most influential of these models is without doubt the Phillips–Thorpe rigidity percolation theory which describes the structure using a single parameter – the mean coordination – defined as the average number of bonds per atom in the network (Thorpe, 1983). This model has proven successful in explaining trends in macroscopic properties such as glass fragility and has recently been the basis for further advances in structural modeling (Tatsumisago *et al.*, 1990; Gupta and Mauro, 2009). This chapter reviews the principles of the mean coordination model, its limitations and success in describing physical properties trends, as well as novel developments accounting for the temperature dependent nature of constraints that lead to the prediction of viscous flow behaviors.

3.2 Mean coordination and topological constraints: the rigidity percolation model

3.2.1 General principles of the rigidity percolation model

In the early 1980s, Phillips and Thorpe developed a topological model of covalent glass structures aimed at establishing a correlation between the degree of network reticulation and physical properties such as elastic moduli and glass forming tendency (Phillips, 1979; Thorpe, 1983). One of the main appeals of this model was its ability to characterize a given glass using a single structural parameter $<r>$ called the average coordination or mean coordination. The mean coordination is a measure of the network connectivity and corresponds to the average number of covalent bonds per atom according to:

$$< r > = \sum r_i a_i \qquad [3.1]$$

where a_i is the molar fraction and r_i is the covalent coordination of atom i. For example, the average coordination of $Ge_2As_3Se_5$ is $<r> = 4 \times 0.2 + 3 \times 0.3 + 2 \times 0.5 = 2.7$. But more importantly, the mean coordination provides a direct estimate of the number of topological constraints in the glassy network following the idea that rigid bonds between two atoms each generate one bonding constraint, and fixed bond angles between three atoms each generate one angular constraint as described in Fig. 3.1 for an $AsSe_3$ trigonal pyramid. An r-coordinated atom therefore generates $r/2$ constraints (since each bond constraint is shared between two atoms) and $2r - 3$ angular constraints (since any additional angle is not independent and would be over-counted). For a network of average coordination $<r>$, the average number of constraints per atom n is then defined as:

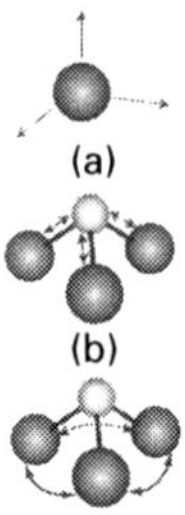

3.1 (a) The three degrees of freedom of an atom in a three-dimensional glassy network. (b) Bond stretching constraints of an $AsSe_3$ trigonal pyramid showing a total of three bond constraints. (c) Bond bending constraints of an $AsSe_3$ pyramid showing three independent angular constraints.

$$n = \frac{<r>}{2} + (2<r> - 3) \qquad [3.2]$$

And since each atom in a three-dimensional glass has $d = 3$ degrees of freedom, the rigidity of the network can be predicted by comparing n and d. When $n < d$, atoms have on average more degrees of freedom than topological constraints and the corresponding network is defined as floppy or underconstrained. When $n > d$, the network becomes overconstrained and is defined as stressed rigid. Finally, if $n = d$, the network is optimally constrained and isostatic.

Solving Equation [3.2] for $n = 3$ show that ideally constrained networks have an average coordination $<r> = 2.4$. Glasses satisfying this topological condition have particular physical properties such as high glass-forming ability and also correspond to the onset of rigidity revealed by an increase in elastic modulus. Indeed, glassy networks with low coordination can undergo deformations at no cost in energy during which the bond angles and bond lengths remain unchanged. The number of such deformations per atom called zero frequency modes is equal to the difference between d and n. The fraction of zero frequency mode f is therefore positive ($f > 0$) in underconstrained glasses as depicted in Fig. 3.2(a) and makes these glasses easily deformable. However, when the number of bonds and constraints increases, f decreases accordingly and rigid domains are created in the structure (Fig. 3.2(b)). For example, it can be shown that an isolated six member ring has no zero frequency modes and is consequently rigid (Thorpe, 1983). When $<r>$ increases, the three-dimensional ring structure of the network decreases and such rigid elements eventually interconnect. This transition occurs at $<r> = 2.4$ when the rigidity is said to percolate through the structure. This process is sketched in Fig. 3.2(c) where the underconstrained network is composed of isolated rigid domains surrounded by a continuous floppy matrix, while

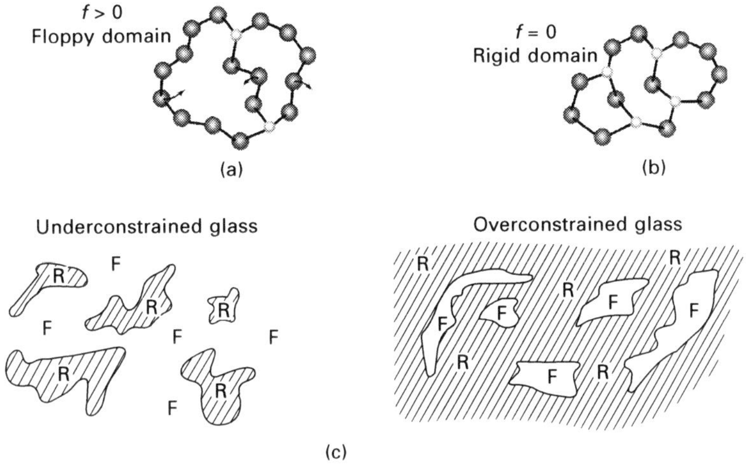

3.2 (a) Topology of a floppy domain containing zero frequency modes of deformation. (b) Topology of a rigid domain where structural constraints outnumber degrees of freedom. (c) Comparative sketch of an underconstrained glass with isolated rigid domains and an overconstrained glass where rigid domains percolate through the structure (reproduced from Thorpe M. F. (1983), 'Continuous deformations in random networks', *J. Non-Cryst. Solids*, 57, 355–370 with permission from Elsevier).

the overconstrained network shows a continuous rigid domain containing floppy inclusions. The transition between these two configurations is often associated with a change in physical properties.

3.2.2 Experimental observation of rigidity percolation in chalcogenide glass

One of the first and most convincing experimental observations of rigidity percolation was provided by Halfpap and Linsay (1986) who measured the elastic constant of two sets of GeAsSe glasses as a function of $<r>$ (Fig. 3.3). Two series of GeAsSe samples with radically different Ge/As ratios but covering the same range of $<r>$ values were used. The results shown in Fig. 3.3(b) indicate that the elastic modulus remains constant until $<r> = 2.4$ at which point it increases exponentially. This result is in good agreement with the numerical prediction of He and Thorpe (1985) who simulated a comparable amorphous network composed of 2, 3 and 4-coordinated atoms by randomly removing bonds from an initial diamond lattice. The fraction of zero frequency modes (f) calculated from this model confirms that f falls to zero at $<r> = 2.4$ (inset in Fig. 3.3(a)). But while the general trend is accurately reproduced between model and experiment, two differences can

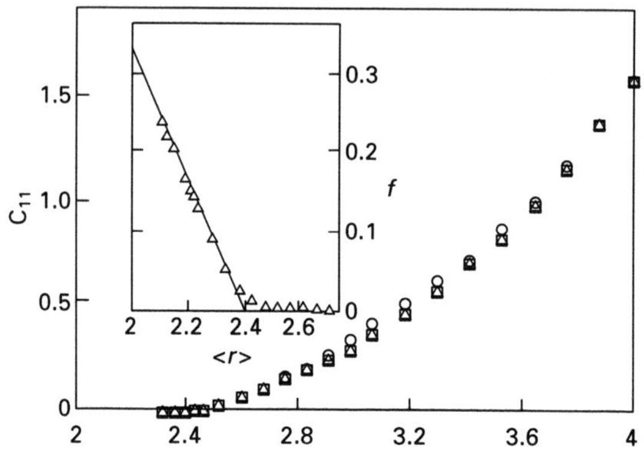

(a)

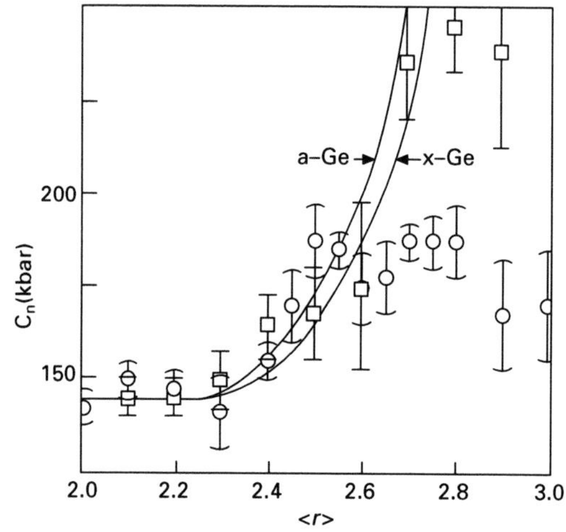

(b)

3.3 (a) Elastic modulus C_{11} calculated from simulated glasses containing various amounts of 2, 3 and 4-coordinated atoms generated by randomly removing bonds from an initial diamond lattice. The inset shows the fraction of zero frequency modes *f* as a function of <*r*> (reproduced from He H. and Thorpe M. F. (1985), 'Elastic properties of glasses', *Phys. Rev. Lett.*, 54, 2107–2110 with permission from American Physical Society). (b) Elastic modulus C_{11} measured by ultrasound pulse echo on two series of GeAsSe glasses containing high and low concentration of As (reproduced from Halfpap B. L. and Lindsay S. M. (1986), 'Rigidity percolation in the germanium-arsenic-selenium alloy system', *Phys. Rev. Lett.*, 57, 847–849 with permission from American Physical Society).

be noticed. First, the elastic constant C_{11} falls to zero in the model while it reaches a finite value for low <*r*> in the experiment. This is a consequence of van der Waals bonding which is not accounted for in the model but clearly plays a significant role in laboratory glasses. Second, the experiment severely deviates from the model for As-rich glasses at high <r> values. This is due to a breakdown of the continuous network due to the formation of isolated molecular species such as As_4Se_4 and As_3Se_4 as is now known (Yang *et al.*, 2010). But overall, the agreement with the model is satisfactory within the limitation of the continuous random network assumption.

Other experimental observation of the percolation threshold involves density measurements (Wang *et al.*, 2009a). The exhaustion of all the degrees of freedom at the rigidity percolation means that the network most efficiently fills space for <r> = 2.4. Hence a maximum in density is reached for isostatic glasses. For stressed rigid glasses, excessive constraints prevent an optimal arrangement of the structure and result in lower densities. This trend is demonstrated in Fig. 3.4a for a series of GeAsSe glasses ranging from <r> = 2.2 to 2.6 (for higher values of <r> the ideal network assumption breaks down). The data show significant scatter but a global trend towards a maximum density at <r> = 2.4 can be observed, as expected, from the rigidity percolation model. Multiple sources of deviation from the model can be invoked to account for the data scatter. In particular, chemical effects have been clearly identified as one of the main factors competing with rigidity effects. An example of chemical effect is shown in Fig. 3.4(b) for a series of $Ge_xAs_ySe_{1-x-y}$ glasses with a fixed value of <r> = 2.5 and an increasing percentage of Ge (Wang *et al.*, 2011). The series of glasses ranges from Se-rich to Se-poor with respect to the stoichiometric composition $Ge_{12.5}As_{25}Se_{62.5}$ (equivalent to $GeAs_2Se_7$ or $(GeSe_2)(As_2Se_3)$). It is known that these glasses respect chemical order (Sen and Aitken, 2002), hence the combination of chemical order and stoichiometry should generate a unique topology where every As and Ge are bridged by a single Se atom. These glasses are therefore expected to display distinct physical properties and indeed Fig. 3.4(b) shows that a clear maximum in density is reached for the stoichiometric glass despite the fact that all samples have the same average coordination.

Topological effects associated with stoichiometry therefore constitute an important factor in addition to the effect of rigidity percolation. This can be further illustrated by comparing the two binary systems As_xSe_{1-x} and Ge_xSe_{1-x} (Fig. 3.5). For the As_xSe_{1-x} system shown in Fig. 3.5(a), the rigidity percolation (<r> = 2.4) is reached at the stoichiometric composition As_2Se_3, hence the density shows a single clear maximum at 2.4. On the other hand, the density of the Ge_xSe_{1-x} system (Fig. 3.5(b)) shows a maximum at the rigidity percolation but also a sharp minimum at the stoichiometric composition $GeSe_2$ (<r> = 2.67), thereby underlining the chemical effect. For ternary systems, the position of these two effects depends on the composition tie line through the ternary phase diagram. Finally, effects of rings and molecular species may further affect the topology as will be discussed in the next section.

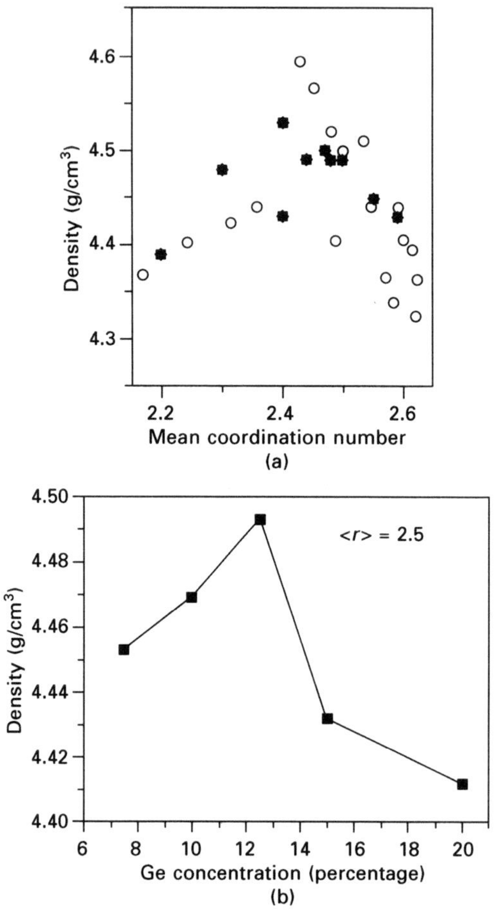

3.4 (a) Density of GeAsSe glasses with mean coordination number <*r*> ranging from 2.2 to 2.6 (adapted from Wang R. P., Smith A., Luther-Davies B., Kokkonen H. and Jackson I. (2009), 'Observation of two elastic thresholds in $Ge_xAs_ySe_{1-x-y}$ glasses', *J. Appl. Phys.*, 105, 056109 with permission from the American Institute of Physics). (b) Density of $Ge_xAs_ySe_{1-x-y}$ glasses with a fixed value of <*r*> = 2.5 as a function of Ge% (adapted from Wang R. P., Bulla D., Smith A., Wang T. and Luther-Davies B. (2011), 'Structure and physical properties of GexAsySe1-x-y glasses with the same mean coordination number of 2.5', *J. Appl. Phys.*, 109, 023517 with permission from the American Institute of Physics).

3.3 Applicability of the rigidity percolation model

3.3.1 Ideal network assumptions of the rigidity percolation model

The covalent nature of chalcogenide glasses makes them ideally suited for theoretical treatment based on modeling network topology. In that respect,

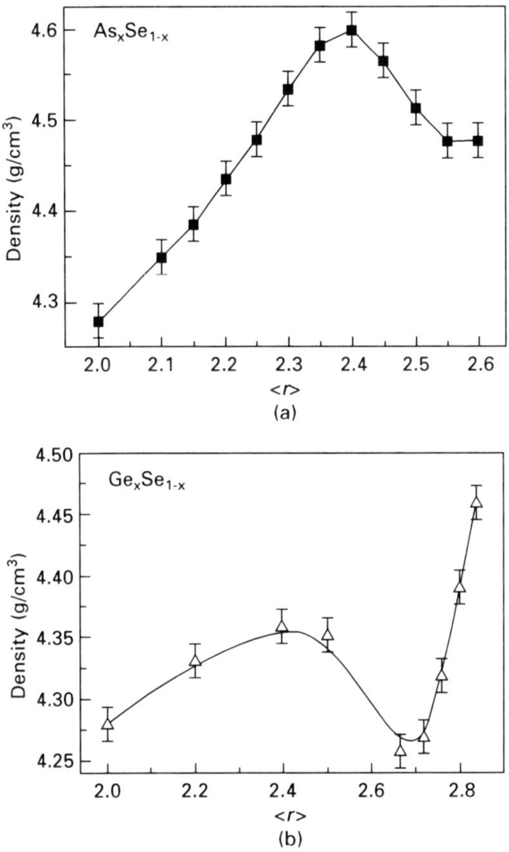

3.5 Density of glass samples from the As_xSe_{1-x} (a) and Ge_xSe_{1-x} (b) systems.

the rigidity percolation has shown much success in describing chalcogenide systems but it is important to point out that this model was built upon several basic assumptions that may limit its suitability for describing certain glasses. Four basic assumptions can be recognized:

1. The glass is described by a continuous random network (CRN) of the type hypothesized by Zachariasen (1932) containing no free-standing clusters.
2. The network is composed of only 2, 3 and 4-coordinated atoms and contains no terminal monovalent atoms or chains with free ends.
3. The ring structure in the network should be large (at least six members) in order to avoid double counting constraints.
4. The model is formulated for a network at absolute zero temperature $T = 0$ K where all constraints are intact.

In the following sections we discuss several cases where experimental observations provide evidence that these assumptions are violated in certain glass systems. Some modified models have been proposed to account for these deviations.

3.3.2 Isolated molecules in chalcogenide glasses

Isolated molecular fragments that are not covalently linked to the backbone network do not contribute to the rigidity percolation since only central forces are accounted for in this model. These molecular species only interact through weak van der Waals forces which are neglected for the purpose of constraint counting. While the rigidity of the remaining backbone network may still be described with the percolation model, its effective average coordination may significantly deviate from that derived from Eq. [3.1]. For example, it is well known that sulfur-rich glasses contain significant amounts of S_8 rings (Lin *et al.*, 2011) which would lead to significant underestimation of the rigidity of the backbone network. On the other hand, arsenic-rich glasses tend to produce molecular species such as As_4Se_4, As_3Se_4 or As_4 (isomorphous to P_4 tetrahedra) that drain arsenic out of the main network and lead to significant overestimation of its effective average coordination (Yang *et al.*, 2010). As mentioned in Section 3.2.2, this may have a significant impact on measured physical properties such as the elastic modulus (Halfpap and Lindsay, 1986). In fact, it can be shown that the dynamic behavior of isolated molecular species is largely decoupled from that of the backbone network (Fig. 3.6). For example, the Raman spectra of As_6Se_4 and As_2Se_3 show that the arsenic-rich glass contains large amounts of molecular species while the stoichiometric glasses only exhibit pyramidal modes from the network (Fig. 3.6(a)). It can be shown that these two different structures result in very distinct structural dynamics when probed by heat capacity spectroscopy. The 'molecular' As_6Se_4 glass shows two distinct dynamic contributions as opposed to a single contribution for the As_2Se_3 glass (Fig. 3.6(b)). The heat capacity spectroscopy measurement championed by Bustin and Descamps (1999) permits the response of the structure to a temperature oscillation to be monitored. Figure 3.6(b) clearly shows two distinct responses for the As_6Se_4 glass, one corresponding to the molecular species and one corresponding to the glassy backbone. On the other hand, the As_2Se_3 glass shows a single narrow response consistent with its uniform structure composed uniquely of pyramidal units. The presence of such molecular species is actually quite common in chalcogenide glasses and has been observed in GeAsS glasses (Soyer-Uzun *et al.*, 2009) as well as PSe glasses (Georgiev *et al.*, 2001). They have also been shown to exhibit decoupled dynamics (Gjersing *et al.*, 2007) and even lead to polyamorphism (Sen *et al.*, 2006).

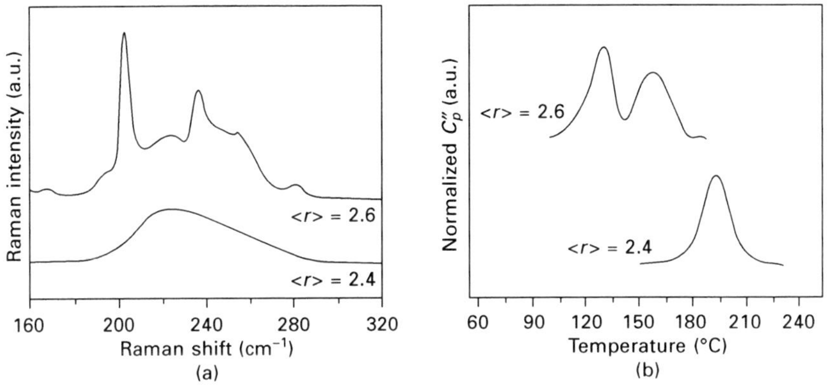

3.6 (a) Raman spectra of an As_6Se_4 sample ($\langle r \rangle = 2.6$) showing sharp molecular Raman modes corresponding to As_4Se_4, As_3Se_4 or As_4 molecules, compared with an As_2Se_3 sample ($\langle r \rangle = 2.4$) showing broad pyramidal network modes. (b) Imaginary heat capacity C_p'' and (increasing temperature and oscillation frequency) of As_6Se_4 and As_2Se_3 samples recorded at increasing temperature oscillation frequency 0.005 Hz with a heating rate of 3°C/min and an amplitude of 3°C, showing a single contribution for the As_2Se_3 sample and two distinct contributions for the As_6Se_4 sample.

3.3.3 Small rings in chalcogenide glasses

Networks with large ring structures are accurately described by the rigidity percolation model and the number of topological constraints is correctly obtained from Eq. [3.2] if the ring size is on the order of 10–12 atoms. However, when the ring size decreases below six atoms, some constraints become linearly dependent and a correction must be applied (Thorpe, 1983). For example, a triangle can be entirely defined by three independent variables (two lengths and one angle), yet Eq. [3.2] would yield $n = 2$ constraints per atom for a total of six constraints thereby overestimating the number of constraints by three. For a four-member ring such as that generated by two edge-sharing tetrahedra (Fig. 3.7), the number of constraints would be overestimated by two. This can easily be demonstrated by enumerating all constraints in edge-sharing tetrahedra (ES) and comparing with corner-sharing tetrahedra (CS) where no rings are present. While the number of bond constraints is the same between ES and CS clusters, the number of angular constraints is actually smaller for ES because a single variable is sufficient to define the bridging angle and because only four outer selenium remain, each contributing one half angular constraint in comparison to six outer selenium for the CS. This leads to a total of 12 angular constraints for the ES as opposed to 14 for the CS. This somewhat counterintuitive result implies that the presence of edge-sharing tetrahedra actually reduces the network rigidity. This is a significant result because the presence of such ES

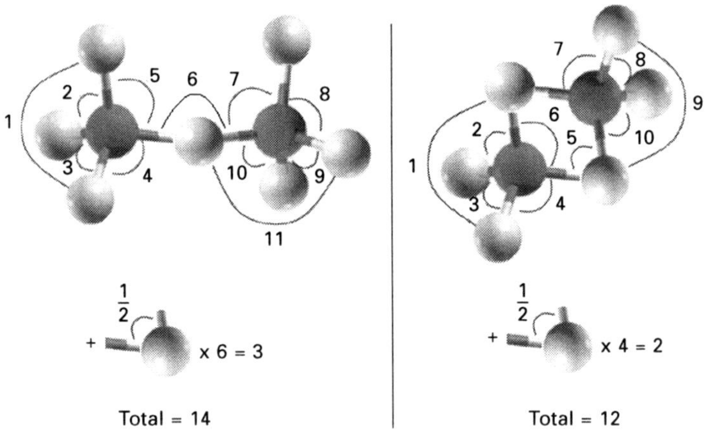

3.7 Angular constraint counting for a pair of corner and edge-sharing tetrahedra. Each outer bridging Se is assigned a half angular constraint.

has been observed in many chalcogenide glasses such as Ge-S and Ge-Se (Jackson *et al*., 1999). For example, Fig. 3.8(a) shows that $GeSe_2$ contains large fractions of ES and it was shown that they remain in the glass even for very low germanium content (Lucas *et al*., 2009). The presence of ES has also been observed in the ternary $Ge_xAs_ySe_{1-x-y}$ system as depicted in Fig. 3.8(b) (Wang *et al*., 2009b). The rigidity percolation threshold being entirely defined by the average number of constraints n (all zero frequency modes vanish when $n = d$), an accurate estimate of the threshold should therefore account for these deviations. In the next section we describe some modified models that incorporate the presence of these short rings.

3.3.4 Modified models of rigidity percolation

The original rigidity percolation theory is based on a random network assumption, where constraints are arbitrarily added throughout the network as $<r>$ increases and where normal probability therefore predicts the formation of stressed domains in coexistence with floppy domains. As more and more constraints are added, the percolation of stressed domains occurs at $<r> = 2.4$. More recently, Thorpe *et al*. (2000) have hypothesized the existence of a second threshold at lower $<r>$ resulting from 'self-organization' of the network. It is suggested that stressed-rigid domains have higher free energy and consequently that the network has a thermodynamic incentive to avoid their formation and produce isostatic-rigid domains instead. In other words, constraints are evenly distributed through the structure and no redundant constraint is created as this would generate stress and increase the free energy. As a consequence, rigidity percolates through the structure at a lower value of

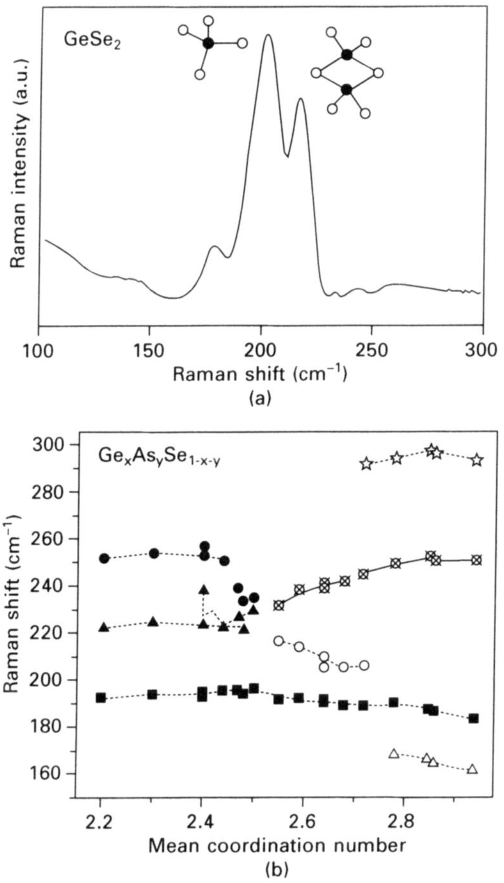

3.8 (a) Raman spectra of $GeSe_2$ glass showing the corner-sharing $GeSe_{4/2}$ tetrahedra breathing mode at 200 cm^{-1} and the edge-sharing tetrahedra mode at 220 cm^{-1}. (b) Raman modes of $Ge_xAs_ySe_{1-x-y}$ glasses showing the appearance of edge-sharing $GeSe_{4/2}$ tetrahedra mode at $\langle r \rangle > 2.5$ (open circles). Remaining mode assignments are: solid squares: corner-sharing $GeSe_{4/2}$ tetrahedra; solid triangles: $AsSe_{3/2}$ pyramids; solid circles: Se–Se bonds; open cross circle: intermetallic bonds, open triangles and open stars: $Ge_2Se_{6/2}$ ethanlike modes (reproduced from Wang R. P., Smith A., Prasad A., Choi D. Y. and Luther-Davies B. (2009), 'Raman spectra of $Ge_xAs_ySe_{1-x-y}$ glasses', *J. Appl. Phys.*, 106, 043520 with permission from the American Institute of Physics).

mean coordination estimated at $\langle r \rangle = 2.375$ (Chubynsky and Thorpe, 2001). Eventually, as more constraints are added, stress becomes unavoidable and a second 'stress' transition is observed at $\langle r \rangle = 2.392$. This results in an 'intermediate phase' between the two transitions where the network is rigid but unstressed as depicted in Fig. 3.9.

Micoulaut and Phillips (2003) later proposed a somewhat different approach

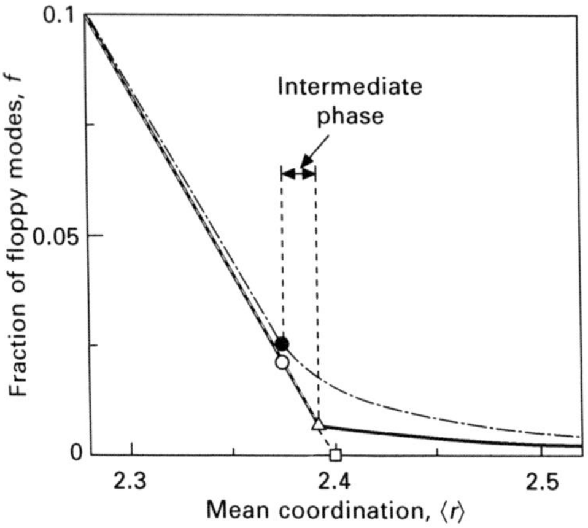

3.9 Depiction of the rigid but unstressed 'intermediate phase' delimited by a rigidity transition at $<r> = 2.375$ and a stress transition at $<r> = 2.392$ (reproduced from Thorpe M. F., Jacobs D. J., Chubynsky M. V. and Phillips J. C. (2000), 'Self-organization in network glasses', *J. Non-Cryst. Solids*, 266–269, 859–866 with permission from Elsevier).

suggesting that the presence of small rings such as edge-sharing tetrahedra would delay the onset of stress build-up and produce a wider intermediate phase with a stress transition at $<r> > 2.4$. The calculated free energy of that network was predicted to decrease with increasing ES content leading to a broad intermediate phase with ES fractions ranging from 30 to 56%. However, this prediction is qualitatively and quantitatively inconsistent with experimental measurements by Edwards and Sen (2011) who observed a positive free energy of formation for ES units near the fictive temperature for ES fractions above 15%. This behavior is also observed for temperatures far below T_g as will be discussed in Section 3.4.2.

So far, the observation of a structural origin of the 'intermediate phase' has been elusive (Shatnawi *et al.*, 2008). The principal experimental observations supporting the existence of the intermediate phase are based on non-reversing enthalpy measurements performed above T_g (Georgiev *et al.*, 2001; Selvanathan *et al.*, 2000). However, these measurements do not satisfy the fourth condition in Section 3.3.1, and their interpretation is subject to debate as will be discussed in the next section.

3.4 The temperature dependence of constraints

3.4.1 Effect of temperature on rigidity percolation: the case of As_xSe_{1-x} glasses

In all previously described models, the number of constraints is tabulated based on the assumption that no constraint is thermally excited or broken.

In other words, these models are formulated for zero temperature or at least for a temperature far below T_g where there is not sufficient thermal energy to overcome central forces. Weaker forces such as van der Waals are not taken into account in the model, and the rigidity prediction would not be affected even if they were broken. It can then be reasonably expected that measurements performed at room temperature satisfy the premises of the rigidity model for most chalcogenide glasses. Multiple room temperature measurements such as the ones described in Section 3.2.2 indeed appear to confirm the presence of a rigidity percolation at <*r*> = 2.4. For example, Ge_xSe_{1-x}, As_xSe_{1-x}, and $Ge_xAs_ySe_{1-x-y}$ all exhibit a maximum in density at the percolation threshold when measured at low temperature.

Additional room temperature measurements are shown in Fig. 3.10 for the As_xSe_{1-x} system, including the linear thermal expansion coefficient, Young's modulus, the shear modulus, the bulk modulus and Poisson ratio. All properties show a single extrema at <*r*> = 2.4 in agreement with the rigidity percolation model (Fig. 3.10(a)–(c). In stark contrast, measurements performed on the same set of sample at temperatures above or across T_g show a radically different trend (Fig. 3.10(d)–(f). Figure 3.10(d) reports the activation energy for viscous flow measured by indentation (Gueguen *et al.*, 2011) while Fig. 3.10(e) reports the activation energy for enthalpy relaxation measured by Moynihan's cool rate method (Moynihan *et al.*, 1976) and Fig. 3.10(f) reports the activation energy for structural relaxation measured by specific heat spectroscopy (Carpentier *et al.*, 2003). All properties show a double minimum near <*r*> = 2.3 and 2.5 and a local maximum at <*r*> = 2.4. This difference in trend is not entirely unexpected considering that these measurements were performed at a temperature where many constraints are already broken. These measurements should therefore be regarded as a reflection of the temperature dependence of constraints in the glasses rather than reflecting the total number of constraints in the ideal network.

These contrasted results indicate that high-temperature properties cannot be correctly used to describe the low-temperature features of glassy networks. In that respect, non-reversing enthalpy measurements performed above T_g have been presented as the main evidence of the presence of an intermediate phase in the As_xSe_{1-x} system (Georgiev *et al.*, 2001). Not surprisingly, the trend of the non-reversing enthalpy closely mimics that of the high-temperature properties reported in Fig. 3.10(d)–(f), with a broad minimum at <*r*> = 2.3 that was interpreted as the intermediate phase. However, this phase is postulated on the basis of constraint counting at low temperature. This may explain why constraint counting compatible with that minimum could only be satisfied by invoking the presence of large quantities (30%) of tetrahedral As that have so far never been observed experimentally (Ahn *et al.*, 2006;

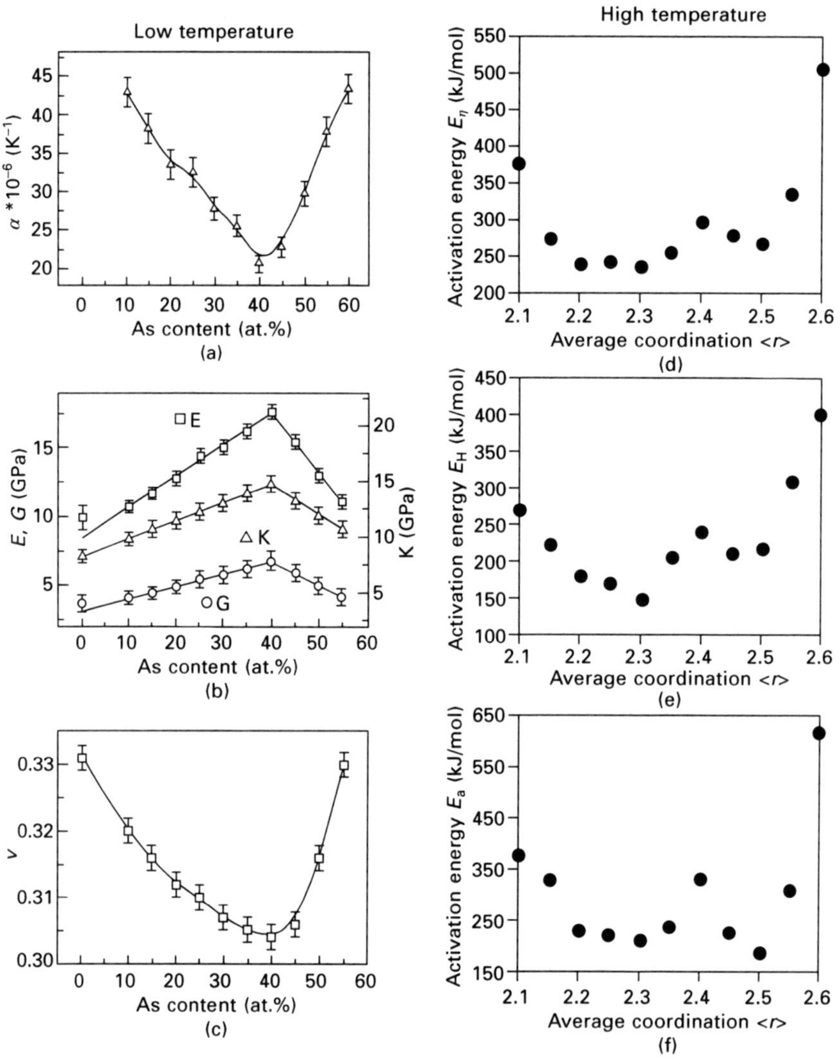

3.10 Comparison of low temperature (below T_g) and high temperature (above T_g) properties of the As_xSe_{1-x} system. (a) Linear coefficient of thermal expansion below T_g. (b) Young's modulus *E*, shear modulus *G*, bulk modulus *K* at room temperature. (c) Poisson ratio at room temperature. (d) Activation energy for viscous flow above T_g. (e) Activation energy for enthalpy relaxation above T_g. (f) Activation energy for structural relaxation obtained by specific heat spectroscopy above T_g.

Golovchak *et al.*, 2007). Hence, the results describe in Fig. 3.10 seem to suggest that the observation of an intermediate phase in the As_xSe_{1-x} system is merely an artifact of high temperature measurements.

3.4.2 Constraints relaxation in the Ge_xSe_{1-x} system

Glasses containing large quantities of ES units such as Ge_xS_{1-x} and Ge_xSe_{1-x} provide an ideal test bed for studying the evolution of constraints with temperature. Indeed, Fig. 3.7 clearly indicates that CS to ES conversion is associated with the loss of two constraints and it can be shown that the ES/CS ratio is also a direct function of temperature (Edwards and Sen, 2011). Hence, the ES/CS ratio should permit the monitoring of the number of constraints with temperature and in turn provide an estimate of entropy creation in the system (Naumis, 2005). In that respect, the Ge_xSe_{1-x} system is of particular interest because the ES and CS Raman features are clearly identifiable and quantifiable as can be seen in Fig. 3.11(a). For example, the ES/CS equilibrium ratio in the supercooled liquid can be monitored as a function of temperature as shown in Fig. 3.11(b). A van't Hoff analysis of these results yields a positive enthalpy $\Delta H_{CS\rightarrow ES} = 0.33\ j{\cdot}g^{-1}{\cdot}K^{-1}$ for the speciation reaction. This is consistent with the idea that the entropy associated with this process should be positive since the formation of ES decreases the number of constraints and increases the number of zero frequency modes in the network. An equilibrium ES/CS ratio should therefore be reached at fixed temperature and then decrease with decreasing temperature. The results shown in Fig. 3.11 for $Ge_{20}Se_{80}$ indicate that no more than 15% ES are created at T_g and that they should have a positive free energy of formation at any lower temperature. This is indeed observed experimentally in Fig. 3.11(a) where the ES/CS ratio is shown to spontaneously decay when the glass is annealed 40° below T_g.

The same behavior is observed when the system is far out of equilibrium down at room temperature. Figure 3.12 compares the ES/CS ratio and the structural enthalpy of $Ge_{20}Se_{80}$ glass fibers aged at room temperature for up to five years. The data show a striking correlation between the ES/CS ratio and the glass enthalpy. Both data sets are fitted with the same stretched exponential function. This indicates that the ES/CS ratio provides a very good measure of the glass fictive temperature. This also shows that ES tetrahedra are not stable structural units at room temperature and that their concentration would ultimately decrease to very small fractions given enough time. While this process can not actually take place within a laboratory time scale, this puts into question the foundation of the intermediate phase in these glasses, whose origin was postulated on the basis of large fractions of ES due to self-organization. If self-organization involves the formation of ES in high concentration, then these results imply that it is not a thermodynamically driven process near T_g. Hence, this raises the question of what would be the driving force for self-organization and for the creation of this intermediate phase.

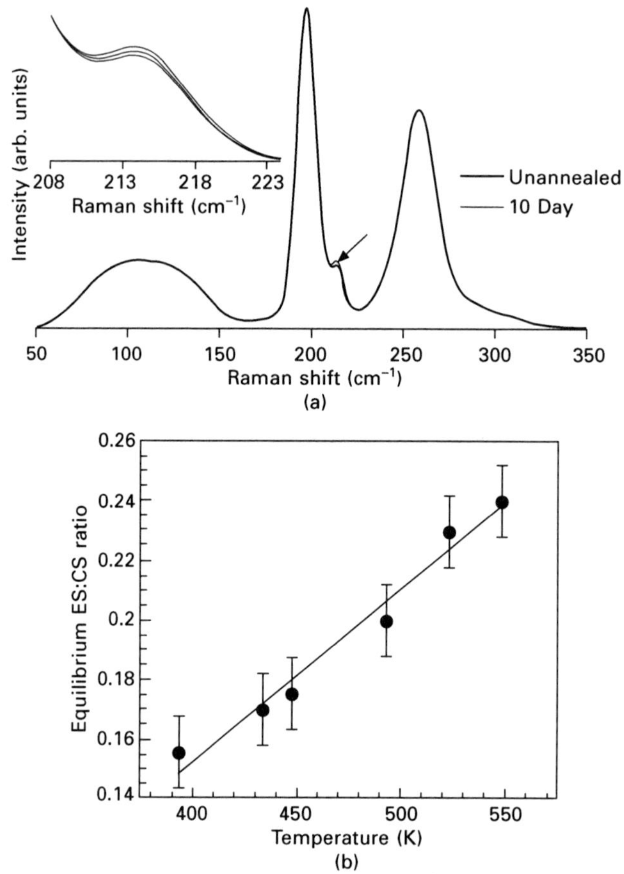

3.11 (a) Raman spectra of a $Ge_{20}Se_{80}$ glass showing the evolution of the CS (200 cm^{-1}) and ES (215 cm^{-1}) peaks during annealing 40° below T_g. The inset show the decay of the ES band over 10 days. (b) Temperature dependence of the equilibrium EC/CS ratio in a $Ge_{20}Se_{80}$ glass obtained from the intensity ratio of ES and CS Raman bands (reproduced from Edwards T. G. and Sen S. (2011), 'Structure and relaxation in germanium selenide glasses and supercooled liquids: a Raman spectroscopic study', *J. Phys. Chem. B*, 115, 4307–4314 with permission from the American Institute of Physics).

3.4.3 Mauro's temperature-dependent constraints model

Viscous flow is a determining factor in controlling glass processing. Modeling high-temperature properties of glasses is therefore of great importance for practical and industrial applications. In that respect, the rigidity percolation model has proven inadequate for accurately describing network behavior at high temperature due to the great fraction of broken constraints. Gupta and Mauro were the first to address this issue and propose a refined model

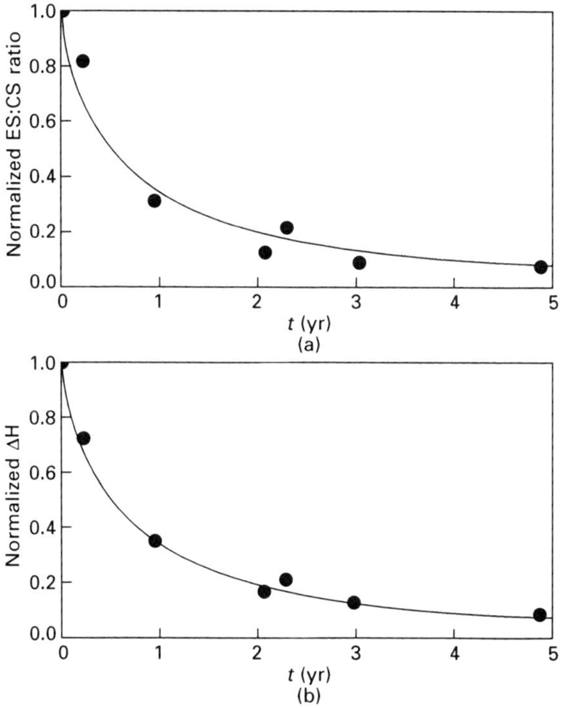

3.12 (a) Decay of the ES/CS ratio of $Ge_{20}Se_{80}$ glass fibers aged at room temperature for 5 years. The data are normalized by the total decay $\Delta = 0.02$. (b) Enthalpy relaxation of the same $Ge_{20}Se_{80}$ glass fibers. The data are normalized by the total relaxed enthalpy $\Delta H = 8$ $j \cdot g^{-1}$. Both data sets are fitted with the same stretched exponential function.

that accounts for the temperature dependence of constraints in an attempt to predict high-temperature properties of chalcogenide glasses (Gupta and Mauro, 2009).

This model builds up from the basic principles of the Phillips and Thorpe theory but includes a characteristic temperature associated with each type of constraint below which the constraint becomes rigid and above which the constraint is broken. The number of constraints is therefore a function of temperature in addition to composition. The number of constraints $n(T)$ may be larger than the number of degrees of freedom d at low temperature, thereby yielding a stressed-rigid glass and become smaller than d at a higher temperature leading to a floppy glass. This added modeling feature then permits the thermal behavior of network glasses to be simulated. Another notable difference is the inclusion of non-central forces which are necessary to enable a quantitative comparison of the model with experiment. This makes this model particularly appealing for its ability to predict the viscous behavior of true laboratory glass-forming systems. It is postulated that the viscosity will diverge at a temperature T_o where $n(T_o) = d$, i.e. when the system becomes globally rigid. This principle permits both the T_g and the fragility parameter m of glass-formers to be estimated as a function of composition.

While this is a powerful method, it must be emphasized that more structural information than just $<r>$ are required to compute T_g and m from $n(T)$. An accurate determination of $n(T)$ requires spectroscopic information about the structure which are necessary to identify and enumerate the number of constraints. Computing $n(T)$ essentially consists in assigning a characteristic temperature to the constraints associated with each structural unit present in the glass. The fraction of each structural units is then derived from the glass composition and $n(T)$ can be tallied for any given temperature. Proper equations giving scaled values of T_g and m from $n(T)$ have been established and tested against experiment with glasses from the Ge_xSe_{1-x} system (Gupta and Mauro, 2009). In that example, five types of constraints are identified: α, the Se-Ge-Se angular constraint; β, the Ge-Se or Se-Se linear constraint; γ, the Se angular constraint; δ, the van der Waals contribution; and λ, the homopolar constraint in Ge_2Se_3 units.

Figure 3.13(a) plots the number of atomic degrees of freedom in the glass ($f(T) = d - n(T)$) as a function of composition and of the characteristic temperature of each constraint. As expected, $f(T)$ increases with temperature at a given composition since more constraints become broken, and $f(T)$ decreases with x at a given temperature because $<r>$ increases. The $f(T) = 0$ line then permits T_o to be determine as a function of composition. For glasses up to 20% Ge, $T_o = T_\delta$ (the characteristic temperature of constraint δ), i.e. the network will be floppy for any temperature sufficient to break van der Waals bonds. For glasses ranging from 20–33% Ge, $T_o = T_\gamma$, i.e. the network will be floppy, for any temperature sufficient to break Se angular constraints. And finally for Ge% > 0.33, $T_o = T_\lambda$. These characteristic temperatures are not known a *priori* and must be obtained by fitting experimental viscosity data. And since T_o corresponds to the temperature at which the viscosity diverged, these values can be used to predict the fragility parameter for all compositions as depicted in Fig. 3.13(b). It is shown that the correlation with experimental data is surprisingly good.

These initial data show great promise but it must be noted that the model is dependent on the quality of the spectroscopic data used to compute the number and type of constraints. It also only yields scaled parameters and requires an initial input of T_g and m to provide the full values for all compositions. However, its ability to correctly predict experimental viscosity parameters is very promising, and more importantly, it constitute a uniquely satisfying viscosity model in that it is derived from structural characteristics as opposed to empirical models such as Vogel-Taumman-Fulcher (VTF).

Finally, it should be mentioned that the temperature-dependent constraints

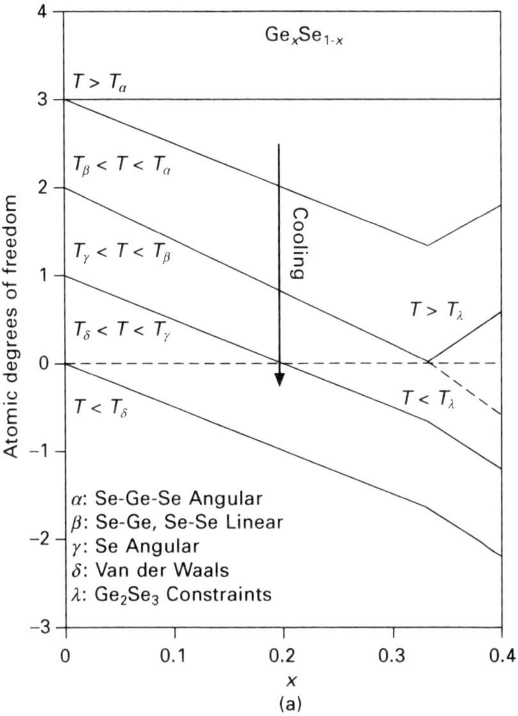

(a)

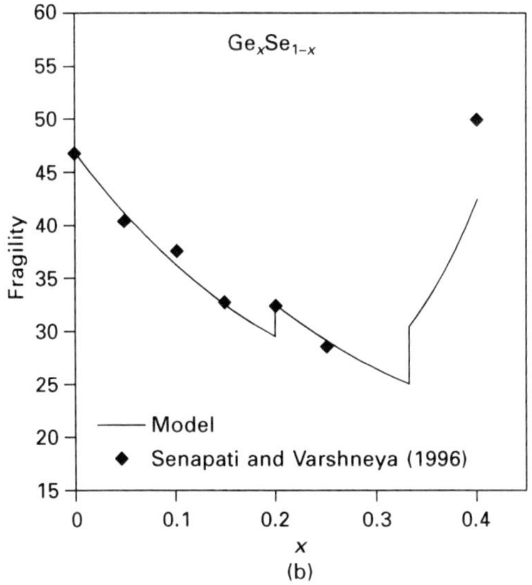

(b)

3.13 (a) Temperature and composition dependence of the number of atomic degrees of freedom $f(T)$ in Ge_xSe_{1-x} glasses. Each line corresponds to a temperature range delimited by the characteristic temperature of topological constraints. (b) Fragility parameter m of glasses from the Ge_xSe_{1-x} system. Diamonds are experimental points (from Senapati and Varshneya, 1996) and solid lines are analytical prediction based on estimation of $n(T)$ and T_0 (reproduced from Gupta and Mauro, 2009).

paradigm has also been successfully applied to oxide glasses (Smedskjaer *et al.*, 2010a, 2010b, 2011). Multiple physical properties such as hardness, fragility, T_g and heat capacity jump were predicted based on nuclear magnetic resonance structural data. Overall, this method is still in its infancy and has much potential for further refinement to improve its predictive power for chalcogenides and other types of glasses.

3.5 Conclusion and future trends

Great advances in glass topology have been made since the early reports of the rigidity percolation theory some 40 years ago. The basic principles laid down by Phillips and Thorpe keep the appeal and elegance of a theory based on a single parameter – the mean coordination – capable of predicting physical properties over entire compositional ranges. The mean coordination $<r>$ is not only a simple and easily predictable parameter, but it is also a universal feature of all chalcogenide glasses and can then be applied to many glassy systems. But while this theory has met with much success, it is also based on an ideal description of glassy network which has been shown to deviate from the complexity of many laboratory glasses. These deviations sometimes lead to a breakdown of the model because rigorous constraint counting is significantly dependent on fine structural details and anomalies which are an intrinsic feature of disordered materials. However, a systematic characterization and quantification of these features is notoriously difficult due to the lack of long-range order inherent to amorphous structures. Indeed, the structure of some simple binary glasses is still not well understood and subject to debate (Lucas *et al.*, 2009; Kibalchenko *et al.*, 2010). Hence, advanced experimental characterization of the structure of glassy systems will be a key element of future improvement and refinement of topological models, for both the construction and validation of these models. Proper understanding of local structural features that enable exact constraint enumeration should then lead to much improved predictive power.

Another key element for improving the predictive power of topological models and to extend their applicability to a wider range of physical properties will involve the temperature dependence of constraints. Initial efforts have already shown much promise, and it is likely that these added features will not only provide a more realistic model of room temperature properties such as mechanical strength, but will also permit the extension of modeling efforts to a wide range of thermal properties. This may lead to a truly fundamental leap forward in our understanding of glass physics since the interconnection between structure and temperature is at the very heart of the glass formation process. Much activity is ongoing in this vibrant field of research which should lead to exciting new discoveries in the near future. Further information on these developments can be found in recent seminal and review papers (Mauro, 2011; Gupta and Mauro, 2009; Smedskjaer *et al.*, 2010a).

3.6 References

Ahn E., Williams G. A. and Taylor P. C. (2006), 'Nuclear quadrupole resonance study of local bonding in glassy AsxSe1-x', *Phys. Rev. B: Condens. Matter Mater. Phys.*, 74, 174206/1-174206/9.

Bustin O. and Descamps M. (1999), 'Slow structural relaxations of glass-forming Maltitol by modulated DSC calorimetry', *J. Chem. Phys.*, 110, 10982–10992.

Carpentier L., Desprez S. and Descamps M. (2003), 'From strong to fragile glass-forming systems: a temperature modulated differential scanning calorimetry investigation', *Phase Transitions*, 76, 787–799.

Chubynsky M. V. and Thorpe M. F. (2001), 'Self-organization and rigidity in network glasses', *Curr. Opin. Solid State Mater. Sci.*, 5, 525–532.

Edwards T. G. and Sen S. (2011), 'Structure and relaxation in germanium selenide glasses and supercooled liquids: a Raman spectroscopic study', *J. Phys. Chem. B*, 115, 4307–4314.

Eggleton B. J., Luther-Davies B. and Richardson K. (2011), 'Chalcogenide photonics', *Nat. Photonics*, 5, 141–148.

Georgiev D. G., Mitkova M., Boolchand P., Brunklaus G., Eckert H. and Micoulaut M. (2001), 'Molecular structure, glass transition temperature variation, agglomeration theory, and network connectivity of binary P-Se glasses', *Phys. Rev. B: Condens. Matter Mater. Phys.*, 64, 134204.

Gjersing E. L., Sen S., Yu P. and Aitken B. G. (2007), 'Anomalously large decoupling of rotational and shear relaxation in a molecular glass', *Phys. Rev. B: Condens. Matter Mater. Phys.*, 76, 214202.

Golovchak R., Kovalskiy A., Miller A. C., Jain H. and Shpotyuk O. (2007), 'Structure of Se-rich As-Se glasses by high-resolution x-ray photoelectron spectroscopy', *Phys. Rev. B: Condens. Matter Mater. Phys.*, 76, 125208.

Gueguen Y., Rouxel T., Gadaud P., Bernard C., Keryvin V. and Sangleboeuf J.-C. (2011), 'High-temperature elasticity and viscosity of GexSe1-x glasses in the transition range', *Phys. Rev. B: Condens. Matter Mater. Phys.*, 84, 064201.

Gupta P. K. and Mauro J. C. (2009), 'Composition dependence of glass transition temperature and fragility. I. A topological model incorporating temperature-dependent constraints', *J. Chem. Phys.*, 130, 094503.

Halfpap B. L. and Lindsay S. M. (1986), 'Rigidity percolation in the germanium-arsenic-selenium alloy system', *Phys. Rev. Lett.*, 57, 847–849.

He H. and Thorpe M. F. (1985), 'Elastic properties of glasses', *Phys. Rev. Lett.*, 54, 2107–2110.

Jackson K., Briley A., Grossman S., Porezag D. V. and Pederson M. R. (1999), 'Raman-active modes of a-$GeSe_2$ and a-GeS_2: a first-principles study', *Phys. Rev. B: Condens. Matter Mater. Phys.*, 60, R14985–R14989.

Kibalchenko M., Yates J. R., Massobrio C. and Pasquarello A. (2010), 'Structural assignments of NMR chemical shifts in $GexSe_{1-x}$ glasses via first-principles calculations for $GeSe_2$, Ge_4Se_9, and GeSe crystals', *Phys. Rev. B: Condens. Matter Mater. Phys.*, 82, 020202.

Lin F.-Y., Gulbiten O., Yang Z., Calvez L. and Lucas P. (2011), 'Mechanism of photostructural changes in mixed-chalcogen As-S-Se glasses investigated by Raman spectroscopy', *J. Phys. D: Appl. Phys.*, 44, 045404.

Lucas J. (1999), 'Infrared glasses', *Curr. Opin. Solid State Mater. Sci*, 4, 181–187.

Lucas P., King E. A., Gulbiten O., Yarger J. L., Soignard E. and Bureau B. (2009), 'Bimodal phase percolation model for the structure of Ge-Se glasses and the existence of the intermediate phase', *Phys. Rev. B: Condens. Matter Mater. Phys.*, 80, 214114.

Mauro J. C. (2011), 'Topological constraint theory of glass', *Am. Ceram. Soc. Bull.*, 90, 31–37.

Micoulaut M. and Phillips J. C. (2003), 'Rings and rigidity transitions in network glasses', *Phys. Rev. B: Condens. Matter Mater. Phys.*, 67, 104204.

Mott N. F. and Davis E. A. (1979), *Electronic Processes in Non-crystalline Materials*, Oxford, Clarendon Press.

Moynihan C. T., Easteal A. J., DeBolt M. A. and Tucker J. (1976), 'Dependence of the fictive temperature of glass on cooling rate', *J. Am. Ceram. Soc.*, 59, 12–16.

Naumis G. G. (2005), 'Energy landscape and rigidity', *Phys. Rev. E: Stat., Nonlinear, Soft Matter Phys.*, 71, 026114.

Phillips J. C. (1979), 'Topology of covalent non-crystalline solids. I. Short range order in chalcogenide alloys', *J. Non-Cryst. Solids*, 34, 153–181.

Sanghera J. S., Shaw L. B., Busse L. E., Nguyen V. Q., Pureza P. C., Cole B. C., Harbison B. B., Aggarwal I. D., Mossadegh R., Kung F., Talley D., Roselle D. and Miklos R. (2000), 'Development and infrared applications of chalcogenide glass optical fibers', *Fiber Integr. Opt.*, 19, 251–274.

Selvanathan D., Bresser W. J. and Boolchand P. (2000), 'Stiffness transitions in Si_xSe_{1-x} glasses from Raman scattering and temperature-modulated differential scanning calorimetry', *Phys. Rev. B: Condens. Matter Mater. Phys.*, 61, 15061–15076.

Sen S. and Aitken B. G. (2002), 'Atomic structure and chemical order in Ge-As selenide and sulfoselenide glasses: an x-ray absorption fine structure spectroscopic study', *Phys. Rev. B: Condens. Matter Mater. Phys.*, 66, 134204.

Sen S., Gaudio S., Aitken B. G. and Lesher C. E. (2006), 'Observation of a pressure-induced first-order polyamorphic transition in a chalcogenide glass at ambient temperature', *Phys. Rev. Lett.*, 97, 025504.

Shatnawi M. T. M., Farrow C. L., Chen P., Boolchand P., Sartbaeva A., Thorpe M. F. and Billinge S. J. L. (2008), 'Search for a structural response to the intermediate phase in Ge_xSe_{1-x} glasses', *Phys. Rev. B: Condens. Matter Mater. Phys.*, 77, 094134.

Smedskjaer M. M., Mauro J. C., Sen S. and Yue Y. (2010a), 'Quantitative design of glassy materials using temperature-dependent constraint theory', *Chem. Mater.*, 22, 5358–5365.

Smedskjaer M. M., Mauro J. C. and Yue Y. (2010b), 'Prediction of glass hardness using temperature-dependent constraint theory', *Phys. Rev. Lett.*, 105, 115503.

Smedskjaer M. M., Mauro J. C., Youngman R. E., Hogue C. L., Potuzak M. and Yue Y. (2011), 'Topological principles of borosilicate glass chemistry', *J. Phys. Chem. B*, 115, 12930–12946.

Soyer-Uzun S., Sen S. and Aitken B. G. (2009), 'Network vs. molecular structural characteristics of ge-doped arsenic sulfide glasses: a combined neutron/x-ray diffraction, extended x-ray absorption fine structure, and Raman spectroscopic study', *J. Phys. Chem. C*, 113, 6231–6242.

Tatsumisago M., Halfpap B. L., Green J. L., Lindsay S. M. and Angell C. A. (1990), 'Fragility of germanium-arsenic-selenium glass-forming liquids in relation to rigidity percolation, and the Kauzmann paradox', *Phys. Rev. Lett.*, 64, 1549–1552.

Thorpe M. F. (1983), 'Continuous deformations in random networks', *J. Non-Cryst. Solids*, 57, 355–370.

Thorpe M. F., Jacobs D. J., Chubynsky M. V. and Phillips J. C. (2000), 'Self-organization in network glasses', *J. Non-Cryst. Solids*, 266–269, 859–866.

Wang R. P., Smith A., Luther-Davies B., Kokkonen H. and Jackson I. (2009a), 'Observation of two elastic thresholds in $Ge_xAs_ySe_{1-x-y}$ glasses', *J. Appl. Phys.*, 105, 056109.

Wang R. P., Smith A., Prasad A., Choi D. Y. and Luther-Davies B. (2009b), 'Raman spectra of $Ge_xAs_ySe_{1-x-y}$ glasses', *J. Appl. Phys.*, 106, 043520.

Wang R. P., Bulla D., Smith A., Wang T. and Luther-Davies B. (2011), 'Structure and physical properties of $Ge_xAs_ySe_{1-x-y}$ glasses with the same mean coordination number of 2.5', *J. Appl. Phys.*, 109, 023517.

Wuttig M. and Yamada N. (2007), 'Phase-change materials for rewriteable data storage', *Nat. Mater.*, 6, 824–832.

Yang G., Bureau B., Rouxel T., Gueguen Y., Gulbiten O., Roiland C., Soignard E., Yarger J. L., Troles J., Sangleboeuf J.-C. and Lucas P. (2010), 'Correlation between structure and physical properties of chalcogenide glasses in the As_xSe_{1-x} system', *Phys. Rev. B: Condens. Matter Mater. Phys.*, 82, 195206.

Zachariasen W. H. (1932), 'The atomic arrangement in glass', *J. Am. Chem. Soc.*, 54, 3841–3851.

4
Thermal properties of chalcogenide glasses

J. D. MUSGRAVES, IRradiance Glass Inc., USA and
S. DANTO and K. RICHARDSON,
University of Central Florida, USA

DOI: 10.1533/9780857093561.1.82

Abstract: Thermal analysis techniques provide powerful tools for both the fundamental understanding and practical applications of chalcogenide glasses. Differential scanning calorimetry, thermogravimetric, and thermomechanical analysis all provide insight not only into those properties of chalcogenide glasses which are critical for hot-forming applications, such as precision molding or fiber drawing, but also into the atomic-scale structural units that give rise to this material behavior. Similarly, a clear understanding of the viscosity–temperature relationship in chalcogenides is necessary for glass processing, from melting to fabrication, but also gives deep insight into fundamental structural differences between sulfide, selenide, and telluride glasses.

Key words: calorimetry, thermomechanical analysis, viscometry, precision molding, fiber drawing.

4.1 Introduction

Thermal analysis is one of the most ubiquitous set of measurement techniques in glass science, and can deliver information about glass structure and properties over a wide range of length scales and application areas. Due to the fundamental differences in chemistry between chalcogenide glasses and their oxide counterparts, the thermal behavior and response of these glass families is markedly different in many important ways. The aim of this chapter is to present a broad overview of the wide variety of thermal analysis techniques and their application to chalcogenide glasses, with a focus on the use of these glasses in hot-forming applications such as extrusion, precision glass molding, and fiber optic drawing. Sections 4.2 through 4.4 will introduce some basic principles of thermal analysis, such as differential scanning calorimetry (DSC), thermogravimetric (TGA), and thermomechanical (TMA) study of the composition dependence of properties in the sulfur, selenium, and tellurium glass families. Section 4.5 opens with an overview of viscometric measurements, followed by a discussion of the important role of the viscosity-temperature curve in the hot-forming of chalcogenide glasses, and concludes with a discussion of viscometry as a tool for fundamental glass science. Section 4.6 discusses thermo-optic behavior, beginning with a description of the measurement and discussion of the impact of thermo-optic properties on device design, and ending with preliminary data on the composition dependence of this important optical design parameter. Section 4.7 will briefly suggest some future directions for the application of thermal analysis to the study of chalcogenide glasses and their various applications. The data presented in the figures in this chapter has been compiled from a large number of literature sources using the SciGlass database [1], and as thus reflect not a single set of experiments, but the best fit of the available data.

4.2 Differential scanning calorimetry (DSC)

Differential scanning calorimetry (DSC) is a thermal analysis technique that measures the heat flow into or out of a sample as a function of temperature, and is used to detect thermodynamic transitions such as glass transition, crystallization or melting. The glass sample, typically powdered in order to increase the surface area available for heat transfer, is heated in an inert sample pan and the heat flux is compared to that of an adjacent, empty, reference pan. Phase transitions are detected through the differential in heat flow supplied to the sample and reference pans; for example, the first-order phase transition of crystallization is an exothermic event, meaning that less heat flow is required from the calorimeter heaters to the sample pan in order to maintain the sample and reference pans at equivalent temperatures. Thus the difference in the heat flow supplied to the reference and sample shows a maximum during the phase change from amorphous to crystalline network structure. Figure 4.1 shows a representative DSC trace for a multi-component chalcogenide glass, indicating the positions of the glass transition (T_g), crystallization onset (T_x), crystallization peak (T_p), and melting peak (T_m).

The glass transition, which is generally considered to be a second-order phase change [2], is visible as a slight endotherm, which may display either a minimum or merely an inflection point depending on the composition and thermal history of the glass. By comparison, the first-order thermodynamic events of crystallization and melting are visible as much larger exo- and endotherms, respectively. As will be discussed in further detail in Section 4.3, crystallization is often accompanied by volatilization of chalcogenide components, thus these measurements are made in hermetically sealed containers. For optical applications, knowledge of the temperature of the crystallization onset is more important than that of the crystallization peak, as the crystalline phases (x) formed in chalcogenides typically have a much higher refractive index (n) than that of the amorphous (a) matrix (e.g. $n(a\text{-}Ge_{50}Te_{50}) = 3.88$ vs. $n(x\text{- }Ge_{50}Te_{50}) = 6.80$ at 1550 nm), meaning that small crystal volume fractions can lead to large scattering losses in these systems [3].

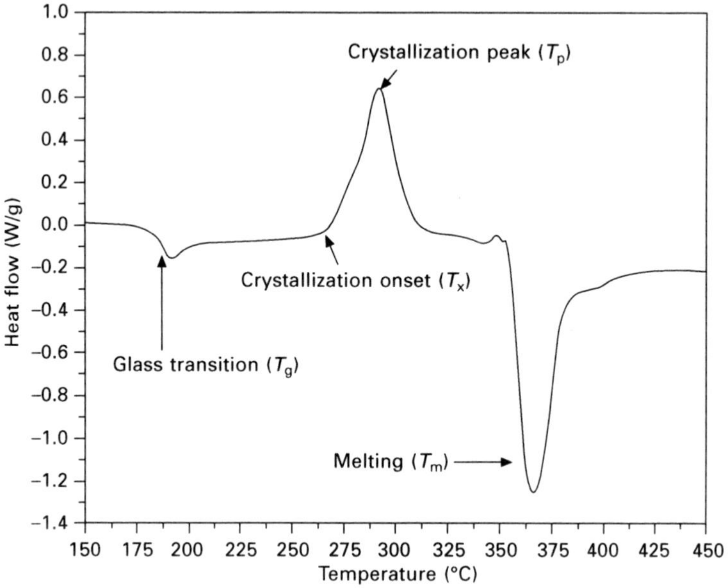

4.1 Example of a DSC trace for a multicomponent chalcogenide glass showing the characteristic temperatures of interest.

4.2.1 Glass transition temperature

The glass transition is perhaps the most important thermal characteristic of an amorphous material in determining its utility in various applications, as it represents the temperature at which the material stops behaving as a rigid solid and begins to show the mechanical response of a deeply supercooled liquid, suitable for hot-forming applications. The temperature of this change in mechanical response and its relationship to the onset of melting or volatilization, discussed in Section 4.3, is critical in hot-forming applications in glasses in general, and chalcogenide glasses in particular, which tend to have much smaller working temperature ranges than oxide glasses. In addition to its utility in optical applications, the glass transition is also profoundly interesting from a fundamental science perspective. Though the existence of a glass transition is one of the defining characteristics of glass as an amorphous solid, the precise atomic-scale nature of the transition is still not completely understood [4].

To better explore some of the physical and chemical factors underlying the glass transition, the values of T_g for arsenic- and germanium-based binary chalcogenide glasses are plotted in Fig. 4.2 with the bottom axis giving the mole percent of the modifier (the As or Ge), and the top axis giving the resulting average coordination number, $<r>$. The average coordination number gives a measure of the mechanical constraint in the network, and is

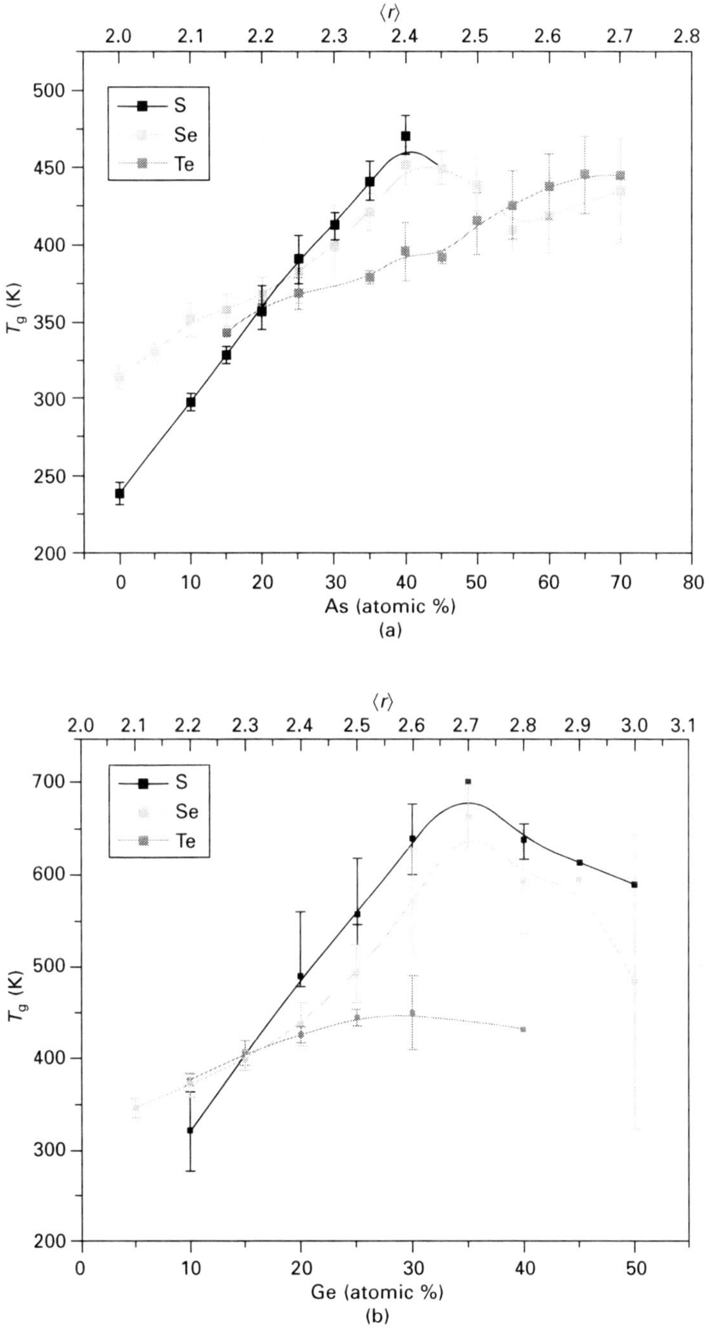

4.2 Variations in T_g for (a) arsenic- and (b) germanium-based binary chalcogenide glasses.

calculated for these binary systems by assuming the coordination number of Ge is 4, that of As is 3, and those of S, Se, and Te are 2, and taking the average weighted by the relative atomic fraction of each species in a given composition. In the definition of Cai and Thorpe and Phillips and Thorpe, amorphous networks with a value of $<r>$ of less than 2.4 are referred to as 'floppy' (meaning they support a population of zero-frequency normal modes) [5], those with $<r>$ greater than 2.4 are referred to as 'stressed rigid' (meaning that the network is overconstrained), and those with $<r>$ equal to 2.4 are called 'isostatic' (meaning stress-free) [6]. By looking at the evolution of various properties both as a function of modifier percent and coordination number, trends which are due to chemistry (modifier) or physical constraint (coordination number) become evident.

As depicted in Fig. 4.2, chalcogen-rich compositions show a roughly linear increase in T_g as a function of modifier. In this low modifier end of the plot, the increase in T_g in the Se- and Te-based glasses in each family appear to proceed at similar rates, while the T_g increase in the S-based glasses proceeds much more rapidly. This change in T_g as a function of modifier concentration was modeled by Gibbs and DiMarzio in the context of crosslinking of polymer chains as [7, 8]:

$$T_g = \frac{T_0}{1 - \kappa X} \qquad [4.1]$$

where X is the fraction of the crosslinking agent, T_0 is the glass transition temperature of the pure polymer chain, and κ is a system constant. The Gibbs–DiMarzio model was developed to describe how the glass transition of long organic polymer chains (i.e., rubbers, plastics, tars) varied as a function of crosslinking agent. This model was later successfully applied to chalcogenide glass systems by replacing X with the average coordination number $<r>$ of the network [9]:

$$T_g = \frac{T_0}{[1 - \beta(\langle r \rangle - 2)]} \qquad [4.2]$$

where β is now a system-dependent constant. This form accurately captures the linear increase in T_g in the low modifier end of the binary chalcogenides shown in Fig. 4.2, where β is left as a fitting parameter, which depends on the composition of the binary. Table 4.1 shows the best fits of Eq. [4.2] to the low modifier (> 20%) data shown in Fig. 4.2, where the values of T_0 for the S, Se, and Te systems are assumed to be 245, 316, and 343 K, respectively [10].

Interestingly, β is more strongly impacted by the identity of the chalcogen (compare 1.679 to 0.272 for AsS to AsTe) than that of the modifier (compare 0.739 to 0.752 for AsSe to GeSe). As may have been expected, the change in the glass transition behavior of the binary chalcogenide glasses is dominated

Table 4.1 Best-fit values of β from Eq. [4.2] to the low modifier (> 20%) data in Fig. 4.2

Glass system	Best-fit β
AsS	1.679
AsSe	0.739
AsTe	0.272
GeS	1.029
GeSe	0.752
GeTe	0.476

more by the network (chalcogen) than the modifier (As or Ge) in the low modifier range of the composition space.

Stochastic Agglomeration Theory attempts, in small part, to derive an analytical expression for β in order to express this constant in terms of more fundamental, network-oriented considerations, namely the coordination number of each atom in the chalcogenide binary [10]:

$$\frac{1}{\beta} = (m' - m)\ln\left(\frac{m'}{m}\right) \qquad [4.3]$$

where m is the coordination number of the chalcogen (i.e., 2) and m' is the coordination number of the modifier (3 for As and 4 for Ge). This expression was arrived at by 'solving' Eq. [4.2] for β, and expressing the average coordination number, $\langle r \rangle$, as a function of the coordination of the network, m, and the modifier, m'. The model predicts a β value of 2.466 for the As-based system and 0.721 for the Ge-based system, which is reasonably close to that shown in Table 4.1 for the GeSe system; clearly, though, this simple relationship linking the system-dependent β parameter to only mechanical constraints in the network has not captured the evolution of β with the chalcogen type, which indicates that there are additional contributions due to the chemistry of the system that have not been accounted for.

Returning to the trends shown in Fig. 4.2, the S- and Se-based glasses in both families exhibit maxima in the glass transition temperature at the stoichiometric compositions: $As_{40}X_{60}$ and $Ge_{33.33}X_{66.66}$ for X = S and Se. In both families there is much less evidence for a clear maximum in variation in T_g with modifier concentration in the Te-based glasses. These maxima in the S and Se binaries occur at $\langle r \rangle = 2.4$ and 2.7 for the As and Ge systems, respectively. Much has been made of the occurrence of the maximum of the glass transition temperature for the AsS and AsSe systems at an average network coordination of 2.4, which is directly in line with the transition from a floppy to stressed-rigid network according to Constraint Theory [11, 12]. However, the shift of this maximum to an average network coordination of 2.7 in the GeS and GeSe compositions, above even the 'intermediate phase' of network structures [13, 14], suggests that this maximum is not strictly determined by network mechanics, but by system chemistry as well. Some composition regions show clear evidence of nanoscale phase separation, which could affect the effective average coordination number of the network and put it more in line with the trends shown in Fig. 4.2, but the exact structural nature of these nanoscale phases is still a topic of debate in chalcogenide glasses [15, 16].

The same maxima shown in the glass transition temperatures of the binary sulfur and selenium glasses are evident in more complicated glass-forming systems based on these elements, as shown in Plate I (between pages 330 and 331), which illustrates the variation in T_g in the As-Ge-S and As-Ge-Se glass-forming regions of composition space.

The authors know of no published model which accurately captures the complexity of the variation in T_g with composition in these ternary systems. Small pockets of elevated T_g in composition space may indicate the presence of the nanoscale phase separation described above. The texture of the contours of T_g variation themselves are interesting in that they are not nearly as regular as those shown by variations in other properties such as the coefficient of thermal expansion (see Plate II between pages 330 and 331).

4.2.2 Crystallization stability

Crystallization stability, defined as the distance in temperature space between the glass transition and the onset of crystallization (i.e., $\Delta T = T_x - T_g$), determines the suitability of a given chalcogenide glass composition for various hot-forming applications. Stability towards crystallization in hot-forming processes represents a balance between the need to deform the glass, which must be above its glass transition temperature in order to flow, and the need to retain a crystal-free network in order to maintain the optical properties and shaping amenability of the amorphous material. A typical 'rule of thumb' suggests a minimum ΔT of 100 K is needed to provide a sufficient temperature window for applications such as fiber drawing, and as such, multi-component systems which show a lower propensity for crystallization are of special interest. It is possible to form chalcogenide glasses with ΔT less than 100 K; however, these typically require extreme forming procedures to prevent crystallization of the network [17]. Table 4.2 shows the evolution of the crystallization stability window as a function of composition in a four-component series of glasses as the chalcogen is gradually changed from sulfur to selenium [18].

Two systems are shown in Table 4.2, $Ge_{28}Sb_{12}S_xSe_{60-x}$ and $Ge_{28}As_{12}S_xSe_{60-x}$, where only the nature of one of the modifying elements has been changed (Sb to As). As both modifiers are threefold coordinated, any discrepancies between the two systems should be considered as chemical rather than

Table 4.2 Glass transition and crystallization temperatures, with corresponding stability windows, for two families of chalcogenide glasses

Composition	T_g(K) [±2K]	T_x(K) [±2K]	T_x(K) [±2K]
$Ge_{28}Sb_{12}Se_{60}$	615	800	185
$Ge_{28}Sb_{12}S_{45}Se_{15}$	596	784	188
$Ge_{28}Sb_{12}S_{30}Se_{30}$	592	766	174
$Ge_{28}Sb_{12}S_{15}Se_{45}$	595	770	175
$Ge_{28}Sb_{12}Se_{60}$	571	764	194
$Ge_{28}As_{12}Se_{60}$	678	698	120
$Ge_{28}As_{12}S_{45}Se_{15}$	660	789	129
$Ge_{28}As_{12}S_{30}Se_{30}$	651	784	134
$Ge_{28}As_{12}S_{15}Se_{45}$	649	782	133
$Ge_{28}As_{12}Se_{60}$	615	778	163

configurational in nature. Though the glass transition temperatures of the antimonide system are lower than those measured in the arsenide system, the antimonide system exhibits a 44 K decrease in T_g and a 36 K decrease in T_x with selenium substitution for sulfur, while the arsenide system exhibits a 63 K decrease in T_g and an 11 K decrease in T_x with the same substitution. These factors combine to give an increasing stability window in both systems with increasing selenium content; however, the antimonide has a higher ΔT for a given sulfur/selenium ratio than the arsenide system.

4.2.3 Heat capacity

The heat capacity of a glass is a measure of the amount of thermal energy, q, required to raise the temperature of the material by 1 K. Thermal energy is converted into normal mode vibrations in the glass network, and the connectivity of the network can be expected to contribute strongly to the measured heat capacity of the glass. Heat capacity is typically measured in ambient laboratory environment using DSC, thus the reported heat capacity is the one measured at a given pressure (i.e., C_p). The heat capacity is calculated as the change in thermal energy delivered by the DSC to the sample over the change in temperature of the system:

$$C_p = \left(\frac{\partial H}{\partial T}\right)_p \approx \frac{q_p}{\Delta T} \qquad [4.4]$$

where the constant pressure heat capacity is defined in thermodynamics as the partial derivative of the enthalpy, H, with respect to temperature at a constant pressure. As enthalpy is a direct measure of heat flow at constant pressure, this can be rewritten as q_p, which is the term measured directly by the DSC.

As it is a reflection of the underlying phonon density of states, the heat capacity of chalcogenide glass, just like any other material, is a function of temperature. Constant pressure heat capacity of a chalcogenide glass exhibits a global maximum in the glass transition range as configurational changes, including structural relaxation, contribute to the heat capacity in this temperature range in addition to the simple vibrational modes discussed above. Below and above the glass transition region, the heat capacity is a weak function of temperature in chalcogenide glasses [19]. To appropriately assess its variation as a function of composition, the heat capacity of two families of glasses below their respective glass transition temperatures is shown in Fig. 4.3.

For both the arsenic- and germanium-modified systems, the heat capacity of the (As,Ge)-Se binary is approximately 300 J/kg.K, regardless of modifier concentration or, alternatively, network constraint. This same phenomenon is exhibited in oxide glasses as well, with the heat capacity of simple soda-lime silicates measuring approximately 900 J/kgK, with very weak dependence on modifier concentration [20]. This observation in the selenide glasses is in stark contrast to that exhibited by the As-S system, which exhibits a clear decrease in sub-T_g heat capacity with increasing arsenic content. However, this apparent trend is an artifact of the units in which the heat capacity is presented: the standard units, J/kg.K, do not account for variation in the molar mass of the glass as a function of arsenic content; a more suitable measure for the heat capacity would be to present it as a molar unit, i.e. J/mol.K. The molar volume of the germanium- and arsenic-selenide glass series is a much weaker function of modifier concentration than that exhibited by the arsenic sulfide series; the molar volume varies from approximately 78.9 to 77.4 g/mol in the arsenic selenides as x varies from 0 to 40, but over the same modifier range, the molar volume changes from 31.6 to 49.2 g/mol in the sulfide system, which gives rise to the apparent variation in heat capacity shown above.

4.3 Thermogravimetric analysis (TGA)

Thermogravimetric analysis (TGA) measures the mass of a glass sample as a function of temperature. In the context of hot-forming applications, this tool is critical in defining the safe upper working temperature of a given composition (here 'safe' means both safety for the experimentalist as well as safety of the fidelity of the original composition). While the crystallization stability window $\Delta T = T_x - T_g$ defines a thermodynamically suitable temperature range, it does not account for the possibility (very high in some compositions) of the glass out-gassing at temperatures above T_g but below T_x, precisely in the working range for forming applications. Figure 4.4 shows a comparison of the viscosity and weight loss profiles of the ternary chalcogenide glass $Ge_{10}As_{40}Se_{50}$. The dashed lines indicate the

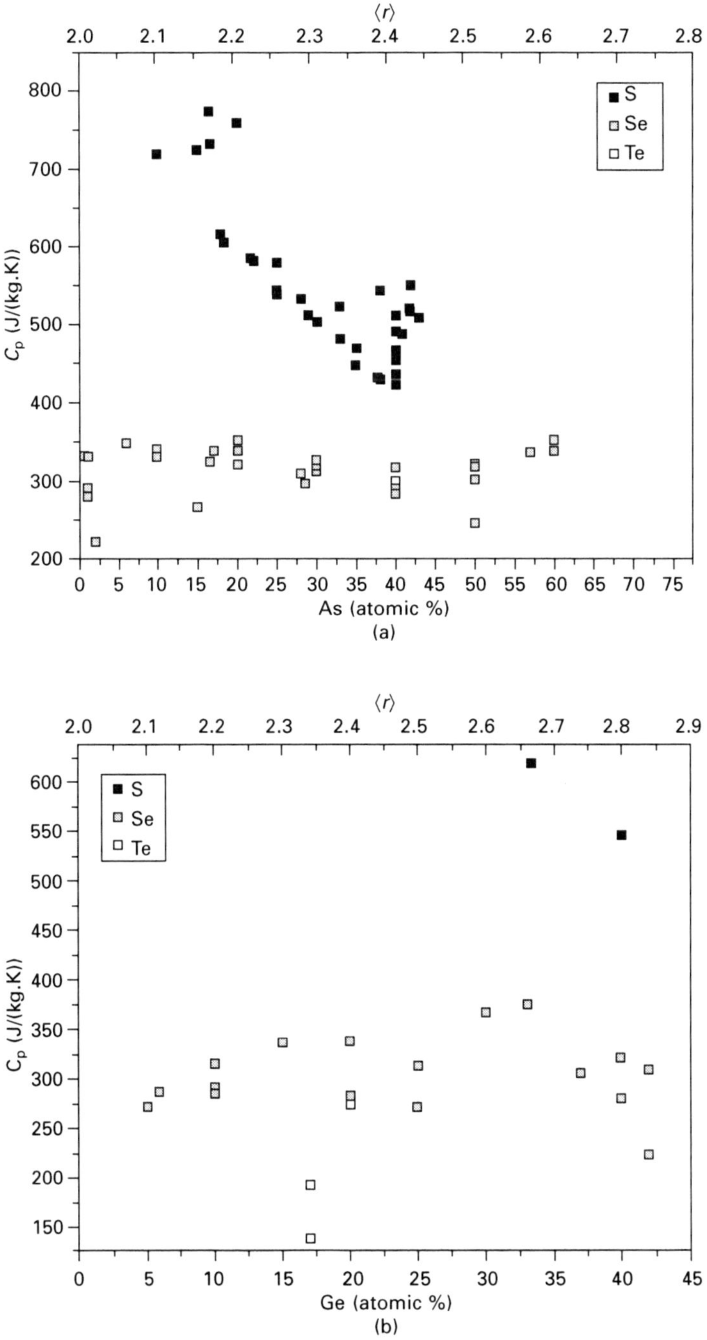

4.3 Variations in C_p for (a) arsenic- and (b) germanium-based binary chalcogenide glasses.

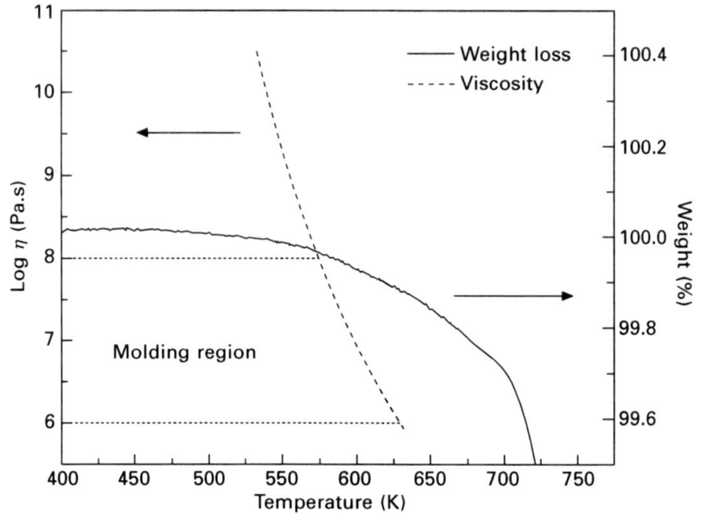

4.4 A comparison of the viscosity and volatilization curves for $Ge_{10}As_{40}Se_{50}$ glass showing evidence of out-gassing in the molding region.

viscosity range (10^6–10^8 Pa.s) necessary for molding or extrusion of the chalcogenide glass, with the associated temperature range (573–628 K). As this figure indicates, by the temperature at which the glass is moldable, there is a non-zero volatilization of glass materials. This change in composition is unacceptable from both a technological and safety perspective, as it will affect the refractive index of the material, and the volatilized components can be hazardous to humans.

If one assumes a maximum tolerable volatilization level of 0.2 wt.% for a given application, the TGA can be used to provide safe working ranges for the hot-forming of chalcogenide glasses. Table 4.3 shows a comparison of the glass transition temperature and temperature of 0.2 wt.% loss for some commercially available chalcogenide glass compositions as determined by TGA measurements made in the authors' laboratory. The heating rate for the glass transition measurement was 10 K/min, and that used for the weight loss measurements was 1 K/min; in both cases the samples consisted of 10–30 mg of powdered glass. Based on these values, a suggested 'safe' working range for each glass is shown at the right of the table. It should be noted that this definition of 'safe working range' is ours as defined here, but will vary depending on application and specific laboratory safety considerations and guidelines.

The largest safe working ranges belong to the $Ge_{30}As_{13}Se_{32}Te_{25}$ and $Ge_{28}Sb_{12}Se_{60}$ compositions, due to their structure containing the heavier (and more difficult to volatilize) Sb and Te elements. The next widest working

Table 4.3 Glass transitions and 0.2 wt% loss temperatures for some commercially available chalcogenide glasses

Composition	T_g(K)	$T_{0.2x}$(K)	Safe working range (K)
$Ge_{33}As_{12}Se_{55}$	611	726	115
$Ge_{30}As_{13}Se_{32}Te_{25}$	565	752	187
$Ge_{10}As_{40}Se_{50}$	516	667	151
$Ge_{28}Sb_{12}Se_{60}$	576	757	181
$As_{40}Se_{60}$	468	621	153

ranges are exhibited by the $Ge_{10}As_{40}Se_{50}$ and $As_{40}Se_{60}$, indicating that low levels of Ge do not significantly impact the volatilization tendency of the arsenic selenide glass family. The lowest safe working range is shown by the $Ge_{33}As_{12}Se_{55}$ composition, not because its 0.2 wt.% loss temperature is lower, but because of its comparatively high T_g. Note that these safe working ranges are not identical to the crystallization stability window, ΔT, for a given glass, and indeed the change in composition may be triggered by the onset of crystallization or vice versa.

4.4 Thermomechanical analysis (TMA)

Thermomechanical analysis (TMA) provides information on the mechanical response of a glass sample to an applied thermal load. Measurement of the expansion or contraction of a glass composition as a function of temperature can be conducted with either a TMA instrument, which operates in a vertical geometry, or a dilatometer, which operates horizontally. In either case, a silica pushrod is placed in contact with one face of the glass sample, and the movement of this pushrod is measured with a linear variable differential transformer (LVDT).

The coefficient of thermal expansion, or CTE, is a measure of the mechanical response of a glass network to an applied thermal load. The linear expansion of a system in response to an increase in temperature is thus written as

$$dL = \alpha_L L dT \quad \text{or} \quad \alpha_L = \frac{1}{L}\frac{dL}{dT} \qquad [4.5]$$

where L is the length of the sample at room temperature, dL is the change in length, dT is the change in temperature and α_L is the linear CTE. Because the CTE measurements are typically performed below the glass transition temperature, there is no contribution to the thermal expansion from conformation or bonding changes in the material; the expansion is a reflection of the aggregate change in interatomic bonding distances. In general, as the temperature of a glass is increased in the region below T_g, the additional thermal energy permits the elongation of the mean bond length as dictated by the shape of the potential energy well representing the bond [21].

As shown in Fig. 4.5, in the arsenic system, the evolution of the measured CTE with As % follows an almost linear trend with both the sulfur- and the selenium-based systems, with an apparent local minimum at the stoichiometric $As_{40}S_{60}$ ($\langle r \rangle = 2.4$) composition, and a similar local minimum at the slightly arsenic-rich composition of $As_{50}Se_{50}$. Following these minima, the linear trend is no longer evident, as the CTE begins to rise with increasing arsenic content before leaving the glass forming region.

This minimum in the CTE in the arsenic selenide series, as well as the seeming lack of minimum in the germanium selenide series, can be seen together in Plate II (between pages 330 and 331). The regularly-banded contours of the CTE variation, indicative of a property which exhibits an orderly variation as a function of composition, are interrupted by a brief excursion where the 20 ppm/K band pushes out to meet the As-Se binary tie line. That this minimum occurs where the arsenic selenide system is overconstrained ($\langle r \rangle = 2.5$) suggests nanoscale phase separation may be occurring in these systems.

4.5 Viscometry

Knowledge of the viscosity of chalcogenide glasses at a given temperature is one of the most fundamental concerns in the formation and processing of these materials, possibly even more critical than the value of the glass transition temperature, T_g, as the glass transition can be estimated from the viscosity curve. The viscoelastic behavior of a glass is critical in all hot-forming processes, as it dictates the time and temperature scales needed to precisely produce optical elements, from slow, low-temperature extrusion to rapid, high-temperature fiber drawing. Aside from its technological importance, the viscosity–temperature relationship has significant value from the perspective of fundamental research into the connections between the atomic structure of amorphous solids and the properties that result from this structural arrangement.

4.5.1 Techniques

A wide variety of viscometric techniques are used in industrial and academic glass science due to the wide range of viscosities exhibited by a glass as a function of temperature. Viscosity of chalcogenide glasses changes over 17 orders of magnitude in less than 500 K, meaning the viscosity ranges appropriate for various hot-forming methods may only be accessible in small (<50 K) temperature windows. Some of these forming regions, as well as an overview of some of the viscometric techniques used are shown in Fig. 4.6.

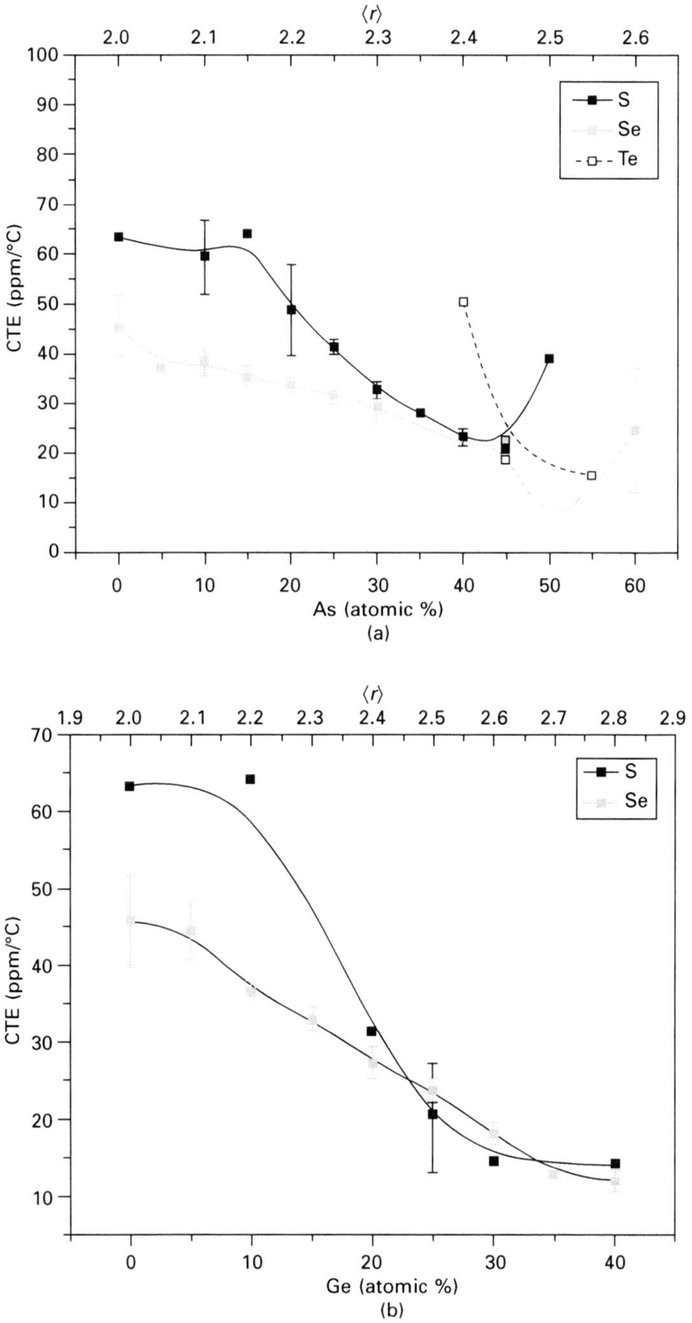

4.5 Variations in CTE for (a) arsenic- and (b) germanium-based binary chalcogenide glasses.

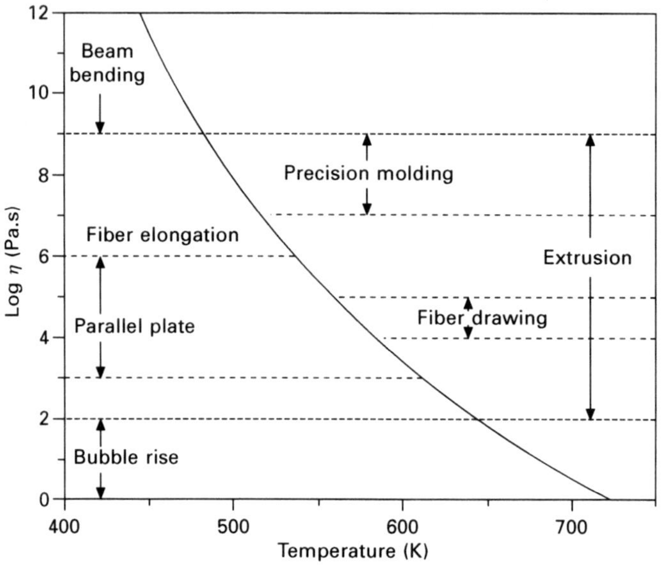

4.6 An overview of the viscosity regions for a chalcogenide glass, showing the ranges for measurements (left) and hot-forming applications (right).

Each measurement technique requires a different sample geometry, some with precise tolerances and others without, and as with many measurements, the quality of the sample dramatically impacts the accuracy of the data obtained. Data from multiple viscometric techniques combined to generate a viscosity–temperature curve for a given material, such as that shown in Figs 4.6 and 4.8.

Parallel plate viscometry

Parallel plate viscometry is used to probe the viscosity of a glass in the range of 10^3–10^6 Pa.s, where the glass is behaving as a viscous supercooled liquid rather than a solid in terms of flow. A right cylinder of the glass is compressed under load at temperature between two parallel plates. From the compression rate, the viscosity of the glass can be calculated using [22]:

$$\eta = 2\pi \frac{Mgh^5}{30V(dh/dt)(2\pi h^5 + V)(1 + \alpha T)} \quad [4.6]$$

where η is the viscosity in Pa.s, M is the applied load, g is the acceleration due to gravity, h is the sample height at time t, V is the sample volume, dh/dt is the compression rate, and α is the CTE of the glass.

Beam bending viscometry

Beam bending viscometry probes the viscosity of a glass in the range of 10^9–10^{13} Pa.s, where the glass's flow behavior resembles that of a solid rather than a viscous liquid. The temperature at which the viscosity is equal to 10^{12} Pa.s is typically equated to the glass transition, T_g, though this is largely an empirical convention rather than a thermodynamic or kinetic distinction. Beam bending viscometry measures the deflection of a beam of glass under load at a defined temperature using a LVDT. From this deflection, the viscosity of the glass at temperature T can be calculated using [23]:

$$\eta = \frac{gL^3}{1440\, I_c \dfrac{dh}{dt}}\left[\frac{M + (\rho AL)}{1.6}\right]\left[\frac{(1 + \alpha_s T)^3}{(1 + \alpha_g T)^4}\right] \qquad [4.7]$$

where η is the viscosity in Pa.s, g is the acceleration due to gravity, L is the support span, I_c is the cross-sectional moment of inertia, dh/dt is the deflection rate, M is the applied load, ρ is the glass density, A is the cross-sectional area of the beam, α_s is the coefficient of thermal expansion (CTE) of the support stand and α_g is the CTE of the test glass.

Fiber elongation

Fiber elongation is a viscometric technique used to measure intermediate viscosities between those probed by the beam bending and parallel plate methods described above. In particular, the fiber elongation technique can be used to precisely calculate the Littleton softening temperature of the glass, where the viscosity is defined as $\eta = 10^{6.6}$ Pa.s [24]. Knowledge of the Littleton softening temperature is critical in the analysis of fiber drawing parameters for chalcogenide glasses, as it describes the temperature at which a glass fiber will deform under its own weight. This viscosity–temperature value is measured by observing the rate of deflection of a sample fiber contained in a furnace. The viscosity can be calculated from this deflection rate using

$$\eta = \frac{\left(Lp + \dfrac{p^2}{2}\right)\rho g - \dfrac{\gamma p}{r}}{3\dfrac{dp}{dt}} \qquad [4.8]$$

where η is the viscosity of the glass in Pa.s, L is the length of fiber outside of the furnace, p is the length of fiber inside the furnace, ρ is the glass density, g is the acceleration due to gravity, γ is the surface tension of the glass, and dp/dt is the elongation rate of the fiber.

4.5.2 Applications

The right-hand side of Fig. 4.6 shows the viscosity ranges for several hot-forming methods commonly used to shape chalcogenide glasses: extrusion, precision glass molding (PGM), and the drawing of optical fibers from bulk chalcogenide glass preforms. Extrusion is the most versatile of these techniques in terms of total possible viscosity range covered by a single technique. However, different applications (i.e., different geometries) require different extrusion viscosities in order to achieve optimal device performance; extrusion can be used to form glass with a viscosity like that of a soft solid being forced through a die (10^9 Pa.s), or like that of a viscous liquid being forced into a channel (10^2 Pa.s). PGM covers the next widest viscosity range of the three techniques, and is a process for molding precision chalcogenide optical lenses while the glass has the viscosity of a slightly softened solid (10^7–10^9 Pa.s). Drawing optical fiber from chalcogenide glasses requires a viscosity below that of the Littleton softening point ($10^{6.6}$ Pa.s) where a glass fiber would deform under its own weight, and is typically done in the viscosity range around 10^5 Pa.s. Each of these techniques is used to obtain precise geometries, dimensions, and surface characteristics by taking advantage of some portion of the wide range of viscosities chalcogenide glasses exhibit in comparatively (to oxides) short temperature windows.

Extrusion

The extrusion of chalcogenide glasses is in many ways no different than the extrusion of polymers: a viscoelastic material is forced into a channel or through a die to form a desired geometry, then the material is cooled back below is glass transition to 'lock in' the solid structure. Depending on the desired viscosity (extrusion can be used to form glasses with viscosities from 10^2 to 10^9 Pa.s), there may or may not be danger of volatilization of the chalcogenide glass during processing; in the high viscosity region, glass can be extruded only 25–50 K above their glass transition temperature. The high CTE of chalcogenide glasses can present challenges in the extrusion of precise geometries. The prediction of the precise volume loss as the material contracts after leaving the hot-forming region is extremely difficult, but efforts are underway using computing techniques such as finite element analysis to simulate the thermal response of chalcogenide glasses in order to define accurate die geometries to obtain a desired final piece geometry. For low viscosity extrusions, there is a risk of crystallizing the glass during processing, which would undermine its shaping ability and the use of the glass in optical applications due to scattering or high index contrast between the two phases.

Extrusion is increasingly used as a tool to produce preform pieces from which to draw microstructured optical fibers, as complex geometries can be extruded that may not be accessible through drilling, stack and draw, or other structure-forming methods [25–27]. In addition to the production of more complex geometries, extrusion can also be used to produce pieces or preforms from complicated material combinations such as chalcogenide glasses paired with oxide glasses, polymers, or metals. An example of such a combination is shown in Fig. 4.7 [28].

When drawn into a fiber, the structure in Fig. 4.7 acts as a one-dimensional photodetector. The insulator material is a polymer, the semiconductor thin film is $As_{40}Se_{60}$ glass, and the metal is tin; if an electrical voltage is applied across the metal electrodes, then an incident photon will excite a charge carrier in the $As_{40}Se_{60}$ glass, which will then register as an increase in current measured in the fiber.

Precision glass molding

Extensive growth of opto-electronic technologies has created a demand for high-quality chalcogenide lenses and has driven the industry toward an inexpensive process for manufacturing of aspheric glass lenses called precision glass molding (PGM). In addition, PGM can achieve more complicated optical element shapes than typically realized via conventional (grinding and polishing) fabrication techniques and where single point diamond turning

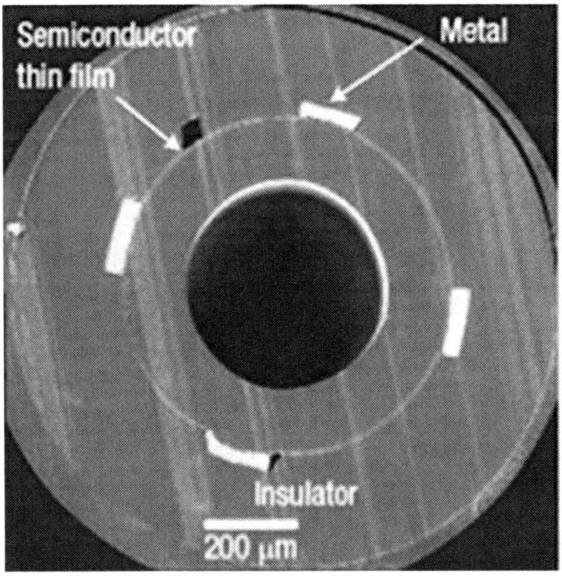

4.7 Cross section of a microstructured optical fiber incorporating chalcogenide glass film, polymeric cladding, and metallic electrodes.

(SPDT) cannot be employed. Aspheric lenses are rapidly replacing spherical lenses in many high-end optical applications, as they eliminate spherical aberrations in the focal element, and quite often, reduce weight. This is especially important in infrared (IR) optical systems where elements are made from high index (high density and weight) materials. PGM is already extensively used to mold oxide glasses and polymers, which have a much shallower viscosity/temperature dependence than chalcogenides, and do not present the same volatilization dangers discussed in Section 4.3

In the PGM process the glass preform, often a precision (mass) gob or with a spherical surface of similar deviation, is raised above its T_g to a viscosity of 10^7–10^9 Pa.s, pressed between two mold surfaces to form the desired geometry, then cooled rapidly back to a solid. A schematic of the heating and pressing process for a chalcogenide lens is shown in Fig. 4.8. After soaking at temperature for 240 s, the chalcogenide is pliant enough to be deformed from its original to its final, desired geometry. Because this is intended to be a single-step process, particular care is given to the surfaces of the molds, which are typically made from a high temperature ceramic such as silicon- or tungsten-nitride. These mold surfaces must remain optically polished, which will be imparted to the chalcogenide lens during the pressing cycle. A major challenge in the pressing of chalcogenide aspheres is preventing sticking between the glass and the mold surface, for which specialized coatings have been developed which serve as an effective barrier to sticking without contaminating the surface of the precision lens [29, 30]

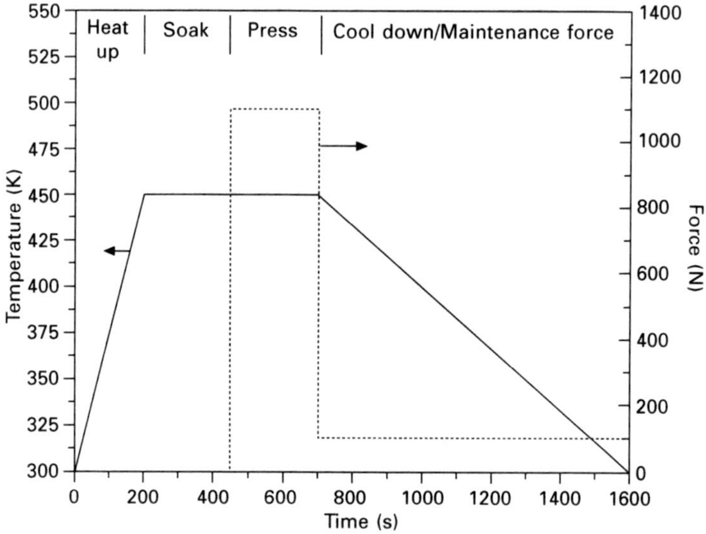

4.8 The temperature and force schedule for the precision molding of a chalcogenide glass.

Precision molding of chalcogenide glasses is non-trivial using the scheme shown in Fig. 4.8 because their thermal properties are much different than those of oxide glasses. For example, the CTE of chalcogenide glasses, discussed in Section 4.4, are much higher than those of visible-optics oxide glasses; an arsenic sulfide lens can have a CTE of 60 ppm/K, as compared to 7.1 ppm/K for commercially available Schott N-BK7 glass. This CTE difference means that the chalcogenide will contract a greater volume during the cool down period than an oxide glass would, and compensations for this change in volume need to be built into the mold geometry. Low thermal conductivity and steep viscosity curve for chalcogenide glasses bring another layer of difficulty, which make precise control of the temperature of the molding system and the glass critical to achieving good parts [31–34]. Thus, to efficiently determine an optimized thermal cycle for a new candidate optical design (component material and shape), it is necessary to employ computational techniques which include contributions due to these various properties along with relaxation properties (structural and stress).

Fiber optic drawing

To date, determination of the forming regions (for both PGM and fiber drawing) for a new chalcogenide glass type has been done using empirical, rather than deterministic methods. Recently, researchers have made large advances in prediction of these thermal regimes using numerical modeling methods [35]. Drawing down a preform of chalcogenide glass to form an optical fiber requires raising the preform's temperature to as much as 100 K or more above its glass transition temperature in order for the glass to flow sufficiently easily ($\eta = 10^5$ Pa.s), allowing it to be pulled to a diameter up to 1000 times smaller than that of the starting bulk. As described in Section 4.3, the volatilization temperature of some chalcogenide compositions, including $Ge_{33}As_{12}Se_{55}$, can be approximately 100 K higher than T_g, so care must be taken to prevent out-gassing of the glass, which would detrimentally change the composition, and could be harmful to humans. Because the viscosity curves of chalcogenide glasses are so much steeper than those of most oxides, the temperature window at which the glass is near a viscosity of 10^5 Pa.s is comparatively smaller, thus precise temperature control is critical in drawing chalcogenides. The crystallization stability window, ΔT discussed in Section 4.2, of a given chalcogenide glass can also limit the available composition options; a $Ge_{20}Te_{80}$ composition may be quenched into an amorphous preform bulk [36], but when it is heated to the draw temperature, it will crystallize completely.

Despite the limitations, chalcogenide glasses have found great success in the application field of fiber optics. Photonic crystal fibers have been demonstrated in chalcogenide glasses [26], which show potential for novel optical properties such as dispersion tailoring [37], supercontinuum generation [38], and single-mode infrared guidance [39]. Figure 4.9 shows a photonic crystal fiber formed from $Ga_5Ge_{20}Sb_{10}S_{65}$ glass. The photonic crystal formed by the holes in the structure effectively guide light through the solid core. The high refractive index of chalcogenide glasses improves the confinement of the guided light, and guiding losses have been demonstrated in the range of 5 dB/km at 1.5 μm [25, 40].

The high optical nonlinearity of chalcogenide glasses is being used in the development of all-optical switching, Raman gain, and infrared signal regeneration applications [41–43]. These nonlinear properties also contribute to the photosensitivity exhibited by chalcogenide glasses, which has been utilized to write waveguides and Bragg gratings into these materials [44]. In many ways, fibers take the best advantage of many of the novel optical properties that define the chalcogenide glasses, as they present a long optical path length over which the nonlinear properties can have the greatest impact.

Recently a new preform-based fiber processing technique has emerged that allows the simultaneous drawing of chalcogenide glasses in combination with metallic electrodes and polymers into tens-of-meters long structures [45]. The incorporation of materials with vastly disparate optical and electrical properties in fibers expands the traditional functionalities of the later ones from optic to acoustic [46], opto-electronic [47] or electronic [48].

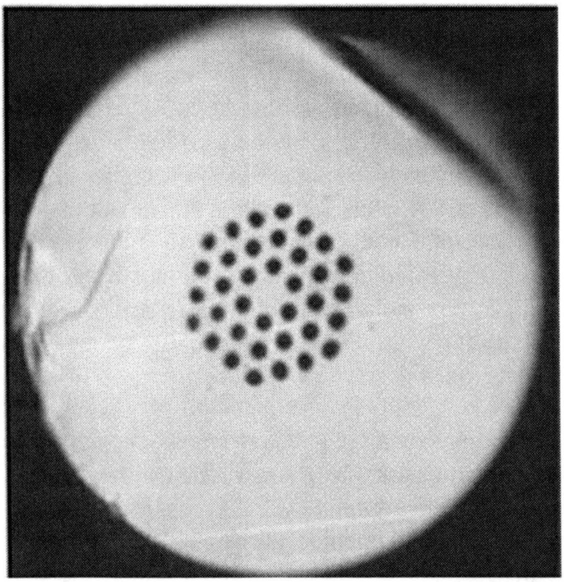

4.9 A photonic crystal fiber formed from $Ga_5Ge_{20}Sb_{10}S_{65}$ glass.

4.5.3 Fundamental science

In addition to the important role of viscosity in the hot-forming applications discussed in Section 4.5.2, scientific study of the viscoelastic behavior of glasses gives powerful insight into the atomic-level structure of amorphous networks. For example, the dimensionality, or connectedness, of an amorphous network of atoms has a direct impact on the viscous flow behavior exhibited by bulk glass: a network composed of intertwined chains of atoms will flow differently than a network composed of sheets or clusters of atoms. Understanding how the viscosity behavior of a glass family evolves with variations in composition can shed light on the underlying network structures and dynamics that give rise to that behavior. The viscosity of the supercooled liquids in the AsS and AsSe families are shown in Fig. 4.10. Again, these data represent a compilation of literature values obtained through the SciGlass database [1].

The available data for each composition has been fit using the empirical model developed by Vogel, Fulcher, and Tamann (known as the VFT model) [49]. The VFT model employs three fitting parameters (A, B, and C in Eq. [4.9] below) to capture the curvature of the viscosity–temperature dependence. Note that while there exist modifications to this simple model, they serve mainly to improve the quality of the extrapolation of the fit to extremely high and low viscosity regimes [50].

$$\log_{10} \eta = A + \frac{B}{T - C} \qquad [4.9]$$

The change in the viscosity–temperature behavior as a function of composition of the chalcogenide glasses in Fig. 4.10 can be evaluated using the fragility parameter, as expressed in Eq. [4.10]. The fragility parameter is a measure of the steepness of the viscosity curve, and is defined as the slope of the curve evaluated at the temperature at which the viscosity is equal to 10^{12} Pa s, T_{12}, which is commonly equated to the glass transition temperature, T_g, of the glass [51]:

$$m \equiv \left. \frac{\partial(\log_{10} \eta)}{\partial\left(\frac{T_{12}}{T}\right)} \right|_{T=T_{12}} = \frac{BT_g}{(T_g - C)^2} \qquad [4.10]$$

The right-hand side of Eq. [4.10] shows the evaluation of the fragility parameter when the VFT formalism of Eq. [4.9] is employed, with B and C taken from the VFT expression. The fragility parameter represents how far from a purely Arrhenius description the viscosity of a glass is: silica, the prototypical Arrhenius glass has a fragility of 17, while more 'fragile' polymeric glass formers have a fragility parameter near 200, meaning that

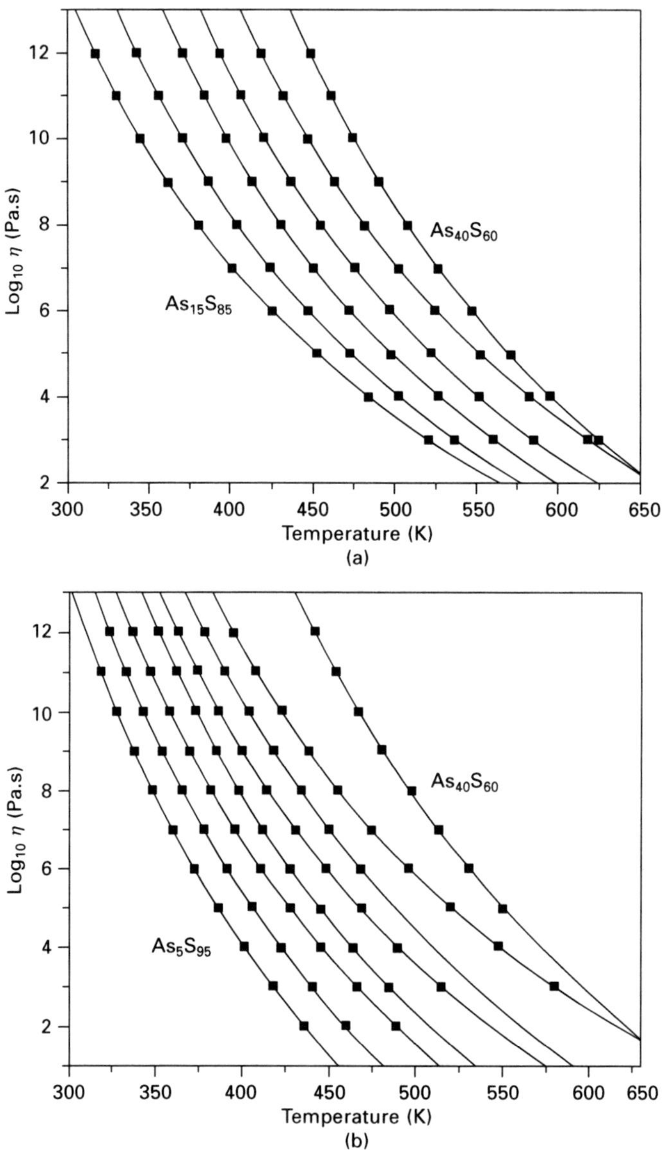

4.10 The temperature–viscosity dependence of the AsS and AsSe glass families.

their behavior is not properly described by the simple single exponential decay of the Arrhenius function [52, 53]. Chalcogenide glasses have been shown to possess 'intermediate' behavior, with fragility parameters typically between 20 and 40, much closer to that of the highly crosslinked silica network than to that of a fragile glassy polymer.

Using Eq. [4.9] to fit the data in Fig. 4.10, the fragility parameter can be calculated for these two families of chalcogenide glasses; these results are shown in Fig. 4.11.

The fragility parameter of the arsenic selenide glass system exhibits an interesting behavior: initially, the fragility of the system decreases (i.e., the viscosity curves become less steep) with the addition of low levels of arsenic, but as the glass approaches 33 molar percent of modifier ($\langle r \rangle$ = 2.33), the fragility changes direction and begins to overlap with that of the arsenic sulfide glass system. Preliminary results indicate that it is the only one of the binary chalcogenide compositions measured to display such a property; the fragility parameters of arsenic sulfide, germanium sulfide, and germanium selenide all vary monotonically with modifier concentration. These results suggest that there is a fundamental change in the atomic-level network topology in the arsenic selenide series that does not occur in the other binary chalcogenides.

4.6 Thermo-optic behavior

The variation in the index of refraction of a chalcogenide glass with a change in temperature is described as thermo-optic behavior, and is an extremely important, and often overlooked, aspect of infrared optical design.

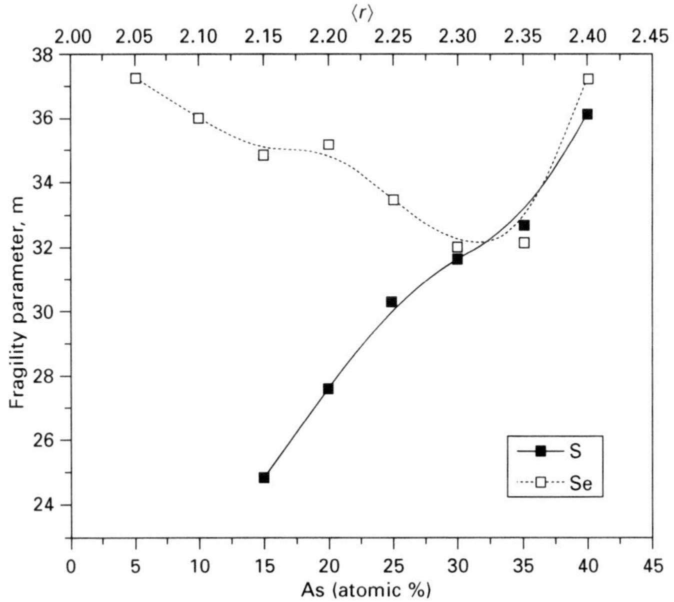

4.11 Variations in fragility parameter for As-Se and As-S binary chalcogenide glasses.

The induced change in refractive index is due to the thermal excitation of phonons (and electrons in some cases), and can be either positive or negative in sign depending on the composition of the glass. A positive index change in response to an applied thermal load is responsible for such phenomena as thermal lensing, self-focusing and laser damage (at high intensities), and spectral instabilities in laser resonators [54]. Despite its importance, this thermal phenomenon has received less study than the properties discussed in preceding sections both because the instrumentation for measuring the thermo-optic coefficient as a function of wavelength is underdeveloped in the infrared, and because, as a weak optical phenomenon, it only begins to impact device design when high-precision optics are necessary [55].

The thermo-optic coefficient (dn/dT) can be measured with a prism coupling setup with an attached hot stage. A schematic of such a prism coupling measurement is shown in Fig. 4.12. The chalcogenide glass sample is used as a substrate, and pressed in optical contact with a prism having a well-characterized index dispersion and dn/dT. The refractive index of the chalcogenide glass at a given wavelength and temperature is measured by launching laser light of appropriate wavelength into the prism. The prism and substrate are then rotated until the incident beam is coupled from the prism into the substrate (seen as a drop in the intensity of the refracted beam); the coupling condition is dependent on the index of the substrate, that of the prism and the angle of refraction, and knowledge of the latter two enables precise calculation of the former.

The attached hot stage can be used to heat the prism and substrate and thus measure the change in index as a function of temperature, dn/dT, at the incident wavelength. A major difficulty in accurately characterizing the thermo-optic coefficient in chalcogenide glasses is in acquiring data at a sufficient number of visible and infrared wavelengths to allow a reasonable fit for the theoretical models [56]. Changes in the refractive index as a

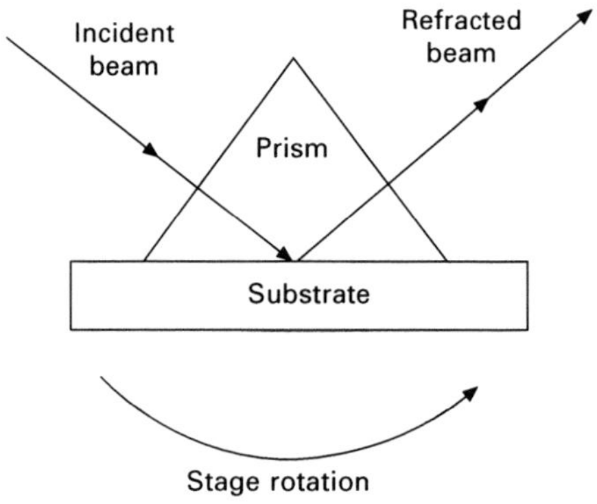

4.12 Schematic of a prism coupling system used to measure dn/dT.

function of both wavelength and temperature for crystalline germanium [57] and a germanium-containing chalcogenide glass are shown in Plate III (between pages 330 and 331). The refractive index of the germanium crystal is a stronger function of temperature than that of the chalcogenide glass; however, this is not the case for all compositions, and in some cases the value of d*n*/d*T* can even be negative.

4.7 Conclusion and future trends

Chalcogenide glasses possess a wide range of properties which make them very attractive for fundamental research and technological development. In this chapter we have presented an overview of their thermal properties, and how these properties emerge from the material's atomic structure and impact their hot-forming ability. We have turned our focus on the standard thermal analysis techniques commonly used in glass science (differential scanning calorimetry, thermogravimetry, thermomechanical analysis, and various viscometric techniques), and discussed briefly the viscoelastic and thermo-optic behaviors of the chalcogenide glasses and their impact on device design and fabrication. Specifically, an attempt was made, in binary and ternary chalcogenide systems, to reveal the composition dependence of thermal properties as a function of modifier element content or average coordination numbers. Though similar trends occasionally arise from direct system-to-system comparisons, no global pattern encompassing the S-, Se-, and Te-based glasses emerges, even in the simplest binary cases. Together, these thermal analysis techniques give precious insight into the response of the chalcogenide glasses with temperature: they contribute to measuring the characteristic thermodynamic transitions of the analytes, and give information on the kinetics underlying crystallization. Ultimately, they allow the prediction, critical from a technological standpoint, of phenomena such as phase separation, vaporization, or deformation. At a fundamental level, the gathered information yields essential understanding of the structural arrangement of chalcogenides, and how it affects their thermo-mechanical, optical and electrical properties.

Thermal analysis of chalcogenide glasses has proceeded from the tools and methodology inherited from the study of other systems, mainly silica. However chalcogenides exhibit specific features (steep viscosity curves, oxidation, vaporization and associated mass changes, shift of the spectral window in the infrared), and span a variety of final product configurations, which yield particular challenges. The thermal drawing of chalcogenides into fibers, for instance, remains largely semi-empirical because of the lack of real-time information on the behavior of the system, especially at the bottom-neck zone, while being simultaneously subjected to heating and tension. The demonstration of the co-drawing of composite structures made of a set of materials (metal, polymer, and chalcogenides) with disparate viscoelastic properties reinforces this assessment [28]. Phase-change materials, which constitute another exciting area of research for chalcogenide glasses [58], encounter similar analytical limitations. These materials, inherently unstable, require multi-layer depositions to reach their target thicknesses. This makes the *in-situ* probing of their glass and crystal phases impractical and leads researchers to rely heavily on indirect measurements and extrapolation. The investigation of chalcogenides would then certainly benefit from the development of dedicated in-depth, real-time analytical instrumentation; however, we expect that the deepest (r)evolution in the future of chalcogenide glass development will proceed from the spectacular progress enabled by computer microprocessors. Already, robust and ergonomic modeling software is spreading in laboratories worldwide to assist in predicting the thermal behavior of materials and to support experimental studies.

Though part of our daily lives, the true nature of the glassy state remains enigmatic. Emblematically, the glass transition temperature, the thermal signature of the glassy state, still lacks a well-accepted theoretical development. Chalcogenide glasses are complex systems where chemistry, kinetic, thermodynamic and atomic-level network mechanics intimately intertwine. It is of the utmost importance, by coupling experimental and modeling efforts, to untangle these microscopic-scale parameters dictating the material's macroscopic properties. Chalcogenides are already extensively used in night vision systems, thermal sensing and monitoring, high-power delivery fibers [44], photonic devices [59], data storage [58], and new applications are constantly emerging (holography [60], metamaterials [61]). Further mastery of their properties would yield the promise of an even brighter future for this fascinating class of materials, an exciting challenge for the glass science community.

4.8 Sources of further information and advice

The following is a list of recommended reading for readers interested in chalcogenide glasses and their applications, and the thermal properties of the amorphous materials:

- B. J. Eggleton, B. Luther-Davies and K. Richardson, 'Chalcogenide photonics', *Nat. Photonics* **5**, 141–148 **(2011)**.
- A. Zakery and S.R. Elliott, 'Optical properties and applications of chalcogenide glasses: a review', *J. Non-Crystal. Solids* **330**, 1–12 (2003).
- D. Lezal, 'Chalcogenide glasses – survey and progress', *J. Optoelectron. Adv. Mater*. **5**, 23–34 (2003).
- B. Bureau, X-H. Zhang, F. Smektala, J-L. Adam, J. Troles, H-L. Ma,

C. Boussard-Pledel, J. Lucas, P. Lucas, D. LeCoq, M. R. Riley and J. H. Simmons, 'Recent advances in chalcogenide glasses', *J. Non-Cryst. Solids* **345**, 276–283 (2004).

- X. Gai, T. Han, A. Prasad, S. Madden, D-Y. Choi, R. Wang, D. Bulla and B. Luther-Davies, 'Progress in optical waveguides fabricated from chalcogenide glasses', *Optics Express* 18, 26635–26646 (2010).
- A large body of work about the glass transition temperature and its modeling has been summarized in the excellent review from Berthier and Biroli [62].

4.9 References

1. SciGlass Infromation System. Available at: http://www.sciglassweb.com/ (accessed May 2013).
2. P. G. Debenedetti and F. H. Stillinger, 'Supercooled liquids and the glass transition,' *Nature* **410**, 259–267 (2001).
3. K. Shportko, S. Kremers, M. Woda, D. Lencer, J. Robertson, and M. Wuttig, 'Resonant bonding in crystalline phase-change materials,' *Nature Materials* **7**, 653–658 (2008).
4. F. Sciortino, W. Kob, and P. Tartaglia, 'Inherent structure entropy of supercooled liquids,' *Phys. Rev. Lett.* **83**, 3214–3217 (1999).
5. Y. Cai and M. F. Thorpe, 'Floppy modes in network glasses,' *Phys. Rev. B* **40**, 10535–10542 (1989).
6. J. C. Phillips and M. F. Thorpe, eds., *Phase transitions and self-organization in electronic and molecular networks* (Dordrecht: Springer, 2001).
7. J. H. Gibbs and E. A. DiMarzio, 'Nature of the glass transition and the glassy state,' *J. Chem. Phys.* **28**, 373 (1958).
8. E. A. DiMarzio and J. H. Gibbs, 'Glass temperature of copolymers,' *J. Polymer Chem.* **40**, 121–131 (1959).
9. A. N. Sreeram, D. R. Swiler, and A. K. Varshneya, 'Gibbs–DiMarzio equation to describe the glass-transition temperature trends in multicomponent chalcogenide glasses,' *J. Non-Cryst. Solids* **127**, 287–297 (1991).
10. R. Kerner and M. Micoulaut, 'On the glass transition temperature in covalent glasses,' *J. Non-Cryst. Solids* **210**, 298–305 (1997).
11. J. C. Phillips and M. F. Thorpe, 'Constraint theory, vector percolation and glass-formation,' *Solid State Commun.* **53**, 699–702 (1985).
12. M. F. Thorpe, D. J. Jacobs, M. V. Chubynsky, and J. C. Phillips, 'Self-organization in network glasses,' *J. Non-Cryst. Solids* **266**, 859–866 (2000).
13. P. Boolchand, X. Feng, and W. J. Bresser, 'Rigidity transitions in binary Ge-Se glasses and the intermediate phase,' *J. Non-Cryst. Solids* **293**, 348–356 (2001).
14. P. Boolchand, D. G. Georgiev, and B. Goodman, 'Discovery of the intermediate phase in chalcogenide glasses,' *J. Optoelectron. Adv. Mater.* **3**, 703–720 (2001).
15. M. Popescu, 'Disordered chalcogenide optoelectronic materials: phenomena and applications,' *J. Optoelectron. Adv. Mater.* **7**, 2189–2210 (2005).
16. K. Tanaka, 'Photoinduced fluidity in chalcogenide glasses,' *C. R. Chim.* **5**, 805–811 (2002).
17. J. Rocca, M. Erazu, M. Fontana, and B. Arcondo, 'Crystallization process on

amorphous GeTeSb samples near to eutectic point Ge(15)Te(85),' *J. Non-Cryst. Solids* **355**, 2068–2073 (2009).
18. G. Guery, 'Influence of iso-structural substitutions on properties of Ge(As,Sb)(S,Se) glasses,' MSc thesis, Clemson University, Clemson (2010).
19. S. Saraswat and S. S. S. Kushwaha, 'Specific heat studies in a-Se and a-Se(90) M(10) (M = In, Sb, Te) alloys,' *J. Therm. Anal. Calorim.* **96**, 923–927 (2009).
20. A. K. Varshneya, *Fundamentals of inorganic glasses* (Waltham, MA: Academic Press, 1993).
21. J. D. Musgraves, P. Wachtel, S. Novak, J. Wilkinson, and K. Richardson, 'Composition dependence of the viscosity and other physical properties in the arsenic selenide glass system,' *J. Appl. Phys.* **110**, 063503 (2011).
22. ASTM C1351M-96(2012) Standard Test Method for Measurement of Viscosity of Glass between 10^4 Pa.s and 10^8 Pa.s by Viscous Compression of a Solid Right Cylinder. West Conshohocken, PA: ASTM International.
23. ASTM C1350M-96(2008) Standard Test Method for Measurement of Viscosity of Glass between Softening Point and Annealing Range Approximately 10^8 Pa.s to Approximately 10^{13} Pa.s) by Beam Bending. West Conshohocken, PA: ASTM International.
24. ASTM C338-93(2008) Standard Test Method for Softening Point of Glass. West Conshohocken, PA: ASTM International.
25. L. Brilland, F. Smektala, G. Renversez, T. Chartier, J. Troles, T. N. Nguyen, N. Traynor, and A. Monteville, 'Fabrication of complex structures of Holey fibers in chalcogenide glass,' *Opt. Express* **14**, 1280–1285 (2006).
26. T. M. Monro, Y. D. West, D. W. Hewak, N. G. R. Broderick, and D. J. Richardson, 'Chalcogenide holey fibres,' *Electron. Lett.* **36**, 1998–2000 (2000).
27. S. D. Savage, C. A. Miller, D. Furniss, and A. B. Seddon, 'Extrusion of chalcogenide glass preforms and drawing to multimode optical fibers,' *J. Non-Cryst. Solids* **354**, 3418–3427 (2008).
28. A. F. Abouraddy, M. Bayindir, G. Benoit, S. D. Hart, K. Kuriki, N. Orf, O. Shapira, F. Sorin, B. Temelkuran, and Y. Fink, 'Towards multimaterial multifunctional fibres that see, hear, sense and communicate,' *Nature Materials* **6**, 336–347 (2007).
29. D. H. Cha, H. J. Kim, H. S. Park, Y. Hwang, J. H. Kim, J. H. Hong, and K. S. Lee, 'Effect of temperature on the molding of chalcogenide glass lenses for infrared imaging applications,' *Applied Optics* **49**, 1607–1613 (2010).
30. H. Niciu, M. Popescu, A. Velea, A. Lorinczi, A. Manea, D. Niciu, and M. Lazarescu, 'Moulding procedure for the preparation of infrared glassy microlenses and prisms based on arsenic sulphide chalcogenide glass,' *Chalcogenide Lett.* **7**, 625–629 (2010).
31. U. Fotheringham, M. Garsche-Andres, and E. Holzel, 'Low temperature relaxation effects: measurement with a double-microscope and relevance for glasses for flat display panels,' *Glass Sci Technol.-Glastech. Ber.* **73**, 400–409 (2000).
32. B. Ananthasayanam, P. Joseph, V. Blouin, D. Cler, S. Gaylord, L. Petit, K. Richardson, M. Stairiker, and M. Tardiff, 'Final shape of precision molded optics: Part I – Computational approach, material definition and the effect of lens shape,' *J. Thermal Stresses* **35**(6), 550–578 (2012).
33. B. Ananthasayanam, P. Joseph, V. Blouin, D. Cler, S. Gaylord, L. Petit, K. Richardson, M. Stairiker, and M. Tardiff, 'Final shape of precision molded optics: Part II – Validation and sensitivity to material properties and process parameters,' *J. Thermal Stresses* **35**(7), 614–636 (2011).

34. S. Gaylord, B. Ananthasayanam, B. Tincher, L. Petit, C. Cox, U. Fotheringham, P. Joseph, and K. Richardson, 'Thermal and structural property characterization of commercially moldable glasses,' J. Am. Ceram. Soc. **93**, 2207–2214 (2010).
35. T. Clopeau, A. Farina, A. Fasano, and A. Mikelic, 'Asymptotic equations for the terminal phase of glass fiber drawing and their analysis,' *Nonlinear Anal.-Real World Appl.* **11**, 4533–4545 (2010).
36. S. Danto, C. Houizot, C. Boussard-Pledel, X. H. Zhang, F. Smektala, and J. Lucas, 'A new family of far IR transmitting glasses in the Ga-Ge-Te system for space applications,' *Adv. Funct. Mater.* **16**, 1847 (2006).
37. T. D. Engeness, M. Ibanescu, S. G. Johnson, O. Weisberg, M. Skorobogatiy, S. Jacobs, and Y. Fink, 'Dispersion tailoring and compensation by modal interactions in OmniGuide fibers,' *Opt. Express* **11**, 1175–1196 (2003).
38. J. Hu, C. R. Menyuk, L. B. Shaw, J. S. Sanghera, and I. D. Aggarwal, 'Maximizing the bandwidth of supercontinuum generation in As(2)Se(3) chalcogenide fibers,' *Opt. Express* **18**, 6722–6739 (2010).
39. J. D. Shephard, W. N. MacPherson, R. R. J. Maier, J. D. C. Jones, D. P. Hand, M. Mohebbi, A. K. George, P. J. Roberts, and J. C. Knight, 'Single-mode mid-IR guidance in a hollow-core photonic crystal fiber,' *Opt. Express* **13**, 7139–7144 (2005).
40. M. El-Amraoui, G. Gadret, J. C. Jules, J. Fatome, C. Fortier, F. Desevedavy, I. Skripatchev, Y. Messaddeq, J. Troles, L. Brilland, W. Gao, T. Suzuki, Y. Ohishi, and F. Smektala, 'Microstructured chalcogenide optical fibers from As(2)S(3) glass: towards new IR broadband sources,' *Opt. Express* **18**, 26655–26665 (2010).
41. M. Asobe, T. Ohara, I. Yokohama, and T. Kaino, 'Low power all-optical switching in a nonlinear optical loop mirror using chalcogenide glass fibre,' *Electron. Lett.* **32**, 1396–1397 (1996).
42. R. E. Slusher, G. Lenz, J. Hodelin, J. Sanghera, L. B. Shaw, and I. D. Aggarwal, 'Large Raman gain and nonlinear phase shifts in high-purity As_2Se_3 chalcogenide fibers,' *J. Opt. Soc. Am. B-Opt. Phys.* **21**, 1146–1155 (2004).
43. S. J. Madden, D. Y. Choi, D. A. Bulla, A. V. Rode, B. Luther-Davies, V. G. Ta'eed, M. D. Pelusi, and B. J. Eggleton, 'Long, low loss etched As_2S_3 chalcogenide waveguides for all-optical signal regeneration,' *Opt. Express* **15**, 14414–14421 (2007).
44. B. Bureau, X. H. Zhang, F. Smektala, J. L. Adam, J. Troles, H. L. Ma, C. Boussard-Pledel, J. Lucas, P. Lucas, D. Le Coq, M. R. Riley, and J. H. Simmons, 'Recent advances in chalcogenide glasses,' *J. Non-Cryst. Solids* **345**, 276–283 (2004).
45. M. Bayindir, F. Sorin, A. F. Abouraddy, J. Viens, S. D. Hart, J. D. Joannopoulos, and Y. Fink, 'Metal-insulator-semiconductor optoelectronic fibres,' *Nature* **431**, 826–829 (2004).
46. S. Egusa, Z. Wang, N. Chocat, Z. M. Ruff, A. M. Stolyarov, D. Shemuly, F. Sorin, P. T. Rakich, J. D. Joannopoulos, and Y. Fink, 'Multimaterial piezoelectric fibres,' *Nature Materials* **9**, 643–648 (2010).
47. A. F. Abouraddy, O. Shapira, M. Bayindir, J. Arnold, F. Sorin, D. S. Hinczewski, J. D. Joannopoulos, and Y. Fink, 'Large-scale optical-field measurements with geometric fibre constructs,' *Nature Materials* **5**, 532–536 (2006).
48. S. Danto, F. Sorin, N. D. Orf, Z. Wang, S. A. Speakman, J. D. Joannopoulos, and Y. Fink, 'Fiber field-effect device via in situ channel crystallization,' *Advanced Materials* **22**, 4162–4166 (2010).
49. G. S. Fulcher, 'Analysis of recent measurements of the viscosity of glass,' *J. Am. Ceram. Soc.* **8**, 339–355 (1925).

50. J. C. Mauro, Y. Yue, A. J. Ellison, P. K. Gupta, and D. C. Allan, 'Viscosity of glass-forming liquids,' *Proc. Natl Acad. Sci. USA* **106**, 19780–19784 (2009).
51. L. Petit, N. Carlie, R. Villeneuve, J. Massera, M. Couzi, A. Humeau, G. Boudebs, and K. Richardson, 'Effect of the substitution of S for Se on the structure and non-linear optical properties of the glasses in the system $Ge_{0.18}Ga_{0.05}Sb_{0.07}S_{0.70}$-Se-x(x),' *J. Non-Cryst. Solids* **352**, 5413–5420 (2006).
52. M. D. Ediger, C. A. Angell, and S. R. Nagel, 'Supercooled liquids and glasses,' *J. Phys. Chem.* **100**, 13200–13212 (1996).
53. V. N. Novikov, Y. Ding, and A. P. Sokolov, 'Correlation of fragility of supercooled liquids with elastic properties of glasses,' *Phys. Rev. E* **71**, 12 (2005).
54. J. Nishii, S. Morimoto, I. Inagawa, R. Iizuka, T. Yamashita, and T. Yamagishi, 'Recent advances and trends in chalcogenide glass-fiber technology – a review,' *J. Non-Cryst. Solids* **140**, 199–208 (1992).
55. J. Mangin, G. Mennerat, and P. Villeval, 'Thermal expansion, normalized thermo-optic coefficients, and condition for second harmonic generation of a Nd:YAG laser with wide temperature bandwidth in $RbTiOPO_4$,' *J. Opt. Soc. Am. B-Opt. Phys.* **28**, 873–881 (2011).
56. N. Cherroret, A. Chakravarty, and A. Kar, 'Temperature-dependent refractive index of semiconductors,' *J. Mater. Sci.* **43**, 1795–1801 (2008).
57. G. Hawkins and R. Hunneman, 'The temperature-dependent spectral properties of filter substrate materials in the far-infrared (6–40 um),' *Infrared Physics & Technology* **45**, 69–79 (2004).
58. M. Wuttig and N. Yamada, 'Phase-change materials for rewriteable data storage', *Nature Materials* **6**, 824–832 (2007); erratum, **6**, 1004 (2007).
59. B. J. Eggleton, B. Luther-Davies, and K. Richardson, 'Chalcogenide photonics,' *Nat. Photonics* **5**, 141–148 (2011).
60. J. Teteris and M. Reinfelde, 'Application of amorphous chalcogenide semiconductor thin films in optical recording technologies,' *J. Optoelectron. Adv. Mater.* **5**, 1355–1360 (2003).
61. E. Semouchkina, D. H. Werner, G. B. Semouchkin, and C. Pantano, 'An infrared invisibility cloak composed of glass,' *Appl. Phys. Lett.* **96**, 233503 (2010).
62. L. Berthier and G. Biroli, 'Theoretical perspective on the glass transition and amorphous materials,' *Rev. Modern Phys.* **83**, 587–645 (2011).

5
Optical properties of chalcogenide glasses and fibers

J. SANGHERA and D. GIBSON, Naval Research Laboratory, USA

DOI: 10.1533/9780857093561.1.113

Abstract: This chapter discusses the optical properties of chalcogenide glasses and fibers for infrared applications. The chapter describes the theory of optical transmission including intrinsic material losses and extrinsic losses due to impurity absorptions. Refractive index and material dispersion of bulk glass are reviewed. Optical fiber transmission loss and laser power delivery data are discussed. Nonlinear optical processes are explored for active devices including Raman amplification, supercontinuum generation, stimulated Brillouin scattering for slow-light applications and four wave mixing.

Key words: chalcogenide glass, optical properties, nonlinear optics.

5.1 Introduction

Chalcogenide glasses, based on the chalcogen elements sulfur, selenium, and tellurium alloyed with group III–V elements such as germanium, arsenic and antimony, form a category of optical materials that have been known and studied since the mid-1960s. These glasses typically possess a broad region of infrared transparency, limited visible transparency, low optical attenuation, large refractive index and high optical nonlinearity, and are generally stable to atmospheric moisture. Interest has been shown in the use of chalcogenide glasses for windows, lenses and fibers for infrared wavelengths beyond 2 μm (Nishii and Yamashita, 1998; Katsuytama and Matsumura, 1989; Sanghera and Aggarwal, 1998). Discussion of chalcogenide glasses and their optical properties is inevitable when investigating infrared materials and are the topics of many collections, chapters, books and review articles (e.g., Sanghera and Aggarwal, 1998; Kokorina, 1996; Popescu, 2000; Hewak, 1998; Harrington, 2004; Bureau *et al.*, 2004). Table 5.1 represents a summary of approximate bulk transmission windows for several chalcogenide glass-forming systems.

5.2 Optical transmission theory

The optical transmission window of chalcogenide glasses, like other dielectric materials, is limited intrinsically by three physical phenomena: electronic

Table 5.1 Approximate transmission windows of some chalcogenide glasses

Glass-forming system	Transmission window (μm)	Reference
Ga–La–S	0.57–9.5	Brady (1998)
As–S	0.6–10	Sanghera and Aggarwal (1999)
Ge–Sb–S	0.64–11	Snopatin *et al.* (2009)
As–S–Se	0.75–12.5	Snopatin *et al.* (2009)
Ge–As–Se	0.84–16	Brady (1998)
As–Se	0.85–17.5	Snopatin *et al.* (2009)
Ge–As–Se–Te	1.0–17	Sanghera and Aggarwal (1999)
As–Se–Te	1.23–18.52	Snopatin *et al.* (2009)
Ge–Se–Te	1.7–17	Brady (1998)
Te–As–Se	2.0–16	Bureau *et al.* (2009)

absorption at short wavelengths, multiphonon absorption at long wavelengths, and Rayleigh scattering in between. Another loss mechanism unique to chalcogenide glasses is the weak absorption tail (WAT) which typically eclipses the Rayleigh scattering loss. Extrinsic losses are often observed within this window and arise from impurities in the glass.

While theoretical predictions indicate minimum attenuation of chalcogenide glass fibers should be very low, 0.1–0.01 dB/km, the practical attenuation is several hundred times higher (Sanghera *et al.*, 1996). The practical attenuation is best understood as the sum of the absorption and scattering losses, defined as:

$$\text{Absorption loss} = A_0 \exp(A/\lambda) + B_0 \exp(-B/\lambda) + \sum c_i C_i + D_0 \exp(D/\lambda) \qquad [5.1]$$

$$\text{Scattering loss} = E/\lambda^4 + F/\lambda^4 + G/\lambda^2 + H \qquad [5.2]$$

where the first two absorption terms represent the intrinsic electronic and multiphonon absorptions, respectively. The third term describes extrinsic absorption due to impurities and the fourth represents the WAT. The scattering losses in optical materials depend on the size of the scatter-center relative to the wavelength of light. Rayleigh, Rayleigh-Gans, and wavelength-independent scattering are caused by scattering centers much smaller than, approximately the same size as, and much larger than the optical wavelength, respectively. Rayleigh scattering can be material intrinsic, arising from density and composition fluctuations frozen into the glass upon cooling from the melt or extrinsic, as impurities and defects. The first and second terms in Eq. [5.2] represent intrinsic and extrinsic Rayleigh scattering, respectively. The third and fourth terms represent Rayleigh-Gans and wavelength-independent scattering, respectively, also arising from extrinsic defects (Sanghera *et al.*, 1996).

The electronic absorption edge governs the short wavelength limit of the optical transmission window (Weaire and Thorpe, 1971; Thorpe and Weaire, 1971). It is useful to consider chalcogenide glasses as amorphous semiconductors with a characteristic energy gap, E_g. Incident photons with energy exceeding this gap, $h\nu > E_g$, will excite an electron from the valence to the conduction band creating an electron-hole pair and are absorbed in the process. Recombination of the electron-hole pair occurs in multiple non-radiative steps generating phonons rather than photons (Brady, 1998). The electronic band gap depends on the electronic configuration and positional arrangement of the atoms in the material. For crystalline materials, the lattice positioning dictates a long-range atomic order suggesting a sharply defined band gap and steep rise in absorption at short wavelengths. However, disorder due to crystal defects, and more importantly temperature, ensures the band gap is not sharply defined and shifts the electronic edge absorption to longer wavelengths and imparts a more gradual dependence of absorption on wavelength, resulting in the Urbach tail (Skettrup, 1978; Urbach, 1953). In the amorphous form of the same material, the crystalline short-range atomic order is typically retained, but the long-range positioning does not adhere to any crystal lattice. The electronic absorption edge therefore is usually situated in approximately the same energy range (wavelength range) as for the crystalline form, but is less sharply defined and takes on a more gradual wavelength dependence (Wood and Tauc, 1972). For many glasses the electronic absorption edge is shifted to longer wavelengths (lower energy). This is demonstrated in Fig. 5.1 by the electronic-edge absorption spectra for amorphous and crystalline As_2S_3 and As_2Se_3 (Zallen *et al.*, 1971).

In general, this peculiar nature of short wavelength absorption behavior is characterized by three regions, labeled A, B, and C in Fig. 5.2. The very high loss region A is the band gap edge and is usually too large to measure in all but the thinnest films. Region B is the broadening of that edge, similar to Urbach tails observed in crystals, and is generally observed as the UV absorption edge. Region C is the weak absorption tail (WAT) and is related to the medium range order in the glass structure and localized energy states in the band gap. The generally weak covalent bonds in chalcogenide glasses is attributed to unsatisfied bonds and network vacancies. This disorder results in a large density of states in the band gap, shown schematically in Fig. 5.3 (Wood and Tauc, 1972) as an overlap of localized states, which permits absorption of lower energy (longer wavelength) photons (Wood and Tauc, 1972). The WAT therefore is sensitive to glass purity and thermal history (Kanamori *et al.*, 1984). Wood and Tauc show the optical absorption near the electronic edge for three chalcogenide glasses, $Ge_{28}Sb_{12}Se_{60}$, $Ge_{33}As_{12}Se_{55}$, and As_2S_3, where the Urbach tail and WAT are evident (Fig. 5.4). The WAT generally extends from the UV edge across the transmission window to the multiphonon edge, and for high-quality chalcogenide glasses, dominates the optical loss spectrum over Rayleigh scattering (Brady, 1998).

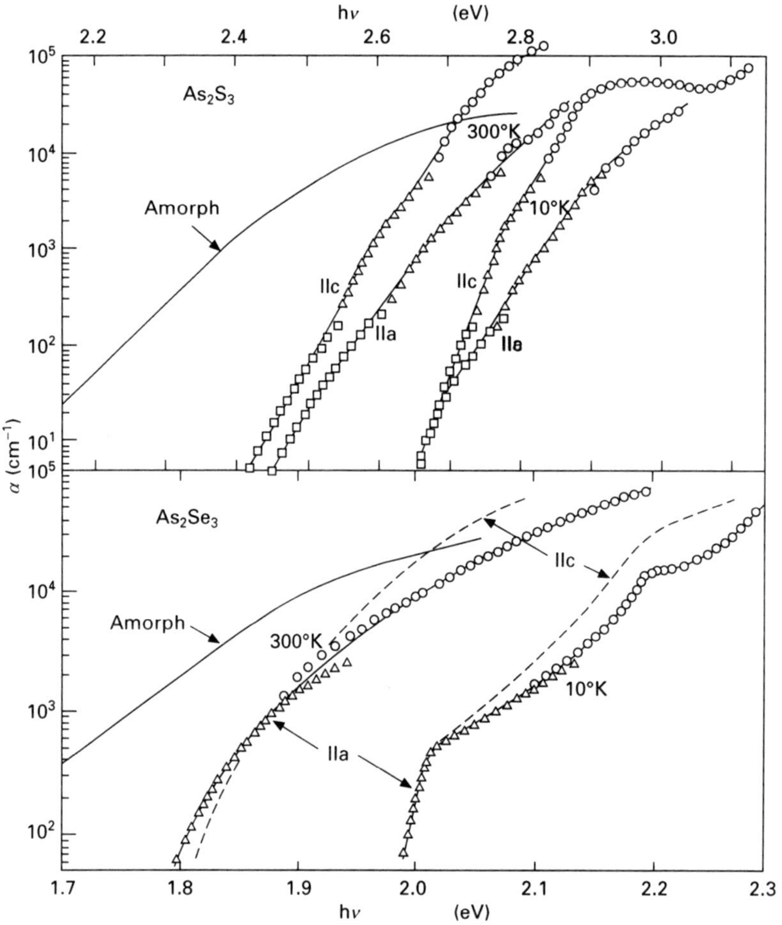

5.1 **Band-edge absorption spectra for crystalline and amorphous As_2S_3 (top) and As_2Se_3 (bottom) (Zallen *et al.*, 1971).**

The infrared cutoff is determined by the multiphonon absorption edge and is governed by vibrational resonances of the atomic network which depend on atomic mass and bond strength. For glasses with larger atoms and weaker bonds, as is the intended case for chalcogenide glasses, this vibrational resonance occurs at lower frequencies, pushing the fundamental absorption infrared cutoff to longer wavelengths. This trend is depicted in Fig. 5.5, where Katsuyama and Matsumura showed that increasing the concentration of Te in the ternary Ge–Se–Te glass shifts the multiphonon edge to longer wavelengths (Katsuyama and Matsumura, 1989). It is important to note that the trends in glass composition which shift the multiphonon edge to longer wavelengths also shift the short wavelength edge to longer wavelengths. For example, in the $xAs_2S_3 - (1\text{-}x)As_2Se_3$ glass system, the substation of S

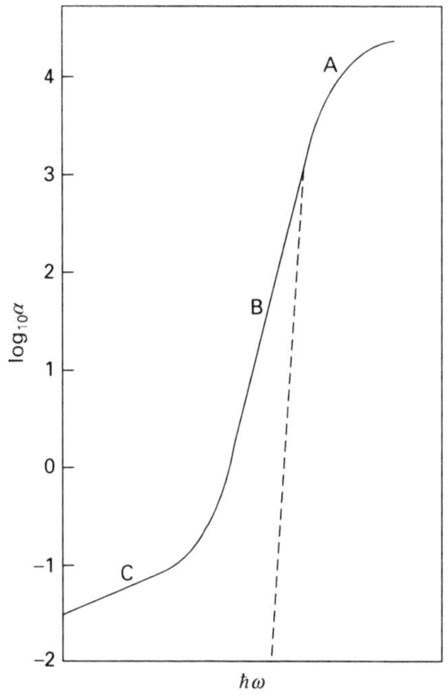

5.2 Three principle regions of optical absorption near the short wavelength cut-off: band gap edge (A), Urbach tail (B) and weak absorption tail (C) (Wood and Tauc, 1972).

for Se shifts both the electronic edge (Fig. 5.6) and the multiphonon edge to shorter wavelengths (Fig. 5.7) (Snopatin *et al.*, 2009).

5.3 Impurity absorptions

The abundance of characteristic absorption peaks in the mid-infrared has led to the prominence of infrared spectroscopy for chemical sensing and identification. These same absorptions are problematic for optical transmission when impurities are present in chalcogenide glasses. Table 5.2 summarizes some common impurity absorption bands in chalcogenide glasses. The most common impurities contained in chalcogenides are also the most difficult to remove: hydrogen, oxygen, and water, which form R–O and R–H bonds (where R is the chalcogen and 4B or 5B elements) in the glass matrix (Nishii and Yamashita, 1998). Extrinsic impurities are especially problematic for chalcogenide optical fibers, where long path lengths ensure that even parts-per-million levels of impurities can have a deleterious effect on the mid-infrared transparency. Sanghera *et al.* (1996) have shown that purification of precursor chemicals can significantly reduce the presence of extrinsic losses. Their fiber attenuation spectra are shown in Fig. 5.8 for As–S–Se fibers

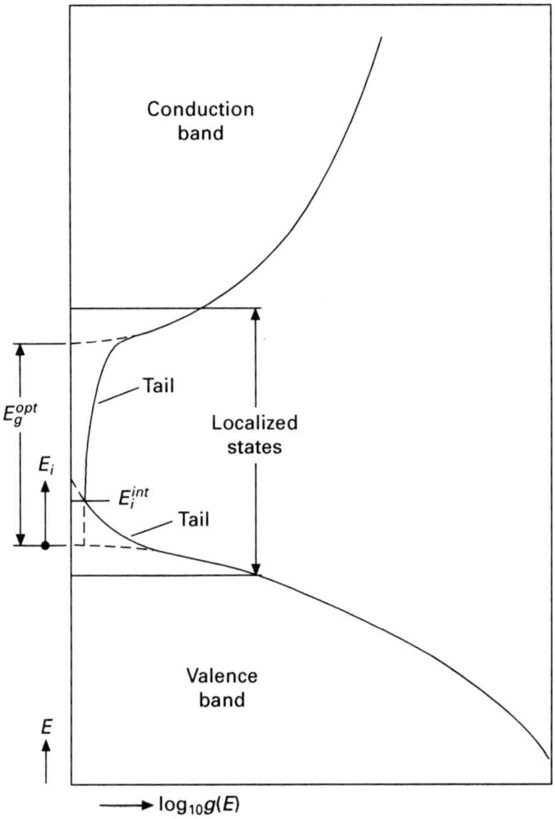

5.3 Density of states diagram for chalcogenide glass depicting localized states, believed to be the origin of the WAT (Wood and Tauc, 1972).

with as-received precursor chemicals and purified chemicals. Removal of water impurities is an area of ongoing research as chalcogenide optics and fibers with low loss in the mid-infrared (3–5 μm) are in increasing demand (Sanghera *et al.*, 1996). Snopatin *et al.* (2009) recently demonstrated high-purity chalcogenide glass optical fibers with losses as low as 12 dB/km for As–S fibers and 60dB/km for As–S–Se fibers. The attenuation spectra are shown in Fig. 5.9 (Snopatin *et al.*, 2009).

5.4 Refractive index, dispersion and *dn*/*dT*

Chalcogenide glasses in general have larger refractive indices than other glasses, due to the considerable degree of polarizability of the chalcogen elements. Refractive index for chalcogenide glasses can range from 2.2 for germanium arsenic sulfides to 3.4 for glasses with large tellurium concentrations. Indices

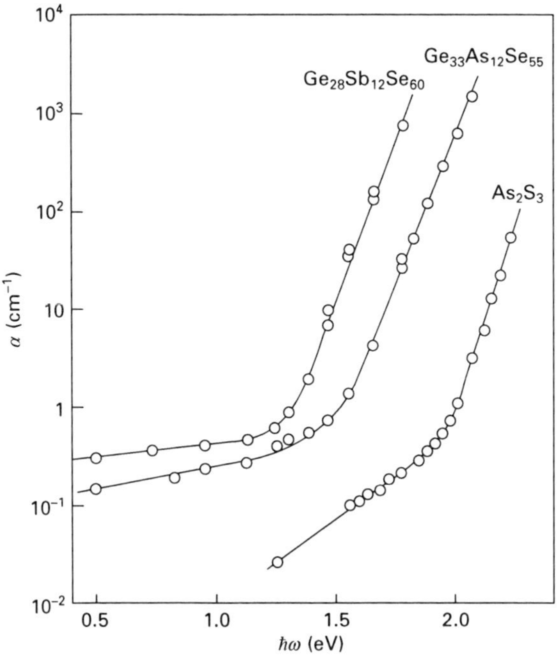

5.4 Optical absorption for three amorphous chalcogenide glasses near the short wavelength cut-off demonstrating the transition from Urbach tail to WAT (Wood and Tauc, 1972).

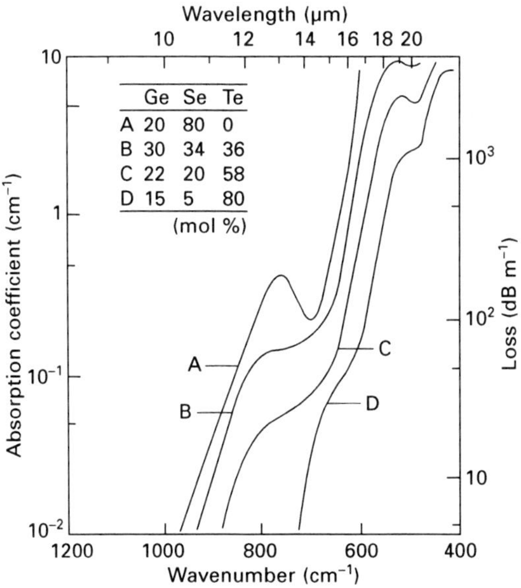

5.5 Effect of increasing Te concentration on multiphonon absorption edge in Ge–Se–Te glass (Katsuytama and Matsumura, 1989).

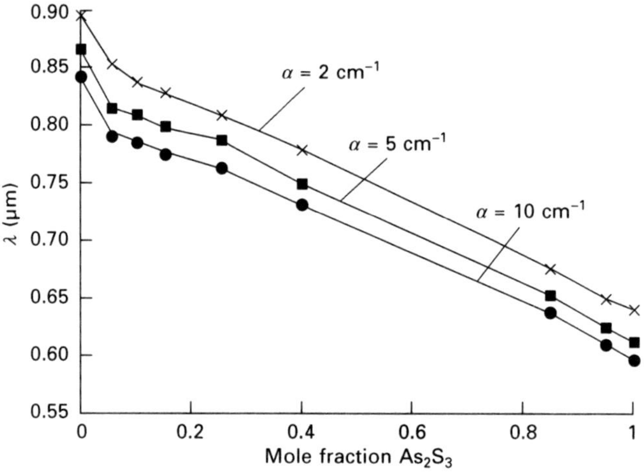

5.6 Spectral position of the electronic edge as a function of As_2S_3 content in the $xAs_2S_3 - (1\text{-}x)As_2Se_3$ glass system. The absorption coefficient corresponding to the electronic edge position is represented by the three curves, from the top down, 2, 5, and 10 cm^{-1} (redrawn from Snopatin *et al.*, 2009).

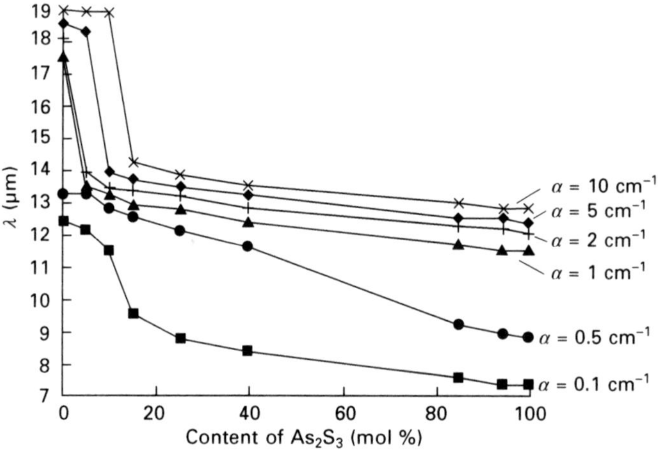

5.7 Spectral position of the multiphonon edge as a function of As_2S_3 content in the $xAs_2S_3 - (1\text{-}x)As_2Se_3$ glass system. The absorption coefficient corresponding to the multiphonon edge position is represented by the six curves, from the top down, 10, 5, 2, 1, 0.5, and 0.1 cm^{-1} (redrawn from Snopatin *et al.*, 2009).

of refraction for several chalcogenide glasses are summarized in Table 5.3. From this data, some general refractive index trends are observed. As the polarizability of the chalcogen elements increases with atomic weight, the refractive index also increases. For example, replacing selenium in As_2Se_3

Table 5.2 Impurity absorptions in chalcogenide glasses

Glass system	Wave number (cm^{-1})	Wavelength (μm)	Bond	Reference
As-S	925	10.8	AsO–H	Kanamori *et al.* (1985)
	1825	5.48	AsO–$H_{(overtone)}$	Kanamori *et al.* (1985)
	3440	2.91	SO–$H_{(fundamental)}$	Kanamori *et al.* (1984)
	4370	2.29	SO–$H_{(combination)}$	Kanamori *et al.* (1985)
	5210	1.92	SO–$H_{(combination)}$	Kanamori *et al.* (1985)
	6950	1.44	SO–$H_{(overtone)}$	Kanamori *et al.* (1985)
	1580	6.32	$H_2O_{(molecular)}$	Kanamori *et al.* (1985)
	3610	2.77	$H_2O_{(molecular)}$	Kanamori *et al.* (1985)
	2480	4.03	S–$H_{(fundamental)}$	Kanamori *et al.* (1984)
	2710	3.69	S–$H_{(combination)}$	Kanamori *et al.* (1985)
	3215	3.11	S–$H_{(combination)}$	Kanamori *et al.* (1985)
	3940	2.54	S–$H_{(combination)}$	Kanamori *et al.* (1985)
	4880	2.05	S–$H_{(overtone)}$	Kanamori *et al.* (1985)
	2025	4.94	Carbon	Kanamori *et al.* (1985)
Ge-S	3570	2.80	SO–H	Shibata *et al.* (1981)
	2500	4.00	S–H	Shibata *et al.* (1981)
	2030	4.92	Ge–H	Shibata *et al.* (1981)
As-Se	3420	2.92	SeO–H	Moynihan *et al.* (1975)
	3600	2.78	OH	Moynihan *et al.* (1975)
	3520	2.84	OH	Moynihan *et al.* (1975)
	1585	6.30	$H_2O_{(molecular)}$	Moynihan *et al.* (1975)
	2830	3.53	Se–H	Moynihan *et al.* (1975)
	2430	4.12	Se–H	Moynihan *et al.* (1975)
	2190	4.57	Se–H	Moynihan *et al.* (1975)
Ge-Se	1280	7.8	Ge–$O_{(combination)}$	Vlček *et al.* (1987)
	800	12.5	Ge–$O_{(stretching)}$	Vlček *et al.* (1987)
	500	20.0	Ge–$O_{(bending)}$	Vlček *et al.* (1987)
Ge-As-Se	3420	2.92	O–H	Kanamori *et al.* (1984)
	2190	4.57	Se–H	Kanamori *et al.* (1984)
	1270	7.90	Oxide	Kanamori *et al.* (1984)
Ge-Se-Te	1230	8.13	Ge–$O_{(combination)}$	Nishii *et al.* (1989)
	765	13.07	Ge–$O_{(stretching)}$	Nishii *et al.* (1989)

with larger, more polarizable tellurium increases index, while the lighter, less polarizable sulfur decreases index. Addition of germanium or silicon lowers the refractive index, while lead, tin and antimony increase it (Seddon, 1995).

For chalcogenide glass lenses, windows, and optical fibers, this large refractive index means optical losses due to Fresnel reflections at surfaces can exceed 31% (17% loss per surface) and are a substantial concern. Anti-reflective (AR) coatings are needed for chalcogenide lenses, windows, and fibers. Commercial AR coatings are routinely applied to silica-based optics and fibers at very high temperatures to ensure full densification. Because of their

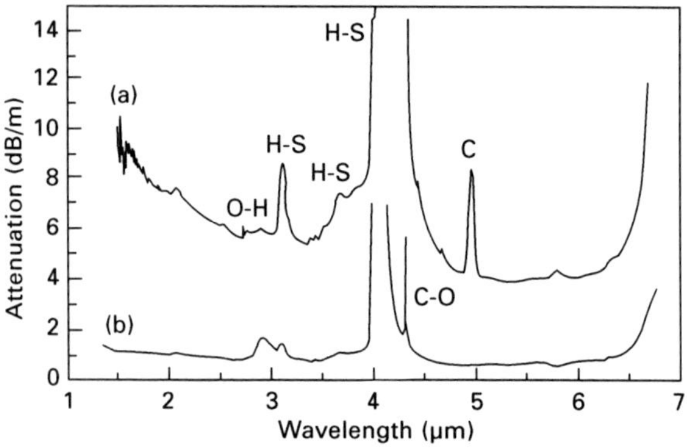

5.8 Optical attenuation spectra of As–S–Se fibers fabricated with commercially available chemical precursors, curve (a), and purified chemical precursors, curve (b), demonstrate a reduction in extrinsic impurity absorption and minimum attenuation (Sanghera *et al.*, 1996).

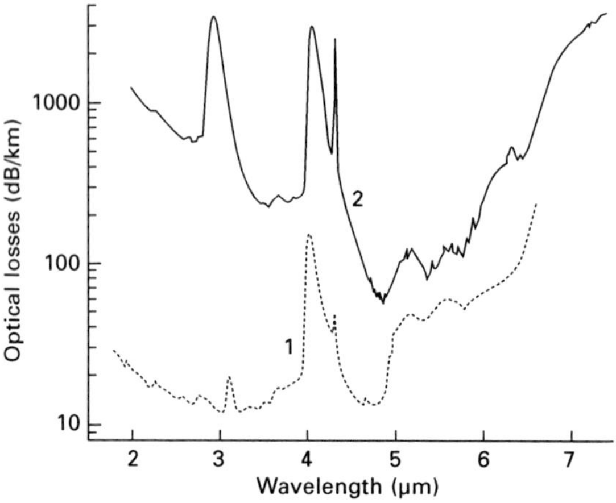

5.9 Attenuation spectra for low-loss As–S (curve 1) and As–S–Se (curve 2) fibers (Snopatin *et al.*, 2009).

comparatively low melt temperatures, chalcogenide glasses are not as easily coated (Graham *et al.*, 2003). Conventional AR coatings, while quite robust when applied to silica-based glasses, have poor adhesion to chalcogenides and are sensitive to humidity. Florea *et al.* (2011) show a novel approach to this problem by imparting fine structural features to the surface, eliminating the need for material coatings. The microstructured fiber surface features a two-dimensional array of ~849 nm tall tapered protrusions, shown in inset of Fig. 5.10, and was applied by a thermal stamping technique. These 'moth eye' structures, so-called as they mimic the antireflective nanostructured

Table 5.3 Optical index, n_0, and nonlinearities for chalcogenide glasses measured at 1.55 μm

Glass	n_0	λ_{gap} (μm)	$n_2/n_2(SiO_2)$	FOM	Reference
$Ge_{25}Se_{75}$	2.7	0.60	120	2	Lenz *et al.* (2000)*
$Ge_{25}Se_{65}Te_{10}$	2.5	0.72	220	1	Lenz *et al.* (2000)*
$Ge_{28}Se_{60}Sb_{12}$	2.61	0.69	360	3	Lenz *et al.* (2000)*
$As_{40}Se_{60}$	2.78	0.70	500	2	Lenz *et al.* (2000)*
$As_{40}Se_{60}$	2.81	0.70	930	11	Harbold *et al.* (2002a)
$As_{39}Se_{61}$	2.81	0.70	660	4	Harbold *et al.* (2002a)
$As_{40}Se_{55}Cu_5$	2.93	0.79	850	5	Harbold *et al.* (2002a)
$As_{25}Se_{55}Te_{20}$	2.52	0.79	470	5	Harbold *et al.* (2002a)
$As_{40}S_{60}$	2.45	0.52	220	>12	Harbold *et al.* (2002a)
$As_{40}S_{50}Se_{10}$	2.49	0.55	380	4	Harbold *et al.* (2002a)
$As_{40}S_{40}Se_{20}$	2.55	0.59	300	8	Harbold *et al.* (2002a)
$As_{40}S_{30}Se_{30}$	2.62	0.62	430	5	Harbold *et al.* (2002a)
$As_{40}S_{20}Se_{40}$	2.70	0.64	460	3	Harbold *et al.* (2002a)
$As_{40}S_{10}Se_{50}$	2.76	0.67	560	7	Harbold *et al.* (2002a)
$Ge_{15.38}As_{30.77}S_{53.85}$		0.49	130	>21	Harbold *et al.* (2002b)
$Ge_{15.38}As_{30.77}S_{32.31}Se_{21.54}$		0.56	250	8.1	Harbold *et al.* (2002b)
$Ge_{15.38}As_{30.77}S_{10.77}Se_{43.08}$		0.61	390	11	Harbold *et al.* (2002b)
$Ge_{20}As_{40}Se_{40}$		0.68	620	4.2	Harbold *et al.* (2002b)
$Ge_{12.5}As_{25}Se_{62.5}$		0.63	450	18	Harbold *et al.* (2002b)
$Ge_{11.11}As_{22.22}Se_{66.67}$		0.63	530	28	Harbold *et al.* (2002b)

*Measured at 1.5 μm.

features on the surface of the eyes of some moths, increase the fiber endface transmission from 83% to 97% (Florea *et al.*, 2011).

Knowledge of the dispersion or wavelength dependence of refractive index is essential for designers of infrared imagers and infrared fibers. The dispersion curves for chalcogenide glasses exhibit a smooth wavelength dependence throughout the transmission window, with increased dispersion near the UV and IR edges. Just as the nominal refractive index can vary widely within the chalcogenide glasses, so can the slope of the dispersion curves. For materials that transmit predominantly in the visible, such as oxide glasses, dispersion is characterized by the Abbe number, v, defined in Eq. [5.3]:

$$v = \frac{n_d - 1}{n_F - n_C} \quad [5.3]$$

where n_d, n_F and n_C are the refractive indices at 587.6 nm (He d-line), 486.1 nm (H_2 F-line) and 656.3 nm (H_2 C-line), respectively. For infrared materials

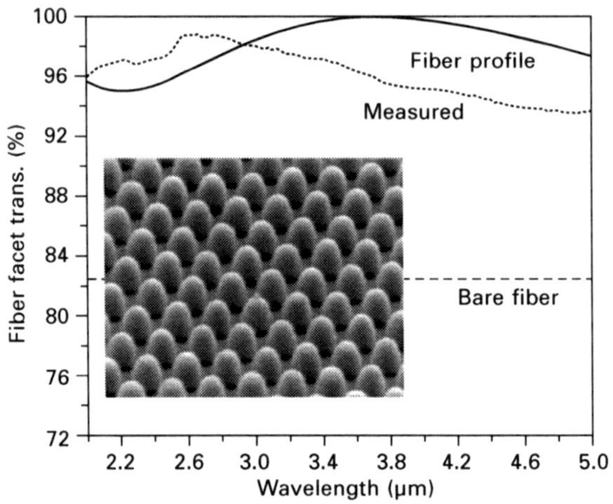

5.10 By applying a microstructured array of ~849 nm tall protrusions to a bare As_2S_3 fiber surface (inset image), Florea *et al.* (2011) reduced the fiber endface reflections and increased the endface transmission from ~83% for bare fiber to 97%.

that do not transmit these wavelengths, the partial dispersion is defined at other wavelengths where the material does transmit,

$$\nu_{S-L} = \frac{n_M - 1}{n_S - n_L} \qquad [5.4]$$

where *S* and *L* are the short and long wavelengths of interest and *M* is the wavelength halfway between them (Riedl, 2001). Two partial dispersion regions are commonly used, $\nu_{3\text{-}5}$ and $\nu_{8\text{-}12}$, for the mid-wave and long-wave infrared. Tables 5.4 and 5.5 show the refractive indices and partial dispersions for some chalcogenide glasses reported by various authors (Savage *et al.*, 1977, 1980; Kokorina, 1996; Aio *et al.*, 1978). It is often more helpful to consider the full dispersion curves for chalcogenide glasses, rather than reducing the dispersion to a single parameter. The dispersion curves for some chalcogenide glasses in Table 5.4 have been plotted in Fig. 5.11.

Thermal lensing due to the thermal change in refractive index, *dn*/*dT* or thermooptic coefficient, is a concern and ultimately limits the power handling capability of chalcogenide glass lenses, windows and fibers. In the case of optical fibers, catastrophic failure due to thermal lensing typically occurs within a few centimeters of the fiber distal end (Sanghera and Aggarwal, 1999). Positive and negative thermooptic coefficients have been measured for chalcogenide glasses (Kokorina, 1996) and As_2S_3 has a thermooptic coefficient comparable to silica (Hilton and Jones, 1967) and thus a high laser damage threshold. The addition of certain elements, including sulfur and iodine, tends to reduce *dn*/*dT* in chalcogenide glasses while the addition of others (Sb, Bi, Te, Th) tends to increase it (Kokorina, 1996).

Table 5.4 Wavelength-dependent refractive index data for chalcogenide glasses measured by Aio *et al.* (1978). Mid-wave and long-wave dispersion figures ($\nu_{3\text{-}5}$ and $\nu_{8\text{-}12}$) are estimates and were calculated from interpolated data (Aio *et al.*, 1978; Kokorina, 1996). Compositions marked with (*) correspond to curves in Fig. 5.11

	Refractive index wavelength, λ (μm)									Estimated dispersion	
	1	1.8	2.2	3	4.6	6.2	8.6	10.2	11	$\nu_{3\text{-}5}$	$\nu_{8\text{-}12}$
$Ge_{20}\,Se_{80}$ *11	2.4932	2.4345	2.4266	2.4190	2.4138	2.4109	2.4058	2.4022	2.4002	240	150
$Ge_{10}\,Se_{65}\,S_{25}$ *13	2.3704	2.3173	2.3103	2.3034	2.2984	2.2954	2.2908	2.2869		226	
$Ge_{17}\,Se_{80}\,Sn_3$ *10		2.4501	2.4420	2.4345	2.4289	2.4256	2.4214	2.4178	2.4158	223	155
$As_{20}\,Se_{80}$ *8	2.6823	2.6013	2.5906	2.5814	2.5745	2.5718	2.5683	2.5664	2.5657	208	360
$As_{30}\,Se_{50}\,Te_{20}$ *3			2.9490	2.9324	2.9211	2.9162	2.9109	2.9074	2.9055	154	210
$As_{30}\,Se_{60}\,Tl_{10}$ *2		3.0376	3.0185	3.0018	2.9899	2.9840	2.9765	2.9711	2.9690	149	164
$As_{30}\,Se_{45}\,Tl_{15}\,Te_{10}$ *1		3.2908	3.2623	3.2369	3.2219	3.2145	3.2062			132	
$Ge_{12.5}\,As_{20}\,Se_{67.5}$	2.6626	2.5900	2.5803	2.5713	2.5650	2.5608	2.5559	2.5523	2.5500	213	155
$Ge_{20}\,As_{20}\,Se_{60}$ *9	2.5983	2.5338	2.5251	2.5168	2.5109	2.5069	2.5010	2.4961	2.4941	219	138
$Ge_{20}\,As_{20}\,Se_{30}$ *12 S_{30} *12	2.4096	2.3629	2.3564	2.3501	2.3439	2.3389				181	
$Ge_{22}\,As_{22}\,Se_{36}\,Te_{20}$ *4		2.8282	2.8129	2.7989	2.7905	2.7861	2.7807	2.7769	2.7745	189	168
$Ge_{17}\,As_{20}\,Se_{58}\,Sn_5$	2.7079	2.6318	2.6214	2.6114	2.6045	2.6002	2.5941	2.5892	2.5870	202	140
$Ge_{17}\,As_{20}\,Se_{28}\,Sn_5\,S_{30}$	2.5312	2.4744	2.4665	2.4586	2.4517	2.4462	2.4366			176	
$Ge_{25}\,As_{20}\,Se_{50}\,Pb_5$ *6		2.7421	2.7299	2.7185	2.7100	2.7050	2.6990	2.6936	2.6910	176	133
$Ge_{12.5}\,As_{10}\,Se_{67.5}\,Sb_{10}$ *7	2.7606	2.6748	2.6637	2.6529	2.6452	2.6407	2.6348	2.6303	2.6272	187	126
$Ge_{12.5}\,Se_{67.5}\,Sb_{20}$ *5	2.8906	2.7888	2.7755	2.7645	2.7547	2.7494	2.7425	2.7368	2.7336	158	119

Table 5.5 Refractive index and dispersion data for chalcogenide glasses in the Ge–As–Se–Te system (Savage *et al.*, 1977, 1980)

	Wavelength (μm)			$\upsilon_{3\text{-}5}$	Wavelength (μm)			
	3	4	5		8	10	12	$\upsilon_{8\text{-}12}$
As_{40} Se_{60}					2.784	2.7789	2.7728	159
Ge_{20} Se_{80}					2.4071	2.4027	2.3973	143
Ge_{10} As_{20} Se_{70}					2.4649	2.4594	2.4526	119
Ge_{10} As_{30} Se_{60}					2.6254	2.6201	2.6135	135
Ge_{10} As_{40} Se_{50}					2.6108	2.6067	2.6016	176
Ge_{20} As_{10} Se_{70}					2.5628	2.5583	2.5528	156
Ge_{30} As_{10} Se_{60}					2.4408	2.4347	2.4271	104
Ge_{30} As_{15} Se_{55}					2.4972	2.4914	2.484	113
Ge_{30} As_{20} Se_{50}					2.569	2.5633	2.556	120
Ge_{33} As_{12} Se_{55}					2.5002	2.4942	2.4867	111
Ge_{30} As_{13} Se_{57}	2.4936	2.4887	2.4859	193	2.4784	2.4724	2.465	110
Ge_{30} As_{13} Se_{47} Te_{10}	2.6118	2.6057	2.6024	171	2.5952	2.5897	2.5829	129
Ge_{30} As_{13} Se_{37} Te_{20}	2.7412	2.7342	2.7305	162	2.7229	2.7178	2.7117	154
Ge_{30} As_{13} Se_{27} Te_{30}	2.8818	2.8732	2.8688	144	2.861	2.8563	2.8509	185

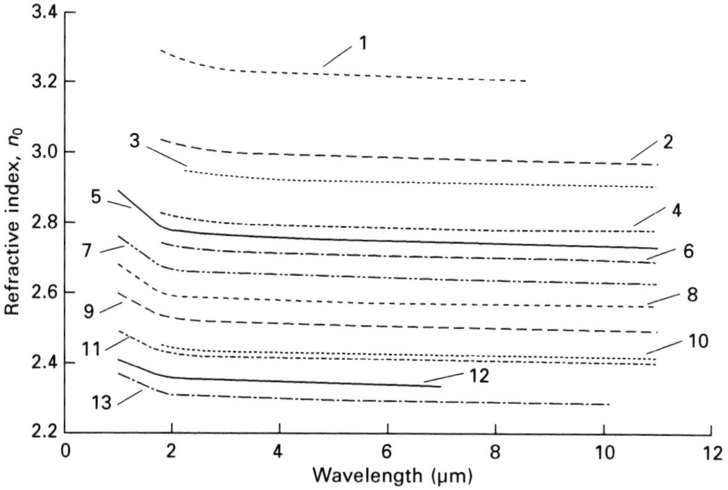

5.11 Dependence of refractive index, n_0, on wavelength, λ (μm), for some chalcogenide glasses reported in *Table* 5.4 and marked with (*). Plotted from data published by Aio *et al.* (1978).

5.5 Transmission and laser power delivery of chalcogenide fibers

Because the optical path lengths are so much longer, the usable spectral transmission windows for chalcogenide glass fibers are significantly limited compared to the bulk form. Depending on composition, the sulfide-, selenide-, and telluride-based fibers transmit between about 0.8-7 μm, 1–10 μm and 2–12 μm, respectively (Sanghera *et al.*, 2010).

The laser power handling capability can vary significantly. High power CO and CO_2 lasers operating at 5.4 and 10.6 μm, respectively, are available and are used for industrial welding and cutting, whereby transmitting the laser power through the fibers enables remote operation. To date, Te-based fibers have demonstrated output powers of 10.7 W for 19.4 W launched power (efficiency of 55.2%) at 10.6 μm (Nishii *et al.*, 1992). The fibers possessed a PbF_2 antireflection (AR) coating and were cooled with water to prevent thermal lensing caused by their increase in absorption coefficient with temperature ($d\alpha/dT$) and increase in refractive index with temperature (dn/dT). In these fibers, damage occurred several centimeters from the input end. On the other hand, arsenic sulfide-based fibers have demonstrated 85 W output power for 169 W launched power (efficiency of 50.3%) without the need for cooling and AR coatings (Nishii *et al.*, 1992). Damage occurred at the input end and is attributed to poor polish quality of the input surface. Unlike the Te-based glasses, the arsenic sulfide-based glasses have smaller $d\alpha/dT$ and dn/dT. Typically the fiber diameters are in excess of 500 μm for high power laser delivery to reduce power density. However, small core diameter (<200 μm) fibers have demonstrated tolerance to power densities of >125 kW/cm^2 at 5.4 μm and >54 kW/cm^2 at 10.6 μm without damage (Busse *et al.*, 1996). Telluride glass fiber losses at 10.6 μm are in the range of 1.5–3 dB/m, depending upon composition and purity (Sanghera *et al.*, 1994; Nishii *et al.*, 1992). The arsenic sulfide fibers transmit in the 2–5 μm region and have demonstrated average power transmission of about 2.69 W. The peak power is 26.9 kW which corresponds to a large input power density, 1.07 GW/cm^2, without fiber damage for $<1.5 \times 10^7$ pulses (Aggarwal *et al.*, 1998). This threshold to damage is close to the predicted threshold of about 3.0 GW/cm^2 due to dielectric breakdown at the surface. Laser damage is prevented by control of fiber polishing; otherwise the fiber end face undergoes damage at lower powers.

5.5.1 Microstructured chalcogenide fibers

Microstructured optical fibers (MOF) feature multiple holes running the entire length of the fiber in a periodic arrangement such that the hole diameter and spacing are on the order of one optical wavelength and are often classified as photonic crystal fibers (PCF) or photonic bandgap fibers (PBG) depending on the light-guiding mechanism and can be solid core (SC) or hollow core (HC). A light guiding core is created by incorporating a defect in the array of holes. In a solid-core photonic crystal fiber (SC-PCF), the core is a solid glass region surrounded by the cladding (an array of holes). Since the diameter and pitch of the holes in the cladding are on the same size scale as the optical wavelength, the cladding has a lower effective index and light is confined to the defect. SC-PCFs can be designed for a variety of optical properties such as very large or small mode areas (Russell, 2003), endlessly single mode operation (Birks *et al.*, 1997), and tunable dispersion (Renversez *et al.*, 2003). PCFs were first fabricated from silica glass in 1996 (Knight *et al.*, 1996) and later from chalcogenide glass in 2000 (Monro *et al.*, 2000), and have been an active area of research ever since (Brilland *et al.*, 2004; Sanghera *et al.*, 2010; Mechin *et al.*, 2010). Le *et al.* (2011) demonstrated an As–Se suspended core or wagon wheel fiber, in which a small solid core is optically isolated from the fiber jacket by thin struts of glass providing a high-NA, air-clad fiber with tight mode confinement for enhanced nonlinearities, with a loss of 0.9 dB/m (Le *et al.*, 2011).

In a PBG fiber the diameter, pitch, and symmetry of the holes are carefully designed to exhibit a photonic band gap, which confines certain optical wavelengths to the defect. A larger hole typically serves this purpose and such fibers are called HC-PBG fibers (Cregan *et al.*, 1999). Since the majority of the optical energy is confined to the air-core rather than the glass, HC-PBG fibers offer the potential for high power handling and low loss. Hollow-core microstructured chalcogenide fibers have been demonstrated by Désévédavy *et al.* (2010) but do not exhibit a photonic band gap. Chalcogenide PBG fibers remain challenging to fabricate and, as of this writing, have not been demonstrated. Kosolapov *et al.* demonstrated hollow core chalcogenide fibers with negative curvature cores and reported transmission of 10.6 μm CO_2 laser radiation with 11 dB/m loss, shown in Fig. 5.12 (Kosolapov *et al.*, 2011).

5.6 Current and future trends

5.6.1 Optical nonlinearities

Chalcogenide glasses have high third-order optical nonlinearities, $\chi^{(3)}$, and consequently high nonlinear refractive indices, n_2, typically two to three orders of magnitude larger than silica (Cerqua-Richardson *et al.*, 1998; Lenz *et al.*, 2000). The sub-picosecond response of these nonlinearities is ideal for high data rate telecommunication devices. Such glasses are being investigated for applications like all optical processing, Raman amplification, parametric amplification, and supercontinuum generation (Sanghera *et al.*, 2006).

For efficient nonlinear devices utilizing the optical Kerr effect, the nonlinearity (n_2) must be high and the nonlinear absorption (i.e., two photon absorption, β) must be low, implying that the wavelength of the energy gap of the material, λ_{gap}, should be less than half the optical wavelength, λ, where β is negligible. For telecommunication applications operating at λ =

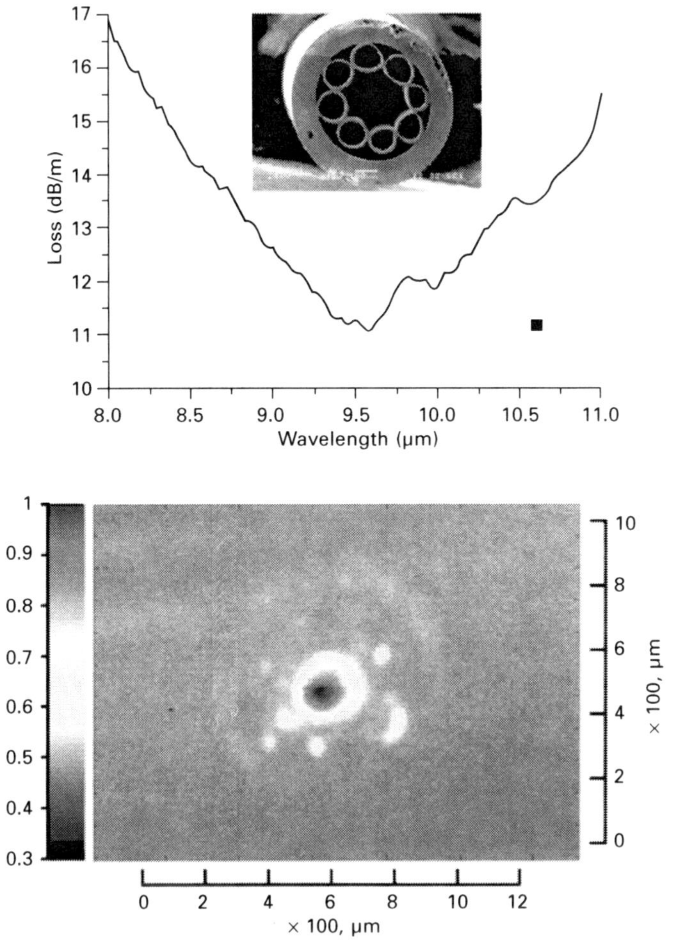

5.12 Optical loss spectrum (left) of negative curvature hollow-core $Te_{20}As_{30}Se_{50}$ (TAS) fiber (micrograph inset) with measured attenuation of 10.6 μm CO_2 laser radiation (square dot) and intensity distribution (right) (composited from figures in Kosolapov *et al.*, 2011).

1.55 μm (photon energy = 0.8 eV), the nonlinear material should have λ_{gap} < 0.77 μm (gap energy > 1.6 eV). A figure of merit (FOM) = $n_2/(\beta\lambda)$ can be defined (Lenz *et al.*, 2000). Table 5.3 lists these properties for several glasses. High speed optical processing has been demonstrated by exploiting these high nonlinearities in chalcogenide glass fiber and waveguides. Earlier work on all optical switching in chalcogenide fiber was performed by Asobe *et al.* (1993), who demonstrated switching of an 80 GHz pulse train in a 2 m length of As_2S_3-based fiber using an optical Kerr shutter configuration. Since then, 640 Gb/s de-multiplexing has been demonstrated in a 5 cm long chalcogenide rib waveguide on silicon by utilizing FWM (Galili *et al.*, 2009). Also, 40 Gb/s all optical wavelength conversion has been demonstrated in chalcogenide tapered fibers (Pelusi *et al.*, 2008).

5.6.2 Raman amplification

Raman amplification at 1.55 μm has been demonstrated in small core As–Se fiber (Thielen *et al.*, 2003b). Over 23 dB of gain was achieved in a 1.1 m length of fiber pumped by a nanosecond pulse of 10.8 W peak power at 1.5 μm. The peak of the Raman gain was shifted by 230 cm^{-1} to 1.56 μm. Subsequent measurements of the Raman gain coefficient showed a value of about 780 times greater than that of silica (Slusher *et al.*, 2004).

A Raman laser has been demonstrated in As–Se fiber by Jackson and Anzueto-Sánchez (2006). They generated 0.64 W of first Stokes at 2062 nm with a slope efficiency of 66% under 2051 nm pumping in a 1 m length fiber with 6 μm core, 0.19 NA. Stimulated Raman scattering (SRS) has been observed in the mid-IR (Aggarwal *et al.*, 2005). Figure 5.13 shows the SRS in a ~ 1 m length of As–Se fiber under CW CO laser pumping at ~ 5.4 μm. The SRS is seen at ~ 6.1 μm. Raman lasers operating in the wavelength range from 6.1 to 6.4 μm would have applications in laser surgery. Calculations show slope efficiencies can approach 80% with moderate threshold powers (Thielen *et al.*, 2003a).

5.6.3 Supercontinuum generation

Supercontinuum generation has been demonstrated between 2 and 3 μm in small core sulfide and selenide fibers as well as photonic crystal selenide

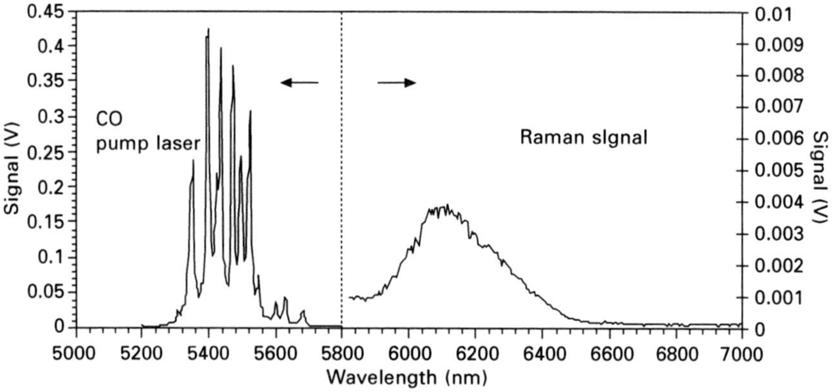

5.13 Stimulated Raman scattering signal observed at 6.1 μm under ~5.4 μm CO laser pumping (Aggarwal *et al.*, 2005).

fibers (PCF) (Shaw *et al.*, 2005). In all cases, pumping was in the normal dispersion region of the fibers and much of the broadening can be attributed to self-phase modulation (SPM) with some broadening to the red due to Raman (Hu *et al.*, 2008). By using chalcogenide glass PCF, the dispersion of the fiber can be controlled and the zero dispersion wavelength can be shifted to the near-IR making it feasible to pump in the anomalous dispersion region of the fiber with conventional near-IR fiber laser pumps. Modeling has shown that very broad supercontinuum bandwidths can be generated with properly designed chalcogenide PCF fiber and proper pump (Hu *et al.*, 2009). Recently, 565 mW of supercontinuum was generated covering the 2–5 μm wavelength region using a COTS all-fiber pumping architecture in a small core, step index fiber, by exploiting Raman as well as self-phase and cross-phase modulation processes (Fig. 5.14) (Gattass *et al.*, 2012).

5.6.4 Brillouin scattering and slow light

Brillouin scattering occurs when an incident or pump photon at a frequency, ω, is converted into a counter-propagating photon with slightly lower frequency, ω_s, and an acoustic phonon at a frequency Ω such that $\omega_s = \omega - \Omega$. While this does occur spontaneously and is a typical mechanism of optical loss, high intensities can cause a stimulated Brillouin scattering (SBS) effect and a strong nonlinear optical gain for the counter-propagating Stokes wave is enhanced exponentially through electrostriction. The counter-propagating wave is typically seeded by a low intensity probe with a frequency near ω_s in order to control the Brillouin gain effect. The measured Brillouin gain

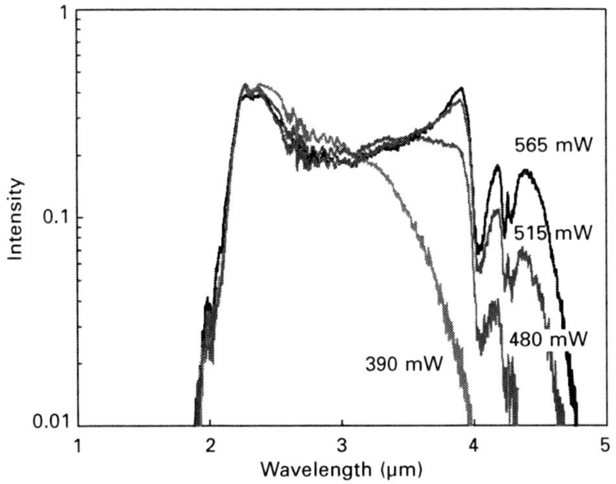

5.14 Spectral evolution of supercontinuum source as a function of measured output power (Gattass *et al.*, 2012).

coefficients are $(5.7 \pm 2.0) \times 10^{-9}$ m.W^{-1} for the As_2S_3 and $(6.75 \pm 0.35) \times 10^{-9}$ m.W^{-1} for As_2Se_3 fibers, respectively. The value for the As_2S_3 fiber, although lower than that for As_2Se_3, is still two orders of magnitude higher than that for fused silica ($\sim 4.4 \times 10^{-11}$m.W^{-1}) (Song *et al.*, 2006; Ogusu *et al.*, 2004). The Brillouin shift is 7.96 GHz, while the linewidth of the Brillouin shift was found to be 31 MHz. The linewidth was measured at low pump powers. Linewidth narrowing was observed for higher powers with linewidths as small as 19 MHz being recorded (Florea *et al.*, 2009).

Another outcome of SBS is a strong dispersion which leads to a slow-light effect (Gehring *et al.*, 2008). The slow-light technique based on SBS in optical fibers enables a tunable optical pulse delay using mostly standard telecom components. Chalcogenide fibers with high nonlinearities inherently require shorter path lengths and lower powers (Sanghera *et al.*, 2010).

A figure of merit (FOM) for slow-light applications in optical fiber accounts for effective mode area (A_{eff}), length (L) and propagation loss (α), refractive index (n_0) and Brillouin gain coefficient (g_B):

$$FOM = 4.34 \frac{g_B k L_{eff}}{n_0 A_{eff} L} \quad [5.5]$$

where k is a constant that reflects whether the polarization is maintained ($k = 1$) or not ($k = 0.5$) and L_{eff} is the effective interaction length $L_{eff} = \frac{1}{\alpha}(1 - e^{-\alpha L})$, in order to compare slow-light capabilities of across fiber materials (Sanghera *et al.*, 2010; Song *et al.*, 2006).

The theoretical gain (G_{th}), expressed in dB, as given by Eq. [5.5], can be used to compare different fibers, if one considers a standard fiber length of 1 m and a standard pump power of 1 mW. Then, the theoretical gain is given by:

$$G_{th}[\mathrm{dB}] = 4.34 \frac{g_B k \times 1\mathrm{mW} \times L_{eff}|_{L=1\mathrm{m}}}{A_{eff}} \quad [5.6]$$

One can use this last, fairly simple expression to compare the most representative fibers considered so far: silica (Song *et al.*, 2005; Ruffin *et al.*, 2005), high-nonlinearity bismuth fiber (Jáuregui *et al.*, 2006; Lee *et al.*, 2005), As_2Se_3 fiber (Song *et al.*, 2006), along with the results reported here. The comparison is provided in Table 5.6, with all the data reported for experiments without polarization control ($k = 0.5$). Also included is the FOM as defined by Song *et al.* (2006). One can easily notice the significant increase in the theoretical gain (or FOM) for the As_2S_3 fiber due to its smaller core, lower loss and slightly reduced refractive index.

Song *et al.* (2006) reported slow-light delays of 0.82 ns/dB gain (~0.58 ns/mW of pump power) in As_2Se_3 fiber and Sanghera *et al.* (2010) demonstrated

Table 5.6 Comparison of figure of merit (FOM) for slow-light based applications at 1.56 μm

	Silica	Bi-HNL	As_2Se_3	As_2Se_3	As_2S_3
n_0	1.47	2.22	2.81	2.81	2.45
A_{eff} (m^2)	6.78×10^{-11}	0.3×10^{-11}	3.94×10^{-11}	6.31×10^{-11}	1.39×10^{-11}
α (dB/m)	0.001	0.91	0.84	0.90	0.57
L (m)	2.0	2.0	5.0	5.0	10.0
L_{eff} (m)	2.0	1.63	3.23	3.1	5.6
g_B (m/W)	4.40×10^{-11}	6.43×10^{-11}	6.10×10^{-9}	6.75×10^{-9}	3.90×10^{-9}
G_{th} (dB)	0.076	0.003	1.084	0.719	3.398
FOM (dB/W/m)	1	17	77	51	139
Reference	Ruffin *et al.* (2005); Song *et al.* (2005)	Jáuregui *et al.* (2006)	Song *et al.* (2006)	Sanghera *et al.* (2010)	Sanghera *et al.* (2010)

slow-light effects in As_2S_3 fiber and reported pulse delays of 0.67 ns/mW of pump power (Fig. 5.15). While delays of 19 ns were demonstrated with only 31 mW of power, delays on the order of 100 ns or more could be obtained for reasonable powers using optimized fibers (Florea *et al.*, 2009).

5.6.5 Degenerate four wave mixing (DFWM)

Nonlinear processes, such as four wave mixing (FWM) that can preserve both amplitude and phase, are uniquely suited for processing of hybrid modulated signals while offering ultrafast response and efficiency, especially in telcom applications. Chalcogenide microstructured fibers are of great interest due to their transparency in the infrared region and nonlinearity 1000 times higher than silica. The ability to dispersion engineer microstructured fiber (e.g., PCF) through control of its geometry could enable phase matching between desired spectral bands for efficient nonlinear interaction in cm-long segments of fibers. Recently, four wave mixing has been demonstrated in a 2.5 cm long segment of chalcogenide microstructured fiber (Fig. 5.16). A continuous wave pump with 345 mW of power enabled the generation of idler waves over more than 50 nm. Ultrafast wavelength conversion of a 4 ps FWHM 8 GHz pulsed signal was also demonstrated (Bres *et al.*, 2011).

5.7 Conclusion

The optical properties of chalcogenide glasses are primarily responsible for their continued use as optical fibers and bulk optical elements for infrared wavelengths. Transmission over a broad wavelength range and low thermo-optic coefficients are inherent in many chalcogenide glasses. Advances in glass

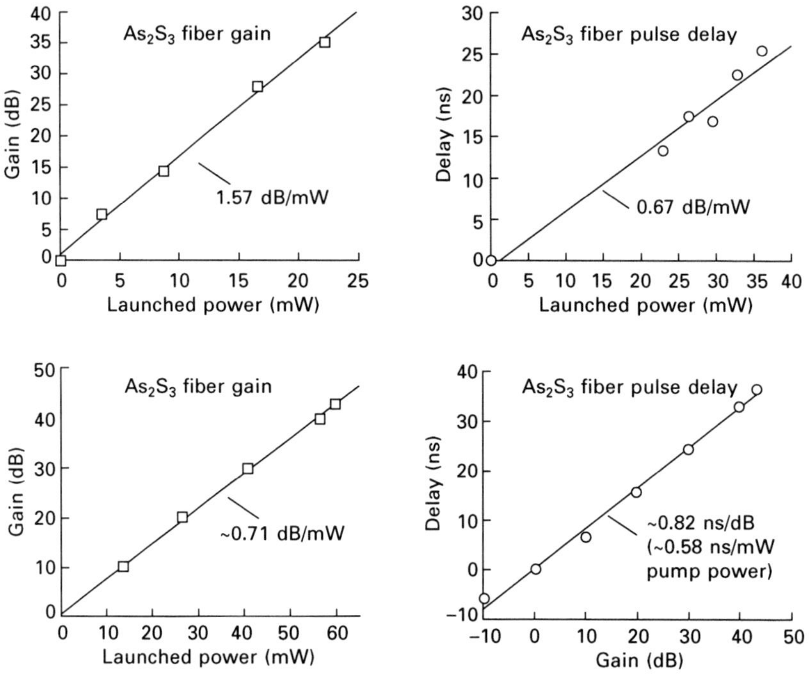

5.15 Gain (top left) and pulse delay (top right) measurements in 10 m long As_2S_3 fiber at 1548 nm (redrawn from data in Sanghera *et al.*, 2010). Gain (bottom left) and pulse delay (bottom right) measurements in 5 m long As_2Se_3 fiber at 1560 nm (redrawn from data in Song *et al.*, 2006). Note pulse delay measurements for As_2S_3 and As_2Se_3 are reported as a function of pump power and gain, respectively, and their slopes are in different units.

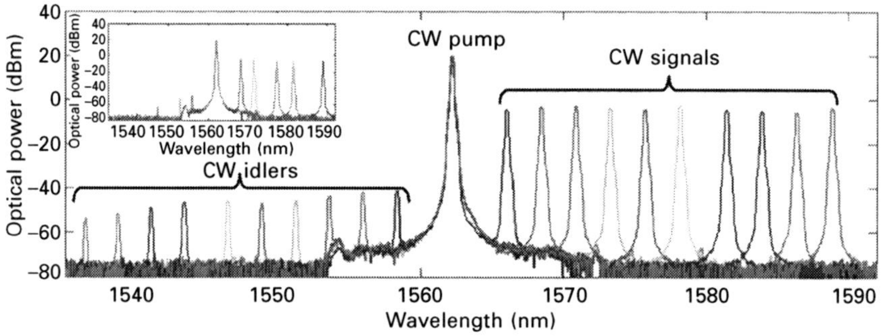

3.16 Superimposed optical spectra at the output of chalcogenide microstructured fiber. Pump positioned at 1562.2 nm and signal swept between 1566 nm and 1589 nm. Idlers are generated between 1559 nm and 1536 nm. Inset: optical spectra at the input of the chalcogenide fiber (Bres *et al.*, 2011).

purification and antireflective surface structures have enabled chalcogenide fibers with low loss and high laser power handling in recent years. Current developments, especially in the non-linear regime, have brought about sustained interest in these materials for novel applications in the infrared including optical switching, broadband sources, wavelength converters, fiber lasers and amplifiers.

5.8 References

Aggarwal, I., Busse, L., Shaw, L., Cole, B. and Sanghera, J., 1998. 'IR transmitting fiber and applications: high-power delivery, sources and amplifiers', in *Proceedings of the Diode Laser Technology Review*, Albuquerque, NM.

Aggarwal, I.D., Shaw, L.B. and Sanghera, J.S., 2005. 'Chalcogenide glass for mid- and long-wave IR fiber lasers', in J.T. Schriempf, ed. *SPIE Proceedings*. SPIE, pp. 242–248.

Aio, L.G., Efimov, A.M. and Kokorina, V.F., 1978. Refractive index of chalcogenide glasses over a wide range of compositions. *Journal of Non-Crystalline Solids*, 27(3), 299–307.

Asobe, M. *et al.*, 1993. Laser-diode-driven ultrafast all-optical switching by using highly nonlinear chalcogenide glass fiber. *Optics Letters*, 18(13), 1056–1058.

Birks, T.A., Knight, J.C. and Russell, P.S.J., 1997. Endlessly single-mode photonic crystal fiber. *Optics Letters*, 22(13), 961–963.

Brady, D.J., 1998. D1.1 Transmission of chalcogenide glass, in D.W. Hewak, ed. *Properties, Processing and Applications of Glass and Rare Earth-Doper Glasses for Optical Fibres*. London: INSPEC, The Institution of Electrical Engineers.

Bres, C.-S. *et al.*, 2011. Demonstration of continuous-wave four-wave mixing in AsSe chalcogenide microstructured fiber. *37th European Conference and Exposition on Optical Communications*, paper Tu.5.LeCervin.4.

Brilland, L. *et al.*, 2004. Fabrication of complex structures of Holey Fibers in chalcogenide glass. *Sensors and Actuators, B: Chemical*, 101, 252–259.

Bureau, B. *et al.*, 2004. Recent advances in chalcogenide glasses. *Journal of Non-Crystalline Solids*, 345–346, 276–283.

Bureau, B. *et al.*, 2009. Chalcogenide glass fibers for infrared sensing and space optics. *Fiber and Integrated Optics*, 28(1), 65–80.

Busse, L.E. *et al.*, 1996. Chalcogenide fibers deliver high IR power. *Laser Focus World*, 32(9), 143–145.

Cerqua-Richardson, K.A. *et al.*, 1998. Comparison of nonlinear optical properties of sulfide glasses in bulk and thin film form. *Optical Materials*, 10(2), 155–159.

Cregan, R. *et al.*, 1999. Single-mode photonic band gap guidance of light in air. *Science*, 285(5433), 1537.

Désévédavy, F. *et al.*, 2010. Chalcogenide glass hollow core photonic crystal fibers. *Optical Materials*, 32(11), 1532–1539.

Florea, C.M. *et al.*, 2009. Slow-light generation through Brillouin scattering in As_2S_3 fibers. *Optical Materials*, 32(2), 358–361.

Florea, C.M. *et al.*, 2011. Reduced Fresnel losses in chalcogenide fibers obtained through fiber-end microstructuring. *Applied Optics*, 50(1), 17–21.

Galili, M. *et al.*, 2009. Breakthrough switching speed with an all-optical chalcogenide glass chip: 640 Gbit/s demultiplexing. *Optics Express*, 17(4), 2182.

Gattass, R.R. *et al.*, 2012. All-fiber chalcogenide-based mid-infrared supercontinuum source. *Optical Fiber Technology*, 18(5), 345–348.

Gehring, G.M. *et al.*, 2008. Fiber-based slow-light technologies. *Journal of Lightwave Technology*, 26(23), 3752–3762.

Graham, A., LeBlanc, R.A. and Hilton, R., Sr, 2003. Low-cost infrared glass for IR imaging applications. In *Window and Dome Technologies VIII*. SPIE, 216–224.

Harbold, J., Ilday, F.Ö., Wise, F., Sanghera, J., *et al.*, 2002a. Highly nonlinear As-S-Se glasses for all-optical switching. *Optics Letters*, 27(2), 119–121.

Harbold, J.M., Ilday, F.Ö., Wise, F.W. and Aitken, B.G., 2002b. Highly nonlinear Ge-As-Se and Ge-As-S-Se glasses for all-optical switching. *Photonics Technology Letters, IEEE*, 14(6), 822–824.

Harrington, J.A., 2004. *Infrared Fibers and Their Applications*, Bellingham, WA: SPIE Press.

Hewak, D.W. (ed.), 1998. *Properties, Processing and Applications of Glass and Rare Earth-Doper Glasses for Optical Fibres* London: INSPEC, The Institution of Electrical Engineers.

Hilton, A.R. and Jones, C.E., 1967. The thermal change in the nondispersive infrared refractive index of optical materials. *Applied Optics*, 6(9), 1513.

Hu, J. *et al.*, 2008. Raman response function and supercontinuum generation in chalcogenide fiber. *Conference on Lasers and Electro-Optics/Quantum Electronics and Laser Science Conference and Photonic Applications Systems Technologies*.

Hu, J. *et al.*, 2009. Generating mid-IR source using As_2S_3-based chalcogenide photonic crystal fibers. *Conference on Lasers and Electro-Optics/International Quantum Electronics Conference*.

Jackson, S.D. and Anzueto-Sánchez, G., 2006. Chalcogenide glass Raman fiber laser. *Applied Physics Letters*, 88(22), 221106.

Jáuregui, C. *et al.*, 2006. Four-fold reduction in the speed of light at practical power levels using brillouin scattering in a 2-m bismuth-oxide fiber, in *Optical Fiber Communication Conference and Exposition and The National Fiber Optic Engineers Conference*.

Kanamori, T. *et al.*, 1984. Chalcogenide glass fibers for mid-infrared transmission. *Journal of Lightwave Technology*, 2(5), 607–613.

Kanamori, T. *et al.*, 1985. Transmission loss characteristics of $As_{40}S_{60}$ and $As_{38}Ge_{5}Se_{57}$ glass unclad fibers. *Journal of Non-Crystalline Solids*, 69(2–3), 231–242.

Katsuyama, T. and Matsumura, H., 1989. *Infrared Optical Fibers*, Philadelphia, PA: IOP Publishing.

Knight, J.C. *et al.*, 1996. All-silica single-mode optical fiber with photonic crystal cladding. *Optics Letters*, 21(19), 1547.

Kokorina, V.F., 1996. *Glasses for Infrared Optics*, Cambridge: Woodhead Publishing Ltd.

Kosolapov, A.F. *et al.*, 2011. Demonstration of CO_2-laser power delivery through chalcogenide-glass fiber with negative-curvature hollow core. *Optics Express*, 19(25), 25723–25728.

Le, S.D. *et al.*, 2011. Efficient four-wave mixing in an ultra-highly nonlinear suspended-core chalcogenide As_{38} Se_{62} fiber. *Optics Express*, 19(26), B653–B660.

Lee, J.H. *et al.*, 2005. Experimental comparison of a Kerr nonlinearity figure of merit including the stimulated Brillouin scattering threshold for state-of-the-art nonlinear optical fibers. *Optics Letters*, 30(13), 1698.

Lenz, G. *et al.*, 2000. Large Kerr effect in bulk Se-based chalcogenide glasses. *Optics Letters*, 25(4), 254–256.

Mechin, D. *et al.*, 2010. Chalcogenide photonic-crystal fibers expand nonlinear applications. *Laser Focus World*, 46(5), 41–47.

Monro, T.M. *et al.*, 2000. Chalcogenide holey fibres. *Electronics Letters*, 36(24), 1998–2000.

Moynihan, C.T. *et al.*, 1975. Intrinsic and impurity infrared absorption in As_2Se_3 glass. *Journal of Non-Crystalline Solids*, 17(3), 369–385.

Nishii, J. and Yamashita, T., 1998. Chalcogenide glass-based fibers. in J.S. Sanghera and I.D. Aggarwal, eds. *Infrared Fiber Optics*. Cambridge: Woodhead Publishing Ltd pp. 143–184.

Nishii, J., Yamashita, T. and Yamagishi, T., 1989. Oxide impurity absorptions in Ge-Se-Te glass fibres. *Journal of Materials Science*, 24(12), 4293–4297.

Nishii, J. *et al.*, 1992. Recent advances and trends in chalcogenide glass fiber technology: a review. *Journal of Non-Crystalline Solids*, 140, 199–208.

Ogusu, K., Li, H. and Kitao, M., 2004. Brillouin-gain coefficients of chalcogenide glasses. *Journal of the Optical Society of America B: Optical Physics*, 21(7), 1302.

Pelusi, M.D. *et al.*, 2008. Applications of highly-nonlinear chalcogenide glass devices tailored for high-speed all-optical signal processing. *IEEE Journal of Selected Topics in Quantum Electronics*, 14(3), 529–539.

Popescu, M.A., 2000. *Non-Crystalline Chalcogenides*, Dordrecht: Kluwer Academic Publishers.

Renversez, G., Kuhlmey, B. and McPhedran, R., 2003. Dispersion management with microstructured optical fibers: ultraflattened chromatic dispersion with low losses. *Optics Letters*, 28(12), 989.

Riedl, M.J., 2001. *Optical Design Fundamentals for Infrared Systems*, Bellingham, WA: SPIE Press.

Ruffin, A. *et al.*, 2005. Brillouin gain analysis for fibers with different refractive indices. *Optics Letters*, 30, 3123–3125.

Russell, P.S.J., 2003. Photonic crystal fibers. *Science*, 299(5605), 358–362.

Sanghera, J.S. and Aggarwal, I.D. (eds) 1998. *Infrared Fiber Optics*. Cambridge: Woodhead Publishing Ltd.

Sanghera, J.S. and Aggarwal, I.D., 1999. Active and passive chalcogenide glass optical fibers for IR applications: a review. *Journal of Non-Crystalline Solids*, 256–257, 6–16.

Sanghera, J.S. *et al.*, 1994. Fabrication of low-loss IR-transmitting $Ge_{30}As_{10}Se_{30}Te_{30}$ glass fibers. *Journal of Lightwave Technology*, 12(5), 737–741.

Sanghera, J.S. *et al.*, 1996. Fabrication of long lengths of low-loss IR transmitting $As_{40}S_{(60-x)}Se_x$ glass fibers. *Journal of Lightwave Technology*, 14(5), 743–748.

Sanghera, J.S. *et al.*, 2006. Nonlinear properties of chalcogenide glass fibers. *Journal of Optoelectronics and Advanced Materials*, 8(6), 2148–2155.

Sanghera, J. *et al.*, 2010. Nonlinear properties of chalcogenide glass fibers. *International Journal of Applied Glass Science*, 1(3), 296–308.

Savage, J.A., Webber, P.J. and Pitt, A.M., 1980. The potential of Ge-As-Se-Te glasses as 3–5 μm and 8–12 μm infrared optical materials. *Infrared Physics*, 20(5), 313–320.

Savage, J.A., Webber, P.J. and Pitt, A.N., 1977. Infrared optical glasses for applications in 8–12-μm thermal imaging systems. *Applied Optics*, 16(11), 2938–2941.

Seddon, A.B., 1995. Chalcogenide glasses: a review of their preparation, properties and applications. *Journal of Non-Crystalline Solids*, 184, 44–50.

Shaw, B. *et al.*, 2005. IR supercontinuum generation in As-Se photonic crystal fiber. *Advanced Solid-State Photonics Conference Proceedings*.

Shibata, S., Terunuma, Y. and Manabe, T., 1981. Sulfide glass fibers for infrared transmission. *Materials Research Bulletin*, 16(6), 703–714.

Skettrup, T., 1978. Urbach's rule derived from thermal fluctuations in the band-gap energy. *Physical Review B: Condensed Matter and Material Physics*, 18(6), 2622–2631.

Slusher, R.E. *et al.*, 2004. Large Raman gain and nonlinear phase shifts in high-purity As_2Se_3 chalcogenide fibers. *Journal of the Optical Society of America B: Optical Physics*, 21(6), 1146–1155.

Snopatin, G.E. *et al.*, 2009. High-purity chalcogenide glasses for fiber optics. *Inorganic Materials*, 45(13), 1439–1460.

Song, K.Y., Herráez, M.G.L. and Thévenaz, L., 2005. Observation of pulse delaying and advancement in optical fibers using stimulated Brillouin scattering. *Optics Express*, 13(1), 82–88.

Song, K.Y. *et al.*, 2006. Highly efficient Brillouin slow and fast light using As_2Se_3 chalcogenide fiber. *Optics Express*, 14(13), 5860–5865.

Thielen, P., Shaw, L., *et al.*, 2003a. Modeling of a mid-IR chalcogenide fiber Raman laser. *Optics Express*, 11(24), 3248.

Thielen, P.A., Shaw, L.B., *et al.*, 2003b. Small-core As-Se fiber for Raman amplification. *Optics Letters*, 28(16), 1406–1408.

Thorpe, M. and Weaire, D., 1971. Electronic properties of an amorphous solid. II. Further aspects of the theory. *Physical Review B: Condensed Matter and Material Physics*, 4(10), 3518–3527.

Urbach, F., 1953. The long-wavelength edge of photographic sensitivity and of the electronic absorption of solids. *Physical Review*, 92(5), 1324.

Vlček, M. *et al.*, 1987. Influence of oxygen traces on physical properties of glassy $GeSe_2$. *Journal of Materials Science*, 22, 2119–2123.

Weaire, D. and Thorpe, M., 1971. Electronic properties of an amorphous solid. I. A simple tight-binding theory. *Physical Review B: Condensed Matter and Material Physics*, 4(8), 2508–2520.

Wood, D. and Tauc, J., 1972. Weak absorption tails in amorphous semiconductors. *Physical Review B: Condensed Matter and Material Physics*, 5(8), 3144–3151.

Zallen, R. *et al.*, 1971. Electronic structure of crystalline and amorphous As_2S_3 and As_2Se_3. *Physical Review Letters*, 26(25), 1564–1567.

6
Photo-induced phenomena in chalcogenide glasses

K. TANAKA, Hokkaido University, Japan

DOI: 10.1533/9780857093561.1.139

Abstract: After briefly reviewing photoinduced phenomena in chalcogenide glasses, we take up two topics: photodarkening and photoinduced anisotropic deformations. The former has been extensively studied due to its simple and unique features, which are considered from new perspectives. We then discuss why the microscopic mechanism remains speculative. The latter topic is attracting growing interest. The deformation, which is macroscopic and surprisingly dramatic, will prelude a paradigm of optical manipulation of viscous materials.

Key words: chalcogenide glass, photodarkening, photoinduced fluidity, vector deformation, optical force.

6.1 Introduction

The photoinduced phenomenon, which emerges from photo-electronically excited atomic motions, is widespread on the earth. The best known examples are undoubtedly biological reactions including plant photosynthesis and animate vision. Also, Toyozawa (2003) proposes that life is created by sunlight through successive photochemical reactions, the idea being consistent with some observations (Ehrenfreund, 1999). For inorganic crystals, we know of much simpler photoinduced processes, such as point-defect formation in Na(K)Cl and photographic reaction in AgCl(Br) (Itoh and Stoneham, 2000). In chalcogenide glasses, photoinduced phenomena have attracted global interest since around 1970, the trigger being the Ovshinskys' discovery of a photo-thermal phase change (Popescu, 2001; Kolobov, 2003; Tanaka and Shimakawa, 2011; Kolobov and Tominaga, 2012).

Research on photoinduced phenomena in non-crystalline chalcogenides has developed through three stages. First, in addition to the phase change, extensive studies had been performed for photoinduced *athermal* processes (photon effects), including photo-crystallization, photodarkening*, and photodoping. De Neufville *et al.* (1974) and other researchers demonstrated that the photoinduced change in As(Ge)-S(Se) consists of irreversible,

*Following De Neufville *et al.* (1974), we restrict this word to the *thermally reversible red-shift* of optical absorption edges, excluding irreversible changes.

reversible, and transitory processes, which are characterized as illustrated in Fig. 6.1. In addition to such isotropic changes, or *scalar* effects, which are not influenced by light polarization, Zhdanov and Malinovsky (1977) discovered so-called *vector* effects, i.e., anisotropic changes induced by polarized light. Then, studies of these phenomena, specifically on induction mechanisms and applications to optical devices, led to exciting topics in international meetings (Lucovsky and Galeener, 1976; Yonezawa, 1981). In the second stage, following this growth period, related research naturally became calmer. Induction mechanisms were continuously explored and disputed. Some researchers investigated photoinduced phenomena in metal-chalcogenide systems (Kolobov and Elliott, 1991), chalcogen liquids (Sakaguchi and Tamura, 2003), and nano-structures (Tanaka, 2004b). Those features have been reviewed extensively (Tanaka, 1990; Pfeiffer *et al.*, 1991; Shimakawa *et al.*, 1995; Fritzsche, 2000), and we will reconsider the scalar change from new perspectives in Sections 6.2–6.4. The third stage seems to begin in around 1995 with three topics (Tanaka and Shimakawa, 2011), namely commercialization of DVDs (digital versatile discs) and new a-Se (amorphous Se) photoconducting devices, extensive studies on nonlinear optical properties, and discoveries of photoinduced macroscopic deformations. The last one, specifically vector photo-deformations, are surprisingly dramatic, and will be discussed in Section 6.5.

Through these studies, we have recognized a variety of aspects of the photoinduced phenomenon. In general, as listed in Table 6.1, a photoinduced

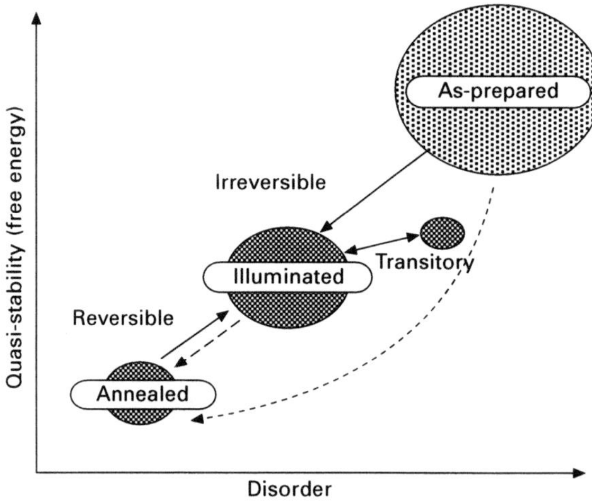

6.1 Characterization of three scalar changes, irreversible, thermally reversible, and transitory processes, in scales of structural disorder and quasi-stability. Solid and dashed arrows indicate the changes induced by illumination and annealing, respectively.

Table 6.1 An energy flow in photoinduced phenomena with macroscopic changes

Light →	Excitation →	Structural change →	Property modification
CW/pulse	Electronic	Atomic change	Density (volume/ shape)
Photon energy	Electro-thermal	Sputtering/explosion	Thermal
Irradiance	Thermal	Relaxation	Mechanical
Polarization	Nonlinear	Chemical/ionic	Optical
Focusing	Plasma	Reaction	Electrical
Temperature			Chemical
			Surface (sticking, etc.)

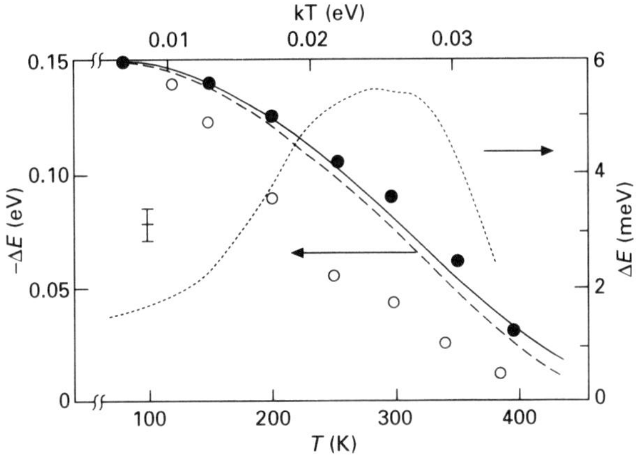

6.2 Dependence of photodarkening (left scale) and photoinduced dichroism (right scale) in As_2S_3 upon temperature at which a sample is exposed to bandgap light.

process proceeds from photo-electronic excitation to some atomic structural changes, which appear as modifications of macroscopic properties.

Naturally, induced changes vary with excitation conditions. For instance, as shown in Fig. 6.2, athermal photoinduced phenomena can be divided in two in respect of temperature variations (Tanaka and Shimakawa, 2011). One includes the purely photoinduced phenomena, such as photodarkening, which become monotonically smaller under illumination at higher temperatures, owing to thermal relaxation. The other is thermally assisted phenomena, such as photo-crystallization in a-Se, photodoping, and vector optical changes, which become the most prominent under illumination at certain temperatures. These different temperature behaviors suggest different scales of responsible structural changes, which may be atomic in the purely photoinduced phenomena and more extended (or diffusional) in the thermally assisted changes.

For these varieties of photoinduced phenomena, our depth of understanding varies with each one, as exemplified in the four cases below (Tanaka and

Shimakawa, 2011). First, the irreversible structural and optical changes (Fig. 6.1), which appear in a-Se and as-evaporated chalcogenide films, are generally gross and related microscopic changes are atomically detectable. The mechanisms have been interpreted as kinds of stabilization processes through crystallization, polymerization (including void collapsing), homo-to-heteropolar bond interchange, oxidation, and vaporization. Second, the photodoping in Ag/As(Ge)-S and other Ag-related phenomena have been understood using electrical and thermodynamical models (Kolobov and Elliott, 1991; Wagner and Frumar, 2003; Kawaguchi, 2003), and a remaining subject is to obtain a unified microscopic (atomic) interpretation, which may be challenging. Third, the photoinduced changes appearing in annealed glassy samples are generally less conspicuous, and the mechanisms remain inconclusive. Specifically, the photodarkening phenomenon (and related refractive index increase) arising from red-shifts of optical absorption edges is simple, seemingly inherent, and unique to covalent chalcogenide glasses, while we cannot yet grasp its atomic mechanism. Finally, we note that recent studies have discovered dramatic vector deformations, which are very surprising, since in optical changes the vector is much smaller than the scalar; e.g., photoinduced refractive index changes in As_2S_3 (with a refractive index of ~2.6) at room temperature are ~0.1 in irreversible, ~0.02 in reversible, ~0.003 in transitory, and ~0.002 in vectoral.

6.2 Scalar changes

6.2.1 Gross features

Considerable studies have been performed over the last 40 years for the thermally reversible scalar change in chalcogenide glasses (De Neufville *et al.*, 1974; Popescu, 2001; Kolobov, 2003; Tanaka and Shimakawa, 2011; Kolobov and Tominaga, 2012). The most appropriate sample for such studies may be bulk As_2S_3 glass, with the Tauc optical gap E_g of ~2.4 eV, in flake forms with a thickness of ~10 μm. We may also employ deposited films of $As_2S(Se)_3$ of a few micron thicknesses which have been stabilized in advance through annealing in Ar atmosphere at just below the glass transition temperature (~450 K). Evaporated film of Se may also be useful for obtaining fundamental insights, due to its elemental constituent, provided that illumination can be provided at substantially lower temperatures than the glass transition temperature of ~310 K. By contrast, we must be careful with non-stoichiometric (specifically, As-rich) compositions and Ge-chalcogenides, the structures seemingly varying with preparation procedures and oxidation. Photoinduced changes appear, as listed in Table 6.1, in optical bandgap (photodarkening), refractive index (increase), electrical resistance, volume (expansion or contraction), elasticity (detected as changes in hardness and sound velocity), glass transition behaviors, and chemical properties. Specifically, as described later, a lot of studies have been performed on the optical change under varied conditions of light exposure, temperature, and pressure.

Resultant atomic changes have been investigated. X-ray diffraction measurements of As_2S_3 reproducibly detect broadening of the so-called first sharp diffraction peak (FSDP) (Tanaka, 1975), which suggests that *disorder increases in 'some' medium-range structures*. The Boson peak reduces upon illumination, which implies similar structural changes. By contrast, to my knowledge, there are no reports demonstrating close connections between the photodarkening and short-range (coordination number, bond length, and bond angle) or defective* structural modifications, for simple glasses such as $As_2S(Se)_3$ and Se. On the other hand, the photoinduced absorption increase and volume expansion in As_2S_3 suggest, under some assumptions, that the density of atomic sites involved is nominally $\sim 10^{20}$ cm^{-3}, corresponding to an atomic fraction of ~1% (Tanaka, 2012).

6.2.2 Remaining problems

Perhaps as many as ten models have been proposed for the reversible photoinduced change (Shimakawa *et al.*, 1995; Popescu, 2001; Kolobov, 2003; Tanaka and Shimakawa, 2011; Kolobov and Tominaga, 2012), while we cannot yet elucidate the mechanism. Why does the mechanism remain inconclusive? At least three possibilities can be put forward.

First, simply because we cannot determine the structure of glasses. We have not yet evolved experimental methods, such as the Bragg-peak analyses of crystals, which can unambiguously identify a disordered structure. For short-range structures, we may follow radial distribution and EXAFS analyses. Such studies have suggested that the structures before and after illumination are similar, although some changes are detected during illumination (Kolobov *et al.*, 1997). Consequently, we seek photoinduced modifications in medium range (as appearing as the FSDP broadening) and/or defective (such as dangling bonds) structures. But, it is difficult to identify these structures in disordered matrices, and hence, the *change* in structures cannot be known. We here note that the (photoinduced) defects in even *crystalline* SiO_2, for instance, cannot be identified using structural analyses, but are suggested or

*Some researchers emphasize density increases in wrong bonds, such as As–As, in $As_2S(Se)_3$ under (super-)bandgap illumination, which can actually be detected using Raman-scattering spectroscopy (for a recent study, see Yannopoulos *et al.*, 2012). However, the increase seems to be irrelevant to the photodarkening, since sub-gap illumination also induces photodarkening in As_2S_3 without appreciable increases in the wrong-bond density. In addition, we cannot envisage the wrong bond in a-Se and a-S, both exhibiting clear photodarkening at low temperatures (Tanaka and Shimakawa, 2011).

determined through electronic and/or optical measurements. Since the initial and the final structure are unknown, we cannot elucidate the dynamics of photo-electro-atomic processes. Under such circumstances, we may consider the mechanism, taking a variety of observations into account.

The second is characteristic complexities. Do all the property changes really arise from a single kind of atomic change? As listed in Table 6.1, several kinds of macroscopic properties are modified by light illumination, and it is questionable if those are produced by single atomic structural changes. The red-shift (photodarkening) and refractive index increase under bandgap illumination can quantitatively be related through Kramers–Kronig relations, and these are ascribable to an enhanced spatial fluctuation of the valence band consisting of lone-pair electron states, leaving the conduction band intact. Such a change in the valence band may arise from a change in intermolecular distances of van der Waals types, for which quantitative analyses will be the next subject. However, the intermolecular interaction may be modified by several kinds of atomic changes, which cannot be identified. In addition, it remains debatable whether such an intermolecular change can cause the observed modifications in macroscopic structural (elastic and thermal) and chemical properties.

Otherwise, as schematically illustrated in Fig. 6.3, it is plausible that upon light illumination, several kinds of atomic changes occur. For instance, some experiments demonstrate structural changes as (i) creation of several kinds of dangling bonds, (ii) some change in medium-range structures, which causes FSDP broadening, and (iii) hetero-to-homopolar bond conversions (Shimakawa *et al.*, 1995; Popescu, 2001; Kolobov, 2003; Tanaka and Shimakawa, 2011). We may then envisage that one and another are responsible for, respectively, optical and structural changes (Tanaka, 2012). In addition, it seems to be common that the atomic change in one material varies with exposure conditions.

The third reason why the mechanism of reversible photoinduced change remains inconclusive is a material variety. There exist notable variations in the photoinduced changes in different materials. It has been demonstrated

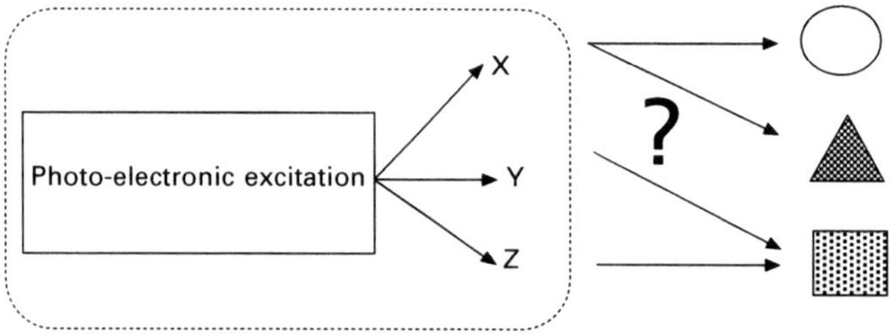

6.3 Possible relationships between photoinduced atomic changes (X, Y, Z) and modifications in macroscopic properites (circle, triangle, square).

that As_2S_3 and GeS_2 expand under illumination. It is also known that Ge-As-S(Se) ternary glasses possess a unique characteristic, photoinduced volume compaction (Hamanaka *et al.*, 1976; Calvez *et al.*, 2009; Arsova and Vateva, 2012). Nevertheless, all these glasses exhibit photodarkening. We may then envisage that illumination produces different kinds of structural disordering in different materials. For instance, the disordering in a-Se(S) may be produced by chain-ring transformations (and/or in chain lengths), in $As_2S(Se)_3$ by fluctuations of van der Waals distances, in As-rich As-Se by hetero-to-homopolar bond conversions, in Ge-S(Se) by modifications of two to three-dimensional structures, and in ternary Ge-As-S by intermixing of Ge- and As-related structural units. As known, in oxide glasses such as $Si(Ge)O_2$, light illumination produces various kinds of defects, which appear as optical absorption peaks, with the absorption edges being intact (Terakado and Tanaka, 2006).

6.3 Enhancement and suppression of photodarkening

It will be fruitful if we can enhance or suppress the photodarkening and other macroscopic changes. The enhancement will afford faster and/or greater optical responses. By contrast, in some applications such as optical amplifiers and nonlinear devices, the photoinduced change may be a nuisance. In fact, in Pr-doped $Ge_{33}As_{12}Se_{55}$ glass, which presents the so-called matrix excitation of rare-earth ions (Pr^{3+}), photodarkening tends to reduce the intensity of photoluminescence from Pr^{3+} (Harada and Tanaka, 1999). We are then interested in chalcogenide glasses giving rise to prominent or negligible photoinduced changes, for which several tactics can be envisaged (Tanaka, 2003a). For spectral effects, see Section 6.4.

6.3.1 Physical

Photodarkening can be enhanced and suppressed by tuning the temperature T_i, at which a sample is illuminated. As exemplified in Fig. 6.4, the photoinduced effect (optical changes and volume expansion) becomes smaller with an increase in T_i/T_g (≤ 1), and ultimately, the effect becomes negligible at $T_i/T_g = 1$ (Tanaka, 1983). In fact, a-Se with $T_g \approx 310$ K does not show appreciable photodarkening at room temperature (Tanaka and Odajima, 1982; Tichý *et al.*, 2002), although it does exhibit photo-expansions at room temperature (Asao and Tanaka, 2007), probably because of longer relaxation time of macroscopic volume changes. Therefore, a simple way to enhance and suppress photodarkening (and related scalar changes) is the tuning of T_i.

Hydrostatic compression can also enhance or suppress photodarkening (Tanaka, 1986; Iketomo *et al.*, 2002), as shown in Fig. 6.5 for a-Se.

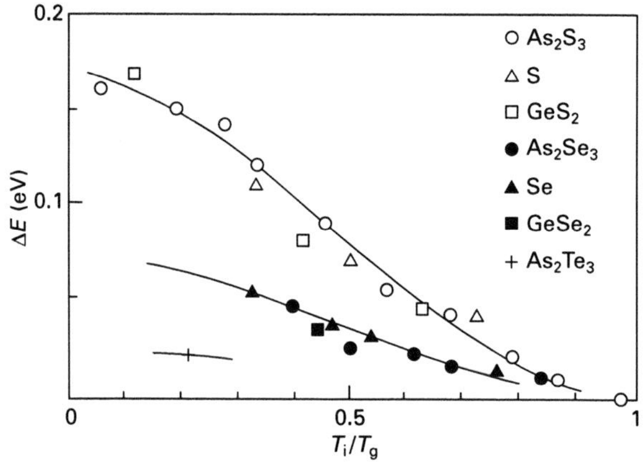

6.4 Photodarkening ΔE in elemental and stoichiometric materials as a function of T_i/T_g, where T_i is the temperature at which illumination is provided and T_g the glass transition temperature (modified from Tanaka, 1983). Measurements of ΔE are performed at (or lower temperatures than) T_i.

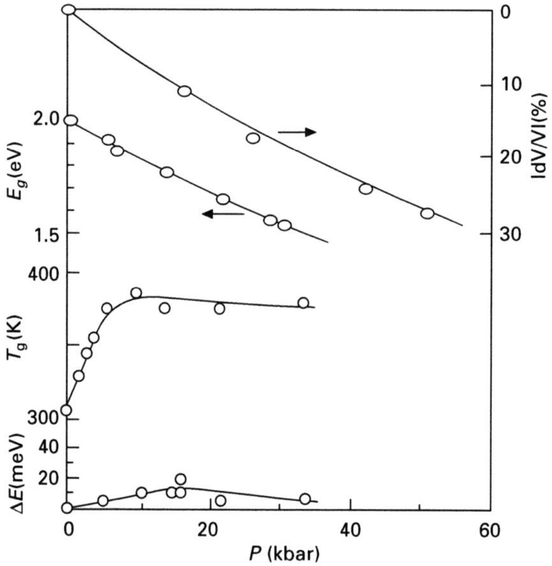

6.5 Variations of volume V, optical gap E_g, glass transition temperature T_g, and photodarkening ΔE at room temperature of free a-Se films as a function of hydrostatic pressure P, which is generated using a diamond-anvil cell (Tanaka, 1986).

The compression enhances the photodarkening through raising the glass transition temperature, which gives rise to a decrease in T_i/T_g. However, at higher pressures than ~15 kbar (= 1.5 GPa), the compression suppresses the photodarkening, which suggests that the existence of free volumes in amorphous networks is a prerequisite to it. Such compression effects may practically be attained by burying (or dispersing) a chalcogenide sample in rigid solids (Indutnyi *et al.*, 1999) and by using piezoelectric materials.

We here mention that, as exemplified on pp. 154-158 and in Section 6.5, a chalcogenide glass tends to photo-deform (scalar and vector) and fluidize. Accordingly, the film on rigid substrates may practically be stressed under illumination, which seems to modify photodarkening characteristics (Tanaka and Mikami, 2012; Abbaszadeh *et al.*, 2012). We must also be careful to consider thermally generated stress effects upon photodarkening.

It has been demonstrated that ultra-thin (~5 nm) films of As_2S_3 do not present photodarkening (Tanaka *et al.*, 1984b; Hayashi and Mitsuishi, 2002), though some irreversible change may exist (Sainov *et al.*, 2007; Abdulhalim *et al.*, 2011). We can speculate at least two possibilities. One is that photo-excited carriers diffuse over distances of ~5/2 nm and they will recombine at sample surfaces, leaving no traces on amorphous structures. Otherwise, some photoinduced structural changes occur, but induced strains may be released at the scale of ~5/2 nm, which is comparable to a typical scale (~2 nm) of the medium-range order in As_2S_3. As mentioned in Section 6.2.1, the number of sites responsible for photodarkening (and photoexpansion) is estimated at 10^{20} cm^{-3}, i.e. one site per 10^2–10^3 atoms (Fig. 6.6(a)), or one site in a volume with a side length of 1–4 nm. It is then tempting to assume that free nanostructured chalcogenide samples undergo little reversible photoinduced changes.

Incidentally, why is the site density responsible for the photoinduced changes ~10^{20} cm^{-3}, ~1 at.%? If we could increase the density, the photodarkening

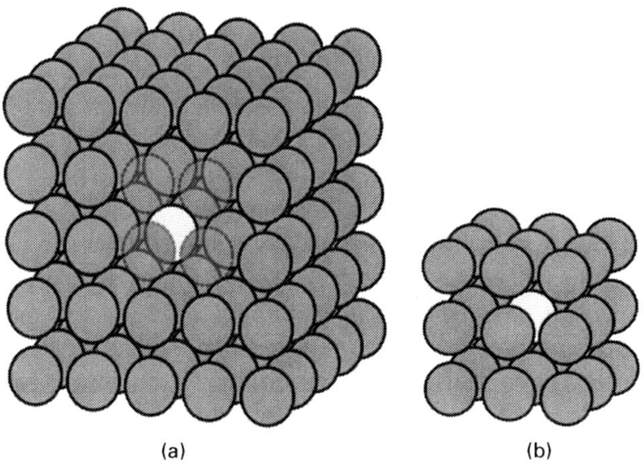

6.6 Schematic illustrations of a defective site in cubic clusters having 125 (a) and 27 (b) atoms.

would become greater (Tanaka, 2012). However, the illustration in Fig. 6.6(b) counters that notion. If the site density were ~10 at.%, which corresponds to a situation of one defective site in every short-range structural unit, a photoinduced atomic motion would strain bond angles and/or lengths, causing cooperative and wider structural transformations, such as crystallization, polymerization, or vaporization. We can then assume that since the responsible atomic site is confined in a medium-range unit, with density of ~1 at.% (Fig. 6.6(a)), the photodarkening can be quasi-stable and thermally recovered. It is mentioned here that, for bi-stable configurations in covalent glassy networks, there seems to exist a universal correlation between the site density and the quasi-stability energy (Tanaka, 1992), the reason being unresolved.

6.3.2 Compositional

Many compositional studies on photodarkening have been reported so far. As shown in Fig. 6.4 for elemental and stoichiometric alloys, the sulfide shows the greatest change, the selenide the next, and the telluride the smallest (Tanaka, 1983). Hence, we may select a sulfide and a telluride for obtaining enhanced and reduced optical changes. Nevertheless, due to fairly metallic characters, binary telluride materials ($E_g \approx 1$ eV) tend to crystallize, and in addition, infrared transparency is likely to be lowered by free-carrier absorption. These drawbacks may be avoided by addition of Se (as As-Se-Te), which enhances compositional disorder and widens the bandgap. On the other hand, in non-stoichiometric binary glasses of As(Ge)-S(Se), the photodarkening tends to become maximal at the compositions with average coordination number Z of 2.67 (Tanaka, 1989). For ternary systems, however, compositional variations are not universal or controversial, as exemplified by the following three kinds of materials.

First is the covalent system, Ge-As-S(Se). Several researchers have demonstrated that photodarkening in Ge-As-S (with Z = 2.3–3.0) induced by bandgap illumination shows a maximum at the composition of $Z \approx 2.67$ (Tanaka *et al.*, 1986; Štábl and Tichý, 2004; Arsova *et al.*, 2005). Or, to my knowledge, stable glasses which exhibit the greatest ($\Delta E \approx 125$ meV) and the least photodarkening at room temperature are $Ge_{0.2}As_{0.25}S_{0.55}$ with $Z \approx 2.67$ and $Ge_{0.4}As_{15}S_{45}$ with $Z \approx 3.0$. By contrast, Calvez *et al.* (2009) recently report for $Ge_xAs_xSe_{1-2x}$ with Z = 2.2–2.6 that the photodarkening and photo-expansion induced by sub-gap illumination becomes smaller at $Z \geq 2.45$. The reasons for this discrepancy are speculative, and we may envisage three possibilities. First is an experimental artifact. In the former reports, the photodarkening is evaluated from red-shifts of optical absorption edges, and in the latter from a transmittance decrease at sub-gap photon energies. The excitation energies (bandgap and sub-gap) are also different. Second, there may be confusion between saturated values and induction speeds (quantum efficiency). I assume that the saturated red-shift ΔE has a peak at $Z \approx 2.67$, while the induction speed becomes slower with an increase in Z from 2.4 to 2.67. Consequently, under incomplete saturation, both the factors tend to yield confusing Z dependence. Third, intrinsic differences may exist between Ge-As-S and Ge-As-Se systems, the latter being constructed by atoms having similar sizes and more covalent bonds. We also recall that, although photo-expansions in As_2S_3 and GeS_2 are well documented, $Ge_1As_4Se_4$ (Z = 2.6) *contracts* upon bandgap illumination (Hamanaka *et al.*, 1976) and also volume change in $Ge_1As_1Se_{13}$ (Z = 2.2) alters (expansion and contraction) with heat treatments and exposure conditions (Calvez *et al.*, 2009).

The second ternary system of interest is the metal–chalcogenide system (Fig. 6.7). Liu and Taylor (1987) have discovered that Cu incorporation, at a concentration of 1–5 at.%, to $As_2S(Se)_3$ destroys the photodarkening. The interpretation, being inconclusive, is twofold. It is demonstrated that Cu is bonded with S(Se), and Cu-S(Se) bonds produce electronic states around the lone-pair electron (valence) band of S(Se), so that the Cu-S(Se) states are likely to mask photoinduced modifications of the lone-pair electron band. Otherwise (or concomitantly), incorporated Cu, being tetrahedrally bonded with the chalcogen, converts flexible structures to crosslinked networks, which suppress photo-structural changes. Similar metal-addition effects have been reported for $x$$Na_2S$-(100–$x$)$GeS_2$ glasses, in which photodarkening disappears in the compositions of $x \geq 10$ mol.%, which are fairly hygroscopic (Tanaka *et al.*, 2003). On the other hand, ternary Ag-chalcogenides such as $AgAsS_2$ are Ag-ion conducting, which causes a photoinduced chemical modification (Yoshida and Tanaka, 1995), making photodarkening undetectable. In short, although studies on incorporation effects of group Ia (1), IIb (2), and IIIb

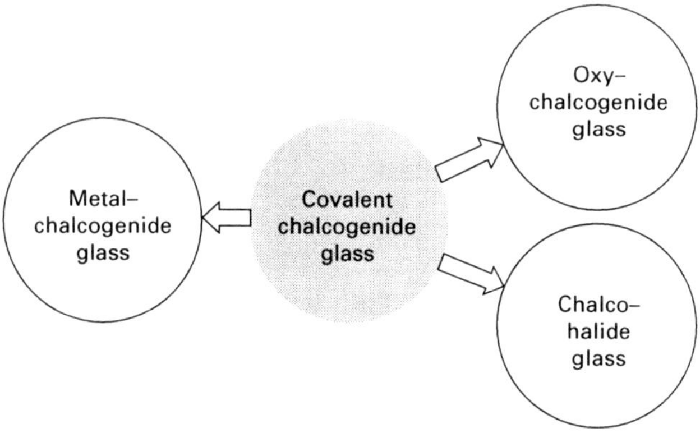

6.7 Development of chalcogenide glasses. The chalcogenide glass such as Se and As_2S_3 is intrinsically covalent, which can be alloyed with metals, oxygen, and halogens. The products become partially ionic or ion-conducting.

(13) metals on the photoinduced phenomena have been limited (Loeffler *et al.*, 1998; Tanaka *et al.*, 2003; Tikhomirov *et al.*, 2003; Ledemi *et al.*, 2009), it seems that the metal–chalcogenide glass exhibits little change. Note that binary systems such as Cu(Na, Ag)-S(Se) cannot be prepared as glassy materials.

Third, behaviors in oxide systems seem to be controversial. Taylor's group has repeatedly asserted that As_2O_3 glass, with an optical gap of ~5 eV, undergoes photodarkening (in Urbach-edge region) upon x-ray irradiation (Hari *et al.*, 2003). By contrast, Terakado and Tanaka (2009) demonstrate that, in an oxy-chalcogenide system xGeO$_2$-(100–x)GeS$_2$, the photodarkening and related structural changes induced by bandgap illumination disappear in the O-rich glasses of $x \geq 50$ mol.%. These different observations may be ascribed to the different excitation energies and/or materials (As- and Ge-systems). It has been well documented that, upon irradiation, SiO_2 glass undergoes volume compaction and also refractive index increase, but the glass presents no photodarkening. We also mention here that few studies have been reported for photoinduced phenomena in chalco-halide glasses (Sramek *et al.*, 2000).

A comment should be added on photobleaching, which refers to a blue-shift of optical absorption edges upon bandgap illumination.* Such phenomena have often been reported for Ge-chalcogenides, specifically in evaporated films (Tanaka *et al.*, 1984a; Kawaguchi and Maruno, 1988; Arsova *et al.*, 2005). We may then envisage that selected compositions can possess both photodarkening and photobleaching, resulting in no photoinduced optical changes. However, illumination in vacuum tends to suppress the photobleaching, and its thermal reversibility has rarely been demonstrated. Photobleaching in bulk glasses is very exceptional (Tanaka *et al.*, 2003), which implies that almost of all photobleaching is caused by irreversible structural changes such as homo-to-heteropolar bond conversion, oxidation, and/or hydration.

6.4 Excitation condition dependent scalar changes

Naturally, the scalar changes vary with excitation conditions. Provided that illumination is given at room temperature, the controllable exposure variables are duration (continuous wave/pulse), light intensity, spectrum (mono- and polychromatic), photon energy, light-spot size, etc. We are interested here in the athermal photoinduced phenomena, and accordingly, intense cw bandgap

*This photobleaching should be distinguished from photon-energy dependence of photodarkening (Fig. 6.9). In fact, blue-shifts occur when a sample, which has been illuminated by bandgap light, is exposed to sub-gap light.

light cannot be employed.* Absorbed light intensity should in practice be weaker than ~1 kW/cm^3, which corresponds to irradiance of ~100 mW/cm^2 for bandgap illumination. Otherwise, pulsed excitations may be preferred, provided that the peak and time-averaged intensities are limited.** Then, a problem occurs if the effects induced by cw and pulsed light are the same.

6.4.1 Continuous wave (CW) and pulse

It seems that cw and pulsed bandgap light provide similar photoinduced effects, at least under moderate average intensities. Tanaka (1988) reports for As_2S_3 ($E_g \approx 2.4$ eV) that cw and pulsed (~500 kW/cm^2, 3 ns, 15 Hz) exposures with the same photon energy of 2.54 eV and time-averaged intensity of ~10 mW/cm^2 give the same red-shift. Lucas *et al.* (2006) demonstrate for $GeSe_9$ ($E_g \approx 2.0$ eV) that sub-gap (~1.6 eV) exposures of cw and pulsed (100 kW/cm^2, 100 fs and 82 MHz) light with average intensities of ~1 W/cm^2 yield similar entropy relaxation, photo-expansion, and darkening. By contrast, Rosenblum *et al.* (1999) report for As_1Se_1 ($E_g \approx 2.0$ eV) that intense pulses (~5 MW/cm^2, 5 ns, ~10 Hz?) of super-gap light (~2.33 eV) provide an increase in photosensitivity by 10^3 times, which they ascribe to a two-photon effect. However, it is not known whether the result is universal to other glasses. Effects of mid-gap pulsed light will be described on pp. 154–158.

6.4.2 Spectral

Spectral studies of photoinduced effects have been performed, mainly for As_2S_3 with $E_g \approx 2.4$ eV under excitations of γ-ray, x-ray, super-gap, band-gap, sub-gap, mid-gap, and infrared photons, with energies covering 1 MeV–0.4 eV (Popescu, 2001). Specifically, sub- and mid-gap excitation effects, the results being summarized in Fig. 6.8, attract considerable interest for two reasons (Tanaka, 2006). The first stems from a fundamental question: why can the sub-gap excitation induce more prominent changes (Tanaka, 2003b), under some circumstances, than those induced by bandgap excitation? The second is related to applications (Zakery and Elliott, 2007; Petit *et al.*, 2008; Eggleton *et al.*, 2011; Ahmad and Rochette, 2011). Extensive studies have demonstrated promising nonlinear optical properties of chalcogenide glasses at telecommunication wavelengths of ~1.5 μm, or photon energies of

*For cw illumination, a maximal temperature rise ΔT can be estimated as ΔT [K] $\approx I$ [W]/(κ[W/cm·K] r[cm]), where I is the absorbed light intensity, κ the thermal conductivity of materials of interest (absorber or substrate for films), and $2r$ is the diameter of light spots.
**For a short pulse, we can estimate a maximal temperature rise, neglecting thermal diffusion, as ΔT [K] $\approx Q$ [J/cm^3]/c [J/cm^3·K], where Q is the absorbed pulsed energy and c the specific heat.

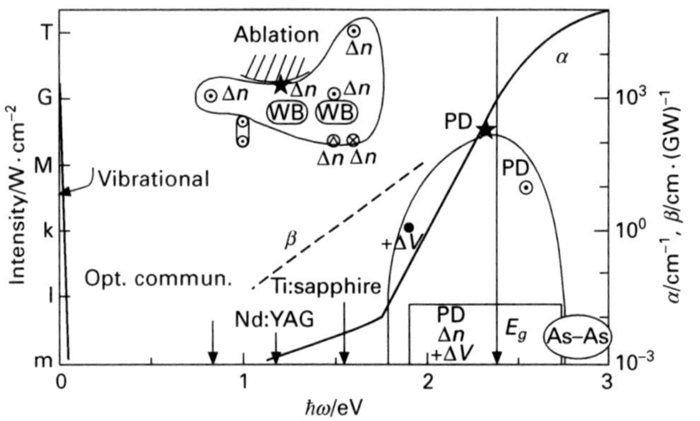

6.8 Photoinduced phenomena and optical absorption spectra (right vertical axis) in As_2S_3 at room temperature (modified from Tanaka, 2006). The photoinduced phenomena depend on photon energy (horizontal axis) and light intensity (left vertical axis). PD denotes photodarkening, Δn refractive-index increase, ΔV volume change, and WB generation of wrong bonds. For the absorption spectra, α (solid line) and β (dashed line) show linear and two-photon absorption coefficients. The linear absorption consists, from high to low photon energy, of square dependence $\alpha\hbar\omega \propto (\hbar\omega - E_g)^2$, Urbach edge $\alpha \propto (\hbar\omega/E_U)$, and weak-absorption tail $\alpha \propto (\hbar\omega/E_{WAT})$, where E_g ($\approx$ 2.4 eV) is the Tauc optical gap, E_U ($\approx$ 50 meV) the Urbach-edge energy, and E_{WAT} ($\approx$ 200 meV) a characteristic energy. Vibrational absorption appears at $\hbar\omega \approx$ 40 meV with a peak value of $\alpha \approx 10^4$ cm^{-1}.

~0.8 eV. It is then necessary to know what kinds of photoinduced phenomena are induced by pulsed mid-gap light with the photon energy of ~0.8 eV.

High-energy (γ-ray, x-ray, super-gap, bandgap) excitation

Substantial studies have been performed for γ-ray irradiation effects upon chalcogenide glasses, as reviewed by Shpotyuk (2004). Since the penetration depth of γ-rays is sufficiently long (≥ 1 cm), we can employ thick bulk samples for the investigations, which makes evaluations of changes in low optical absorptions and bulky properties relatively easy. It has been demonstrated that the γ-ray (~1 MeV) from ^{60}Co gives rise to increases in optical absorption and micro-hardness. However, for optical changes, unlike photodarkening, induced absorptions occur not only in the Urbach edges but also in the weak absorption tails, which suggests that several kinds of disordered structures, including defects such as wrong bonds, are produced. Nevertheless, it is reported that no electron-spin resonance signals emerge under irradiation at room temperature.

Studies on x-ray excitation effects are fewer. To my knowledge, no comprehensive studies have been performed for hard x-rays of ~10 keV

(Apling *et al*., 1975; Popescu, 2001). For soft x-rays, Hayashi *et al*. (1996) demonstrate that when annealed As_2S_3 films are exposed to monochromatic 44 eV photons, which can resonantly excite 3d core-electrons of As, a red-shift (photodarkening) of ~40 meV in the optical absorption edge ($\hbar\omega \approx 2.6$ eV) occurs, with an apparent quantum efficiency higher than that obtained with bandgap illumination by two orders of magnitude. No structural analyses have been performed for the irradiated films. It seems as if any kinds of structural disordering could cause the red shift, probably reflecting enhanced interaction between lone-pair electrons.

Illumination of super-gap light (~3.0 eV) to As_2S_3 tends to segregate As, which is likely to crystallize to As_2O_3 (Berkes *et al*., 1971; Tanaka, 1999; Popescu, 2001; Ogusu *et al*., 2005). Super-gap light most probably excites the bonding electrons, directly breaking =As–S– covalent networks, which may assist the segregation of As atoms through some mechanism presently unspecified. As a result, photo-enhanced vaporization (Janai *et al*., 1978) occurs, which can be employed for producing relief patterns (Tanaka and Shimakawa, 2011).

Many experimental studies using bandgap illumination have been reported for As_2S_3 (Popescu, 2001; Kolobov, 2003; Tanaka and Shimakawa, 2011; Kolobov and Tominaga, 2012). The illumination gives rise to photodarkening, refractive index increase, volume (density) changes, mechanical changes, and chemical durability changes. Dependences of a quantum efficiency and the red-shift ΔE of the photodarkening upon the excitation photon energy manifest that the photon with $\hbar\omega \geq E_g$ appears to be effective (Fig. 6.9(a)). On the other hand, Kimura *et al*. (1981) have found intensity dependence

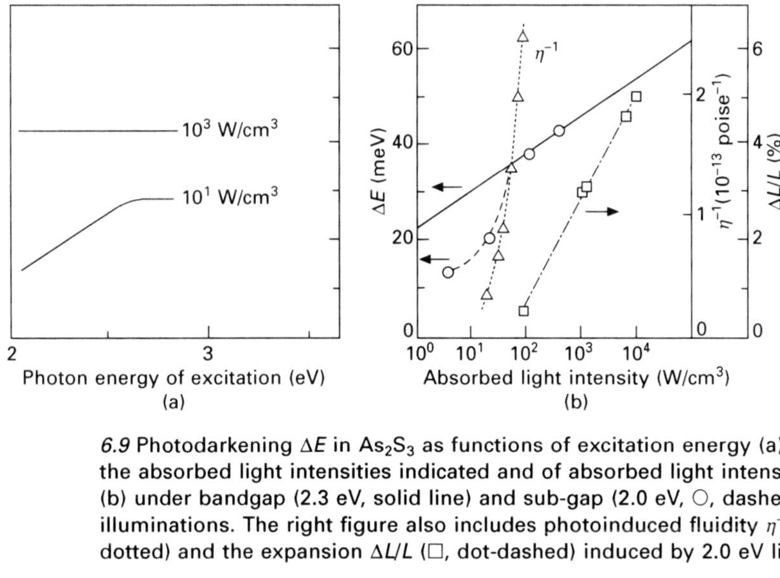

6.9 Photodarkening ΔE in As_2S_3 as functions of excitation energy (a) at the absorbed light intensities indicated and of absorbed light intensity (b) under bandgap (2.3 eV, solid line) and sub-gap (2.0 eV, ○, dashed) illuminations. The right figure also includes photoinduced fluidity η^{-1} (△, dotted) and the expansion $\Delta L/L$ (□, dot-dashed) induced by 2.0 eV light.

of $\Delta E \propto I$ (Fig. 6.9(b)), which can be interpreted taking relaxational effects into account.

However, induced photo-electro atomic processes remain speculative. Since the photon energy is comparable to the optical gap and the top of the valence band in chalcogenide glasses is composed by lone-pair electron states, we may straightforwardly envisage photo-excitation of lone-pair electrons, leaving the bonding electrons intact. Chalcogen atoms will be ionized, which may be viewed as generation of electron-hole pairs, and under such excited states, some atoms will move as polarons, and the trace may remain after recombination of the pairs.

Here, theoretical analyses may give some insight. *Ab initio* computer simulations for S_8 rings (Shimojo *et al.*, 1998), a-Se (Zhang and Drabold, 1999), and so forth (Simdyankin and Elliott, 2007) delineate electro-structural dynamics, in which breaking and interchange of the covalent bonds occurs after hybridization of the lone-pair electron and covalent-bonding states. In these simulations, however, photo-excitation processes cannot be rigorously simulated, and instead, these are replaced by addition of electronic charges. Such treatments necessarily disregard the photon energy, light polarization, etc., and accordingly, the pictures obtained should be taken as useful ideas.

Low-energy (sub-gap, mid-gap, vibrational) excitation

Despite the excitation-energy dependence of photodarkening mentioned above, prominent sub-gap excitation effects have been demonstrated for structural changes. For instance, using As_2S_3 ($E_g \approx 2.4$ eV) and 2.0 eV photons (Tanaka, 2003b), we can produce giant volume expansions (Hisakuni and Tanaka, 1994) and bulky fluidity (Hisakuni and Tanaka, 1995). Note that the excitation lies in the Urbach-edge region (Fig. 6.8), i.e. it is *Urbach-edge light*, the energy shortage (~0.4 eV from the Tauc optical gap) of which is much greater than the thermal energy (~25 meV) at room temperature.

We also note that the sub-gap illumination has peculiar intensity dependence. As shown in Fig. 6.9(a), under weaker excitations than ~10 W/cm^3, ΔE varies with the photon energy. However, when intense ($\geq 10^2$ W/cm^3), sub-gap illumination exerts the same photodarkening (red-shift) and volume expansion with those induced by bandgap illumination. That is, *intense 2.0 eV light behaves as if it were bandgap light*, both giving rise to increases in optical absorption at $\hbar\omega \approx 2.4$ eV. In addition, as shown in Fig. 6.9(b), the photoinduced fluidity (η^{-1}) becomes prominent under such excitations. How can we interpret such spectral intensity dependence?

Tanaka (2000, 2003a,b) has proposed a model, which is consistent with marked photocurrents (Tanaka, 1998) and structural modifications (Tikhomirov *et al.*, 2006) in As_2S_3 under intense Urbach-edge light. We assume that the Urbach edge is governed by localized states, reflecting spatial fluctuation of inter-molecular (layer) distances, above the valence band. Then, as illustrated in Fig. 6.10(a), weak Urbach-edge light (~2.0 eV) cannot excite free holes to the valence band, giving rise to no hole currents (electrons being immobile in chalcogenide glasses). However, when the sub-gap excitation is intense ($\geq 10^2$ W/cm^3), photo-excited holes fill all the localized states (with density of 10^{16}–10^{17} cm^{-3}) lying above 0.4 eV (= 2.4–2.0) from the valence-band edge, and hence, a subsequently-excited hole can no longer be trapped. The hole will be excited through multi-phonon (Fig. 6.10(b)) and/or two-step (Fig. 6.10(c)) transitions to the valence band, and it acts as if it were excited by bandgap light. Geminate recombination may be suppressed under such situations, which is required for photoconduction and probably for photodarkening (Fig. 6.12). A similar idea may apply to photoinduced fluidity.

However, the existence of the threshold intensity has recently been debated. Gueguen *et al.* (2010) demonstrate that, upon sub-gap illumination, $GeSe_9$ glass presents no intensity threshold in the photoinduced fluidity. I assume that the characteristic may depend upon the glass composition, since the band-edge and tail states are known to critically vary with glass compositions (Tanaka and Shimakawa, 2011). Further compositional studies on the threshold will be valuable.

Recently, much interest is aroused by mid-gap excitation (~$E_g/2$) effects. At such photon energies the absorption coefficient is less than 1 cm^{-1}, so that studies on athermal photoinduced phenomena using cw mid-gap light are limited (Aleksandrov *et al.*, 2001). For pulsed exposures, a problem is the distinction between photoinduced phenomena and permanent damage, which may be accompanied by thermal effects (Masselin *et al.*, 2011), micro-void formation, and ablation. On this subject, several studies have demonstrated,

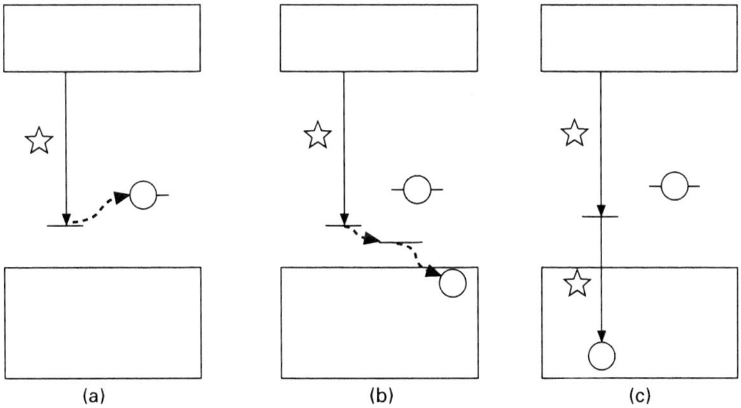

6.10 Hole excitation by sub-gap (Urbach-edge) light from the conduction band to (a) a trapping state under weak and intense excitations and to the valence band under (b) thermally-assisted one-photon and (c) two-step excitations.

as summarized in Fig. 6.8, that moderate peak-intensity (~1 GW/cm^2) pulsed (100 fs – 5 ns) mid-gap (0.8 – 1.5 eV) excitations of As_2S_3 give rise to refractive index increases *without accompanying the photodarkening* (Asobe *et al.*, 1995; Tanaka, 2004a). However, it is not known whether the refractive index increase can be removed by annealing at around the glass transition temperature, since the annealing makes irradiated samples frosty. On the other hand, as shown in Table 6.2, Raman-scattering spectroscopy of As_2S_3 demonstrates a density increase of wrong bonds (As-As, S-S) under such mid-gap excitations (Tanaka, 2004a). This result suggests that irradiated samples are heterogeneous, which may cause the index increase and the frosty appearance.

Photo-electro-atomic processes induced by mid-gap photons are highly speculative. For the mid-gap excitation, as illustrated in Fig. 6.11, we should consider the effects of (a) one-photon (linear) absorption to a state producing the weak absorption tail (see Fig. 6.8), (b) two-photon absorption, and (c) two-step absorption, which is a successive one-photon absorption

Table 6.2 Photoinduced changes in absorption edge ΔE (photodarkening), refractive index Δn, and Raman scattering spectrum in As_2S_3 by pulsed 1.17 and 2.33 eV light (the peak intensitites of 10^9 and 10^7 W/cm^2) at absorbed photon numbers of 10^{23}–10^{24} cm^{-3}, in comparison with those induced by cw bandgap light. Linear and two-photon absorption coefficients α and β are also listed

$\hbar\omega$ (eV)	ΔE (meV)	Δn	Raman scattering	α (cm^{-1})	β (cm/W)
1.17	0	0.005	Wrong bond	10^{-3}	10^{-10}
2.33	20	0.003	No change	300	–
CW	50	0.02			

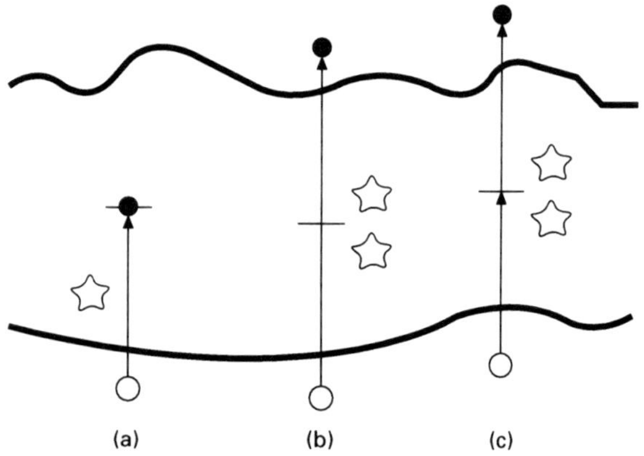

6.11 Three excitation processes in disordered semiconductors having mid-gap states by photons with energy of ~$E_g/2$; (a) one-photon, (b) resonant two-photon, and (c) two-step absorptions.

via mid-gap state. (Note that, in an ideal crystalline semiconductor, only the two-photon absorption can take place.) Here, the one- and two-photon absorption coefficients α and β can be written approximately as:

$$\alpha(\hbar\omega) \propto |\langle\varphi_f|H|\varphi_i\rangle|^2 \int D_f(E + \hbar\omega)\, D_i(E)\, dE \qquad [6.1]$$

and

$$\beta(\hbar\omega) \propto |\Sigma_n \langle\varphi_f|H|\varphi_n\rangle\langle\varphi_n|H|\varphi_i\rangle/(E_{ni} - \hbar\omega)|^2 \int D_f(E + 2\hbar\omega)\, D_i(E)\, dE \qquad [6.2]$$

where H is a light-electron interaction Hamiltonian, φ a related electron wavefunction, D a density-of-state, E an electron energy, for which the subscripts i, n, and f represent initial, intermediate, and final states, $E_{ni} = E_n - E_i$, and Σ_n denotes summation for all possible n states.

It is thus clear that, to understand the mid-gap photon process, the origin of the weak absorption tail (see Fig. 6.8) is a prerequisite. I assume that the tail in (pure) As_2S_3 arises from electronic tail states below the conduction band, which are produced by anti-bonding states of As-As and S-S homopolar bonds, i.e. wrong bonds (Tanaka, 2002a)*. This idea straightforwardly suggests that the wrong bond is responsible for the one-photon and two-step excitations by mid-gap photons. In addition, we see in Fig. 6.8 that the two-photon absorption (β) has a similar spectral slope to that of the weak absorption tail (Tanaka, 2002b). This feature implies that the two-photon absorption is resonated also with the wrong bond. Such a transition is very plausible, since in Eq. [6.2], a resonance occurs when $E_{ni} = \hbar\omega$, in which the n state may originate from a wrong bond.

We therefore envisage that all the (one, two-photon, two-step) mid-gap excitation processes occur at around wrong bonds, and as the result, the glass network becomes more heterogeneous, through unknown processes, which may be similar to that induced by super-gap illumination. The heterogeneous structure is likely to cause a higher refractive index, through photoelastic and/or interfacial effects, without giving rise to photodarkening. Since the shape of weak absorption tails in As-S glasses varies with the atomic ratio (Tanaka *et al.*, 2002), it is plausible that the mid-gap excitation effect also varies with the glass composition. In short, as illustrated in Fig. 6.12, we can assume that (sub-)bandgap photons modify the top of the valence band and mid-gap photons affect the wrong bonds.

Finally, we refer to a pioneering work for $As_2S(Se)_3$ by Hari *et al.* (2000). They report that, upon pulsed irradiation of infrared light with photon energies of ~0.4 eV ($< E_g/2$), vibrational absorptions of impurities such as C-H_x are

*A recent calculation suggests that As atoms with sp^2 electron configuration also produce tail states (unpublished).

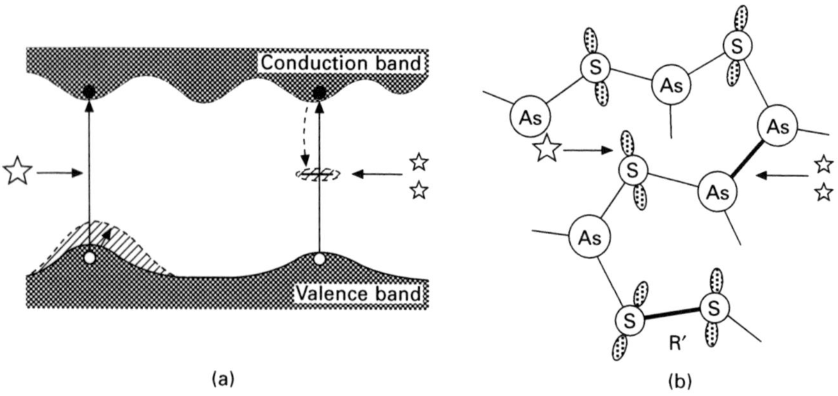

6.12 Effects of (sub-)bandgap (a) and mid-gap (b) excitations in a band diagram (a) and a real atomic structure (b).

modified. The modification is wavelength-selective and non-thermal, which evokes some structural changes induced through direct photon–phonon (localized vibrations) interaction. Further studies will be valuable for this untouched process.

6.5 Vector deformations

6.5.1 Gross features

Zhdanov and Malinovsky (1977) discovered vector changes in optical properties: dichroic red-shift of absorption edges and birefringence. For instance, $n(//) - n(+) \approx -0.002$ in As_2S_3, where $n(//)$ and $n(+)$ are the refractive indices monitored by probe light having the electric field parallel and perpendicular to that of linearly-polarized excitation light. These absorptive and refractive changes are induced by bandgap illumination, in a similar way to those of the scalar changes (photodarkening and refractive index increase), with the magnitudes being ~1/10 of the scalars. We may then envisage that ~10% of the atomic sites, which give the scalar change, participate in the vector change.

However, this idea has been disputed by some observations (Tanaka, 2001; Lyubin and Klebanov, 2003). For instance, variations of the optical changes with excitation-light intensity and illuminating temperature are clearly different between the scalar and the vector. As shown in Fig. 6.2, the scalar change becomes monotonically smaller at higher illumination temperatures, but the vector shows a maximum at a certain temperature. Such differences strongly suggest different origins of the two effects. On the other hand, to get insight into the vector change, some researchers performed microscopic structural investigations (Matsuishi *et al.*, 2002; Antoine *et al.*, 2009), while conclusive results have not been obtained. It is also noted here that, before around 2000, we had naively believed that the vector change was smaller than the scalar, as actually demonstrated in the optical properties.

Surprisingly, however, several kinds of marked vector photo-deformations have been discovered (Yannopoulos and Trunov, 2009; Tanaka and Shimakawa, 2011). Specifically, the deformations of *semi-freely fixed* samples are very prominent (Tanaka, 2008; Tanaka and Mikami, 2009), as exemplified in Fig. 6.13 for an As_2S_3 flake laid on viscous grease. Upon illumination of linearly polarized bandgap light, the flake first curls and deforms as a U-shape, the U direction always being parallel to the electric field. (The photoinduced birefringence occurs before this deformation.) After the U-deformation, the flake elongates (as if it were a snake) orthogonal to the electric filed, sometimes accompanying screwing deformations. Under prolonged illumination, the elongation grows to a size (~0.5 mm) of the light spot. To my knowledge, no other abiotic solids undergo such a dramatic photoinduced athermal shape change, although the deformation speed may be slower than those appearing in azo-polymers (Barrett *et al.*, 2007). Anyhow, these deformations in semi-freely fixed As_2S_3 flakes have manifested that the substrate exerts decisive mechanical constraints upon deposited chalcogenide films. Why do semi-free chalcogenide samples undergo such anomalous deformations as shown in Fig. 6.13?

6.5.2 Deformation mechanisms

As mentioned above, the dye-doped organic polymer undergoes deformation resembling vector photo-deformations (Barrett *et al.*, 2007), so that it seems fruitful to comparatively consider these deformation mechanisms. Figure 6.14 gives a model for dye polymers, in which the deformation is motivated by trans-cis photo-isomerization of azobenzene molecules, which triggers cooperative motions of polymer networks, resulting in host orientation (Barrett *et al.*, 2007; Koshiba *et al.*, 2009). Oriented atomic structures are and seem

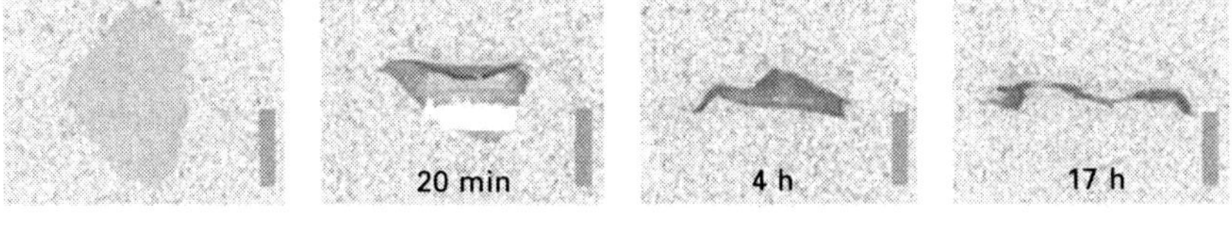

6.13 Vector deformation of an As_2S_3 flake (~0.1 mm in lateral dimension and 4 μm in thickness) laid on frosted glass as a function of exposure time of linearly (vertically) polarized unfocused laser light with photon energy and intensity of 2.3 eV and 40 mW (Tanaka, 2008). Video footage of this process can be viewed at http://www.youtube.com/watch?v=sHeYBHBCEWQ

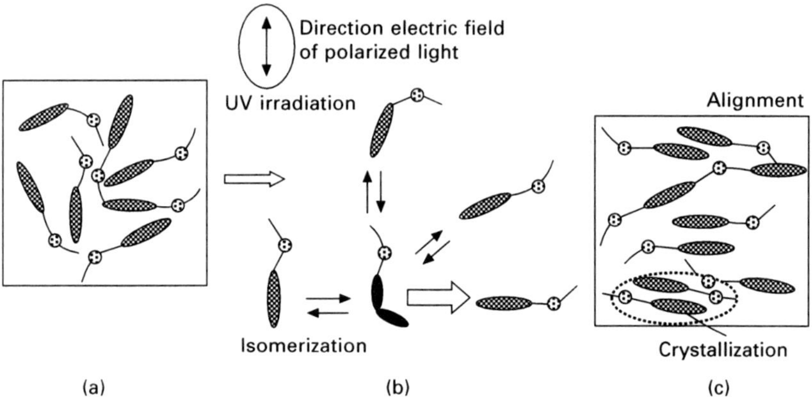

6.14 Schematic illustrations of azobenzene-attached polymers under illumination of linearly (vertically) polarized ultraviolet light (modified from Koshiba *et al.*, 2009): (a) a disordered isotropic structure before illumination, (b) selective trans-cis isomerizations under illumination, and (c) a partially crystallized aligned structure after illumination. Single and doubly-connected ellipsoids represent, respectively, *trans* and *cis* forms of the dye molecules.

to be produced, respectively, also in vectrally (by linearly polarized light) photo-crystallized Se films (Ishida and Tanaka, 1997) and M-engraved As_2S_3 films (Tanaka and Asao, 2006). Hence, it is straightforward to conceive similar atomic motions for the vector deformations in Fig. 6.13. However, despite the prominent deformations, we could detect no traces of anisotropic crystallization or oriented amorphous structures using x-ray diffraction and micro Raman scattering spectroscopy.

This fact has advocated a macroscopic deformation model (Tanaka *et al.*, 2009). We assume that the deformation occurs viscously, being produced by some macroscopic (not atomic) force and the photoinduced fluidity ($\propto$ 1/viscosity) (Tanaka and Shimakawa, 2011), in which 'deformation speed = force/viscosity'. If a non-crystalline solid is sufficiently viscous (fluidal), atomic changes (containing strain, orientation, etc.) will be relaxed, as illustrated in Fig. 6.15. And after illumination, no *atomically-detectable* traces possibly remain, which is consistent with the x-ray and Raman scattering observations. Then, what is the motive force?

We can nominate several candidates for the force. In the present context, photo-thermal forces are dismissed because of the clear anisotropic deformation which is governed by light polarization. Photon drag effects may be difficult to interpret overall features. In addition, it is difficult to envisage electric forces such as Coulombic repulsions being responsible, since the deformation has appeared also in samples laid on metallic plates.

I assume an optical force to be the most plausible (Tanaka *et al.*, 2009). As known, there are several kinds of optical forces, in which the gradient

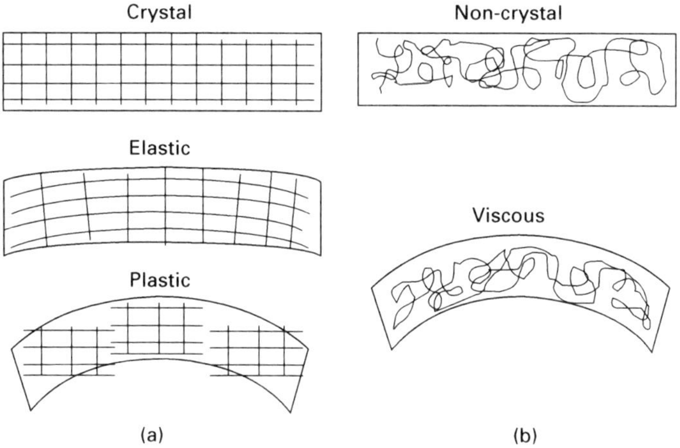

6.15 Deformations in a single crystal (a) and a viscous non-crystal (b).

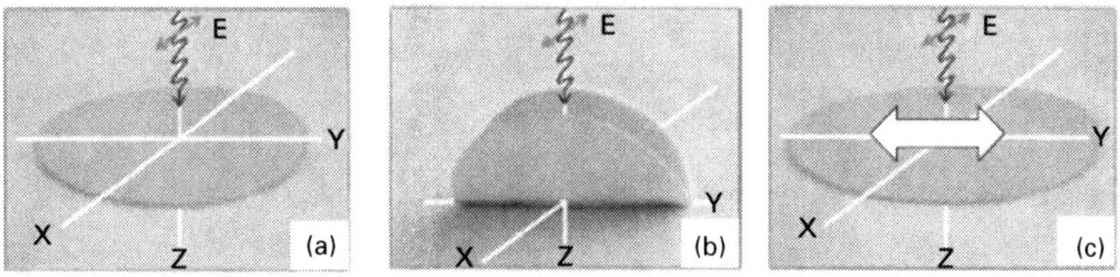

6.16 Deformation of an As_2S_3 disc under linearly polarized bandgap light: (a) shows the initial state, (b) U-shape deformation under optical torque and photoinduced fluidity, and (c) elongation orthogonal to the electric field by light pressure of scattered light.

force, being responsible for laser tweezers, can be discarded, since the laser beam is unfocused in Fig. 6.13 and the flake is smaller than the light spot. We assume that the U-shape deformation is caused by *optical torque* (photon momentum and spin), the idea being inferred from *rotation* of c-As_2S_3 (orpiment, a layer-type crystal having birefringence) fragments under illumination of linearly polarized light (Tanaka, 2008). Orpiment fragments rotate so that the layer plane becomes perpendicular to the electric field of light, which can be ascribed to the optical torque. It is then plausible that the optical torque deforms glassy As_2S_3 flakes to U (or L) shapes through photoinduced fluidity, as illustrated in Fig. 6.16(b). A common feature to these behaviors, i.e. the rotation of crystalline fragments and the U deformation of glassy flakes, is that (effective) sample thicknesses become minimal in the direction parallel to the electric field. On the other hand, as illustrated in Fig. 6.16(c), *optical pressure* of Mie-scattered light may cause elongation of As_2S_3 flakes orthogonal to the electric field (4 and 17 h in Fig. 6.13). It is mentioned here that the pressure is assumed to be responsible also for wrinkling deformations of As-S(Se) films on grease (Tanaka and Mikami, 2009, http://www.youtube.com/watch?v=9EAOBSyK3Ts). Note that deformation by optical forces has been detected in fluids, and related studies are growing to a scientific field called 'opto-fluidics' (Jonáš and Zemánek, 2008; Emile and Emile, 2011).

However, the optical force model is confronted with two big problems. One is a quantitative problem (Tanaka *et al.*, 2009). Naturally, the optical force is very small. But, an observed deformation speed (= force/viscosity) is faster by ~10 orders(!) than that estimated using a theoretical optical force and a viscosity of $\sim 10^{12}$ P, the value appearing in the photoinduced fluidity (Hisakuni and Tanaka, 1995; Tanaka and Shimakawa, 2011). The optical force may be amplified and/or the sample becomes much more fluidal, through some mechanisms presently unknown. The other problem is the behavior of a-Se. It has recently been demonstrated that a-Se also undergoes photoinduced elongation, while its direction is *parallel* to the electric field, in contrast to the orthogonal direction in As-S(Se) (Tanaka and Mikami, 2011). A force having a different origin seems to govern the deformation in a-Se, which is totally speculative at present. In addition, a recent study demonstrates more complicated deformation features of Se and As_2S_3 films and flakes under varied exposure conditions (Tanaka and Mikami, 2012), the understanding of which will be valuable for advancing the light-matter science.

6.6 Conclusion

After taking a bird's-eye view of the photoinduced phenomenon in covalent chalcogenide glasses, we have considered two topics. The first is a scalar change including photodarkening, which has been studied for a long time after a comprehensive study by De Neufville *et al.* (1974). The phenomenon appears to be simple, unique, and inherent to the chalcogenide glass, while its microscopic mechanism remains speculative in labyrinths. The reason is ascribable to difficulties in identification of subtle atomic changes in disordered networks.

The second topic is vector deformation. Specifically, deformation appearing in semi-freely fixed As(Ge)-S(Se) flakes under illumination of linearly polarized bandgap light is surprisingly dramatic. Despite that, no photoinduced changes in atomic scales have been detected, which prompts a new idea taking optical forces into account. We can interpret almost all the observations by the idea that the deformation arises from viscous flows (photoinduced fluidity) of the glass under optical forces. Nevertheless, the idea must surmount quantitative and compositional problems. Finally, we note that the optical-force deformation may occur in all viscous materials such as oxide glasses at softening temperatures, in a similar way to those observed in the opto-fluidics, which will be a challenging subject for future studies.

6.7 References

Abbaszadeh S, Rom K, Bubon O, Weinstein B A, Karim K S, Rowlands J A and Reznik A (2012), 'The effect of the substrate on transient photodarkening in stabilized amorphous selenium', *J. Non-Cryst. Solids* 358, 2389–2392.

Abdulhalim I, Gelbaor M, Klebanov M and Lyubin V (2011), 'Photoinduced phenomena in nano-dimensional glassy As_2S_3 films', *Opt. Mater. Express* 1, 1192–1201.

Ahmad R and Rochette M (2011), 'Photosensitivity at 1550 nm and Bragg grating inscription in As_2Se_3 chalcogenide microwires', *Appl. Phys. Lett.* 99, 061109.

Aleksandrov A P, Babin A A, Kiselev A M, Kulagin D I, Lozhkarev V V and Stepanov A N (2001), 'Formation of microstructures in As_2S_3 by a femtosecond laser pulse train', *Quantum Electron.* 31, 398–400.

Antoine K, Jain H, Vlcek M, Senanayake S D and Drabold D A (2009), 'Chemical origin of polarization-dependent photoinduced changes in an $As_{36}Se_{64}$ glass film via *in situ* synchrotron x-ray photoelectron spectroscopy', *Phys. Rev. B* 79, 054204.

Apling A J, Daniel M F and Leadbetter A J (1975), 'The oxidation of amorphous films of arsenic chalcogenides', *Thin Solid Films* 27, L11–L13.

Arsova D and Vateva E (2012), 'Dual action of light in photodarkened Ge-As-S films', *Phys. Stat. Sol.* 249, 153–157.

Arsova D, Skordeva E, Pamukchieva V and Vateva E (2005), 'Photoinduced changes in Ge-As-S thin films with various network rigidities', *J. Optoelectron. Adv. Mater.* 7, 1259–1264.

Asao H and Tanaka K (2007), 'Polarization-dependent photoinduced mechanical deformations in covalent chalcogenide glasses', *J. Appl. Phys.* 102, 043508.

Asobe M, Kanamori T, Naganuma K, Itoh H and Kaino T (1995), 'Third-order nonlinear spectroscopy in As_2S_3 chalcogenide glass-fibers', *J. Appl. Phys.* 77, 5518–5523.

Barrett C J, Mamiya J, Yanger K G and Ikeda T (2007), 'Photo-mechanical effects in azobenzene-containing soft materials', *Soft Matter* 3, 1249-1261.

Berkes J S, Ing S W and Hillegas W J (1971), 'Photodecomposition of amorphous As_2Se_3 and As_2S_3', *J. Appl. Phys.* 42, 4908–4916.

Calvez L, Yang Z and Lucas P (2009), 'Reversible giant photocontraction in chalcogenide glass', *Opt. Express* 17, 18581–18589.

De Neufville J P, Moss S C and Ovshinsky S R (1974), 'Photostructural transformations in amorphous As_2Se_3 and As_2S_3 films', *J. Non-Cryst. Solids* 13, 191–223.

Eggleton B J, Luther-Davies B and Richardson K (2011), 'Chalcogenide photonics', *Nature Photonics* 5, 141–148.

Ehrenfreund P (1999), 'Molecules on a space odyssey', *Science* 283, 1123–1124.

Emile O and Emile J (2011), 'Low-power laser deformation of an air-liquid interface', *Phys. Rev. Lett.* 106, 183904.

Fritzsche H (2000), 'Light induced effects in glasses', in Boolchand P, *Insulating and Semiconducting Glasses*, Singapore, World Scientific, 653–690.

Gueguen Y, Sangleboeuf J G, Keryvin V, Lépine E, Yang Z, Pouxel T, Point C, Bureau B, Zhang X-H and Lucas P (2010), 'Photoinduced fluidity in chalcogenide glasses at low and high intensities: a model accounting for photon efficiency', *Phys. Rev. B* 82, 134114.

Hamanaka H, Tanaka K, Matsuda A and Iizima S (1976), 'Reversible photo-induced volume changes in evaporated As_2S_3 and $As_4Se_5Ge_1$ films', *Solid State Commun.* 19, 499–501.

Harada H and Tanaka K (1999), 'Photoluminescence from Pr^{3+}-doped chalcogenide glasses excited by bandgap light', *J. Non-Cryst. Solids*. 246, 189–196.

Hari P, Cheney C, Luepke G, Singh S, Tolk N, Sanghera J S and Aggarwal I D (2000), 'Wavelength selective materials modification of bulk As_2S_3 and As_2Se_3 by free electron laser irradiation', *J. Non-Cryst. Solids* 270, 265–268.

Hari P, Guzel S, Su T, Taylor P C, Kuhns P L, Moulton W G and Sullivan N S (2003), 'Photodarkening effect in glassy As_2S_3 and As_2O_3', *J. Non-Cryst. Solids* 326–327, 199–204.

Hayashi K and Mitsuishi N (2002), 'Thickness effect of the photodarkening in amorphous chalcogenide films', *J. Non-Cryst. Solids* 299, 949–952.

Hayashi K, Kato D and Shimakawa K (1996), 'Photoinduced effects in amorphous chalcogenide films by vacuum ultra-violet light', *J. Non-Cryst. Solids*. 198–200, 696–699.

Hisakuni H and Tanaka K (1994), 'Giant photoexpansion in As_2S_3 glass', *Appl. Phys. Lett.* 65, 2925–2927.

Hisakuni H and Tanaka K (1995), 'Optical microfabrication of chalcogenide glasses', *Science* 270, 974–975.

Iketomo H, Tsuzuki T, Inui M, Yao M and Endo H (2002), 'Photodarkening of amorphous selenium under high pressure', *Z. Phys. Chem.* 216, 1107–1121.

Indutnyi I Z, Shepeliavyi P E and Indutnyi V I (1999), 'Relaxation of photodarkening in SiO-As2$(S,Se)_3$ composite layers', *Semicond. Phys., Quant. Electron. Optoelectron.* 2, 59–62.

Ishida K and Tanaka K (1997), 'Photoinduced anisotropic crystallization of amorphous Se', *Phys. Rev. B* 56, 206–209.

Itoh N and Stoneham M (eds) (2000), *Material Modification by Electronic Excitation*, Cambridge, Cambridge University Press.

Janai M, Riess I and Rudman P S (1978), 'Substrate dependence of photooxidation of amorphous As_2S_3 thin films', *Surf. Sci.* 74, 13–20.

Jonáš A and Zemánek P (2008), 'Light at work: The use of optical forces for particle manipulation, sorting, and analysis', *Electrophoresis* 29, 4813–4851.

Kawaguchi T (2003), Photo-induced deposition of silver particles on amorphous semiconductors', in Kolobov A V, *Photoinduced Metastability in Amorphous Semiconductors*, Weinheim, Wiley-VCH, 182–198.

Kawaguchi T and Maruno S (1988), 'Photoinduced and thermally-induced bleaching of amorphous Ge-S film', *Jpn J. Appl. Phys.* 27, 2199–2204.

Kimura K, Nakata H, Murayama K and Ninomiya T (1981), 'Interrelation between glass transition and reversible photostructural change in a-As_2S_3', *Solid State Commun.* 40, 551–554.

Kolobov A V (2003) (ed.), *Photoinduced Metastability in Amorphous Semiconductors*, Weinheim, Wiley-VCH.

Kolobov A V and Elliott S R (1991), 'Photodoping of amorphous chalcogenides by metals', *Adv. Phys.* 40, 625–684.

Kolobov A V and Tominaga J (2012), *Chalcogenides: Metastability and Phase Change Phenomena*, New York, Springer.

Kolobov A V, Oyanagi H, Tanaka K and Tanaka K (1997), 'Structural study of amorphous selenium by *in situ* EXAFS: observation of photoinduced bond alternation', *Phys. Rev. B* 55, 726–734.

Koshiba Y, Yamamoto M, Kinashi K, Misaki M, Ishida K, Oguchi Y and Ueda Y (2009), 'Photo-induced alignment behavior of azobenzene compound in thin film', *Thin Solid Films* 518, 805–809.

Ledemi Y, Messaddeq S H, Skhripachev I, Ribeiro S J L and Messaddeq Y (2009), 'Influence of Ga incorporation on photoinduced phenomena in Ge-S based glasses', *J. Non-Cryst. Solids*. 355, 1884–1889.

Liu J Z and Taylor P C (1987), 'Absence of photodarkening in bulk, glassy As_2S_3 and As_2Se_3 alloyed with copper', *Phys. Rev. Lett.* 59, 1938–1941.

Loeffler P, Schwarz T, Sautter H and Lezal D (1998), 'Structural changes of GeGaS bulk glasses induced by UV exposure', *J. Non-Cryst. Solids*. 232–234, 526–531.

Lucas P, King E A and Doraiswamy A (2006), 'Comparison of photostructural changes induced by continuous and pulsed laser in chalcogenide glass', *J. Optoelecton. Adv. Mater*. 8, 776–779.

Lucovsky G and Galeener F L (eds) (1976), *Structure and Excitations of Amorphous Solids*, New York, AIP.

Lyubin V M and Klebanov M L (2003), 'Photo-induced anisotropy in chalcogenide glassy semiconductors', in Kolobov A V, *Photoinduced Metastability in Amorphous Semiconductors*, Weinheim, Wiley-VCH, 91–108.

Masselin P, Le Coq D and Bychkov E (2011), 'Refractive index variations induced by femtosecond laser direct writing in the bulk of As_2S_3 glass at high repetition rate', *Opt. Mater*. 33, 872–876.

Matsuishi K, Kagota K and Onari S (2002), 'Anisotropic structural change induced by sub-bandgap light in As-S glasses studied by quasi-elastic light scattering', *J. Non-Cryst. Solids* 299–302, 929–934.

Ogusu K, Hosokawa Y, Maeda S, Minakata M and Li H (2005), 'Photo-oxidation of As_2Se_3, Ag-As_2Se_3, and Cu-As_2Se_3 chalcogenide films', *J. Non-Cryst. Solids* 351, 3132–3138.

Petit L, Carlie N, Anderson T, Choi J, Richardson M and Richardson K C (2008), 'Progress on the photoresponse of chalcogenide glasses and films to near-infrared femtosecond laser irradiation: a review', *IEEE J. Selected Topic Quant. Electron.* 14, 1323–1334.

Pfeiffer G, Paesler M A and Agarwal S C (1991), 'Reversible photodarkening of amorphous arsenic chalcogens', *J. Non-Cryst. Solids* 130, 111–143.

Popescu M (2001), *Non-Crystalline Chalcogenides*, Dordrecht, Kluwer Academic.

Rosenblum G, Sfez B G, Kotler Z, Lyubin V and Klebanov M (1999), 'Nonlinear optical effects in chalcogenide photoresists', *Appl. Phys. Lett.* 75, 3249–3251.

Sainov S, Sainov V, Dikova J and Beev K (2007), 'Optical and holographic properties of nano-sized amorphous semiconductor', *J. Optoelectron. Adv. Mater.* 9, 3125–3127.

Sakaguchi Y and Tamura K (2003), 'Photo-induced changes in liquid sulfur and selenium', in Kolobov A V, *Photo-Induced Metastability in Amorphous Semiconductors*, Weinheim, Wiley-VCH, 199–219.

Shimakawa K, Kolobov A and Elliott S R (1995), 'Photoinduced effects and metastability in amorphous semiconductors and insulators', *Adv. Phys.* 44, 475–588.

Shimojo F, Hoshino K and Zempo Y (1998), 'Photo-induced bond breaking in the S_8 ring: an *ab initio* molecular-dynamics simulation', *J. Phys.: Condens. Matter.* 10, L177–L182.

Shpotyuk O (2004), 'Radiation-induced effects in chalcogenide vitreous semiconductors', in Fairman R and Ushkov B, *Semiconducting Chalcogenide Glass I*, Amsterdam, Elsevier, 215–260.

Simdyankin S I and Elliott S R (2007), 'Light induced effects in amorphous and glassy solids', in Drabold D A and Estreicher S K, *Theory of Defect in Semiconductors* (*Topics Appl. Phys.* 104), Berlin, Springer-Verlag, 269–287.

Sramek R, Smektala F, Xie W X, Douay M and Niay P (2000), 'Photoinduced surface expansion of fluorozirconate glasses', *J. Non-Cryst. Solids*. 277, 39–44.

Štábl M and Tichý L (2004), 'On the photo-darkening in some Ge-As-S glasses', *J. Optoelectron. Adv. Mater.* 6, 781–786.

Tanaka K (1975), 'Reversible photoinduced change in intermolecular distance in amorphous As_2S_3 network', *Appl. Phys. Lett.* 26, 243–245.

Tanaka K (1983), 'Mechanisms of photodarkening in amorphous chalcogenides', *J Non-Cryst. Solids*. 59–60, 925–928.

Tanaka K (1986), 'Configurational and structural models for photodarkening in glassy chalcogenides', *Jpn J. Appl. Phys.* 25, 779–786.

Tanaka K (1988), 'Light intensity dependence of photodarkening in amorphous As_2S_3 films', *Thin Solid Films* 157, 35–41.

Tanaka K (1989), 'Structural phase transitions in chalcogenide glasses', *Phys. Rev. B* 39, 1270–1279.

Tanaka K (1990), 'Photoinduced structural changes in chalcogenide glasses', *Rev. Solid State Sci.* 4, 641–659.

Tanaka K (1992), 'Defect densities in glasses', *J. Non-Cryst. Solids* 139, 179–181.

Tanaka K (1998), 'Free-carrier generation in amorphous semiconductors by intense subgap excitation', *Appl. Phys. Lett.* 73, 3435–3437.

Tanaka K (1999), 'Spectral dependence of photoexpansion in As_2S_3 glass', *Philos. Mag. Lett.* 79, 25–30.

Tanaka K (2000), 'Sub-gap excitation effects in As_2S_3 glass', *J. Non-Cryst. Solids* 266–269, 889–893.

Tanaka K (2001), 'Photoinduced anisotropy in chalcogenide glass', in Nalwa H S, *Handbook of Advanced Electronic and Photonic Materials and Devices Vol. 5*, San Diego, Academic Press, 120–140.

Tanaka K (2002a), 'Wrong bond in glasses: a comparative study on oxides and chalcogenides', *J. Optoelectron. Adv. Mater.* 4, 505–512.

Tanaka K (2002b), 'Two-photon absorption spectroscopy of As_2S_3 glass', *Appl. Phys. Lett.* 80, 177–179.

Tanaka K (2003a), 'Chalcogenide glass: its optical properties and related applications to photonics', *Kogaku* 32, 294–299 (in Japanese).

Tanaka K (2003b), 'Sub-gap photo-induced phenomena in chalcogenide glasses', in Kolobov A V, *Photoinduced Metastability in Amorphous Semiconductors*, Weinheim, Wiley-VCH, 69–90.

Tanaka K (2004a), 'Midgap photon effects in As_2S_3 glass', *Philos. Mag. Lett.* 84, 601–606.

Tanaka K (2004b), 'Nanostructured chalcogenide glasses', in Nalwa H S, *Encyclopedia of Nanoscience and Nanotechnology Vol. 7*, Valencia, Am Sci Pub, 629–640.

Tanaka K (2006), 'Photo-induced phenomena in chalcogenide glasses: comparison with those in oxide and polymer', *J. Non-Cryst. Solids* 352, 2580–2584.

Tanaka K (2008), 'Visible anisotropic deformation of chalcogenide glass by illumination of linearly polarized light', *Appl. Phys. Express*. 1, 012006.

Tanaka K (2012), 'Relation between photodarkening and photoexpansion in As_2S_3 glass', *Phys. Stat. Sol. B* 249, 2019–2023.

Tanaka K and Asao H (2006), 'Photoinduced anisotropic deformation in As_2S_3 glass', *Jpn. J. Appl. Phys.* 45, 1668–1674.

Tanaka K and Mikami M (2009), 'Laser-induced wrinkling in chalcogenide glass films on viscous layers', *Appl. Phys. Express* 2, 081301.

Tanaka K and Mikami M (2011), 'Photoinduced vector deformations of semi-freely fixed glassy As-S(Se)', *Phys. Stat. Sol. C* 8, 2756–2760.

Tanaka K and Mikami M (2012), 'Photoinduced macroscopic vector deformations in chalcogenide glasses', *J. Non-Cryst Solids* 357, 2385–2388.

Tanaka K and Odajima A (1982), 'Photodarkening in amorphous selenium', *Solid State Commun.* 12, 961–964.

Tanaka K and Shimakawa K (2011), *Amorphous Chalcogenide Semiconductors and Related Materials*, New York, Springer.

Tanaka K, Kasanuki Y and Odajima A (1984a), 'Physical properties and photoinduced changes of amorphous Ge-S films', *Thin Solid Films* 117, 251–260.

Tanaka K, Kyohya S and Odajima A (1984b), 'Anomaly of the thickness dependence of photodarkening in amorphous chalcogenide films', *Thin Solid Films* 111, 195–199.

Tanaka K, Nakagawa T and Odajima A (1986), 'Photodarkening in glassy chalcogenide systems', *Philos. Mag.* B54, L3–L7.

Tanaka K, Gotoh T, Yoshida N and Nonomura S (2002), 'Photothermal deflection spectroscopy of chalcogenide glasses', *J. Appl. Phys.* 91, 125–128.

Tanaka K, Nemoto N and Nasu H (2003), 'Photoinduced phenomena in Na_2S-GeS_2 glasses', *Jpn. J. Appl. Phys.* 42, 6748–6752.

Tanaka K, Terakado N and Saitoh A (2009), 'Visible anisotropic deformation of chalcogenide glass by optical force', *Phys. Stat. Sol. A* 206, 892–897.

Terakado N and Tanaka K (2006), 'Photoinduced phenomena in GeO_2 glass', *J. Non-Cryst. Solids* 352, 3815–3822.

Terakado N and Tanaka K (2009), 'Ag-photodoping in nano-structured GeO_2-GeS_2 glasses', *Phys. Stat. Sol. C* 6, 575–578.

Tichý L, Tichá H, Nagels P and Mertens R (2002), 'Photo-induced shifts in the optical gap of a-Se thin films', *J. Optoelectron. Adv. Mater.* 4, 785–791.

Tikhomirov V K, Seddon A B, Asatryan K, Galstian T V and Vallee R (2003), 'The role of van der Waals bonding in photosensitivity of chalcogenide glasses', *J. Non-Cryst. Solids* 326–327, 205–208.

Tikhomirov V K, Barj M, Turrell S, Kobelke J, Idrissi N, Bouazaoui M, Capoen B and Seddan A B (2006), 'Non-linear Raman effects and photodarkening in chalcogenide glass As_2S_3', *Europhys. Lett.* 76, 312–317.

Toyozawa Y (2003), *Optical Processes in Solids*, Cambridge, Cambridge University Press.

Wagner T and Frumar M (2003), 'Optically-induced diffusion and dissolution of metals in amorphous chalcogenides', in Kolobov A V, *Photoinduced Metastability in Amorphous Semiconductors*, Weinheim, Wiley-VCH, 160–181.

Yannopoulos S N and Trunov M L (2009), 'Photoplastic effects in chalcogenide glasses: a review', *Phys. Stat. Sol B* 246, 1773–1785.

Yannopoulos S N, Andrikopoulos K S, Kastrissios D Th and Papatheodorou G N (2012), 'Origin of photoinduced defects in glassy As_2S_3 under band gap illumination studied by Raman scattering: a revisory approach', *Phys. Stat. Sol B* 249, 2005–2012.

Yonezawa F (ed.) (1981), *Fundamental Physics of Amorphous Semiconductors*, Berlin, Springer-Verlag.

Yoshida N and Tanaka K (1995), 'Photoinduced Ag migration in Ag-As-S glasses', *J. Appl. Phys.* 78, 1745–1750.
Zakery A and Elliott S R (2007), *Optical Nonlinearities in Chalcogenide Glasses and Their Applications*, New York, Springer.
Zhang X and Drabold D A (1999), 'Direct molecular dynamics simulation of light-induced structural change in amorphous selenium', *Phys. Rev. Lett.* 83, 5042–5045.
Zhdanov V G and Malinovsky V K (1977), 'Photoinduced birefringence and dichroism in As_2S_3 films', *Sov. Tech. Phys. Lett.* 3, 387–388.

7
Ionic conductivity of chalcogenide glasses

A. PRADEL and M. RIBES, University of Montpellier, France

DOI: 10.1533/9780857093561.1.169

Abstract: Owing to the high polarisability of the chalcogens, sulphur or selenium, chalcogenide glasses show higher ionic conductivity than their oxide counterparts. In this chapter, methods for preparing ion conducting chalcogenide glasses and techniques used to characterise ion diffusion in these glasses will first be described. Academic research aiming at getting an insight on both composition/structure/property relationships and ion dynamics in these glasses will then be developed. Finally, the last section will be devoted to a brief summary of the main applications of ion conducting chalcogenide glasses which includes development of solid state batteries, development of ionic memories for data storage and development of sensors for environmental control.

Key words: modifier, dc conductivity, conductivity spectroscopy, mixed alkali effect, mixed former effect, ion dynamics, model, battery, sensor, memory device.

7.1 Introduction

In the last decades, a great deal of interest has been devoted to investigation of ionic conductivity of chalcogenide glasses for both academic and application purposes. Due to the high polarisability of the chalcogen, sulphur or selenium, the chalcogenide glasses show higher conductivity than their oxide counterparts. Alkali (mainly lithium) conducting chalcogenide glasses were then studied as potential candidates for the development of solid state batteries while silver-doped chalcogenide films were considered for the development of ionic memories for data storage. Ag-doped chalcogenide membranes are proposed for the development of sensors for environmental control. Academic research involved investigation into new glasses, their characterisation with emphasis on the structure/property relationship in order to gain insight into the ion dynamics of these glasses.

First, we will describe briefly the preparation of ion conducting chalcogenide glasses as bulk materials or in the thin film form. The techniques to measure the main parameters (ionic conductivity, ion diffusion coefficient, ion transport number, redox stability domain) will then be described. The two main sections will concern the experimental data obtained for ion conducting chalcogenide glasses over the years along with a description of the main models developed to account for them. Finally, the last section will be devoted to a brief summary of the main applications of ion conducting chalcogenide glasses.

7.2 Preparation of ionic conductive chalcogenide glasses

Just like their oxide counterparts, chalcogenide glasses can be prepared as bulk materials or in the form of (thin or thick) films. On the other hand, and unlike the oxides, their preparation requires an oxygen-free atmosphere. Moreover, alkali-based chalcogenide glasses must be handled in a moisture-free glove box since they are highly hygroscopic.

7.2.1 Bulk glasses

Most of the ionic conductive chalcogenide glasses can be obtained by direct synthesis of the constituting elements or more commonly by synthesis of the constituting modifiers, formers and eventually dopants mixed in stoichoimetric quantities and placed in quartz tubes sealed under vacuum. The mixture is then melted at high temperature (1,000–1,350 K). Depending on the ease of vitrification, the tube is quenched in air or in various liquids that increase the heat dissipation (water, salted water, liquid nitrogen). Many glasses in the systems M_2X-YX_n-MZ, where M = Li, Na, Ag; X = S, Se; Y = Ge, Si, As, B, P; Z = Cl, Br, I; n = 2, 3/2, 5/2, were prepared in this way.[1–26]

When preparing alkali conducting chalcogenide glasses, vitreous carbon crucibles are often inserted in the tube in order to avoid contact of the reacting mixture with the quartz tube and thus contamination of the glass by silica.[11,16,27] The vitreous carbon crucible can be replaced by a layer of carbon obtained by pyrolysis of an organic compound such as acetone.[21,28]

Sulphur and, to a lesser extent, selenium have high vapour pressures. A slow heating ramp of ~30°/h at the beginning of the reaction process is therefore required in order to avoid any risk of explosion of the quartz tube. The heating treatment can include steps with constant temperature for several hours to optimise the reaction.

7.2.2 Fast quenching method

The higher the modifier content in a glass is, the higher is the conductivity. Unfortunately, the glasses are then more difficult to obtain. The quenching rate must be increased. Several techniques can be considered: from the simple crush of a droplet between two plates to more sophisticated techniques such as the twin roller quenching. This technique, shown in Fig. 7.1, allows quenching rates of about 10^6K/s to be obtained.[29] Glasses of the system

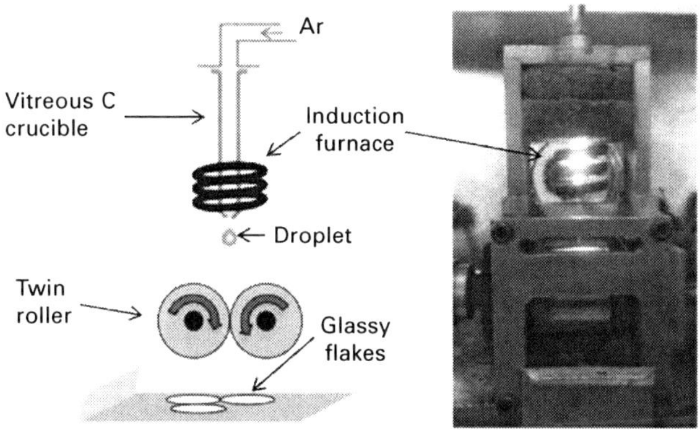

7.1 Twin roller quenching system: droplets of molten material are laminated between the rotating twin rollers, giving 50–80 μm thick flakes.

xLi_2S-$(1-x)SiS_2$ were prepared in this way up to large modifier content ($x = 0.6$, $\sigma_{298K} = 5 \times 10^{-2}$ Sm^{-1}).[27] Several series of chalcogenide glasses were obtained in this way.[27,29–36]

7.2.3 Mechanical alloying

Mechanical alloying, or high energy milling, is not only a way to prepare nanoparticles. The energy brought by the balls allows the preparation of stable but out-of-equilibrium compounds at low temperature. Mechanical alloying has been used for metals, alloys and refractory ceramics and to a lesser extent for the materials of interest here, i.e. ion conductive glasses or glass-ceramics.[37–42] For example, glasses of the system Li_2S-P_2S_5 were prepared by ball milling using a planetary ball mill apparatus. Alumina pots and balls were used for milling using a rotation speed of about 370 rpm.[43,44]

7.2.4 Thin film deposition

Ion conducting chalcogenide films were deposited either by thermal evaporation under vacuum (classical, flash, co-evaporation, e-beam evaporation)[45,46] or various sputtering techniques (RF sputtering, reactive sputtering, magnetron sputtering).[45–49] Silver was often introduced afterwards by thermal or photo-dissolution.[50–54]

Thin films of chalcogenide glassy membranes for heavy metal analysis were also elaborated by means of pulse laser deposition.[55,56] A schematic of the technique is shown in Fig. 7.2.[55] This technique was also used for deposition of silver chalcogenide films[56,57] and recently, Li_2S-P_2S_5 films.[58] Chemical deposition techniques were used to obtain thin films of Ag^+

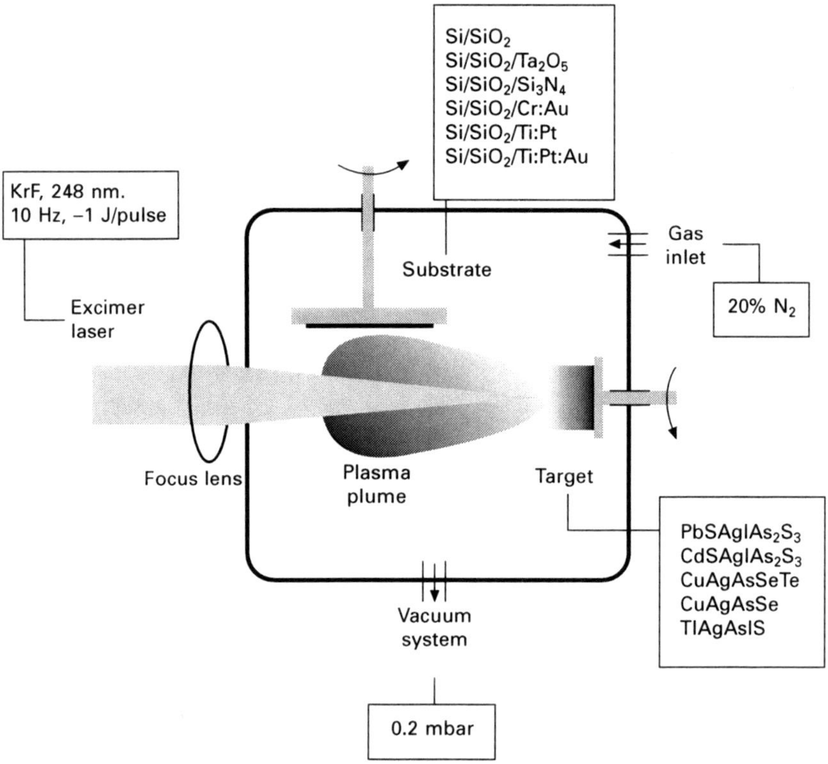

7.2 Schematic of the pulsed laser deposition process (off-axis geometry) and parameters used for depositing the chalcogenide glass materials.[55]

conductive chalcogenide glasses.[45,59] In this case, the indirect way, with silver being introduced in the chalcogenide layers, further deposition by thermal- or photo-dissolution was used. These layers were used for applications as photoresists of submicronic resolution[60,61] or as components for programmable metallisation cells[52–54] or electrolyte for microbatteries.[46–49,58]

7.3 Electrical and electrochemical characterisations

The techniques used to measure the electrical and electrochemical properties of an ion conducting material are similar whatever the material considered, glassy or crystalline, oxide or chalcogenide. In the last case, some additional difficulties may arise when the material is hygroscopic.

On the whole, it is important to know the conductivity of the material but also to measure the contribution of ions and/or electrons to the total conductivity. Important information is the voltage range the material can be used in without being damaged. Finally, measurements of the electrical relaxation at different frequency is a good way to obtain information on the ion dynamics which is of utmost interest for gaining insight into the mechanism of ion transport in the glasses. The different techniques to obtain this information are described in the next section with some examples taken from Ag^+ conducting chalcogenide glasses.

7.3.1 Conductivity measurement

First, it should be noted that dc measurements cannot be carried out successfully to measure the conductivity of an ion conductor. In fact, most electrodes will (at least partially) block the ions at the interface 'electrode/ion conductor' and the current flux will decrease over time and finally reach zero. To overcome the problem, ac impedance measurements over a wide range of frequencies have to be used.

If an ac voltage $\hat{U}$ is applied to the sample (e.g., a cylinder of thickness d and area A), an ac current $\hat{I}$ is created through the sample with $\hat{U} = U_0 \exp(i\omega t)$ and $\hat{I} = I_0 \exp(i(\omega t + \phi))$, U_0 and I_0 are the magnitude, ω is the angular frequency and ϕ the dephasing between voltage and current. The complex impedance, $\hat{Z}$, is given by $\hat{Z} = \hat{U}/\hat{I}$ and the complex conductivity $\hat{\sigma} = (d/A)(1/\hat{Z})$.

Different ways to exploit the experimental data can be used. The Nyquist diagram is commonly used for the measurement of dc conductivity, $\sigma_{dc} = \sigma(\omega = 0)$. In the Nyquist diagram, the opposite of the imaginary part of Z, $-\mathrm{Im}(Z)$, is plotted against the real part of Z, $\mathrm{Re}(Z)$. A good approximation for an equivalent electrical circuit for an ion conducting glass is a RC parallel circuit. In this case, $Z_R(\omega) = R$ and $Z_C(\omega) = 1/iC\omega$, i.e., $\hat{Z} = R/(1 + R^2C^2\omega^2) - iR^2C\omega/(1 + R^2C^2\omega^2)$. The Nyquist diagram of such a circuit, shown in Fig. 7.3, is a semi-circle, whose intercept with the abscissa is the total resistance of the system $\lim_{\omega \to 0}[\mathrm{Re}(Z)] = R$. Dc conductivity can be extracted from R knowing that $\sigma_{dc} = d/A(1/R)$.

For those who are concerned about the ion dynamics of the system, plots giving the real part of conductivity with frequency, $\sigma(\omega)$, are more commonly used. In this case, measurements can be extended to higher frequencies in

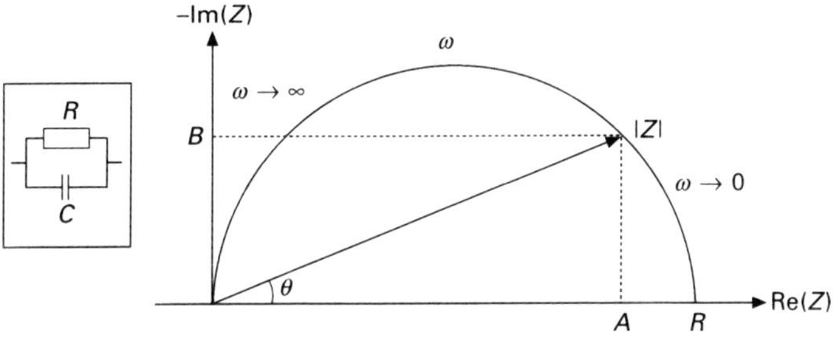

7.3 Nyquist diagram of a RC parallel circuit.

order to have the 'complete' conductivity spectrum of the material. The conductivity spectroscopy covers large frequency ranges from 10^{-3} Hz to more than 10^{14}Hz. Four different ranges have to be considered, which correspond to different measurement techniques sketched in Fig. 7.4.[62] The impedance regime covers the lowest frequency range, below a few MHz; the radiofrequency regime lies in between a few MHz up to a few GHz. The microwave regime covers higher frequencies from a few GHz to 150 GHz, and finally above 150 GHz lays the far infrared regime. Interested readers are invited to refer to Ref.[62] where all the details on measurements of electrical relaxation at high frequency can be found.

Very few materials have been measured in such an extended range.[62] The complete conductivity spectrum of the $0.5Ag_2S$-$0.5GeS_2$ glass is shown in Fig. 7.5.[63] At low frequency, a plateau is first observed (its extrapolation at

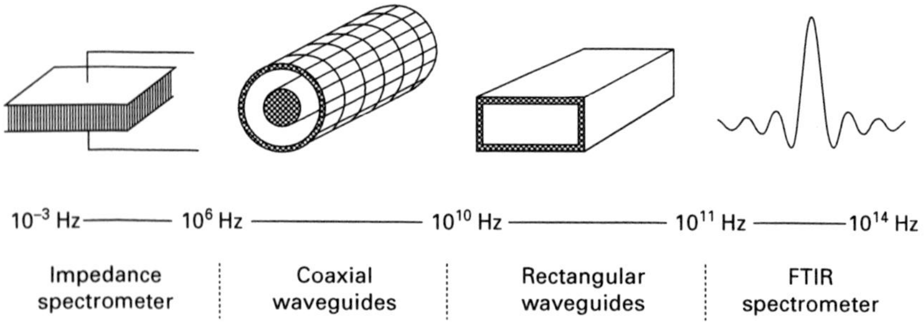

7.4 Schematic overview of different techniques for the measurement of frequency-dependent conductivities (redrawn from Ref. 62).

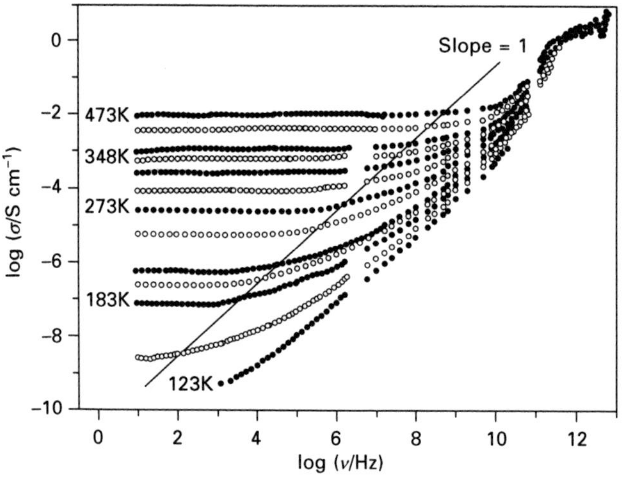

7.5 Frequency dependent conductivity spectra of $0.5Ag_2S$-$0.5GeS_2$ glass at various temperatures (from Ref. 63).

frequency 0 gives σ_{dc}) and a dispersive part $A\omega^n$ occurs at higher frequency. This representation was used to validate models to explain the ion dynamics and will be discussed in Section 7.5.2.

Other types of representations were also used, but less often, to study the ion dynamics, in particular in chalcogenide glasses: the complex permittivity $\hat{\varepsilon}$ (ω) with $\hat{\sigma}$ (ω) = $i\omega\varepsilon_0\hat{\varepsilon}$ (ω) and the complex electrical modulus, $\hat{M}(\omega)$ with $\hat{\sigma}$ (ω) = $i\omega\varepsilon_0/\hat{M}(\omega)$.[64,65]

Before concluding this conductivity measurement section, let us note that very recent papers report on the investigation of local conductivity. Measurements rely on the use of an atomic force microscope (AFM).[66,67] In particular, Piarristeguy *et al.*[66] carried out AFM-based current–voltage measurements on phase-separated Ag-Ge-Se glasses in order to measure the conductivity of the silver-rich and silver-poor phases separately. This pioneering work will probably develop in the future and should be of great help for the characterisation of heterogeneous systems.

7.3.2 Transport number measurements

The electrochemical semi-permeability method[68] can be used to measure the electronic transport number, $t_e = \sigma_e/\sigma_T$, which helps in knowing the part of ionic and electronic contributions to the total conductivity $\sigma_T = \sigma_e + \sigma_i$. It aims at the determination of the electronic part of the conductivity of the glass by measuring the variation, with time, of the electromotive force, E, of a concentration cell comprising the glass placed in between two amalgams containing the mobile ion, Ag^+ for example, at two different electrochemical activities, ${a_{Ag}}^{(1)}$ and ${a_{Ag}}^{(2)}$. The variation of E with time is due to the flux of matter resulting from the internal short circuit due to the electronic conductivity, i.e. due to the electrochemical semi-permeability. The part of electronic conductivity in two silver chalcogenide glasses was measured using this method.[69] The following cell was realized:

Ag-saturated Hg/Ag amalgam [2]	Ag^+conducting vitreous electrolyte	Hg/Ag amalgam [1]

with $\quad a_{Ag}^{(1)} \ll a_{Ag}^{(2)} = 1.$

The relation between E, the electronic resistance, R_e, and the silver concentration in compartment 1, $x_{Ag}^{(1)}$ was calculated in Ref.[68] as:

$$R_e = \frac{RT}{F} E \frac{1}{n_T} \frac{1}{x_{Ag}^{(1)}} \frac{1}{dE/dt}$$

where n_T is the total number of Ag moles in the glass.

As shown in Fig. 7.6[69] dE/dt shifts of 5 and 11 mV per month were observed for the 0.6 (GeS_2-Ag_2S), 0.4 AgI ($n_T = 5.5 \times 10^{-6}$) and 0.7 $GeSe_3$-0.3 Ag_2Se ($n_T = 5.9 \times 10^{-6}$) glasses, respectively. When taking the maximum value at saturation for $x_{Ag}^{(1)}$, i.e. 6×10^{-4}, it was possible to estimate the minimal electronic resistances, $R_e > 1.27 \times 10^9\ \Omega$ and $R_e > 4 \times 10^8\ \Omega$, and therefore the maximal electronic conductivities, $\sigma_e < 5.9 \times 10^{-9}\ \Omega^{-1}\,cm^{-1}$ and $\sigma_e < 1.8 \times 10^{-10}\ \Omega^{-1}cm^{-1}$ for the sulphide and selenide glasses, respectively. From these values and the total conductivity of the glasses, σ_T, the electronic transference numbers, $t_e = \sigma_e/\sigma_T$, could be estimated. Very low numbers were obtained, $t_e \leq 1.5 \times 10^{-8}$ for the sulphide glass and $t_e \leq 3 \times 10^{-6}$ for the selenide glass, indicating that these glasses can be considered as nearly-pure Ag^+ conductors.

7.3.3 Radioactive tracer diffusion coefficient measurements

Radioactive tracer diffusion measurement is the way to measure the auto diffusion measurement D_{Ag}. Measurements are usually carried out using the thin-layer geometry method.[70] It is a very long process. A drop of radioactive solution, a radioactive $^{110m}AgNO_3$ solution in the case of Ag^+ ion conducting glass for example, is deposited onto one face of each sample. The samples are then annealed at different temperatures for different periods of time ranging from hours to months, depending on the value of the diffusion coefficient. Diffusion anneals are terminated by quenching the samples in air. Sections are then removed from the diffusion face using a parallel

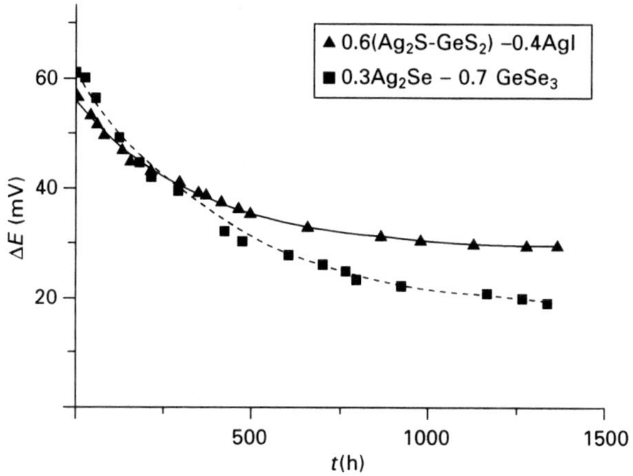

7.6 Variations of the emf of a concentration cell 'Ag-saturated Hg/Ag amalgam //Ag^+ conducting vitreous electrolyte // Hg/Ag amalgam' as function of time for two chalcogenide glasses.

grinder. A scintillation counter is used to measure the gamma activity. The penetration depth x obtained by sectioning of the glass is described by the usual equation for the thin-layer geometry[70]

$$1 - \frac{A(x,t)}{A_0} = \mathrm{erf}\,(q)$$

where $q = x/(2\sqrt{D_{Ag} \cdot t})$, A (x, t) is the corrected residual activity of the sample after a thickness x was removed, t is the diffusion anneal time, A_0 is the initial residual activity and D_{Ag} is the silver tracer coefficient. Using this type of measurements, Bychkov *et al.* measured the diffusion coefficients of several Ag conducting chalcogenide glasses.[71,72] Cu conducting chalcogenide glasses were also investigated.[72–74] These data were further used to propose mechanisms of ion transport in these glasses as described in Section 7.4.1.

7.3.4 Electrochemical stability

The redox stability domain of a solid electrolyte is the range lying between the oxidation voltage and the reduction voltage of the electrolyte. It can be measured by triangular voltammetry. This technique aims at measuring the change in faradic current that flows through the glass while sweeping potentials are applied.[75] During a cycle, anodic and cathodic sweeps are performed in the same potential range. Data obtained for a 0.5 Ag_2S-0.5GeS_2 glass are shown in Fig. 7.7.[9,68] During the anodic sweep, a slight increase in current

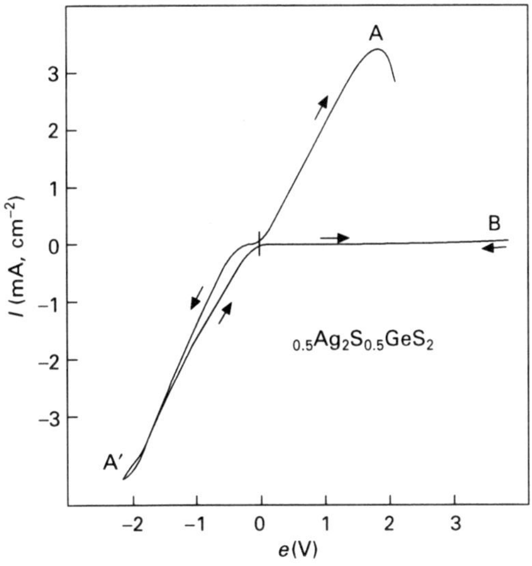

7.7 Triangular voltammetry curve for a 0.5Ag_2S-0.5GeS_2 glass (experiment carried outwith a gold working electrode at 25°C and a sweeping rate of 200 mV/s) (from Ref. 9).

is observed at B. It corresponds to slow electro-active species and can be attributed to the oxidation of sulphur in the glass: $S^{2-} \rightarrow -S^0 + 2^{e-}$. During the cathodic sweep, a drastic increase in current (A) is due to the reduction of silver ions which are the main charge carriers. Silver deposited in this way can be re-oxidised during the reverse sweep at A. The glass possesses a redox stability domain larger than 3V.

7.4 Conductivity versus composition

7.4.1 Modifier content

The alkali conducting glasses, essentially Li^+ and Na^+, were studied mainly as potential solid electrolytes for the development of solid state batteries. In this case, the search for high conductivity and therefore for glasses containing large amounts of modifier was the main goal. The very first paper reporting on the synthesis of these highly hygroscopic ion conducting chalcogenide glasses was published in 1978 by Barrau *et al.*[6] The authors aimed at checking a recently reported model, the weak electrolyte theory (described in Section 7.5.2), that predicted that the ion diffusion in a glass should be facilitated in a more polarisable environment. The replacement of oxygen by sulphur in the glassy network led to producing $x$$Na_2S$-(1-$x$)$GeS_2$ glasses, $0 < x < 0.6$ whose highest conductivity was 5×10^{-4} Sm^{-1} at room temperature, more than an order of magnitude higher than the conductivity of the oxide counterpart. In the following years, this group and others prepared other sulphide glasses including lithium conducting glasses with different formers: GeS_2,[8,9] SiS_2,[76,77] P_2S_5,[15] B_2S_3,[16] and As_2S_3.[18,78] The glasses were prepared by air or water quenching and also, in order to enlarge the vitreous domain towards high modifier contents, by fast quenching using the twin roller quenching method.[27,29] The vitreous domain depends upon the investigated system, but on the whole it extends from 30 mol% in modifier up to 50–60 mol% in modifier. Beyond these limits, high tendency to crystallization is observed. Later, the preparation and characterization of alkali conducting selenide glasses was investigated.[30]

Conductivity measurements in temperature indicated that, for temperatures lower than the vitreous transition temperature T_g, the conductivity of the glasses (and with the exception of a few glasses discussed later on in this section) obeys the Arrhenius law: $\sigma = \sigma_0 \exp(-E_\sigma/k_B T)$, where σ_0 is the pre-exponential factor, E_σ is the activation energy of conductivity, k_B is the Boltzmann factor and T the temperature.

Figure 7.8 shows the electrical characteristics of a series of $x$$Li_2X$-(1-$x$)$SiX_2$, X = O, S, Se, glasses.[79] The electrical characteristics of sulphide and selenide glasses are very similar. The chalcogenide glasses show much higher conductivity than their oxide counterparts by a factor larger than 100, as predicted by the weak electrolyte theory.

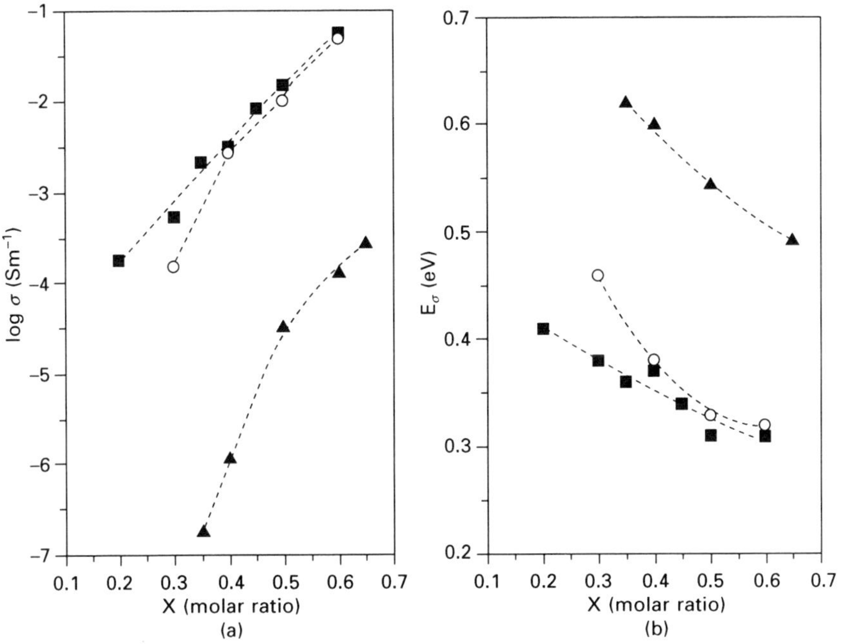

7.8 Variation of (a) conductivity, σ_{dc}, at room temperature and (b) activation energy, E_σ, with composition for xLi$_2$X-(1-x)SiX$_2$ glasses (X= ○ (▲), S (○), Se (■)) (redrawn from Ref 79).

A small increase in the modifier content leads to a huge (exponential) increase in electrical conductivity. For example, to double the lithium concentration in lithium thiosilicate glasses leads to an increase in conductivity by a factor larger than 100. These variations in conductivity are ruled by the variations in activation energy, the pre-exponential factor being almost constant. Its logarithm, log σ_0, is close to 2 whatever the chalcogenide system considered.[80]

In the case of silver (and to a much less extent copper) doped glasses, stable in air and easier to handle than alkali glasses, the aim was mainly to acquire information on the mechanism of ion transport in the glasses, often by investigating the relationships between electrical properties and glassy structure. Most of the work concerned $As_2S(Se)_3$[1–5,13,72] and $GeS(Se)_x$, x = 2, 3[2,3,9–11,71,81,82] glass networks even though a number of papers dealt with other formers (e.g., SiS_2,[34] B_2S_3,[26] P_2S_5[3,19]).

Experimental data covering a large range of mobile cation concentration (several orders of magnitude) are available for a few systems, including Ag_2S-GeS_2,[83] Ag_2S-As_2S_3,[2,4,84] $x$$Ag_2S$-60GeS-(40 − x)GeS_2,[2,71] $Ag_x(Ge_{0.25}Se_{0.75})_{(1-x)}$[85] and $Ag_xGe_{28}Sb_{12}Se_{60}$.[71] Depending on the system, the conductivity data showed a continuous increase or a sudden break that could reach 4–5 orders of magnitude for a silver content of about 5–8 at.% (Fig. 7.9).[83,86] Field-emission scanning electron microscopy (FE-SEM) or electric field microscopy (EFM) show clearly the role of the macroscopic structure in this behaviour.[87,88] The sudden break in conductivity occurred in those systems, which are phase separated with a silver-rich phase embedded in a silver-poor phase at low silver content and vice versa at high silver content (Fig. 7.10).[86] The jump in conductivity occurred at the percolation threshold when the silver-rich phase starts connecting throughout the glassy matrix. The glasses which did not show the break in conductivity were shown to be homogeneous at least down to a few nanometres.[86] Nonetheless, even in these glasses a change in the conductivity regime did occur at about 5 at.% of silver as observed in Fig. 7.9. Therefore, it is clear that two drastically different transport regimes exist at low silver concentration (< 5–8 at.% of Ag) and high silver concentration (> 8 at.% of Ag).

In order to obtain information on the mechanism of ion transport in these glasses, Bychkov *et al.*[71,89–93] measured the tracer diffusion coefficient D_{Ag} and compared it with the diffusion coefficient, D_σ, derived from the ionic conductivity σ_i via the Nernst–Einstein relation $D_\sigma = \sigma_i kT/N(Ze)^2$, where N is the concentration of the mobile species, Ze is their electric charge and k and T are as defined above. The calculated values of D_σ and the experimentally-determined D_{Ag} were further compared using the Haven ratio: $H_R = D_{Ag}D_\sigma$. Typical data for the xAg$_2$S-60GeS-(40−x)GeS$_2$ glasses are shown in Figs 7.11[93] and 7.12.[71] A close relationship between the variations of conductivity and ion diffusion coefficients with a change occurring at about 5 at.% Ag

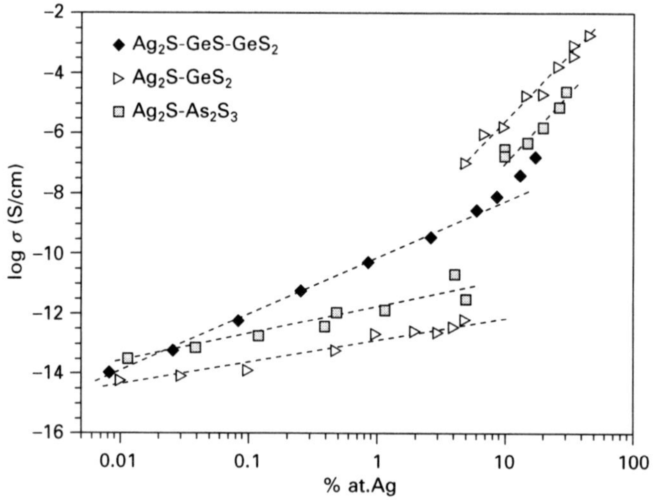

7.9 Variation of conductivity at room temperature with silver content (in at.%) for xAg$_2$S-(100-x) GeS$_2$[83] xAg$_2$S-(100-x)As$_2$S$_3$[84] and xAg$_2$S-60GeS - (40-x) GeS$_2$[71] glasses.

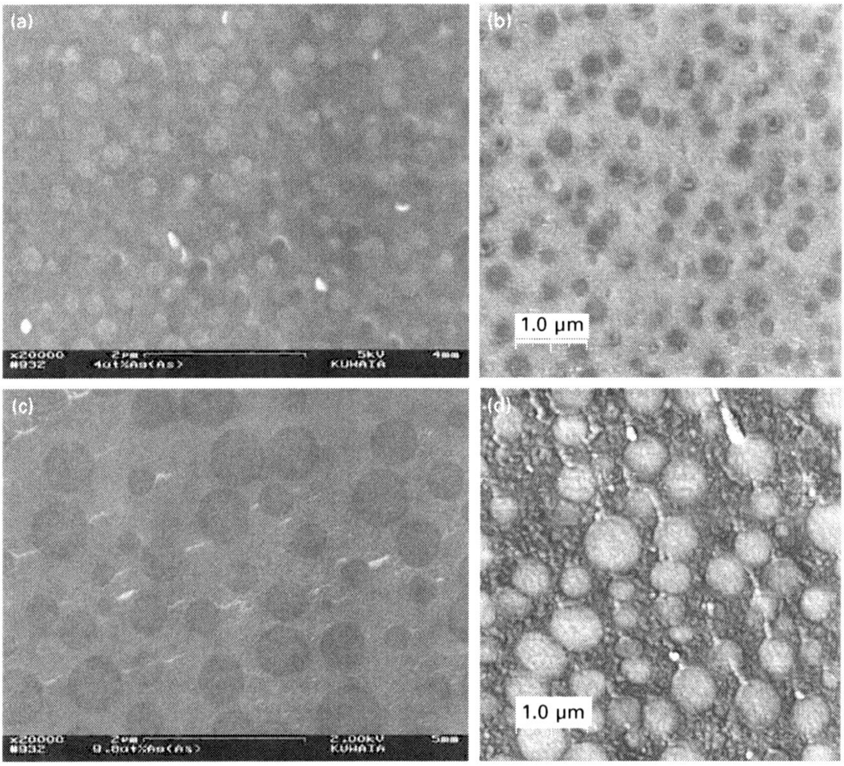

7.10 FE-SEM micrographs of $x\text{Ag}_2\text{S}$-(100-x)As_2S_3 glasses containing 4 at.% Ag (a) and 9.8 at.% Ag (c); EFM micrographs of the same glasses, i.e $x\text{Ag}_2\text{S}$-(100-x)As_2S_3 with 4 at.% Ag (b) and 9.8 at.% Ag (d) (applied voltage = –3V) (redrawn from Ref. 83).

is observed. The variation of both conductivity and diffusion coefficients with silver concentration ([Ag] in at.%) can be fitted by a power law in the low conductivity domains with the equations $\sigma([\text{Ag}])\ \mu\ [\text{Ag}]^{t\sigma}$ and $D([\text{Ag}])\ \alpha\ [\text{Ag}]^{tD}$ leading to consistent exponents, i.e. $t_D = t_\sigma - 1$ according to the Nernst–Einstein equation. Such behaviour is consistent with a transport regime due to percolation. The available composition range in the high conductivity domain is not large enough to allow a reliable fitting of the σ and D_{Ag} curves, but Haven ratio values in this domain along with additional structural investigation using mainly neutron diffraction experiments helped Bychkov in proposing a microscopic mechanism based upon strongly correlated motion of silver at the basis of ion transport in this 'modifier-controlled' domain.[92,94,95] In fact, the short Ag-Ag correlations at ~ 3 Å indicated direct contact of the AgS_3 trigonal pyramids forming the first coordination sphere of silver, and the low H_R values = 0.3 are in agreement with the presence of the correlated motion of the mobile ions. Extensive work has been carried

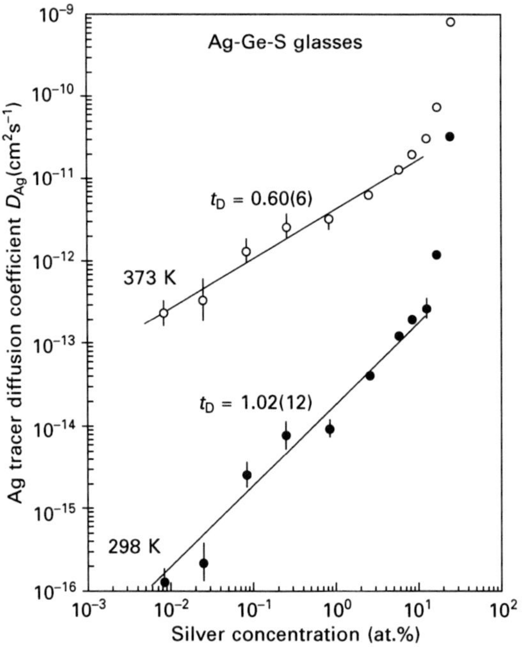

7.11 Diffusion coefficient $D_{Ag}(x)$ isotherms at 298 K and 373 K for Ag_2S-60GeS-(40-x) GeS_2 glasses plotted on a log-log scale (from Ref. 93).

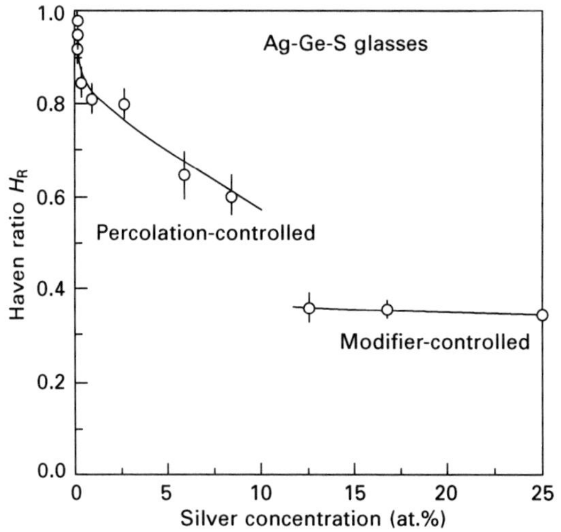

7.12 Haven ratio, H_R, versus the mobile ion content (at.% Ag) for Ag_2S-60GeS-(40-*x*) GeS_2 glasses (redrawn from Ref. 71).

out on Ag(Cu) doped thin films.[45,50–57,59–61] In this case, silver was often introduced in the amorphous chalcogenide layer by photodiffusion, more rarely by thermal diffusion. Both academic and applicative research was carried out. A review of early work has been published by Kolobov and Elliott in 1991.[96] This is described in detail in Chapter 6 of this book dealing with photo-induced phenomena.

7.4.2 Dopant content

A classical idea for improving ionic conductivity is to dissolve a salt of the cation responsible for the conduction in the glass. Generally the dissolved salts are alkaline/silver halides.[15,27,76,77] About 40–50 mol% of halide can be introduced in a glass.[10,11,13,15,78,97–99] It has been proposed that the halide acts as a plasticizer, facilitating the cooperative movements in the glass, which results in a significant decrease in T_g.[10] The increase in conductivity and related decrease in activation energy, resulting from the introduction of a salt in a glass, makes it all the more important that the conductivity of the base glass is low, as shown for a series of Ag^+ conductive sulphide and oxide glasses in Fig. 7.13.[11] On the whole, whatever the base glass used, the conductivity tends to have a limiting value of about 10^{-1} Sm^{-1} for Li^+ conducting glasses[80] and 1 Sm^{-1} for Ag^+ conducting glasses.[11] In both cases, the activation energy tends towards the limiting value of 0.3 eV. In the Ag^+ conducting glasses, two assumptions to explain this behaviour were proposed: either the existence of αAgI microdomains[100,101] or the migration of Ag^+ by simple rotation around the iodide anions when the I^- ions come close together at high silver content.[11] The second assumption is more consistent with an activation energy of 0.3 eV, much larger than that of αAgI (0.05 eV). In the same way, the second assumption alone can apply to Li^+ conducting glasses.[80] In fact, pure LiI has very different electrical characteristics than the doped glasses ($\sigma_{LiI} \sim 10^{-5}$ Sm^{-1} and $E_\sigma \sim 0.45$ eV).

Oxysalts were also used to improve the conductivity of Li^+ conductive chalcogenide glasses.[36,102,103] For example, small amounts of lithium oxysalts, Li_3PO_4, Li_4SiO_4, Li_3BO_3 and Li_4GeO_4 doped into the Li_2S-SiS_2 glass system helped not only to increase the conductivity (sometimes by a factor of 10) but also the stability of the glasses toward Li metal, a crucial parameter for an electrolyte to be selected for use in a battery.[104]

7.4.3 Mixed alkali effect/mixed cation effect

Nonlinear variations of several physical properties including electrical ones are observed in a glass when the relative proportion of two alkali (mobile) ions is varied while their total concentration is maintained constant. This is the well-known ‘mixed alkali (cations) effect’ (MAE).[105,106] This intriguing phenomenon has been reported many times for oxide glasses, e.g. silicate,[107] phosphate,[108,109] borate,[110] but few studies were reported that dealt with chalcogenide glasses. Whenever investigation was carried out, as in the glasses Na-Rb silicon sulphide glasses,[111] $0.5[(1\text{-}x)Li_2S\text{-}xNa_2S]\text{-}0.5SiS_2$[79] and $0.5[(1\text{-}x)Rb_2S\text{-}xAg_2S]\text{-}0.5GeS_2$, $0 < x < 1$,[81] clear manifestations of MAE were observed. Diffusivity crossover, i.e. a decrease in mobility of a cation upon addition of the second cation, was observed by radioactive diffusion coefficients for Na-Rb silicon sulphide glasses.[111] As shown in Fig. 7.14 for the thiosilicate glasses,[79] both the vitreous transition temperature and conductivity display a high minimum for $x \sim 0.5$, while the activation energy displays a maximum for the same value of x. Many suggestions were proposed to explain this phenomenon:[109,112] cations of different sizes are not interchangeable at their respective sites and/or interaction between alkali cations and glassy network gives rise to 'complexes', mixed interstitial pairs of low

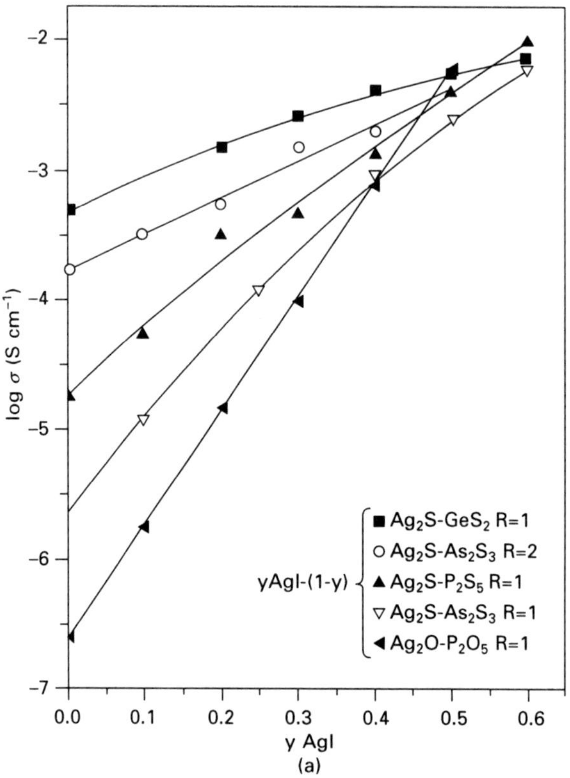

7.13 (a) Electrical conductivity and (b) activation energy for various oxide and sulfide based glasses versus AgI content (*R* is the ratio glass former/glass modifier): ■, 11 ○ and ▽, 13 ▲ and ◀ 99 (redrawn from Ref 11).

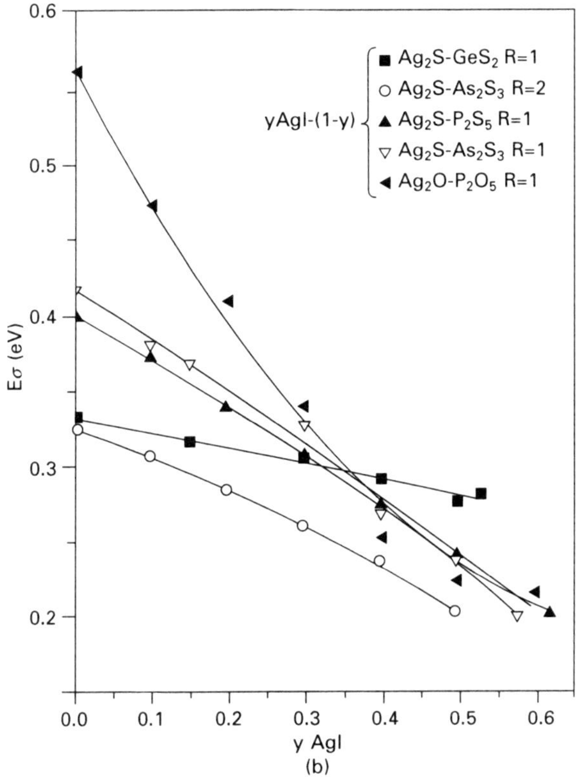

7.13 Continued

mobility would exist. Models based upon computer simulations and focusing on cation-induced relaxations of the network in mixed cation glasses[113] were supported by the structural investigation of the Rb-Ag thiogermanate glasses. Small-angle x-ray scattering, extended x-ray absorption fine structure, Raman and infrared spectroscopy showed that: (i) the glasses were homogeneous in the mesoscopic domain, (ii) each mobile cation formed its own specific environment and was not influenced by the presence of the dissimilar cation, and (iii) nonlinear structural changes of the glassy matrix occurred upon cation mixing.[81] The maximum deviation from structural additivity was observed for compositions giving the extreme values for conductivity and T_g, and was found to result from cation-triggered rearrangements of the local germanate polyhedra towards a more homogeneous glass structure.

A so-called 'degenerated mixed cation effect' was observed in the CuI-AgI-As_2S_3 by Bolotov *et al.*[72] In disagreement with the MAE, the diffusivity crossover was not present in the system and, as a result, the ionic conductivity increased linearly upon silver addition. To our knowledge, this is the only system that shows such behaviour.

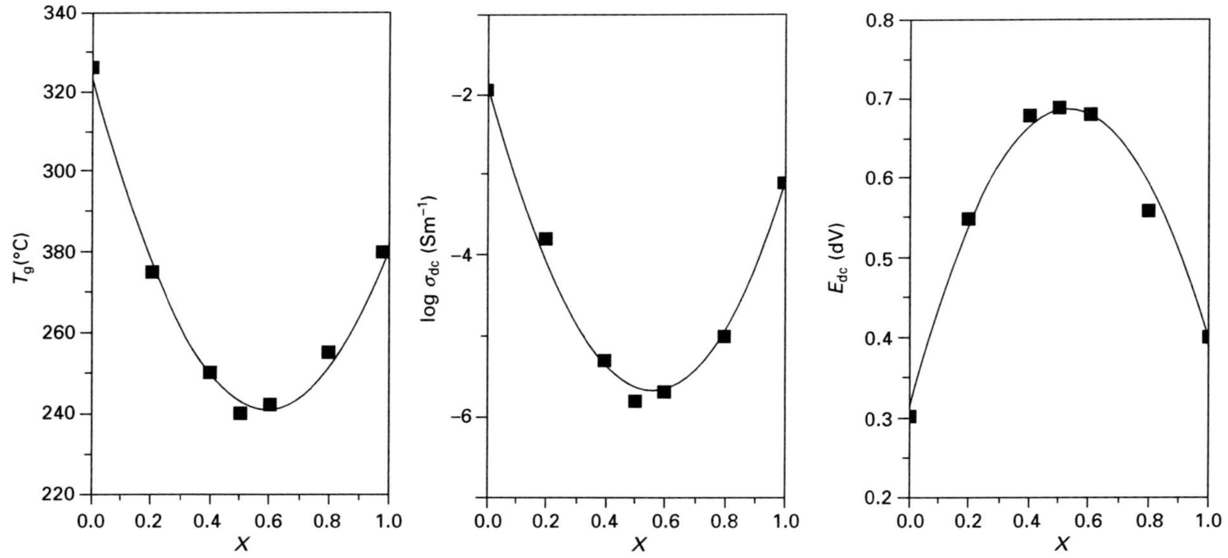

7.14 Variation of glass transition temperature, T_g, conductivity at room temperature, σ_{dc}, and activation energy of conductivity, E_{dc}, with x for 0.5 [$x$$Na_2S$-(1-$x$)$Li_2S$]-0.5$SiS_2$ glasses (redrawn from Ref. 79).

7.4.4 Mixed former effect

Another intriguing effect that exists in ion conducting glasses is the 'mixed glass former effect' (MGFE). Nonlinear increase in both electrical conductivity and vitreous transition temperature can be achieved by mixing two glass formers at constant fraction of the mobile cation.[114,115]

MGFE has been known for many years and was first reported in borophosphate glasses by Tsuchiya and Moriya in the 1980s.[114] A structural origin has been suggested with the proposed existence of BPO_4 entities. Controversies arose over time and investigation is still going on.[116–118]

Pradel *et al.*[31] were the very first to observe a MGFE in chalcogenide glasses, i.e. in Li_2S-GeS_2-SiS_2 glasses where, depending upon the modifier content, the chalcogenide glasses show (low) or do not show (high) a MGFE (Fig. 7.15).[80] Such a difference between glasses with different modifier contents but belonging to the same family has never been observed in oxide glasses.

Structural investigation of $0.3Li_2S$-$0.7[yGeS_2$-$(1$-$y)SiS_2]$ glasses by small angle X-ray scattering has shown that phase separation exists for $0.5 \leq y \leq 0.64$ and accounts for the large increase in conductivity in the central composition region.[119] Nevertheless, one can observe in Fig. 7.15 that deviation of conductivity from linearity also exists in regions where the glasses are homogeneous, $y < 0.5$ and $y > 0.64$.

Zhang and Kennedy studied the $0.33[(1-y)B_2S_3-yP_2S_5]-0.67Li_2S$[28] and

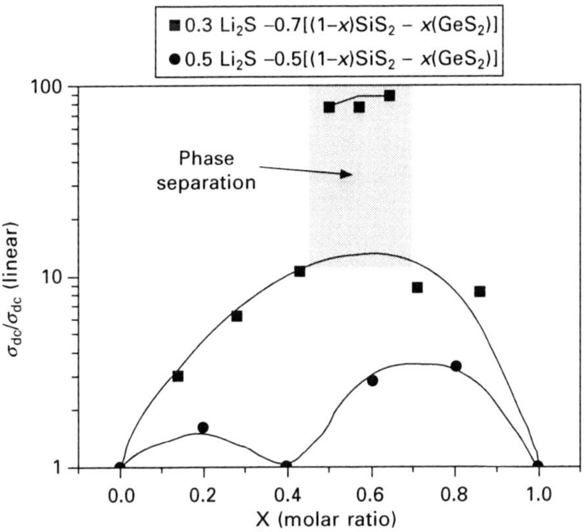

7.15 Evolution of ratio 'conductivity/theoretical linear conductivity' with x for yLi_2S-$(1$-$y)[xGeS_2$-$(1$-$x)SiS_2]$ glasses (y = 0.3, 0.5) (experimental data from Ref. 80).

$0.4[(1-y)SiS_2\text{-}yP_2S_5]\text{-}0.6Li_2S$[120] systems. They reported a higher Li^+ ionic conductivity for the coformer sulfide glasses than for the single sulfide network former glasses. An investigation of two series of $Na_2S\text{-}B_2S_3\text{-}P_2S_5$ glasses was carried out by Haynes *et al.*[121] While the conductivity of the glasses was not studied, the glass transition temperatures showed a pronounced maximum for both series, in agreement with a MGFE. In contrast, no MGFE was observed in $Li_2S\text{-}B_2S_3\text{-}As_2S_3$ glasses.[122]

Li^+ conducting oxy-sulphide glasses of composition $0.5Li_2S\text{-}0.5[(1\text{-}x)GeS_2\text{–}xGeO_2]$, where $0.1 < x < 0.8$, were prepared by Kim *et al.*[123] In this case an increase in conductivity of $0.5Li_2S\text{-}0.5GeS_2$ by a factor 4–5 was observed for an addition of 5 mol% of GeO_2. Further addition of GeO_2 monotonically decreased the conductivity while increasing the activation energy.

7.4.5 Very fast ion conducting glasses: non-Arrhenius behaviour

While it is generally reported that the conductivity of ion conducting glasses obeys the Arrhenius law below T_g, it has been shown that deviation from this law could occur in some very fast ion conducting glasses at high temperature when the conductivity approaches 1 Sm^{-1}.[124] Owing to their high intrinsic conductivity, many examples of this behaviour are found in chalcogenide glasses as shown in Fig. 7.16.[124] It has also been reported in a $0.5Ag_2S\text{-}0.5GeS_2$ glass.[90,125] Kincs and Martin[124] explained this behaviour by a saturation effect. The activation energy, usually the sum of electrostatic and mobility terms (see models in Section 7.5), would be associated with the mobility term alone at these high temperatures, where all ions would participate equally in the conductivity events. Ribes *et al.*[125,126] gave a different interpretation of the phenomenon and associated it with a manifestation of the configurational entropy associated with the cation disordering.

7.5 Direct current (dc) conductivity models

Most of the models used to describe ion transport in glasses are not specific to chalcogenide glasses. In most cases, oxide glasses, and in much fewer cases, chalcogenide glasses, were used as examples to validate the proposed model. The space available here does not allow a complete description of the models. The main basis and predictions only will be reported.

Before describing the models, let summarize briefly the main characteristics of ion transport in glasses.

The ionic conductivity in glasses is a thermally activated phenomenon. Below the vitreous transition temperature T_g, the conductivity obeys the Arrhenius law: $\sigma = \sigma_0 \exp(-E_\sigma/k_B T)$, where σ_0 is the pre-exponential factor,

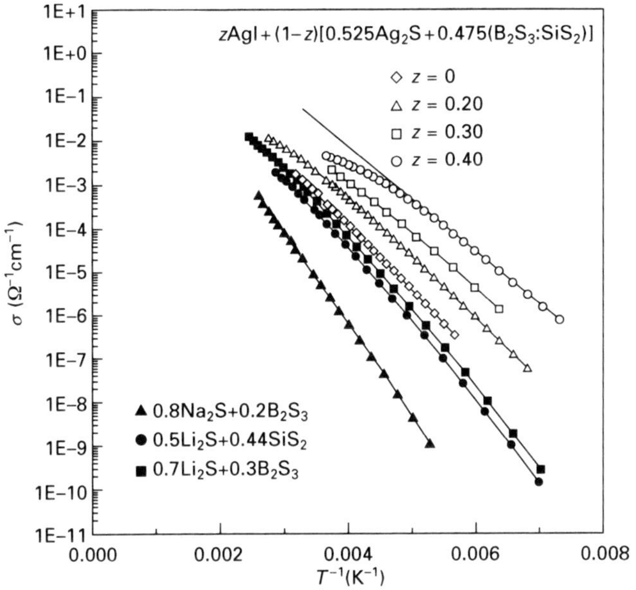

7.16 Arrhenius plots of the ionic conductivity for optimized Ag conducting FIC glasses compared to those for other Li and Na conducting glasses. Notice that for the poorer conducting glasses, the Arrhenius plots have a straight slope, whereas the optimized Ag conducting FIC glasses have significant curvature at highest temperatures (from Ref. 124).

E_σ is the activation energy of conductivity, k_B is the Boltzmann factor and T the temperature. Above T_g, the ion motion is made easier thanks to the cooperative motion of the macromolecular chains constituting the glass. The conductivity can then be described by a free volume law, a general law that accounts for many other properties of liquids and overcooled liquids above T_g: $\sigma = \sigma_0\exp\text{-}E_\sigma/k_B\ (T-T_0)$, where T_0 is the ideal vitreous temperature.

The ionic conductivity σ can be written as $\sigma = nq\mu$, where n is the concentration in mobile carriers of charge q and μ is the mobility of these carriers. It is very difficult to measure separately the mobility and the number of mobile carriers. Therefore it is not known which of the two factors is the main contributor to the huge variation of σ with the modifier content (Remember that to double the modifier content results in an increase in conductivity by a factor of 10 to 100! (Fig. 7.8).)

In the case of a model involving independent thermally activated jumps of carriers between equivalent sites, the mobility can be expressed as $\mu = (\alpha\lambda^2\nu_0 q/k_B T)\exp(\text{-}E_m/k_B T)$, where α is the degree of freedom, λ the average jump distance, ν_0 the attempt frequency of the ions and E_m the migration energy. The number of charge carriers is also thermally activated $n = n_0 \exp(-E_c/k_BT)$, where n_0 is the total number of modifier cations and E_c the energy of creation of mobile carriers.

The energy of activation of conductivity is therefore the sum of two terms, the creation energy of mobile carriers E_c (directly linked to n) and the migration energy of the ions E_m (directly linked to μ). Since the pre-exponential factor does not change much when changing the composition in chalcogenide glasses,[80] the conductivity changes are ruled by changes in activation energy.

7.5.1 Anderson and Stuart model (A.S.)

The pioneering Anderson and Stuart model, published in 1954, was proposed to account for the conductivity of ion conducting silicate glasses.[127] Nevertheless, it can apply to chalcogenide glasses with the adjustment of a few parameters.

According to the A.S. model, the activation energy of conductivity is the result of two contributions: an electrostatic one accounts for the $M^+ ... O^-$ binding energy E_b and the other one is the strain energy E_s, arising from the elastic bending of the glass when the ion passes through.

$$E_\sigma = E_b + E_s = 1/\gamma(ZZ_0e^2/(r + r_0) - ZZ_0e^2/(\lambda/2)) + 4\pi G R_D(r + r_d)$$

where γ is a covalence parameter arbitrarily taken to be equal to the relative permittivity ε_r of the glass; Z_0 and r_0 are the charge and the radius of the oxygen cation, λ is the jump distance, r is the mobile cation radius, r_d is the doorways radius and G is the elastic modulus. Anderson and Stuart calculated the activation energy for several compositions of ion conducting silicate glasses. For small cations (Li^+, Na^+, Ag^+), $E_b > E_s$. In these cases, the variation of E_b rules over the variation of E_σ with composition.

Note that through the covalence parameter which scales with the relative permittivity of the glass (higher for a chalcogenide glass than for an oxide one) it was possible to predict the lower activation energy of conductivity (by a factor 2 on average) for chalcogenide glasses.

7.5.2 Weak electrolyte (W.E.) theory

Ravaine and Souquet's model was proposed in 1977. It was built on the basis of data obtained by thermodynamic activity measurements on a series of sodium silicate glasses[128] and later, confirmed on silver phosphate glasses by Reggiani *et al.*[129] and on lithium borates by Levasseur *et al.*[130] These data indicated that the glasses behaved as weak electrolytes. It led Ravaine and Souquet to consider a glass as an electrolyte with the glassy network being the solvent and the modifier being the solute. As in any weak electrolyte, the solute would be weakly dissociated, leading to the conclusion that only a small fraction of the cations would be mobile at time t.

While the original model was a macroscopic one, Souquet *et al.* proposed a microscopic version later on.[131] The basic hypothesis assumes that ionic displacement results from the migration of cationic pairs formed by a partial dissociation, which would correspond to the creation of a Frenkel defect in an ionic crystal. An ion would leave a 'normal' position close to a non-bridging oxygen (eventually chalcogen) and jump in an already occupied position then forming a cationic pair. Owing to the energy needed to succeed in such a jump, the concentration in cationic pairs is small. Expressions were then formulated for the variation in ionic conductivity as a function of temperature. Below T_g, $\sigma = n\,(e^2\lambda^2\nu/6k_BT)\exp((\Delta S_f/2 + \Delta S_m)/k_BT)\exp(-(\Delta H_f/2 + \Delta H_m)/k_BT)$. Fitting the expressions with experimental data measured on several alkali disilicate glasses allowed the calculation of a mobility of effective charge carriers close to 10^{-4} $cm^2s^{-1}V^{-1}$ at room temperature and a ratio between the number of effective charge carriers and the total number of alkali cations comprised between 10^{-8} and 10^{-10}, comparable to the concentration of intrinsic defects in an ionic crystal or dissociated species from the weak electrolyte solution.[131] Recent data on $AgPO_3$-AgI glasses led to similar conclusions.[132]

The weak electrolyte theory is at the basis of the preparation of the very first ion conductive chalcogenide glasses.[6,133] In fact, the model predicts that the conductivity would increase by increasing the dielectric constant of the glassy network, i.e. by replacing oxygen by a more polarisable ion, for instance sulphur.

Finally, both the A.S. model and the W.E. theory predict the increase in conductivity by increasing the dielectric constant of the medium.

7.5.3 Dynamic structure model

In the early 1990s, computer simulation had strongly progressed. It was then possible to produce 'simulated' (virtual) glasses whose behaviour could be compared to 'real' materials. Based on such possibilities, Maass *et al.* developed the dynamic structure model based on the concept of site memory effect.[134,135] The basic idea is that the mobile ions create well-defined sites and their own ion-specific pathways.[113] Let us consider an A cation occupying an $\overline{A}$ site, and leaving this site behind as it moves through the glassy matrix. This empty $\overline{A}$ site then acts as a 'stepping stone' for the other A cations to move through the glass. However, after a time τ, the empty $\overline{A}$ site will relax and lose the memory of the A cation passing through. It will become a 'less good' $\overline{C}$ site. It will be energetically less favourable for an A cation to enter a 'wrong' $\overline{C}$ site than an $\overline{A}$ one. The jump probability of A cation to a vacant nearest-neighbour $\overline{A}$ site is defined as $\omega_{A\overline{A}} = \nu_{A\overline{A}}\exp(-E_{A\overline{A}}/k_BT)$. Similarly, its jump probability to a neighbouring $\overline{C}$ site is $\omega_{A\overline{C}} = \nu_{AC} \exp(-E_{A\overline{C}}/k_B T)$. Since the environment of the $\overline{C}$ site must distort in order to accommodate the arrival of A cation and the lattice distortion requires appropriate elastic energy $\Delta E_{A\overline{C}}$, it can be written as $E_{A\overline{C}} = E_{A\overline{A}} + \Delta E_{AC}$ with the 'mismatch energy' $\Delta E_{AC} > 0$.

In the same way, glasses containing two types of cations A and B can be considered with the presence of two types of sites $\overline{A}$, favourable to A, and $\overline{B}$, favourable to B, along with sites $\overline{C}$ neither adjusted to cations A nor cations B. The main consequence of this site memory effect, as shown by Monte Carlo simulations, is the emergence of fluctuating pathways which facilitate the migration of particular ions in the glass. The general occurrence of the mixed cation effect in glassy materials can be explained on the basis of the dynamic structure model. As reported in Section 7.4.3, the structural investigation of the Rb-Ag thiogermanate glasses by Rau *et al.*[33] indeed supports the assumption of the dynamic structure model. Finally, the model is also consistent with the dependence of conductivity on the modifier content in single alkali glasses following a simple power-law relation.

7.6 Frequency-dependent conductivity models

The evolution of conductivity with frequency exhibits some particular characteristics. In the earlier work, when measurements by complex impedance were carried out in the frequency range from a few mHz up to a few KHz, it was proposed that after the dc plateau at low frequency, a dispersive part appeared that showed a power law dependency with frequency, $A\omega^n$ with $n \sim 0.5$. This is Jonscher's universal law.[136] However further measurements exploring wider frequency domains indicated that it was somewhat more complicated with the exponent n apparently increasing with increasing frequency to finally reach (or even exceed) unity.[64,65,137–139] Much controversy a rose at the time that led to conductivity measurements in the whole frequency range from a few mHz up to the IR regime. Funke relied upon this experimental data, in particular upon data on the chalcogenide $0.5Ag_2S$-$0.5GeS_2$ glass[63,140] to develop and further improve his theoretical models.[141,142] At the same time, measurements at very low temperature were carried out.[139,143,144] In this region the dc plateau could no longer be observed and the frequency dependence of conductivity confirmed the presence of a regime where the conductivity changed (nearly) linearly with frequency and showed a weak (if any) temperature dependence (see the iso-frequency representation of log σ versus $1/T$, for a chalcogenide glass, i.e. $0.4\ Ag_2S$-$0.6GeS_2$ glass, shown in Fig. 7.17.[139] This is the domain of nearly constant loss (NCL) discovered by Nowick.[143,144] On the whole, it appeared that a NCL region that emerged both at low temperature and high frequency is present in addition to the dc region and dispersive Jonscher's region.

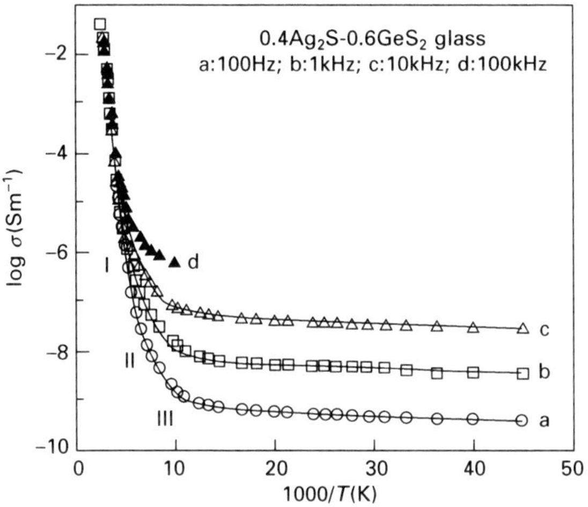

7.17 Temperature dependence of conductivity for glass $0.4Ag_2S$-$0.6GeS_2$. Solid lines are guide for the eyes. Region I = dc regime, region II = dispersive Jonscher's regime, region III = nearly constant loss regime (redrawn from Ref. 139).

The models developed to account for ion transport in glasses had to reproduce these features. The readers can refer to the review proposed by Dieterich and Maass[145] for a description of the proposed models. The three main models that were much discussed over the years will be briefly described in the following sections.

7.6.1 Coupling model

The coupling model developed by Ngai[146] is based on general concepts and aims at dealing with all kind of relaxations in complex systems including those resulting from electrical excitations. The coupling model is based on the assumption that the primitive relaxation of a microscopic unit at short time, $t << t_c$, is independent of other relaxing units, while, after the critical time t_c, interactions between the relaxing units cannot be neglected and the relaxation occurs in a cooperative way. With this assumption, the time dependent relaxation rate $W(t)$ behaves as:

$$W(t) = W_0 \text{ for } t << t_c$$

$$W(t) = W_0 (t/t_c)^{-n}, 0 < n < 1, \text{ for } t >> t_c$$

Ngai and co-workers[147–149] assumes that the time t_c is temperature independent while the short-time relaxation rate is thermally activated, $W_0 = \nu\exp(-E_0/k_BT)$.

The associated normalized relaxation function $\phi(t)$ can then be calculated.

Knowing that $d\phi(t)/dt = -W(t)\,\phi(t)$, we get:

$$\phi(t) = \exp(-W_0 t) \text{ for } t \ll t_c$$

$$\phi(t) = \exp(-(t/\tau_\sigma)^{1-n}) \text{ for } t \gg t_c$$

with $\tau_\sigma = (t_c^{-n}/W_0)^{1/(1-n)}$. For short intervals, a Debye relaxation is observed, while a non-Debye, Kolrausch-Williams-Watts-type relaxation occurs over longer times.

From the above relaxation function, the evolution of conductivity with frequency can be calculated. Typical plots are reported in Fig. 7.18.[150,151]

The coupling model predicts a constant conductivity for $t > \tau_\sigma$ and $\tau < t_c$, i.e. a plateau at low and high frequencies and a dispersive behaviour $\sigma = A\omega^n$ $(0 < n < 1)$ for $\tau_\sigma < t < t_c$.

7.6.2 Counter-ion model

The counter-ion model (CM) was proposed by Dieterich and co-workers[152,153] in the 1990s. In this model, the mobile ions move among immobile counterions that are placed randomly among the centres of cubic lattice cells. The site energies encountered by the mobile ions are attributed to the Coulomb fields of the immobile counterions. Monte Carlo simulations were performed to analyse the consequence of these assumptions.

A typical conductivity spectrum $\sigma(\omega)$ which results from the CM model is shown in Fig. 7.19.[145] Four regimes of conductivity are observed. At

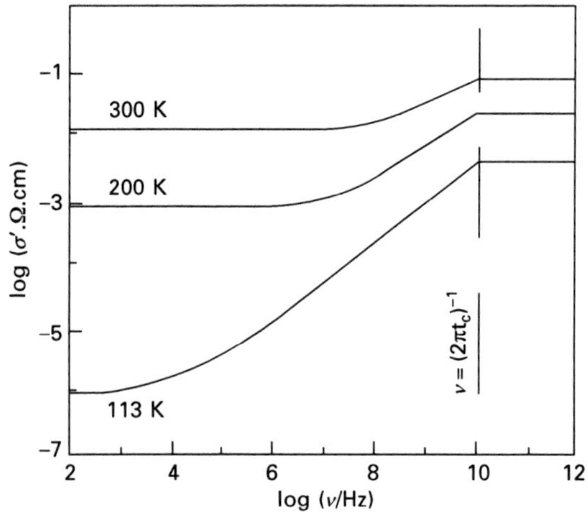

7.18 Conductivity spectra at different temperatures as derived from the coupling model. The parameters have been chosen to fit the data obtained for Na-βalumina in Ref. 150 (redrawn from Ref. 151).

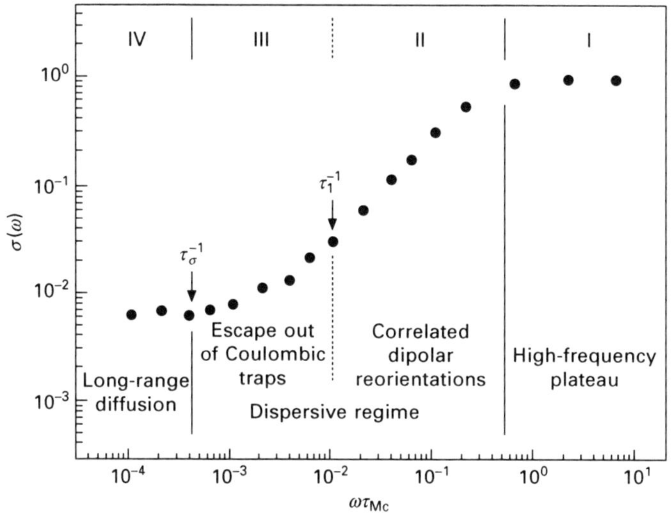

7.19 Illustration of different frequency regimes in the dynamic of the counterion model. The two intermediate regimes II and III pertain to 'nearly constant loss' and Jonscher-type behaviour, respectively (redrawn from Ref. 153).

high frequency when every hop contributes to conductivity, a plateau is observed. At lower frequency, the dispersive region occurs ($\sigma(\omega) \propto \omega^n$) which comprises two regimes, a NCL regime when only correlated dipolar reorientations are observed (n equal to or even slightly larger than 1) and the Jonscher regime with the ions being able to escape out of Coulombic trap ($n \sim 0.5$–0.6). Finally, the dc plateau occurs when long-range diffusion becomes possible.

7.6.3 Jump relaxation model and further developments

The jump relaxation model (JRM) was proposed by Funke and Riess.[154] In this model, the dispersion in the conductivity results from the strong forward-backward jump correlations in the motion of an ion. When an ion jumps from its previously relaxed position into a new site, it is no longer at equilibrium. Equilibrium will be recovered owing to the relaxation of the system in one of the two following ways: the surrounding can relax due to the motion of other ions or the ion itself can perform a backward jump and go back to its former position. The probability of the first or second process occurring changes with time. The longer the ion stays in the new position, the smaller becomes the probability for a backward jump to occur. The physical treatment of the concept led to calculating the time dependent relaxation rate $W(t)$ and finally the frequency-dependent conductivity. The resulting spectrum is shown in Fig. 7.20.[151] The model predicts high and low frequency plateaus separated by a dispersive region at intermediate frequencies.

One of the further developments of the model aimed at integrating the concepts developed in the dynamic structure model. While the initial JRM considered that all sites were equivalent, the new unified site relaxation model (USRM)[155] introduced the presence of two types of sites, an $\overline{A}$ site adapted to the A cation and a less favourable $\overline{C}$ site (see Section 7.5.3 for more detail). The probability for a backward jump is larger for an ion A jumping to a $\overline{C}$ site than for its jumping to an $\overline{A}$ site. In the framework of USRM, a superlinear frequency dependence of conductivity is possible which was not the case for the JRM.

Two other steps in the development of the JMR led to the concept of mismatch and relaxation (CMR)[141,156] and the MIGRATION concept.[157–159] In the CMR, the neighbouring ions, rearranging in the mismatch induced dipole field, had the same dynamic properties as the 'central' hopping one. The assumption led to the development of a new formulation for the relaxation rate $W(t)$ and the resulting conductivity spectra displayed the property of a continuously increasing slope. The last improvement is the MIGRATION concept, which considers a time-dependent shielding of the dipole-field associated with the mismatch of the central ion. Better fitting of the experimental conductivity spectra was then achieved.[142]

The developments to Funke's model in the most recent years are described in Ref.[62]

7.7 Applications

7.7.1 Batteries

The development of solid state batteries which would help in overcoming the main problems of batteries containing liquid electrolytes, i.e. leakage and/or

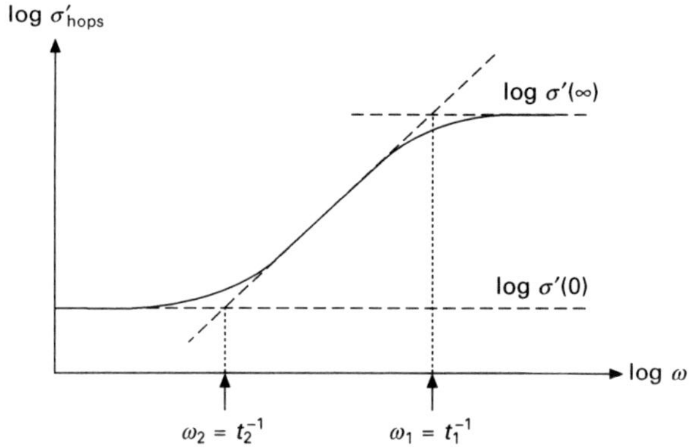

7.20 Dynamic conductivity as derived from the jump relaxation model (from Ref. 151).

corrosion at the electrodes, required the use of solid electrolytes with high ionic conductivity in order to limit the ohmic drop at the electrodes. In this sense, lithium conducting chalcogenide glasses were excellent candidates with conductivity 10–100 times larger than that of their oxide counterparts. Several solid state batteries comprising lithium conducting chalcogenide glass or glass-ceramic as the solid electrolyte have been developed.[38,44,97,160–172] Thin film technology was also considered to develop rechargeable miniaturised Li batteries with Li^+ conducting amorphous chalcogenide films as the electrolyte.[47–49,163]

This work is described in more detail in Chapter 19.

7.7.2 Ionic memories

Due to an ever increasing demand for information storage, investigation into new memory devices has expanded dramatically in the last decades. The requirements are complex: non-volatility, fast response, large scalability, low power consumption. From the late 1990s and further developed over the years, a new type of resistive-random access memory (ReRAM) device based upon a silver/copper conductive chalcogenide layer, e.g. a Ag-photodissolved As_2S_3 or $Ge_{0.3}Se_{0.7}$ layer, as the active layer was proposed by Kozicki *et al.*[164–167] In such a device, as shown in Fig. 7.21,[166,168,169] the chalcogenide layer of ~ 30 nm in thickness, is sandwiched between an active silver/copper electrode and an inert one, e.g. W, Ni. Under the application of a positive bias voltage to the oxidisable Ag/Cu electrode, ions migrate through the chalcogenide layer and are reduced at the negative electrode. An Ag/Cu filament is formed throughout the layer, creating a conductive path. The application of a reverse bias leads to the dissolution of the metal filament and a resistive state is restored. This type of device, referred to as a programmable metallisation cell (PMC)[167] or conductive-bridge random access memory (CB-RAM),[170] presents many interesting characteristics.[168,169,171,172] A fast switching of a few ns from resistive to conductive states is observed. Switching has been tested for over than 10^{10} cycles.[167] The power consumption is low since the write currents lay in the microampere range and the switching voltages are found in the few hundred millivolts range.[173] It has been shown that the resistive and conductive states are stable in the absence of voltage application.[174] Therefore PMC belongs to the category of non-volatile memories. Finally, the feasibility of multi-bit data storage by variation of the write current was demonstrated, in particular by Schindler *et al.*,[175] which proved the potential high scalability of these devices. The main drawback in the use of chalcogenide glasses to develop PMC is the lack of temperature stability of these materials.[53] In order to successfully develop PMC at an industrial stage, temperatures of several hundred degrees Celsius usually used in back end of line processes might have to be tolerated

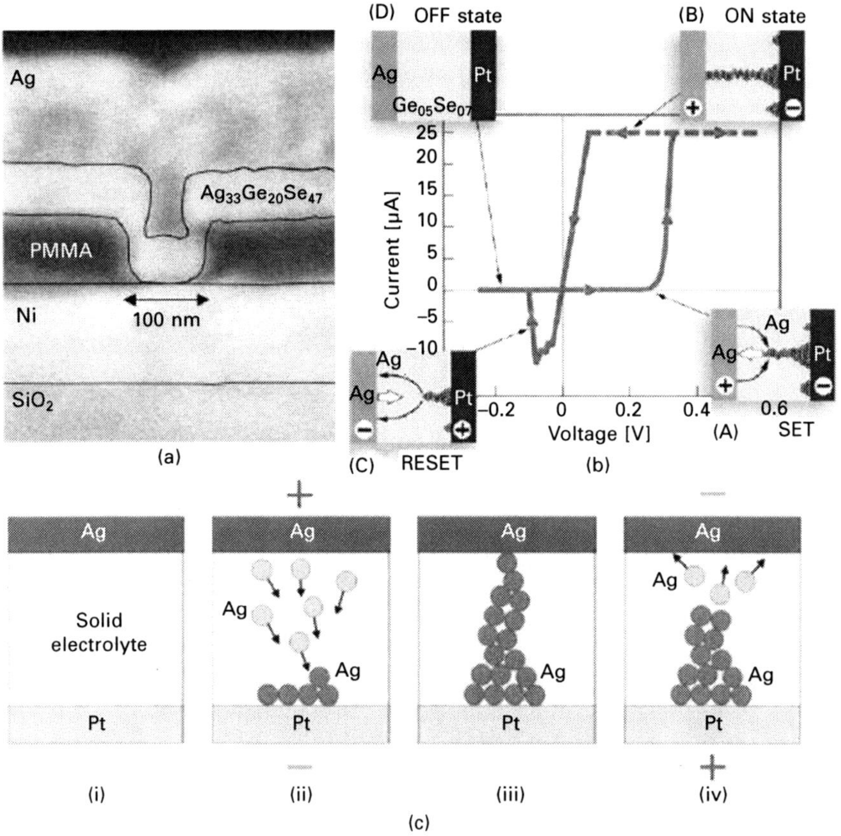

7.21 (a) SEM micrograph of an electrochemical metallisation cell Ag/Ag-Ge-Se/Ni (PMC memory) from Ref. 166. (b) Typical current-voltage characteristic of an Ag/Ag-Ge-Se/Pt electrochemical metallisation cell using a triangular voltage sweep. The ON conductance is limited by compliance current of 25 mA. The insets A to D show the different stages of the switching procedure (from Ref. 168). (c) Schematic of the switching process in an electrochemical memory cell with an oxidable Ag electrode and an inert Pt electrode; (i) off state; (ii) switching on; (iii) on state; (iv) switching off (from Ref. 169).

by the different parts of the device, a difficult challenge with chalcogenide as the active layer.[170]

From an academic point of view, the investigation mainly concerned the mechanism of the filament formation. In order to explain the fast switching from resistive to conductive states, and owing to the relatively low Ag^+ ion mobility at room temperature (~ 10^{-3}–10^{-4} $cm^2\ V^{-1}\ s^{-1}$),[82,131] the presence of pre-existing silver-rich domains was proposed[176] and investigated by Raman spectroscopy.[177] The need for an electroforming process prior to normal device operation was also demonstrated.[178,179] The generally accepted idea is that a first filament is formed during a first cycle operating at a voltage larger than the 'normal' switching voltage. This filament is not completely dissolved during the reverse process and therefore it will require a lower voltage and shorter time for this filament to be restored in further cycles. Conductive atomic force microscopy was used to prove this concept.[178,179] Finally, while a dendritic type of filament growth was sometimes proposed, a growth of the filament with a conical shape is probably more realistic as shown in the case of copper sulphide layer by Kaeriyama *et al.*[180]

7.7.3 Sensors

Environmental control is becoming a major requirement. It implies constant analysis of natural and industrial waters. While laboratory analysis is very reliable and precise, its cost is very high and the need for sampling makes the amount of checking quite low. Sensors which could be used for continuous *in-situ* measurements are therefore under active investigation. For many years, ion selective electrodes (ISE) based upon chalcogenide glasses as sensitive membranes have been considered for the detection of heavy metals (e.g., Cu^{2+}, Cd^{2+}, Pb^{2+}, Hg^{2+})[181–196] or toxic anions (e.g., S^{2-}...).[197,198] The ISE is based upon the following electrochemical cell:

reference electrode | solution || chalcogenide sensitive membrane | current collector

While the potential of the reference electrode E_r is constant, the potential E_m of the sensitive membrane depends on the activity of the ion to be detected in the solution, a_i, according to the Nernst equation: $E_m = E_0 + RT/z_i \log a_i$, with E_0 standard potential of the membrane and F Faraday constant.[199] The continuity of the electrochemical cell is required for the ISE to give a stable response.[192] The electromotive force (emf) measured between the reference electrode and the ISE can then be used to obtain the concentration of the ions, c_i, in the solution (as long as the concentration is low and therefore the approximation $a_i \sim c_i$ is valid). A typical measurement curve is shown in Fig. 7.22.[192]

Based on this principle, ISE comprising crystalline chalcogenide membranes have been developed and commercialised. Compared to these crystalline membranes, chalcogenide glasses show easier shaping, higher chemical durability and stability over time, a major advantage for continuous *in-situ* measurements. Another major advantage is the possibility for a glass to have its composition adjusted by changing the quantity or nature of its components. It is then possible to monitor the properties of the membrane and therefore the performance of the sensors. Generally, the ion to be detected is one of the glass components.[181] On the other hand, a membrane, being sensitive to any ion that it contains, any other type of potentially interfering ions is

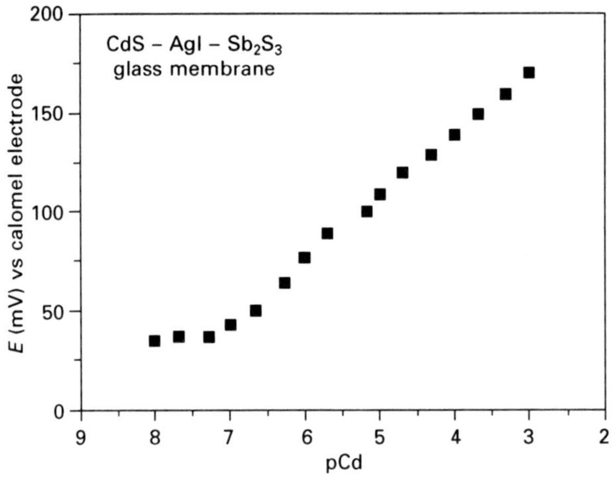

7.22 Cd^{2+} ion response of a CdS-AgI-Sb_2S_3 microsensor in $Cd(NO_3)_2$ solutions. The sensitive membrane is a thin film elaborated by thermal evaporation of a 3CdS-97(AgI-Sb_2S_3) glass (redrawn from Ref. 192).

prohibited in the glass composition. There is one major exception to this rule: most of the membranes contain a certain amount of silver ions. Their presence helps in increasing the conductivity of the membrane and has a drastic positive influence on both the response time and stability.[184,200] While an ion-chemical exchange process is the simplest mechanism to explain how an ISE works, the formation of a modified active layer at the membrane surface has been proposed in the case of chalcogenide glass membranes, in particular to account for the conditioning time of several hours required for a chalcogenide membrane to get its sensing properties.[184,185,201,202]

From the pioneering work to present, many different chalcogenide glasses have been investigated, different types of sensors, mainly ISE but also ISFET comprising chalcogenide glasses either in bulk or film forms, have been studied,[55,203,204] with the most sophisticated systems proposing arrays of sensors to measure several ions simultaneously.[55,205–207]

7.8 Conclusion

A review of the work carried out over the years to prepare ion conducting chalcogenide glasses and to gain an insight into the ion dynamics in these materials has been presented in this chapter. Techniques for preparing and characterizing ion mobility in theses glasses have been described. It has been shown that, while these glasses show higher conductivity than their oxide counterparts, they exhibit similar intriguing features: exponential increase of conductivity with modifier content, mixed alkali effect, mixed glass former effect.

Models proposed to account for ion diffusion in glasses and experimental investigation carried out for their validation have been described. Finally, it has been shown that several applications based upon the use of these glasses have been proposed. They include development of solid state batteries, development of ionic memories for data storage and development of sensors for environmental control.

7.9 References

1. S. Maruno, T. Yamada, M. Noda and Y. Kondo, *Japanese Journal of Applied Physics* **1972**, *11*, 116–117.
2. Y. Kawamoto, N. Nagura and S. Tsuchiha, *Journal of the American Ceramic Society* **1974**, *57*, 489–491.
3. Y. Kawamoto and M. Nishida, *Journal of Non-Crystalline Solids* **1976**, *20*, 393–404.
4. Z. U. Borisova, *Glassy Semiconductors*, Plenum Press, New York, **1981**.
5. P. Patel and N. J. Kreidl, *Journal of the American Ceramic Society* **1975**, *58*, 263–268.
6. B. Barrau, A. Kone, M. Ribes, J. L. Souquet and M. Maurin, *Comptes Rendus Hebdomadaires des Seances de L'Academie des Sciences Serie C* **1978**, *287*, 43–46.
7. M. Ribes, D. Ravaine, J. L. Souquet and M. Maurin, *Revue De Chimie Minerale* **1979**, *16*, 339–348.
8. M. Ribes, B. Barrau and J. L. Souquet, *Journal of Non-Crystalline Solids* **1980**, *38–39*, 271–276.
9. J. L. Souquet, E. Robinel, B. Barrau and M. Ribes, *Solid State Ionics* **1981**, *3–4*, 317–321.
10. B. Carette, E. Robinel and M. Ribes, *Glass Technology* **1983**, *24*, 157–160.
11. E. Robinel, B. Carette and M. Ribes, *Journal of Non-Crystalline Solids* **1983**, *57*, 49–58.
12. B. Carette, M. Ribes and J. L. Souquet, *Solid State Ionics* **1983**, *9–10*, 735–737.
13. J. P. Malugani, A. Saida, A. Wasniewski and G. Robert, *Comptes Rendus Hebdomadaires des Seances de L'Academie des Sciences Serie C* **1979**, *289*, 69–71.
14. J. P. Malugani and G. Robert, *Comptes Rendus Hebdomadaires des Seances de L'Academie des Sciences Serie C* **1980**, *290*, 251–253.
15. R. Mercier, J. P. Malugani, B. Fahys and G. Robert, *Solid State Ionics* **1981**, *5*, 663–666.
16. A. Levasseur, R. Olazcuaga, M. Kbala, M. Zahir and P. Hagenmuller, *Comptes Rendus de L'Academie des Sciences Serie Ii* **1981**, *293*, 563–565.
17. H. Wada, M. Menetrier, A. Levasseur and P. Hagenmuller, *Materials Research Bulletin* **1983**, *18*, 189–193.
18. M. C. R. Shastry, M. Menetrier and A. Levasseur, *Solid State Communications* **1993**, *85*, 887–889.
19. Z. M. Zhang and J. H. Kennedy, *Journal of the Electrochemical Society* **1993**, *140*, 2384–2390.

20. Z. M. Zhang, J. H. Kennedy and H. Eckert, *Journal of the American Chemical Society* **1992**, *114*, 5775–5784.
21. J. H. Kennedy, *Materials Chemistry and Physics* **1989**, *23*, 29–50.
22. J. H. Kennedy, C. Schaupp, H. Eckert and M. Ribes, *Solid State Ionics* **1991**, *45*, 21–27.
23. J. H. Kennedy and Z. M. Zhang, *Solid State Ionics* **1988**, *28*, 726–728.
24. S. W. Martin and D. R. Bloyer, *Journal of the American Ceramic Society* **1991**, *74*, 1003–1010.
25. J. Cho and S. W. Martin, *Journal of Non-Crystalline Solids* **1994**, *170*, 182–189.
26. A. E. Burns, M. Royle and S. W. Martin, *Journal of Non-Crystalline Solids* **2000**, *262*, 252–257.
27. A. Pradel and M. Ribes, *Solid State Ionics* **1986**, *18–19*, 351–355.
28. Z. M. Zhang and J. H. Kennedy, *Solid State Ionics* **1990**, *38*, 217–224.
29. A. Pradel, T. Pagnier and M. Ribes, *Solid State Ionics* **1985**, *17*, 147–154.
30. V. Michellledos, A. Pradel and M. Ribes, *European Journal of Solid State and Inorganic Chemistry* **1992**, *29*, 301–310.
31. V. K. Deshpande, A. Pradel and M. Ribes, *Materials Research Bulletin* **1988**, *23*, 379–384.
32. V. K. Deshpande, A. Pradel and M. Ribes, *Solid State Ionics* **1988**, *28*, 756–761.
33. C. Rau, P. Armand, A. Pradel, C. P. E. Varsamis, E. I. Kamitsos, D. Granier, A. Ibanez and E. Philippot, *Physical Review B* **2001**, *63*, 184–204.
34. A. Pradel, G. Taillades, M. Ribes and H. Eckert, *Journal of Non-Crystalline Solids* **1995**, *188*, 75–86.
35. N. Aotani, K. Iwamoto, K. Takada and S. Kondo, *Solid State Ionics* **1994**, *68*, 35–39.
36. M. Tatsumisago, K. Hirai, T. Hirata, M. Takahashi and T. Minami, *Solid State Ionics* **1996**, *86–88*, 487–490.
37. A. Hayashi, S. Hama, H. Morimoto, M. Tatsumisago and T. Minami, *Journal of the American Ceramic Society* **2001**, *84*, 477–479.
38. T. Minami, A. Hayashi and M. Tatsumisago, *Solid State Ionics* **2006**, *177*, 2715–2720.
39. Y. Kim and S. W. Martin, *Solid State Ionics* **2006**, *177*, 2881–2887.
40. K. Minami, A. Hayashi, S. Ujiie and M. Tatsumisago, *Solid State Ionics* **2011**, *192*, 122–125.
41. K. Minami, A. Hayashi and M. Tatsumisago, *Solid State Ionics* **2010**, *181*, 1505–1509.
42. S. S. Berbano, I. Seo, C. M. Bischoff, K. E. Schuller and S. W. Martin, *Journal of Non-Crystalline Solids* **2012**, *358*, 93–98.
43. F. Mizuno, T. Ohtomo, A. Hayashi, K. Tadanaga and M. Tatsumisago, *Solid State Ionics* **2006**, *177*, 2753–2757.
44. A. Hayashi, K. Minami and M. Tatsumisago, *Journal of Solid State Electrochemistry* **2010**, *14*, 1761–1767.
45. M. Ribes, *Proc. of the International Seminar on Soid State Ionic Devices* (Eds B. Chowdari and S. Radhakrishna) **1988**, 135.
46. R. Creus, J. Sarradin and M. Ribes, *Solid State Ionics* **1992**, *53*, 641–646.
47. R. Creus, J. Sarradin, R. Astier, A. Pradel and M. Ribes, *Materials Science and Engineering B – Solid State Materials for Advanced Technology* **1989**, *3*, 109–112.

48. I. Seo and S. W. Martin, *Journal of the Electrochemical Society* **2011**, *158*, A465–A470.
49. S. D. Jones and A. Jr, *Solid State Ionics* **1992**, *53*, 628–634.
50. T. Wagner, M. Frumar and V. Suskova, *Journal of Non-Crystalline Solids* **1991**, *128*, 197–207.
51. M. Frumar and T. Wagner, *Current Opinion in Solid State & Materials Science* **2003**, *7*, 117–126.
52. D. Brauhaus, C. Schindler, U. Bottger and R. Waser, *Thin Solid Films* **2008**, *516*, 1223–1226.
53. M. Mitkova, M. N. Kozicki, H. C. Kim and T. L. Alford, *Journal of Non-Crystalline Solids* **2004**, *338*, 552–556.
54. M. Mitkova, M. N. Kozicki, H. C. Kim and T. L. Alford, *Thin Solid Films* **2004**, *449*, 248–253.
55. M. J. Schoning and J. P. Kloock, *Electroanalysis* **2007**, *19*, 2029–2038.
56. M. Frumar, B. Frumarova, P. Nemec, T. Wagner, J. Jedelsky and M. Hrdlicka, *Journal of Non-Crystalline Solids* **2006**, *352*, 544–561.
57. M. Erazu, J. Rocca, M. Fontana, A. Urena, B. Arcondo and A. Pradel, *Journal of Alloys and Compounds* **2010**, *495*, 642–645.
58. A. Sakuda, A. Hayashi, S. Hama and M. Tatsumisago, *Journal of the American Ceramic Society* **2010**, *93*, 765–768.
59. M. Ribes, Cros B., Julien P., *SPIE: Optical microlithographic Technology for Integrated Circuit Fabrication and Inspection* (Eds H.L. Stover and S. Wittecoek **1987**, *811*, 202.
60. P. G. Huggett, K. Frick and H. W. Lehmann, *Applied Physics Letters* **1983**, *42*, 592–594.
61. *Proceedings of the Symposium on Inorganic Resist Systems, vol. 82–9 (Eds D. Doane and A. Heller) The Electrochemical Society*, **1982**.
62. K. Funke, R. D. Banhatti, D. M. Laughman, L. G. Badr, M. Mutke, A. Santic, W. Wrobel, E. M. Fellberg and C. Biermann, *Zeitschrift für Physikalische Chemie – International Journal of Research in Physical Chemistry & Chemical Physics* **2010**, *224*, 1891–1950.
63. R. Belin, G. Taillades, A. Pradel and M. Ribes, *Solid State Ionics* **2000**, *136*, 1025–1029.
64. B. Durand, G. Taillades, A. Pradel, M. Ribes, J. C. Badot and N. Belhadjtahar, *Journal of Non-Crystalline Solids* **1994**, *172*, 1306–1314.
65. M. Cutroni, A. Mandanici, A. Piccolo, C. Fanggao, G. A. Saunders and P. Mustarelli, *Philosophical Magazine B* **1996**, *73*, 349–365.
66. A. A. Piarristeguy, M. Ramonda, N. Frolet, M. Ribes and A. Pradel, *Solid State Ionics* **2010**, *181*, 1205–1208.
67. J. Kruempelmann, M. Balabajew, M. Gellert and B. Roling, *Solid State Ionics* **2011**, *198*, 16–21.
68. E. Robinel, A. Kone, M. J. Duclot and J. L. Souquet, *Journal of Non-Crystalline Solids* **1983**, *57*, 59–70.
69. M. Ribes, Unpublished result.
70. G. H. Frischat, Mass transport in Solids, *Angewandte Chemie International Edition*, 1974, 13(6), 384–398.
71. E. Bychkov, V. Tsegelnik, Y. Vlasov, A. Pradel and M. Ribes, *Journal of Non-Crystalline Solids* **1996**, *208*, 1–20.
72. A. Bolotov, E. Bychkov, Y. Gavrilov, Y. Grushko, A. Pradel, M. Ribes, V. Tsegelnik and Y. Vlasov, *Solid State Ionics* **1998**, *113*, 697–701.

73. E. Bychkov, A. Bolotov, Y. Grushko, Y. Vlasov and G. Wortmann, *Solid State Ionics* **1996**, *90*, 289–294.
74. Y. G. Vlasov, E. A. Bychkov, A. M. Bolotov, V. S. Tsegelnik and Y. A. Gavrilov, *Radiochemistry* **1996**, *38*, 316–321.
75. A. Kone, M. Armand and J. L. Souquet, *Electrochimica Acta* **1982**, *27*, 653–656.
76. J. H. Kennedy and Y. Yang, *Journal of the Electrochemical Society* **1986**, *133*, 2437–2438.
77. S. Sahami, S. W. Shea and J. H. Kennedy, *Journal of the Electrochemical Society* **1985**, *132*, 985–986.
78. S. J. Visco, P. J. Spellane and J. H. Kennedy, *Journal of the Electrochemical Society* **1985**, *132*, 1766–1770.
79. A. Pradel and M. Ribes, *Journal of Non-Crystalline Solids* **1994**, *172*, 1315–1323.
80. A. Pradel and M. Ribes, *Materials Chemistry and Physics* **1989**, *23*, 121–142.
81. C. Rau, P. Armand, A. Pradel, C. Varsamis, E. Kamitsos, D. Granier, A. Ibanez and E. Philippot, *Physical Review B* **2001**, *63*, 184–204.
82. M. A. Urena, A. A. Piarristeguy, M. Fontana and B. Arcondo, *Solid State Ionics* **2005**, *176*, 505–512.
83. A. Pradel, N. Kuwata and M. Ribes, *Journal of Physics – Condensed Matter* **2003**, *15*, S1561–S1571.
84. E. Bychkov, A. Bychkov, A. Pradel and M. Ribes, *Solid State Ionics* **1998**, *113*, 691–695.
85. A. Piarristeguy, J. M. C. Garrido, M. A. Urena, M. Fontana and B. Arcondo, *Journal of Non-Crystalline Solids* **2007**, *353*, 3314–3317.
86. V. Balan, A. Piarristeguy, M. Ramonda, A. Pradel and M. Ribes, *Journal of Optoelectronics and Advanced Materials* **2006**, *8*, 2112–2116.
87. A. Piarristeguy, M. Ramonda, N. Kuwata, A. Pradel and M. Ribes, *Solid State Ionics* **2006**, *177*, 3157–3160.
88. A. Piarristeguy, M. Ramonda, A. Urena, A. Pradel and M. Ribes, *Journal of Non-Crystalline Solids* **2007**, *353*, 1261–1263.
89. Y. Drugov, V. Tsegelnik, A. Bolotov, Y. Vlasov and E. Bychkov, *Solid State Ionics* **2000**, *136*, 1091–1096.
90. E. Bychkov, *Solid State Ionics* **2000**, *136*, 1111–1118.
91. M. Ribes, E. Bychkov and A. Pradel, *Journal of Optoelectronics and Advanced Materials* **2001**, *3*, 665–674.
92. E. Bychkov, D. L. Price, C. J. Benmore and A. C. Hannon, *Solid State Ionics* **2002**, *154*, 349–359.
93. E. Bychkov, *Solid State Ionics* **2009**, *180*, 510–516.
94. E. Bychkov and D. L. Price, *Solid State Ionics* **2000**, *136*, 1041–1048.
95. E. Bychkov, D. L. Price and A. Lapp, *Journal of Non-Crystalline Solids* **2001**, *293*, 211–219.
96. A. V. Kolobov and S. R. Elliott, *Advances in Physics* **1991**, *40*, 625–684.
97. J. P. Malugani, B. Fahys, R. Mercier, G. Robert, J. P. Duchange, S. Baudry, M. Broussely and J. P. Gabano, *Solid State Ionics* **1983**, *9–10*, 659–665.
98. A. Levasseur, *Materials for Solid State Batteries* (Eds B. V. R. Chowdari and S. Radhakrishna) World Scientific, Singapore, **1986**.
99. J. P. Malugani, G. Robert and R. Mercier, *Materials Research Bulletin* **1980**, *15*, 715–720.

100. T. Minami, H. Nambu and M. Tanaka, *Journal of the American Ceramic Society* **1977**, *60*, 467–469.
101. A. Schiraldi, *Electrochimica Acta* **1978**, *23*, 1039–1043.
102. S. Kondo, K. Takada and Y. Yamamura, *Solid State Ionics* **1992**, *53*, 1183–1186.
103. A. Hayashi, *Solid State Ionics* **2002**, *148*, 381–389.
104. K. Hirai, *Solid State Ionics* **1995**, *78*, 269–273.
105. D. E. Day, *Journal of Non-Crystalline Solids* **1976**, *21*, 343–372.
106. M. D. Ingram, *Glastechnische Berichte – Glass Science and Technology* **1994**, *67*, 151–155.
107. R. M. Hakim and D. R. Uhlmann, *Physics and Chemistry of Glasses* **1967**, *8*, 174–177.
108. R. Chen, R. Yang, B. Durand, A. Pradel and M. Ribes, *Solid State Ionics* **1992**, *53*, 1194–1199.
109. P. W. S. K. Bandaranayake, C. T. Imrie and M. D. Ingram, *Physical Chemistry Chemical Physics* **2002**, *4*, 3209–3213.
110. S. Voss, A. W. Imre and H. Mehrer, *Physical Chemistry Chemical Physics* **2004**, *6*, 3669–3675.
111. M. P. Thomas, N. L. Peterson and E. Hutchinson, *Journal of the American Ceramic Society* **1985**, *68*, 99–104.
112. A. Vegiri, C. P. E. Varsamis and E. I. Kamitsos, *Physical Review B* **2009**, *80*, 184202.
113. M. D. Ingram and A. H. J. Robertson, *Solid State Ionics* **1997**, *94*, 49–54.
114. T. Tsuchiya and T. Moriya, *Journal of Non-Crystalline Solids* **1980**, *38–39*, 323–328.
115. A. Magistris, G. Chiodelli and M. Villa, *Journal of Power Sources* **1985**, *14*, 87–91.
116. D. Zielniok, C. Cramer and H. Eckert, *Chemistry of Materials* **2007**, *19*, 3162–3170.
117. S. Elbers, W. Strojek, L. Koudelka and H. Eckert, *Solid State Nucl Magn Reson* **2005**, *27*, 65–76.
118. B. Raguenet, G. Tricot, G. Silly, M. Ribes and A. Pradel, *Journal of Materials Chemistry* **2011**, *21*, 17693–17704.
119. A. Pradel, C. Rau, D. Bittencourt, P. Armand, E. Philippot and M. Ribes, *Chemistry of Materials* **1998**, *10*, 2162–2166.
120. J. H. Kennedy and Z. M. Zhang, *Journal of the Electrochemical Society* **1989**, *136*, 2441–2443.
121. M. J. Haynes, C. Bischoff, T. Kaufmann and S. W. Martin, *Physics and Chemistry of Glasses – European Journal of Glass Science and Technology Part B* **2009**, *50*, 144–148.
122. D. Y. Seung, A. Levasseur and M. Couzi, *Physics and Chemistry of Glasses* **1998**, *39*, 295–300.
123. Y. Kim, J. Saienga and S. W. Martin, *J Phys Chem B* **2006**, *110*, 16318–16325.
124. J. Kincs and S. W. Martin, *Phys Rev Lett* **1996**, *76*, 70–73.
125. M. Ribes, G. Taillades and A. Pradel, *Solid State Ionics* **1998**, *105*, 159–165.
126. F. Mizuno, J. P. Belieres, N. Kuwata, A. Pradel, M. Ribes and C. A. Angell, *Journal of Non-Crystalline Solids* **2006**, *352*, 5147–5155.
127. O. L. Anderson and D. A. Stuart, *J. Am. Ceram. Soc.* **1954**, *37*, 573–580.

128. D. Ravaine and J. L. Souquet, *Physics and Chemistry of Glasses* **1977**, *18*, 27–31.
129. J. C. Reggiani, J. P. Malugani and J. Bernard, *Journal de Chimie Physique et de Physico-Chimie Biologique* **1978**, *75*, 849–854.
130. A. Levasseur, J. C. Brethous, M. Kbala and P. Hagenmuller, *Solid State Ionics* **1981**, *5*, 651–654.
131. J. L. Souquet, M. L. Nascimento and A. C. Rodrigues, *J Chem Phys* **2010**, *132*, 034704.
132. A. C. Rodrigues, M. L. Nascimento, C. B. Bragatto and J. L. Souquet, *J Chem Phys* **2011**, *135*, 234504.
133. B. Barrau, J.-M. Latour, D. Ravaine and M. Ribes, *Silicates Industriels* **1979**, *12*, 275.
134. P. Maass, A. Bunde and M. D. Ingram, *Phys Rev Lett* **1992**, *68*, 3064–3067.
135. A. Bunde, M. D. Ingram and P. Maass, *Journal of Non-Crystalline Solids* **1994**, *172*, 1222–1236.
136. A. K. Jonscher, *Nature* **1977**, *267*, 673–679.
137. A. Burns, G. D. Chryssikos, E. Tombari, R. H. Cole and W. M. Risen, *Physics and Chemistry of Glasses* **1989**, *30*, 264–270.
138. C. Cramer and K. Funke, *Berichte der Bunsen-Gesellschaft – Physical Chemistry Chemical Physics* **1992**, *96*, 1725–1727.
139. A. Pradel, G. Taillades, C. Cramer and M. Ribes, *Solid State Ionics* **1998**, *105*, 139–148.
140. C. Cramer, S. Bruckner, Y. Gao, K. Funke, R. Belin, G. Taillades and A. Pradel, *Journal of Non-Crystalline Solids* **2002**, *307*, 905–912.
141. K. Funke, R. D. Banhatti, S. Bruckner, C. Cramer, C. Krieger, A. Mandanici, C. Martiny and I. Ross, *Physical Chemistry Chemical Physics* **2002**, *4*, 3155–3167.
142. K. Funke and R. D. Banhatti, *Solid State Sciences* **2008**, *10*, 790–803.
143. W. K. Lee, J. F. Liu and A. S. Nowick, *Phys Rev Lett* **1991**, *67*, 1559–1561.
144. A. S. Nowick, A. V. Vaysleyb and W. Liu, *Solid State Ionics* **1998**, *105*, 121–128.
145. W. Dieterich and P. Maass, *Chemical Physics* **2002**, *284*, 439–467.
146. K. L. Ngai, *Journal de Physique IV* **1992**, *2*, 61–73.
147. K. L. Ngai and K. Y. Tsang, *Phys Rev E Stat Phys Plasmas Fluids Relat Interdiscip Topics* **1999**, *60*, 4511–4517.
148. K. Y. Tsang and K. L. Ngai, *Phys Rev E Stat Phys Plasmas Fluids Relat Interdiscip Topics* **1996**, *54*, R3067–R3070.
149. K. L. Ngai and R. W. Rendell, *Journal of Non-Crystalline Solids* **1991**, *131*, 233–237.
150. K. L. Ngai and U. Strom, *Phys Rev B Condens Matter* **1988**, *38*, 10350–10356.
151. K. Funke, *Progress in Solid State Chemistry* **1993**, *22*, 111–195.
152. D. Knodler, P. Pendzig and W. Dieterich, *Solid State Ionics* **1996**, *86–88*, 29–39.
153. P. Pendzig and W. Dieterich, *Solid State Ionics* **1998**, *105*, 209–216.
154. K. Funke and I. Riess, *Zeitschrift für Physikalische Chemie Neue Folge* **1984**, *140*, 217–232.
155. A. Bunde, K. Funke and M. D. Ingram, *Solid State Ionics* **1996**, *86–88*, 1311–1317.
156. K. Funke and D. Wilmer, *Solid State Ionics* **2000**, *136*, 1329–1333.

157. K. Funke and R. D. Banhatti, *Solid State Ionics* **2004**, *169*, 1–8.
158. K. Funke, R. D. Banhatti and C. Cramer, *Phys Chem Chem Phys* **2005**, *7*, 157–165.
159. K. Funke and R. D. Banhatti, *Solid State Ionics* **2006**, *177*, 1551–1557.
160. K. Takada, N. Aotani, K. Iwamoto and S. Kondo, *Solid State Ionics* **1996**, *86–88*, 877–882.
161. J. R. Akridge and H. Vourlis, *Solid State Ionics* **1986**, *18–19*, 1082–1087.
162. J. R. Akridge and H. Vourlis, *Solid State Ionics* **1988**, *28*, 841–846.
163. S. D. Jones and J. R. Akridge, *Journal of Power Sources* **1993**, *44*, 505–513.
164. M. N. Kozicki, M. Yun, S. J. Yang, J. P. Aberouette and J. P. Bird, *Superlattices and Microstructures* **2000**, *27*, 485–488.
165. M. Mitkova and M. N. Kozicki, *Journal of Non-Crystalline Solids* **2002**, *299*, 1023–1027.
166. M. N. Kozicki, M. Mitkova, M. Park, M. Balakrishnan and C. Gopalan, *Superlattices and Microstructures* **2003**, *34*, 459–465.
167. M. N. Kozicki, M. Park and M. Mitkova, *IEEE Transactions on Nanotechnology* **2005**, *4*, 331–338.
168. R. Waser, R. Dittmann, G. Staikov and K. Szot, *Advanced Materials* **2009**, *21*, 2632–2663.
169. C. Schindler, *http://darwin.bth.rwth-aachen.de/opus3/volltexte/2009/2704/pdf/Schindler_Christina.pdf* **2009**.
170. R. Bruchhaus, M. Honal, R. Symanczyk and M. Kund, *Journal of the Electrochemical Society* **2009**, *156*, H729–H733.
171. I. Valov, R. Waser, J. R. Jameson and M. N. Kozicki, *Nanotechnology* **2011**, *22*, 254003.
172. F. Wang and X. L. Wu, *Non-volatile Memory Devices Based on Chalcogenide Materials*, IEEE, New York, **2009**.
173. M. N. Kozicki, M. Balakrishnan, C. Gopalan, C. Ratnakumar and M. Mitkova, *Programmable Metallization Cell Memory based on Ag-Ge-S and Cu-Ge-S Solid Electrolytes*, **2005**, 83–89.
174. R. Symanczyk, R. Bruchhaus, R. Dittrich and M. Kund, *IEEE Electron Device Letters* **2009**, *30*, 876–878.
175. C. Schindler, X. Guo, A. Besmehn and R. Waser, *Zeitschrift für Physikalische Chemie – International Journal of Research in Physical Chemistry & Chemical Physics* **2007**, *221*, 1469–1478.
176. M. N. Kozicki and M. Mitkova, *Journal of Non-Crystalline Solids* **2006**, *352*, 567–577.
177. M. Mitkova, M. N. Kozicki, H. C. Kim and T. L. Alford, *Journal of Non-Crystalline Solids* **2006**, *352*, 1986–1990.
178. C. Schindler, I. Valov and R. Waser, *Phys Chem Chem Phys* **2009**, *11*, 5974–5979.
179. A. Pradel, N. Frolet, M. Ramonda, A. Piarristeguy and M. Ribes, *Physica Status Solidi A–Applications and Materials Science* **2011**, *208*, 2303–2308.
180. S. Kaeriyama, T. Sakamoto, H. Sunamura, M. Mizuno, H. Kawaura, T. Hasegawa, K. Terabe, T. Nakayama and M. Aono, *IEEE Journal of Solid-State Circuits* **2005**, *40*, 168–176.
181. C. T. Baker and I. Trachten, *Journal of the Electrochemical Society* **1971**, *118*, 571–576.
182. N. Tohge and M. Tanaka, *Journal of Non-Crystalline Solids* **1986**, *80*, 550–556.

183. Y. G. Vlasov, E. A. Bychkov, A. D. Safarov, P. P. Antonov and M. S. Miloshova, *Journal of Analytical Chemistry of the USSR* **1985**, *40*, 1131–1138.
184. Y. G. Vlasov and E. A. Bychkov, *Ion-Selective Electrode Reviews* **1987**, *9*, 5–93.
185. Y. G. Vlasov, E. A. Bychkov and A. M. Medvedev, *Analytica Chimica Acta* **1986**, *185*, 137–158.
186. Y. G. Vlasov, E. A. Bychkov and A. V. Legin, *Talanta* **1994**, *41*, 1059–1063.
187. A. V. Legin, Y. G. Vlasov, A. M. Rudnitskaya and E. A. Bychkov, *Sensors and Actuators B – Chemical* **1996**, *34*, 456–461.
188. M. Miloshova, E. Bychkov, V. Tsegelnik, V. Strykanov, H. Klewe-Nebenius, M. Bruns, W. Hoffmann, P. Papet, J. Sarradin, A. Pradel and M. Ribes, *Sensors and Actuators B – Chemical* **1999**, *57*, 171–178.
189. B. Pejcic and R. De Marco, *Applied Surface Science* **2004**, *228*, 378–400.
190. R. De Marco, B. Pejcic, K. Prince and A. van Riessen, *Analyst* **2003**, *128*, 742–749.
191. R. De Marco, B. Pejcic and X. D. Wang, *Laboratory Robotics and Automation* **1999**, *11*, 284–288.
192. A. Guessous, J. Sarradin, P. Papet, K. Elkacemi, S. Belcadi, A. Pradel and M. Ribes, *Sensors and Actuators B – Chemical* **1998**, *53*, 13–18.
193. A. Guessous, P. Papet, J. Sarradin and M. Ribes, *Sensors and Actuators B – Chemical* **1995**, *24*, 296–299.
194. F. O. Mear, M. Essi, M. F. Guimon and A. Pradel, *Chalcogenide Letters* **2008**, *5*, 117–124.
195. F. O. Mear, M. Essi, P. Sistat, M. F. Guimon, D. Gonbeau and A. Pradel, *Applied Surface Science* **2009**, *255*, 6607–6611.
196. M. Kassem, PhD Thesis, ULCO, Dunkerque **2010**.
197. C. Cali, G. Taillades, A. Pradel and M. Ribes, *Sensors and Actuators B – Chemical* **2001**, *76*, 560–564.
198. M. Miloshova, D. Baltes and E. Bychkov, *Water Sci Technol* **2003**, *47*, 135–140.
199. W. E. Morf, *The Principles of Ion-selective Electrodes and of Membrane Transport*, Akademiai Kiado, Budapest, **1981**.
200. E. Bychkov, Y. Tveryanovich and Y. Vlasov, *Semiconducting Chalcogenide Glass III: Applications of Chalcogenide Glasses* **2004**, *80*, 103–168.
201. Y. G. Vlasov and E. A. Bychkov, *Sensors and Actuators* **1987**, *12*, 275–283.
202. C. Cali, D. Foix, G. Taillades, E. Siebert, D. Gonbeau, A. Pradel and M. Ribes, *Materials Science & Engineering C – Biomimetic and Supramolecular Systems* **2002**, *21*, 3–8.
203. G. Taillades, O. Valls, A. Bratov, C. Dominguez, A. Pradel and M. Ribes, *Sensors and Actuators B – Chemical* **1999**, *59*, 123–127.
204. J. P. Kloock, L. Moreno, A. Bratov, S. Huachupoma, J. Xu, T. Wagner, T. Yoshinobu, Y. Ermolenko, Y. G. Vlasov and M. J. Schoning, *Sensors and Actuators B – Chemical* **2006**, *118*, 149–155.
205. Y. G. Mourzina, J. Schubert, W. Zander, A. Legin, Y. G. Vlasov, H. Luth and M. J. Schoning, *Electrochimica Acta* **2001**, *47*, 251–258.
206. J. P. Kloock, Y. G. Mourzina, Y. Ermolenko, T. Doll, J. Schubert and M. J. Schoning, *Sensors* **2004**, *4*, 156–162.
207. A. Legin, A. Rudnitskaya, Y. Vlasov, C. Di Natale, E. Mazzone and A. D'Amico, *Electroanalysis* **1999**, *11*, 814–820.

8
Physical ageing of chalcogenide glasses

O. SHPOTYUK, Jan Dlugosz University of Częstochowa, Poland,
R. GOLOVCHAK, Austin Peay State University, USA and
A. KOZDRAS, Opole University of Technology, Poland

DOI: 10.1533/9780857093561.1.209

Abstract: This chapter covers state of the art in physical ageing of chalcogenide glasses. The thermodynamic origin of this phenomenon, experimental possibilities for its investigation, up-to-date results, general phenomenology and observed regularities are critically reviewed. The influence of various external factors, such as elevated temperatures, light exposure, high-energy irradiation, on the physical ageing effect in chalcogenide glasses are considered and analyzed from the constraints theory point of view. Possible mechanisms of structural relaxation caused by these factors are discussed thoroughly.

Key words: physical ageing, chalcogenide glass, structural relaxation, constraints theory, glass transition.

8.1 Introduction

8.1.1 Thermodynamic origin of physical ageing in disordered solids

If the cooling rate of glass-forming liquid $q = dT/dt$ is high enough to prevent the formation of crystal seeds and their further growth, the thermodynamic system emerges into a supercooled liquid regime (Feltz, 1993; Varshneya, 2006). Usually, viscosity of the supercooled liquid still allows the internal parameters of the considered thermodynamic system to follow their equilibrium values determined by a current bath temperature (Feltz, 1993; Varshneya, 2006; Donth, 2001; Ngai, 2011). The corresponding structural changes involve breaking of bonds, their switching, conformation of steady structural units, changes in their dimensions, as well as translational, vibrational and rotational movements. At further cooling, the time needed by the considered thermodynamic system to achieve a new equilibrium state at a particular bath temperature becomes greater (because of temperature dependence of the viscosity) and at some point (the glass transition temperature) it exceeds the time allowed by a cooling rate $q = dT/dt$ to attain equilibrium. As a result, the system departs from equilibrium and enters the glassy regime, being stuck in one of the local potential minima of the energy/enthalpy landscape (Ngai, 2011; Angell *et al.*, 2000). The internal parameters that describe a thermodynamic system in equilibrium are no longer in time with the cooling rate. As a consequence, the achieved glassy state is characterized by higher values of configurational entropy S_c, enthalpy H and free volume V available for relaxation, in comparison to what can be expected at a certain temperature in equilibrium (Feltz, 1993; Varshneya, 2006; Donth, 2001; Ngai, 2011; Angell *et al.*, 2000; Zallen, 1983). Contrary to liquid and supercooled liquid regimes, in a glassy state the bond breaking or switching mechanisms are not expected to be significant, the only possible structural perturbations seem to rely on the conformations of structural units.

Figure 8.1 illustrates the above glass formation paradigm. Cooling of the high-temperature liquid (path AB) through the melting point T_m can result in two different products. If the cooling rate permits formation of crystal seeds and their further growth, we have a crystallization process (path BC) resulting in the thermodynamic equilibrium of crystalline state (path CD). If the cooling rate is high enough at point B, crystallization does not occur and we achieve thermodynamically equilibrium supercooled liquid states (path BG). At further cooling, the viscosity increases dramatically (up to ~10^{12}–10^{13} Poises) at some critical point F and the system departs from the equilibrium line BG forming a glass (path FI). The corresponding temperature is called the glass transition temperature T_F (coincides with the fictive temperature, as defined by Moynihan *et al.*, 1974). Strictly speaking, it does not coincide

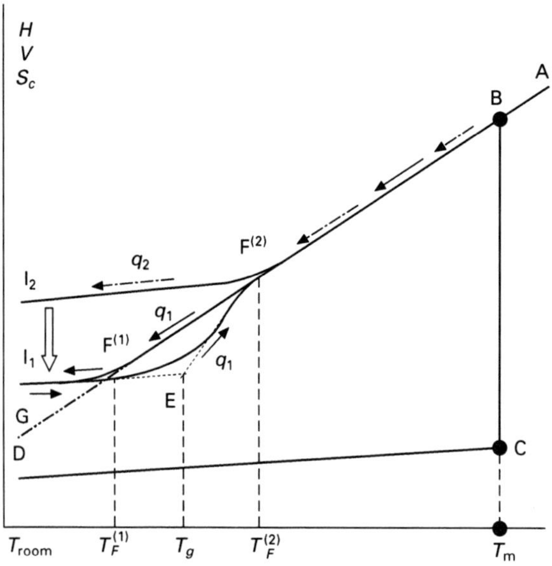

8.1 Schematic diagram, showing temperature behavior of enthalpy (*H*), configurational entropy (S_c), and free volume available for relaxation (*V*) during glass formation and reheating of the glass with rates $q_1 < q_2$ (see text for more details).

with the softening point, determined from the heating curve (path I_1EB, for example) and often used in the literature as the glass transition temperature T_g. The reason is the hysteresis, which is usually observed between heating and cooling paths in the glass transition interval (Feltz, 1993; Varshneya, 2006; Donth, 2001; Ngai, 2011; Angell *et al.*, 2000; Zallen, 1983; Moynihan *et al.*, 1974). The slower the cooling rate, the lower the T_F value and the closer to thermodynamic equilibrium would be the final glassy state. The infinite slow cooling will result in the formation of an ideal glass at the Kauzmann temperature T_K. However, the real glasses (I_1 or I_2 in Fig. 8.1) obtained by conventional melt quenching always are in a thermodynamic non-equilibrium metastable state, characterized by an excess of configurational entropy ΔS_c, enthalpy ΔH and free volume ΔV available for relaxation over the crystalline state (line CD) or extrapolated states of supercooled liquid (line BG). Therefore, thermodynamically driven forces tend such system with time towards thermodynamic equilibrium (transition from I_2 to I_1 state in Fig. 8.1, for example), which is represented by supercooled liquid states extrapolated into the low-temperature region (line BG). This occurs via structural relaxation, leading to spontaneous densification of glass. The related processes are called *physical ageing* (Struik, 1978; Golovchak *et al.*, 2006a; Saiter *et al.*, 2005; Shieh and La Course, 1993; Nemilov, 2000; Kurchan, 2005; Hutchinson, 1995; Drozdov, 1999).

We will distinguish here the *natural physical ageing* occurring at room temperature and *induced physical ageing* occurring under external influences (light, high-energy radiation, elevated temperatures, etc.), even if such classification is somewhat artificial. Note, that ageing due to the chemical reactions with the atmosphere (chemical ageing), crystallization, and phase separation are not included in this categorization.

Physical ageing is known to be an important concern for organic polymers, because it results in time dependence of important properties, such as density, elastic modulus, brittleness, permeability, impact strength, fracture energy, deformation, etc. (Struik, 1978; Golovchak *et al.*, 2006a; Saiter *et al.*, 2005; Shieh and La Course, 1993; Nemilov, 2000; Kurchan, 2005; Hutchinson, 1995; Drozdov, 1999; Churbanov *et al.*, 2002). However, up to now there seems to be no consensus on the microstructural mechanisms of this effect. The typical difficulties with polymers are due to the complicated nature of most monomer units, the large variety of their possible conformations, and the high concentration of impurities ultimately brought into their polymeric matrices due to general peculiarities of preparation procedures (catalysts, etc.). In this sense, the investigation of physical ageing in more structurally simple inorganic polymers like chalcogenide glasses can be considered as a model to reveal general regularities and microstructural mechanisms. These materials can be obtained of a high purity (Feltz, 1993; Borisova, 1981), their networks are built of covalent bonds and structural parameters have been evaluated for many years, even if some current controversies remain (Feltz, 1993; Borisova, 1981; Fairman and Ushkov, 2004).

On the other hand, the effect of physical ageing in silicate glasses was first documented by J.P. Joule (Joule, 1867; 1884). To date, the kinetics of changes in physical-chemical properties of many silicate glasses has been studied on timescales from several hours to several decades (Nemilov, 2000, 2001). As such, silicate glasses are characterized by an over-constrained network, whose ageing kinetics are very slow under normal conditions. Therefore, the physical ageing in these materials under normal conditions is far from complete, which can be estimated using the fictive temperature of the aged glasses. As a consequence, the exponential law, which is assumed for ageing kinetics in silicate glasses under ambient conditions (Nemilov, 2000, 2001), describes only a part of the relaxation process. To develop a complete picture and understand the fundamental nature of this phenomenon, the glasses with fast kinetics of physical ageing should be studied, such as the chalcogenide glasses (ChG). These materials have lower glass transition temperatures and faster kinetics of physical ageing under ambient conditions in comparison to silicate glasses, which allows the observation of the almost complete picture of physical ageing within acceptable experimental timescales (~tens of years) (Golovchak *et al.*, 2006a; Saiter *et al.*, 2005; Shieh and La Course, 1993). Furthermore, extended glass-forming regions of typical chalcogenide systems permit wide ranges of fragility and connectivity within the same glass family (Feltz, 1993; Borisova, 1981; Vinogradova, 1984), that is nearly impossible to realize in silicate glasses.

8.2 Experimental characterization of physical ageing in glasses using thermal analysis

Structural relaxation of polymeric glass network associated with physical ageing usually results in a characteristic endothermic peak superimposed on the endothermic step of experimental heating curves recorded by differential scanning calorimetry (DSC) or temperature-modulated DSC (TMDSC) in the vicinity of glass-to-supercooled liquid transition (Ngai, 2011; Angell *et al.*, 2000; Struik, 1978; Golovchak *et al.*, 2006a; Saiter *et al.*, 2005; Shieh *and* LaCourse, 1993). It is associated with regaining of the entropy/enthalpy lost during natural storage (Golovchak *et al.*, 2006a; Saiter *et al.*, 2005), or photo- (Lucas *et al.*, 2005, 2006) or high-energy radiation influences (Golovchak *et al.*, 2006b; 2009a). In fragile glass-formers, development of the endothermic peak as a result of physical ageing is usually accompanied by T_g changes (Golovchak *et al.*, 2006a; Saiter *et al.*, 2005). Additionally, for highly quenched (hyperquenched) glasses, the decrease in area of exothermic pre-T_g peak is observed over time (Yue, 2004). It is, presumably, associated with self-annealing under the ambient conditions of elastic strains that appear in the glass as a result of quenching.

In the case of chalcogen-rich glasses obtained by conventional quenching, the typical DSC curves and corresponding behavior of configurational enthalpy (H) in the as-prepared and aged states can be schematically represented as in Fig. 8.2. The difference in the area under this endothermic peak in DSC (or TMDSC) signal of the aged and rejuvenated ChG is directly proportional to the enthalpy losses ΔH. Along with the ΔH value, the changes in softening point T_g (usually designated as onset T_g^{onset}, midpoint T_g^{mid}, endset T_g^{end} or inflection T_g^{infl} glass transition temperatures) and partial area A (determined as shown in Fig. 8.2 taking into account an appropriate q value), as well as fictive temperature T_F could be analyzed as the main parameters describing physical ageing in ChG. Here, the A and T_g^{onset} values will be used as control parameters for the qualitative/quantitative description of physical ageing along with ΔH and T_F values (insets to Fig. 8.2). It should be noted here that the value of T_g as determined from DSC patterns in a heating mode depends significantly not only on the heating rate during DSC scans, but also on the cooling rate through the glass transition region during the preparation procedure (Moynihan *et al.*, 1974; Hutchinson, 1992). In particular, it was demonstrated that the T_g value for inorganic glasses (determined in heating mode) exhibited U-like dependence on cooling rate q, increased with high q (more than 15 K/min) and very low q (less than 1 K/min) (Hutchinson, 1992). The latter effect is attributed to annealing of the glass during slow cooling through T_g, while the former corresponds to usual T_g behavior established by Moynihan *et al.* (1974). Despite these difficulties and ambiguities in

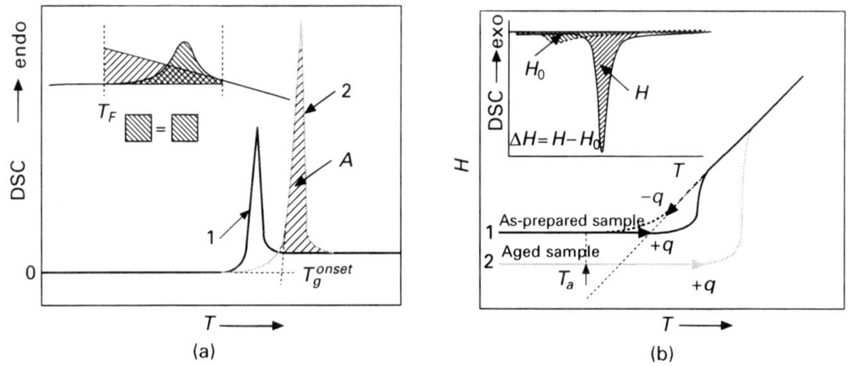

8.2 Schematic representation of DSC curves (a) and deduced from them configurational enthalpy H curves (b) for Se-rich samples: 1 – just after rejuvenation procedure, 2 – after physical ageing. The dotted line corresponds to the extrapolated glass equilibrium line of supercooled liquid. Insets show Moynihan's graphical method for T_F determination (Moynihan *et al.*, 1974) (a) and method for ΔH calculation (b) from DSC curves.

T_g determination from DSC heating measurements, we will be using this parameter in our further discussion on physical ageing.

If T_a is the annealing (ageing) temperature, the maximum energy that a glass can lose after an infinite annealing can be approximated as (Saiter *et al.*, 2005):

$$\Delta H_\infty \approx \Delta C_{p(T_g)} \, (T_g^c - T_a) \tag{8.1}$$

where $\Delta C_{p(T_g)} = C_{pl} - C_{pg}$ measured at T_g (C_{pl} is the thermal capacity in supercooled liquid state and C_{pg} is the thermal capacity in glassy state); and T_g^c is the glass transition temperature, when the glass is cooled through the transition region at a given rate. In the first approximation, it is equal to the fictive temperature T_F of the rejuvenated glass.

According to Eq. [8.1], the increase in the annealing temperature T_a leads to a decrease in ΔH_∞ values. Therefore, if T_a became very close to T_g, the enthalpy losses ΔH vanish because of negligible ΔH_∞ value at $T_a = T_g$ and the beginning of the rejuvenation process. So, the excess of enthalpy which a glass can lose during physical ageing depends considerably on the $T_g - T_a$ departure.

However, not only an excess of enthalpy determines the ability of the system to undergo relaxation; a connectivity of glass backbone should additionally be taken into account (Wang and Boolchand, 2004), which is related to the viscosity and fragility of the glass-forming liquid (Feltz, 1993; Varshneya, 2006; Ngai, 2011) and determines the time constraints of the physical ageing kinetics (Ngai, 2011; Angell *et al.*, 2000). The higher is the connectivity of glass backbone, the higher is the viscosity and more constraints for a structure to relax. On the other hand, the lower the ($T_g - T_a$) value, the faster the glassy system approaches equilibrium because of temperature dependence of time relaxation constants (Ngai, 2011; Angell *et al.*, 2000). The fictive temperature T_F of the glass approaches T_a during physical ageing (Feltz, 1993; Nemilov, 2000), such as:

$$\lim_{t \to \infty} T_F \approx T_a \tag{8.2}$$

Changes in DSC parameters caused by physical ageing are well correlated with corresponding changes in optical transmission/absorption characteristics of ChG (Golovchak *et al.*, 2010a), volumetric/density (Hach *et al.*, 1997), and stress relaxation (McEnroe and LaCourse, 1989) measurements.

8.3 Physical ageing effects in chalcogenide glasses

8.3.1 Natural physical ageing

The first comprehensive investigations of natural physical ageing in chalcogenide glasses using thermal analysis techniques appeared a short time ago (Saiter, 2001; Saiter *et al.*, 1994, 1995, 2004, 2005; Echeverria *et al.*, 2003; Lima *et al.*, 2001; Tonchev and Kasap, 2002; Cortes *et al.*, 1998; Chakravarty *et al.*, 2005; Bohmer and Angell, 1992). Long-term physical ageing in ChG was first documented by Saiter and co-authors for Se-rich Ge-Se glasses, stored more than 10 years under room conditions (Saiter *et al.*, 2005; Saiter, 2001). These authors have applied to chalcogenide glasses the protocol, which was usually used for thermal analysis of physical ageing in organic polymers (Donth, 2001; Struik, 1978; Saiter *et al.*, 2004). In particular, it was shown that the thermal history of ChG can be erased by a rejuvenation procedure like in the case of organic polymers, bringing them into a state close to the as-prepared one (Saiter *et al.*, 2005; Saiter, 2001). The rejuvenation procedure implied heating of the samples above their glass transition temperatures (T_g), waiting equilibrium there and subsequent cooling in the chosen regime. It has opened a broad possibility for studies of physical ageing phenomena using either forward or backward chronology. Short-term physical ageing (up to a year) was also reported in glassy Se (Echeverria *et al.*, 2003; Lima *et al.*, 2001) and some ChG of binary (Hach *et al.*, 1997; Tonchev and Kasap, 2002; Saiter *et al.*, 1994; Cortes *et al.*, 1998) and ternary (Chakravarty *et al.*, 2005; Saiter *et al.*, 1995; Bohmer and Angell, 1992) systems.

By definition, the natural physical ageing is a spontaneous transition of the glass into a more thermodynamically favorable state closer to the equilibrium state of supercooled liquid under ambient conditions (Struik, 1978; Hodge, 1995). It is revealed by conventional DSC through the development of strong endothermic peak and shift in the glass transition temperature, measured during heating of a glass sample through the glass-to-supercooled liquid transition domain (Struik, 1978; Golovchak *et al.*, 2006a, 2006b, 2009a, 2009b; Saiter *et al.*, 1994, 1995, 2004, 2005; Lucas *et al.*, 2005, 2006, Saiter, 2001; Echeverria *et al.*, 2003; Lima *et al.*, 2001; Tonchev and Kasap, 2002; Cortes *et al.*, 1998; Chakravarty *et al.*, 2005; Bohmer and Angell, 1992; Hodge, 1995). The greater the ageing duration, the greater is the energy loss and endothermic peak at T_g, whose area after subtracting the non-aged curve for the same material is directly proportional to the enthalpy lost during ageing (ΔH).

Typical DSC behavior during natural physical ageing of As_xSe_{100-x} and Ge_xSe_{100-x} glasses is shown in Fig. 8.3, the DSC curves being recorded with q = 5 K/min heating rate (Golovchak *et al.*, 2006a, 2009b). As expected, a large endothermic peak of structural relaxation accompanied by a shift toward higher temperatures of the apparent glass transition is observed for Se-rich ChG. Analogous behavior is also characteristic of S-rich ChG (Fig. 8.4) (Shpotyuk *et al.*, 2011). Sometimes, even double peak relaxation is observed after a long period of natural storage (Fig. 8.4(a), Ge_4Se_{96} composition) (Saiter *et al.*, 2005; Golovchak *et al.*, 2009b), which is assumed to be dependent on the quenching rate during ChG synthesis (Golovchak *et al.*, 2009b).

For a given composition, the value of $\Delta C_{p(T_g)}$ should not depend on the sample's age, which is to confirm that the glass undergoes only physical ageing, not associated with any crystallization or phase separation effects (Saiter *et al.*, 2004, 2005; Saiter, 2001). Then, from the area of the measured

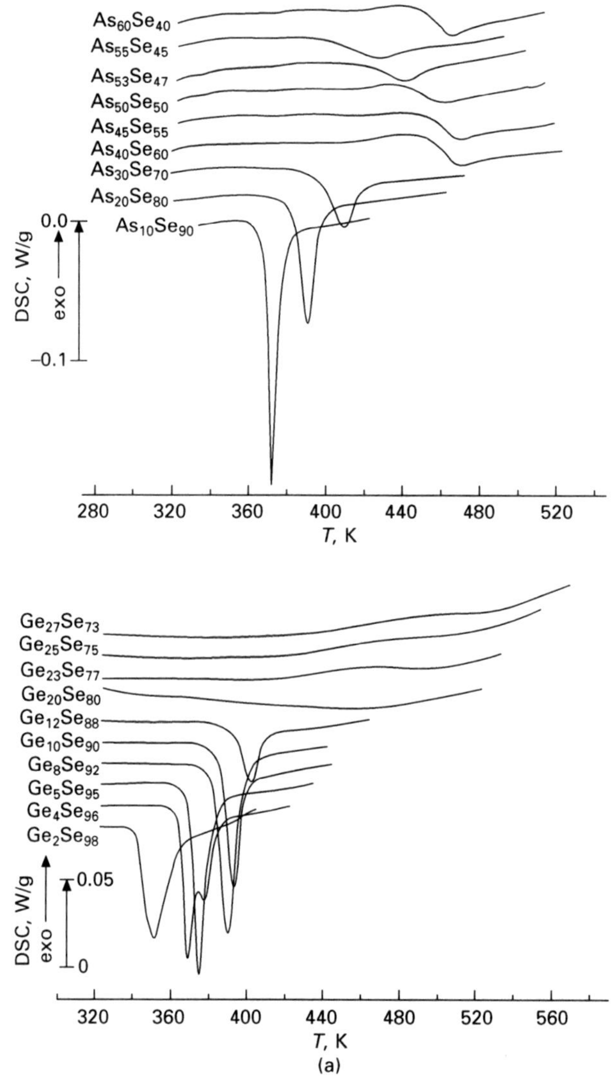

8.3 DSC curves of vitreous As_xSe_{100-x} and Ge_xSe_{100-x} samples subjected to long-term (~two decades) physical ageing (a) and subsequent rejuvenation procedure (b).

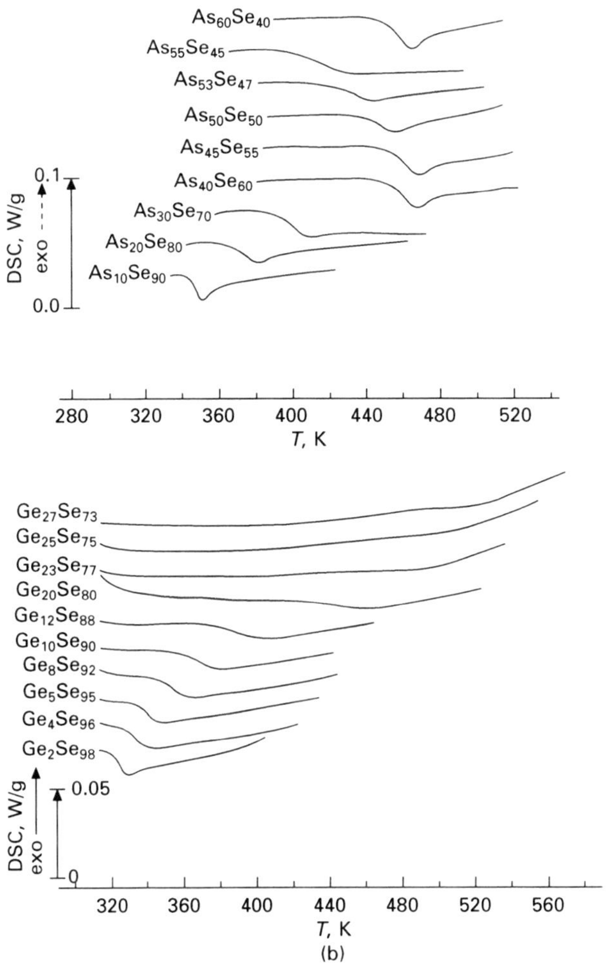

8.3 Continued

endothermic peak, after subtracting the non-aged (rejuvenated) curve, and using Eq. [8.1], the value of ΔH_∞ can be estimated for each glass composition. Doing this, completeness of physical ageing for each particular glass composition in addition to T_F can be also evaluated by comparison of ΔH_∞ values with those determined from DSC heating curves of long-term aged ChG. Contrary to ΔH and T_F values, the T_g^{onset} parameter for aged and rejuvenated ChG depends significantly on the heating rate during DSC experiments, as demonstrated in Fig. 8.4 for 25 years aged and rejuvenated $As_{25}S_{75}$ glassy samples (Shpotyuk *et al.*, 2011). Therefore, this quantity should be used with some caution during quantitative analysis of physical ageing and its kinetics.

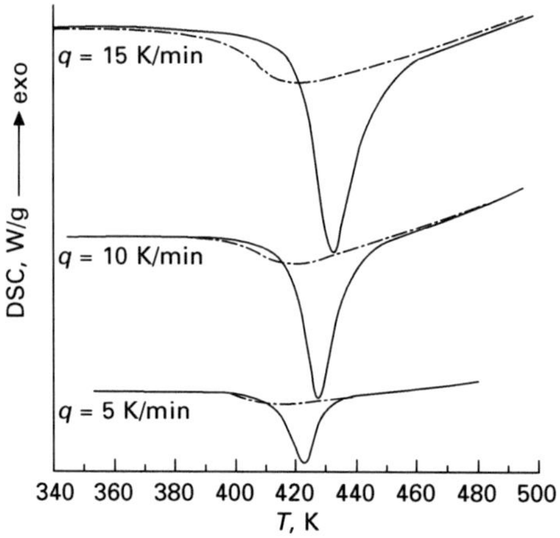

8.4 DSC curves of 25-years aged (solid) and rejuvenated (dash-dot) $As_{25}S_{75}$ glass at various heating rates *q*.

In the case of TMDSC signal, the physical ageing results in a significant non-zero out-of-phase (non-reversible) component of heat flow (Golovchak *et al.*, 2008a; Chen *et al.*, 2010). These components of complex heat flow are shown in Fig. 8.5 for typical representatives of aged and rejuvenated Se-rich glasses. The area under the peak denoted as ΔH_{nr} may be used for description of physical ageing. This term provides a measure of how different a glass going through the softening process is from the liquid in a configurational sense at the moment of measurement. In addition, the ΔH_{nr} depends on a lot of other factors: the conditions of TMDSC experiment (heating ramp, modulation parameters), glass composition and prehistory, completeness of physical ageing, etc. (Tonchev and Kasap, 2002; Golovchak *et al.*, 2008a; Hutchinson, 2003). It also depicts the average cooperative rearranging region size distribution during transition from glassy to supercooled liquid state (Saiter *et al.*, 2009). Behavior of non-reversible heat flow component ΔH_{nr} in TMDSC experiments can be roughly compared to that of *A* values determined from DSC measurements (Fig. 8.6). Thus, according to either ΔH_{nr} or *A* parameters the compositional dependence of isothermal physical ageing at room temperature for As_xSe_{100-x} and Ge_xSe_{100-x} glasses (Fig. 8.6) suggests significant effect for chalcogen-rich compositions and an almost negligible one for glasses with higher concentration of high-coordinated atoms (As or Ge) (Golovchak *et al.*, 2006a, 2009b). This result testifies that physical ageing can be adequately probed either by conventional DSC or TMDSC techniques with the same reliability in good agreement with Tonchev and Kasap (2002).

In general, the enthalpy losses ΔH follow a sigmoidal time dependence, which is defined by a choice of ageing temperature T_a in respect to T_g (T_g –

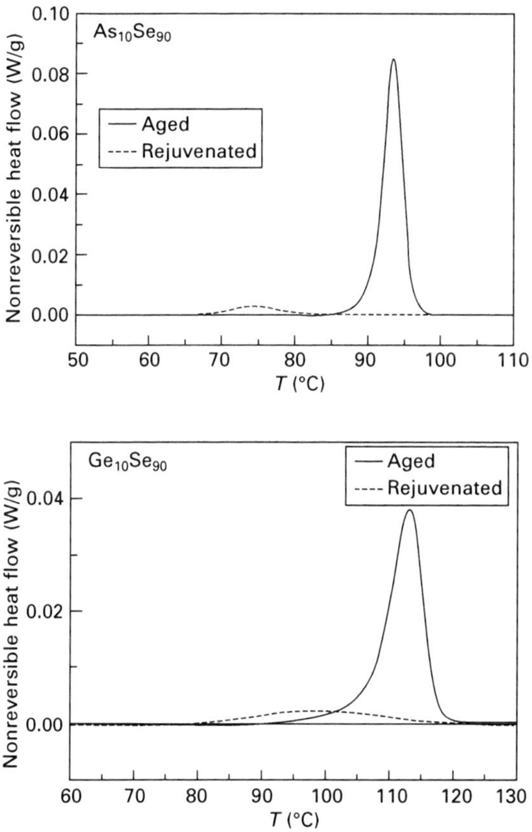

8.5 Out-of-phase component (or 'non-reversible' heat flow) of TMDSC signal for typical ~two decades aged (solid) and rejuvenated (dash) glasses.

T_a difference) (Ngai, 2011; Angell *et al.*, 2000). Therefore, for some ChG compositions with fast component of natural physical ageing (for example, with lower T_g), a significant enthalpy loss can be detected within a short experimental timescale, while for glass compositions with more extended kinetics (higher T_g), their marginal value just after synthesis (or short-term periods of physical ageing) can lead to misleading conclusions (Fig. 8.6(a)). The longer periods of storage depending on specific kinetics for chosen glass composition are needed to conclude on the existence of the ageing process (Golovchak *et al.*, 2006a, 2008a, 2009b; Shpotyuk *et al.*, 2011; Chen *et al.*, 2010). Thus, for $As_{30}Se_{70}$ composition (Fig. 8.6(a)), the first year of natural storage does not lead to any observable changes in A or ΔH_{nr} parameters, while further storage induces significant physical ageing in this specimen (Golovchak *et al.*, 2006a, 2008a).

This circumstance should be taken into account to study correctly physical

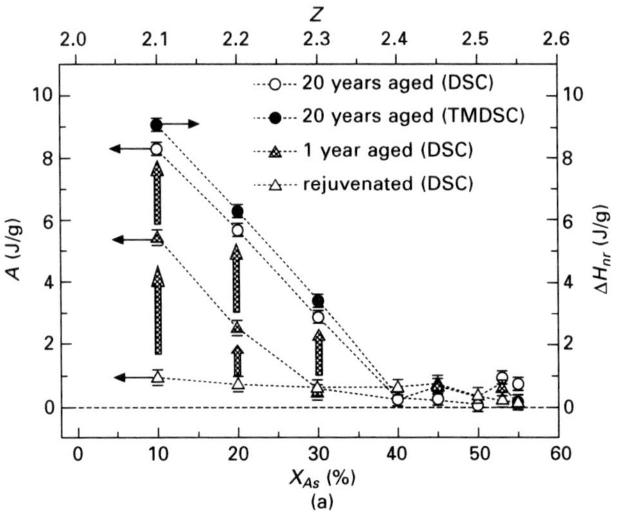

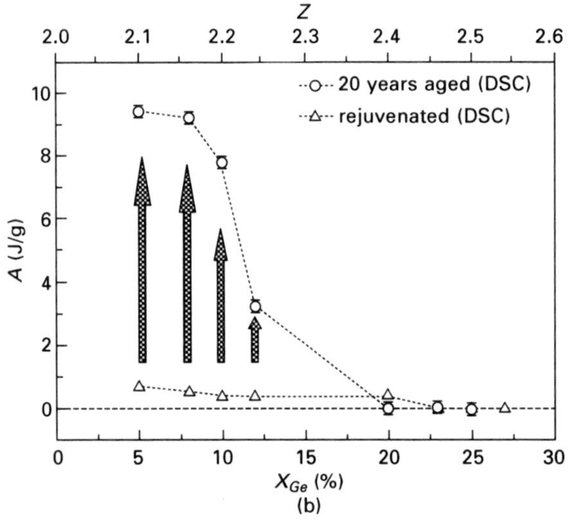

8.6 Areas *A* under the endothermic peaks of DSC curves recorded with 5 K/min heating rate for aged and rejuvenated As_xSe_{100-x} (a) and Ge_xSe_{100-x} (b) glasses. TMDSC ΔH_{nr} values are shown for As_xSe_{100-x} glasses (Golovchak *et al.*, 2008a) for comparison.

ageing in network glasses. We suggest that only DSC experiments, whether temperature-modulated or conventional, performed in a real-time scale chronology from earliest stages of as-prepared (rejuvenated) up to long-term aged state can be used to draw strict conclusions on physical ageing ability and quantification of physical ageing effect in ChG.

8.3.2 Radiation-induced physical ageing

The first announcements about the changes of mechanical properties of As_2S/Se_3 ChG caused by ^{60}Co γ-irradiation appeared in the former Soviet Union in the early 1960s (I.A. Domoryad, Institute of Nuclear Physics, Tashkent, Uzbekistan) (Starodubcev *et al.*, 1961). In particular, it was established that γ-irradiation of 10^5–10^6 Gy doses led to the experimentally detectable radiation-induced changes of microhardness, Young's modulus, internal friction and geometrical dimensions of various ChG (Starodubcev *et al.*, 1961; Kolomiets *et al.*, 1971).

Then, in the mid-1980s, a lot of experimental studies appeared, concerning radiation-induced effects in optical properties, photoluminescence, photoconduction, dissolution of ChG, as well as their compositional, dose, and temperature dependences (Fairman and Ushkov, 2004). The mechanism of radiation-induced structural transformations was proposed on the basis of FT-IR spectroscopy, EPR and mass-spectrometry data for glassy As_2S_3 – the model compound with high radiation sensitivity and well-studied structural properties (Fairman and Ushkov, 2004; Shpotyuk *et al.*, 1994). At the same time, thermal analysis studies of γ-induced effects, which could be treated in terms of physical ageing phenomenon, were lacking. The pioneering work in this direction was performed by Calemczuk and Bonjour (1981). They showed a γ-induced enhancement of structural relaxation in amorphous Se associated with physical ageing. Since that time, the problem of radiation-induced physical ageing was not raised for the next two decades.

Only recently it was shown that γ-irradiation can accelerate physical ageing in a greater variety of ChG (Golovchak *et al.*, 2006b, 2006c, 2009a, 2011a; Imran, 2008; Lucas *et al.*, 2011). Thus, the increased rate of structural relaxation towards thermodynamic equilibrium was recorded for chalcogen-rich fragile As-Se (Golovchak *et al.*, 2006b), Ge-Se (Golovchak *et al.*, 2009a), As-Ge-Se (Golovchak *et al.*, 2006c), Sn-Se (Imran *et al.*, 2008) and As-S (Golovchak *et al.*, 2011a; Lucas *et al.*, 2011) ChG under the influence of γ-radiation. The DSC curves recorded after isochronal periods of natural storage and γ-irradiation are shown in Fig. 8.7 for the example of typical representatives of sulfide and selenide ChG (Golovchak *et al.*, 2006b, 2006c, 2009a). The physical ageing under γ-radiation field conditions causes greater changes in T_g and endothermic peak area A values for Se-rich glasses than an equal period of natural storage does. Increase in the concentration of As or Ge leads to decrease in the value of radiation-induced physical ageing as detected by DSC (Golovchak *et al.*, 2006b, 2009a, 2011a; Lucas *et al.*, 2011). Finally, no detectable γ-induced changes (within an accuracy error) in DSC curves were recorded for ChG close to the Phillips–Thorpe rigidity percolation threshold at the average coordination number $Z = 2.4$ (Golovchak *et al.*, 2006b, 2009a, 2011a). Thus, we can speak of the acceleration of physical

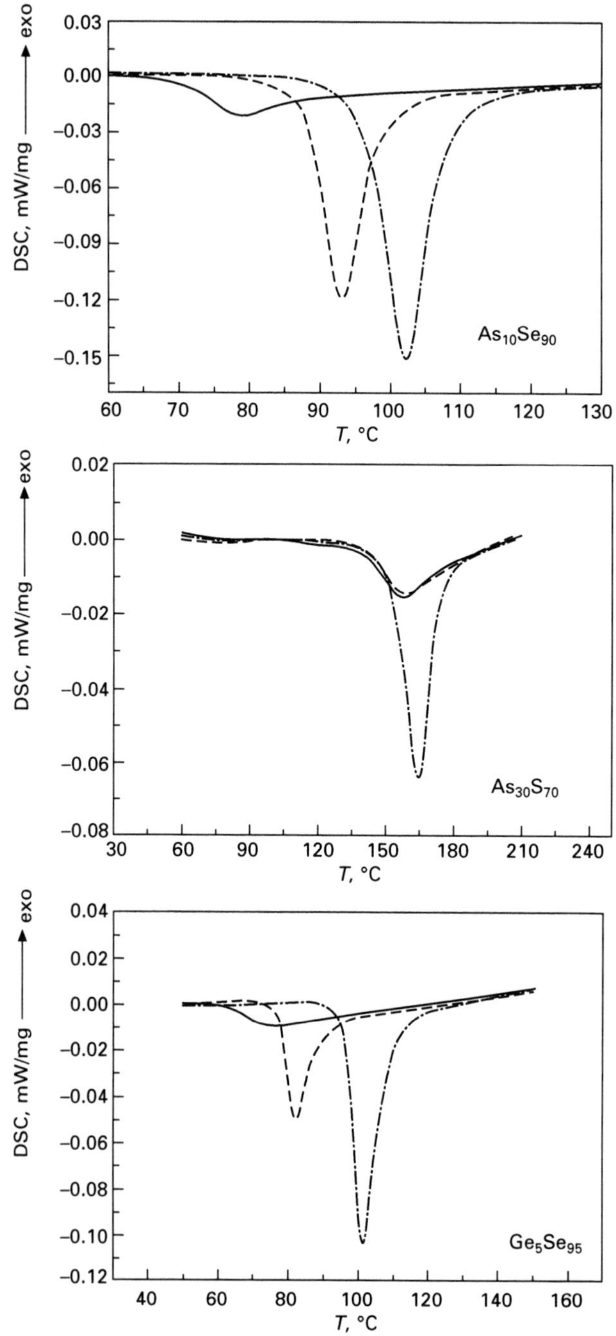

8.7 DSC curves recorded for typical representatives of selenide and sulfide ChG after equal periods of natural storage in dark (dashed line) and within γ-radiation field (dash-dot line). Solid black curves correspond to DSC signal from rejuvenated (as-prepared) samples.

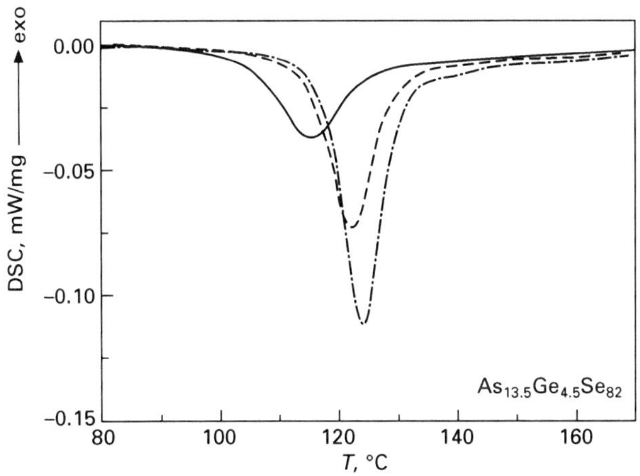

8.7 Continued

ageing due to radiation treatment only in floppy chalcogen-rich glasses. As far as the similar behavior in DSC scans after γ-irradiation was recorded for γ-irradiated vitreous Se, As-Se, As-Ge-Se, As-S, and Sn-Se glasses (showing increase in T_g and endothermic peak area *A* values) (Golovchak *et al.*, 2006b, 2009a, 2011a; Imran *et al.*, 2008; Lucas *et al.*, 2011), the conclusion about acceleration of physical ageing and related structural relaxation far below glass transition interval can be generalized to all chalcogen-rich fragile ChG. In other words, radiation treatment of these materials modifies their glass networks towards thermodynamically equilibrium extrapolated states of supercooled liquid as shown in Fig. 8.1.

Recent comprehensive studies on γ-induced physical ageing in binary As-S and As-Se ChG (Golovchak *et al.*, 2011a) have shown that, despite structural similarity of the main constituent building blocks ($As(S)Se_{3/2}$ pyramids and chalcogen chains), the compositional dependence of radiation-induced physical ageing significantly depends on the nature of the chalcogen atom (Fig. 8.8). In particular, a considerable radiation-induced physical ageing was recorded in As_xS_{100-x} glasses within $30 \le x \le 42$ compositional range, while no effect was observed for analogous ChG of As_xSe_{100-x} system (Fig. 8.8) (Golovchak *et al.*, 2011a). This result correlates with previous investigations of γ-induced changes in mechanical properties (shear modulus, microhardness) of binary As-based ChG, showing decay of γ-induced changes in the compositional row $As_2S_3 \rightarrow As_2Se_3 \rightarrow As_2Te_3$ (Starodubcev *et al.*, 1962; Domoryad and Kolomiets, 1971).

To explain the different behavior of S- and Se-based ChG, which are expected to have similar structure of covalent networks, the processes under γ-radiation influence should be considered in more detail. As a rule, three main mechanisms of the interaction of γ-rays with matter are considered:

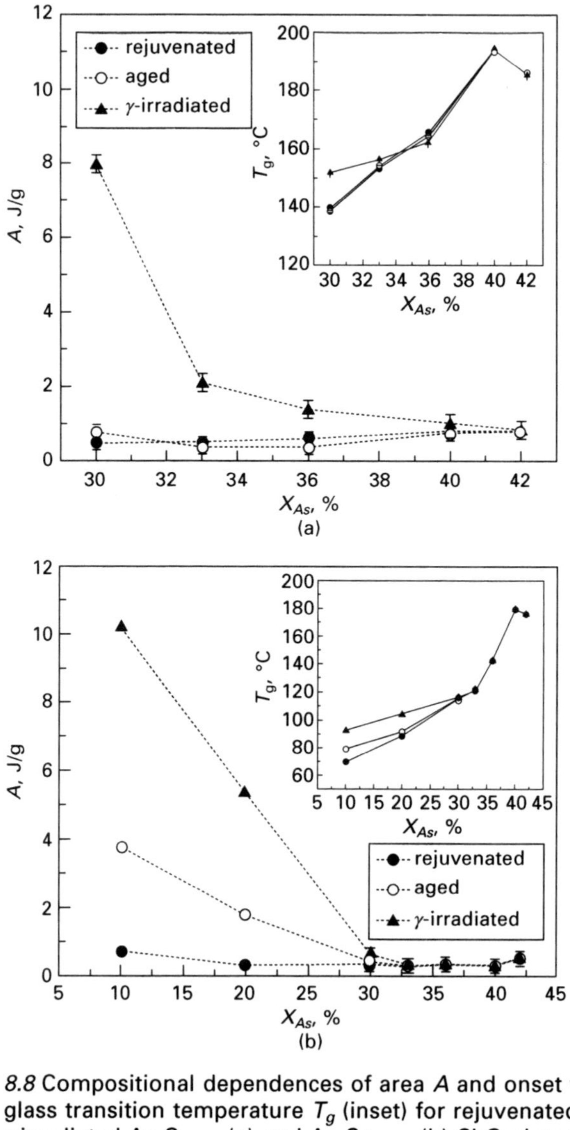

8.8 Compositional dependences of area *A* and onset value of glass transition temperature T_g (inset) for rejuvenated, aged and γ-irradiated As_xS_{100-x} (a) and As_xSe_{100-x} (b) ChG, determined from corresponded DSC curves recorded with q = 5°C/min heating rate.

γ-quanta can lose their energy as a result of absorbing, Compton scattering and creation of electron-positron pairs (Pikaev, 1985). In the case of ChG and γ-quanta energies usually used (1.25 MeV of ^{60}Co isotope), the main mechanism is related to Compton scattering (Pikaev, 1985). Because of strong electron-phonon coupling character for ChG (Feltz, 1993; Mott and Davis, 1971), the Compton photoelectron produces specific destructive structural transformations. They are associated with the breaking/switching of chemical bonds, which can be accompanied, in part, by the formation of coordination defects (Golovchak and Shpotyuk, 2005). In the case of selenide ChG, these processes decay rapidly without conservation of any coordination defects, which is evident from the absence of significant changes in optical properties of these materials under γ-irradiation influence (Fairman and Ushkov, 2004; Shpotyuk, 1990). In the case of sulfide ChG, it is shown that γ-induced covalent bond switching is accompanied by formation of metastable topological coordination defect pairs, like diamagnetic positively-charged, over-coordinated and negatively-charged under-coordinated atoms (Fairman and Ushkov, 2004; Golovchak and Shpotyuk, 2005).

So, it can be assumed that γ-stimulated structural transformations (broken or switched covalent chemical bonds, topological coordination defect pairs, in the first approximation) are more stable and appropriate to sulfide networks giving additional possibility for reorientation of constituent structural units and more extended structural fragments under irradiation. These processes provide an additional effective channel to release free volume frozen during melt-quenching and, thus, contribute towards general shrinkage of a glass backbone, identified as physical ageing with DSC. In other words, destruction-polymerization transformations are more effective in sulfide ChG than selenide ones, giving additional possibility for shrinkage of glassy backbone under γ-radiation conditions. Under γ-radiation, the relaxation of network sites can be achieved even if it would never occur during very long natural storage because of significant spatial constraints. This process is possible due to destruction of covalent bonds under γ-irradiation (Fairman and Ushkov, 2004; Golovchak and Shpotyuk, 2005), which makes it different from the natural physical ageing, where only conformations of the network are considered to be significant. So, with γ-radiation we can achieve a more thermodynamically favorable state (Fig. 8.1) than it is achievable after many years of natural storage, provided γ-stimulated excitations do not decay rapidly in the glass network. Therefore, the use of γ-radiation is an alternative way to achieve a thermodynamic quasi-equilibrium state of glass for a short period of time (days) at much lower (room) than T_g temperatures, at which relaxation rate normally is very slow (tens of years).

8.3.3 Light-assisted physical ageing

High photosensitivity of ChG has been known right from the earliest stages of the discovery of their semiconducting properties in 1950s by Kolomiets and Goryunova (1955). Since that time, different effects induced by bandgap, sub-bangap and above-bandgap light have been reported (Kolobov, 2003; Zakery and Elliott, 2007; Krecmer *et al.*, 1997; Stuchlik *et al.*, 2001; Ganjoo *et al.*, 1999; Kuzukawa *et al.*, 1998; Tanaka, 2002; Trunov *et al.*, 2009;

Poborchii *et al.*, 1999; Tallman *et al.*, 2008; Kovalskiy *et al.*, 2008), and numerous models of photoinduced effects have been developed (Kolobov, 2003; Shimakawa *et al.*, 1995; Frumar *et al.*, 1983, 1997; Tanaka, 1983; Tanaka *et al.*, 2009). Photoinduced darkening/bleaching (Kolobov, 2003; Zakery and Elliott, 2007), optomechanical effect (Krecmer *et al.*, 1997; Stuchlik *et al.*, 2001), photoexpansion/photocontraction (Ganjoo *et al.*, 1999; Kuzukawa *et al.*, 1998), photofluidity (Tanaka, 2002; Trunov *et al.*, 2009), photomelting/photocrystallization (Poborchii *et al.*, 1999; Tallman *et al.*, 2008), and photodiffusion (Kovalskiy *et al.*, 2008) are only a few examples of the progress achieved in the field.

Investigations on light-assisted physical ageing in Se-based ChG, induced by sub-bandgap light (laser sources), revealed its strong compositional dependence, as well as dependence on the intensity of sub-bandgap light and thermal prehistory of ChG samples (Lucas *et al.*, 2005, 2006; Calvez *et al.*, 2008; Lucas, 2006). A summative bond-breaking mechanism within a concept of strong/fragile glass-formers (Lucas, 2006) was assumed to explain the observed phenomena. Physical ageing effects produced by bandgap or above-bandgap light exposure were not considered to be significant in bulk ChG for a long time, because of too low penetration depth of corresponding photons (< 0.1 μm). The first to question this common assumption were. Larmagnac *et al.* (1982). Using conventional DSC, they recorded for amorphous Se films (6 μm thickness) an unusual increase in the relaxation rate under above-bandgap light exposure (404, 546, 643 nm wavelengths): the highest value of photorelaxation was achieved with 404 nm light, which had the lowest penetration depth into the material (~0.04 μm) (Larmagnac *et al.*, 1982). It was assumed that photoinduced excitations in the form of valence alternation pairs diffused from the surface, where they had been created by illumination, into the remaining bulk interior of the film, effectively promoting physical ageing as probed by DSC. Later on, similar wavelength dependence of photostructural relaxation induced by above-bandgap light exposure (404, 546, 581, 643 nm wavelengths) was reported for Ge_xSe_{100-x} films (5 μm thickness) (Vautier *et al.*, 1991). The conclusion was made about the significant role of Se chains, which again were considered as a main source of valence alternation pairs created by the illumination. More recent studies revealed also a considerable athermal physical ageing effect in Se-rich bulk As-Se ChG caused by the above-bandgap light exposure (Golovchak *et al.*, 2008b). The mechanism of the observed photoinduced enthalpy losses was assumed to be connected with dissipation into bulk of elastic strains produced at the surface layers of the sample by light excitations (Golovchak *et al.*, 2008b).

More comprehensive investigations of light-assisted physical ageing in broader spectral range, using for excitation discrete wavelengths from above- and sub-bandgap regions, are performed in Kozdras *et al.* (2011)

for typical sulfide and selenide ChG. The DSC curves recorded for ChG samples after equal time intervals of dark storage and irradiation with light of different wavelengths are shown in Fig. 8.9 (Kozdras *et al.*, 2011). Their behavior is similar to radiation-induced effects – the light exposure causes greater changes in T_g and endothermic peak area A values than equal period of natural storage does. A measurable light-assisted physical ageing effect at room temperature is recorded for all ChG with $Z < 2.4$ regardless of choice of chalcogen (anion) or cation atoms. The samples with $Z = 2.4$ ($As_{40}S_{60}$,

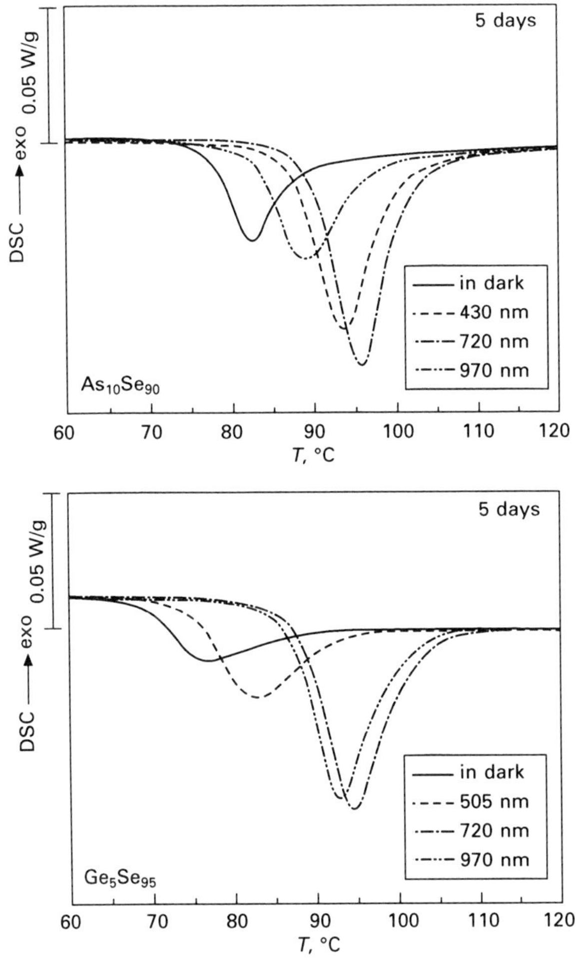

8.9 Typical DSC curves of the investigated ChG recorded after same periods of dark storage and light exposure at different wavelengths. The method for A determination (cross-hatched area) is shown at the example of DSC curve for $As_{10}Se_{90}$ sample aged during 5 days in darkness (Kozdras *et al.*, 2011).

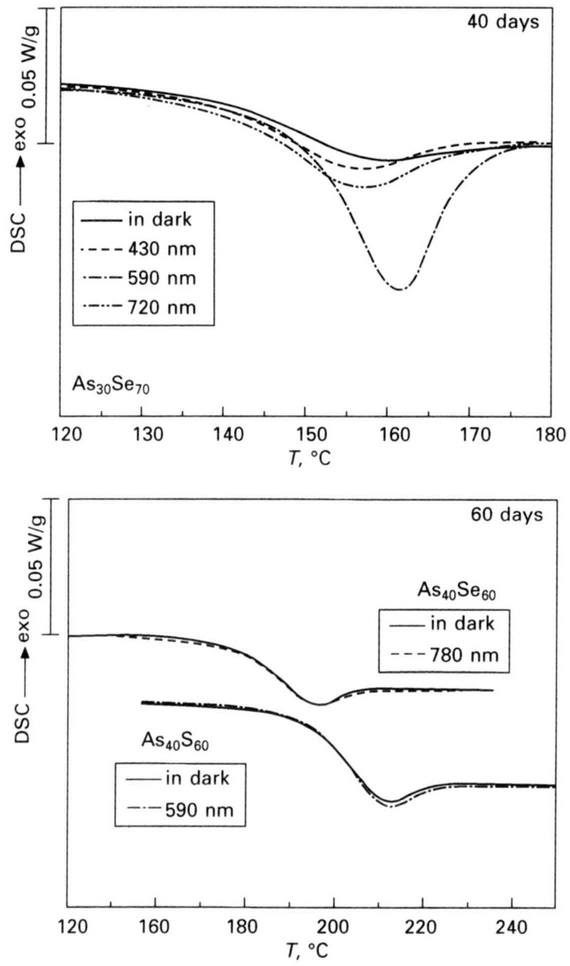

8.9 Continued

$As_{40}Se_{60}$ or $Ge_{20}Se_{80}$) did not exhibit any significant changes in DSC curves either under a dark storage or under light exposure at any of the wavelengths used (Fig. 8.9) (Kozdras *et al.*, 2011).

The influence of the exposure with light of different discrete wavelengths on the physical ageing of ChG with $Z < 2.4$ (expressed through ΔA values) is presented in Fig. 8.10. The effect of physical ageing in the same ChG caused by isochronal dark storage is indicated by horizontal lines. It is shown that for chalcogen-rich selenide ChG, ΔA values show almost threshold increase (Fig. 8.10) at the energies lower than the optical gap of the corresponding material (E_g) (Kozdras *et al.*, 2011). After that, ΔA values slightly decrease in the transparent region of ChG spectra with further decrease in the energy of incident photons (Fig. 8.10). The overall ΔA magnitude in selenide ChG decreases significantly with increasing connectivity (Z) or As(Ge) concentration in the samples. On the other hand, in sulfide ChG, the light-assisted physical

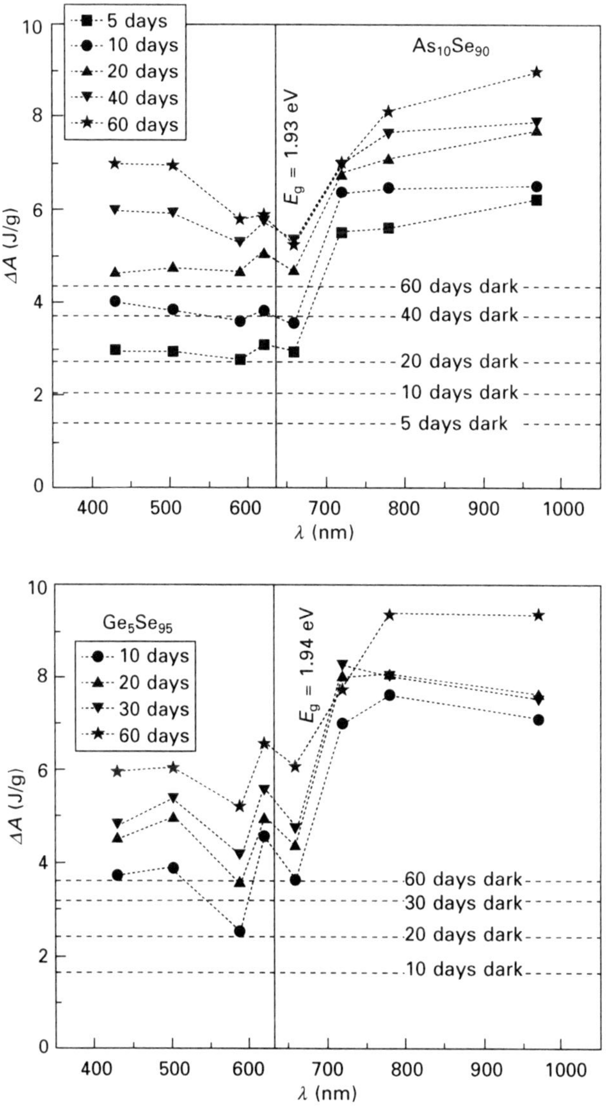

8.10 Dispersion of light-assisted physical ageing in typical representatives of ChG expressed through ΔA values (calculated as the difference between A values of the aged and rejuvenated samples). The connection lines are drawn as guides for the eyes. The value of the physical ageing effect caused by isochronal dark storage is given by the dashed lines (Kozdras *et al.*, 2011).

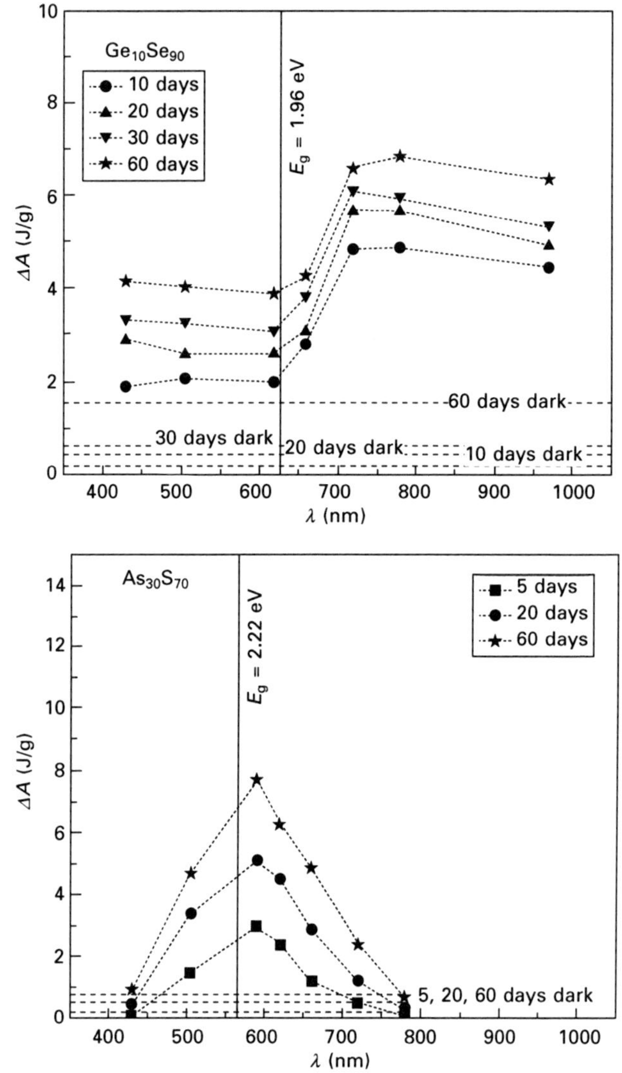

8.10 Continued

ageing is almost one order of magnitude greater than in selenide ChG. It shows maximum in spectral dependence at the illumination wavelengths comparable to the optical gap of these ChG (Fig. 8.10), almost vanishing with deviation of the incident photon energies from E_g in both directions. So, from these results it is evident that dispersion of light-assisted physical ageing differs for S- and Se-based ChG, while influence of cation type (Ge or As) is not so essential. Therefore, type and number of chalcogen atoms play a decisive role in the light-assisted physical ageing.

The mechanisms of photostructural transformations in ChG are usually considered in respect to three phenomena: photodarkening, photoinduced volume changes, and photoinduced carriers/defects creation (related to photoconductivity) (Shimakawa, 2007). Recent *in situ* studies and simultaneous kinetic measurements of these effects suggest rather the absence of direct relationships between them (Shimakawa, 2007). Therefore, it is interesting to compare the wavelength dispersion of light-assisted physical ageing with spectral dependences of photoconductivity, photodarkening, and photoinduced volume expansion in ChG (El Gharras, 1989; Averyanov, 1972; Nagels *et al.*, 2000; Oheda, 1979; Derrey *et al.*, 1986). Of the considered effects, the maximum on spectral dependence of photoconductivity is observed at the highest energies of the incident photons above the ChG optical gap, which is demonstrated by the example of vitreous $As_{20}Se_{80}$ and Ge_5Se_{95} samples (Fig. 8.11) (El Gharras, 1989; Averyanov, 1972). It is assumed that light-assisted physical ageing under above-bandgap light exposure is connected with generation of charge carriers. Indeed, when the energy of incident photons exceeds the bandgap value, we have direct excitation of the electrons from top energetic levels in the valence band (in the case under consideration they are formed by lone-pair and bonding electrons of chalcogen atoms) (Golovchak *et al.*, 2007a, 2009c) into the conduction band, producing electron-hole pairs and broken covalent bonds. Owing to a low penetration depth of the above-bandgap light and low mobility of electrons in ChG (usually these materials possess *p*-type of conductivity) (Feltz, 1993; Mott and Davis, 1971), these excitations should be effective only in the surface layer, where above-bandgap incident light is directly absorbed. They cause transient and metastable displacements of atoms from their most energetically favorable positions in a glass structure due to thermalization of excited hot electrons through inelastic collisions with surrounding atoms (note, ChG are characterized by strong electron-phonon coupling; Mott and Davis, 1971), switching or restoration of the destroyed covalent bonds and possible valence alternation pair formation. In turn, the disturbed atomic sites ultimately cause elastic stresses at their nearest surroundings in the covalent network, whose further relaxation promotes physical ageing into the remaining volume of the glass, as happens in surface-modified Se-rich fibers (McEnroe and LaCourse, 1989). In other words, the surface under above-bandgap light excitation plays the role of 'mechanical hummer' for the rest of the bulk volume. In such a way, the changes produced by above-bandgap photoexposure near the surface of the samples can diffuse into their depth, producing a measurable DSC effect (which is known to be an integral method sensitive to bulk properties of materials) (Golovchak *et al.*, 2008b; Kozdras *et al.*, 2011). The effect produced by above-bandgap light exposure (in the region of 400–600 nm) decreases significantly with increasing connectivity of glass backbone (or

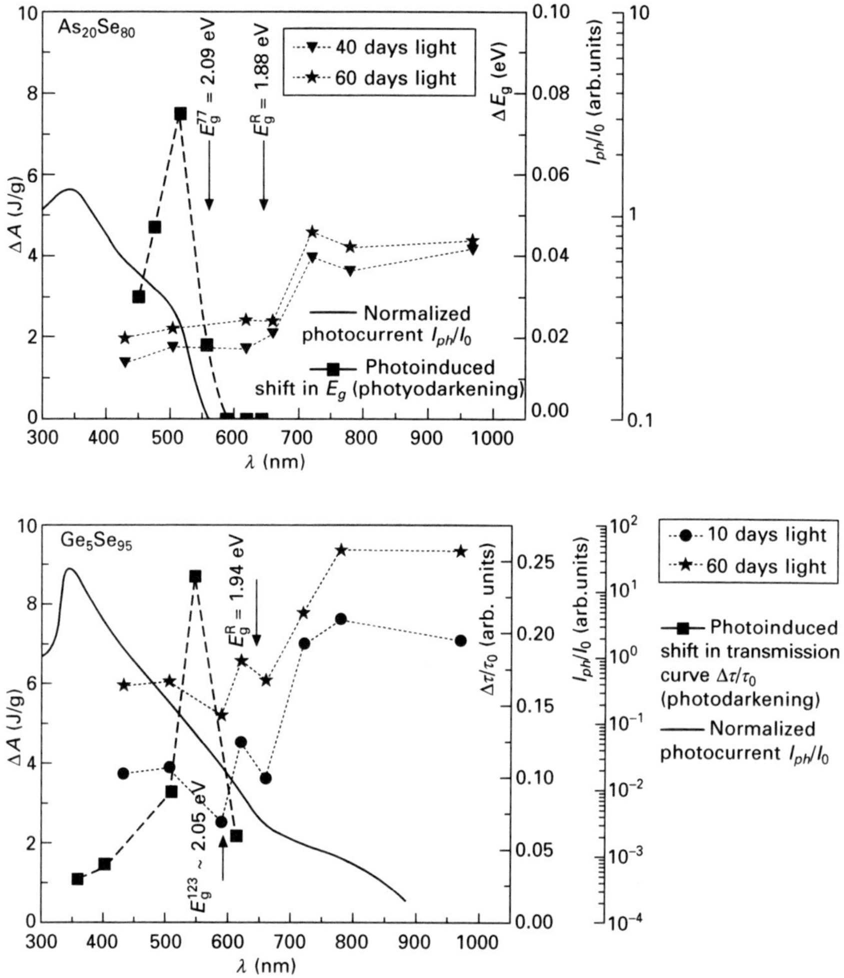

8.11 Comparison of spectral dependences of photoconductivity, photodarkening, and light-assisted physical ageing in $As_{20}Se_{80}$ and Ge_5Se_{95} ChG (El Gharras, 1989; Averyanov, 1972; Nagels *et al.*, 2000; Oheda, 1979). Optical gap values measured at room and lower temperatures (indicated by superscript E_g index) are shown by arrows (Nagels *et al.*, 2000; Oheda, 1979).

Z). It fully disappears in $As_{40}Se_{60}$ ($As_{40}S_{60}$, $Ge_{20}Se_{80}$) glass with $Z = 2.4$ (Kozdras *et al.*, 2011).

Spectral studies of photodarkening effects in ChG, associated with long-wave shift of their optical absorption edge, have provided important insights into the understanding of microstructural mechanisms of these phenomena (Tanaka, 1986, 2004, 2006; Ho *et al.*, 2003). It has been demonstrated that photodarkening can be induced by bandgap illumination with photon energy $h\nu \sim E_g$ (Nagels *et al.*, 2000; Oheda, 1979; Tanaka, 1986, 2006), as well as sub-bandgap illumination with photon energy $h\nu < E_g$, but much higher intensities (~100 W/cm^2) (Tanaka, 2004; Ho *et al.*, 2003). If the intensity of light is fixed, then the maximum of photodarkening is observed at the photon energies comparable with ChG bandgap ($h\nu \sim E_g$) (Nagels *et al.*, 2000; Oheda, 1979; Derrey *et al.*, 1986; Tanaka, 1986). Position of this maximum is slightly shifted from corresponding E_g values towards the high-energy side of the optical spectrum, as is shown by the examples of $As_{20}Se_{80}$ and Ge_5Se_{95} glasses (Fig. 8.11) (Kozdras *et al.*, 2011). The developed microstructural mechanism of photodarkening in ChG films is based on chemical bond switching processes, presumably also accompanied by valence alternation pair formation (Shimakawa *et al.*, 1995; Frumar *et al.*, 1983, 1997; Tanaka, 1983, 1986; Tanaka *et al.*, 2009). It is also known, that the magnitude of the induced darkening decreases in the row S → Se → Te and with increasing temperature (reversible component of the effect can be erased by near-T_g annealing at $T/T_g \sim 0.75$) (Tanaka, 1986; Tanaka, 1983). Therefore, one order increase in the magnitude of light-assisted physical ageing in As-S ChG observed around the optical gap can be correlated with greater efficiency of the mechanisms responsible for the photodarkening in sulfide ChG compared to selenide ones. On the other hand, the photodarkening is observed even in stoichiometric compositions (like $As_{40}S_{60}$ and $As_{40}Se_{60}$) with $Z = 2.4$ (Kolobov, 2003), while light-assisted physical ageing is not (Fig. 8.9). Owing to the reversible nature of photodarkening, which is typically erased when approaching T_g (Kolobov, 2003; Tanaka, 1983), the irradiated (after light illumination) $As_{40}S_{60}$ and $As_{40}Se_{60}$ samples lose the reversible effect produced by light exposure when approaching temperature $0.75T_g$ during DSC scans. As a result, the signature of reversible photodarkening cannot be seen in DSC curves during structural relaxation through T_g. The DSC measurements can reveal only photostructural rearrangements, which can permanently affect a general connectivity of glass backbone or facilitate (through valence alternation pair formation, bond switching) the light-assisted physical ageing.

The maximum of light-assisted physical ageing in selenide ChG is shifted towards photon energies lower than corresponding E_g values (Fig. 8.11). In this case, the energy of the incident photons is not enough to excite electrons from valence to conduction bands directly. The valence alternation pair formation is also doubtful due to a lower probability of bond breaking. So, only weak two-photon absorption processes, phonon-assistant electronic transitions or inelastic photon scattering, can be considered as a reason for light-assisted physical ageing under sub-bandgap illumination of relatively low intensity. However, owing to a much greater penetration depth of sub-bandgap light, these processes can occur in the whole volume of the sample, not only at the surface layer as in the case of above-bandgap photoexposure (Kozdras *et al.*, 2011). Therefore, sub-bandgap light can produce greater effect of light-assisted physical ageing in comparison to above-bandgap light, because of greater amount of photoinduced perturbation events (more network sites of the bulk are involved).

8.3.4 Thermally-induced physical ageing

Physical ageing of ChG at elevated temperatures is relatively well studied only for stoichiometric $As_{40}Se_{60}$ (Cernoskova *et al.*, 2001; Cernosek *et al.*, 2003; Golovchak *et al.*, 2011b) and a few Se-rich compositions, including glassy Se (Echeverria *et al.*, 2003; Svoboda *et al.*, 2006, 2007, 2010). Like in the case of natural or induced physical ageing, the endothermic peak superimposed on the endothermic step of the glass transition signal is observed in DSC curves as a result of prolonged annealing at the elevated temperatures. Considering $As_{40}Se_{60}$ glass as typical example, the ΔH values determined from DSC curves after annealing of this material during 1 and 5 days at various temperatures (starting from room up to near T_g) are plotted in Fig. 8.12 (Golovchak *et al.*, 2011b). The results obtained in a heating mode show a clear threshold around annealing temperature $T_a \sim 90°C$ for the onset of structural relaxation, revealed through the increase in ΔH values (Fig. 8.12). Enthalpy losses achieve maximum at temperatures 20–40 K below the glass transition temperature T_g^{onset} and decrease significantly when approaching $T_g^{onset} = 181°C$ (Fig. 8.12). The maximum of the enthalpy losses occurs exactly at the temperatures that are usually used to anneal as-prepared $As_{40}Se_{60}$ glass during synthesis procedure (annealing of glasses at temperatures ~20–40 K below their T_g is a well-known technological operation to homogenize and stabilize glass structure; Feltz, 1993; Borisova, 1981). This behavior is in good agreement with Eq. [8.1], where the maximum of energy that a glass can lose during ageing depends on the $T_g - T_a$ value. Indeed, the maximum energy that $As_{40}Se_{60}$ glass can lose after an infinite annealing duration at room temperature can be estimated at the level of $\Delta H_\infty \approx 28$ J/g, while at $T_a = 140°C$ at the level of only $\Delta H_\infty \approx 6$ J/g, which is quite close to ΔH values observed experimentally after 5 days annealing at $T_a = 140–160°C$ (Fig. 8.12) (Golovchak *et al.*, 2011b). It testifies to near saturation of physical ageing in $As_{40}Se_{60}$ glass under these annealing conditions. Further increase in the annealing temperature towards T_g leads to vanishing of the enthalpy losses ΔH because of vanishing ΔH_∞ at $T_a = T_g$ according to Eq. [8.1], and the beginning of the rejuvenation process.

To exclude from consideration the redistribution of significant chemical bonds or phase separation at higher annealing temperatures, which also could be a reason for strong endothermic peak in the vicinity of glass-to-supercooled liquid transition, the structural investigations were performed for $As_{40}Se_{60}$ glass using *in-situ* EXAFS measurements at As and Se K-edges (Golovchak

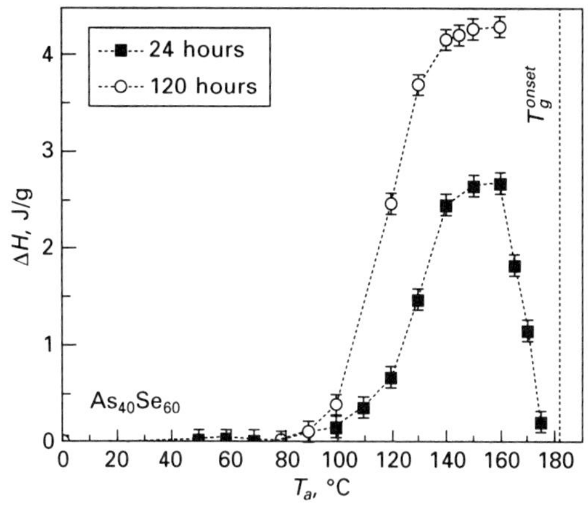

8.12 Enthalpy losses ΔH caused by isochronal thermal annealing of $As_{40}Se_{60}$ glass at different temperatures (Golovchak *et al.*, 2011b).

et al., 2011b). According to these results, the coordination number did not change for both As and Se atoms either during annealing at T_a = 140°C or thereafter. However, at this high-temperature stage, a slight increase in the nearest neighbor distance (~0.003 Å) and Debye-Waller factor (~0.0008 Å^2) around both atoms was recorded (Golovchak *et al.*, 2011b). These values did not change during an annealing period of 24 hours at T_a = 140°C and retained their initial values after cooling the glass to room temperature. Therefore, the observed significant enthalpy losses at T_a = 140°C (Fig. 8.12) are not associated with changes in the first coordination spheres around Se and As atoms. So, phase separation effects or significant bond redistribution during the annealing process should be excluded from consideration. Thus, the only possible way to explain the observed changes in DSC curves is structural relaxation of $As_{40}Se_{60}$ glass at the elevated annealing temperatures.

8.4 Phenomenological description of physical ageing

8.4.1 Topological consideration of covalent-bonded glassy networks

Consider an *ideal* covalent network with N atoms ($N \to \infty$). Let it be constructed by n_r atoms each having r bonds in respect to their valence (Thorpe, 1983, 1995):

$$\sum_r n_r = N \qquad [8.3]$$

An *ideal* means that there are no surfaces, voids, defects, dangling bonds or any phase-separated regions in this network and every atom has saturated its valence by covalent bonds with nearest neighbors. So, the mean coordination number Z of each atom can be calculated as:

$$Z = \left(\frac{\sum_r r n_r}{\sum_r n_r} \right) \qquad [8.4]$$

In covalently bonded networks, the bond lengths and angles are well defined and small displacements from equilibrium are described by a potential of small vibrations (Keating, 1966):

$$U = \frac{1}{2}\sum_{i,j} \alpha_{ij}(\Delta r_{ij})^2 + \frac{1}{2}\sum_{i,j,k} \beta_{ijk}(\Delta\Theta_{ijk})^2 \qquad [8.5]$$

where Δr_{ij} is change in nearest-neighbor bond length between atoms $<ij>$, $\Delta\Theta_{ijk}$ is change in bond angle between two adjustment $<ij>$ and $<jk>$ bonds, and α_{ij} and β_{ijk} are bond stretching and bond bending force constants.

The rank of dynamical matrix formed from potential [8.5] gives a number of linearly independent constraints N_c (Frazer *et al.*, 1938).

Another way to estimate the number of constraints is as follows (Thorpe, 1983):

1. one constraint per bond is associated with bond stretching forces (first term in Eq. [8.5]) $n_\alpha = r/2$ for each r-coordinated atom;
2. the number $n_\beta = \frac{(D-1)(2r-D)}{2}$ of angular constraints, where D is space dimensionality, is associated with bending forces (second term in Eq. [8.5]): if $D = 3$, then $n_\beta = 2r - 3$. (It is clear, that atoms with $r \leq D - 2$ should be excluded from consideration.)

According to this analysis, the total number of Lagrangian constraints can be estimated as:

$$N_c = \sum_r n_r(n_\alpha + n_\beta) = \sum_r n_r\left[\frac{r}{2} + (2r-3)\right] \qquad [8.6]$$

Consequently, taking into account Eqs [8.3] and [8.4], the number of Lagrangian constraints per atom is:

$$N_c = \frac{Z}{2} + (2Z - 3) \qquad [8.7]$$

Now, let's consider two deviations typical for ChG from an *ideally* composed covalent network. The first one is associated with dangling bonds and the second with ring-like fragments. If dangling bonds are present in a glass network, then $n_\beta = -1$ instead of zero for $r = 1$. Therefore, to account for the total number of Lagrangian constraints, Eq. [8.6] should be corrected for a number of singly coordinated atoms (dangling bonds) n_1 (Thorpe, 1995):

$$N_c = \sum_r n_r \left[\frac{r}{2} + (2r - 3) \right] + n_1 \qquad [8.8]$$

The existence of small rings in the structure of ChG means that not all the constraints specified above are linearly independent. It is easy to show that a six-sided ring is just rigid (Thorpe, 1983). Triangles, quadrilaterals and pentagons are also rigid, but the number of constraints calculated using Eq. [8.6] is overcounted by three for a triangle, two for a quadrilateral and one for a pentagon (Thorpe, 1983). All the remaining isolated rings have seven or more sides and are floppy (Thorpe, 1983). So, three constraints should be removed from the total number of constraints calculated using Eq. [8.6] for three-sided rings, two for four-sided rings, one for five-sided rings and none for lager rings. Therefore, Eq. [8.8] should be rewritten as:

$$N_c = \sum_r n_r \left[\frac{r}{2} + (2r - 3) \right] + n_1 - n^{ring} \qquad [8.9]$$

where n^{ring} is a corresponding ring-correction parameter, which depends on the number of rings and their type.

Finally, the number of Lagrangian constraints per atom for ChG should be calculated as follows:

$$N_c = \frac{Z}{2} + (2Z - 3) + \frac{n_1}{N} - \frac{n^{ring}}{N} \qquad [8.10]$$

It was assumed by Phillips (1979) that optimal mechanical stability of the network can be achieved when $n_c = D$ ($n_c = 3$ in the case of a three-dimensional network). According to Eq. [8.7], $n_c = 3$ when $Z = 2.4$ – a well-known floppy-to-rigid transition, in which extrema in compositional dependence of many physical-chemical properties are observed (Feltz, 1993; Borisova, 1981; Vinogradova, 1984). At this point, the number of floppy modes becomes zero (Eq. [8.8]) dividing the glass formation region into sections with preferential floppy or under-coordinated/constrained ($Z < 2.4$, $n_c < 3$) and stressed rigid or over-coordinated/constrained ($Z > 2.4$, $n_c > 3$) networks. The network with $n_c = 3$ is called optimally constrained (Phillips, 1979; Boolchand *et al.*, 2005; Thorpe *et al.*, 2000).

The classic Maxwell constraints counting algorithm predicts a solitary transition from floppy to rigid networks at $Z = 2.4$ with increasing backbone connectivity by changing glass composition (Z) (Thorpe, 1983, 1995; Phillips, 1979). However, for networks that have a possibility to avoid stresses by structural variation at low entropy cost, this solitary transition splits into two points: the second-order transition from floppy to unstressed-rigid and first-order transition from unstressed-rigid to stressed-rigid networks (Boolchand *et al.*, 2005; Thorpe *et al.*, 2000). In other words, the topological self-organization occurs when a structure is able to keep the optimal number of constraints per atom ($n_c = 3$, for instance) with changing composition as long as possible, avoiding creation of over-constrained stressed regions (Thorpe *et al.*, 2000; Chubynsky *et al.*, 2006). The range of ChG compositions with optimally constrained networks forms a so-called self-organized phase (the terms *self-organized* and *intermediate* are equivalent), also according to Eqs [8.9] and [8.10]. It is believed that optimally constrained networks of self-organized phases do not undergo any structural relaxation during storage or under the influence of external factors (Phillips, 1979; Boolchand *et al.*, 2005). This remarkable feature makes ChG with $n_c = 3$ very appealing for application in highly reliable optoelectronics.

However, within the Phillips–Thorpe approach (Thorpe, 1983, 1995; Phillips, 1979), the number of constraints for a given ChG composition is assumed to be constant with respect to temperature. A mathematically equivalent approach, which accounts for temperature dependence of constraints, has been derived by Gupta and Mauro (2009) and Smedskjaer *et al.* (2010a, 2010b). According to their approach, a constraint is considered floppy above an onset temperature and rigid below this temperature (Smedskjaer *et al.*, 2010a, 2010b):

$$n_c(T) = \theta(T_{th} - T) \qquad [8.11]$$

where $\theta(T_{th} - T)$ is a unit step function and T_{th} represents the temperature below which a particular constraint becomes rigid.

Using temperature-dependent constraints theory (Gupta and Mauro, 2009; Smedskjaer *et al.*, 2010a), it was possible to predict compositional dependence of glass transition temperature in Ge-Se ChG (Gupta and Mauro, 2009), as well as fragility and hardness in some oxide glasses (Smedskjaer *et al.*, 2010a, 2010b). An attempt to develop an atomistic understanding of the dynamical processes responsible for viscous flow and shear relaxation in Ge-Se glasses was made using the basics of temperature-dependent constraints theory and temperature-dependent nuclear magnetic resonance data (Gjersing *et al.*, 2010). Molecular dynamics simulations also provided some evidence for temperature degeneracy of constraints in sodium silicate glasses (Bauchy and Micoulaut, 2011). Thermally induced physical ageing of optimally constrained $As_{40}Se_{60}$ glass (Fig. 8.12) can be also explained by temperature removal of some of constraints (their degeneracy) (Golovchak *et al.*, 2011b).

To distinguish the influence of temperature degeneracy of constraints and temperature dependence of structural relaxation time constant ($\tau = f(T)$) on the degree and kinetics of physical ageing, the isothermal annealing of ChG with almost the same T_g (same $T_g - T_a$ value), but different n_c should be considered. If $\tau = f(T)$ alone determines the structural relaxation and n_c does not have any influence, then such glass should demonstrate the same value of physical ageing effect, otherwise n_c also governs the structural relaxation. Long-term physical ageing under normal conditions in $As_{30}Se_{70}$ and $As_{55}Se_{45}$ ChG, having almost the same T_g (~110°C) and comparable fragility, but different n_c ($As_{30}Se_{70}$ has $n_c < 3$, $As_{55}Se_{45}$ has $n_c > 3$), can be considered as typical examples of such isothermal annealing studies (Golovchak *et al.*, 2006a, 2008a). The noticeable effect of long-term (tens of years) natural physical ageing was recorded for $As_{30}Se_{70}$ glass, while this effect was negligible for $As_{55}Se_{45}$ glass (Fig. 8.6(a)) (Golovchak *et al.*, 2006a, 2008a). So, structural relaxation of optimally constrained glasses at the elevated temperatures during relatively short time periods (~days) can be well explained by temperature removal of some of the constraints rather than $\tau = f(T)$ dependence. So, the initial optimally constrained network behaves as under-constrained with $n_c < 3$ at higher temperatures, resulting in pronounced physical ageing (Fig. 8.12) (Golovchak *et al.*, 2011b).

8.4.2 Role of topological self-organization in physical ageing

Compositional trends in physical ageing of ChG can be satisfactorily explained within mean-field constraints theory (Thorpe, 1983, 1995; Phillips, 1979; Boolchand *et al.*, 2005; Thorpe *et al.*, 2000; Chubynsky *et al.*, 2006; Gupta and Mauro, 2009; Smedskjaer *et al.*, 2010a, 2010b; Gjersing *et al.*, 2010; Bauchy and Micoulaut, 2011). This explanation is grounded on a supposition that the ability to attain physical ageing of a specific covalently bonded network is fully determined by its constraints number n_c. The under-constrained ChG ($n_c < 3$) are characterized by a pronounced physical ageing, revealed through significant drift in their physical properties. The over-constrained ChG ($n_c > 3$) can be affected by ageing too, while ChG with $n_c = 3$ should not age at all. Consequently, ChG belonging to a self-organized phase should be distinguished by ideal non-ageing ability.

Mean-field topological consideration predicts self-organized phase for network glasses within a very narrow range of compositions close to $Z \approx 2.4$ (in the case of 3D networks) (Thorpe *et al.*, 2000; Chubynsky *et al.*, 2006). To much surprise, it was claimed on the basis of TMDSC data that the width of self-organized phase in real glasses could be about one order of magnitude greater (Chakravarty *et al.*, 2005; Chen *et al.*, 2010; Selvanathan *et al.*, 1999, 2000; Georgiev *et al.*, 2000, 2003; Wang *et al.*, 2000, 2005; Boolchand *et al.*, 2001a, 2001b, 2002a, 2002b; Mamedov *et al.*, 2003; Qu *et al.*, 2003; Gump *et al.*, 2004; Vempati and Boolchand, 2004; Qu and Boolchand, 2005; Cernoskova *et al.*, 2005). In this case, the marginality of non-reversible heat flow component in TMDSC experiments performed on

ChG samples aged for a few weeks (or as-prepared) was used as the universal criterion to determine compositional limits of the self-organized phase. The range of compositions with marginal non-reversible heat flow component was called the *reversibility window* (Table 8.1). On the other hand, many reports on a collapse of reversibility windows in ChG after long-term natural storage (Golovchak *et al*., 2006a; Chen *et al*., 2010) or under the influence of external factors (Wang *et al*., 2007), supported by lack of direct experimental evidence for structural signature of a self-organized phase (Shatnawi *et al*., 2008), have brought into question the validity of this criterion for correct identification of the compositional boundaries of self-organized phases. Furthermore, the enormously wide reversibility windows, especially those extended far below $Z = 2.4$ such as for As-containing ChG (Table 8.1), have

Table 8.1 Reversibility windows (x_1, Z_1 = onset composition, x_2, Z_2 = end point) identified by TMDSC for as-prepared or short-term aged glasses

Glass composition	x_1	x_2	Z_1	Z_2	References
Si_xSe_{100-x}	20	27	2.40	2.54	Selvanathan *et al.* (1999, 2000); Boolchand *et al.* (2002a)
Si_xSe_{100-x}	20	26	2.40	2.52	Boolchand *et al.* (2002b)
Si_xSe_{100-x}	20	26.5	2.40	2.53	Wang and Boolchand (2004)
As_xSe_{100-x}	29	37	2.29	2.37	Georgiev *et al.* (2000); Boolchand *et al.* (2001b); Wang and Boolchand (2004)
As_xSe_{100-x}	27	38	2.27	2.38	Boolchand *et al.* (2002b)
As_xS_{100-x}	22.5	29.5	2.225	2.295	Boolchand *et al.* (2001a, 2002b)
Ge_xSe_{100-x}	20	26	2.40	2.52	Gump *et al.* (2004)
Ge_xSe_{100-x}	19	–	2.38	–	Wang and Boolchand (2004); Wang *et al.* (2005)
Ge_xSe_{100-x}	20	25	2.40	2.50	Wang and Boolchand (2004); Georgiev *et al.* (2003)
P_xSe_{100-x}	28	40	2.28	2.40	Wang *et al.* (2000)
$Ge_xAs_xSe_{100-2x}$	9	14	2.27	2.42	Boolchand *et al.* (2002b)
$Ge_xAs_xSe_{100-2x}$	10	14	2.30	2.42	Mamedov *et al.* (2003)
$Ge_xAs_xSe_{100-2x}$	9	16	2.27	2.48	Qu *et al.* (2003)
$Ge_xAs_xSe_{100-2x}$	11	15	2.33	2.45	Wang and Boolchand (2004)
$Ge_xAs_xSe_{100-2x}$	9	15.5	2.27	2.46	Qu and Boolchand (2005)
$Ge_xAs_xS_{100-2x}$	9.3	15.7	2.28	2.47	Cernoskova *et al.* (2005)
$Ge_{25}Se_{75-x}I_x$	15.5	16.4	2.345	2.336	Wang and Boolchand (2004)
$Ge_{25}S_{75-x}I_x$	15.5	16.4	2.345	2.336	Wang and Boolchand (2004)
$Ge_xP_xSe_{100-2x}$	9	14.5	2.27	2.43	Wang and Boolchand (2004); Chakravarty *et al.* (2005)
$Ge_xP_xSe_{100-2x}$	9	13.5	2.270	2.405	Vempati and Boolchand (2004)
$Ge_{7.5}As_xTe_{92.5-x}$	10	20	2.35	2.45	Wang and Boolchand (2004)

led to significant discrepancies with Phillips–Thorpe rigidity percolation theory (Chubynsky, 2009). Thus, to overcome these obstacles for As-based ChG, the existence of quasi-tetrahedral $Se{=}As(Se_{1/2})_3$ structural units, having double-bonded Se atoms and optimal number of Lagrangian constraints per atom $n_c = 3$, was admitted by Georgiev *et al.* (2000). Another approach was based on 'repulsive interactions of lone pair electrons of As atoms' to introduce additional constraints (Lucovsky *et al.*, 2007). Whichever the case, to explain the expansion of the intermediate phase into the $Z < 2.4$ region, the four-fold coordinated As atoms, was necessarily introduced (Georgiev *et al.*, 2000; Lucovsky *et al.*, 2007). In contrast to P-(S)Se ChG, having their crystalline counterparts in the form of corresponding penta-chalcogenides $P_2S(Se)_5$ based on quasi-tetrahedral units (Feltz, 1993), the As-based ChG are not the same, as most of the reliable experimental data speak in favor of exceptionally three-fold coordinated As atoms (Feltz, 1993; Borisova, 1981; Vinogradova, 1984; Golovchak *et al.*, 2008a; Saiter *et al.*, 2009; Renninger and Averbach, 1973; Yang *et al.*, 2010).

Collapse of wide reversibility windows into the narrow regions near $Z \approx 2.4$ for As-based ChG after a long-term period of natural storage (Fig. 8.6(a) and 8.13) (Golovchak *et al.*, 2006a, 2008a; Elabbar and Abu-Sehly, 2011) has shown that all glasses, whose backbone is under-constrained (n_c<3), are characterized by a pronounced physical ageing, removing the discrepancy

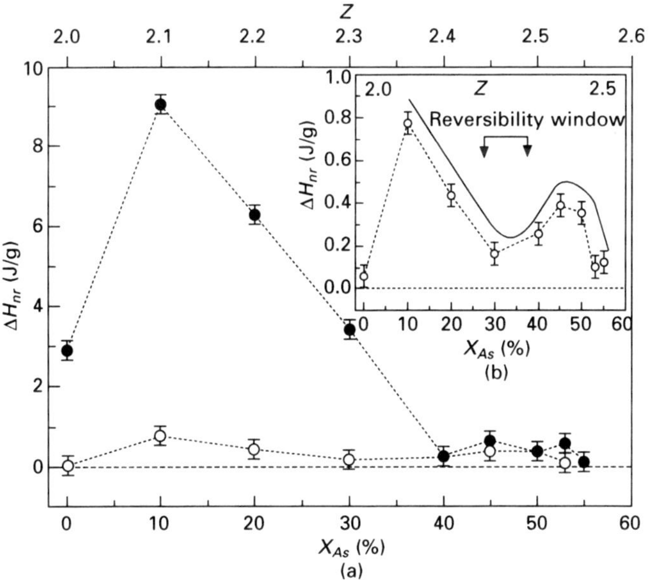

8.13 Compositional dependence (a) of ΔH_{nr} term for aged (solid) and rejuvenated (open) As_xSe_{100-x} (Golovchak *et al.*, 2008a). Inset (b) shows the compositional trend (solid line) of ΔH_{nr} published by Georgiev *et al.* (2000).

with rigidity percolation theory. In this case, the non-reversible component of heat flow in TMDSC experiments (as well as the endothermic peak area in DSC experiments) only indicates temporary compositional boundaries of natural physical ageing, determined by a time interval passed between ChG preparation and current DSC/TMDSC measurements. The possibility for long-term kinetics of physical ageing should be taken into account (Fig. 8.13). Therefore, the compositional boundaries of reversibility windows indicated in Table 8.1 can be accepted only as the boundaries of short-term physical ageing rather than the range of self-organized topological phases in ChG.

However, the absence of long-term physical ageing itself is a necessary, but not sufficient, criterion for identification of topologically self-organized phases in ChG. Sometimes, the over-constrained ($n_c > 3$) networks can be also characterized by non-ageing ability, as in the case of the pseudo-self-organization phenomenon of the binary Ge_xSe_{100-x} system (Golovchak *et al.*, 2009b; Shpotyuk and Golovchak, 2011). In this case, the range of non-ageing ability (both short-term and long-term) was identified by DSC and TMDSC techniques within $20 \leq x \leq 25$, accepted also as the reversibility window in this system (Wang and Boolchand, 2004; Georgiev *et al.*, 2003).

However, careful structural investigations show that the network of Ge_xSe_{100-x} glasses within this domain contains structural fragments proper to high-temperature modification of crystalline $GeSe_2$ (Golovchak *et al.*, 2009c). The latter consist of two edge-shared $GeSe_{4/2}$ tetrahedra connected with four corner-shared $GeSe_{4/2}$ ones (so-called *outrigger raft* clusters), which are over-constrained with $n_c > 3$ (Fig. 8.14(a)). Quantum mechanics calculations show that all glass compositions of binary Ge_xSe_{100-x} system within $20 < x < 28$ range are more energetically preferable in their atomic configurations,

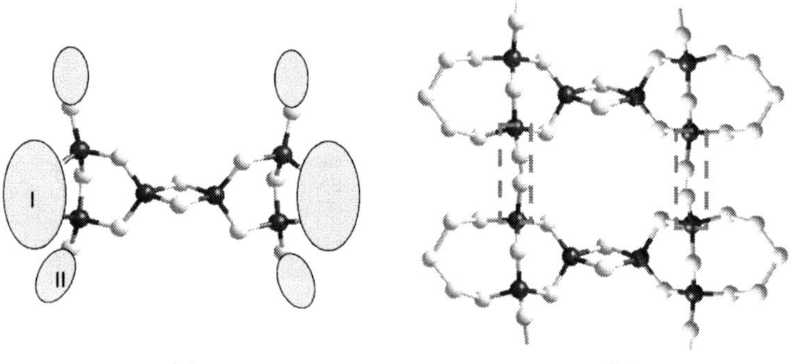

(a) (b)

8.14 Outrigger raft structural motive in Ge_xSe_{100-x} glasses within $20 < x < 28$ compositional range (a) and most energetically favorable linking for optimally constrained $Ge_{21.5}Se_{78.5}$ composition (b). Area I denotes a place of initial Se-Se dimers, area II shows a place of inter-cluster bridges. Dark four-fold coordinated spheres denote Ge atoms, while light two-fold coordinated spheres denote Se atoms.

which involve two Se atoms incorporated between separate outrigger raft network-forming clusters in the form of optimally constrained ≡Ge-Se-Se-Ge≡ inter-cluster bridges (Fig. 8.14(b)) (Shpotyuk and Golovchak, 2011; Shpotyuk *et al.*, 2010, Golovchak *et al.*, 2009c). Other Se atoms, which do not participate in the backbone connectivity, form Se-rich fragments attached to place I in Fig. 8.14(a), as is arbitrarily shown in Fig. 8.14(b). According to this scheme of structural evolution, only one glass composition $Ge_{21.5}Se_{78.5}$ (Fig. 8.14(b)) possesses $n_c = 3$ in a global sense, e.g. for separate out-rigger raft clusters themselves and inter-cluster bridges. Thus, the real self-organized phase in a strong sense is expected to exist in these glasses only in a narrow domain around $x \approx 20$–22, in good agreement with Thorpe *et al.* (2000), Chubynsky *et al.* (2006) and Chubynsky (2009). All the remaining ChG in the range of reversibility window ($20 \leq x \leq 25$) (Wang and Boolchand, 2004; Georgiev *et al.*, 2003) possess glassy networks built of locally over-constrained ($n_c > 3$) outrigger raft clusters interconnected via optimally constrained inter-cluster bridges. This evolution trend, restricted by $Ge_{27.25}Se_{72.75}$ and $Ge_{21.5}Se_{79.5}$ glass compositions (determined with the accuracy of x-step in the modelling procedure; Shpotyuk and Golovchak, 2011; Shpotyuk *et al.*, 2010), forms a so-called *pseudo-self-organized phase*.

So, in addition to known experimental observations of long-term physical ageing, the energetic characteristics of optimally constrained network-forming units should be examined by theoretical calculations. If one could show the energetic preference of units with $n_c = 3$ over the remaining possible units (or, at least, they should coexist at low entropy cost) in a specific range of ChG compositions where long-term physical ageing is absent, then this range can be accepted as a *topologically self-organized phase*. Nevertheless, the existence of these units in real glass networks still needs to be verified experimentally.

So, compositional trends in long-term physical ageing of different ChG systems can also demonstrate the existence of pseudo-self-organized topological phases. The *self-organized optimally constrained intermediate topological phases* are possible in the glasses whose networks allow formation of a variety of optimally constrained network-forming clusters ($n_c = 3$) at low entropy cost. In contrast, *the pseudo-self-organized topological phases* occur when rigid structural units (like stressed over-constrained with $n_c > 3$) are linked via optimally constrained inter-cluster bridges with $n_c = 3$.

8.4.3 Kinetics of physical ageing in chalcogen-rich glasses

In general, time evolution of the departure from equilibrium of any physical quantity, δ, during isothermal ageing can be represented by Kohlrausch's stretch-exponential law (Ngai, 2011; Angell *et al.*, 2000):

$$\delta(t) = \delta_0 \exp\left\{-\left(\int_0^t \frac{dt}{\tau}\right)^{\beta}\right\} \qquad [8.12]$$

where δ_0 is the initial departure of physical quantity from equilibrium; τ is the relaxation time constant; and $0 \leq \beta \leq 1$ is the non-exponentiality index.

Then, non-linearity of structural recovery can be accounted for by an appropriate model like the Tool–Narayanaswamy–Moynihan (TNM) (Narayanaswami, 1971), Kovacs–Aklonis–Hutchinson–Ramos (KAHR) (Kovacs *et al.*, 1979) or Hodge–Scherer (Hodge, 1991) models, which are considered as equivalent. These models, however, do not address the microscopic nature of structural relaxation, describing only the behavior of macroscopic physical quantities. The most frequently used is the TNM model (Kovacs *et al.*, 1979; Pustkova *et al.*, 2005):

$$\tau = \tau_0 \exp\left\{y\frac{E_a}{RT} + \frac{(1-y)E_a}{RT_F}\right\} \qquad [8.13]$$

where y is structural factor ($0 \leq y \leq 1$); E_a is activation energy and R is gas constant. Nevertheless, this model cannot account for the temperature or structural dependence of the β parameter, if any (Ngai, 2011).

At the same time, it is also shown that various relaxing properties including enthalpy, volume, stress, strain and refractive index behave differently in the glass transition region and reach equilibrium at widely different times (Ngai, 2011; Struik, 1978; Moynihan *et al.*, 1976; Simon *et al.*, 1997). The β and τ values depend also on the waiting time t below the glass transition temperature (Ngai, 2011; Angell *et al.*, 2000). In particular, a large increase in τ accompanied by decrease in β upon ageing is testified by studying colloidal suspensions (Ngai, 2011; Leonardo *et al.*, 2006). On the other hand, increase in β is expected with increasing of non-linearity y of the system (Hodge, 1991) and temperature (in the case of thermorheologically complex systems) (Ngai, 2011; Angell *et al.*, 2000; Stickel, 1995). Again, the lower the $(T_F - T_a)$ value, the faster the glass system approaches an equilibrium because of $\tau(T)$ dependence (Ngai, 2011; Angell *et al.*, 2000). All these correlations make a lot of complications for developing a *master empirical equation* capable of describing ageing kinetics in complex glassy systems.

Physical ageing kinetics can be studied with DSC/TMDSC, using changes in the glass transition region (Plate IV between pages 330 and 331). Enthalpy losses ΔH, changes in T_g (onset T_g^{onset}, midpoint T_g^{mid}, inflection T_g^{infl} or endset T_g^{end} values), partial area A or fictive temperature T_F (Moynihan *et al.*, 1974) can be used for quantitative analysis of physical ageing kinetics. As a rule, the $\Delta H(t_a, T_a)$ dependence in semilogarithmic scale has a sigmoidal trend (Fig. 8.15), approaching saturation value determined by a current ageing temperature T_a, while the fictive temperature T_F is known to approach the

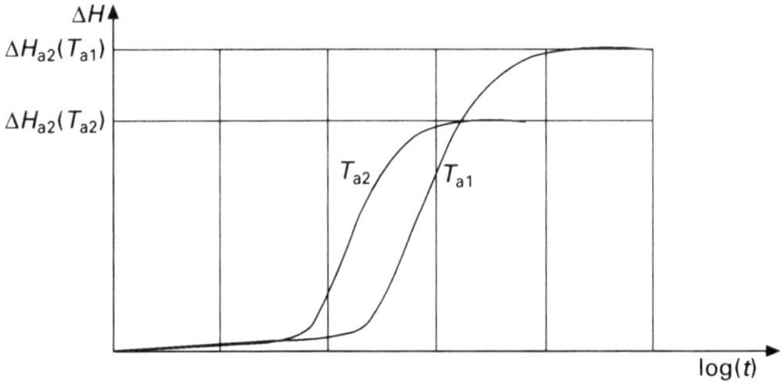

8.15 Schematic view of expected variation of energy recovery associated with energy lost during ageing for two different annealing temperatures $T_{a1} < T_{a2} < T_g$.

ageing temperature ($T_a \approx 300$ K, in case of room storage) over time according to Eq. [8.2] (Ngai, 2011; Angell *et al.*, 2000; Saiter *et al.*, 2005; Nemilov, 2000; Hodge, 1994). Typical time dependences of enthalpy losses ΔH and fictive temperature T_F values, obtained from DSC measurements for some representative As-Se glasses, are shown in Fig. 8.16. The completeness of the physical ageing process can be estimated by T_F values. Thus, in the ChG considered, the physical ageing is fully completed for vitreous Se ($T_F \approx 300$ K) and almost completed for the $As_{10}Se_{90}$ sample ($T_F \approx 305$ K) after two decades of ambient storage in the dark (Fig. 8.16) (Golovchak *et al.*, 2010b). Contrary to these glasses, the physical ageing for $As_{20}Se_{80}$ and, more pronounced, for $As_{30}Se_{70}$ ChG is far from its completeness, since corresponding T_F values are far from $T_a \approx 300$ K even after long-term ageing (Golovchak *et al.*, 2010b). So, different stages of physical ageing kinetics can be captured: the saturation exemplified by vitreous Se, beginning (glassy $As_{30}Se_{70}$), intermediate (glassy $As_{20}Se_{80}$) and almost full kinetics of physical ageing (glassy $As_{10}Se_{90}$). This circumstance should be taken into account during theoretical description of kinetic dependences, because different functions can fit various stages of physical ageing kinetics with different quality. At present, the physical ageing kinetics in ChG are not fully established.

8.5 On the origin of physical ageing in chalcogenide glasses

8.5.1 Physical ageing by structure-sensitive probes

Structural investigations of physical ageing with Raman spectroscopy allowed the identification of the mutual re-conformations of bent-deformed (Se atoms in *cis* configurations) and chain-like (Se atoms in *trans* configurations) Se fragments (Golovchak *et al.*, 2007b), as well as small changes in bond

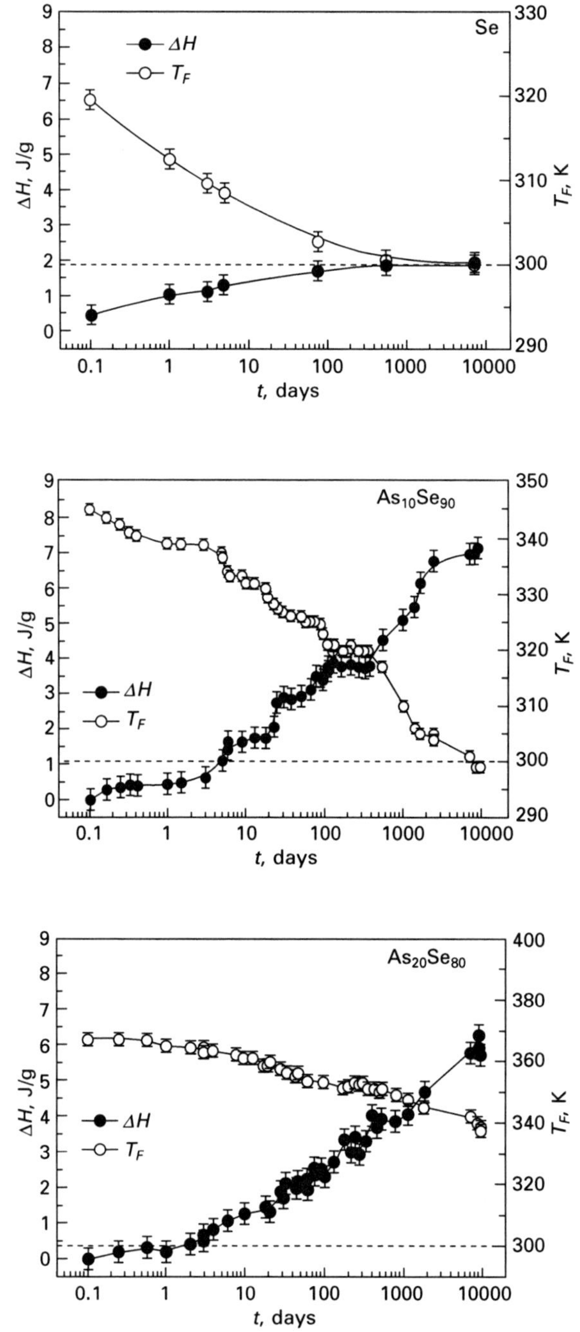

8.16 Physical ageing kinetics probed by DSC measurements for typical representative As_xSe_{100-x} glasses.

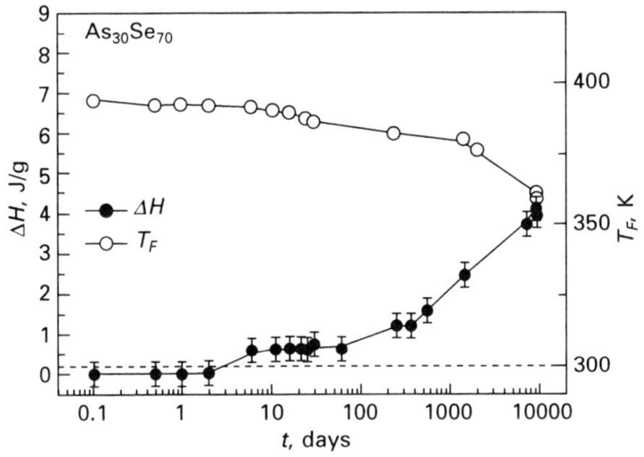

8.16 Continued

statistics (Golovchak *et al.*, 2008c). The Raman spectra of 20-years aged and rejuvenated $As_{10}Se_{90}$, $As_{20}Se_{80}$ and $As_{30}Se_{70}$ ChG as well as their differences are shown in Fig. 8.17, as common examples. The observed band components (ξ) in Fig. 8.17(a) and (b) for glassy $As_{10}Se_{90}$ and $As_{20}Se_{80}$ have been assigned to $AsSe_3$ pyramidal units (~230 cm^{-1}), Se_n chains (230–250 cm^{-1}) and Se ring-like fragments (250–260 cm^{-1}) (Nakamura and Ikawa, 2002; Bogomolov *et al.*, 1985; Kovanda *et al.*, 2003; Yannopoulos and Andrikopoulos, 2004). As a result of prolonged natural storage, the changes in Raman spectra of these glasses can be conveniently represented in the form of difference spectra (insets to Fig. 8.17) (Golovchak *et al.*, 2007b, 2008c). The bands of negative intensity (Δξ) at ~260 cm^{-1} in the insets to Fig. 8.17(a) and (b) can be associated with the disappearance of Se ring-like fragments (*cis* configurations of Se atoms in chains), while the appearance of straightened/aligned Se_n chains instead of the former should cause the band of positive intensity at Δξ ~240 cm^{-1}. The observed peaks in the Raman spectra of aged and rejuvenated $As_{30}Se_{70}$ glass (Fig. 8.17(c)) have been assigned to bond-stretching vibrations of directly corner-shared (~230 cm^{-1}) or Se-Se shared (~240 cm^{-1}) $AsSe_{3/2}$ pyramids and Se_3 segments (~250–260 cm^{-1}) (Bogomolov *et al.*, 1985; Kovanda *et al.*, 2003; Yannopoulos and Andrikopoulos, 2004). The weak changes after prolonged natural storage are seen in the difference spectrum between aged and rejuvenated samples (inset to Fig. 8.17(c)) (Golovchak *et al.*, 2008c). The band of positive intensity (Δξ) at ~260 cm^{-1} is assigned to the appearance of Se-Se-Se fragments and one at ~230 cm^{-1} to the formation of As-Se-As fragments or corner-shared $AsSe_{3/2}$ pyramids as a result of long-term physical ageing. Analysis of broad features in the 80–150 cm^{-1} range corresponding to bond-bending vibrations is more complicated because of overlapping of different vibrational modes.

The NMR data, recorded for rejuvenated and ~20-years aged $As_{10}Se_{90}$

glass (Fig. 8.18(a)) (Golovchak *et al.*, 2007b) show the general invariance of the chain-crossing model during physical ageing. The lines corresponding to Se-Se-Se (~860 ppm), As-Se-As (~380 ppm) and Se-Se-As (~580 ppm) fragments, can be identified from NMR data obtained for a number of samples of the same ChG series (Fig. 8.18(b)), rich in Se-Se-Se, Se-Se-As and As-Se-As fragments (Bureau *et al.*, 2003). According to such identification, a slight increase of Se-Se-Se (~860 ppm), As-Se-As (~380 ppm) and decrease of Se-Se-As (~580 ppm) sites in the structure of aged $As_{30}Se_{70}$ glass can be found by appropriate fitting of ^{77}Se NMR spectra (Fig. 8.18(b)) (Golovchak

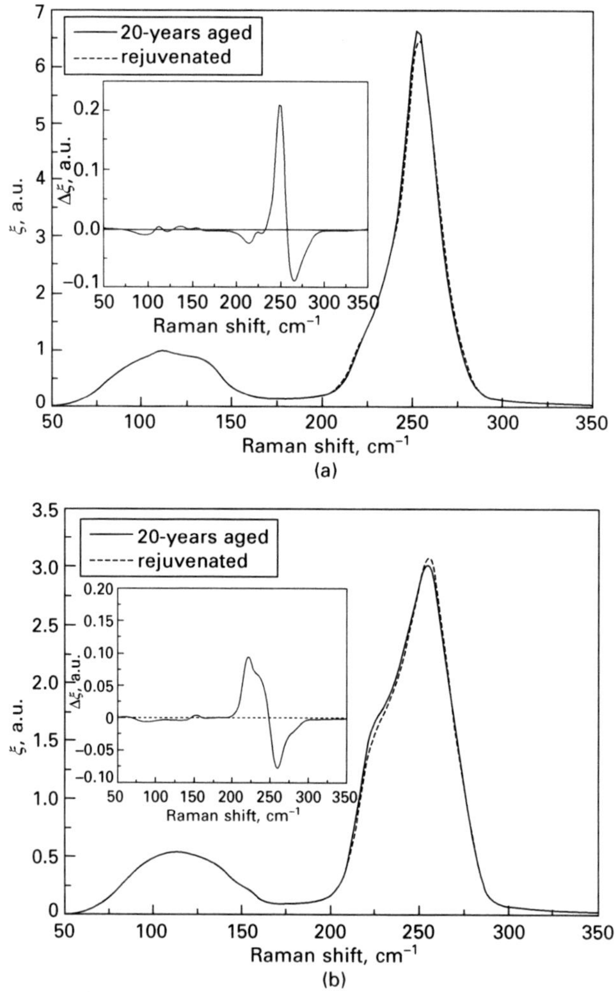

8.17 Raman spectra of 20-years aged and rejuvenated glassy $As_{10}Se_{90}$ (a) $As_{20}Se_{80}$ (b) and $As_{30}Se_{70}$ (c) samples. Insets show difference between aged and rejuvenated curves.

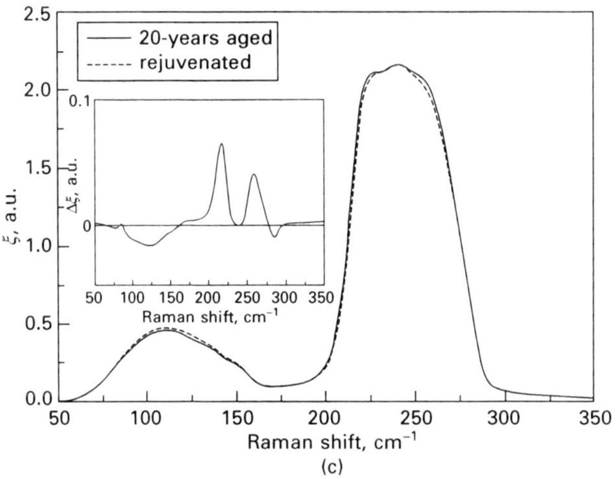

8.17 Continued

et al., 2008c). This conclusion is supported also by high resolution XPS data for Se and As core levels of glassy $As_{30}Se_{70}$ sample, which allowed relative fractions of transformed complexes to be determined (Golovchak *et al.*, 2008c). So, although the changes observed for $As_{30}Se_{70}$ ChG after 20 years of physical ageing are weak in their magnitude (in the case of NMR, they only slightly exceed a noise level), the Raman scattering, NMR and XPS support the idea that two As-Se-Se-As structural units transform into As-Se-As and As-Se-Se-Se-As fragments during prolonged natural storage.

In general, we can conclude from Raman, XPS and NMR results that long-term physical ageing does not cause drastic changes in short-range ordering of ChG. The same conclusion is also supported by extended X-ray absorption fine structure (EXAFS) spectroscopy, which is sensitive to changes in the local coordination number around the target atom, to the distances in the coordination shells, as well as to the average deviations in bond lengths and angles (mean square relative displacement or the so-called Debye–Waller factor). Thus, the differences between 20-years aged and rejuvenated glassy $As_{10}Se_{90}$ samples are shown in Fig. 8.19 in the form of Fourier transformed (FT) EXAFS spectra (Golovchak *et al.*, 2007b). The well-pronounced peak in partial radial distribution functions is associated with the first coordination shell, and minor peaks at longer distances are attributed to second and further coordination shells. Comparative analysis of the EXAFS data for $As_{10}Se_{90}$ glass shows only slight changes (possible small increase in the local coordination number N_j for Se K-edge, distance R_j for As K-edge and Debye–Waller factor σ^2 for both of them) after 20 years of natural storage (Golovchak *et al.*, 2007b), which can be well explained by shrinkage of glass backbone and re-conformation of some Se-based structural

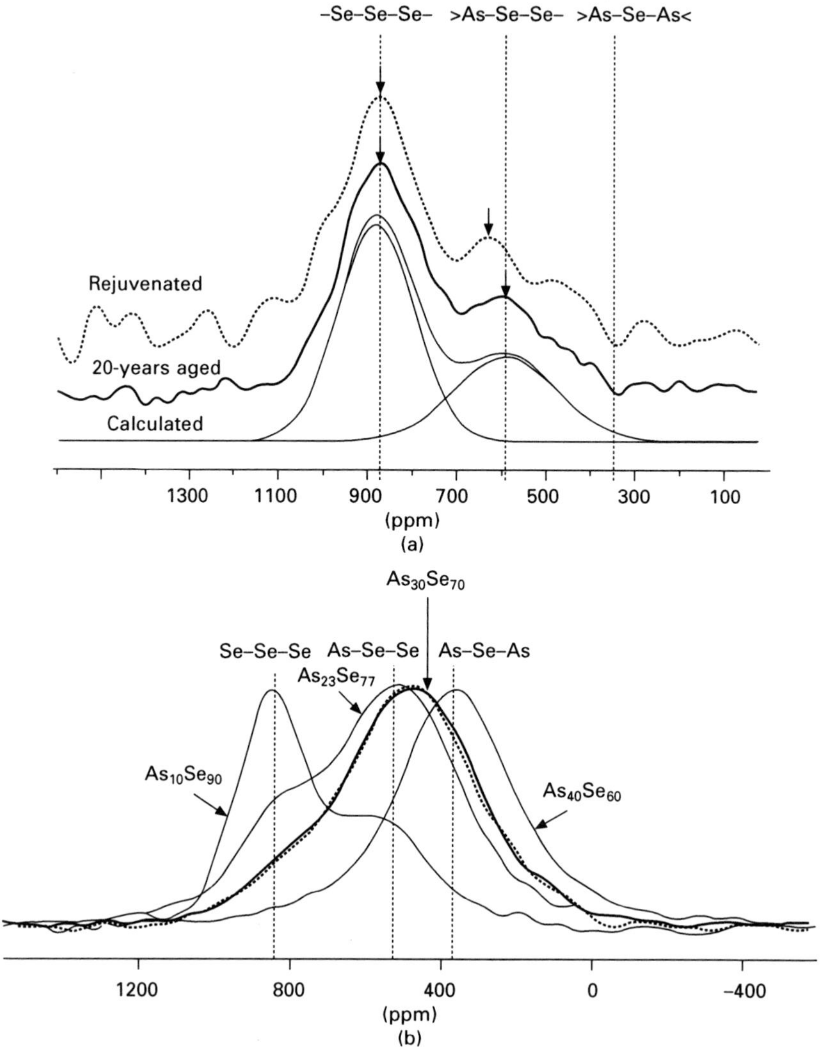

8.18 Solid state Se^{77} NMR spectra of 20-years aged and rejuvenated glassy $As_{10}Se_{90}$ (a) and $As_{30}Se_{70}$ (b) samples (Golovchak *et al.*, 2007b, 2008c).

fragments. With increase in As content, even these changes become less prominent (Golovchak *et al.*, 2008c). Thus, fitting parameters of EXAFS spectra for $As_{30}Se_{70}$ and $As_{40}Se_{60}$ ChG do not show any significant changes in the average interatomic distances (R_j) or coordination number (N_j) after 20 years of natural storage or ageing at near T_g temperature for both As and Se elements (Golovchak *et al.*, 2008c, 2011b). There is also no evidence for

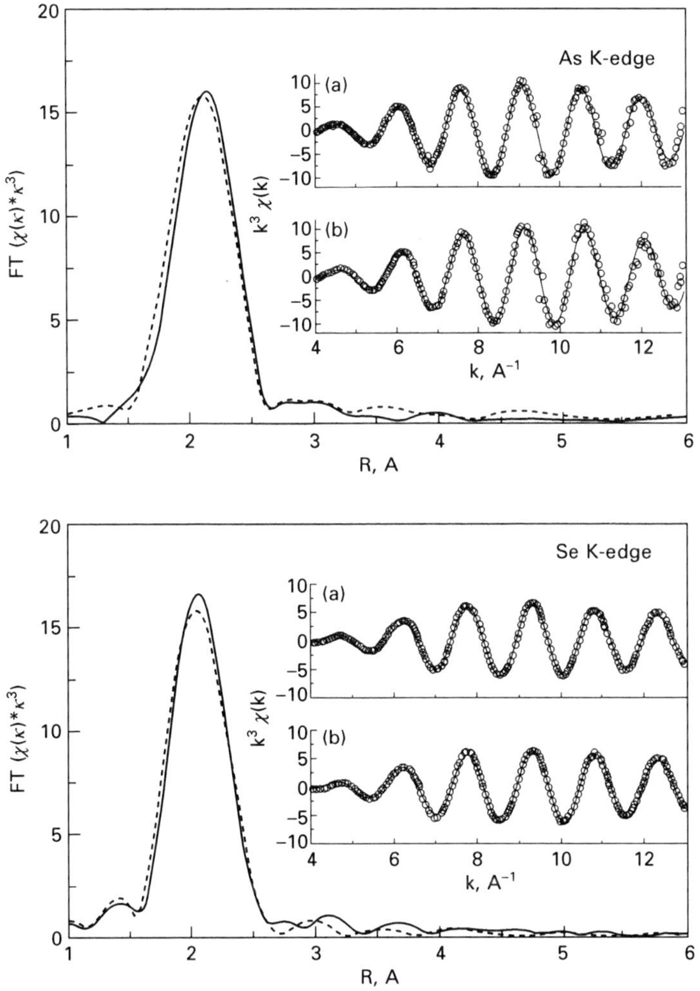

8.19 FT-EXAFS spectra of g-$As_{10}Se_{90}$ taken at As and Se K-edges in rejuvenated (dash) and 20-years aged (solid) states. k^3-weighted EXAFS oscillations $\chi(k)$ (open circles) and their best fits (solid lines) are shown as insets for 20-years aged (a) and rejuvenated (b) sample (Golovchak *et al.*, 2007b, 2008c).

coordination defects or broken bonds in significant concentrations formed during physical ageing of Se-based ChG.

More significant changes are expected at a so-called intermediate range ordering level of these materials. However, the corresponding experimental investigations have not been performed to date.

8.5.2 A unified structural model of physical ageing in chalcogenide glasses

Recent experimental investigations and theoretical achievements suggest that physical ageing well below T_g occurs through the Johari–Goldstein (JG) β-relaxation mechanism (Ngai, 2011), which can be considered as the initiating process (precedent) for α-relaxation events (Ngai, 2011; Nemilov and Johari, 2003). However, the microstructural nature of these processes is not yet confidently established for any of the materials.

The nanostructural mechanism proposed for physical ageing in ChG (Golovchak *et al.*, 2007b, 2008c) consists of two stages of structural perturbations (Fig. 8.20). The first one is grounded on the elementary relaxation events (twisting) of inner Se atoms within double-well potentials associated with high flexibility of chalcogen bonds (Fig. 8.20) (Golovchak *et al.*, 2007b). It causes spontaneous densification of some local regions (identified as the alignment stage in Golovchak *et al.*, 2007b), which ultimately leads to the formation of loosely packed regions and elastic strains in their immediate surrounding (as was suggested for physical ageing of oxide glasses; Nemilov and Johari, 2003) under the condition that total macroscopic volume of the glass is still the same (Stage I in Fig. 8.20). Such situation leads to a lowering of the general connectivity strength of the glass backbone (due to elastic strains and partitioning of the system into loosely and more densely packed regions, in the first approximation), which is partially confirmed by simultaneous positron annihilation lifetime spectroscopy (PALS) and DSC measurements (Fig. 8.21) (Ingram *et al.*, 2012). Indeed, PALS is known to be a powerful tool for characterization of local free volumes (open-volume holes, inner pores, vacancies and vacancy-like agglomerations) in crystals, liquids, polymers and glasses on a sub-nanometer scale (Krause-Rehberg and Leipner, 1999; Jean *et al.*, 2003). The average positron lifetime (τ_{av}) is sensitive to the total amount of nanoscale volume free of the electron density (Krause-Rehberg and Leipner, 1999; Jean *et al.*, 2003). In general, the greater the amount of this volume, the greater the value of τ_{av} expected. A well-expressed correlation between time depenences of τ_{av}, onset value of T_g (T_g^{onset}) and ΔA is emphasized for $As_{20}Se_{80}$ glass (Fig. 8.21) (Ingram *et al.*, 2012), testifying the appearance of local free volume and collapsed regions (increased density fluctuations) during the initial stage of physical ageing (Stage I in Fig. 8.20). The same conclusion was drawn for glassy poly(methylmethacrylate) by Takahara *et al.* (1999), showing formation of loosely packed and densified local regions at the initial stage of physical ageing by means of light scattering and oscillating DSC studies.

Then, if the accumulated elastic energy became sufficient to overcome the energetic barrier of inter-metabasin transition in the potential energy/enthalpy landscape (Golovchak *et al.*, 2011c), the cooperative rearrangement

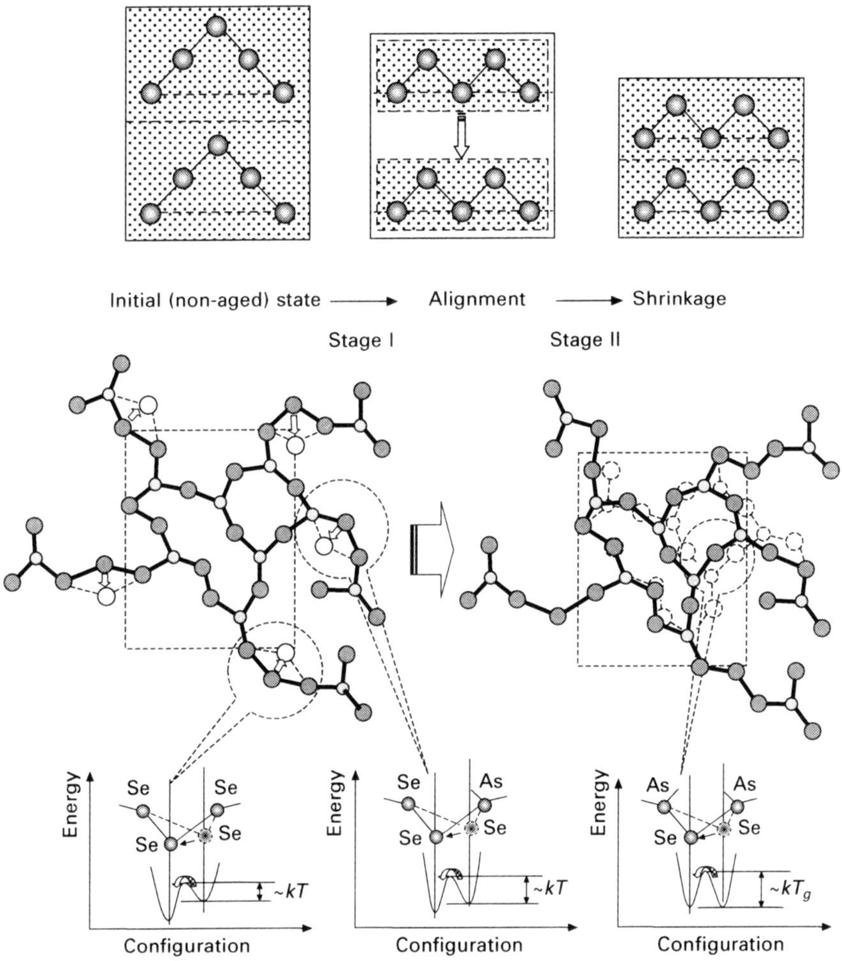

8.20 Schematic illustration of possible structural rearrangements in g-As-Se during physical ageing. In the first stage, the individual transitions of Se atoms within DWP occur resulting in the alignment of some regions, appearance of elastic strains and open volume nanovoids in their nearest surrounding. In the second stage, the cooperative atomic rearrangements and dissipation of elastic strains lead to reorientation of more extended structural groups, which result in shrinkage of glass.

occurs during next stage of physical ageing (Stage II in Fig. 8.20), eliminating redundant free volume from the glass. This leads to the lowering of potential energy of a whole system and, thus, to a decrease in the macroscopic volume of a glass (shrinkage) through elimination of strained regions and free volume from the bulk. Therefore, the second stage of physical ageing was identified as shrinkage of ChG network (Golovchak *et al.*, 2007b, 2008c; Ingram *et al.*,

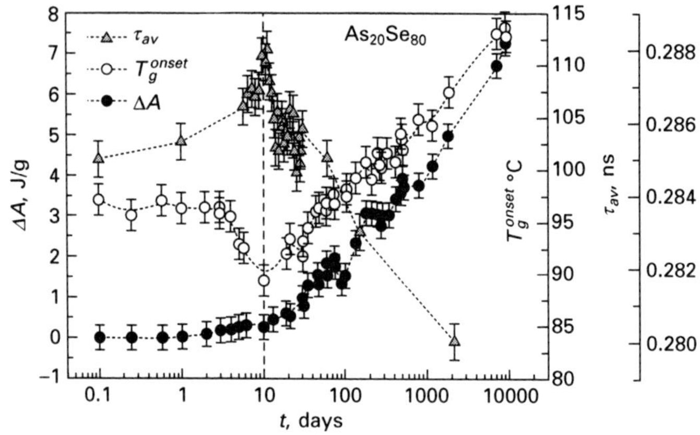

8.21 Time evolution of endothermic peak area difference ΔA, T_g^{onset} and τ_{av} during physical ageing in $As_{20}Se_{80}$ glass (Ingram *et al.*, 2012).

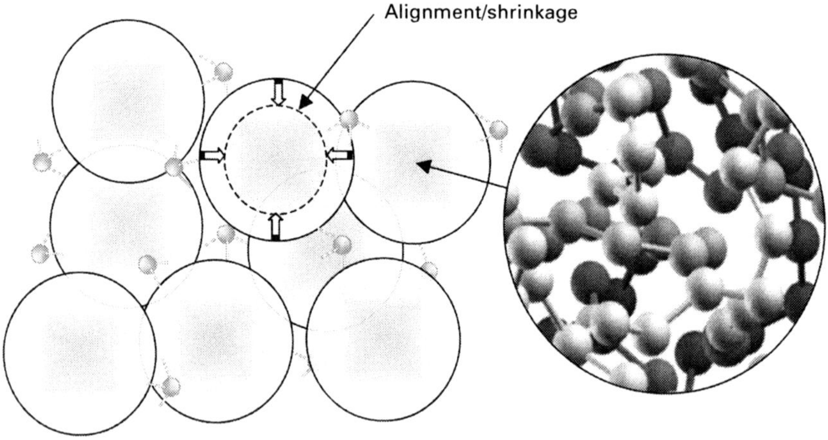

8.22 Schematic representation of physical ageing in g-$As_{10}Se_{90}$ through subsequent alignment shrinkage processes. The simulated structure of Se chains is taken from Golovchak *et al.* (2007b).

2012), leading to general densification of glassy bulk, as well as observed increase in T_g^{onset}, ΔA and decrease in τ_{av} values (Fig. 8.21).

In the case of very Se-rich samples (like glassy $As_{10}Se_{90}$), the Monte Carlo simulations (Mauro and Varshneya, 2005) allow one to schematically represent the cross-linked Se chains between $AsSe_3$ pyramids as spheres of ~15–20 Å diameter interlinked by As atoms (Fig. 8.22). When aligning shrinkage processes occur, they lead to a decrease in the volume of these virtual spheres and distortion of some chemical bonds (presumably seen with EXAFS and NMR; Golovchak *et al.*, 2007b). Taking place independently in each individual step, the alignment shrinkage processes have attained complex hierarchical behavior (cyclic repeating of Stages I and II in Fig. 8.20). The latter can be a reason for step-wise kinetics of physical ageing (like ΔH in Fig. 8.16), resulting in non-uniform shrinkage of glassy network.

At the same time, much lower dissociation energies of chalcogen bonds (within ~1.5–2.8 eV range) (Feltz, 1993; Varshneya, 2006) compared to that of the Si–O bond in SiO_2 glass (~4.6 eV) (Varshneya, 2006) together with the possibility of chalcogen–chalcogen bond formation (in the case of silicate glasses, the homopolar O–O bonds do not exist), introduce an additional possibility for covalent bond switching accompanying physical ageing. This is evident also from structural studies of physical ageing phenomena in $As_{30}Se_{70}$ glass (see previous section) (Golovchak *et al.*, 2008c). When the pyramid I in a floppy configuration starts rotating (Fig. 8.23) due to some relaxation processes, re-conformation of some Se-Se shared $AsSe_{3/2}$ pyramids (As-Se-Se-As structural fragments) into directly corner-shared (via As-Se-As bridges) pyramids and Se-Se-Se structural fragments can occur (Golovchak *et al.*, 2008c). This cooperative many-body relaxation process is expected to be more significant for As and/or Ge-rich ChG approaching rigidity transition (e.g., $As_{40}Se_{60}$ composition in the case of As-Se ChG, possessing optimally constrained network of directly corner-shared $AsSe_{3/2}$ pyramids).

8.6 Conclusion and future trends

Despite extensive experimental and theoretical works on the problem of physical ageing in network glasses, our understanding of this phenomenon is still in its infancy. The existing problems can be divided into three categories.

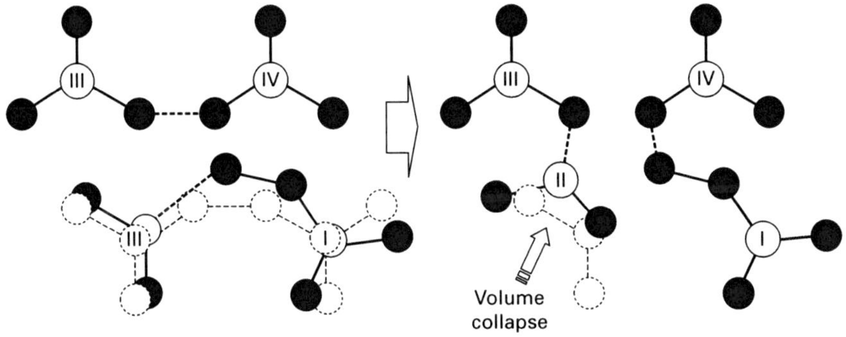

8.23 Schematic representation of bond-changing structural transformations in g-$As_{30}Se_{70}$ associated with long-term physical ageing (the switched bonds are highlighted; the previous positions of structural complexes are shown by dashed lines) (Golovchak *et al.*, 2008c).

First of all, the full kinetics of physical ageing at various temperatures should be investigated by DSC-TMDSC methods depending on glass fragility and network dimensionality (connectivity). Then, the influence of various external fields on the parameters of structural relaxation should be examined to understand the role of the chemical nature of the constitutive elements in physical ageing. Finally, the changes at medium range ordering of glassy structure should be emphasized to correlate them with theoretical models. Successful resolution of these problems will allow in the future the effect of network self-organization to be used in practice, to predict the stability and durability of ChG-based devices, as well as to develop high-reliable glassy media for modern optoelectronics.

8.7 References

Angell C A, Ngai K L, McKenna G B, McMillan P F and Martin S W (2000), 'Relaxation in glassforming liquids and amorphous solids', *J. Appl. Phys.*, 88, 3113.

Averyanov V L (1972), 'Investigations of local states in vitreous semiconductors of Se-As system', *Fiz. Tehn. Polupr.*, 6, 9, 1709.

Bauchy M and Micoulaut M (2011), 'Atomic scale foundation of temperature-dependent bonding constraints in network glasses and liquids', *J. Non-Cryst. Solids*, 357, 2530.

Bogomolov V N, Poborchy V V, Romanov S G and Shagin S I (1985), 'Raman spectra of chalcogen chains isolated in zeolite matrixes', *J. Phys. C: Solid State Phys.*, 18, L313.

Bohmer R and Angell C A (1992), 'Correlations of ninexponentiality and state dependence of relaxations with bond connectivity in Ge-As-Se supercooled liquids', *Physical Review B*, 45, 17, 10091–10094.

Boolchand P, Feng X and Bresser W J (2001a), 'Rigidity transitions in binary Ge-Se glasses and the intermediate phase', *Journal of Non-Crystalline Solids*, 293–295, 348–356.

Boolchand P, Georgiev D G and Goodman B (2001b), 'Discovery of the intermediate phase in chalcogenide glasses', *J. Optoelectron. Adv. Mater.*, 3, 703–720.

Boolchand P, Georgiev D G and Micoulaut M (2002a), 'Nature of glass transition in chalcogenides' *Journal of Optoelectronics and Advanced Materials*, 4, 4, 823–836.

Boolchand P, Georgiev D G, Qu T, Wang F, Cai L and Charkravarty S (2002b), 'Nanoscale phase separation effects near $r = 2.4$ and 2.67, and rigidity transitions in chalcogenide glasses', *C.R. Chimie*, 5, 713–724.

Boolchand P, Lucovsky G, Phillips J C and Thorpe M F (2005), 'Self-organization and the physics of glassy networks', *Phil. Mag.* 85, 3823.

Borisova Z U (1981), *Glassy Semiconductors*, New York: Plenum Press.

Bureau B, Troles J, Le Floch M, Smektala F, Silly G and Lucas J (2003), 'Solid state ^{77}Se NMR investigations on arsenic-selenium glasses and crystals', *Solid State Sciences*, 5, 219.

Calemczuk R and Bonjour E (1981), 'Gamma-ray induced relaxation in selenium glass', *J. Non-Cryst. Solids*, 43, 427–432.

Calvez L, Yang Z. and Lucas P (2008), 'Light-induced matrix softening of Ge-As-Se network glasses', *Phys. Rev. Lett.*, 101, 177402.

Cernosek Z, Holubova J and Cernoskova E (2003), 'Non-isothermal structural relaxation and the glass transition temperature', *Solid State Sci.*, 5, 1087–1093.

Cernoskova E, Cernosek Z, Holubova J and Frumar M (2001), 'Structural relaxation near the glass transition temperature', *J. Non-Cryst. Solids*, 284, 73–78.

Cernoskova E, Qu T, Mamedov S, Cernosek Z, Holubova J and Boolchand P (2005), 'Reversibility window in as-quenched Ge-As-S glasses', *Journal of Physics and Chemistry of Solids*, 66, 185–189.

Chakravarty S, Georgiev D G, Boolchand P and Micoulaut M (2005), 'Ageing, fragility and the reversibility window in bulk alloy glasses', *J. Phys.: Condens. Matter*, 17, L1–L7.

Chen P, Boolchand P and Georgiev D G (2010), 'Long-term aging of selenide glasses: evidence of sub-T_g endotherms and pre-T_g exotherms', *J. Phys.: Condens. Mat.*, 22, 065104.

Chubynsky M (2009), *Characterizing of intermediate phases through topological analysis*, in: *Rigidity and Boolchand Intermediate Phases in nanomaterials*, Bucharest: INOE Publishing House.

Chubynsky M V, Briere M-A and Mousseau M (2006), 'Self-organization with equilibration: a model for the intermediate phase in rigidity percolation', *Phys. Rev. E*, 74, 016116.

Churbanov M F, Shiryaev V S, Gerasimenko V V, Pushkin A A, Skripachev I V, Snopatin G E and Plotnichenko V G (2002), 'Stability of the optical and mechanical properties of chalcogenide fibers', *Inorganic Materials*, 38, 1063.

Cortes P, Montserrat S, Ledru J and Saiter J M (1998), 'Enthalpy relaxation in Ge_xSe_{1-x} chalcogenide glasses', *J. Non-Cryst. Solids*, 235–237, 522–526.

Derrey T, Saiter J M and Vautier C (1986), 'Photodarkening of amorphous $Se_{100-x}Ge_x$ alloys ($2 < x < 16$)', *Materials Letters*, 4, 8–9, 365.

Domoryad I A and Kolomiets B T (1971), 'Changes in the elastic properties of As_2S_3-As_2Se_3 glasses under the influence of ionizing radiation', *Izv. AN SSSR: Inorg. Mat.*, 7, 9, 1620–1621.

Donth E (2001), *The Glass Transition Relaxation Dynamics in Liquids and Disordered Materials*, Berlin: Springer.

Drozdov A D (1999), 'Physical aging in amorphous polymers far below the glass transition temperature', *Computational Materials Sci.*, 15, 422.

Echeverria I, Kolek P L, Plazek D J and Simon S L (2003), 'Enthalpy recovery, creep and creep–recovery measurements during physical aging of amorphous selenium', *J. Non-Cryst. Solids*, 324, 242.

Elabbar A A and Abu-Sehly A A (2011), 'Structural relaxation and rigidity transition in aged and rejuvenated As_xSe_{100-x} glasses', *Physica B*, 406, 4261–4265.

El Gharras Z (1989), Contribution to the role of defects in the transport phenomena in the amorphous Ge-Se, Thesis, Faculty of Science, University of Rouen.

Fairman R and Ushkov B, eds (2004), *Semiconducting Chalcogenide Glasses 1: Glass Formation, Structure, and Simulated Transformations in Chalcogenide Glasses*, Amsterdam: Elsevier Academic Press.

Feltz A (1993), *Amorphous Inorganic Materials and Glasses*, New York: VCH Publishers.

Frazer R A, Dunkan W J and Collar A R (1938), *Elementary Matrices*, Cambridge: Cambridge University Press.

Frumar M, Firth A P and Owen A E (1983), 'A model for photostructural changes in the amorphous As-S system', *J. Non-Cryst. Solids*, 59–60, 921.

Frumar M, Vlcek M, Cernosek Z, Polak Z and Wagner T (1997), 'Photoinduced changes of the structure and physical properties of amorphous chalcogenides', *J. Non-Cryst. Solids*, 213–214, 215.
Ganjoo A, Ikeda Y and Shimakawa K (1999), '*In situ* photoexpansion measurements of amorphous As_2S_3 films: role of photocarriers', *Appl. Phys. Letters*, 74, 2119.
Georgiev D G, Boolchand P and Micoulaut M (2000), 'Rigidity transitions and molecular structure of As_xSe_{1-x} glasses', *Phys. Rev. B*, 62, R9228.
Georgiev D G, Boolchand P, Eckert H, Micoulaut M and Jackson K (2003), 'The self-organized phase of bulk P_xSe_{1-x} glasses', *Europhysics Letters*, 62, 1, 49–55.
Gjersing E L, Sen S and Youngman R E (2010), 'Mechanistic understanding of the effect of rigidity percolation on structural relaxation in supercooled germanium selenide liquids', *Phys. Rev. B*, 82, 014203.
Golovchak R and Shpotyuk O (2005), 'Radiation-induced bond switching in mixed Ge-As sulphide glasses', *Phil. Mag.*, 85, 2847–2860.
Golovchak R, Gorecki Cz, Kozdras A and Shpotyuk O (2006a), 'Physical ageing effects in vitreous arsenic selenides', *Sol. State Commun.*, 137, 67.
Golovchak R, Kozdras A, Gorecki Cz and Shpotyuk O (2006b), 'Gamma-irradiation-induced physical ageing in As–Se glasses', *J. Non-Cryst. Solids*, 352, 42–49, 4960.
Golovchak R, Kozdras A and Shpotyuk O (2006c), 'Physical ageing in vitreous $As_{13.5}Ge_{4.5}Se_{82}$ modified by high-energy gamma-irradiation', *Physica B: Condensed Matter*, 371, 2, 323–326.
Golovchak R, Kovalskiy A, Miller A C, Jain H and Shpotyuk O (2007a), 'Structure of Se-rich As-Se glasses by high-resolution x-ray photoelectron spectroscopy', *Physical Rev. B*, 76, 125208.
Golovchak R, Shpotyuk O, Kozdras A, Bureau B, Vlcek M, Ganjoo A and Jain H (2007b), 'Atomistic model of physical ageing in Se-rich As–Se glasses', *Phil. Mag.*, 87, 28, 4323.
Golovchak R, Jain H, Shpotyuk O, Kozdras A, Saiter A and Saiter J-M (2008a), 'Experimental verification of the reversibility window concept in binary As-Se glasses subjected to a long-term physical aging', *Phys. Rev. B*, 78, 014202.
Golovchak R, Kozdras A and Shpotyuk O (2008b), 'Physical ageing in glassy As–Se induced by above-bandgap photoexposure', *Solid State Commun.*, 145, 423.
Golovchak R, Shpotyuk O, Kozdras A, Vlcek M, Bureau B, Kovalskiy A and Jain H (2008c), 'Long-term physical ageing in As–Se glasses with short chalcogen chains', *J. Phys.: Cond. Matter*, 20, 245101.
Golovchak R, Kozdras A, Kozyukhin S and Shpotyuk O (2009a), 'High-energy γ-irradiation effect on physical ageing in Ge–Se glasses', *Nuclear Instr. Methods in Phys. Res. B*, 267, 2958.
Golovchak R, Kozdras A, Shpotyuk O, Kozyukhin S and Saiter J-M (2009b), 'Long-term ageing behaviour in Ge–Se glasses', *J. Mater. Sci.*, 44, 15, 3962.
Golovchak R, Shpotyuk O, Kozyukhin S, Kovalskiy A, Miller A C and Jain H (2009c), 'Structural paradigm of Se-rich Ge-Se glasses by high-resolution X-ray photoelectron spectroscopy', *J. Appl. Physics*, 105, 103704.
Golovchak R, Kozdras A and Shpotyuk O (2010a), 'Optical signature of structural relaxation in glassy $As_{10}Se_{90}$', *J. Non-Cryst. Solids*, 356, 23–24, 1149–1152.
Golovchak R, Kozyukhin S, Kozdras A, Shpotyuk O and Novotortsev V (2010b), 'Physical aging of chalcogenide glasses', *Inorganic Materials*, 46, 8, 919–921.
Golovchak R, Shpotyuk O, Kozdras A, Riley B J, Sundaram S K and McCloy J S (2011a), 'Radiation effects in physical aging of binary As-S and As-Se glasses', *J. Thermal Analysis and Calorimetry*, 103, 1, 213–218.

Golovchak R, Kozdras A, Shpotyuk O, Gorecki Cz, Kovalskiy A and Jain H (2011b), 'Temperature-dependent structural relaxation in $As_{40}Se_{60}$ glass', *Phys. Letters A*, 375, 3032–3036.

Golovchak R, Kovalskiy A, Shpotyuk O and Jain H (2011c), 'In search of energy landscape for network glasses', *Appl. Phys. Lett.*, 98, 171905.

Gump J, Finkler I, Xia H, Sooyakumar R, Bresser W J and Boolchand P (2004), 'Light-induced giant softening of network glasses observed near the mean-field rigidity transition', *Phys. Rev. Lett.*, 92, 24, 245501.

Gupta P K and Mauro J C (2009), 'Composition dependence of glass transition temperature and fragility. I. A topological model incorporating temperature-dependent constraints', *J. Chem. Phys.*, 130, 094503.

Hach C T, Cerqua-Richardson K, Varner J R and LaCourse W C (1997), 'Density and microhardness of As-Se glasses and glass fibers', *J. Non-Cryst. Solids*, 209, 159.

Ho N, Laniel J M, Vallee R and Villeneuve A (2003), 'Photosensitivity of As_2S_3 chalcogenide thin films at 1.5 μm', *Opt. Lett.* 28, 965.

Hodge I M (1991), 'Adam-Gibbs formulation of non-linear enthalpy relaxation', *J. Non-Cryst. Solids*, 131–133, 435–441.

Hodge I M (1994), 'Enthalpy relaxation and recovery in amorphous materials', *J. Non-Cryst. Solids*, 169, 211.

Hodge I M (1995), 'Physical ageing in polymer glasses', *Science*, 267, 1945–1947.

Hutchinson J M (1992), 'Interpretation of structural recovery of amorphous polymers from DSC data', *Progr. Colloid. Polym. Sci.*, 87, 69–73.

Hutchinson J M (1995), 'Physical ageing of polymers', *Prog. Polym. Sci.* 20, 703–760.

Hutchinson J M (2003), 'Studying the glass transition by DSC and TMDSC', *J. Thermal Analysis and Calorimetry*, 72, 619–629.

Imran M M A, Al-Hamarneh I F, Awadallah M I and Al-Ewaisi M A (2008), 'Physical ageing in $Se_{94}Sn_6$ glass induced by gamma irradiation', *Physica B*, 403, 2639–2642.

Ingram A, Kozdras A, Shpotyuk O and Golovchak R (2012), 'PAL signature of physical ageing in chalcogenide glasses', *Phys. Status Solidi B*, 249, 1017–1019.

Jean Y C, Mallon P E and Schrader D M, eds (2003), *Principles and Application of Positron and Positronium Chemistry*, Singapore: World Scientific.

Joule J (1884), *The Scientific Papers of J.P. Joule*, London: Physical Soc., vol. 1, 558.

Joule J P (1867), Mem. Manchr Literary Philos. Soc., 3rd ser., 292.

Keating P N (1966), 'Effect of invariance requirements on the elastic strain energy of crystals with application to the diamond structure', *Phys. Rev.*, 145, 637–645.

Kolobov A V, ed. (2003), *Photo-induced Metastability in Amorphous Semiconductors*, Weinheim: Wiley-VCH.

Kolomiets B T and Goryunova N A (1955), 'Properties and structure of ternary semiconductor systems: 2. Electrical properties and structure of materials in thallium, antimony, and arsenic selenide systems', *Zh. Tekh. Fiz.*, 25, 12, 2069.

Kolomiets B T, Mamontova T N, Domoryad I A and Babaev A A (1971), 'Photoluminescence in γ-irradiated vitreous and monocrystalline As_2S_3 and As_2Se_3', *Phys. Stat. Sol.*, A7, K29.

Kovacs A J, Aklonis J J, Hutchinson J M and Ramos A R (1979), 'Isobaric volume and enthalpy recovery of glasses. II. A transparent multiparameter theory', *Polym. Sci. Polym. Phys. Ed.*, 17, 1097–1162.

Kovalskiy A, Miller A C, Jain H and Mitkova M (2008), '*In situ* measurements of X-ray-induced silver diffusion into a $Ge_{30}Se_{70}$ thin film', *J. Am. Ceram. Soc.*, 91, 3, 760.

Kovanda V, Vlcek M and Jain H (2003), 'Structure of As–Se and As–P–Se glasses studied by Raman spectroscopy', *J. Non-Cryst. Solids*, 326–327, 88.

Kozdras A, Golovchak R, Shpotyuk O, Szymura S, Saiter A and Saiter J-M (2011), 'Light-assisted physical ageing in chalcogenide glasses: dependence on the wavelength of incident photons', *J. Materials Research*, 26, 18, 2420–2427.

Krause-Rehberg R and Leipner H (1999), *Positron Annihilation in Semiconductors*, Heidelberg: Springer.

Krecmer P, Moulin A M, Stephenson R J, Rayment T, Welland M E and Elliott S R (1997), 'Reversible nanocontraction and dilatation in a solid induced by polarized light', *Science*, 277, 1799.

Kurchan J (2005), 'In and out of equilibrium', *Nature*, 433, 222–225.

Kuzukawa Y, Ganjoo A and Shimakawa K (1998), 'Photoinduced structural changes in obliquely deposited As- and Ge-based amorphous chalcogenides: correlation between changes in thickness and band gap', *J. Non-Cryst. Solids*, 227–230, 715.

Larmagnac J P, Grenet J and Michon P (1982), 'Photodependence of sub-T_g relaxation in a-Se thin films', *Phil. Mag.* B45, 6, 627.

Leonardo R Di, Gentilini S, Ianni F and Ruocco G (2006), 'Aging and flow in a complex fluid', *J. Non-Cryst. Solids*, 352, 4928–4933.

Lima J C, Grandi T A and Biasi R S (2001), 'Influence of aging on the thermal and structural properties of amorphous selenium prepared by ball milling', *J. Non-Cryst. Solids*, 286, 93–98.

Lucas P (2006), 'Energy landscape and photoinduced structural changes in chalcogenide glasses', *J. Phys.: Cond. Matter*, 18, 5629.

Lucas P, King E A, Doraiswamy A and Jivaganont P (2005), 'Competitive photostructural effects in Ge-Se glass', *Phys. Rev. B*, 71, 104207.

Lucas P, King E A, Horner A D, Johnson B R and Sundaram S K (2006), 'Photostructural relaxation in As–Se–S glasses: effect of network fragility', *J. Non-Cryst. Solids*, 352, 2067.

Lucas P, King E A, Erdmann R G, Riley B J, Sundaram S K and McCloy J S (2011), 'Thermal and gamma-ray induced relaxation in As–S glasses: modeling and experiment', *J. Phys. D: Appl. Phys.*, 44, 395402.

Lucovsky G, Baker D A, Peasler M A, Phillips J C and Thorpe M F (2007), 'Intermediate phases in binary and ternary alloys. How far can we go with a semi-empirical bond-constraint theory?', *J. Optoelectron. Adv. Mater.*, 9, 2979–2988.

Mamedov S, Georgiev D G, Qu T and Boolchand P (2003), 'Evidence for nanoscale phase separation of stressed-rigid glasses', *J. Phys.: Cond. Matter*, 15, S2397–S2411.

Mauro J C and Varshneya A K (2005), 'Monte Carlo simulation of Se_xTe_{1-x} glass structure with *ab initio* potentials', *Phys. Rev. B*, 72, 024212.

McEnroe D J and LaCourse W C (1989), 'Tensile strengths of Se, As_2S_3, As_2Se_3, and $Ge_{30}As_{15}Se_{55}$ glass fibers', *J. Am. Ceram. Soc.*, 72, 1491.

Mott N F and Davis E A (1971), *Electronic Processes in Non-crystalline Materials*, Oxford: Clarendon Press.

Moynihan C T, Easteal A J and Wilder J (1974), 'Dependence of the glass transition temperature on heating and cooling rate', *J. Phys. Chem.*, 78, 2673.

Moynihan C T, Macedo P B, Montrose C J, Gupta P K, DeBolt M A, Dill J F, Dom B E, Drake P W, Eastel A J, Elterman P B, Moeller R P, Sasabe H and Wilder J A (1976), 'Structural relaxation in vitreous materials', *Ann. N.Y. Acad. Sci.*, 279, 15.

Nagels P, Tichy L, Mertens R and Callaerts R (2000), 'Low-temperature photodarkening of the As_xSe_{100-x} system prepared by PECVD', *Materials Letters*, 46, 234.

Nakamura K and Ikawa A (2002), 'Vibrational properties of a regular helical Se chain', *Phys. Rev. B*, 66, 024306.

Narayanaswami O S (1971), 'A model of structural relaxation in glass', *J. Am. Ceram. Soc.*, 54, 491–498.

Nemilov S V (2000), 'Physical ageing of silicate glasses at room temperature: general regularities as a basis for the theory and the possibility of *a priori* calculation of the ageing rate', *Glass Phys. and Chem.*, 26, 511.

Nemilov S V (2001), 'Physical ageing of silicate glasses at room temperature: the choice of quantitative characteristics of the process and the ranking of glasses by their tendency to ageing', *Glass Phys. and Chem.*, 27, 214.

Nemilov S V and Johari G P (2003), 'A mechanism for spontaneous relaxation of glass at room temperature', *Phil. Mag.*, 83, 3117.

Ngai K L (2011), *Relaxation and Diffusion in Complex Systems*, New York: Springer.

Oheda H (1979), 'The exponential absorption edge in amorphous Ge-Se compounds', *Jap. J. Appl. Phys.*, 18, 1973.

Phillips J C (1979), 'Topology of covalent non-crystalline solids. I: short-range order in chalcogenide alloys', *J. Non-Cryst. Sol.*, 34, 153–181.

Pikaev A K (1985), *Modern Radiation Chemistry. Main Statements, Experimental Technique and Methods*, Moscow: Nauka.

Poborchii V, Kolobov A and Tanaka K (1999), 'Photomelting of selenium at low temperature', *Appl. Phys. Letters*, 74, 2, 215.

Pustkova P, Shanelova J, Malek J and Cicmanec P (2005), 'Relaxation behavior of selenium based glasses', *J. Thermal Analysis and Calorimetry*, 80, 643–647.

Qu T and Boolchand P (2005), 'Shift in elastic phase boundaries due to nanoscale phase separation in network glasses: the case of $Ge_xAs_xS_{1-2x}$', *Phil. Mag.*, 85, *8*, 875–884.

Qu T, Georgiev D G, Boolchand P and Micoulaut M (2003), 'The intermediate phase in ternary $Ge_xAs_xSe_{1-2x}$ glasses', *Mat. Res. Soc. Symp. Proc.*, 754, CC8.1.1–CC8.1.12.

Renninger A L and Averbach B L (1973), 'Atomic radial distribution functions of As-Se glasses', *Phys. Rev. B*, 8, 1507.

Saiter A, Oliver J M, Saiter J M and Gomez Ribelles J L (2004), 'Entropic model for the relaxation in vitreous systems. Estimation of uncertainty in the calculation of the conformational relaxation times', *Polymer*, 45, 2743–2750.

Saiter A, Saiter J-M, Golovchak R, Shpotyuk M and Shpotyuk O (2009), 'Cooperative rearranging region size and free volume in As–Se glasses', *J. Phys.: Condens. Matter*, 21, 075105.

Saiter J M (2001), 'Physical ageing in chalcogenide glasses', *J. Optoelectron. Adv. Mater.*, 3, 685.

Saiter J M, Hamou A and Vautier C (1994), 'Physical aging kinetics in vitreous Se-based alloys', *J. Non-Cryst. Solids*, 172–174, 580–583.

Saiter J M, Benazeth S, Ma Quing, Belhadji M and Vautier C (1995), 'Physical aging in vitreous Ge-Te-Sb calorimetric and X-ray absorption studies', *J. Non-Cryst. Solids*, 192–193, 369–371.

Saiter J M, Arnoult M and Grenet J (2005), 'Very long physical ageing in inorganic polymers exemplied by the Ge_xSe_{1-x} vitreous system', *Physica B*, 355, 370.

Selvanathan D, Bresser W J, Boolchand P and Goodman B (1999), 'Thermally reversing window and stiffness transitions in chalcogenide glasses', *Solid State Comm.*, 111, 619–624.

Selvanathan D, Bresser W J and Boolchand P (2000), 'Stiffness transitions in Si_xSe_{1-x} glasses from Raman scattering and temperature-modulated differential scanning calorimetry', *Phys. Rev. B*, 61, 22, 15061–15076.

Shatnawi M, Farrow C, Chen P, Boolchand P, Sartbaeva A, Thorpe M and Billinge S (2008), 'Search for a structural response to the intermediate phase in Ge_xSe_{1-x} glasses', *Phys. Rev. B*, 77, 094134.

Shieh S H M and LaCourse W C (1993), 'Ageing and sub-T_g relaxation in arsenic selenide glass fibers', *Materials Chem. Phys.*, 35, 160.

Shimakawa K (2007), 'Photon effects in chalcogenide glasses', *J. Optoelectr. Adv. Materials*, 9, 10, 2973.

Shimakawa K, Kolobov A and Elliott S R (1995), 'Photoinduced effects and metastability in amorphous semiconductors and insulators', *Adv. in Physics*, 44, 6, 475.

Shpotyuk O I (1990), 'Gamma-induced changes in optical properties of chalcogenide glass semiconductors', *Ukr. Phys. J.*, 35, 10, 1545–1549.

Shpotyuk O and Golovchak R (2011), 'Pseudo-self-organized topological phases in glassy selenides for IR photonics', *Phys. Stat. Solidi C*, 8, 9, 2572–2576.

Shpotyuk O I, Kovalsky A P, Vakiv M M and Mrooz O Ya (1994), 'Reversible radiation effects in vitreous As_2S_3. I. Changes of physical properties', *Phys. Stat. Sol. A*, 144, 277–283.

Shpotyuk O, Golovchak R, Boyko V and Kozyukhin S (2010), 'Cluster modeling of quasi-adaptive phases in vitreous germanium selenides', *Physica Status Solidi C*, 7, 3–4, 921–924.

Shpotyuk O, Kozdras A, Golovchak R and Iovu M (2011), 'Is the marginality of non-reversible heat flow in MDSC experiments a sufficient criterion for self-organization in network glass-formers?', *Phys. Stat. Solidi C*, 8, 11–12, 3043–3046.

Simon S L, Plazek D J, Sobieski J W and McGregor E T (1997), 'Physical aging of a polyetherimide: volume recovery and its comparison to creep and enthalpy measurements', *J. Polym. Sci. B Polym. Phys.*, 35, 929–936.

Smedskjaer M M, Mauro J C, Sen S and Yue Y (2010a), 'Quantitative design of glassy materials using temperature-dependent constraint theory', *Chem. Mater.*, 22, 5358–5365.

Smedskjaer M M, Mauro J C and Yue Y (2010b), 'Prediction of glass hardness using temperature-dependent constraint theory', *Phys. Rev. Lett.*, 105, 115503.

Starodubcev S V, Domoryad I A and Khiznichenko L P (1961), 'Gamma-induced changes of mechanical properties in amorphous selenium', *Dokl. AN SSSR*, 139, 3, 594–595.

Starodubcev S V, Azizov S A, Domoryad I A, Peshykov E V and Khiznichenko L P (1962), 'Changes of the mechanical characteristics of some solids under gamma-irradiation', in: *Influence of Nuclear Radiation on the Materials*, Moscow: AN SSSR.

Stickel F (1995), *Untersuchung der Dynamik in niedermolecularen Flussigkeiten mit dielektrischer Spektroskopie*, Aachen: Shaker.

Struik L C E (1978), *Physical Ageing in Amorphous Polymers and other Materials*, New York: Elsevier.

Stuchlik M, Krecmer P and Elliott S R (2001), 'Opto-mechanical effect in chalcogenide glasses', *J. Opt. Adv. Materials*, 3, 361.

Svoboda R, Pustkova P and Malek J (2006), 'Volume relaxation of a-Se studied by mercury dilatometry', *J. Non-Cryst. Solids*, 352, 4793–4799.

Svoboda R, Pustkova P and Malek J (2007), 'Relaxation behavior of glassy selenium', *J. Phys. Chem. Solids*, 68, 850–854.

Svoboda R, Honcova P and Malek J (2010), 'Relaxation in Ge_2Se_{98} and As_2Se_{98} glasses', *J. Non-Cryst. Solids*, 356, 447–455.

Takahara K, Saito H and Inoue T (1999), 'Physical aging in poly(methyl methacrylate) glass: densification via density fluctuation', *Polymer*, 40, 3729–3733.

Tallman R E, Weinstein B A, Reznik A, Kubota M, Tanioka K and Rowlands J A (2008), 'Photo-crystallization in a-Se imaging targets: Raman studies of competing effects', *J. Non-Cryst. Solids*, 354, 4577.

Tanaka K (1983), 'Mechanisms of photodarkening in amorphous chalcogenides', *J. Non-Cryst. Solids*, 59–60, 925.

Tanaka K (1986), 'Configurational and structural models for photodarkening in glassy chalcogenides', *Japanese J. Appl. Phys.*, 25, 6, 779.

Tanaka K (2002), 'Photoinduced fluidity in chalcogenide glasses', *C.R. Chemie*, 5, 805.

Tanaka K (2004), 'Midgap photon effects in As_2S_3 glass', *Phil. Mag. Lett.*, 84, 9, 601.

Tanaka K (2006), 'Photo-induced phenomena in chalcogenide glass: comparison with those in oxide glass and polymer', *J. Non-Cryst. Solids*, 352, 2580.

Tanaka K, Saitoh A and Terakado N (2009), 'Photoinduced phenomena in group VIb glasses', *J. Mater. Sci: Mater. Electron.*, 20, S38.

Thorpe M F (1983), 'Continuous deformations in random networks', *J. Non-Cryst. Solids.*, 57, 355–370.

Thorpe M F (1995), 'Bulk and surface floppy modes', *J. Non-Cryst. Solids.*, 182, 135–142.

Thorpe M F, Jacobs D J, Chubynsky M V and Phillips J C (2000), 'Self-organization in network glasses', *J. Non-Cryst. Solids*, 266–269, 859–866.

Tonchev D and Kasap S O (2002), 'Effect of aging on glass transformation measurements by temperature modulated DSC', *Mater. Sci. Engineer.*, A328, 62–66.

Trunov M L, Nagy P M, Takats V, Lytvyn P M, Kokenyesi S and Kalman E (2009), 'Surface morphology of as-deposited and illuminated As–Se chalcogenide thin films', *J. Non-Cryst. Solids*, 355, 1993.

Varshneya A K (2006), *Fundamentals of Inorganic Glasses*, Sheffield: Society of Glass Technology.

Vautier C, Derrey T, Saiter J M and Lebaudy Ph (1991), 'Photorelaxation in amorphous Ge_xSe_{1-x} ($0 < x < 0.08$) thin films', *J. Non-Cryst. Solids*, 131–133, 520.

Vempati U and Boolchand P (2004), 'The thermally reversing window in ternary $Ge_xP_xS_{1-2x}$ glasses', *J. Phys.: Cond. Matter*, 16, S5121–S5138.

Vinogradova G Z (1984), *Glass Formation and Phase Equilibria in Chalcogenide Systems*, Moscow: Nauka.

Wang F and Boolchand P (2004), 'Photo-structural transformations and global connectedness of network glasses', in: *Non-Cryst. Materials for Optoelectronics* (ed. G. Lucovsky and M. A. Popescu), Bucharest: INOE Publishing House.

Wang F, Mamedov S, Boolchand P, Goodman B and Chandrasekhar M (2005), 'Pressure Raman effects and internal stress in network glasses', *Phys. Rev. B*, 71, 174201.

Wang F, Boolchand P, Jackson K A and Micoulaut M (2007), 'Chemical alloying and light-induced collapse of intermediate phases in chalcohalide glasses', *J. Phys.: Condens. Matter*, 19, 226201.

Wang Y, Boolchand P and Micoulaut M (2000), 'Glass structure, rigidity transitions and the intermediate phase in the Ge-As-Se ternary', *Europhys. Lett.*, 52, 6, 633–639.

Yang G, Bureau B, Rouxel T, Gueguen Y, Gulbiten O, Roiland C, Soignard E, Yarger J L, Troles J, Sangleboeuf J C and Lucas P (2010), 'Correlation between structure and physical properties of chalcogenide glasses in the As_xSe_{1-x} system', *Physical Review B*, 82, 195206.

Yannopoulos S N and Andrikopoulos K S (2004), 'Raman scattering study on structural and dynamical features of noncrystalline selenium', *J. Chem. Phys.*, 121, 4747.

Yue Y (2004), 'Influence of physical ageing on the excessive heat capacity of hyperquenched silicate glass fibers', *J. Non-Cryst. Solids*, 348, 72–77.

Zakery A and Elliott S R (2007), *Optical Nonlinearities in Chalcogenide Glasses and their Applications*, Berlin: Springer-Verlag.

Zallen R (1983), *The Physics of Amorphous Solids*, New York: John Wiley and Sons.

9
Deposition techniques for chalcogenide thin films

J. ORAVA, University of Cambridge, UK and Tohoku University, Japan, T. KOHOUTEK, Toyota Technological Institute, Japan and University of Pardubice, Czech Republic and T. WAGNER, University of Pardubice, Czech Republic

DOI: 10.1533/9780857093561.1.265

Abstract: The chapter describes details of deposition techniques, namely thermal evaporation, sputtering, pulsed-laser deposition, chemical vapour deposition and spin-coating for chalcogenide thin films. We discuss key issues, advantages and influence of different deposition techniques and experimental conditions on properties of as-prepared amorphous chalcogenide thin films. The final chemical composition, structure and physico-chemical properties of the prepared films predetermine their application and vice versa. Their applications are in many fields such as (nano)electronics, (nano)optics, information storage, security, health protection and alternative energy sources.

Key words: chalcogenide glasses, thin films deposition, physical vapour deposition, chemical vapour deposition.

9.1 Introduction

Chalcogenide glasses (CHGs) are very well known for their unique optical properties (high refractive index infra-red (IR) transparency and optical non-linearity, see corresponding chapters in this book) and for their reversible amorphous-to-crystal phase-change transition (Wuttig and Yamada, 2007). The optical and other properties of thin films (thermal, mechanical, resistive, chemical, photo-sensitive, etc.) are inevitably very dependent on the deposition technique (from vapour, solution, or via reaction), which controls the stoichiometry, structure and defect states in as-prepared and relaxed thin films. Thus, a particular deposition technique will make chalcogenide films attractive for certain applications, while other deposition techniques fail to produce films of desirable properties. In this chapter, we will identify and discuss the capabilities, the main issues and future key challenges for a variety of deposition techniques. The coverage is far from comprehensive, and we will focus on the dependence of the physical-chemical properties of as-prepared chalcogenide films on the deposition technique and experimental conditions. We will show that the primary process of thin-film deposition has not been well understood and even not studied in detail in many cases. Deposition is generally recognized as the most crucial step for any real or potential applications of chalcogenide thin films.

We will give an overview of the major physical and chemical vapour deposition techniques, their main attributes and limitations using CHGs as target materials. Detailed descriptions are given of: vacuum *thermal evaporation* – TE (including glancing-angle TE, and flash evaporation – FE) (Section 9.2.1), *sputtering* (Section 9.2.2), *pulsed-laser deposition* – PLD (Section 9.2.3), *chemical vapour deposition* – CVD (Section 9.2.4) and *spin-coating* – SC (Section 9.2.5). The influence of different deposition conditions on the properties of as-prepared films is discussed. In Section 9.3, probable future trends in thin-film deposition are discussed. Recommended further reading is given in Section 9.4.

9.2 Thin-film deposition

The aim of any deposition technique is the preparation of thin films in reproducible, controllable and predictive ways. There are many reasons why thin films may be preferred to bulk glass, and therefore there is a clear need for effective thin-film deposition techniques. In any case, some compositions cannot be prepared in the glassy state in bulk form. A clear example is the phase-change memory material $Ge_2Sb_2Te_5$ (GST), which is assumed to be supercooled liquid of high 'fragility' (Fig. 9.1) (Orava *et al.*, 2012), and which is a very poor glass-former with a critical cooling rate $\sim 10^{10}$ K/s. This is much beyond the cooling rates possible for the preparation of bulk samples, but such an effective rate can be achieved readily in thin-film deposition. Good glass-formers with 'strong' supercooled liquids, like As_2S_3 (critical cooling rate $\sim 2.4 \times 10^{-6}$ K/s (Snopatin *et al.*, 2009), close to that for silica glass) can easily be prepared in bulk form. Thin films are required for scaling-down devices, new structural states, or new and advanced functionalities that are difficult to achieve in a bulk form (Bragg reflectors, photoresists, functional coatings, optical memory, etc.). For amorphous chalcogenides, the disadvantages of thin films (compared to bulk analogues) are: worse thermal stability (lowered glass-transition temperature T_g, more evident release of volatile species, e.g. S at high temperatures); degraded mechanical resilience (the film-substrate interface may induce high internal stress especially when rough; films are easily scratchable); and lowered resistance to laser power density.

For amorphous solids, it is often very difficult to achieve consistent optical, electrical, chemical, structural, mechanical or thermal properties across the variety of available deposition techniques. The crucial steps of target-to-substrate mass transfer, and condensation on the substrate, are unlikely to be the same for different deposition techniques, and these steps can strongly

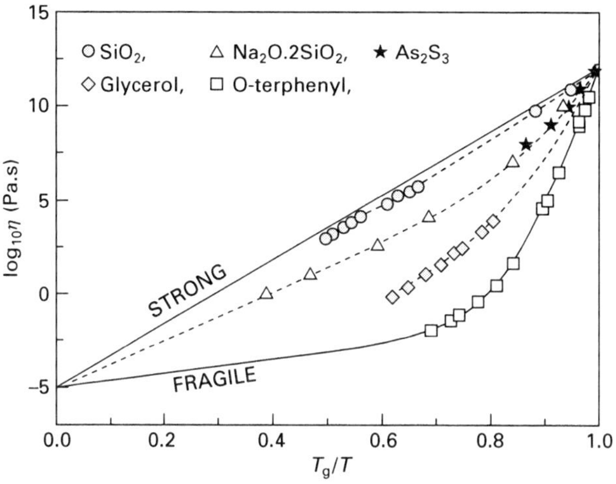

9.1 An Angell plot showing the temperature dependence of viscosity $\eta(T)$ for variety of glass-forming liquids plotted on an inverse temperature scale normalized to the glass-transition temperature T_g. 'Strong' liquids show an Arrhenius $\eta(T)$, while 'fragile' liquids show non-Arrhenius behaviour. Glass-forming ability tends to worsen from strong (e.g., As_2S_3, fragility is 35) to fragile liquids. (Redrawn from Angell, 1995, published with permission.)

influence the stoichiometry, the short- and the medium-range order and the defect states (homopolar pnictides and chalcogen 'wrong' bonds, valence alternation pairs and dangling bonds) in as-prepared amorphous thin films. The variations in structure and properties may be reduced by annealing the thin films or exposing them to light. Both annealing and exposure of as-prepared films can lead to more stable glasses (lower photo-sensitivity), but the products of these external stimuli are generally different. For example, it has been shown that $As_{50}Se_{50}$ thin films prepared by PLD have a better response to external stimuli (near band-gap illumination and annealing) than the corresponding bulk glass or thin films prepared by TE (Kalyva *et al.*, 2007). In the same binary system, an induction time for optically induced dissolution and diffusion of silver is measurable in TE films, but not detectable for PLD films. The lack of induction time was attributed to a lower concentration of As-As homopolar bonds in PLD films, suppressing the surface formation of arsenic oxide, through which the Ag^+ ions would have to diffuse. In contrast, an induction time is observed in TE thin films (Kalyva *et al.*, 2008, 2009), which have a high As-As concentration (as-prepared TE As-S films may contain more than 32% of As-As bonds from the total number of As-neighbourhood bonds; Kosek *et al.*, 1987). General trends can be observed: for example the formation of homopolar As-As bonds in As-S glass leads to an increase in refractive index (n) and a decrease in optical band-gap energy (E_g^{opt}), and vice versa when breaking As-As bonds. The implications are not only that precise control of experimental conditions is crucial during deposition (as discussed in the following sections), but also that the correct deposition technique must be chosen to achieve the desired properties of thin films in particular applications.

Whether we deposit a single film or a multilayer structure, it is necessary to consider the effects of the substrate (BK7, SiO_2, silicon) and any sandwich layers (TiN, Si_3N_4, metals – e.g., Au, W, Ti, Pt) on the surface roughness and adhesion of the chalcogenide. These substrate and multilayer influences are difficult to predict.

There are some unusual deposition techniques that are not yet used widely. Thick films of Ga-La-S can be deposited by *hot-dip* SC, i.e. by spinning (at ~1250°C) of molten Ga-La-S glass, which has very good stability against crystallization. Buried waveguides with optical losses <0.2 dB/cm at λ = 1064 nm were inscribed in films using direct laser writing (Mairaj *et al.*, 2005). Electrodeposition from aqueous or non-aqueous solutions has been used to deposit crystalline chalcogenide thin films mainly for phase-change memory (PCM) and solar-cell applications. At present the technique lacks precise control of film composition and the incorporation of contaminants from the solutions is an issue (Lincot, 2005). Micrometre-size lenses (~11 μm in length and ~0.8 μm in height) of As_2S_3 have been prepared directly by inkjet printing from As_2S_3 in *n*-propylamine solution (Sanchez *et al.*, 2011). Inkjet printing is not generally considered to be a deposition technique, but direct printing of optical elements may play an important role in future applications, even though the combinations of glasses and proper solvents are very limited (see Section 9.2.5). In the following sections we present details of the thin-film deposition techniques most commonly applied to chalcogenides.

9.2.1 Vacuum thermal evaporation

Over the last few decades, amorphous thin films have predominantly been prepared by physical vapour deposition techniques. Vacuum thermal evaporation (TE) (Fig. 9.2) is the simplest of these. The vapour is formed by heating of a solid (*sublimation*, e.g. as used for As purification, when removing AsO_x oxides) or of a liquid (melt) phase (*evaporation*). We describe TE in some detail, limiting ourselves to resistive (Joule) heating; other techniques such as electron-beam heating are described in Mattox (2010). With the introduction of more versatile deposition techniques such as sputtering and CVD (Section 9.2.2 and 9.2.4), TE is now less widely used than it once was. But TE remains attractive for good glass-forming binary chalcogenides like As-S(Se), Ge-S(Se), because it is low-cost, relatively simple, and reproducible, and capable of deposition over a large substrate area. TE is the favoured technique for deposition of amorphous selenium (a-Se) for flat-panel X-ray detectors (Belev *et al.*, 2008), nano-ionic memories (programmable metallization cell, using Ag-doped Ge-S, or Ge-Se) (Valov *et al.*, 2011), thin-film waveguides (Madden *et al.*, 2007; Ting *et al.*, 2010), hard and soft stamp-imprinting techniques (Lian *et al.*, 2009; Han *et al.*, 2010)

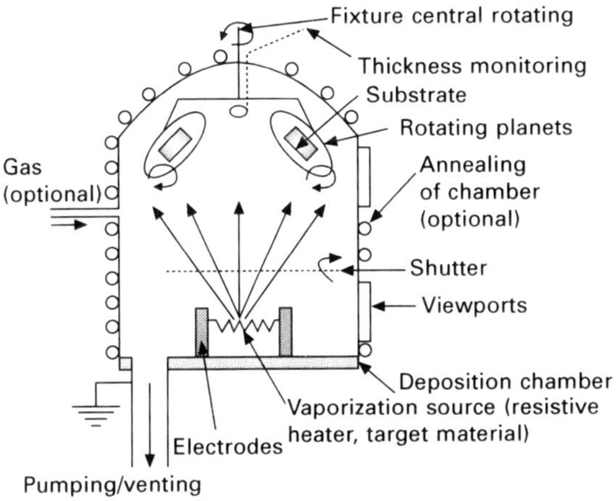

9.2 Schematics (not to scale) of two TE systems, with the substrate mounted (a) on a planetary rotation system (to give films of uniform thickness), and (b) directly above the evaporating source with a temperature stage allowing heating/cooling of the substrate during deposition. The heating stage increases mobility of species condensed on the substrate, permitting more relaxation to produce low-stress films. Cooling increases the effective quench rate of the incident species. Part (c) shows a simplified cosine distribution of evaporated species from a point source (bottom), and the corresponding thickness distribution of the film deposited on the substrate (top), where d*m*/d*A* is the mass *m* deposited per unit *A*, and *r* is the distance of the substrate from the source (the angles θ and ϕ are shown in the figure). The distribution depends on the form of the source (wire, spiral, cylinder, dimpled sheet, Al_2O_3 boat, SiO_2 crucible, etc.); for further details, see Holland and Steckelmacher (1952). Co-evaporation from two boats is possible, but is not commonly used for chalcogenide glasses. Co-evaporation has, however, been used for deposition of some crystalline analogues (Goncalves *et al.*, 2011). Thickness monitoring is predominantly done by measuring a change in oscillation frequency of a quartz-crystal resonator. It is also common to use *in-situ* optical techniques (e.g., spectroscopic ellipsometry) for simultaneous monitoring of thickness and optical properties. The source-substrate distance is much less than the mean free path l of the evaporated molecules: for example, for a molecule of typical 0.3 nm in diameter, $l \approx 100$ m at 10^{-4} Pa (Smith, 1995). (Part (c) is published from Mattox (2010), with permission.)

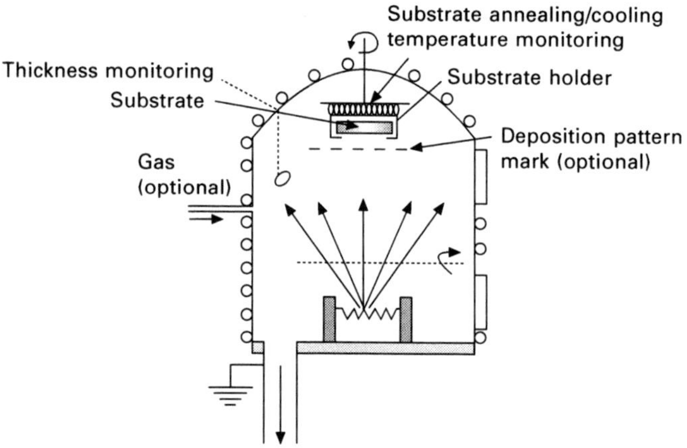

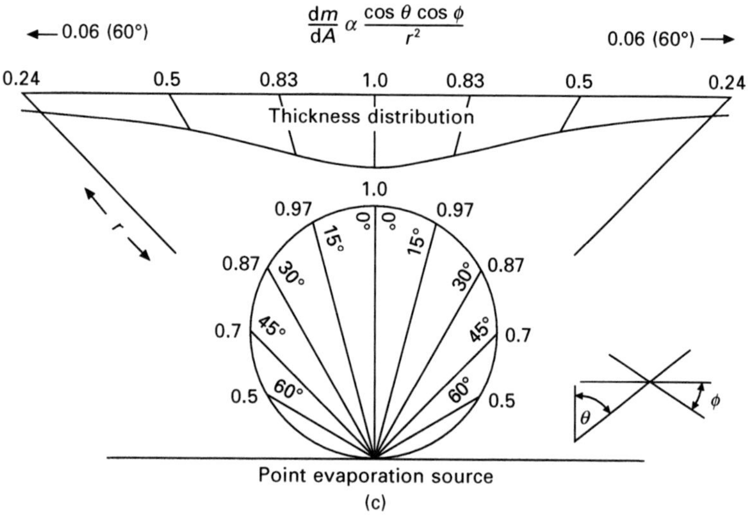

9.2 Continued

and, for example, Bragg reflectors (DeCorby *et al.*, 2005; Kohoutek *et al.*, 2009). Film thicknesses are usually from tens of nanometres to hundreds of micrometres.

We can derive the *Clausius-Clapeyron* equation from the equality of the chemical potentials of two phases in thermodynamic equilibrium (Moore, 1972), describing the equilibrium vapour pressure P^* at the evaporant surface, dependent on temperature T:

$$dP^*/dT = \Delta H_v/[T(v_g - v_c)], \quad [9.1]$$

where ΔH_v is the latent heat of the phase transformation (vaporization), and v_g

and v_c are molar volumes of the gaseous and condensed phases, respectively. The quantities ΔH_v, v_g and v_c are functions of temperature (T). The P^*–T dependences are known (see Kasap and Capper, 2006, and references therein) for the elements in solid and liquid form (Fig. 9.3). The vapour pressure P^* increases sharply with temperature. For common metals (e.g., Pb, Ag, Au, Al, Cu), the temperature required for evaporation is as high as 1500 K. Materials such as Pt, Mo, Al_2O_3, fused silica, etc., have very low vapour pressures and high melting temperatures; they can hardly be vaporized by conventional TE, and are therefore used as materials for the boats containing the evaporant sources. At high temperature, alloying between the material of the boat and the evaporant is a potential concern, but can be avoided. For example, a boat made of W can withstand several depositions of Al, which otherwise alloys easily with many metals at high temperatures. From the kinetic theory of gases, the impingement rate N_i ($m^{-2}.s^{-1}$) of gas molecules on a surface is given by (Totten *et al.*, 2004):

$$N_i = (2\pi m k_B T)^{-1/2} P, \qquad [9.2]$$

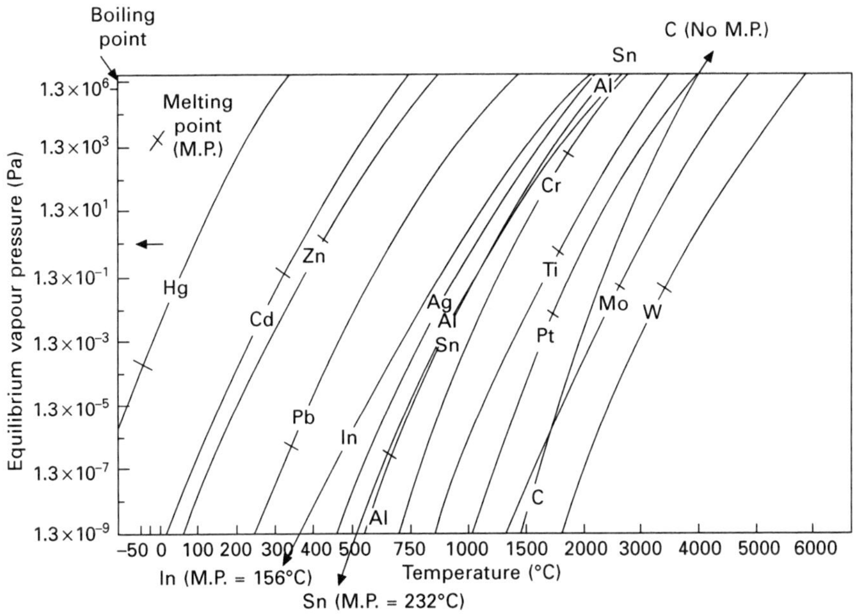

9.3 The dependence of equilibrium vapour pressure (P^*) on temperature (T) for various metals. The slashes on the curves mark the melting temperature. The temperature dependences are, for example ~1.3 kPa/100°C for Cd and 1.3 kPa/250°C for W. Vapour pressures of 1.3 Pa, if above a liquid, indicate a good evaporating material, while if above a solid, indicate a good subliming material. (From Mattox, 2010, the *y*-axis recalculated from torr to pascal, published with permission.)

where m is the molecular weight, k_B is the Boltzmann's constant and P is the gas pressure. Equation [9.2] applies for the residual gases and the evaporant molecules. Hertz (1882) and then Knudsen (1915) measured the evaporation rates of mercury and postulated an equation describing the net molecular evaporation rate, N_e nowadays known as the Hertz–Knudsen equation:

$$N_e = \alpha_v (2\pi m k_B T_e)^{-1/2}(P^* - P), \qquad [9.3]$$

where T_e is the evaporation temperature, P is the ambient pressure of species in the evaporation chamber, and α_v is the evaporation coefficient (Fedkin *et al.*, 2006). The quantity N_e can also be regarded as the impingement rate (Eq. [9.3]) of gas molecules based on the kinetic theory of gases, corrected for the returning molecules corresponding to gas pressure P and for the evaporation coefficient α_v. We can see that the maximum evaporation rate corresponds to the condition where $P = 0$. If $P = P^*$, i.e. under equilibrium conditions, there is no net evaporation. Evaporation is favourable only if $P \ll P^*$, i.e. under non-equilibrium conditions. The evaporation coefficient α_v (tabulated in Kasap and Capper, 2006; Fedkin *et al.*, 2006) is defined as the ratio of the real evaporation rate in the evaporation chamber to the theoretically calculated values. The vacuum chamber in TE usually operates at base pressures less than 10^{-4} Pa.

It is obvious that the relation between the solid material (amorphous or crystalline in the form of a powder or fractionalized sample), the liquid melt and the vapour flux can be very complex, especially when binary, ternary or multi-component compounds are evaporated. The Ge-S solid-liquid binary diagram (Fig. 9.4), for example, clearly shows complex melting behaviour. At a *eutectic* (local minimum in the melting temperature where the solid consists of two different phases), and at a *congruent* composition, the solid is in thermal equilibrium with a melt of the same composition, and the vapours have the composition of that starting point. Melting at other compositions is *incongruent*, resulting in vapours having a different composition from the solid and leading to vapour fractionation (Elliott, 1998). A compositional mismatch between the deposited thin film and the original bulk material can also arise from the presence in the vapour of molecular species with composition different from the target material; the presence of such species is difficult to control. Because the substrate is relatively cold, the vapour species have a very short relaxation time on the surface (~10^{-12} s at room temperature) and thus the preferential evaporation in the vapour is conserved in as-prepared thin film. The dissociation in vapours (almost unavoidable in TE) is very complex, but can be schematically expressed using Ge-S and As-S as examples:

$$2GeS_2 \leftrightarrow Ge_2S_3 + {}^1/_2S_2 \rightarrow 2GeS + S_2, \qquad [9.4]$$

$$2(As_2S_3)_b \leftrightarrow bAs_4S_4 + 2S_b, \qquad [9.5]$$

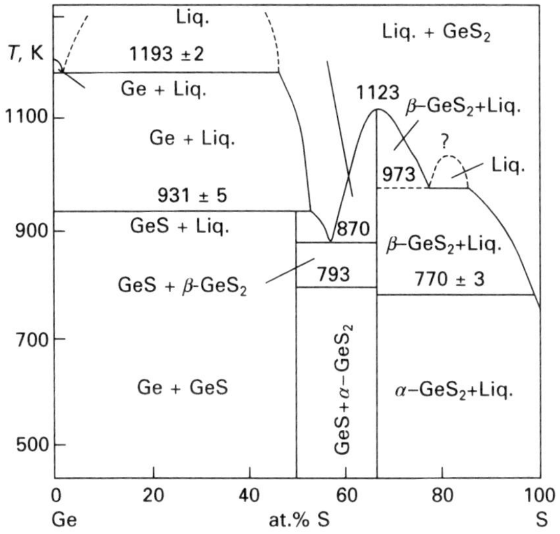

9.4 Equilibrium phase diagram of the Ge-S binary system, showing the compounds GeS (which melts incongruently at ~931 K) and GeS_2 (which melts congruently at 1123 K). Between lies a eutectic either with 57.3 at.% of S, T_E = 870 K (as shown) or 60 at.% S, T_E = 883 K. The eutectic parameters are in dispute (Bletskan, 2005). (From Bletskan (2005), published with permission.)

where *b* can be much greater than one. Thus inhomogeneities in thin films arise from species such as S_n, As_4S_3, As_4S_4, GeS_x and many combinations (Frumar *et al.*, 2003, 2011). As vapours condense on a substrate, 'wrong' homopolar bonds are formed in pnictides (e.g., As-As) and/or in chalcogens (e.g., S-S or Se-Se). Usually those molecular fragments with crystallographic analogues are more stable in vapours (Mikhailov *et al.*, 2000). These as-prepared films are prone to a variety of photo-induced phenomena (Frumar *et al.*, 2011). Such changes are induced by using proper light exposure or alternatively by annealing of as-prepared films (this can also be referred to as irreversible photo-induced polymerization, giving products that are thermodynamically more stable). The optical properties are then changed accordingly due to different local bonding. In TE films, it can be observed that the refractive index is *non-uniform* through the thickness: the bottom part of the film is enriched with the more volatile species and therefore has a lower *n* than nearer to the surface. Preferential evaporation effects of this kind are particularly pronounced for Ge-Sb-Te and Ag-In-Sb-Te alloys, where TE leads to an enormous shift in composition between target and films, the films being enriched in the more volatile species. The problems of preferential evaporation can be reduced (though not completely avoided) using higher evaporation rates up to 10 nm/s (normal rate ~1 nm/s) (González-Leal *et al.*, 2003), special Knudsen cells (Kozicki *et al.*, 2005) or deposition of thin films from targets of different composition demonstrated for As-S-Se and Ge-Se thin films (Orava *et al.*, 2009). Bearing in mind the complex processes during TE, it is surprising that thermally evaporated as-prepared $Ge_{10}As_{35}Se_{55}$ films are the only photo-stable samples in the $Ge_xAs_{45-x}Se_{55}$ glass series, making them useful for photo-stable optical devices (Yang *et al.*, 2008). Vapour flux-shadowing has to be considered during film deposition onto pre-prepared patterns (Fig. 9.5) (e.g., onto etched photoresist).

Glancing-angle thermal evaporation is an oblique-deposition technique that allows growth of films with a columnar structure (Fig. 9.6(a)). The density ρ of the resulting as-prepared films is given by $\rho = \rho_0\, 2\cos\alpha/(1 + \cos\alpha)$, where ρ_0 is the density of the bulk material, and α is the incidence angle of arriving vapour species with respect to the substrate normal. Over a wide range of α, the refractive index n of the thin films (e.g., Ge-Sb-Se) can vary by more than one (Martín-Palma *et al.*, 2008) (Fig. 9.6(b)). It has been shown that the composition is almost unchanged with changing α and with or without substrate rotation. This invariance is not true for the sample density (directly related to the effective refractive index) or for the thickness distribution. In production of thin films, oblique-angle deposition and columnar growth lead to reproducibility problems, which can be eliminated by choosing a proper geometry for the deposition chamber (Fig. 9.2). On the other hand, the controlled variation of n in columnar structures of selected α can be exploited for the fabrication of, e.g., one-dimensional photonic chemical sensors (Martín-Palma *et al.*, 2009).

Thermal *flash evaporation* (FE) is a modified form of TE, mostly similar to the arrangement in Fig. 9.2, but where the vaporization source (i.e., the boat including a feeder for the source material) is designed to eliminate preferential evaporation (Fig. 9.7(a)), for example in a-Ge-Te, a-Ge-Sb-Te, a-Se-Sn-Te, a-In-Se and other compositions (Prikryl *et al.*, 2009; Kahnt and Feltz, 1982) (Fig. 9.7(b)). Because these chalcogenides are usually poor glass formers, the target material is crystalline. FE as-prepared GST thin films have the same optical properties as as-prepared GST films sputtered at an argon pressure of 1.2 Pa (Dieker and Wuttig, 2005). The sheet resistance of FE as-prepared GST is very reproducible and in good agreement with that of as-prepared sputtered films (Wagner *et al.*, 2009). To ensure rapid evaporation of the powdered sample being fed into the boat, it is resistively heated at higher temperatures than during normal TE. The as-prepared FE films are effectively '*self-annealed*' and, if the deposition is prolonged, the as-prepared thin films (especially when they are poor glass-forming compositions) can be partially crystalline.

We have discussed the most common attributes of TE. In laboratories all around the world there are many modifications of the technique. These modifications are appropriate to achieve the deposition conditions needed for the special needs of particular samples.

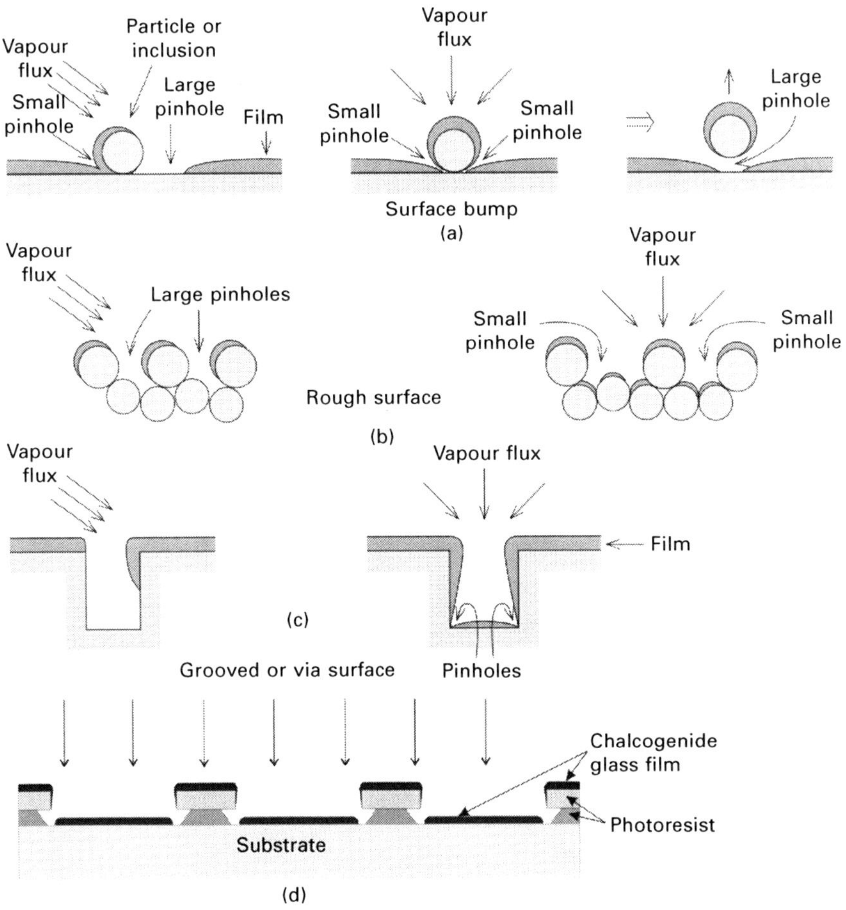

9.5 During physical vapour deposition with different directions of incident flux, there is a vapour flux-shadowing due to variety of features on the substrate (a) bumps, (b) a rough surface and (c) a grooved surface (this represents a traditional problem when depositing CHG film throughout the open windows of a single layer of photoresist, where $d_{photoresist} >> d_{chalcogenide\ film}$). Flux shadowing is a problem common to all physical vapour deposition techniques. (d) An example of one of the possible masking solution (for deposition of a waveguide) using two layers of photoresist (not to scale). (Parts (a)–(c) are from Mattox (2010), published with permission.)

9.2.2 Sputtering

Sputtering (together with chemical vapour deposition – CVD) is the most versatile technique for depositing chalcogenide thin films. It can be used for deposition of any binary (e.g., a-As-S, a-Ge-S, a-Ge-Te) (Tan *et al.*, 2010), ternary (e.g., a-Ge-Ga-S, a-Ge-Sb-Te) (Simpson *et al.*, 2010), multi-component

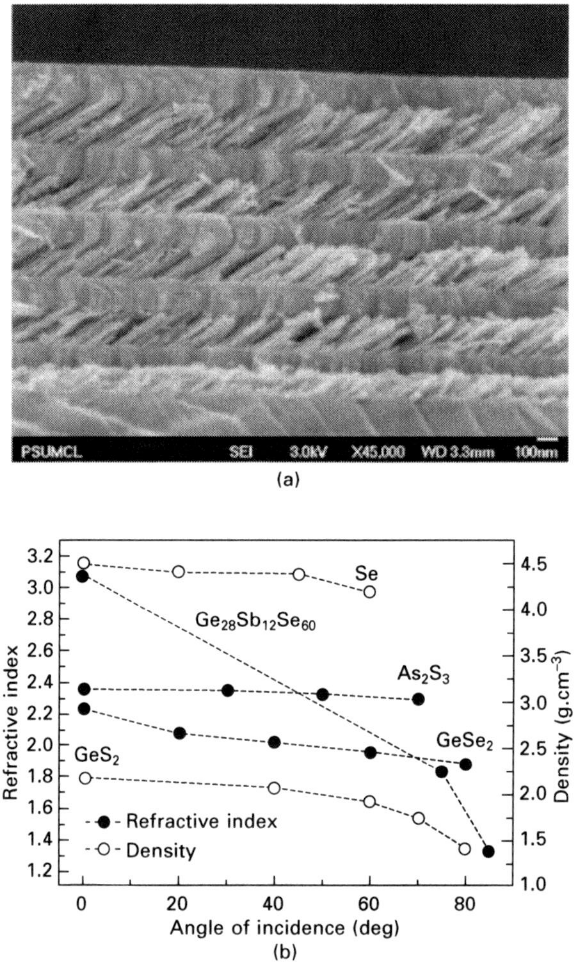

9.6 (a) SEM picture of an oblique-angle-deposited Ge-Sb-Se Bragg reflector used as an optical chemical sensor. Depending on the vapour's angle of incidence, the photonic band-gaps were tuned to 0.8 and 1.5 μm. (From Martín-Palma *et al.* (2009), published with permission). (b) The columnar structures show refractive index n and density ρ that vary with angle for each given composition or the TE CHG thin films deposited under otherwise identical conditions. The absolute values of n and ρ (e.g., the low density of GeS_2 for perpendicular deposition) may vary with different deposition conditions and also the true film composition can differ from nominal. The values for a-Se are from Lukács *et al.* (2008) and are calculated based on a molecular-dynamics simulation. Lines, only to guide the eye, are plotted through data for a-$Ge_{28}Sb_{12}Se_{60}$ (Martín-Palma *et al.*, 2008), a-As_2S_3 (Dikova *et al.*, 2005), a-$GeSe_2$ (Bhardwaj *et al.*, 2007), and a-GeS_2 (Rajagopalan *et al.*, 1982). (Published with permission.)

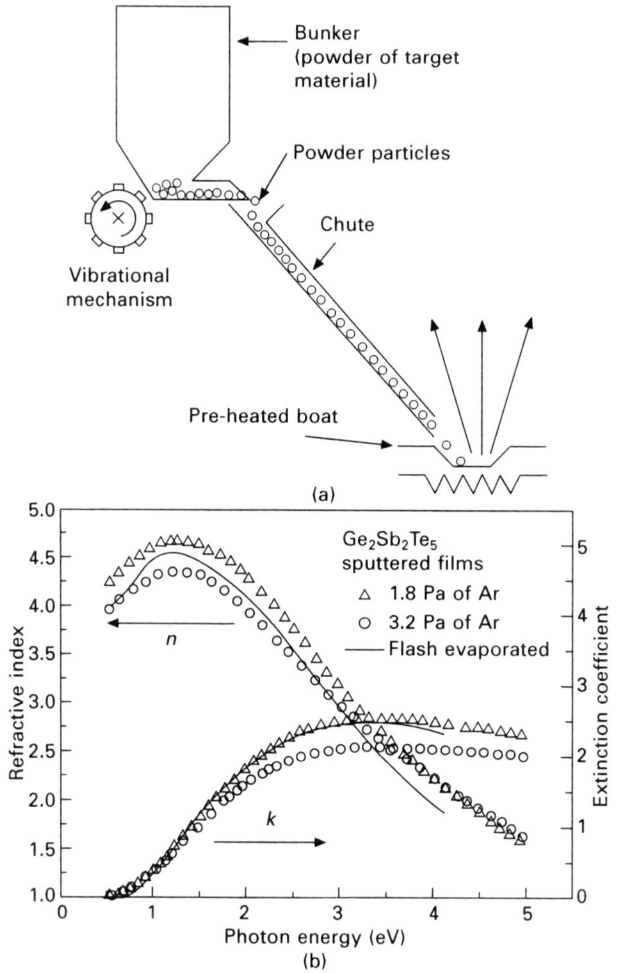

9.7 (a) A schematic (not to scale) showing the operating principle of thermal flash evaporation. The bunker with the powder sample is vibrated and the powder falls in small continuous amounts into the pre-heated boat, where the particles quickly evaporate, i.e. the equilibrium between solid and liquid melt is not achieved. The important parameters are particle size, rate and also the shape of the boat (often with complex inner shape to guide the vapour). (b) Comparison of the spectral dependence of refractive index and extinction coefficient of $Ge_2Sb_2Te_5$ thin films prepared by FE (thickness, $d \sim 200$ nm) and sputtering ($d \sim 100$ nm) technique (at different argon pressures). The values of optical band-gap energy E_g^{opt} and Urbach's energy E_U are 0.65 eV and ~70 meV, respectively, for as-prepared FE amorphous films. The values of E_g^{opt} and E_U for sputtered as-prepared amorphous films depend on the deposition conditions and are ~0.7 eV and 50–90 meV, respectively (Orava *et al.*, 2008). No change in optical properties of as-prepared $Ge_2Sb_2Te_5$ sputtered thin film was observed with increasing power from 20 to 100 W. (Part (b) is from Wagner *et al.* (2009), published with permission.)

(e.g., a-Ag-In-Sb-Te, a-Ge-As-Se-Te) (Seddon *et al.*, 2006), or rare-earth doped glasses (e.g., a-Er-Ga-La-S) (Frantz *et al.*, 2006). Sputtering is the dominant industrial technique in depositing chalcogenide films for optical (DVD, Blu-ray) and electrical data-storage (FLASH, RAM) applications, but is now being challenged by promising new techniques (PLD, CVD, solution-based processes) (Milliron *et al.*, 2009).

In most forms of sputtering there is an electrical potential difference (can be of the order of 1500 V) between the target and the substrate, that (via generation of plasma) is the source of the high-energy ions. Sputtering is a complex process where the material in a solid target is vaporized by bombardment with high-energy ions (predominantly Ar^+ for CHGs) (Fig. 9.8). There are many variants of sputtering, and here we limit ourselves to

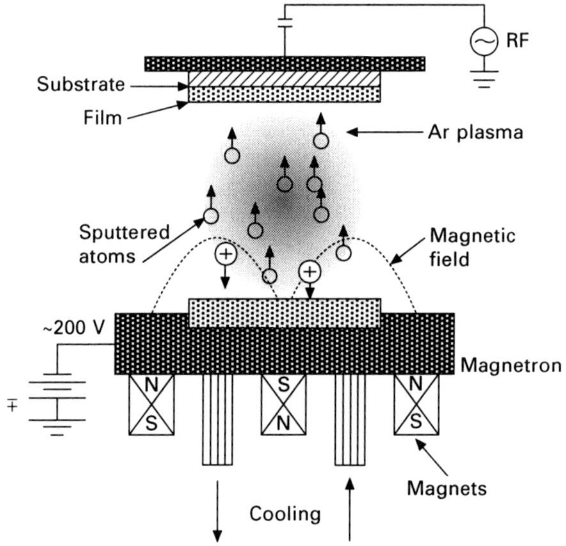

9.8 Schematic (not to scale) of RF (radio-frequency) magnetron sputtering showing target species projected towards the substrate. The atoms of, e.g., Ar low-pressure gas are ionized in the chamber due to the large electric field and are excited to create a stable plasma. Magnets are used to control volume and position of the plasma. Water cooling is necessary as thermal energy is released upon collision of Ar^+ ions with the target material, which could fracture due to an uneven temperature if left uncooled. This could be an issue for typical low-thermal conductivity bulk CHGs. The rotating substrate (a heating stage can be employed) can be much closer to the target than in TE as there is no risk of radiant heating from the target. By adding reactive gas such O_2 or N_2, reactive sputtering can successfully deposit, e.g., SiO_2, Al_2O_3 and ZrN. RF sputtering is used to sputter materials of poor electrical conductivity, where RF is in the range 0.5–30 MHz, with 13.56 MHz being the most standard. (Redrawn from Smith (1995) and Rossnagel (2002), published with permission.)

only the techniques most commonly used for CHG thin-film deposition. Further details of sputtering can be found in Smith (1995), Rossnagel (2002), Mattox (2010) and references therein. The target material is in the form of elemental (multi-target deposition) or multi-component amorphous/crystalline discs made from high-purity elements. When high-energy ions arrive at a target surface, ablation of target atoms is just one of many possible events which can happen to the target material (Fig. 9.9). Based on time-scale, these can be classified as *prompt effects* ($<10^{-12}$ s), e.g. lattice collisions, physical sputtering, reflection from the surface; *cooling effects* (from $>10^{-12}$ s to $<10^{-10}$ s), e.g. thermal spikes along collision cascades; *delayed effects* ($>10^{-10}$ s), e.g. atomic diffusion in the target material, strain-induced diffusion, segregation, etc.; and *persistent effects*, e.g. gas incorporation, compressive stress due to recoil implantation, etc. (Mattox, 2010).

Our aim during sputtering is to adjust different parameters (chamber geometry, target-to-substrate distance and orientation, magnetron (Fig. 9.8), flow rate and pressure of incident ions, power, and substrate temperature) to obtain as-prepared thin films that are adherent to the substrate, and have dense uniform structures with near-zero internal stress. There is a transition pressure at which intrinsic stress in as-prepared thin films goes through zero and changes from thin films being under compression to tension with increasing argon pressure. A similar transition can be observed with substrate bias variation. Further the product of pressure × target-to-substrate distance, in dependence on atomic mass of films, controls the degree of *thermalization*

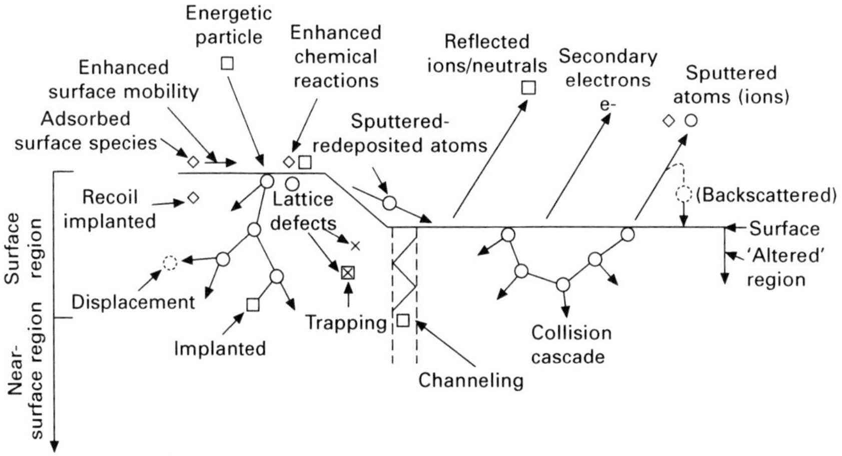

9.9 Schematic of events that can occur upon bombarding the surface of the target material with high-energy ions. Sputtered particles originate only from the top surface region (few 10s of nm) of target material, which helps stoichiometric ablation. When the target is polycrystalline, some crystallographic directions can be sputtered faster than others. (From Mattox (2010), published with permission.)

of the species arriving at the substrate, and thereby controls the structure and stress in thin film (Windischmann, 1992). This stress control is not possible in TE. Even though there are many parameters to be adjusted, the sputtering conditions can be easily reproduced over different runs. Films with excellent thickness uniformity can be sputtered across substrates of large area. Sputtering rates are low (<1 nm/s) and film thicknesses are usually ($d < 1$ μm). The low deposition rates allow precise control of film thickness down to the Å level. Most of the energy of the incident ions is dissipated within the target material; the targets heat up and must be cooled (Fig. 9.8). The high energy of species arriving at the substrate can cause preferential sputtering of light species from the thin film being deposited. We have already shown that, for example, the value of the argon pressure during sputtering has a very pronounced effect on the final optical properties of as-deposited thin films (Fig. 9.7(b)).

Prior to sputtering there is a *threshold energy* of ions, below which deposition does not occur. An important quantity is the *sputtering yield*, which is the number of atoms leaving a target per incident ion (Fig. 9.10). The momentum transfer from an incident ion to the atoms in the target is most efficient when the ions and the atoms have the same atomic mass. Argon ions (40 amu) have high sputtering yields for targets atoms of roughly similar mass. For heavier target atoms, heavier incident ions give better yields: for example, for Au (197 amu) the sputtering yield from 500 eV ions is 2.4 for Ar^+ (40 amu) and 3.06 for Kr^+ (84 amu) (Mattox, 2010). However, argon is still the most common sputtering gas because of the high cost of the heavier noble gases. The sputtering yield is dependent on the angle of incidence (of the Ar^+ ions on the target); it increases with off-normal incidence, and peaks at 50–60°, where the yield is increased by 20–50% (Fig. 9.10).

DC (direct-current) sputtering is feasible only for electrically conducting targets (e.g., polycrystalline Ge-Sb-Te). Most CHG targets, however, have poor conductivity, and accordingly deposition is predominantly by RF (radio-frequency) magnetron sputtering (Fig. 9.8). Chalcogenide targets (containing Ge, As, S) are prone to oxidation and gas contamination. The advantage of sputtering is that multiple targets can be positioned in the chamber so that, in one deposition run without breaking vacuum, complex structures can be produced such as multilayered DVD discs or superlattice consisting of thin (few Å) alternating layers of $GeTe/Sb_2Te_3$ (Tominaga *et al.*, 2008). Unfortunately, a-GST thin films are very often reported in the literature to be deposited by simple TE rather than sputtering, and studies on TE GST films can then be misleading.

By adding a reactive gas (N_2 or O_2, or a mixture) into the deposition chamber, these species can be incorporated into the thin films. This *reactive sputtering*, for example in the deposition of a-Ge-Se layers in the presence of added N_2, can be represented according to the scheme: $Ge_xSe_{1-x}(s) + N_2(g)$

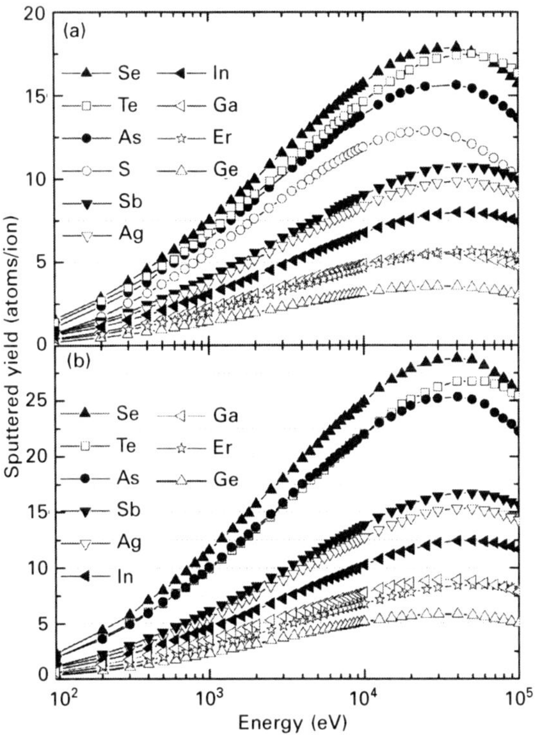

9.10 Calculated dependence of the sputtering yield of chosen elements on (a) normal incidence with respect to target, and (b) 45° angle of incidence, as a function of the energy of the incident of Ar^+ ions. The yield can be increased by a factor of five for certain elements (S, Sb, Te, As) due to polyatomic sputtering (Seah, 2005). At very low incident energies there is little sputtering. Then there is the 'knock-on' sputtering regime (the usual one for deposition in practice), where the sputtering yield scales roughly linearly with ion energy (up to ~10^4). This is followed by a 'collision-cascade' sputtering, where ions have enough energy to break any bonds in the target, and the 'high-energy implantation' regime, where the ions can penetrate well into the target, get trapped or implanted, sputtering does not occur near to the surface and the sputtering yield decreases (Rossnagel, 2002). The lines are guides to the eye only. (Data are from National Physical Laboratory (2011), calculated according to Seah (2005) and Seah *et al.*, 2005).

+ Ar(g) → Ge_xSe_{1-x}:N(s) + Ar(g) (Pinnow and Happ, 2010). Finding the proper deposition conditions to maintain the exact composition is not easy, bearing in mind the many adjustable parameters. In Er-doped Ge-Ga-Se thin films, the photoluminescence (PL) intensity increases with increasing RF power during sputtering (Maeda *et al.*, 2008), and PL lifetime is shorter than in the corresponding bulk glass, probably as a consequence of shifted composition. Another PL study (Nazabal *et al.*, 2010) on Tm-doped Ga-Ge-Sb-S glass showed that with increasing argon pressure, the as-prepared films are enriched with sulphur and antimony. Under optimum deposition conditions, however, a PL lifetime of 50 μs measured in the thin films is close to that of the original bulk sample (120 μs). During sputtering the growing films may undergo photo-annealing by the visible light emitted by the Ar^+ plasma near the face of the target. For As_2S_3 thin films, it has been shown that as-prepared (photo-annealed) films have a higher refractive index than as-deposited films prepared by TE or PLD. Further thermal annealing of as-prepared As_2S_3 leads to an increase of ~0.01 in n, a change that is much smaller than is found on thermal annealing TE as-deposited films (Tan *et al.*, 2010). Such '*self-annealed*' as-prepared films are suitable for the fabrication of stable optical devices. Metal-doped a-chalcogenides can easily be deposited by sputtering. Silver-doped Ge-Te or Sb-Te sputtered layers, for example, can be used as the solid-electrolyte in nano-ionic memory (Lee *et al.*, 2006; Park *et al.*, 2007). An ion-selective electrode membrane based on co-sputtered Cu-Ge-Sb-Se shows high detection sensitivity for Cu^{2+} ions over a wide concentration range with a lower limit of 10^{-6} mol/l (Méar *et al.*, 2008). State-of-the-art chalcogenide planar waveguide sensors based on sputtered Ga-Ge-Sb-S have been successfully demonstrated (Fig. 9.11) (Anne *et al.*, 2009). Ion-beam sputtering was reported for As_xS_{1-x} glasses, where no As_4S_4 and As_4S_3 molecular fragments were observed contrary to samples prepared by TE evaporation (Mikhailov *et al.*, 2000).

As we have seen, sputtering is a widespread and versatile technique. Even so, there is still a lack of detailed studies concerning the influence of different experimental conditions on the structure, stability, optical properties, and photo-sensitivity of sputtered chalcogenide thin films.

9.2.3 Pulsed-laser deposition

Pulsed-laser deposition (PLD) (Eason, 2007) is a relatively new physical vapour deposition technique, which has nevertheless been used for deposition of a vast range of materials including CHG thin films. In PLD, laser pulses of high-energy density are focused onto a target, the material of which is vaporized after the laser power reaches the ablation threshold and forms a plasma plume. This plume of ablated species is projected towards substrate, where the species condense and form a thin film. Laser ablation is terminated immediately on closing optical pathway, so no shutters are needed. Figure 9.12 shows a schematic of the PLD process and the time evolution of a plume, in which a variety of high-energy charged species can be detected (Fig. 9.13) (Pangavhane *et al.*, 2010). Interaction of the laser pulse with the target is a highly non-equilibrium process, where the absorption of the laser energy is

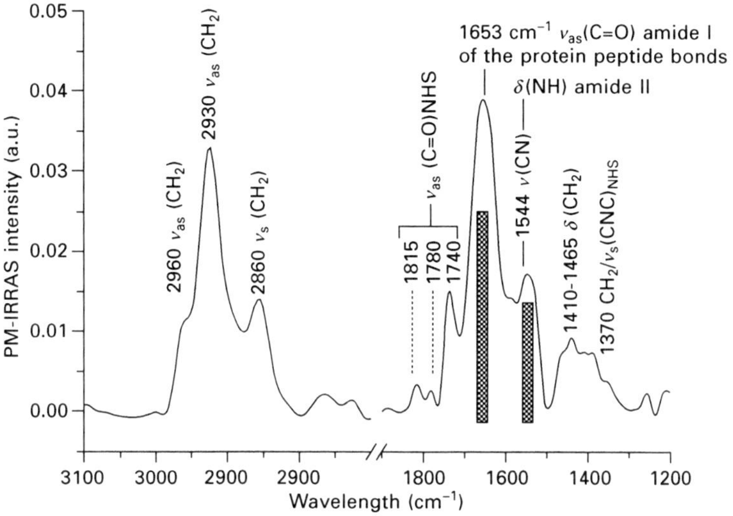

9.11 Polarization-modulation infrared reflection absorption spectroscopy (PM-IRRAS) of a functionalized chalcogenide Ga-Ge-Sb-S based sensor. The technique of self-assembled monolayers (SAMs) (Ulman, 1996) was used to introduce anchor molecules (1,1 mercaptoundecanoic acid), which were attached to the Au film followed by introducing antibody SAMs, activated by *N*-hydroxysuccinimide, 1-ethyl-3-carbodiimide hydrochloride). The spectra confirm the presence of amide groups originating from avidin bounded to antibody. (From Anne *et al.* (2009), published with permission.)

confined to very small volume. The stoichiometry of the target is then well preserved in the deposited film even for multi-component or rare-earth doped compositions. Chalcogenide films deposited by PLD can often be obtained containing structural units similar to those in the corresponding bulk glass (target material) (Frumar *et al.*, 2006). PLD is a complex deposition technique with many technological aspects reviewed in detail elsewhere (Eason, 2007; Frumar *et al.*, 2006; Gamaly *et al.*, 2002, 2006).

There are two PLD approaches for chalcogenide films: conventional PLD (Eason, 2007) and ultrafast PLD (Gamaly *et al.*, 2002, 2006). The techniques make use of different pulsed-laser sources, resulting in different predominant ablation mechanisms. In conventional PLD, ultraviolet excimer gas lasers (F_2* operating at 157 nm, ArF* – 193 nm, KrF* – 248 nm, XeCl* – 308 nm, XeF* – 351 nm) provide nanosecond laser pulses of average energy ~0.01–10 J, intensity ~10^8–10^9 W/cm^2 and repetition rates ~1–50 Hz (Eason, 2007). The laser spot area on the target is typically 0.5–5 mm^2 but can be larger and non-circular in shape (Delmdahl and Pätzel, 2008). Penetration depths are up to several micrometres, and depend strongly on laser power

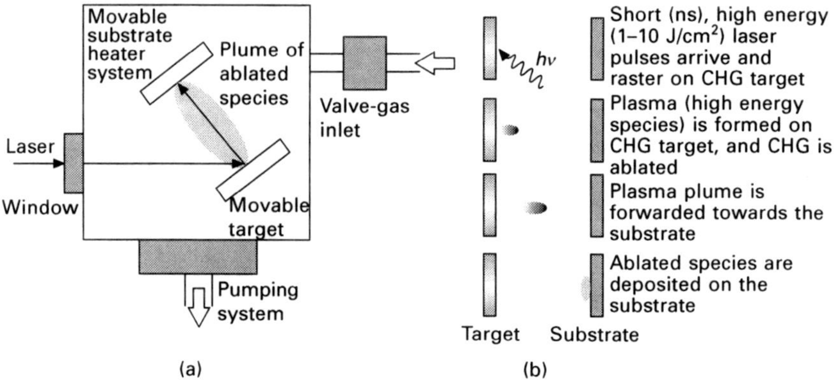

9.12 (a) Schematic of a PLD set-up with a movable target. Rastering over the target can also be achieved by laser movement. Because of the high energy of the incident laser, the silica window has to be replaced regularly due to its ablation. *In-situ* thickness is usually monitored by optical techniques (e.g. ellipsometry). (b) Time evolution of the deposition processes in conventional PLD. (From Eason (2007), published with permission.)

density and material absorption coefficient. Similarly, deposition rates can reach up to tens of nm per pulse (10–24 nm per pulse (Rode *et al.*, 2002) or even more), and are likely difficult to control.

These conditions are associated with the thermal ablation mechanism (Gamaly *et al.*, 2006), in which species are ejected from the target mostly by local thermal overheating, i.e. melting and vaporization. In contrast, in ultrafast PLD, electrostatic ablation dominates (Gamaly *et al.*, 2006), in which species (ions) are ejected from the target due to the strong electric fields. These fields are formed through charge separation by ionization, associated with electrons being emitted from the target. Ultrafast PLD uses ps or fs pulsed lasers, e.g. higher harmonic generation in Nd:YAG laser (natural frequency at $\lambda = 1064.1$ nm) at 532, 355, 266 nm. The repetition rates reach ~50–100 MHz, the intensities above ~10^{13} W/cm^2 and the average energy is in the μJ range. The penetration depths in the target, up to 1–2 μm, are typically smaller than in conventional PLD. Deposition rates can be controlled easily and are 1–15 nm per pulse (Luther-Davies *et al.*, 2005). Ultrafast PLD has the advantages of: (1) forming of a continuous flow of ablated species, (2) changing the ablation mechanism from thermal to electrostatic, and (3) smoothing the effective intensity distribution, allowing the creation of a single homogeneous phase (suppressing preferential ablation which can occur in conventional PLD) in the ablated plume.

PLD has been used in the deposition of a variety of chalcogenide thin films, for example sulphides (As-S (Rode *et al.*, 2002), Ge-S (Ogura *et al.*, 2000), Ge-Ga-S (Martino *et al.*, 2003), Ga-La-S (Youden *et al.*, 1993), Ge-

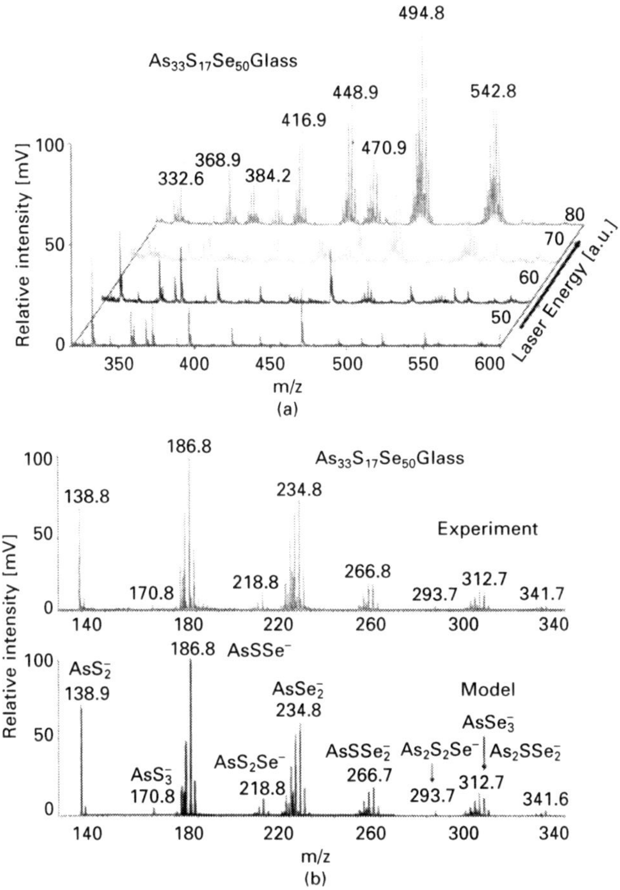

9.13 (a) MALDI-TOF (laser-desorption-assisted mass spectroscopy) linear positive ion mode, laser energy at 50, 60, 70 and 80 a.u., relative to 180 a.u. corresponding to full laser power 6 mW at λ = 337 nm – N_2 laser, repetition 10 Hz and pulse width of 3 ns. Analysis of charged ablated species was performed in the plume during laser ablation of $As_{33}S_{17}Se_{50}$ glass, and demonstrates the effect of relative laser power density on charged cluster formation. The *m*/*z* values (Da units) given in the spectrum at highest power density correspond to maxima in the corresponding cluster isotopic envelope. (b) Comparison of experimental and theoretical mass spectra of $As_{33}S_{17}Se_{50}$ glass (reflectron negative ion mode, laser power density corresponding to 80 a.u. in (a)). Model of the isotopic envelopes assuming formation of AsS_2^-, AsS_3^-, $AsSSe^-$, AsS_2Se^-, $AsSe_2^-$, $AsSSe_2^-$, $As_2S_2Se^-$, $AsSe_3^-$, $As_2SSe_2^-$. So far it is not fully understood how MALDI-TOF species relate to practical PLD conditions. Firstly, the geometry (angle of incidence, target-to-detector distance) and conditions (lasers, power density, wavelength and matrix effect) are different from PLD, and secondly, the positive and negative species probably represent only a small portion of all particles, most of which do not have any charge in the plume in conventional PLD. (From Pangavhane *et al.* (2010), published with permission.)

Ga-S-Cs-I (Caricato *et al.*, 2003)), selenides (As-Se (Kalyva *et al.*, 2008) Ge-Se (Gmachl *et al.*, 2001), Ge-Ga-Se (Nemec *et al.*, 2000)) and telluride (Ge-Sb-Te (Wagner *et al.*, 2007)) films. PLD also permits the deposition of rare-earth doped chalcogenide amorphous films, which exhibit near-infrared PL at λ = 1.335–1.54 μm, such as Ge-Ga-Sb-S:Dy_2S_3 (Nazabal *et al.*, 2006), Ge-Ga-Se:Er_2S_3 (Dwivedi *et al.*, 2005) and Ge-Ga-S-Cs-I:Pr_2S_3 (Martino *et al.*, 2003). Due to very good control of composition, PLD is very effective for fabrication of functional devices like ion-selective thin-film membranes based on complex materials, for example Pb-Ag-As-I-S, Cd-Ag-As-I-S, Cu-Ag-As-Se or Cu-Ag-As-Se-Te, Cu-Ag-As-Se, Tl-Ag-As-I-S and Ag-As-S (Kloock *et al.*, 2002; Kloock and Schoning, 2008). These are used in sensors for the detection of heavy metal ions such as Pb^{2+}, Cd^{2+}, Cu^{2+}, Tl^{+} and Ag^{+} in the concentration range 10^{-5}–10^{-7} mol/l. By using PLD, As_2S_3 and As_2Se_3 planar optical waveguides have been fabricated with attenuation losses as low as ~0.3 dB/cm at λ = 1064 nm (Luther-Davies *et al.*, 2005; Zakery, 2002). Active optical elements such as photonic crystal membranes and quantum-cascade laser devices have been prepared using PLD Ge-Se and Ge-As-Se films (Gmachl *et al.*, 2001; Freeman *et al.*, 2005), respectively.

PLD of chalcogenide films also has several technological issues which curtail its wider application. It is still difficult to control film deposition rates precisely, especially in conventional PLD. To achieve good thickness uniformity (this could be an issue even for small samples) and composition homogeneity of films over a large substrate area, optimization of the laser power and optical path are necessary, as well as control of the process geometry and target-to-substrate positioning (by substrate rotation/translation). To avoid crater formation on the target, programmed raster patterns are employed. Another problem of PLD is droplets. These can be deposited on the substrate as result of fast condensation of the supersaturated vapours of the ablated species expanding from the plume towards the substrate. The arrival of droplets on the substrate can be restricted by mechanical and geometrical means (off-axis deposition, velocity filters), but it is most effective to use short-pulse lasers with MHz repetition rates and lower power density. Laser ablation of glassy materials can also be improved by using *plasma-assisted* PLD (De Giacomo *et al.*, 2002), where the plume is biased by an additional plasma discharge between the target and the substrate. This gives an improved plume with higher temperature and higher degree of vaporization. Another disadvantage of PLD is its relatively high cost, mainly associated with the lasers; picosecond and femtosecond pulsed lasers are still quite far from widespread use in laboratories or in industry.

Although PLD chalcogenide films usually reproduce closely the composition of the target, deviations of composition can be observed, e.g. dependent on pulse power density as shown in Fig. 9.14(a). The structures of PLD films also reflect different deposition conditions (Fig. 9.14(b)). As-prepared films exhibit good adhesion to the substrate (Nemec *et al.*, 2002) because of the high energy of the ablated species. The density of PLD films is high, due to the high kinetic energy of species in the plume; the high-energy incident ions result in thin films under compressive stress on the substrate (Fig. 9.14(c)). High kinetic energy of the plume species enables chemical synthesis or polymeric reactions, and as-deposited films can become *self-annealed*. The reduced photo-sensitivity of self-annealed as-prepared films can be an advantage in optical-device (optical-circuit) fabrication, where formation of defects and cracks could otherwise occur during film annealing.

The ease with which multi-component CHG films can be deposited makes

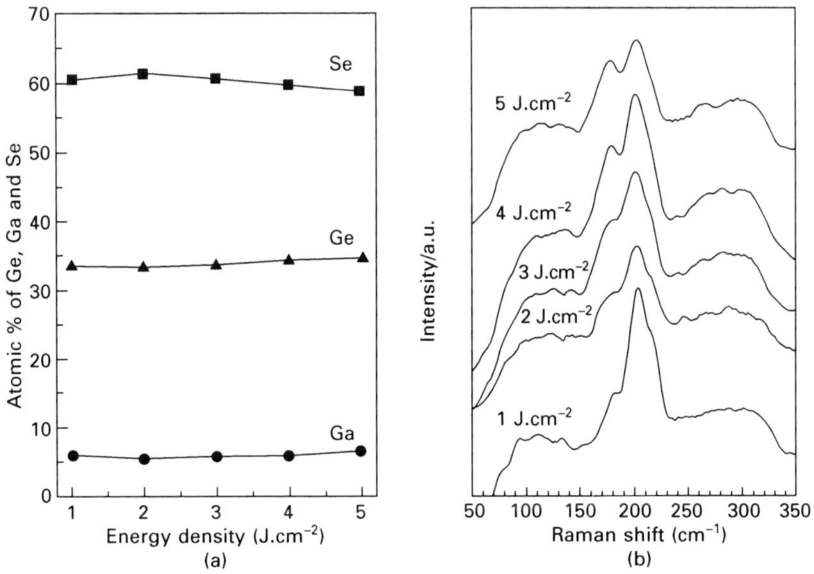

9.14 (a) Slight variation in the composition of PLD $Ge_{35}Ga_5Se_{60}$ as-prepared films (Nemec *et al.*, 2002) on changing the pulsed laser energy density. (b) depicts the corresponding influence on film structure: Raman spectroscopy shows the relative change in concentration of molecular units 'building blocks' in the amorphous network. Units such as $GeSe_{8/2}$ near 180 cm^{-1}, a mixture of $Ge_2Se_{6/2}$, corner-shared $GeSe_{4/2}$ units near 201 cm^{-1}, edge-shared $Ga_2Se_{6/2}$ or $GeGaSe_{6/2}$ units near 216 cm^{-1}, Se-Se bond vibration near 245 and 265 cm^{-1}, and $GeSe_4$ units near 300 cm^{-1}, can all be identified in the spectra. An increase in bond concentrations was observed for units assigned to the bands at 180 cm^{-1} and 265 cm^{-1}, assuming slightly higher chemical disorder in films with increasing pulse power density. (c) The refractive index dispersion is shown for As_2Se_3 films deposited under similar conditions as in (a). An increase in *n* reflects the higher density of films ablated at higher laser power densities (Frumar *et al.*, 2006). (From Frumar *et al.* (2006) and Nemec *et al.* (2002), published with permission.)

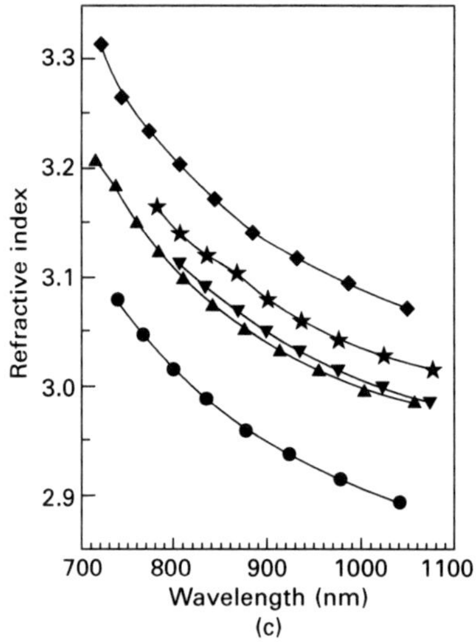

9.14 Continued

the PLD method worth studying, and PLD is certainly promising for future device fabrication. The challenges are mainly to improve film-deposition quality, e.g. by optimization of target-to-substrate geometry, by using tunable lasers, and by optimization of pulse duration, power density, repetition rates, etc. There are further challenges and opportunities associated with a direct synthesis of new materials, nano-particles or functional nano-structures by using reactive gas backgrounds during the PLD.

9.2.4 Chemical vapour deposition

Chemical vapour deposition (CVD) is an important technique, particularly promising for conformal deposition of chalcogenide thin films into embedded 1D, 2D and 3D structures. CVD is a method where gaseous precursors react either in gas phase or at the substrate–gas interface, producing thin film on the substrate. The composition of the thin film is different from the gases being used. The thickness of the conformal coatings can be more uniform and controlled more precisely than in physical vapour deposition. CVD is well known for the deposition of crystalline films (ceramics), but it is rather new for thin films of amorphous chalcogenides. The key advantages of CVD using gaseous sources are: (1) uniform thickness (and composition) can be achieved in coatings of very complex spatial features, reaching remote areas of substrates; (2) the correct purification processes can give high-purity precursors; and (3) high deposition rates (~1 μm/min). Optimizing the deposition of CHGs involves a strong and complex competition between the various techniques, and is emerging as a preferred option (Anscombe, 2011).

CVD reactions typically start and proceed only at quite high temperatures of 400–600°C, too high for pre-deposited temperature-sensitive structures (memory, sensors, etc.) to be used as substrates. High temperatures are necessary to achieve large values of equilibrium constants and a good conversion rate of precursor's reactions at the substrate. The required temperature can be reduced by employing a plasma in the reaction chamber to enhance (i.e., accelerate) the chemical reactions (PECVD) (González-Leal *et al.*, 2004b), or by using metal-organic precursors (MOCVD) (Ahn *et al.*, 2010). MOCVD is the technique of choice for depositing crystalline chalcogenide films for application in solar-cells (Irvine *et al.*, 2008). Because of lower deposition temperatures, PECVD and MOCVD are feasible for the deposition of, e.g., Ge-Sb-Te based thin films onto active device structures such as those for phase-change memory. The pressure in the deposition chamber may vary from atmospheric-pressure CVD to low-pressure CVD depending on the required deposition conditions. CVD can be performed: (1) in a hot-wall reactor, in which all parts of the CVD system are heated to the desired temperature and the reaction occurs everywhere, not only on the substrate; or (2) in a cold-wall reactor, where only the substrate is heated, and the gaseous precursors react only on the substrate (other parts of the system, including the input gases are kept cold and apart from the substrate, other parts are free of any deposits). A schematic of atmospheric-pressure CVD used for deposition of amorphous Ge-S thin films is shown in Fig. 9.15(a) (different reactor types are shown in detail in Fig. 9.15(b), (c) and (d)).

The complex thermodynamics and flow dynamics in CVD are beyond the scope of this chapter; readers should refer to Smith (1995), Pierson (1999), Yan and Xu (2010) and references therein. The sequence of process events in CVD is shown in Fig. 9.16(a). The slowest event in this scheme determines the deposition rate. At high pressures, boundary layer dynamics apply, but in low-pressure CVD this is not so (Pierson, 1999). The reactions during CVD are complex, but can often be presented in terms of a simple overall equation. The best known and most studied example is probably the thermal decomposition of silane: $SiH_4(g) \rightarrow Si(s) + H_2(g)$ (Fig. 9.16(b)). The thermodynamics and kinetics of similar reactions for multi-component materials (e.g., quaternary and rare-earth doped glasses) can be extremely complex and further studies are required.

Atmospheric-pressure, hot-wall CVD has been used successfully to deposit a-Ge-S thin films from gaseous precursors, via the overall reaction: $GeCl_4(g) + 2H_2S(g) \leftrightarrow GeS_2(s) + 4HCl(g)$ in the temperature range 400–600°C, and with argon gas used as the $GeCl_4$ carrier. For this reaction, the Gibbs free energy change is –49 kJ/mol at 527°C. The deposition (reaction) rate

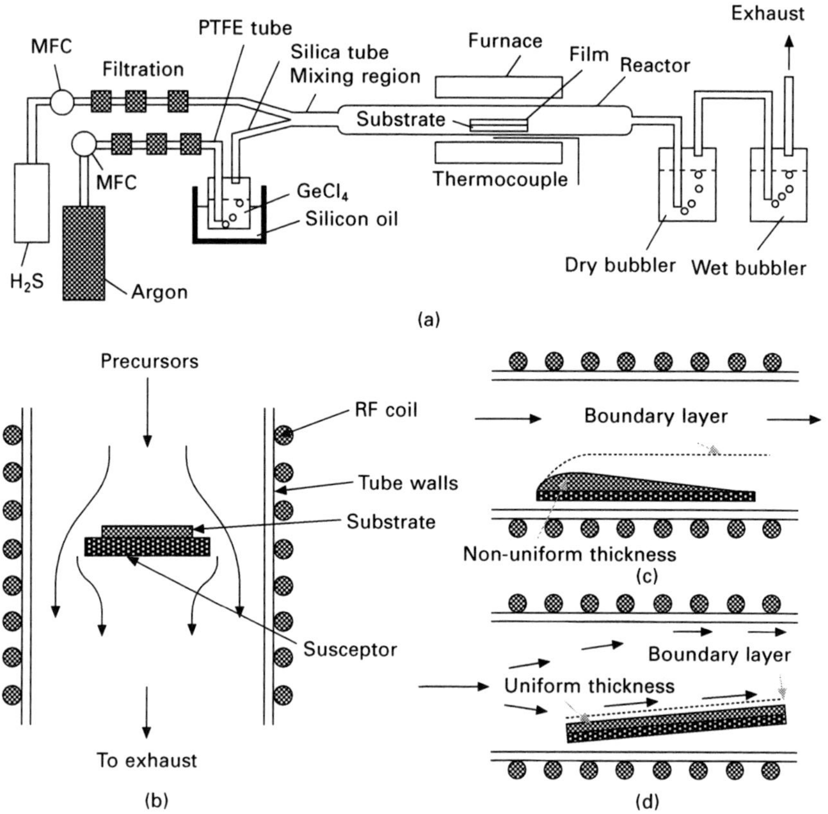

9.15 (a) A schematic of hot-wall CVD (atmospheric-pressure) used for Ge-S deposition at 450–600°C. Argon serves as the $GeCl_4$ carrier (MFC: mass flow controller). Typical reactor layouts include: (b) cold-wall axisymmetric reactor (for best uniformity of thickness); (c) cold-wall tube reactor with uncompensated substrate for gas flow, which gives non-uniform thickness on the substrate, largest near the gas inputs and lowest on the gas output side; (d) cold-wall tube reactor with compensated tilting of substrate, resulting in a uniform film thickness. All three configurations can also be used in hot-wall set-up. (a) is from Huang *et al.* (2004), published with permission.)

increases with increasing temperature. Crystallization occurs in films grown above 500°C, so amorphous films were prepared around this temperature, and deposition rates up to 12 μm/h were achieved (Curry *et al.*, 2005). The demonstrated optical losses in CVD GeS_2 amorphous waveguides were 2.1 ± 0.3 dB/cm at $\lambda = 632.8$ nm (Huang *et al.*, 2004). By adjusting the CVD deposition conditions (temperature, gas flow) and introducing silver into the GeS_2 amorphous films, a channel waveguide with optical losses of 0.67 ± 0.05 dB/cm at $\lambda = 632.8$ nm was demonstrated (Huang and Hewak, 2006). Thin films of a-Ge-As-Se have been deposited by CVD at 190–330°C and

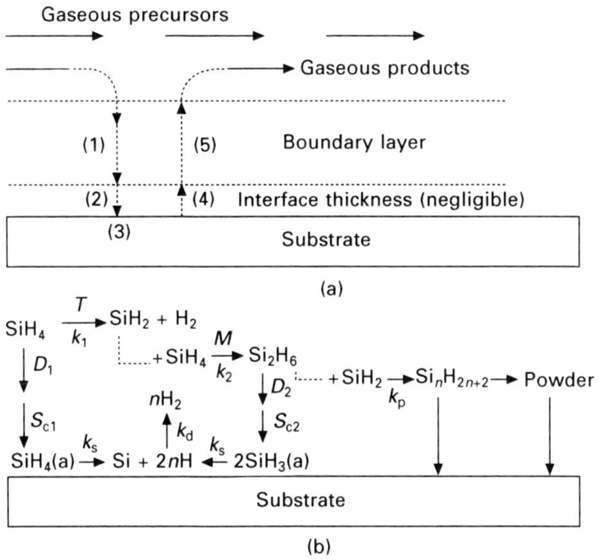

9.16 (a) Schematic of process events during CVD deposition. Under forced or free convection, gaseous precursors are transported into the reaction tube. 1) Gas-phase reactants diffuse through the boundary layer, 2) followed by their adsorption on substrate surface, where 3) chemical reactions take place (decomposition, oxidation, reduction, hydrolysis, displacement) forming a solid continuous layer of film. 4) There is a desorption of adsorbed species, and 5) the gaseous products diffuse from the surface going to exhaust. (From Pierson (1999), published with permission). (b) Schematic of different steps believed to be important during gas-phase thermal decomposition of SiH_4: k_i is a reaction rate constant, M is a third body in a reactive collision, D_i is a gas diffusivity, and S_{ci} is a sticking coefficient. (From Smith (1995), published with permission.)

at 670–1330 Pa using volatile chlorides as precursors (Shurgalin *et al.*, 2005). A variation in film composition can be easily achieved by controlling the experimental parameters, i.e. precursor flow rate (carrier-gas flow rate), precursor type (different starting composition), pressure, temperature (constant or graded across the reaction tube), substrate orientation and type of reactor (Fig. 9.17) (see details in Reso *et al.*, 2011a, 2011b). When the rate of CVD is limited by mass transport, the composition of as-prepared films varies with the ratio of the partial pressures of the precursors. In contrast, when the rate is limited by the surface reaction, the film composition can easily be tuned by selection of pressure and substrate temperature (Milliron *et al.*, 2009). Crystalline thin films of Ge-Sb-Te have been deposited using a CVD configuration similar to that in Fig. 9.15(a) using preheated Te, $GeCl_4$ and $SbCl_5$ (co-reactant H_2, argon carrier) at 250°C (Huang *et al.*, 2011). PECVD has been used to deposit thin films of a-As_2S_3 (precursors AsH_3/

H_2S at 20 Pa and 15 W RF power), a-As_2Se_3 (AsH_3/H_2Se at 25 Pa and 60 W RF power) (González-Leal *et al.*, 2004b), a-Se (H_2Se at 10 Pa and 30 W RF power), a-Ge_xS_{1-x} (GeH_4/H_2S at 10 Pa and RF power 20 W), a-Ge_xSe_{1-x}

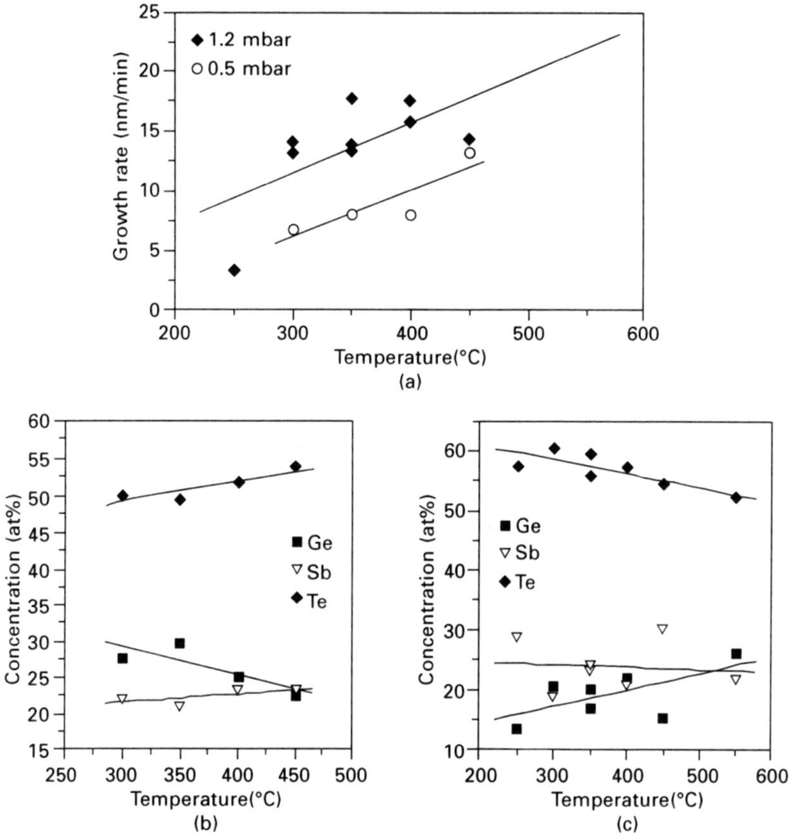

9.17 The influence of hot-wire MOCVD deposition conditions on the final properties of crystalline as-prepared Ge-Sb-Te thin films using $Ge(C_3H_5)_4$, $Sb(C_3H_7)_3$ and $Te(C_3H_7)_2$ precursors, N_2 as carrier gas and H_2 as a co-reactant. The composition $Ge_2Sb_2Te_5$ was achieved at 350°C and 25% of hydrogen flow. (a) Growth rate dependence on deposition temperature using a gas mixture of 25% of H_2. The corresponding compositions of as-prepared thin films are plotted for (b) 0.5 mbar and (c) 1.2 mbar. (d) and (e) depict the dependence of growth rate and composition on relative hydrogen flow. The thin films were deposited at 350°C and 1.2 mbar. The experimental conditions also have a very pronounced effect on grain size and thin-film roughness. Thin films deposited at 0.5 mbar have smaller grains (~30 nm) and lower mean square roughness (2–3 nm) than samples deposited at 1.2 mbar with grain size 80 nm and roughness (5–8 nm) (Reso *et al.*, 2011a, 2011b). The lines are guides to the eye only. (From Reso *et al.* (2011a), published with permission.)

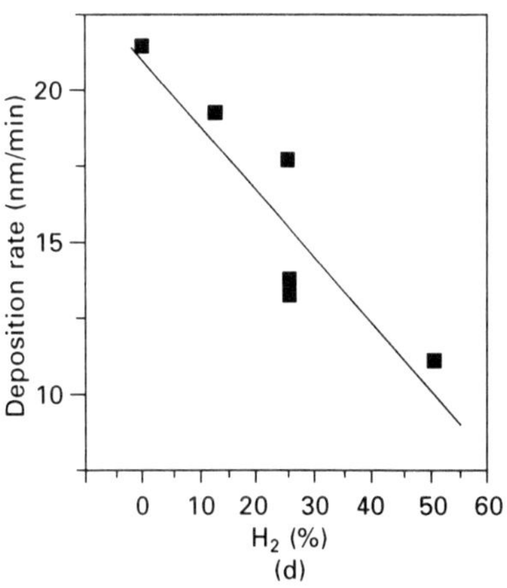

9.17 Continued

(GeH_4/H_2Se at 10 Pa and 80 W RF power) (Nagels, 1998), all nominally at room temperature (noting that the substrate temperature can rise due to the nearby plasma), and, e.g., a-Ge_xSe_{1-x} ($GeCl_4/(CH_3)_2Se/H_2$ at 125–140°C at ~25 Pa) (Whitham *et al.*, 2011).

MOCVD uses metal-organic precursors rather than inorganic ones. It is typically used to deposit crystalline thin films of chalcogenides for phase-change data recording or solar-cell applications. Common compositions are based on Ge-Sb-Te, Ge-Te, Sb-Te, $CuInS_2$, CdS, or ZnS. Mostly MOCVD takes place at $T < 400°C$, which is lower than would be typical for conventional CVD. Typical precursors for different elements are: for germanium – tetrakis(dimethylamino)germanium ($Ge[N(CH_3)_2]_4$), tetra(allyl)germanium ($Ge(C_3H_5)_4$), tetra(isobutyl)germanium ($Ge(i\text{-}C_4H_9)_4$), isobutylgermane (H_3-Ge-(i-C_4H_9); for antimony – tris(dimethylamino)antimony ($Sb[N(CH_3)_2]_3$), tri(isopropyl)antimony ($Sb(C_3H_7)_3$); and for tellurium – di(isopropyl)tellurium ($Te(C_3H_7)_2$) (Longo *et al.*, 2008; Milliron *et al.*, 2009). In the case of Ge-Sb-Te compounds, H_2 is used as a co-reactive gas. CdS can be grown from $CdCl_2/SC(NH_2)_2$ through the intermediate complex state $Cd(SC(NH_2)_2]_2Cl_2$, and, e.g., $CuInS_2$ using single-molecule precursors $[(C_4H_9)_2In(S(i\text{-}C_3H_7))Cu(S_2CN(C_4H_9)_2)]$, $[C_4H_9In(S(i\text{-}C_3H_7))_2Cu(S_2CN(C_4H_9)_2)]$, or $[(C_4H_9)_2In(S(i\text{-}C_3H_7))Cu(S_2CN(i\text{-}C_3H_7)_2)]$ and many more (Kemmler *et al.*, 2002; Gleizes, 2000). As already noted, very good thickness uniformity over large substrate areas and uniform infilling of, e.g., trenches, can be achieved by CVD (Fig. 9.18).

CVD has already been demonstrated to be promising for the deposition of amorphous and crystalline chalcogenide thin films for waveguides, phase-change RAM and photovoltaic applications. It is still challenging to find appropriate precursors for rare-earth metals in the deposition of

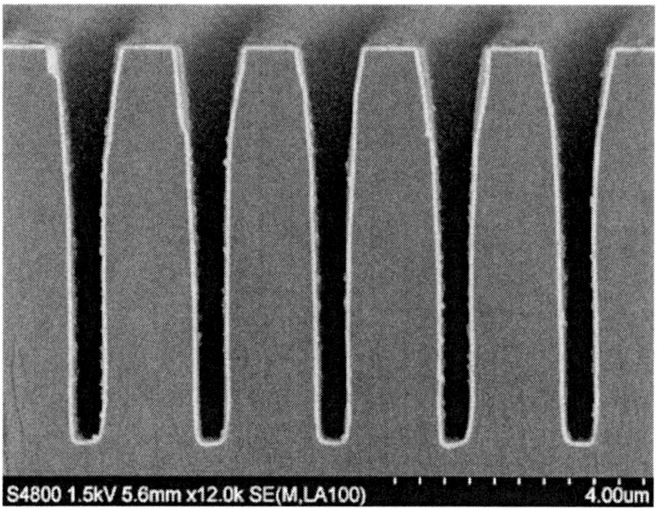

9.18 Uniform thickness coverage of micrometre-scale trenches (aspect ratio 52:1) by $Ge_2Sb_2Te_5$ layer. The GST films were deposited by atomic layer deposition, which is a variant of CVD, where the thin film is grown via self-limiting surface reactions (Ritala and Ninisto, 2009). The GST was deposited at 90°C using $GeCl_2.C_4H_8O_2$, $SbCl_3$ and $(Et_3Si)_2Te$ as precursors. (From Ritala *et al.* (2009), published with permission.)

rare-earth doped chalcogenide films. Single-crystal nano-wires (~30 nm in diameter) of GST have been successfully prepared, which demonstrates nm-scale controlled growth in a new application of CVD (Lee *et al.*, 2007). It is, however, difficult to make CVD deposition fully predictable as too many parameters and complex processes have to be considered. Toxicity of precursors represents risk to health. In principle, though, CVD should be widely useful for CHGs for different applications, especially where thin films of very high purity are required.

9.2.5 Spin-coating

Spin-coating (SC) is a solution-based process developed for low-cost deposition of thin films of materials ranging from polymers (as organic photoresists) to functional inorganic films including amorphous (Chern and Lauks, 1982) and crystalline chalcogenides (Mitzi, 2009). SC films are formed by the centrifugal flow of a liquid solution of dissolved material on a rotating planar substrate, as schematically shown in Fig. 9.19. With increasing spinning speed, the liquid spreads over the substrate forming a liquid layer. The layer thickness depends on the rotational forces (dependent on spinning speed), the viscosity of the solution (dependent on the concentration of glass material in

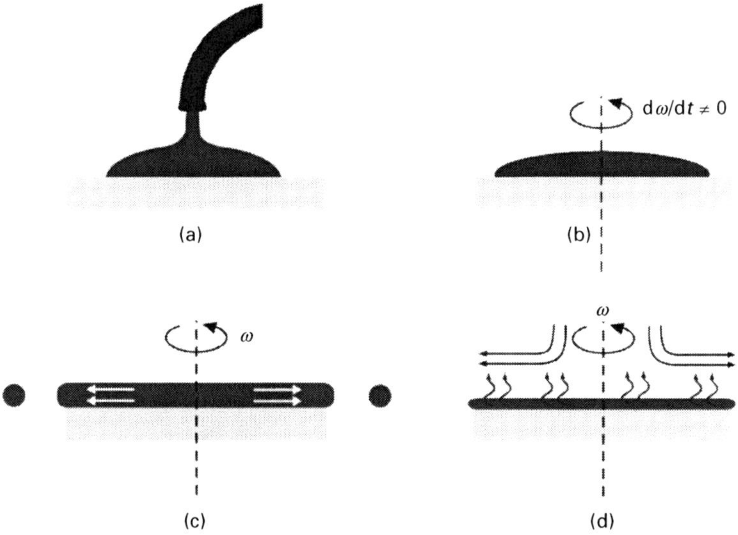

9.19 Schematic of the spin-coating technique: (a) a solution of a material is dropped onto the surface of the planar substrate; (b), (c) substrate rotation applies a centrifugal force on the solution and spreads it over the surface; (d) a solid thin film is formed by a solvent evaporation while substrate rotates at a constant speed. (Redrawn from Bornside *et al.* (1987), published with permisson.)

solution), and on the adhesion of the solution to the substrate. Acceleration usually lasts for only a few seconds and is followed by substrate rotation at constant speed (typically for 30–120 s). In this second step, most of the solvent evaporates from the solution, leaving a solid thin film behind. The spinning dynamics is well described theoretically (see, e.g., Flack *et al.*, 1984). SC films, especially chalcogenide ones, have to undergo post-annealing to release residual solvent. This annealing is crucial in ensuring thin films of good quality, with desirable properties (e.g., optical, thermal, mechanical) and with sufficient durability.

SC was first applied to CHG thin films in the early 1980s by Chern and Lauks (1982) for coating of As-S photoresist thin films. They suggested the mechanism of dissolution of chalcogenide As-S glasses in amine solvents (Chern and Lauks, 1983), which are used to dissolve CHGs and make a solution applicable in SC. The composition, structure, optical and thermal properties, surface morphology, roughness and nano-colloidal behaviour of CHG solutions (Chern and Lauks, 1982, 1983; Chern *et al.*, 1983; Hajto *et al.*, 1989; Norian *et al.*, 1984; González-Leal *et al.*, 2004a; Song *et al.*, 2010; Kohoutek *et al.*, 2008) have been reported mainly for As_2S_3 or As-S thin films that are the amorphous systems most suitable for SC. Spin-coating of amorphous selenide (Shtutina *et al.*, 1995) or telluride films has rarely been reported, since their solubility is either limited in amines or requires the use of higher molecular weight amines (or diamines) that cannot be removed completely from the film's amorphous networks. A special case of SC is *hot-dip spin-coating* (Mairaj *et al.*, 2005) (already mentioned in the introduction to Section 9.2), where the melt of a glass is spun, instead of a liquid or nano-colloidal glass solution.

SC of amorphous chalcogenide thin films involves specific features, summarized below. We take as an example As-S films deposited from propylamine solvent.

1. The dissolved chalcogenide glass is present in amine solution in the form of clusters, with a diameter of several nanometres (Norian *et al.*, 1984; Kohoutek *et al.*, 2008), surrounded by molecules of solvent bound to the cluster surfaces. There is a significant excess of free amine solvent.
2. In the process of dissolution, the nano-clusters are formed by break-up of the glass when the free-electron pairs present on nitrogen attack the As-S bonds in the glass leading to the formation of As-N bonds and the quaternary ammonium salt $R\text{-}NH_3^+$ (Chern *et al.*, 1983):

$$As_2S_3 + 6C_3H_7NH_2 \rightarrow (C_3H_7NH_3)_3AsS_3^- + As(C_3H_7NH)_3 \quad [9.6a]$$

$$C_3H_7NH + S_n \rightarrow C_3H_7NH\,^+S_n^- \quad [9.6b]$$

$$C_3H_7NH^+S_n^- + As_2S_{4/2} \rightarrow (C_3H_7NH^+S_n^-)AsS_{2/2} + (C_3H_7N)AsS_{2/2}. \quad [9.6c]$$

3. A thin film is formed on the substrate from the solution by aggregation of nano-clusters during spinning under inert atmosphere (e.g., nitrogen, argon). Evaporation of free solvent is promoted by the spinning. Some macroscopic morphological defects such as radial stripes, 'comets', and edge defects, can occur in films when the spinning conditions have not been optimized (Birnie, 2001).
4. Before annealing, the amorphous network in the as-deposited SC chalcogenide films consists of aggregates formed from the nano-clusters present in the original glass solution. The diameter of the aggregates is tens of nanometres, some 15–20 times larger than the diameter of the nano-clusters themselves (Norian *et al.*, 1984; Kohoutek *et al.*, 2008). The amount of free solvent trapped in the network is reduced, but is still not negligible. Bound solvent is not released during spinning.
5. Heating (*soft annealing*) of as-deposited SC chalcogenide films leads to removal of the residual free solvent and removal of the solvent bound to the amorphous network. However, a distinct fraction of residual solvent remains in the films even at annealing temperatures far above the boiling point of the solvent. For example, the boiling point of propylamine is 48°C, but its residues were detected even in As_2S_3 films annealed at

T = 150°C (Chern and Lauks, 1983; Song *et al.*, 2010). Solvent release from the As-S amorphous network is illustrated in Fig. 9.20.

6. Long-term annealing of SC films (usually 10–20°C below the 'threshold' for their thermal damage) under vacuum should lead to essentially complete release of the solvent from the amorphous network of the thin film. There is cross-linking of the glass aggregates, followed by significant film densification and reduction in film thickness (Song *et al.*, 2010). After appropriate annealing, the film structure, physical and chemical properties should be close to those of the original bulk glass (Song *et al.*, 2010; Kohoutek *et al.*, 2011) (Table 9.1). The conservation of the structural building units maintains the bulk glass in SC CHG films, a characteristic unique to this low-temperature solution-based process.

To produce good-quality films by SC, it is crucial to have a suitable chalcogenide glass/solvent pair. This should enable proper dissolution of glass, and provide good film-forming ability with reasonable viscosity, but the solvent has to have a high volatility to be effectively released from the

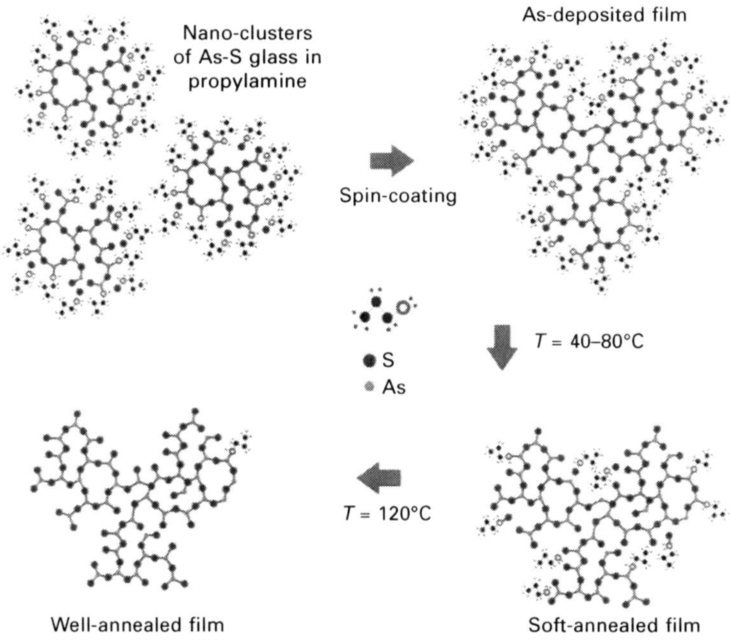

9.20 The aggregation of nano-clusters during spin-coating of an As-S film, and solvent release on annealing the film are shown. Annealing promotes cross-linking of the aggregates, leading to densification. For a SC As_2S_3 film, thickness reduction as great as 20% has been observed during annealing, together with increase in density (27%) and refractive index (15%) (Song *et al.*, 2010).

Table 9.1 The composition *x*, refractive index *n*, 'free' volume *V*, thickness *d*, and surface roughness R_q determined for a SC $As_{30}S_{70}$ film on long-term (20 hours) annealing under vacuum. An increase in refractive index reflects increasing annealing temperature, reaching $n = 2.209$ at $\lambda = 1.55$ μm, which is comparable with $n = 2.231$ at $\lambda = 1.55$ μm for an annealed TE $As_{30}S_{70}$ film whose deposition was solvent-free. The thin films were spin-coated from propylamine solution.

$As_{30}S_{70}$ films	$x_{As:S}$ (at.%)	$n(\lambda = 1.55$ μm)	*V* (vol.%)	*d* (nm)	R_q (nm)
T = 20°C	28.7:71.3	1.944 ± 0.001	35.2 ± 0.2	1071 ± 7	0.21 ± 0.05
T = 40°C	28.7:71.3	2.029 ± 0.001	26.2 ± 0.1	877 ± 5	0.30 ± 0.05
T = 80°C	28.9:71.1	2.110 ± 0.001	18.2 ± 0.1	747 ± 5	0.42 ± 0.05
T = 120°C	29.9:70.1	2.209 ± 0.001	9.1 ± 0.2	632 ± 5	2.79 ± 0.20
TE film	29.9:70.1	2.231 ± 0.001	–	800 ± 7	1.79 ± 0.10
Bulk glass	29.9:70.1	2.278 ± 0.001	–	–	–

amorphous thin film network by annealing at relatively low temperatures. The annealing temperature is limited by the low temperature limit of chalcogenide films, usually in the range 120–250°C, depending on the composition. Insufficiently annealed films undergo unwanted *ageing* through hydrolysis of the residual amine by reaction with water vapour in the atmosphere. Solvent residues reduce the resistance of SC chalcogenide films to attack by acidic and alkaline solutions. The presence of amine residues also significantly influences the optical, thermal and mechanical properties of SC thin films. In practice, the use of SC chalcogenide thin films is thus limited to the very few chalcogenide glass/amine solvent combinations that satisfy all the above criteria.

The application of SC chalcogenide thin films in functional devices is still rare (Zha *et al.*, 2013). Recently, however, significant progress was achieved, especially in the fabrication of planar waveguides. Tsay *et al.* (2010) have reported that planar optical waveguides operating at mid-IR wavelengths (based on SC As_2S_3 films) can be fabricated either by infilling polymeric (PDMS) moulds with As_2S_3 chalcogenide glass solution, or by spin-coating the solution onto the mould followed by micro contact transfer. The characteristics of such waveguides are shown in Fig. 9.21. Capillary-force lithography was reported (Carlie *et al.*, 2010) to produce waveguides directly in as-deposited SC As_2S_3 films, where solvent removal from the moulded SC films was through the PDMS mould. Also poly- or nano-crystalline SC chalcogenide films such as SnS, Sn-Se (Mitzi *et al.*, 2005), $CuInTe_2$ (Mitzi *et al.*, 2006) or ZnTe (Mitzi, 2005) have been shown to be applicable in high-mobility transparent conductive layers in thin-film transistors. These films can be fabricated by a chemical route exploiting hydrazine and hydrazinium salt precursors (highly toxic but undergoing decomposition at attractively low temperatures).

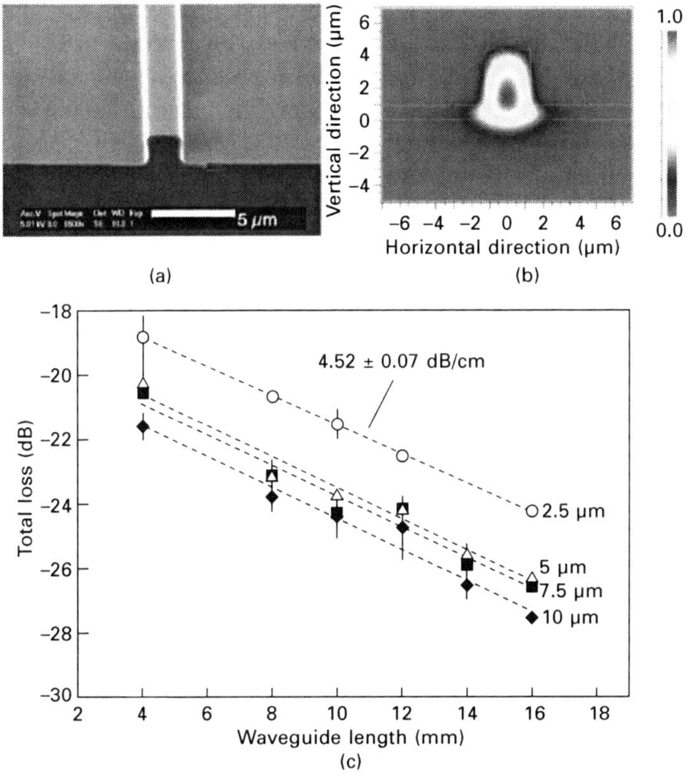

9.21 (a) SEM image (45° inclination) of a 2.5 μm wide by 4.5 μm high rib waveguide on a $LiNbO_3$ substrate, (b) simulated mode field profile in the rib waveguide at λ = 5 μm, and (c) cut-back measurement of losses in the chalcogenide rib. Losses of rib waveguides of varying widths are shown, with height of 4.5 μm, on a $LiNbO_3$ substrate. The waveguides are aligned to and end-fired coupled to a QCL emitting at λ = 4.8 μm. Each point represents averaged data from at least five different waveguides. Propagation loss of the 2.5 μm wide waveguide is given. (From Tsay *et al.* (2010), published with permission.)

9.3 Conclusion and future trends

Over the last decades there has been significant improvement of deposition techniques in terms of their reliability, reproducibility and final thin-film quality having desired physical-chemical properties. We have 'mastered' a variety of physical and chemical deposition techniques for a variety of materials, but with some limitations with chalcogenide single thin films (1D). We are able to deposit thin films (few Å) and fabricate functional multi-layered devices in one deposition go, using multiple targets without a need to break the vacuum. *In-situ* analysis and diagnostic tools are widely used, allowing precise and accurate monitoring of all possible properties, providing useful feedback for reproducible thin film deposition. Nevertheless with all the heavy loads of sophisticated *in-situ* and *off-situ* analysis tools, we are still lacking detailed knowledge of film formation under different experimental conditions (e.g., different types of substrates, sandwich layers, annealing/cooling of the substrate, different target-to-sample geometry, etc.), especially in the new deposition techniques such as PLD or CVD (atomic layer deposition; Ritala and Ninisto, 2009). It is quite exciting for basic research how the short- and medium-range orders and the defects in as-prepared chalcogenide films vary with change in deposition conditions. Without this detailed knowledge it is unlikely that thin films of a good optical quality can be deposited in a reproducible way. As an example, the thin-film quality in PLD has a number of problems (presence of droplets and thickness non-uniformity) and much has to be done to improve PLD performance.

Laser-assisted reactive deposition techniques can help in complex sample deposition. Possibilities of such deposition are barely studied. Molecular beam epitaxy employing wedge shutters has been demonstrated to be feasible for a-chalcogenide thin film deposition with graded composition across the substrate area. This allows much faster characterization of different film compositions covering large parts of phase diagram almost from 0–100 at.% in some cases (Hewak *et al.*, 2011). Growth of nano-crystalline structures (e.g., wires, tubes, etc.) represents a modern trend of down-scaling memory cells and fabrication of low power devices (Jung *et al.*, 2011a, 2011b). Recently, a solution-based deposition has been used in device fabrication processes (Milliron *et al.*, 2009). We are now looking for new chalcogenide glass/solvent combinations and selective solvents, which could extend the applicability of the solution-based depositions. It is demanding to deposit ultra thin (<50 nm), conformal thick (several microns) and multi-layered structures by SC. Also possibilities of reactive spin-coating have to be studied in detail.

The direct inkjet printing of chalcogenide glass optical elements (combining deposition and optical element fabrication into one, as opposed to hard (Lian *et al.*, 2009) and soft stamp imprinting (Orava *et al.*, 2011), where both steps have to be done individually), as has been demonstrated by Sanchez *et al.* (2011), will certainly be on the radar of many scientists, but limited to a few chosen chalcogenide (mainly binary)/solution systems. At the moment, we are transferring our knowledge of single chalcogenide thin-film deposition to functional device fabrication, where chalcogenide deposition represents only one step in a very complex process. Deposition into the 2D and 3D embedded structures is more challenging, and many detail studies have to be done to meet industrial requirements.

9.4 Sources of further information and advice

General features of physical and chemical vapour deposition techniques can be found in many books and papers (see e.g., Smith, 1995; Martin, 2009). Further details can also be found for: physical vapour deposition (Mattox, 2010; Powell and Rossnagel, 1998; Hopwood, 2000); sputtering (Wasa *et al.*, 2011); pulsed-laser deposition (Eason, 2007; Chrisey and Hubler, 1994) (similarities can be found in PLD of polymers (Bityurin, 2005) or carbon-like diamond films (Voevodin and Donley, 1996)); chemical vapour deposition (Pierson, 1999; Jones and Hitchman, 2009), and spin-coating (Mitzi, 2009; Yuan and Mitzi, 2009). The reader may like to consider additional further readings about chalcogenide glasses (Tanaka and Shimakawa, 2011; Bhattacharya *et al.*, 2011), the mechanisms of nucleation and growth during thin-film deposition (Venablesk, 2010; Ohring and Gall, 2012), heat transfer at atomic level (Carey *et al.*, 2008), and top-down and bottom-up fabrication techniques (Mijatovic *et al.*, 2005). There are also many on-line tutorials, which can be found on the internet.

9.5 Acknowledgements

Professor A. L. Greer is thanked for helpful discussion. JO acknowledges support from the Engineering and Physical Sciences Research Council (UK). TW acknowledges EU Operational Program Education for Competitiveness (OP EfC) projects 'ReAdMat' and MEYS bilateral cooperation KONTAKT II, SOLARMAT. All correspondence should be addressed to T. Wagner (tomas.wagner@upce.cz).

9.6 References

Ahn J K, Park K W, Hur S G, Seong N J, Kim C S, Lee J Y and Yoon S G (2010), 'Metalorganic chemical vapor deposition of non-GST chalcogenide materials for phase change memory applications', *J. Mater. Chem.*, 20, 1751–1754.

Angell C A (1995), 'Formation of glasses from liquids and biopolymers', *Science*, 267, 1924–1935.

Anne M-L, Keirsse J, Nazabal V, Hyodo K, Inoue S, Boussard-Pledel C, Lhermite H, Charrier J, Yanakata K, Loreal O, Le Person J, Colas F, Compère C and Bureau B (2009), 'Chalcogenide glass optical waveguides for infrared biosensing', *Sensors*, 9, 7398–7411.

Anscombe N (2011), 'The promise of chalcogenides', *Nature Photon.*, 5, 474.

Belev G, Kasap S, Rowlands J A, Hunter D and Yaffe M (2008), 'Dependence of the electrical properties of stabilized a-Se on the preparation conditions and the development of a double layer X-ray detector structure', *Curr. Appl. Phys.*, 8, 383–387.

Bhardwaj P, Shishodia P K and Mehra R M (2007), 'Optical and electrical properties of obliquely deposited a-$GeSe_2$ films', *J. Mater. Sci.*, 42, 1196–1201.

Bhattacharya P, Fornari R and Kamimura H (2011), *Amorphous and glassy semiconducting glasses*, Amsterdam, Elsevier.

Birnie D P (2001), 'Rational solvent selection strategies to combat striation formation during spin coating of thin films', *J. Mater. Res.*, 16, 1145–1154.

Bityurin N (2005), '8 studies on laser ablation of polymers', *Annu. Rep. Prog. Chem., Sect. C: Phys. Chem.*, 101, 216–247.

Bletskan I (2005), 'Phase equilibrium in the systems A^{IV} - B^{VI}', *J. Ovonic Res.*, 1, 53–60.

Bornside D E, Macosko C W and Scriven L E (1987), 'On the modeling of spin coating', *J. Imag. Tech.*, 12, 122–130.

Carey V P, Chen G, Grigoropoulos C, Kaviany M and Majumdar A (2008), 'A review of heat transfer physics', *Nanosc. Microsc. Thermophys. Eng.*, 12, 1–60.

Caricato A P, De Sario M, Fernández M, Ferrari M, Leggieri G, Luches A, Martino M, Montagna M, Prudenzano F and Jha A (2003), 'Chalcogenide glass thin film waveguides deposited by excimer laser ablation', *Appl. Surf. Sci.*, 208–209, 632–637.

Carlie N, Musgraves J D, Zdyrko B, Luzinov I, Hu J, Singh V, Agarwal A, Kimerling L C, Canciamilla A, Morichetti F, Melloni A and Richardson K (2010), 'Integrated chalcogenide waveguide resonators for mid-IR sensing: leveraging material properties to meet fabrication challenges', *Opt. Express*, 18, 26728–26743.

Chern G C and Lauks I (1982), 'Spin-coated amorphous chalcogenide films', *J. Appl. Phys.*, 53, 6979–6982.

Chern G C and Lauks I (1983), 'Spin coated amorphous chalcogenide films: structural characterization ', *J. Appl. Phys.*, 54, 2701–2705.

Chern G C, Lauks I and McGhie A R (1983), 'Spin coated amorphous chalcogenide films: thermal properties', *J. Appl. Phys.*, 54, 4596–4601.

Chrisey D B and Hubler G K (1994), *Pulsed laser deposition of thin films*, New York, Wiley-Blackwell.

Curry R J, Mairaj A K, Huang C C, Eason R W, Grivas C, Hewak D W and Badding J V (2005), 'Chalcogenide glass thin films and planar waveguides', *J. Am. Ceram. Soc.*, 88, 2451–2455.

De Giacomo A, Shakhatov V A, Senesi G S, De Pascale O and Prudenzano F (2002), 'Plasma-assisted pulsed laser deposition for the improvement of the film growth process', *Appl. Surf. Sci.*, 186, 533–537.

DeCorby R, Nguyen H, Dwivedi P and Clement T (2005), 'Planar omnidirectional reflectors in chalcogenide glass and polymer', *Opt. Express*, 13, 6228–6233.

Delmdahl R and Pätzel R (2008), 'Pulsed laser deposition with excimer lasers', *Phys. Stat. Sol. C*, 5, 3276–3279.

Dieker H and Wuttig M (2005), 'Influence of deposition parameters on the properties of sputtered GeSbTe films', *Thin Solid Films*, 478, 248–251.

Dikova J, Babeva T and Sharlandjiev P (2005), 'Photoinduced scalar and vector effects in obliquely deposited a-As_2S_3 thin films', *J. Optoelectron. Adv. Mater.*, 7, 361–364.

Dwivedi P K, Sun Y W, Tsui Y Y, Tonchev D, Munzar M, Koughia K, Haugen C J, Decorby R G, McMullin J N and Kasap S O (2005), 'Rare-earth doped chalcogenide thin films fabricated by pulsed laser deposition', *Appl. Surf. Sci.*, 248, 376–380.

Eason R (2007), *Pulsed laser deposition of thin films: applications-led growth of functional materials*, Hoboken, NJ, Wiley.

Elliott S (1998), *The physics and chemistry of solids*, Chichester, Wiley.

Fedkin A V, Grossman L and Ghiorso M S (2006), 'Vapor pressures and evaporation coefficients for melts of ferromagnesian chondrule-like compositions', *Geochim. Cosmochim. Acta*, 70, 206–223.

Flack W W, Soong D S, Bell A T and Hess D W (1984), 'A mathematical model for spin coating of polymer resists', *J. Appl. Phys.*, 56, 1199–1206.

Frantz J A, Shaw L B, Sanghera J S and Aggarwal I D (2006), 'Waveguide amplifiers in sputtered films of Er^{3+}-doped gallium lanthanum sulfide glass', *Opt. Express*, 14, 1797–1803.

Freeman D, Madden S and Luther-Davies B (2005), 'Fabrication of planar photonic crystals in a chalcogenide glass using a focused ion beam', *Opt. Express*, 13, 3079–3086.

Frumar M, Frumarova B, Wagner T and Nemec P (2003), 'Photo-induced phenomena in amorphous and glassy chalcogenides', in Kolobov A V, *Photo-induced metastability in amorphous semiconductors*, Weinheim, Wiley, 23–44.

Frumar M, Frumarova B, Nemec P, Wagner T, Jedelsky J and Hrdlicka M (2006), 'Thin chalcogenide films prepared by pulsed laser deposition – new amorphous materials applicable in optoelectronics and chemical sensors', *J. Non-Cryst. Solids*, 352, 544–561.

Frumar M, Frumarova B and Wagner T (2011), 'Amorphous and glassy semiconducting chalcogenides', in Bhattacharya P, Fornari R and Kamimura H, *Comprehensive semiconductor science and technology*, Amsterdam, Elsevier, 206–261.

Gamaly E G, Rode A V, Luther-Davies B and Tikhonchuk V T (2002), 'Ablation of solids by femtosecond lasers: ablation mechanism and ablation thresholds for metals and dielectrics', *Phys. Plasmas*, 9, 949–957.

Gamaly E G, Rode A V and Luther Davies B (2006), 'Ultrafast laser ablation and film deposition', in Eason R, *Pulsed laser deposition of thin films: applications-led growth of functional materials*, Hoboken, NJ Wiley, 99–129.

Gleizes A N (2000), 'MOCVD of chalcogenides, pnictides, and heterometallic compounds from single-source molecule precursors', *Chem. Vapor Depos.*, 6, 155–173.

Gmachl C, Hwang H Y, Paiella R, Sivco D L, Baillargeon J N, Capasso F and Cho A Y (2001), 'Quantum cascade lasers with low-loss chalcogenide lateral waveguides', *IEEE Photonic Tech. Lett.*, 13, 182–184.

Goncalves L M, Alpuim P, Rolo A G and Correia J H (2011), 'Thermal co-evaporation of Sb_2Te_3 thin-films optimized for thermoelectric applications', *Thin Solid Films*, 519, 4152–4157.

González-Leal J M, Prieto-Alcón R, Angel J A and Márquez E (2003), 'Optical properties of thermally evaporated amorphous $As_{40}S_{60-x}Se_x$ films', *J. Non-Cryst. Solids*, 315, 134–143.

González-Leal J M, Prieto-Alcón R, Stuchlik M, Vlcek M, Elliott S R and Márquez E (2004a), 'Determination of the surface roughness and refractive index of amorphous $As_{40}S_{60}$ films deposited by spin coating', *Optical Mater.*, 27, 147–154.

González-Leal J, Prieto-Alcón R, Vlcek M and Marquez E (2004b), 'Structural and optical characterization of amorphous AsS and AsSe films prepared by plasma-enhanced chemical vapor deposition', *J. Non-Cryst. Solids*, 345–346, 88–92.

Hajto E, Ewen P J S, Hill P G and Owen A E (1989), 'Spectroscopic investigation of spin-deposited As-S amorphous films', *Phys. Stat. Sol. A*, 114, 587–597.

Han T, Madden S, Bulla D and Luther-Davies B (2010), 'Low loss chalcogenide glass waveguides by thermal nano-imprint lithography', *Opt. Express*, 18, 19286–19291.

Hertz H (1882), 'Ueber die Verdunstung der Flüssigkeiten, insbesondere des Quecksilbers, im luftleeren Raume', *Ann. Phys.*, 17, 177–193.

Hewak D, Huang C C, Gholipour B, Al-Saab F, Guerin S, Anastasopoulos A, Hayden B and Purdy G (2011), 'Thermally stable, low current consuming galium and germanium chalcogenides for consumer and automotive memory applications', *European Symposium on Phase Change and Ovonic Science*, 4–6 Sept. 2011, Switzerland, 60–67.

Holland L and Steckelmacher W (1952), 'The distribution of thin films condensed on surfaces by the vacuum evaporation method', *Vacuum*, 2, 346–364.

Hopwood J A (2000), *Ionized physical vapor deposition*, San Diego, CA, Academic Press.

Huang C and Hewak D (2006), 'Silver-doped germanium sulphide glass channel waveguides fabricated by chemical vapour deposition and photo-dissolution process', *Thin Solid Films*, 500, 247–251.

Huang C C, Hewak D and Badding J (2004), 'Deposition and characterization of germanium sulphide glass planar waveguides', *Opt. Express*, 12, 2501–2506.

Huang C C, Gholipour B, Ou J Y, Knight K and Hewak D W (2011), 'Electrical phase change of CVD-grown Ge-Sb-Te thin-film device', *Electron. Lett.*, 47, 288–289.

Irvine S J C, Barrioz V, Lamb D, Jones E W and Rowlands-Jones R L (2008), 'MOCVD of thin film photovoltaic solar cells – next-generation production technology?', *J. Cryst. Growth*, 310, 5198–5203.

Jones A C and Hitchman M L (2009), *Chemical vapour deposition: precursors, processes and applications*, Cambridge, RSC Publishing.

Jung Y, Agarwal R, Yang C-Y and Agarwal R (2011a), 'Chalcogenide phase-change memory nanotubes for lower writing current operation', *Nanotechnology*, 22, 254012.

Jung Y, Nam S-W and Agarwal R (2011b), 'High-resolution transmission electron microscopy study of electrically-driven reversible phase change in $Ge_2Sb_2Te_5$ nanowires', *Nano Letters*, 11, 1364–1368.

Kahnt H and Feltz A (1982), 'Modified flash evaporation technique for the preparation of thin amorphous semiconductor films', *Thin Solid Films*, 98, 211–214.

Kalyva M, Siokou A, Yannopoulos S N, Nemec P and Frumar M (2007), 'Electronic and structural changes induced by irradiation or annealing in pulsed laser deposited $As_{50}Se_{50}$ films: an XPS and UPS study', *J. Phys. Chem. Solids*, 68, 906–910.

Kalyva M, Siokou A, Yannopoulos S N, Wagner T, Krbal M, Orava J and Frumar M (2008), 'Soft X-ray induced Ag diffusion in amorphous pulse laser deposited $As_{50}Se_{50}$ thin films: an X-ray photoelectron and secondary ion mass spectroscopy study', *J. Appl. Phys.*, 104, 043704.

Kalyva M, Siokou A, Yannopoulos S N, Wagner T, Orava J and Frumar M (2009), 'Ag diffusion in amorphous $As_{50}Se_{50}$ films studied by XPS', *J. Non-Cryst. Solids*, 355, 1844–1848.

Kasap S and Capper P (2006), *Springer handbook of electronic and photonic materials*, New York, Springer.

Kemmler M, Lazell M, O'Brien P, Otway D J, Park J H and Walsh J R (2002), 'The growth of thin films of copper chalcogenide films by MOCVD and AACVD using novel single-molecule precursors', *J. Mater. Sci.*, 13, 531–535.

Kloock J P and Schoning M J (2008), 'Heavy metal detection with semiconductor devices based on PLD-prepared chalcogenide glass thin films', *Armen. J. Phys.*, 1, 95–98.

Kloock J P, Mourzina Y G, Schubert J and Schoning M J (2002), 'A first step towards a microfabricated thin-film sensor array on the basis of chalcogenide glass materials', *Sensors*, 2, 356–365.

Knudsen M (1915), 'Die maximale Verdampfungsgeschwindigkeit des Quecksilbers', *Ann. Phys.*, 47, 697–708.

Kohoutek T, Wagner T, Frumar M, Chrissanthopoulos A, Kostadinova O and Yannopoulos S N (2008), 'Effect of cluster size of chalcogenide glass nanocolloidal solutions on the surface morphology of spin-coated amorphous films', *J. Appl. Phys.*, 103, 063511.

Kohoutek T, Orava J, Prikryl J, Mistrik J, Wagner T and Frumar M (2009), 'Near infrared quazi-omnidirectional reflector in chalcogenide glasses', *Optical Mater.*, 32, 154–158.

Kohoutek T, Orava J, Sawada T and Fudouzi H (2011), 'Inverse opal photonic crystal of chalcogenide glass by solution processing', *J. Coll. Inter. Sci.*, 353, 454–458.
Kosek F, Cimpl Z, Tulka J and Chlebny J (1987), 'New analytic method for investigation of the distribution of bonds in As-S system', *J. Non-Cryst. Solids*, 90, 401–404.
Kozicki M N, Balakrishnan M, Gopalan C, Ratnakumar C and Mitkova M (2005), 'Programmable metallization cell memory based on Ag-Ge-S and Cu-Ge-S solid electrolytes', *Non-Volatile Memory Technology Symp.*, 83–89.
Lee S-H, Jung Y and Agarwal R (2007), 'Highly scalable non-volatile and ultra-low-power phase-change nanowire memory', *Nature Nanotechnol.*, 2, 626–630.
Lee S J, Yoon S G, Choi K J, Ryq S O, Yoon S M, Lee N Y and Yu B G (2006), 'Characterization of *in situ* diffusion of silver in Ge-Te amorphous films for programmable metallization cell memory applications', *J. Vac. Sci. Technol. B*, 24, 2312–2316.
Lian Z G, Pan W, Furniss D, Benson T M, Seddon A B, Kohoutek T, Orava J and Wagner T (2009), 'Embossing of chalcogenide glasses: monomode rib optical waveguides in evaporated thin films', *Opt. Lett.*, 34, 1234–1236.
Lincot D (2005), 'Electrodeposition of semiconductors', *Thin Solid Films*, 487, 40–48.
Longo M, Salicio O, Wiemer C, Fallica R, Molle A, Fanciulli M, Giesen C, Seitzinger B, Baumann P K, Heuken M and Rushworth S (2008), 'Growth study of $Ge_xSb_yTe_z$ deposited by MOCVD under nitrogen for non-volatile memory applications', *J. Cryst. Growth*, 310, 5053–5057.
Lukács R, Hegedüs J and Kugler S (2008), 'Structure and photoinduced volume changes of obliquely deposited amorphous selenium', *J. Appl. Phys.*, 104, 103512.
Luther-Davies B, Rode A V, Madsen N R and Gamaly E G (2005), 'Picosecond high-repetition-rate pulsed laser ablation of dielectrics: the effect of energy accumulation between pulses', *Opt. Eng.*, 44, 051102.
Madden S J, Choi D Y, Bulla D A, Rode A V, Luther-Davies B, Ta'eed V G, Pelusi M D and Eggleton B J (2007), 'Long, low loss etched As_2S_3 chalcogenide waveguides for all-optical signal regeneration', *Opt. Express*, 15, 14414–14421.
Maeda K, Imai T, Fujita S and Saito N (2008), 'Preparation and optical properties of erbium doped chalcogenide glass films by magnetron sputtering', *J. Vac. Soc. Jpn.*, 51, 675–677.
Mairaj A K, Curry R J and Hewak D W (2005), 'Inverted deposition and high-velocity spinning to develop buried planar chalcogenide glass waveguides for highly nonlinear integrated optics', *Appl. Phys. Lett.*, 86, 094102.
Martín-Palma R J, Redondo-Cubero A, Gago R, Ryan J V and Pantano C G (2008), 'Rutherford backscattering spectrometry characterization of nanoporous chalcogenide thin films grown at oblique angles', *J. Anal. Atomic Spectrom.*, 23, 981–984.
Martín-Palma R J, Torres-Costa V and Pantano C G (2009), 'Distributed Bragg reflectors based on chalcogenide glasses for chemical optical sensing', *J. Phys. D*, 42, 055109.
Martin P M (2009), *Handbook of deposition technologies for films and coatings: science, applications and technology*, Norwich, William Andrew.
Martino M, Caricato A P, Fernández M, Leggieri G, Jha A, Ferrari M and Mattarelli M (2003), 'Pulsed laser deposition of active waveguides', *Thin Solid Films*, 433, 39–44.
Mattox D M (2010), *Handbook of physical vapor deposition (PVD) processing*, Oxford, Elsevier.
Méar F O, Essi M, Guimon M F and Pradel A (2008), 'Processing and characterization

of thin film $Cu_x(Ge_{28}Se_{60}Sb_{12})_{1-x}$ ion-selective electrode membrane', *Chalcogenide Lett.*, 5, 117–124.

Mijatovic D, Eijkel J C T and Berg A V D (2005), 'Technologies for nanofluidic systems: top-down vs. bottom-up – a review', *Lab. Chip*, 5, 492–500.

Mikhailov M D, Kryzhanowsky I I and Petcherizin I M (2000), 'Structure and properties of ion-beam sputtered As_xS_{1-x} films', *J. Non-Cryst. Solids*, 265, 1–8.

Milliron D J, Huang Q and Zhu Y (2009), 'Novel deposition methods', in Raoux S and Wuttig M, *Phase change materials, science and applications*, New York, Springer, 227–244.

Mitzi D B (2005), 'Polymorphic one-dimensional $(N_2H_4)_2ZnTe$: soluble precursors for the formation of hexagonal or cubic zinc telluride', *Inorg. Chem.*, 44, 7078–7086.

Mitzi D B (2009), *Solution processing of inorganic materials*, Hoboken, NJ, John Wiley.

Mitzi D B, Copel M and Chey S J (2005), 'Low-voltage transistor employing a high-mobility spin-coated chalcogenide semiconductor', *Adv. Mater.*, 17, 1285–1289.

Mitzi D B, Copel M and Murray C E (2006), 'High-mobility p-type transistor based on a spin-coated metal telluride semiconductor', *Adv. Mater.*, 18, 2448–2452.

Moore W J (1972), *Physical chemistry*, 5th edition, London, Longman.

Nagels P (1998), 'Plasma-enhanced chemical vapor deposition and structural characterization of amorphous chalcogenide films', *Semiconductors*, 32, 855–860.

National Physical Laboratory (2011), *Sputter yield values* Available at: http://www.npl.co.uk/nanoscience/surface-nanoanalysis/products-and-services/sputter-yield-values (accessed 5 December 2011).

Nazabal V, Nemec P, Jedelsky J, Duverger C, Le Person J, Adam J L and Frumar M (2006), 'Dysprosium doped amorphous chalcogenide films prepared by pulsed laser deposition', *Optical Mater.*, 29, 273–278.

Nazabal V, Jurdyc A M, Nemec P, Brandily-Anne M L, Petit L, Richardson K, Vinatier P, Bousquet C, Cardinal T, Pechev S and Adam J L (2010), 'Amorphous Tm^{3+} doped sulfide thin films fabricated by sputtering', *Optical Mater.*, 33, 220–226.

Nemec P, Frumar M, Frumarova B, Jelinek M, Lancok J and Jedelsky J (2000), 'Pulsed laser deposition of pure and praseodymium-doped Ge-Ga-Se amorphous chalcogenide films', *Optical Mater.*, 15, 191–197.

Nemec P, Frumar M, Jedelsky J, Jelinek M, Lancok J and Gregora I (2002), 'Thin amorphous chalcogenide films prepared by pulsed laser deposition', *J. Non-Cryst. Solids*, 299–302, 1013–1017.

Norian K H, Chern G C and Lauks I (1984), 'Morphology and thermal properties of solvent-cast arsenic sulfide films', *J. Appl. Phys.*, 55, 3795–3798.

Ogura H, Matsuishi K and Onari S (2000), 'Raman scattering and photodarkening of amorphous $Ge_{1-x}S_x$ ($0 \le x \le 0.62$) films', *J. Non-Cryst. Solids*, 270, 147–153.

Ohring M and Gall D (2012), *Materials science of thin films: deposition and structure*, New York, Academic Press.

Orava J, Wagner T, Sik J, Prikryl J, Frumar M and Benes L (2008), 'Optical properties and phase change transition in $Ge_2Sb_2Te_5$ flash evaporated thin films studied by temperature dependent spectroscopic ellipsometry', *J. Appl. Phys.*, 104, 043523.

Orava J, Kohoutek T, Wagner T, Cerna Z, Vlcek M, Benes L, Frumarova B and Frumar M (2009), 'Optical and structural properties of Ge-Se bulk glasses and Ag-Ge-Se thin films', *J. Non-Cryst. Solids*, 355, 1951–1954.

Orava J, Kohoutek T, Greer A L and Fudouzi H (2011), 'Soft imprint lithography of a bulk chalcogenide glass', *Opt. Mater. Express*, 1, 796–802.

Orava J, Greer A L, Gholipour B, Hewak D W and Smith C E (2012), 'Characterization of supercooled liquid $Ge_2Sb_2Te_5$ and its crystallization by ultrafast-heating calorimetry', *Nature Mater.*, 11, 279–283.

Pangavhane S D, Houska J, Wagner T, Pavlista M, Janca J and Havel J (2010), 'Laser ablation of ternary As-S-Se glasses and time-of-flight mass spectrometric study', *Rapid. Commun. Mass. Sp.*, 24, 95–102.

Park Y S, Lee S Y, Yoon S M, Jung S W, Yu B G, Lee S J and Yoon S G (2007), 'Nonvolatile programmable metallization cell memory switching element based on Ag-doped SbTe solid electrolyte', *Appl. Phys. Lett.*, 91, 162107.

Pierson H O (1999), *Handbook of chemical vapor deposition (CVD): principles, technology, and applications*, Park Ridge, NJ, Noyes Publications.

Pinnow C U and Happ T (2010), *Reactive sputtering process for optimizing the thermal stability of thin chalcogenide layers*. US patent application 7,692,175 B2, 6 April.

Powell R and Rossnagel S M (1998), *PVD for microelectronics: sputter deposition to semiconductor manufacturing*, London, Academic Press.

Prikryl J, Hrdlicka M, Frumar M, Orava J, Benes L, Vlcek M, Kostal P, Hromadko L and Wagner T (2009), 'Structure, electrical, optical and thermal properties of $Ge_4Sb_4Te_x$ ($x = 8$, 9 and 10) thin films', *J. Non-Cryst. Solids*, 355, 1998–2002.

Rajagopalan S, Harshavardhan K S, Malhotra L K and Chopra K L (1982), 'Photo-optical changes in Ge-chalcogenide films', *J. Non-Cryst. Solids*, 50, 29–38.

Reso D, Silinskas M, Kalkofen B, Lisker M and Burte E P (2011a), 'Hot wire chemical vapor deposition of $Ge_2Sb_2Te_5$ thin films', *J. Electrochem. Soc.*, 158, D187–D190.

Reso D, Silinskas M, Lisker M, Gewalt A and Burte E P (2011b), 'The role of hydrogen in hot wire chemical vapor deposition of Ge-Sb-Te thin films', *Thin Solid Films*, 519, 2150–2154.

Ritala M and Ninisto J (2009), 'Atomic layer deposition', in Jones A C and Hitchman M L, *Chemical vapour deposition: precursors, processes and applications*. Cambridge, RSC Publishing, 158–206.

Ritala M, Pore V, Hatanpää T, Heikkilä M, Leskelä M, Mizohata K, Schrott A, Raoux S and Rossnagel S M (2009), 'Atomic layer deposition of $Ge_2Sb_2Te_5$ thin films', *Microelectronic Engineering*, 86, 1946–1949.

Rode A V, Zakery A, Samoc M, Charters R B, Gamaly E G and Luther-Davies B (2002), 'Laser-deposited As_2S_3 chalcogenide films for waveguide applications', *Appl. Surf. Sci.*, 197–198, 481–485.

Rossnagel S (2002), 'Sputtering and sputter deposition', in Seshan K, *Handbook of thin-film deposition processes and techniques: principles, methods, equipment and applications*, 2nd edn, Norwich, William Andrew, 319–348.

Sanchez E A, Waldmann M and Arnold C B (2011), 'Chalcogenide glass microlenses by inkjet printing', *Appl. Optics*, 50, 1974–1978.

Seah M P (2005), 'An accurate semi-empirical equation for sputtering yields, II: for neon, argon and xenon ions', *Nucl. Instrum. Meth. B*, 229, 348–358.

Seah M P, Clifford C A, Green F M and Gilmore I S (2005), 'An accurate semi-empirical equation for sputtering yields I: for argon ions', *Surf. Interf. Anal.*, 37, 444–458.

Seddon A B, Pan W J, Furniss D, Miller C A, Rowe H, Zhang D, McBrearty E, Zhang Y, Loni A, Sewell P and Benson T M (2006), 'Fine embossing of chalcogenide glasses – a new fabrication route for photonic integrated circuits', *J. Non-Cryst. Solids*, 352, 2515–2520.

Shtutina S, Klebanov M, Lyubin V, Rosenwaks S and Volterra V (1995), 'Photoinduced phenomena in spin-coated vitreous As_2S_3 and AsSe films', *Thin Solid Films*, 261, 263–265.

Shurgalin M, Fuflyigin V N and Anderson E G (2005), 'The structural heterogeneity and optical properties in chalcogenide glass films', *J. Phys. D*, 38, 4037–4047.

Simpson R E, Krbal M, Fons P, Kolobov A V, Tominaga J, Uruga T and Tanida H (2010), 'Toward the ultimate limit of phase change in $Ge_2Sb_2Te_5$', *Nano Letters*, 10, 414–419.

Smith D L (1995), *Thin-film deposition: principles and practice*, New York, McGraw-Hill.

Snopatin G E, Shiryaev V S, Plotnichenko V G, Dianov E M and Churbanov M F (2009), 'High-purity chalcogenide glasses for fiber optics', *Inorg. Mater.*, 45, 1439–1460.

Song S, Dua J and Arnold C B (2010), 'Influence of annealing conditions on the optical and structural properties of spin-coated As_2S_3 chalcogenide glass thin films', *Opt. Express*, 18, 5472–5480.

Tan W C, Solmaz M E, Gardner J, Atkins R and Madsen C (2010), 'Optical characterization of a-As_2S_3 thin films prepared by magnetron sputtering', *J. Appl. Phys.*, 107, 033524.

Tanaka K and Shimakawa K (2011), *Amorphous chalcogenide semiconductors and related materials*, New York, Springer.

Ting H, Madden S, Bulla D and Luther-Davies B (2010), 'Low loss chalcogenide glass waveguides fabricated by hot embossing', *35th Australian Conference on Optical Fibre Technology*, 5–9 December 1–3.

Tominaga J, Fons P, Kolobov A, Shima T, Chong T C, Zhao R, Lee H K and Shi L (2008), 'Role of Ge switch in phase transition: approach using atomically controlled GeTe/Sb_2Te_3 superlattice', *Jpn. J. Appl. Phys.*, 47, 5763–5766.

Totten G E, Funatani K and Xie L (2004), *Handbook of metallurgical process and design*, New York, Marcel Dekker.

Tsay C, Zha Y and Arnold C B (2010), 'Solution-processed chalcogenide glass for integrated single-mode mid-infrared waveguides', *Opt. Express*, 18, 26744–26753.

Ulman A (1996), 'Formation and structure of self-assembled monolayers', *Chem. Rev.*, 96, 1533–1554.

Valov I, Waser R, Jameson J R and Kozicki M N (2011), 'Electrochemical metallization memories – fundamentals, applications, prospects', *Nanotechnology*, 22, 254003.

Venablesk J A (2010), *Introduction to surface and thin-film processes*, Cambridge, Cambridge University Press.

Voevodin A A and Donley M S (1996), 'Preparation of amorphous diamond-like carbon by pulsed laser deposition: a critical review', *Surface and Coatings Technology*, 82, 199–213.

Wagner T, Havel J, Houska J, Pena-Mendez E M, Prikryl J, Pavlista M and Frumar M (2007), 'Thin films of $Ge_2Sb_2Te_5$ prepared by pulsed laser deposition, properties and plume particles analysis', *European Symposium on Phase Change and Ovonic Science*, 1–4 September, part F2, 1–9.

Wagner T, Orava J, Prikryl J, Kohoutek T, Bartos M and Frumar M (2009), 'Medium-term thermal stability of amorphous $Ge_2Sb_2Te_5$ flash-evaporated thin films with regards to change in structure and optical properties', *Thin Solid Films*, 517, 4694–4697.

Wasa K, Kitabatake M and Adachi H (2011), *Thin films material technology: sputtering of compound materials*, Heidelberg, Springer.

Whitham P J, Strommen D P, Lau L D and Rodriguez R G (2011), 'Thin film growth of germanium selenides from PECVD of $GeCl_4$ and dimethyl selenide', *Plasma Chem. Plasma Proc.*, 32, 251–256.

Windischmann H (1992), 'Intrinsic stress in sputter-deposited thin films', *Crit. Rev. Solid State Mat. Sci.*, 17, 547–596.

Wuttig M and Yamada N (2007), 'Phase-change materials for rewriteable data storage', *Nature Mater.*, 6, 824–832.

Yan X T and Xu Y (2010), *Chemical vapour deposition: an integrated engineering design for advanced materials*, London, Springer.

Yang G, Jain H, Ganjoo A, Zhao D, Xu Y, Zeng H and Chen G (2008), 'A photo-stable chalcogenide glass', *Opt. Express*, 16, 10565–10571.

Youden K E, Grevatt T, Eason R W, Rutt H N, Deol R S and Wylangowski G (1993), 'Pulsed laser deposition of Ga-La-S chalcogenide glass thin film optical waveguides', *Appl. Phys. Lett.*, 63, 1601–1603.

Yuan M and Mitzi D B (2009), 'Solvent properties of hydrazine in the preparation of metal chalcogenide bulk materials and films', *Dalton Trans.*, 31, 6078–6088.

Zakery A (2002), 'Low loss waveguides in pulsed laser deposited arsenic sulfide chalcogenide films', *J. Phys. D*, 35, 2909–2913.

Zha Y, Waldmann M and Arnold C B (2013), 'A review on solution processing of chalcogenide glasses for optical components, *Opt. Mater. Express*, 3, 1259–1272.

10
Transparent chalcogenide glass-ceramics

L. CALVEZ, University of Rennes, France

DOI: 10.1533/9780857093561.1.310

Abstract: Glass-ceramics are defined as composite materials constituted of crystals in a glassy matrix. This combination between amorphous and crystalline states leads to a new type of material having unique properties that can be adjusted. Chalcogenide glass-ceramics are a recent target of research and appear very promising to make more efficient materials for passive and active applications working from the visible to the mid-infrared ranges. In this chapter, the theoretical and practical ways to make reproducible glass-ceramics containing nanoparticles will be described. Their efficiency in making materials with improved mechanical properties or with exacerbated nonlinear or luminescence phenomena will be described.

Key words: glass-ceramics, nucleation and growth, scatterings, optical and mechanical properties.

10.1 Introduction

Glass-ceramics are defined as composite materials constituted of crystals in a glassy matrix. They now form their own family of materials. This combination between amorphous and crystalline states leads to a new type of material that has unique properties that can be adjusted depending on the intended application. The first glass-ceramics were discovered in the 1950s by S.D. Stookey with alumino-silicate glasses (Stookey, 1957, 1959). Only a few years later, glass-ceramics presenting very low coefficient of thermal expansion were commercialized under the name of Corningware® (Stookey, 1964). Since this date, a large panel of glass-ceramic materials has been developed. The theoretical and applied research, based mainly on controlled crystallization of oxide-based glasses during the last 40 years, have produced rapid technological developments making this material accessible for the greatest number of applications. Indeed, real-world applications of glass-ceramics are numerous and diversified, including mirrors, telescopes, lasers, dental surgery, hot plates for electric cookers, mechanical workability materials, bioceramics, construction materials and laser hosts (Strnad, 1986; Holand and Beall, 2002). The main advantage of glass-ceramics compared to conventional ceramics lies in their ability to create complex shapes by casting without any machining.

From a thermodynamical point of view, a glass, whatever the constituent materials, is out of its equilibrium state. The high free enthalpy of amorphous materials confers to them a metastable state. By heat treating a glass close to its glass transition temperature (T_g), where the glass viscosity allows a slow modification of the structure, some more stable state can be attained. The phenomenon which consists in an extremely localized aggregation of atoms is called the nucleation step. The nucleation phenomenon is directly linked to the decrease of the free enthalpy in order to reach an equilibrium state. As shown in Fig. 10.1, the crystalline phase produced is thermodynamically more stable than the initial phase.

Controlled isothermal treatment leads to the birth of crystallites within the glassy matrix, the size typically ranging from several nanometers to 1 μm. These particles allow modifications of properties such as its mechanical (e.g., hardness, toughness), thermal, electrical properties or modifies its optical properties by crystallization of rare earth or no centrosymmetric crystalline phase. It then becomes possible to design a glass-ceramic with the desired properties by controlling the size, the ratio and the nature of the crystalline phase.

The chalcogenide glasses, as the name implies, contain at least one chalcogen element, S, Se or Te. They are the subject of many studies for passive or active applications due to their large transmission range, extending from the visible to the mid-infrared depending on the chalcogen element. These glasses are commercialized in particular for thermal imaging used, for example, for driving assistance at night or in foggy conditions, but they also present a high potential for various fields of applications (e.g., chemical sensing, IR sources and lasers, optical switches) (Bureau *et al.*, 2004).

Chalcogenide glass-ceramics are a recent subject of research and appear very promising to make more efficient materials for wide fields of applications

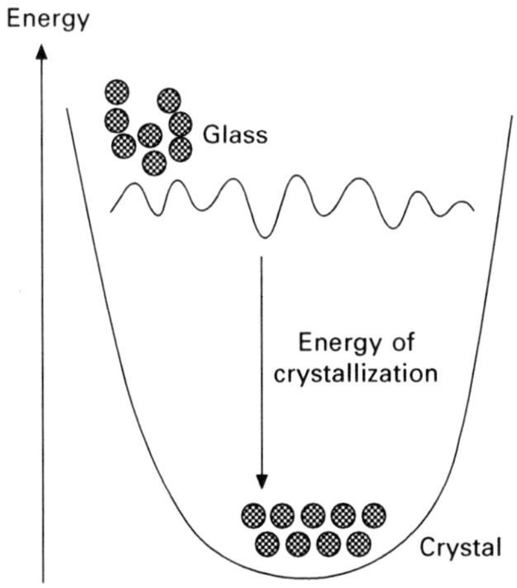

10.1 Schematic diagram of transformation between glass and glass-ceramic.

which will be discussed in this chapter. The main interest in chalcogenide glasses or glass-ceramics is their potential to cover simultaneously a part of the visible range, the first atmospheric window (1000–2500 nm), the second (3–5 μm) and the third ones (8–12 μm).

Associated with their wide transmission range, chalcogenide glasses have relatively weak chemical bonds, leading to poor thermo-mechanical properties compared to oxide glasses, for example. One technique for improving these properties is to fabricate composite materials such as glass-ceramics. The first experimental studies on infrared glass-ceramics were performed in order to improve their thermo-mechanical properties.

To understand the evolution of research dedicated to chalcogenide glass-ceramics, their recent history will be described in Section 10.2. Section 10.3 will give some indications to select the glass composition potentially appropriate for the production of controllable and reproducible glass-ceramics.

Section 10.4 will be devoted to the evolution of mechanical properties (hardness, toughness, thermal expansion coefficient, etc.) and active optical properties such as luminescence (rare earth doping) or permanent second-order nonlinear phenomena by generating nanoparticles.

Finally, the direction of future trends will be discussed in Section 10.5, and novel efficient techniques to make IR glass-ceramics will be described then.

It is important to note that this chapter will be focused on chalcogenide glass-ceramics for optical applications only. Thus many applications concerning non-transparent composite materials such as phase change materials or battery materials containing amorphous and crystalline states will not be treated. For further information, two separate chapters are devoted to these areas of research. Furthermore, in this chapter the theory related to the nucleation and growth of crystals in glass-ceramics will not be described in detail as it has already been the subject of many chapters in other books. Consequently, I advise the reader to consult Strnad (1986) and Holand and Beall (2002).

10.2 The recent history of chalcogenide glass-ceramics

Mecholsky started to work in 1973 on the $PbSeGe_{1}.{_5}As_{0.5}Se_3$ glass composition obtaining for the first time a relatively transparent glass-ceramic between 8 and 12 μm. The heat treatments led to a crystallization of 60% of PbSe, $PbSe_2$ and $GeSe_2$ of about 0.5 μm. Modulus of rupture of the glass-ceramic has doubled and the hardness increased by 30% (Mecholsky 1973; Mecholsky *et al.*, 1976). The 1973 paper was the first approach on the way to make chalcogenide glass-ceramics still transparent in the infrared range.

Later, Cheng worked on the As-Ge-Se-Sn system with different types of nucleating agents ($ZrSe_2$, $ZrSe_2$ + SnSe), a high proportion of crystals of about 1 μm has been achieved while maintaining a moderate transparency in the infrared range. The generation of $ZrSe_2$ crystals led to an increase in certain mechanical properties including resistance to crack propagation (Cheng, 1982, 1986).

In the last decade, much work has been performed on thermodynamic studies of crystallization of chalcogenide glasses. However, the majority of these reports demonstrate the difficulty to control the crystallization of chalcogenide glasses to generate submicron crystals uniformly (Song *et al.*, 1997; Cernoskova *et al.*, 1998; Vázquez *et al.*, 1998). The two methods to produce glass-ceramics using homogeneous nucleation or heterogeneous nucleation with the addition of nucleant agent, have not led so far to the production of glass-ceramics keeping a good transmission in the infrared range.

More recently, studies on Ga-Ge-Sb-Se glasses made by Ma *et al.* have shown that it seems necessary to have stable glasses against crystallization to obtain transparent glass ceramics (Ma *et al.*, 2003). A unique step of nucleation combined with crystal growth slightly higher than the T_g (about 30°C higher) led to transparent glass-ceramics from 2 to 15 μm with low fraction of crystals. A higher temperature of annealing led to the rapid growth of crystals, reducing the transparency of the material in the IR region. It was also shown that it is possible to obtain optical precision molding (Zhang *et al.*, 2004a) which is essential for the production of IR optics.

However, all the ceramics mentioned above are difficult to reproduce. Using stable glasses to induce a controllable crystallization has been reconsidered recently with the latest results on chalcogenide glass-ceramics (Ma *et al.*, 2003; Calvez *et al.*, 2007b).

The first glass-ceramics transparent in the infrared up to 11.5 μm, obtained in a reproducible manner, were discovered in 2004, through the development of novel vitreous compositions, containing both chalcogenides and halide elements. In the GeS_2-Sb_2S_3-CsCl ternary diagram, the simultaneous presence of covalent and ionic compounds creates competition preventing a rapid and uncontrollable crystallization (Zhang *et al.*, 2004b).

An adequate heat treatment of a selected glass composition has led to the generation of a significant proportion of quasi alkali halide microcrystals in chalcogenide glassy matrix, which allows an improvement of mechanical properties, especially an increase of the resistance to fracture propagation (Zhu *et al.*, 2007) which should provide a better resistance to thermal shocks.

As observed, the first papers mentioned above were solely oriented toward the improvement of mechanical properties. More recently, novel aspects of research have been investigated. Even if the possibility to make glass-ceramics containing rare-earth ions has for several decades, been largely related to fluoride glass-ceramics (Auzel *et al.*, 1975), the first attempts on chalcogenide matrices were very innovative.

As is well known, chalcogenide glasses are excellent hosts for the rare earth ions and can be of great interest as materials for active optical applications because of their low phonon energy. Seznec *et al.* have firstly shown that it is possible to generate nanocrystals from a Nd^{3+} containing glassy matrix Ge-Ga-Sb-S-CsCl (Seznec *et al.*, 2006a). They have demonstrated that the emissions of this ion are much more intense in the glass ceramic than in the glass with exactly the same composition, paving the way for a new field of research. Moreover, since 2005, the generation of non-centro-symmetric crystalline particles within the glass led to glass-ceramics presenting permanent second-order nonlinear properties.

As can be seen in Fig. 10.2, with these new routes of investigation, the number of published items concerning chalcogenide glass-ceramics has increased during the last decade, showing the activity around this area of research.

10.3 Synthesis of transparent chalcogenide glass-ceramics

10.3.1 Controlled nucleation and crystal growth

Nucleation in a glassy matrix is related to the decrease of the Gibbs energy ΔG to achieve a steady state. The crystalline phase formed is thermodynamically more stable than the initial amorphous phase. It is recognized that the number and size of particles formed from a given quantity of matter should be related to relative rates of nucleation and growth. To obtain particles of uniform size throughout the glass, it is necessary that the nucleation and growth

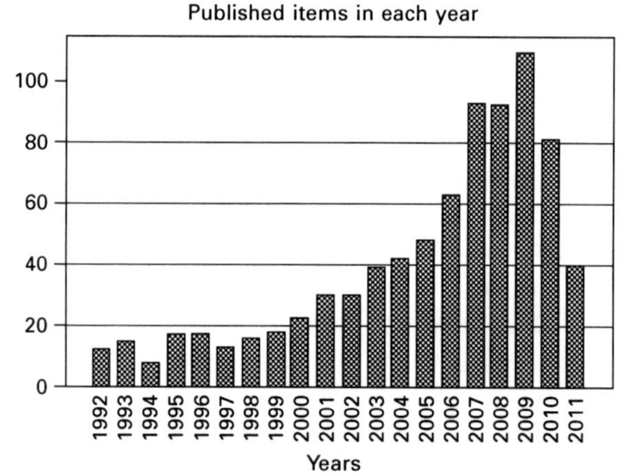

10.2 Evolution of published papers when searching for chalcogenide glass-ceramics in Isiweb of knowledge.

steps are clearly separated, so that a single forming step takes place first, and then seeds grow by accumulation of the rest of the available material (Fig. 10.3).

When both nucleation and growth phenomena appear at similar temperatures, a fast and uncontrollable crystallization occurs leading to a strong and fast decrease of optical transparency. In most cases, the formation of germs occurs on the glass surface or around impurities which are sites where the nucleation is facilitated. In these two cases, the energy necessary for the formation of a germ is strongly reduced. This nucleation phenomenon, called heterogeneous nucleation, causes a fast growth of particles, deteriorating the optical properties, leading rapidly to the whole opacity of the glass-ceramic.

The crystal growth can be represented by two successive steps. According to the diffusion theory, the ions migrate to the interface crystal/solution and are inserted into the crystal lattice by surface reaction. Also, if the diffusion component is fast, growth is controlled by the reaction at the interface. However, if the surface reaction is predominant, then the overall reaction is limited by the transport of ions to the surface. We direct the reader wishing to address the theoretical approach to crystal growth controlled by the interface to the writings of Volmer and Marder (1931) and Turnbull and Cohen (1960). In this chapter most of the thermodynamic calculations leading to the number of nuclei formed or the overall speed of nucleation have not been described. Readers are invited to read reports published by Turnbull (1949) and Weinberg and Zanotto (1991).

The progressive crystallization of the melt bath during the quenching could be an alternative way to make glass-ceramics. However, the low thermal conductivity of silica tubes used during the synthesis of chalcogenide glasses induces a gradient of cooling rate from the outside to the inner part of the tube. Therefore, a crystalline part in the center of the batch surrounded by a glassy matrix can be observed in unstable glasses when the cooling rate is too low. In fact, it is difficult to obtain a homogeneous crystallization with uniform crystal size in the entire volume of a glass.

The most commonly used method in industry to make oxide-based glass-

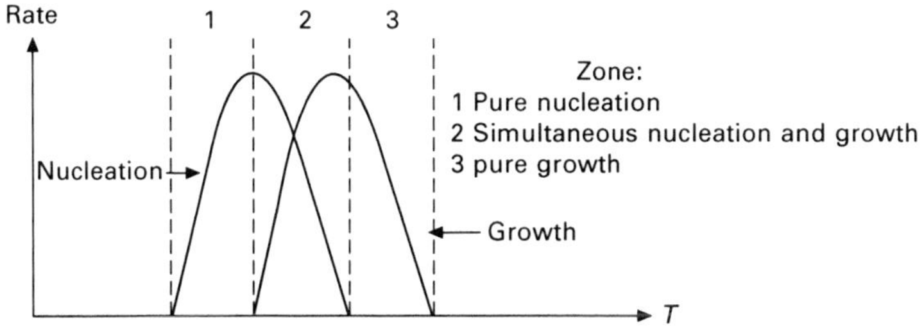

10.3 Nucleation and growth intensity as a function of temperature.

ceramics is to achieve a level of nucleation at a temperature close to T_g, followed by a second level of growth. This technique is possible when the temperatures of nucleation of glass and crystal growth are well separated and have been previously determined by kinetic studies of crystallization. Contrary to what is usually done with silicate glasses, chalcogenide glass-ceramics are mostly obtained using one crystallization step (Fig. 10.4). From a practical point of view, glassy samples were heat treated at temperatures slightly above T_g (from T_g to T_g + 50°C) in ventilated furnaces and then slowly cool down to room temperature. Another way, consisting in heating the glass sample progressively up to its T_x, has been tested also to go through the nucleation step and then the growth step but without conclusive results (Calvez *et al.*, 2006). The use of temperatures close to the glass transition temperature permits high enough viscosity to be reached to allow movement of the atoms leading to a rearrangement of the structure, but also low enough to avoid a fast crystallization of the whole glassy matrix.

The methods reported by Ray and Day are usually used to determine the optimal temperature and time of nucleation in chalcogenide glasses (Ray and Day, 1990; Ray *et al.*, 1996, 2000). The experimental procedure is as follows. To determine the best time of nucleation, the glass is annealed in a ventilated furnace at a temperature slightly above T_g with a variable duration. The glass is then rapidly cooled to room air. Thermal analysis (DSC) is performed at a rate of 10°C/min with 10 mg of glass. The height of the crystallization peak is considered to be proportional to the number of nuclei created during the previous annealing thermal analysis. This study is then performed at different temperatures to determine the optimum temperature.

Many papers reporting the kinetics of isothermal and non-isothermal crystallization of chalcogenide glasses are based on two different theories. Both of them permit calculation of the activation energy. The first method was established by Johnson–Mehl–Avrami (JMA) according to the following equation. It is possible to determine the crystalline fraction (x) of a glass as a function of time (Johnson and Mehl, 1939; Avrami, 1939).

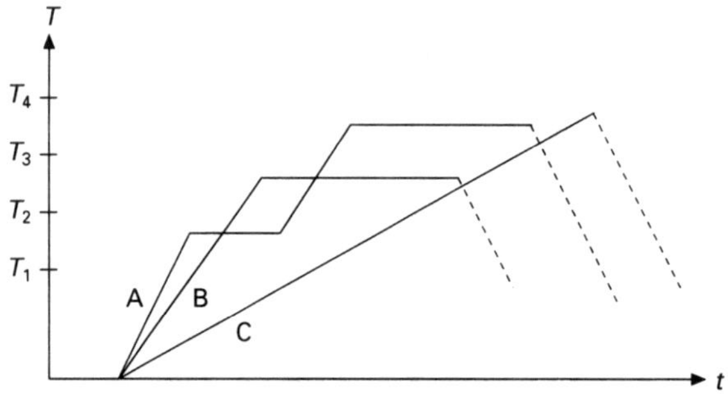

10.4 Dependence of the temperature versus time for heat treatment at two levels (A), at one level (B), continuous (C).

$$x = 1 - \exp(-kt)^n \qquad [10.1]$$

where t is the time, n the Avrami exponent (which allows determination of the nucleation dimension rate and the way the crystal growth is controlled) and K is the reaction rate constant according to:

$$k = k_0 \exp\left(-\frac{E_c}{RT}\right) \qquad [10.2]$$

with k_0 the frequency factor (s-1), R the Boltzmann constant and E_c is the activation energy for the crystallization reaction.

Ozawa and Chen proposed the following equations based on the variation of the onset crystallization temperature to determine the activation energy parameter (Ozawa, 1965; Chen, 1965):

- According to Chen:

$$\frac{d\left(\dfrac{\ln T_p^2}{\alpha}\right)}{d\left(\dfrac{1}{T}\right)_x} = \frac{E_c}{R} \qquad [10.3]$$

- According to Ozawa's method:

$$\frac{d(\ln \alpha)}{d\left(\dfrac{1}{T_p}\right)} = \frac{E_c}{R} \qquad [10.4]$$

where α is the heating rate and T_p is the apex of the crystallization peak.

The value of E_c is the value of the slope obtained by plotting ln α versus $1000/T_p$. Then, Kissinger (1957) proposed a modified Ozawa's equation to calculate E_c based on the following equation:

$$\left(\frac{\ln \alpha}{T_p^2}\right) = -\frac{E_C}{RT_p} + \text{constant} \qquad [10.5]$$

A prerequisite for the success of this study is the need to always have the same phase that crystallizes during the various heat treatments which unfortunately often leads to uncontrollable crystal growth as mentioned before.

However, while many papers deal with the study of the kinetics of crystallization of chalcogenide glasses by using differential scanning calorimtry (DSC) (Ivanova and Cernoskova, 2004; Zhao *et al.*, 2005; Fayek *et al.*, 2007; Khan, 2000), few of them clearly distinguish the nucleation and growth steps in order to make glass-ceramics with controllable crystal size and proportion (Song *et al.*, 1997; Calvez *et al.*, 2007a; Lin *et al.*, 2011b).

10.3.2 Systems of transparent infrared glass-ceramics

In order to make controllable and reproducible chalcogenide glass-ceramics transparent in the infrared range, the glass compositions have to satisfy certain requirements. First, the selected glass must present a rather high viscosity at a temperature close to its glass transition temperature to minimize the diffusion of atoms in the glassy matrix, allowing a slow aggregation of particles. Knowing the fact that particle size is the main parameter to lower scatterings, the growing step must be controlled to avoid fast and uncontrollable crystallization growth.

The best way to control the nucleation and growth step should be to perform preliminary thermal experiments in order to calculate the kinetic of crystallization in glasses. As mentioned earlier, very few papers deal with dissociation of the nucleation and growth steps in order to make controllable glass-ceramics transparent in the infrared range.

Recently, two methods of crystallization have been clearly identified to produce chalcogenide glass-ceramics. The first one uses a phase separation as spinodal crystallization, while the second one places in competition the crystallization of two crystalline phases. These two routes to make infrared transparent glass-ceramics will be described in more detail in the next section.

It seems that in most of the studied chalcogenide glass compositions leading to controllable crystal growth, a phase separation occurs. One of the phases presents a higher liquidus temperature; this would result in a significant decrease in the thermodynamic barrier to cross to start the nucleation phenomenon.

Zhang *et al.* introduced ionic compounds such as alkali halide in a covalent glassy matrix such as germanium-based chalcogenide glasses to make reproducible and controllable glass-ceramics (Zhang *et al.*, 2004b). This work has been followed by several papers reporting the crystallization of various alkali halide components in Ge-Sb-S/Se ternary systems. More recently, new families of glass-ceramics were obtained in the Ge-(Ga/In)-(S/Se) systems. These results are described in the following sections.

Ge-Sb-(S/Se)-MX (MX: alkali halide)

The first way used to make controllable and reproducible glass-ceramics has been to consider the incorporation of alkali halide or metal halide in well-known ternary systems such as Ge-Sb-S. The idea was here to introduce ionic compounds in a highly covalent glassy matrix to induce a phase separation when heating at higher temperature than T_g (T_g + 30°C). Cesium chloride was first added in this ternary system using a systematic study of crystallization process on various glass compositions, the $62.5GeS_2$-$12.5Sb_2S_3$-25CsCl glass composition having been demonstrated to be the best candidate to make controllable glass-ceramics (Zhang, 2004b). As shown in Plate V (between pages 330 and 331), this glass composition is situated really close to the limit of the vitreous domain. This low stability against crystallization (ΔT < 100°C) does not allow bulk glasses with diameter higher than 10 mm to be made. In fact, the low thermal conductivity of silica strongly decreases the quenching rate inducing a crystallization of the melting bath.

The $62.5GeS_2$-$12.5Sb_2S_3$-25CsCl glass composition was annealed 30°C above T_g (260°C) for various durations. The progressive glass opacity is observable in Fig. 10.5. The loss of transparency in the visible range is due to scatterings of growing size of crystals inducing stronger scatterings. However, the infrared transmittance remains unchanged in the second and third atmospheric windows, keeping a high transparency up to 11 μm.

The mechanism of crystallization in alkali halide-based chalcogenide glasses is described as a progressive phase separation between the alkali halide and the residual glassy matrix (Fig. 10.6). In this case, the demixion from the glassy matrix and crystallization of CsCl has been determined by X-ray diffraction and NMR spectra. The rejection of CsCl from the glass network induces its progressive crystallization forming crystals of about 100 nm.

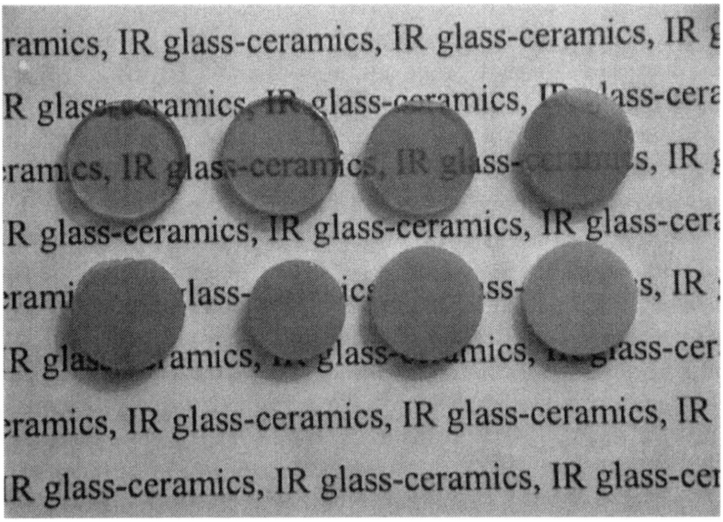

10.5 Glass-ceramics obtained with an annealing treatment at 290°C for various durations from the $62.5GeS_2$-$12.5Sb_2S_3$-25CsCl glass composition.

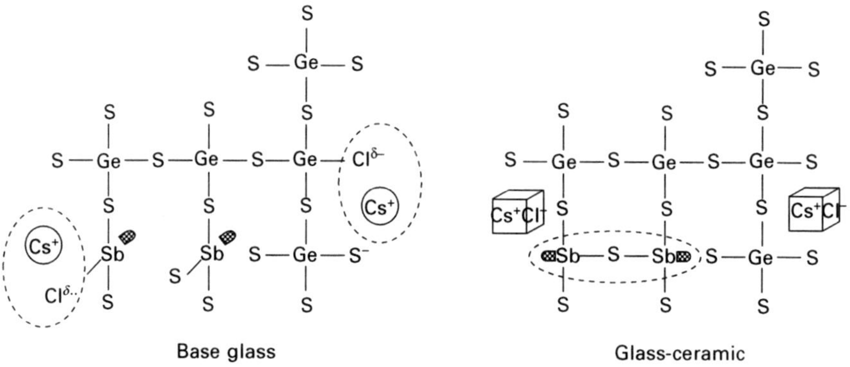

10.6 Hypothesis of the structural evolution during crystallization for the 62.5GeS_2-12.5Sb_2S_3-25CsCl glass composition.

Thus, the choice of glass compositions mixing alkali halide within a covalent glassy matrix size has demonstrated its efficiency to make controllable crystals with various systems: Ge-Sb-Se-(CsCl/RbI/CsI), allowing a larger transmittance in the mid-IR up to 16 μm (Calvez *et al.*, 2006; Liu *et al.*, 2009). Some of the glass compositions have only demonstrated crystallization of the alkali halide compound when this last one is added in a high amount close to supersaturation. The various alkali halide crystals obtained from glasses close to the boundary line of the vitreous domain are presented in Fig. 10.7.

The other advantage of the introduction of alkali halide is to clearly shift the beginning of transmission toward shorter wavelength by localizing the free electron pairs of selenium. In order to obtain chalcogenide glasses transparent in the visible range and up to the mid-IR, CsCl was introduced in the Ge-Ga-Se system. In fact, gallium-based glasses are known to present a broad range of transmission extended toward the visible range compared to antimony-based glasses. Following this thinking, other research has been carried out, adding various alkali halide (e.g., CsBr, RbI) in matrices containing antimony, gallium or indium with sulfur or selenium (Calvez *et al.*, 2007a, b, Liu *et al.*, 2009). The first glasses slightly transparent in the visible range and up to 16 μm were prepared by introducing CsCl in this system (Calvez *et al.*, 2007a,b). Experiments of controlled crystallization were also performed. Unlike antimony-based glasses, the controlled crystallization of Ga-based glasses demonstrated the germination of Ga_2Se_3 or Ga_4GeSe_8 crystalline phases instead of halide.

The introduction of CsCl was also studied in the sulfur-based system. The introduction of high amounts of cesium chloride in the Ge-Ga-S system has permitted the beginning of transmission to be shifted strongly toward shorter wavelengths leading to fully transparent glasses in the visible range. But even with 40mol% incorporated in the glassy matrix, the crystallization

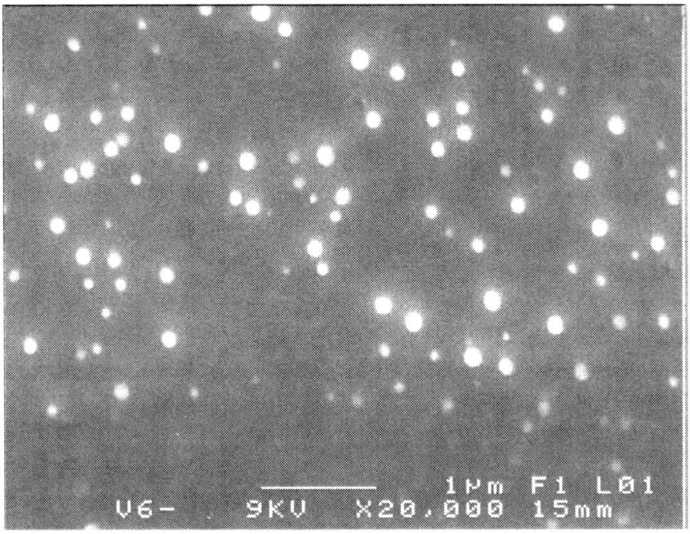

(a)

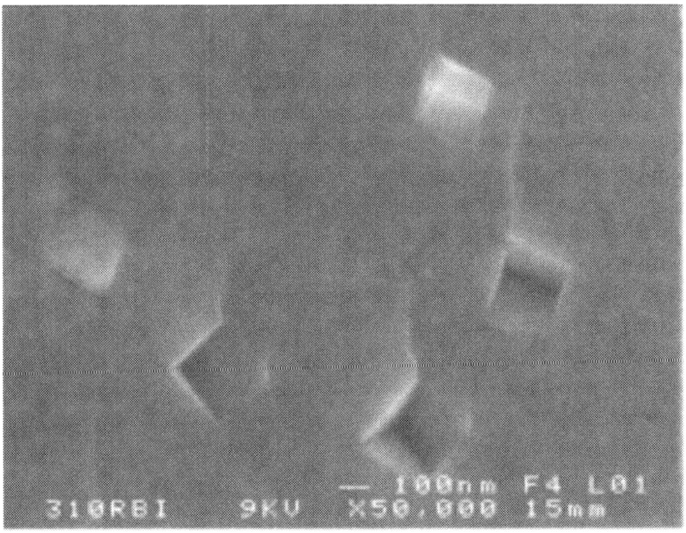

(b)

10.7 Different shapes of crystals obtained from various Ge-Sb-S/Se based glasses containing alkali halide: CsCl crystals (a), RbI crystals (b) (Zhang *et al.*, 2004b; Calvez *et al.*, 2007a).

mechanism remains unchanged and is still due to a phase separation of a gallium mixed phase. In fact, the same phenomenon as observed in Ge-Ga-Se systems appeared with sulfur-based glasses, in which the crystallization of Ga_2S_3 nanosize particles was observed (Ledemi *et al.*, 2009).

Ge-(Ga/In)-(S/Se)

However, as halides are placed at the end of chains, they progressively break bonds of the glass network. Thus, alkali halide leads to a decrease of the glass network rigidity and consequently to a lowering of mechanical properties such as hardness (Calvez *et al.*, 2007a). Thus, recent studies have been oriented toward compositions without halide.

Recently, Liu and Lin have considered the chemical structure of Ge-Ga-Se glassy matrices to understand the limit of glass composition leading to controllable glass-ceramics (Liu *et al.*, 2009; Lin *et al.*, 2011a). These papers give the key to understand the progressive crystallization of the glass network. In fact, a nucleation-rate-like curve can be determined by thermal analysis to promote fine phase separation in the glassy matrix. The control of phase separation acts as a slow nucleation, then the crystallization of the nodules can be performed at higher temperature (Plate VI (between pages 330 and 331)). This work has recently been followed by a new ternary diagram involving In instead of Ga in the Ge-In-Se/S glassy systems, leading to similar results, i.e. the crystallization of $In_{2.67}S_4$ nanoparticles (Xu *et al.*, 2008; Yang *et al.*, 2011).

In these last papers, a common point of base glasses leading to reproducible glass-ceramics is the presence of at least two close peaks of crystallization in the DSC curves. It seems that one of these peaks has to be broad and with a low maximum compared to the second one (Fig. 10.8). In these glasses, a competition between two phases having approximately the same activation

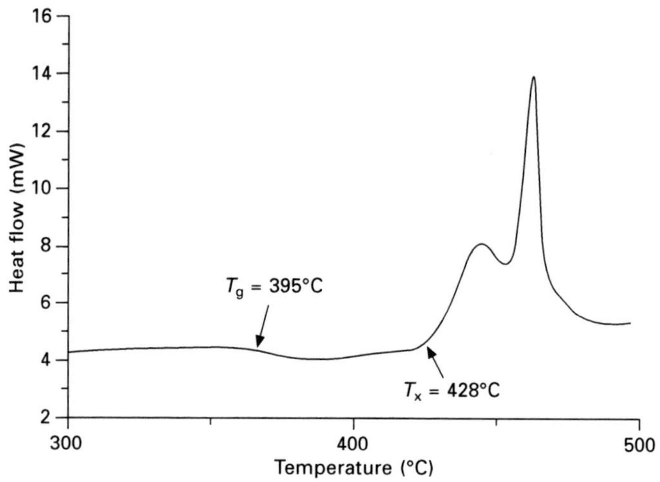

10.8 Typical DSC curve of a glass composition leading to reproducible and controllable glass-ceramics ($80GeSe_2$-$20Ga_2Se_3$ in this case).

energy occurs, and the phase corresponding to the broad peak separates from the glassy matrix before crystallizing.

More efforts have been dedicated to the Ge-Ga-S/Se glass systems. Controllable glass-ceramics were obtained from several glass compositions belonging to these systems. Whatever the glass, the crystallization process appears to be similar. In fact, nanoparticles of about 5 nm start to aggregate under larger agglomerates which progressively grow. Depending on the crystallization time, particles having a size from 10 to 100 nm can be observed by SEM (scanning electron microscopy) or TEM (transmission electron microscopy) pictures (Fig. 10.9). XRD patterns have demonstrated

10.9 Scanning electron microscopy (a) and transmission electron microscopic patterns (b) of the $80GeSe_2$-$20Ga_2Se_3$ glass sample 43 h at 380° in the bright-field mode (left) and dark-field mode (right).

the precipitation of cubic Ga_2S_3 in sulfide base-glasses and of Ga_4GeSe_8 or Ga_2Se_3 in selenide base-glasses.

It is known that a glass is considered stable against crystallization if the difference of temperature ΔT between its glass transition temperature (T_g) and the onset crystallization temperature (T_x) is higher than 100°C for chalcogenide glasses. Meanwhile, all the reproducible and controllable glass-ceramics obtained in the last decade present a ΔT ranging from 80°C to 120°C. Up to now, the chalcogenide glass-ceramics dimensions have been strongly limited because of the low quenching rate which prevents large bulk glasses.

Other IR transparent glass-ceramics

While the glass-ceramic families detailed above are the most common, other chalcogenide glass-ceramics have to be mentioned in this chapter. First, Tver'yanovich *et al.* demonstrated in 2005 the possibility to generate $CuCr_2Se_4$ ferromagnetic spinel microcrystals in a Cu_2Se-As_2Se_3-Cr_2Se_3 glassy system (Tver'yanovich *et al.*, 2005). While no information is given on the infrared transparency of the glass-ceramics, it is interesting to note that when samples are placed in a magnetic field and irradiated by an IR source, an increase of the magnetization is observed.

In 2006, Xia *et al.* generated controllable glass-ceramics in highly covalent GeS_2-Sb_2S_3-PbS glasses (Xia *et al.*, 2006), in which PbS can be introduced in high proportion (higher than 50 mol.%). By annealing glasses slightly above T_g, several glass compositions led to the crystallization of Pb-based crystals ($PbGeS_4$, $PbGeS_3$, $PbSb_2S_4$) having a size lower than 100 nm. Authors have demonstrated that PbS acts both as a glass former but also as a nucleant agent. The glass-ceramics obtained present higher resistance to thermal shocks and an enhanced resistance to fracture propagation. These data will be discussed in Section 10.4.2.

10.4 Properties of glass-ceramics

10.4.1 Optical properties evolution

MIE and Rayleigh scatterings

The possibility to generate a high fraction of crystals without modifying significantly the transmission can be explained by the turbidity equation. As determined by Hendy, light scatterings are mainly controlled by the particle radius R, the volume fraction of crystals φ, the wave vector k and Δn, the difference of refractive index between the glassy matrix and the crystallized phase, according to the following equation (Hendy, 2002):

$$\tau = \frac{14}{15\pi}\varphi(1-\varphi)k^8R^7\left(\frac{\Delta n}{n}\right)^2$$

Following this approach, a closer refractive index of crystals to that of the residual glassy matrix permits glass-ceramics with higher crystallized fraction, lower scattering losses and better mechanical properties to be made most of the time.

The particle radius must be considered as the critical parameter to keep the glass-ceramics transparency in the infrared range. Going beyond crystals of few microns induces Rayleigh scatterings. These scatterings have a strong impact on the whole infrared transparency which progressively decreases as the particle size increases up to the whole opacity of the glass-ceramics. Usually, this phenomenon is observed when a heterogeneous and uncontrollable crystallization occurs, for example on the glass surface or around impurities, places where the crystallization process is facilitated.

However, by maintaining a size at least ten times lower than the wavelength used, the transmission will be affected only by MIE scatterings. As already observed in several chalcogenide glass-ceramics, the generation of nanosize particles ranging from 5 to 100 nm provokes a shift of the beginning of transmission toward longer wavelength without any changes in the optimal transparency in the near and mid-infrared (Plate VII between pages 330 and 331).

So whatever the crystalline phase generated within the glassy matrix, the glass-ceramic can keep a broad transparency, without disturbing the maximum of transmission if the size of crystals can be controlled at a dimension about 100 times lower than the wavelength used (Fig. 10.5).

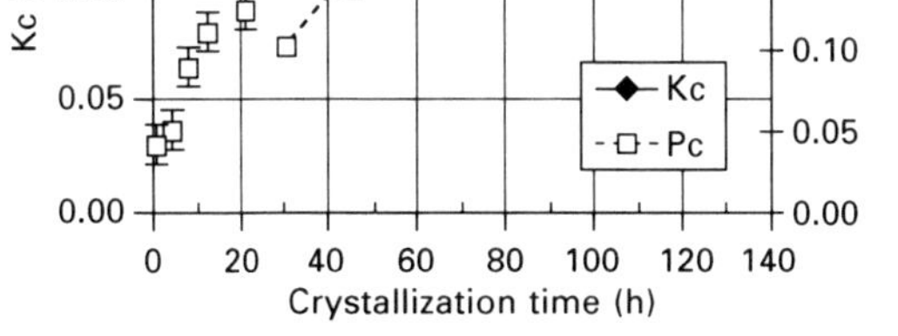

10.10 Evolution of the fracture toughness (*Kc*) and critical load for crack initiation (*Pc*) of the $62.5GeS_2$-$12.5Sb_2S_3$-25CsCl glass-ceramics (heat treatment at T_g + 30°C) (Zhu *et al.*, 2007).

Permanent nonlinear phenomenon

Because of their wide range of transmission in the infrared, the generation of second-order nonlinear effects in chalcogenide glasses is of great interest. In fact, the second harmonic generation (SHG) could pave the way to various fields of applications that can be cited here: electro-optics modulators, frequency convertors, as well as biological imaging. However, all glasses are isotropic materials which do not present second-order non-linear properties. Until recently, the SHG effects were only induced by thermal poling, which results in disturbing the centro-symmetrical media of the glass system (Guignard *et al.*, 2005). Even if high second-order nonlinear susceptibility (χ^2) can reach high values (8 pm/V), these effects are not permanent as they strongly decrease with time and are not stable under normal conditions. Recently, more attention has been paid to the crystallization of nonlinear nanoparticles. In order to generate nonlinearity, crystals must not be centro-symmetric so that all the cubic phases are prohibited.

In 2007, the first experiments demonstrating permanent SHG were performed by Guignard *et al.* (2007). From a glass composition belonging to the Ge-Sb-S glass system doped with 1mol.% of CdS, the authors produced glass-ceramics containing Cd_4GeS_6 and β-GeS_2 microcrystals (1–6 μm) with appropriate heat treatment time and temperature. A second-order nonlinear intensity of about 0.01 that of α-quartz was measured using a process based on Maker fringes set-up and it was ascribed to the presence of β-GeS_2. However, due to the crystal size and the high difference of refractive index between the residual glassy matrix and the crystalline phase as well, strong scatterings occur in the short wavelength.

One year later, further research has confirmed the efficiency of chalcogenide glass-ceramics for obtaining permanent SHG phenomenon. Dong's paper explored the crystallization in the GeS_2-Ga_2S_3-CdI system with a glass composition reaching 10 mol.% of CdI_2. Heat treatment performed 15°C above T_g led to a homogeneous crystallization of α-$CdGa_2S_4$ crystalline phase. Using the same method as previously, second harmonic intensity of 0.8 times larger than that of the Z-cut quartz was determined (Dong *et al.*, 2007).

In 2008, using cadmium sulfide instead of iodine sulfide, Gu *et al.* obtained a slight crystallization on the glass surface of highly nonlinear $CdGa_2S_4$ phase. Particles from 150 to 600 nm have permitted a high value of χ^2 of 38.85 pm/V to be reached. As mentioned earlier, the surface crystallization is a hard parameter to control, reducing the reproducibility of glass-ceramics.

The same year, Lin *et al.* have demonstrated that the homogeneous crystallization of $AgGaGeS_4$ is efficient for inducing second-order optical nonlinearity up to 5.79 pm/V (Lin *et al.*, 2008). In this last paper, a fast decrease of the maximum of transmittance is also observed with increasing heat treatment time showing a rapid growth of crystals.

The evolution of the third-order optical nonlinearity has been studied by Shen *et al.* (2011), who demonstrated that the formation of Ga_2S_3 and GeS_2 nanosize particles ranging from 2 to 80 nm has a strong influence on the nonlinear refractive index n_2 which has been increased four-fold (4.3 $\times 10^{-11}$esu) compared to the base glass as well as the nonlinear coefficient absorption β (Shen *et al.*, 2011). This result has been confirmed by Lin *et al.* in transparent $90GeS_2$-$10Ga_2S_3$ chalcogenide glass-ceramics containing β-GeS_2 nanocrystals. A high value figure of merit (FOM $\approx$ 11.64) was obtained in the crystallized sample using the conventional Z-scan method (Lin *et al.*, 2011b). All these results are promising to make novel efficient active materials with highly nonlinear properties for various fields of applications (e.g., optical convertors, frequency doubling, waveguides, super continuum sources).

Enhancement of luminescence

Chalcogenide glasses are of great interest as host for rare earth doping compared to oxide or fluoride glasses, as they present low phonon energy which is favorable for obtaining high quantum efficiency for wavelength conversion. Several spectroscopic studies concerning sulfide-based glasses doped with rare earth elements have been published (Balda *et al.*, 2001; Dai *et al.*, 2010; Guimond *et al.*, 1999; Tverjanovich *et al.*, 2001; Heo, 2003). It has already been demonstrated that the presence of gallium in a chalcogenide glass allows the incorporation of a larger amount of rare earth (Scheffler *et al.*, 1999). The rare earth incorporation can be explained by the extended hole structure formed by Ga-based glasses.

Furthermore, since the discovery of the ease of making transparent Ga-based chalcogenide glass-ceramics both in the visible and mid-IR ranges, many papers dealing with the incorporation of rare earth ions within these glassy systems have been reported. Thus various rare earths (e.g., Er^{3+}, Nd^{3+}, Tm^{3+}), excitable in the visible and near infrared range and emitting in the same range of wavelength, were incorporated in glassy matrices such as Ge-Ga-S or Ge-Ga-Sb-S (Seznec, 2006a, 2006b). Many attempts to produce rare earth nanocrystals embedded in that low phonon energy matrix have been performed to enhance the wavelength conversion.

Seznec *et al.* (2006a, 2006b) were the first to study the influence of rare earth ions embedded in crystals in controllable chalcogenide glass-ceramics. They have demonstrated the structuration of emission peaks after crystallization assuming the progressive incorporation of Er^{3+} ions in the Ga_2S_3 nanoparticles. In addition, the intensity of emission bands (at 670 nm with an excitation at 980 nm) is increased up to 100 times compared to the glass without heat treatment, showing a strong increase of luminescence by up-conversion. The erbium cations seem to be the best candidate to reach such narrowing of the emission peak because of their smaller ionic radius allowing their diffusion in the nanoparticles.

This work has been followed by Guillevic et al. with the introduction of Nd^{3+} in Ge-Ga-S-CsCl glassy matrix which has been heat treated (Guillevic *et al.*, 2010). As observed in the Plate VIII (betwen pages 330 and 331), they pointed out that although the ceramization process increases the luminescence efficiency significantly (emission at 583 nm, 600 nm, 660 nm and 677 nm under an excitation at 532 nm), the fact that no peak structuration is observed indicates that the neodymium ions having a bigger ionic radius than the Er^{3+} and are not incorporated in the nanocrystals.

In fact, Dai *et al.* paved the way for new applications by doping sulfide glass-ceramics with Tm^{3+}. Enhanced mid-infrared emission spectra were observed at 2.3 and 3.8 μm in a $65GeS_2$–$25Ga_2S_3$–10CsI (mol.%) glass-ceramic composition doped with 0.6 wt.% Tm^{3+} ions (Dai *et al.*, 2011).

By replacing erbium with other rare earths such as Tm^{3+} (Dai *et al.*, 2010) or by co-doping with two rare earths (Tm^{3+}/Ho^{3+}, Pr^{3+}/Er^{3+}) (Hewak, 1998; Park *et al.*, 2000; Li *et al.*, 2011), these materials could be promising candidates for the making of new laser sources working in the second atmospheric window ranged between 2 and 5 μm. This work will now be extended to selenide-based glass-ceramics, as Hubert *et al.* (2013) have demonstrated that the same phenomenon can be observed in selenide-based glass-ceramics doped with Er^{3+}.

10.4.2 Mechanical and physical properties

Chalcogenide glasses are interesting because of their good optical transmittance in the mid-IR region, relatively low cost compared with the single crystalline germanium or polycrystalline ZnSe, and the possibility to produce complex optics by molding. However, their weak mechanical properties slow down their full exploitation. To date, most chalcogenide glass-ceramics studies have tackled ways to increase the mechanical properties. This tendency is progressively countered by the increasing number of papers dealing with active chalcogenide glass-ceramics containing rare earth or other active ions. Three main aspects are studied to strengthen chalcogenide glasses: the resistance to fracture propagation, the hardness and the thermal expansion coefficient.

Resistance to fracture propagation

From a practical point of view, the fracture toughness (K_I) is estimated from the length of the cracks that emanate from the corners of a Vickers indentation. The measurement of the lengths of the cracks and the indentation half-diagonal size permit the toughness to be assessed using analytical expressions which involve hardness and elastic modulus. Even if the first work claimed that the accuracy of the toughness value is within 30–40%, the ability to perform Vickers indentation with identical force (N) and time (s) allows comparison to be made between glasses and glass-ceramics and glasses of various families.

Thus, using this technique, many papers have reported a strong correlation of the resistance to fracture propagation with the crystallization rate of the glass-ceramics. The first significant study dealing with the evolution of mechanical properties of controllable and reproducible infrared glass-ceramics was reported by Zhu *et al.* (2007). They have put forward the idea that the increase of toughness and of the critical load for crack initiation is a function of the crystallization time (Fig. 10.10).

These observations have also been reported in Ge-Ga-Se-based glass-ceramics (Rozé *et al.*, 2008). One can see a progressive strengthening of the glass by generating Ga_4GeSe_8 nanoparticles inside the glassy matrix. The usual flakes observed after Vickers indentations tend to disappear with increasing heat-treatment time (Plate IX between pages 330 and 331).

Hardness

A detrimental aspect for chalcogenide applications is their mechanical weakness. In fact, compared to materials commonly used for IR applications (e.g., Ge, ZnSe) and even compared to oxide base glasses, chalcogenide glasses present a much weaker hardness. The hardness of a common borosilicate glass is around 650 kg/mm^2 while the AMTIR and GASIR, which are two of the most used glasses, have a hardness of 170 kg/mm^2. However, the hardness evolution in IR transparent glass-ceramics is difficult to apprehend. As already mentioned by Guin *et al.* (2002a, 2002b), the toughness and hardness measurements by Vickers indentation experiments are time dependant due to the viscoelastic behavior of chalcogenide glasses. Thus, it is of a paramount importance to keep the time and load of indentation constant in order to draw comparisons.

Depending on the crystallized fraction in the glassy matrix, the hardness can be controlled mainly by the residual glassy matrix. As shown by Ledemi *et al.* (2009) the crystallization of Ga_2S_3 nanoparticles in a GeS_2-Ga_2S_3-CsCl glass composition can lead to a lowering of hardness due to a decrease of the residual glass network connectivity. In contrast Lin *et al.* (2009) have demonstrated that in a GeS_2-Ga_2S_3 glass composition, the progressive formation of Ga_2S_3 induces an increase of hardness. So a crystallization of nanoparticles of the same nature does not always lead to a strengthening of mechanical properties. The final hardness value clearly depends on the residual glass network viscoelastic property at room temperature. So the hardness evolves in the same way as the glass network connectivity when the glass-ceramics contain a low fraction of crystals. For a high amount of crystals, several parameters can influence the hardness: internal stresses induced by crystal growth, crystal size and the mean length between crystals (Miyata and Jinno, 1982).

In the future, further work will have to be devoted to the mechanical properties in order to reach a better understanding of the hardness evolution as a function of the glass-ceramic structure.

Coefficient of thermal expansion

The most famous glass-ceramic is the Zerodur® due to its coefficient of thermal expansion (CTE) being close to zero in a large range of temperatures. It is made from aluminosilicate glass and its CTE is close to 0×10^{-6} K^{-1} depending on the crystalline content (Bach and Krause, 2005). This parameter is a good indicator of the glass-ceramic resistance to thermal shocks.

The coefficient of thermal expansion, α, is a very important parameter as it is known that the ratio α/thermal conductivity is used to characterize the resistance of materials to thermal shocks. Low CTE and high thermoconductivity are favorable to highly resistant materials.

Xia *et al.* (2006) first firstly demonstrated the progressive decrease of the thermal expansion coefficient according to the annealing time in GeS_2-Sb_2S_3-PbS system (Fig. 10.11) leading to the crystallization of Pb-based phases.

What is most striking is observed in Ge-Ga-S/Se-based chalcogenide glass-ceramics. Lin *et al.* (2009) have also put forward the decrease of the CTE in glass-ceramics made from the $80GeS_2$-$20Ga_2S_3$, reaching a very low value of $8.2 \times 10^{-6}K^{-1}$ after controlled crystallization, when the hardness

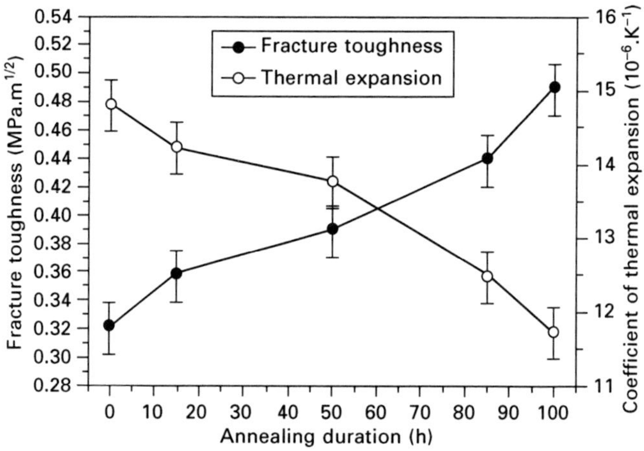

10.11 Fracture toughness and coefficient of thermal expansion as a function of annealing duration (Xia *et al.*, 2006).

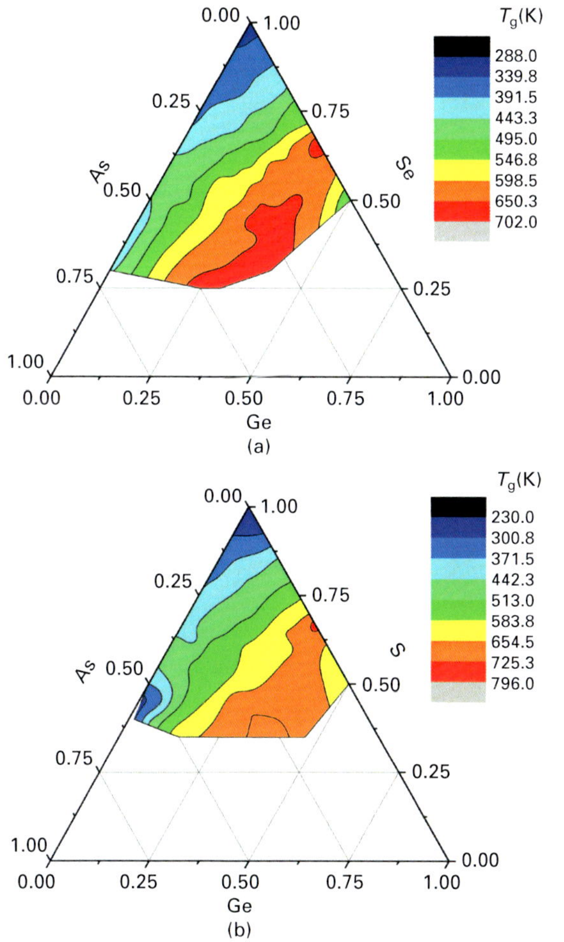

Plate I (Chapter 4) Variations in T_g for As-Ge(S,Se) ternary chalcogenide glasses.

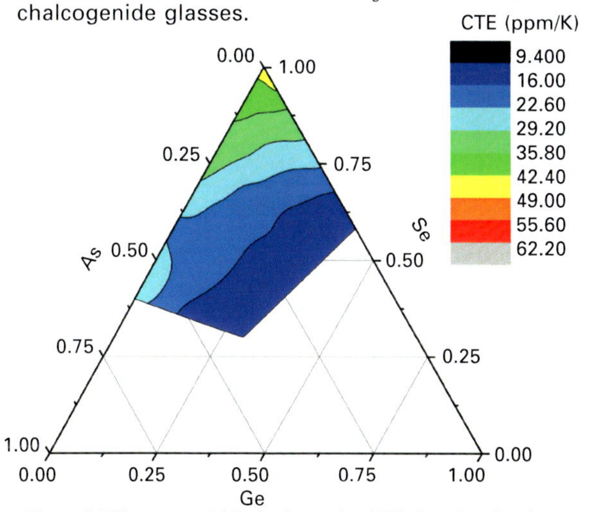

Plate II (Chapter 4) Variations in CTE for As-Ge-Se ternary chalcogenide glasses.

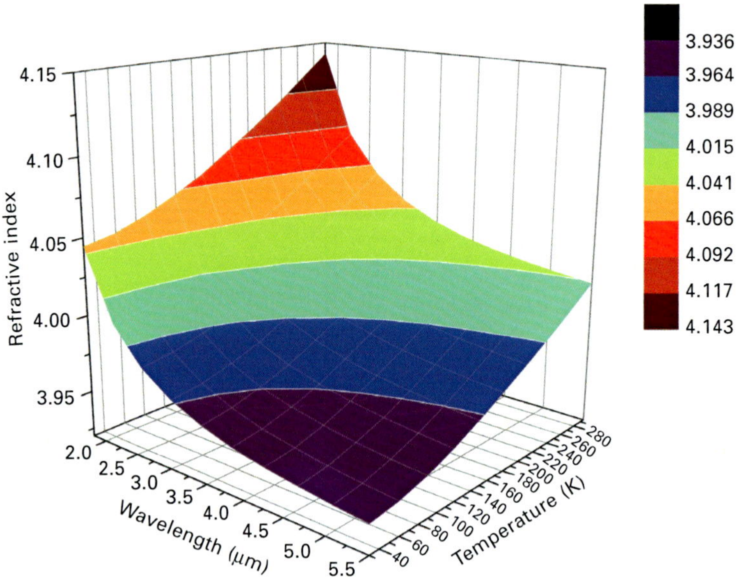

(a)

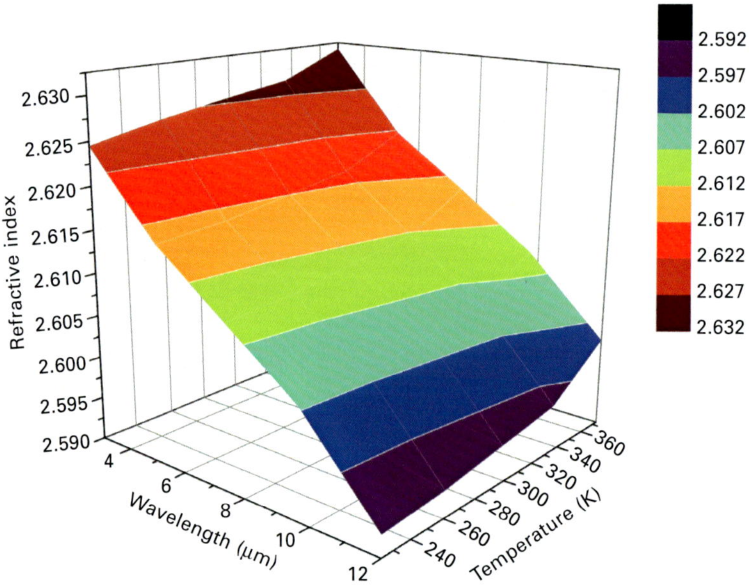

(b)

Plate III (Chapter 4) Variation in refractive index with wavelength and temperature for crystalline germanium (a) and a germanium-containing chalcogenide glass (b).

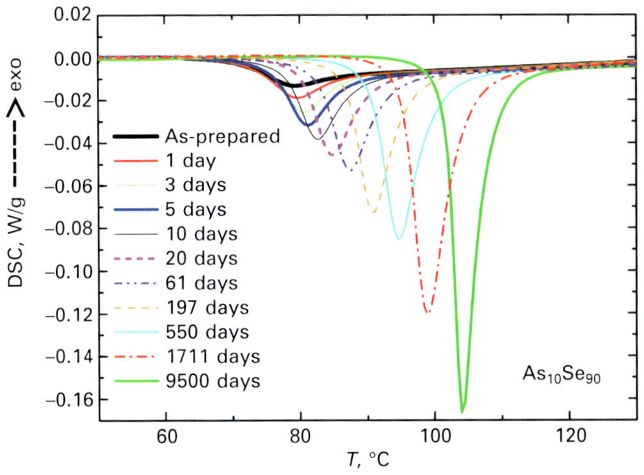

Plate IV (Chapter 8) DSC curves recorded at room temperature for g-$As_{10}Se_{90}$ after certain periods of dark storage.

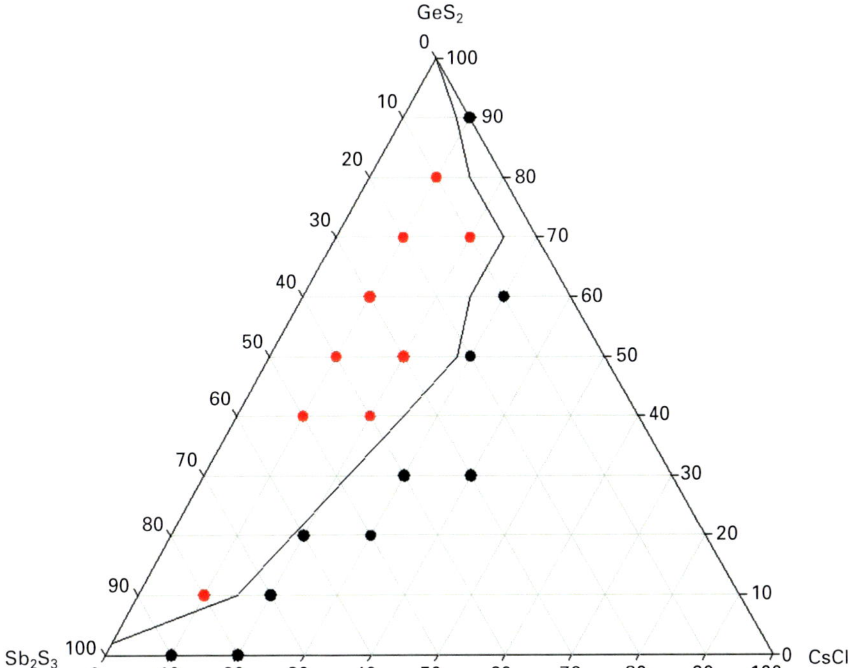

Plate V (Chapter 10) GeS_2-Sb_2S_3-CsCl ternary system and the 62.5GeS_2-12.5Sb_2S_3-25CsCl glass composition.

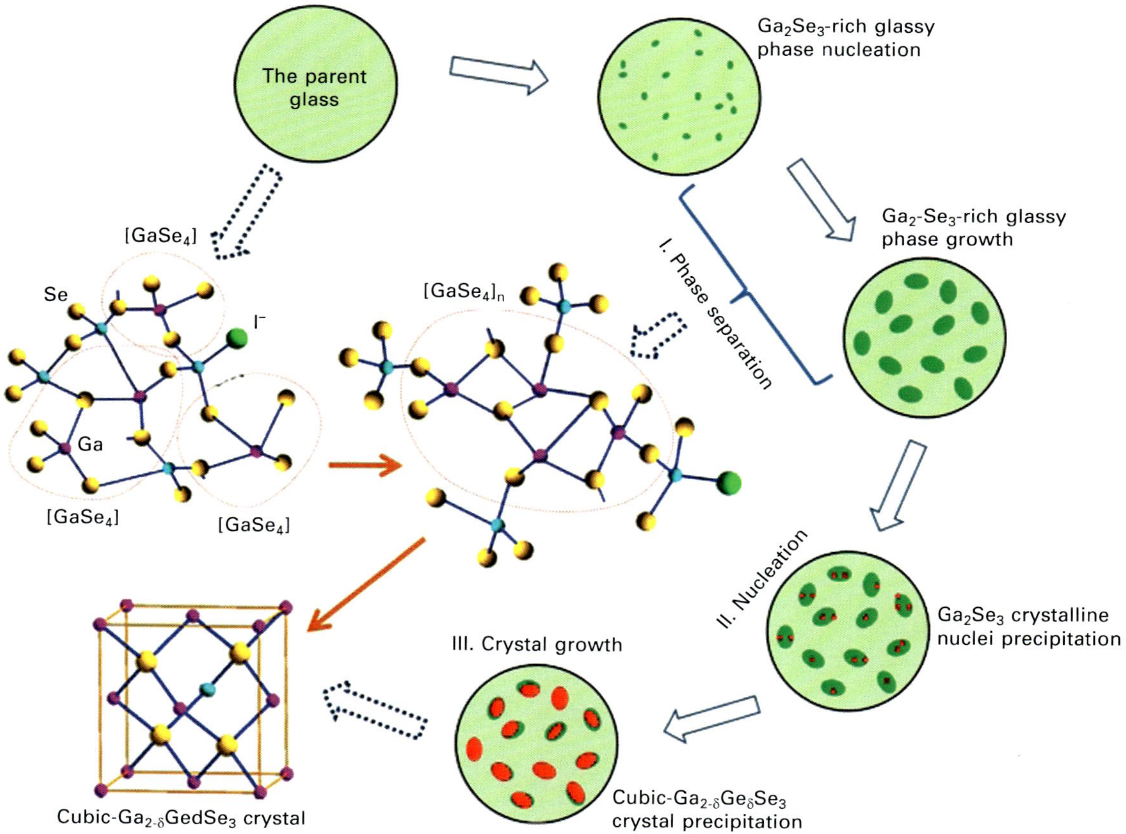

Plate VI (Chapter 10) Crystallization process in Ge-Ga-Se based chalcogenide glass-ceramics (Liu *et al.*, 2009).

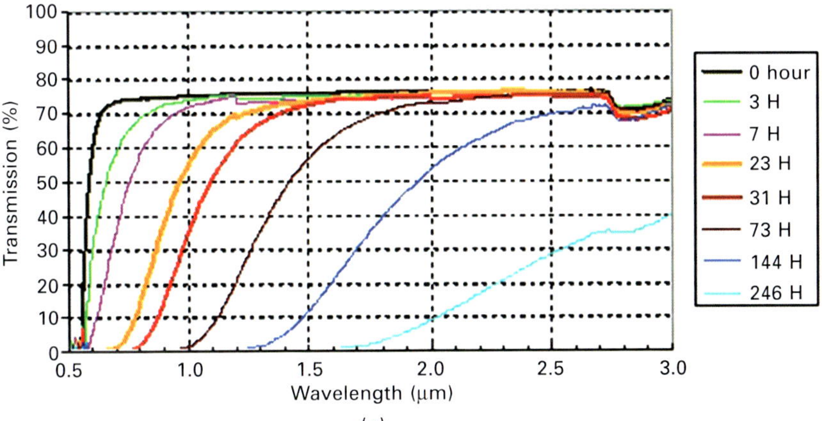

(a)

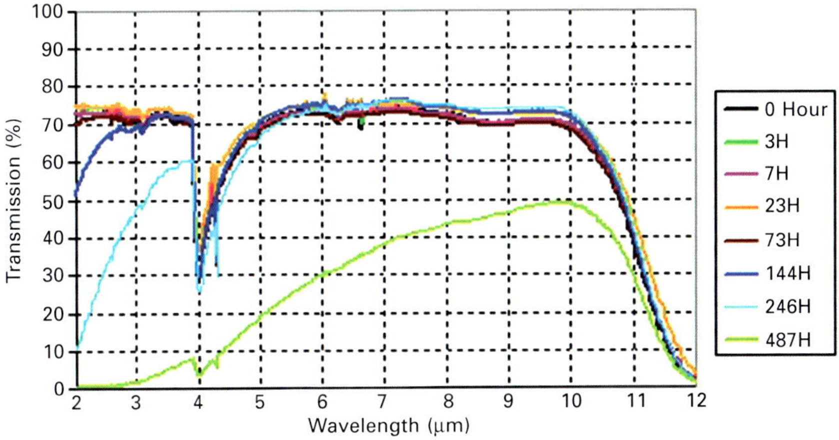

(b)

Plate VII (Chapter 10) (a) Influence of MIE diffusion due to nanoparticles on the transmission in the short wavelength. (b) Influence of Rayleigh scattering due to microcrystals on the transmission in the mid-infrared range.

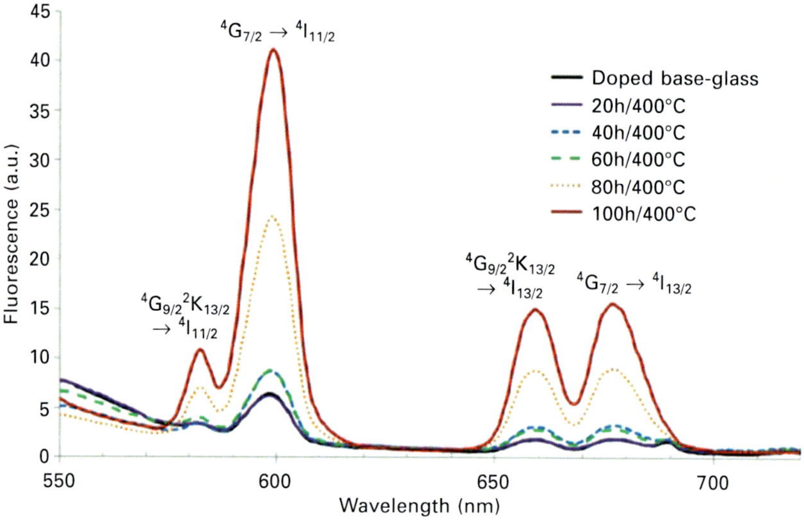

Plate VIII (Chapter 10) Photoluminescence spectra with excitation at 532 nm of the $(GeS_2)70$–$(Ga_2S_3)20$–(CsCl)10 0.1 wt.% doped glass-ceramic series.

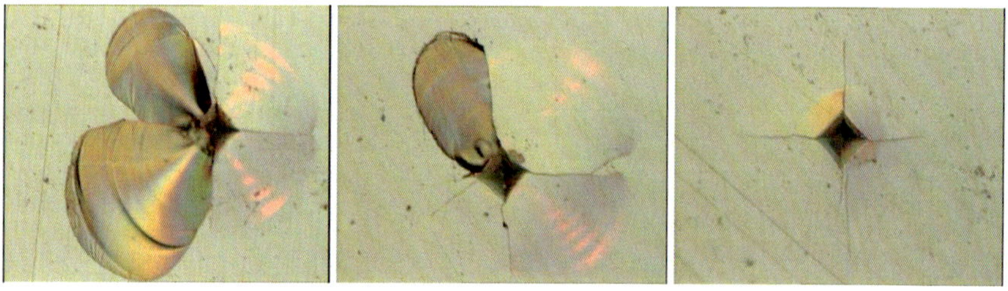

Plate IX (Chapter 10) Optical microscope pictures of the $80GeSe_2$-$20Ga_2Se_3$ base glass and glass-ceramics heat treated at 380°C for different times (0 h, 15 h, 33 h from left to right) after Vickers indentation (300 g) (Rozé *et al.*, 2008).

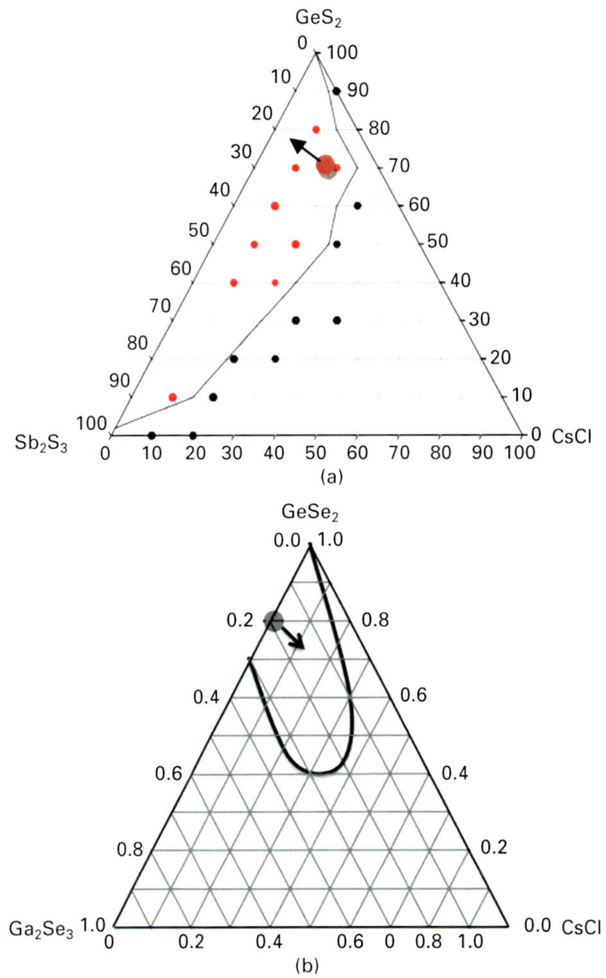

Plate X (Chapter 10) Vitreous domains of the GeS_2-Sb_2S_3-CsCl (a) and $GeSe_2$-Ga_2Se_3-CsCl systems (b).

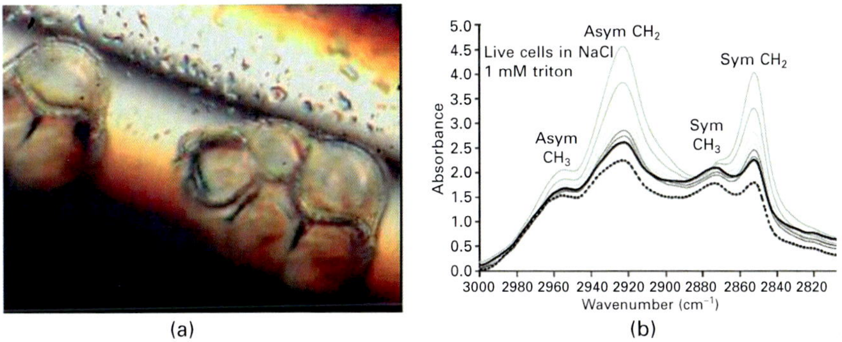

Plate XI (Chapter 12) (a) Human live cells forming strong attachment on the surface of a TAS fibre. (b) IR signatures of the cell membrane upon introduction of micromolar quantities of surfactant in the cell environment.

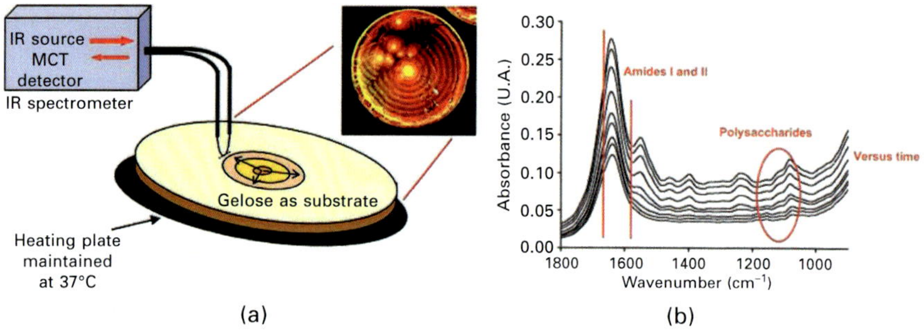

Plate XII (Chapter 12) (a) IR fibre set-up used to investigate the colonization of a gelose surface by a bacteria biofilm. (b) Progression of the IR signature of the *Proteus mirabilis* bacteria membrane as the biomass propagates through the substrate.

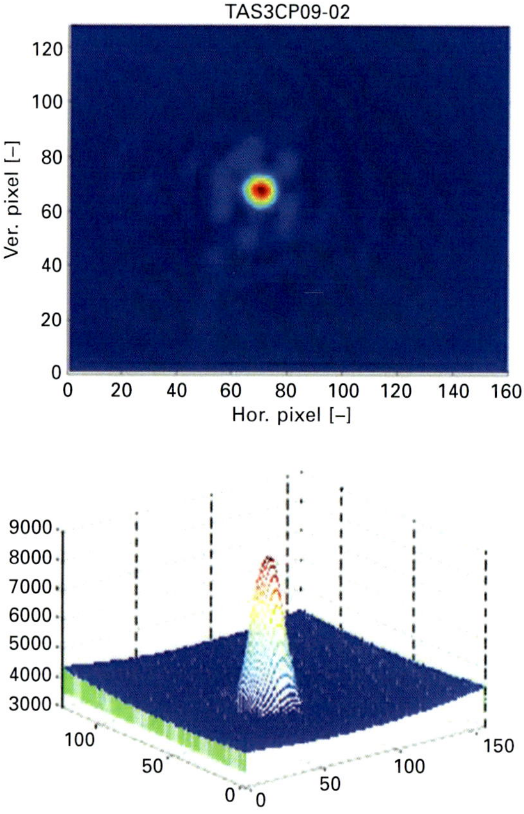

Plate XIII (Chapter 12) Far-field intensity distribution at 10.6 μm of a TAS step index fiber showing single-mode propagation.

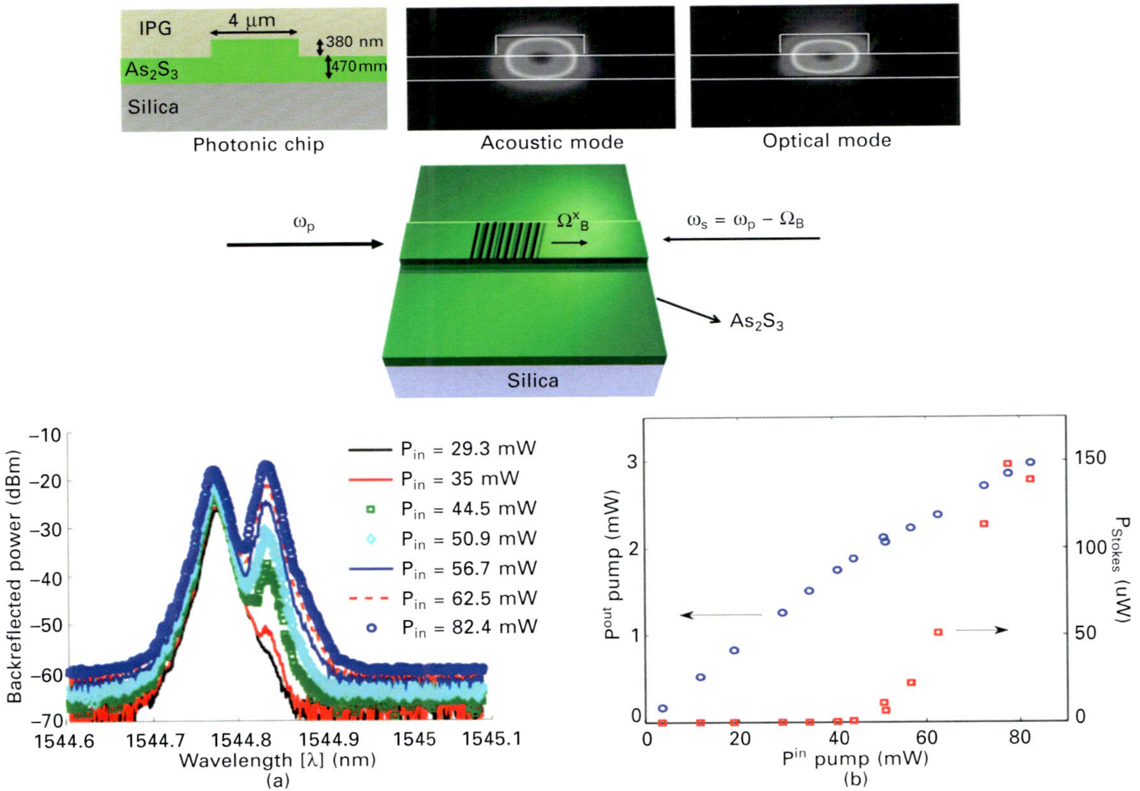

Plate XIV (Chapter 14) Characterization of on-chip stimulated Brillouin scattering in a chalcogenide rib waveguide [2] using (a) backscattered signal at different pump powers showing an increase in the Stokes power as the pump power is increased and (b) plot of the output pump power in the forward direction and filtered Stokes signal with increase in input pump power showing that as the Stokes signal starts to increase at 50 mW the slope of output pump power reduces.

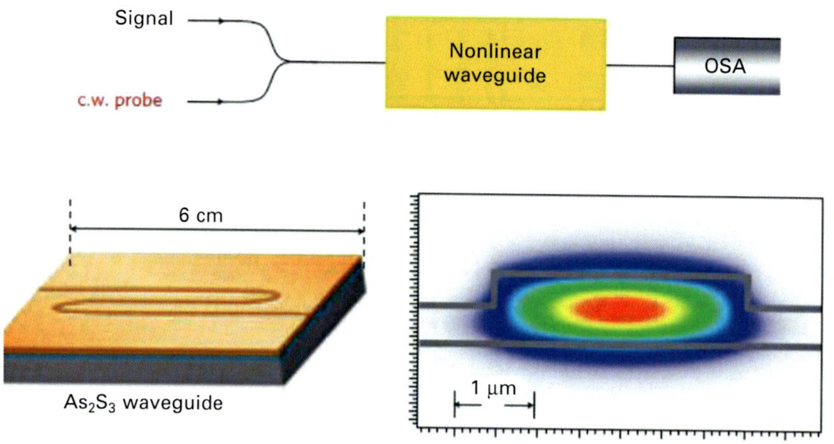

(a)

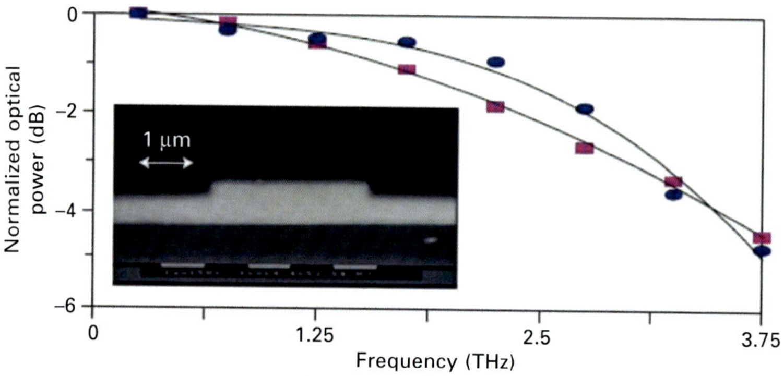

(b)

Plate XV (Chapter 14) (a) Schematic of photonic chip based implementation of radio-frequency spectrum analyser based on XPM showing the optical mode profile for the chip (b) measured bandwidth of the RF spectrum analyser for the 4 μm wide chalcogenide rib waveguides demonstrating a 3 dB bandwidth of ~ 3 THz and 3.2 THz for the probe wavelengths of 130 nm and 1569 nm respectively demonstrating that there is no significant change in the RFSA bandwidth with probe wavelength showing the potential for colourless operation [3].

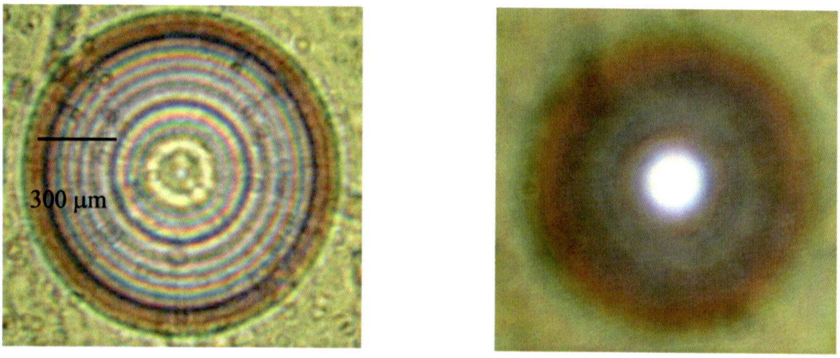

(a) (b)

Plate XVI (Chapter 17) Optical characterization of individual Fresnel lens (400 μm diameter, 600 nm thickness) on glass substrate: (a) microscope lens focused on the surface, and (b) at the focal point that is 5.38 mm from the surface (Kovalskiy *et al.*, 2009b).

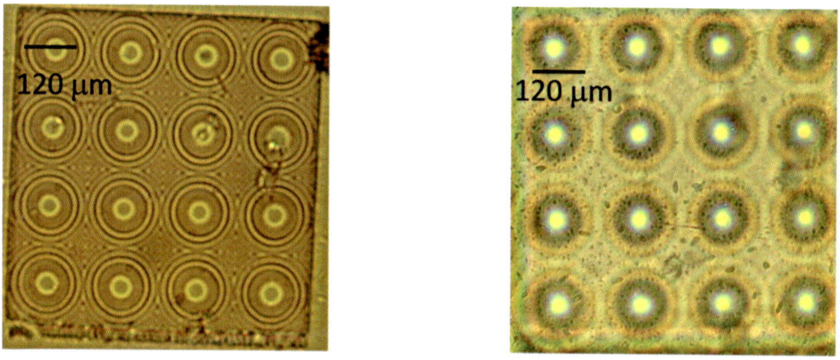

(a) (b)

Plate XVII (Chapter 17) Optical characterization of Fresnel lens array on glass substrate: lens focused on the surface (a) and at the focal point (b) (Kovalskiy *et al.*, 2009b).

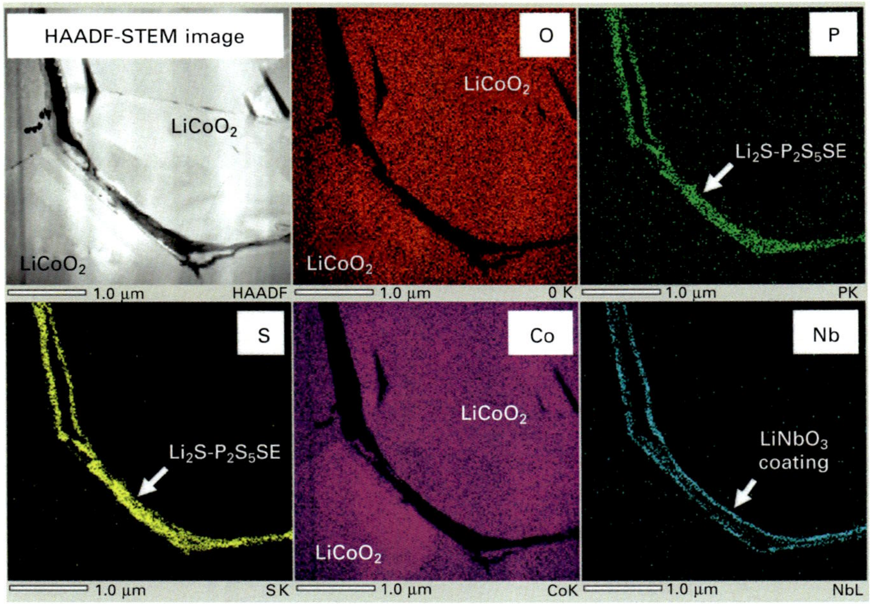

Plate XVIII (Chapter 19) Cross-sectional high-angle annular dark field (HAADF)-STEM image and EDX mappings for the composite electrode consisting of only SE-coated LiCoO2 after charge-discharge measurements for 450 cycles.

and toughness are strongly improved also Rozé *et al.* studied the same glass-ceramics composition using selenium instead of sulfur. While the base glasses present an average CTE of $11.8 \times 10^{-6}K^{-1}$, much lower than that of GASIR ($17 \times 10^{-6}K^{-1}$), the crystallization of nanoparticles still reduces the CTE reaching up to $10.9 \times 10^{-6}K^{-1}$ (Rozé *et al.*, 2011). These values are the lowest coefficient of chalcogenide glasses or glass-ceramics ever seen for glasses having a relatively low content of Ge.

10.4.3 Shaping of glass-ceramics

Molding ability

As demonstrated, the nucleation and growth of particles within chalcogenide glassy matrix occur at temperatures slightly above the glass transition temperature. Most of the glasses or glass-ceramics designed for optical applications in the infrared range are shaped by molding or fiber drawing. Owing to their intrinsic viscoplastic properties when heated above their T_g, chalcogenide glasses can be shaped into fibers or complex lenses around the Littleton softening point at a viscosity of $10^{7.6}$ Poises (Zhang *et al.*, 2003). However, the decrease in glass viscosity due to an increase of temperature is offset by the progressive crystallization inducing a reticulation of the glassy network. This phenomenon, which could be dramatic for a fine-shaping process, has been well described on glasses belonging to the $GeSe_2$-Ga_2Se_3 glassy system either with alkali halide (Yang *et al.*, 2008) or without (Rozé *et al.*, 2011).

As can be seen in Fig. 10.12, the shaping process efficiency of the $80GeSe_2$-$20Ga_2Se_3$ glass composition (T_g: 380°C) is optimum around 435°C, i.e. only

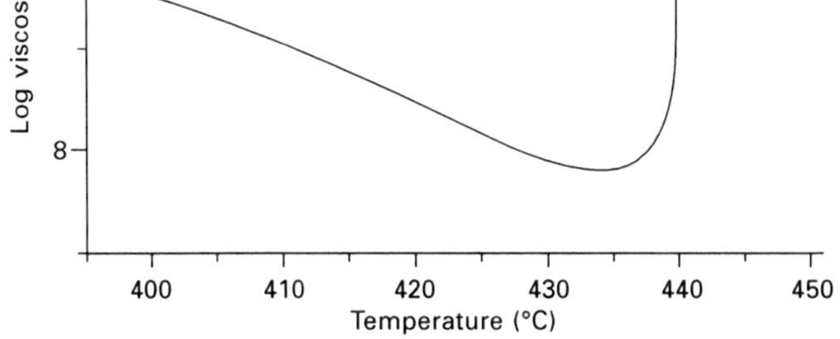

10.12 Viscosity as a function of temperature under a 150 g weight of the $GeSe_2$-$20Ga_2Se_3$ glass (Rozé *et al.*, 2011).

55°C above T_g. This temperature corresponds to the lowest viscosity at which the glass should be molded. However, a slow crystallization occurs which requires achieving a rapid shaping of the glass. At higher temperatures, the curve shows that a fast and uncontrollable crystallization of the whole glassy matrix leads to an increase of viscosity, preventing the glass from being molded. As can be understood, the drawing of fibers, which necessitates reaching a temperature of at least 100°C above T_g, is hardly compatible with the growth of controllable and homogeneously dispersed nanoparticles. At present, no paper has reported a controllable crystalline growth into chalcogenide fibers.

However, very recently, Mironov *et al.* have demonstrated the possibility of incorporating nanoparticles in chalcogenide glasses without using heat treatment (Mironov *et al.*, 2010). They have introduced ZnSe or ZnS nanoparticles (0.1–4 wt.%) doped with Cr^{2+} (1wt.%) in a silica tube containing the pure As-S raw materials necessary to make the As_2S_3 base glass. A conventional way of synthesis was then used to obtain a rod of a composite material containing a chalcogenide glass and doped nanoparticles. The rods were drawn into As_2S_3 fibers of 170–300 μm diameter and 2–20 m length. Luminescence spectrums of doped fibers were recorded showing an efficient luminescence from 1.8 to 3 μm with fibers presenting optical losses ranging from 2 to 4 $dB.m^{-1}$. This innovative method to fabricate composite materials containing glass and crystalline phase paves the way for making new families of chalcogenide glass-ceramics.

Etching ability

Recently, a new way to tailor oxide-based glass-ceramics to various applications (e.g., bioactive materials, dental materials, sensors, membranes) consists in selectively etching the crystalline fraction or the residual glass. Research on porous glass-ceramics started in the 1970s with borosilicate such as Vycor® from Corning (Hammel, 1974). In 1989 Hosono *et al.* reported the first experiments on bioactive porous $CaO-TiO_2-P_2O_5$ glass-ceramics. The materials obtained offer the advantages of porosity and ease of shaping.

In 2007, Delaizir *et al.* were the first to report the possibility of etching selectively the particles from the chalcogenide glassy matrix. They selected the $GeS_2-Sb_2S_3-CsCl$ glass-ceramic (see Section 10.3.2) as the crystalline ratio of hygroscopic CsCl can be easily controlled with heat treatment time. The use of strong acid H_2SO_4 mixed with a strong oxidant H_2O_2 solution has permitted the nanoparticles to be etched rapidly to a depth of about 100 nm within the glass. The progressive dissolution of nanoparticles can be observed in Fig. 10.13. A specific surface 7,000 times higher than the base surface is attained. While crystalline particles are first dissolved, the pore

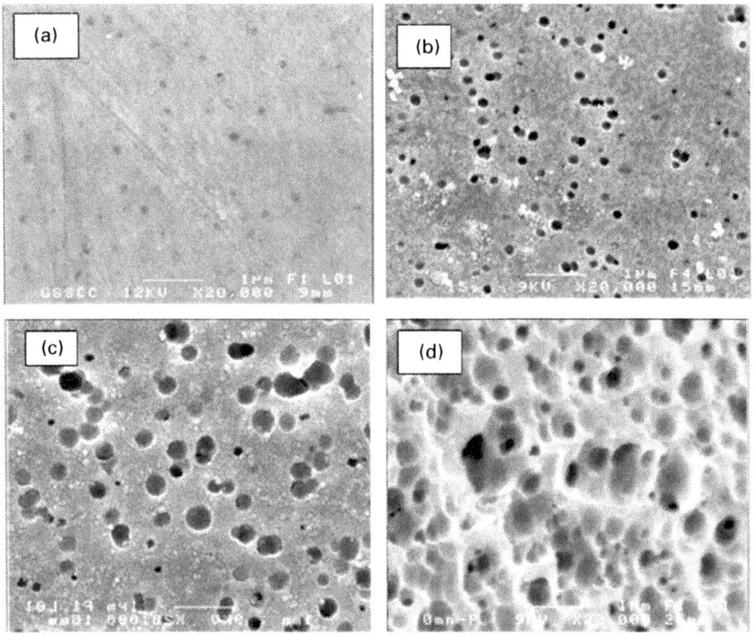

10.13 Scanning electron micrograph of a glass-ceramic etched in acidic solution for 1 min after different heat treatments: (a) no treatment, (b) 20 h, (c) 40 h, and (d) 60 h at 300°C (Delaizir, *et al.* 2007).

size still increases with longer treatment time as the solution attacks the glassy matrix, thus decreasing the surface area.

The as-made porosity does not affect the transmittance measured on slices of bulk glass-ceramics. Delaizir *et al.* took advantage of the increase of the specific area on the surface making an ATR (attenuated total reflexion) plate in order to experiment with it as an optical IR sensor. The detection of an organic analyte and silanol on the etched glass-ceramic surface was performed showing a strong increase of the infrared signal (Fig. 10.14).

10.5 Future trends

10.5.1 Lower the scatterings losses

As mentioned in the previous section, the glass-ceramic transparency is determined mainly by several parameters, of which the main ones are the size and number of particles, but also the difference of refractive index between the generated crystalline phase and the remaining glassy matrix (Hendy, 2002).

The first way to decrease optical losses in the near and mid-IR is to control the generation of particles with size much smaller than the wavelength used.

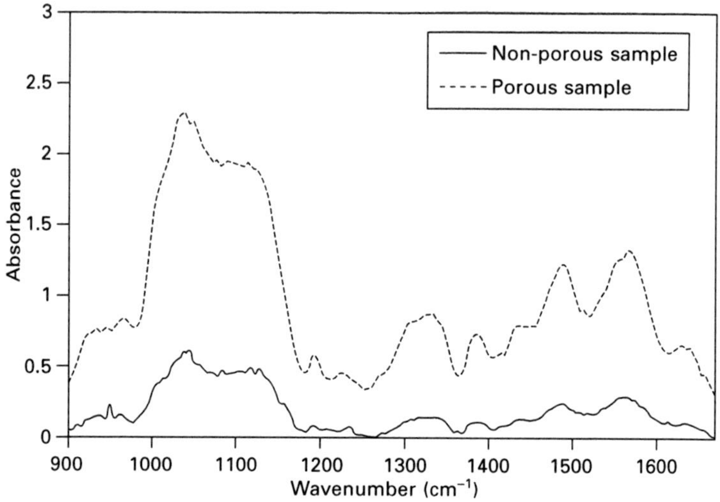

10.14 Transmission spectra in the region of Si–O–Si and amino group vibrations of silanol adsorbed at the surface of a porous and a flat glass-ceramic disk (Delaizir *et al.*, 2007).

While the nucleation and growth step can be easily distinguished in silicate glasses allowing excellent control of the volumic crystalline ratio and of the particles size, the low glass transition temperature combined with lower energy bonds of chalcogenide glasses gives a narrow difference of temperature between both steps. Until now, only glass compositions presenting a difference between the crystallization temperature and the glass transition temperature not exceeding 150°C led to controllable and reproducible glass-ceramics. This low ΔT makes it difficult to control the size and fraction of crystals separately. A low temperature of heat treatment permits the diffusion speed of atoms to be limited, emphasizing the nucleation, meaning the crystalline fraction instead of the particle size. However, the improvement of mechanical properties or active optical properties usually needs crystals having a size and a fraction high enough to induce significant improvement. Thus, another step at higher temperature has to be performed to induce crystal growth.

The third parameter influencing the optical losses by scatterings is the difference of refractive index between the glass and the crystalline phase. Thus, the nature of the crystalline phase could be mastered by directly introducing known crystalline particles in the base glass (see Section 10.4.3). It is well known that nanoparticles such as CdS and ZnS can be made by controlled precipitation. Knowing their interesting optical properties, it would be of great interest to incorporate them in a base glass having a very close refractive index. By introducing nanosize particles and lowering the difference of the refractive indices of glass and crystalline part, scatterings will be minimized. As a result, high fraction of crystals could be added in the glassy matrix. The main barrier lies in the difference in density between the amorphous and the crystalline phase, which may lead to an inhomogeneous glass-ceramic during quenching.

10.5.2 Anticipate the glass-ceramic forming ability

The major theoretical challenge in the coming years will consist in determining which glass system, and within this vitreous system which glass composition, will be the most appropriate to form nanosize particles. As far as we know, chalcogenide glass-ceramics have always been generated by heat treating several samples and then choosing the glass composition presenting the best ability to form glass-ceramic.

Up to now, there is still not any publication predicting the crystalline phase from a defined composition when heat treated. However, by coupling the results from the latest publications concerning both passive and active glass-ceramics working in the infrared range, a hypothesis can be put forward.

If we carefully examine the various ternary diagrams studied (Plate X between pages 330 and 331), one can observe that by rejecting the crystalline phase from the initial glassy matrix, the remaining glass composition progressively reaches a higher stability. In fact, the remaining glass tends to move into the middle of the glass-forming region. Taking this into account, the base glass can be chosen in a suitable zone of the glassy domain to force the desired crystalline phase to appear.

10.5.3 New ways to make chalcogenide glass-ceramics

Photo-induced crystallization

The photo-fluidity effects are mainly observed in low coordinated systems presenting a coordination number z usually lower than 2.4 according to the Phillips theory (Phillips, 1979). In contrast, chalcogenide glasses presenting a good ability to form nanocrystals are among highly coordinated composition with $z > 2.4$. Recent studies have demonstrated the possibility to give rise to microstructural changes by using laser illumination with highly coordinated systems which suggests that these bond rearrangement effects can also lead to the generation of nanocrystals in chalcogenide glassy matrix. These phenomena have already been observed with oxide glasses in order to reduce ionic metallic particles (Qiu *et al.*, 2004), but few papers report such results with chalcogenide glasses.

The first study dealing with photo-induced crystallization in chalcogenide glasses was reported by Griffiths *et al.* in 1981. Raman spectroscopy of irradiated $GeSe_2$ thin films in the band tail demonstrated the crystallization of $GeSe_2$ microparticles (Griffiths *et al.*, 1981; Haro *et al.*, 1985). Other attempts were performed on amorphous sulfide and selenium-based glasses (Griffiths *et al.*, 1983; Sugai 1986; Sakai *et al.*, 2000). The crystallization process is considered to be both athermal and thermal. The former is due to the Se-Se lone pair of electrons excited by photons which provoke the movement of atomic position to a more stable position (relaxation) while the latter is correlated with the electron excitation. So Tanaka preferred to use the term 'photo-enhanced crystallization' instead of 'photo-induced crystallization', as the absorption of light increases the crystallization rate (Tanaka and Shimakawa, 2011).

The As-Se and As-S glasses were explored by Mikla and Brazkhin, respectively, showing photo-induced crystallization (Mikla, 1996; Brazkhin *et al.*, 2007), while the reverse phenomenon, called photo-induced amorphization, has been observed as well (Frumar *et al.*, 1995). Since these studies, few aspects of the photo-crystallization process in chalcogenide glasses have been tackled, leaving the door open for further investigations.

We can work out that by focusing the laser beam onto planar waveguides, the nanocrystallization of rare earth particles could highly exacerbate the nonlinear effect or the luminescence to make efficient optical systems for active applications.

Mechanical milling and hot pressing

Few aspects of the potential of chalcogenide glass-ceramics have been investigated if we consider the wide range of applications of oxide-based glass-ceramics. Some recent developments have paved the way for new routes of investigation.

In a recent work, Hubert *et al.* (2011) report an innovative approach for making transparent chalcogenide glasses and glass-ceramics. The combination of amorphization by mechanical milling and densification by spark plasma sintering (SPS) was described. The SPS technique has demonstrated the possibility to make chalcogenide glasses and glass-ceramics with desired dimensions without using silica tubes (Delaizir *et al.*, 2010). The advantages of this approach are associated with the synthesis of amorphous chalcogenide powders near room temperature and with the possibility of fast precision molding of infrared optics by SPS. Furthermore, this technique permits unstable infrared glasses to be obtained without limitation of dimensions as it depends only on the mold size. They demonstrated the possibility to make an amorphous powder of the $80GeSe_2$-$20Ga_2Se_3$ glass composition from metallic Ge, Ga and Se elements at room temperature using mechanical milling (see Fig. 10.15). The amorphous powder was then sintered by SPS around T_g for several durations varying from 2 minutes to 1 hour, about 50 times faster than a usual heat treatment performed on this glass composition. With increasing SPS sintering duration, a controlled crystallization of nanoparticles

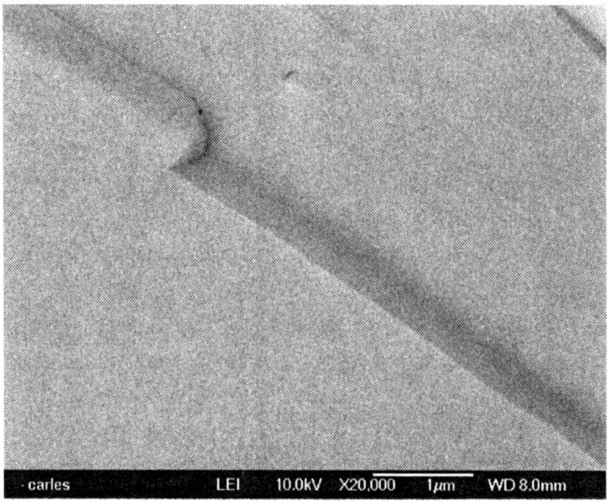

(a)

(a)

10.15 SEM pictures of the $80GeSe_2$-$20Ga_2Se_3$ base glass (a) and glass-ceramic (b) of glass-ceramics obtained from SPS treatment of 2 and 15 min respectively.

progressively occurred. The as-made glass-ceramics by combining the ball milling and SPS sintering show identical properties as glass-ceramics made with classical heat treatment of bulk glasses obtained by melt-quenching.

Furthermore, due to its low stability against crystallization, the $80GeSe_2$-$20Ga_2Se_3$ glass composition can only be fabricated in silica tubes with a diameter smaller than 9 mm. By using mechanical milling combined with SPS, much larger (36 mm in diameter) transparent glass ceramics have been obtained.

10.6 Conclusion

This chapter gives several keys to understand the evolution of research in the field of chalcogenide glass-ceramics. The first controllable and reproducible transparent glass ceramics were discovered very recently. The first motivations to make IR glass-ceramics are provided in this chapter irrespective of the targeted applications: increase of mechanical properties, enhancement of luminescence, nonlinear phenomena, etc.). Many papers report the kinetic study of crystallization using thermal analysis such as differential scanning calorimetry. However, few of them have presented both the theoretical analysis to obtain the activation energy or the characteristic temperature of nucleation and growth, and the experimental results to make controllable bulk glass-ceramics.

To date, the most efficient glass-ceramics are obtained from relatively unstable glasses against crystallization. The low difference between the glass transition temperature and the crystallization temperature makes it difficult to synthesize massive glass-ceramics or to shape them into lenses or fibers.

The formation of nanoparticles within the glassy matrix has demonstrated several important advantages concerning mechanical properties (hardness, toughness) without deteriorating the infrared transmittance. Moreover, the incorporation of rare earth within crystalline phases significantly increases the luminescence efficiency compared to the base glass. The last and most recent aspect is the crystallization of non-centro-symmetric nanosized particles which lead to permanent second-order nonlinear phenomenon in infrared glassy matrix.

All these results are the first steps toward some very promising and exciting future research. Also, compared to oxide glass-ceramics, few domains of investigation have been explored with chalcogenide elements, leaving the door open for numerous future areas of research.

10.7 References

Auzel F., Pecile D. and Morin D. (1975), 'Rare earth doped vitroceramics: new efficient blue and green emitting materials for infrared up-conversion', *J. of the Electrochem. Soc.*, 122, 101–107.

Avrami M. (1939), 'Kinetics of phase change: I General Theory', *Journal of Chemical Physics*, 7, 12, 1103.

Bach H. and Krause D. (2005), 'Low thermal expansion glass-ceramics', *Schott Series on Glass and Glass Ceramics*, 121–235.

Balda R., Sanz M., Mendioroz A., Fernández J., Griscom L.S. and Adam J.-L. (2001), 'Infrared-to-visible upconversion in Nd^{3+}-doped chalcohalide glasses', *Phys. Rev. B*, 64, 144101.

Brazhkin V.V., Gavrilyuk A.G., Lyapin A.G., Timofeev Yu.A., Katayama Y. and Kohara S. (2007), 'AsS: bulk inorganic molecular-based chalcogenide glass', *Appl. Phys. Lett.*, 91, 031912.

Bureau B., Zhang X.H., Smektala F., Adam J.L., Troles J., Ma H.L., Boussard-Plèdel C., Lucas J., Lucas P., Le Coq D., Riley M.R. and Simmons J.H. (2004), 'Recent advances in chalcogenide glasses', *Journal of Non-Crystalline Solids*, 345–346, 276–283.

Calvez L., Ma H-L., Lucas J. and Zhang X.H. (2006), 'Preparation and properties of glasses and glass-ceramics based on $GeSe_2$-Sb_2Se_3 and halides', *Physics and Chemistry of Glasses: European Journal of Glass Science and Technology*, 47 (2), 142–145.

Calvez L., Ma H.L., Lucas J., Glouannec P. and Zhang X.H. (2007a), 'Thermal analysis of RbI crystallization in the $GeSe_2$-Sb_2Se_3-RbI system', *Journal of Non-Crystalline Solids*, 353(52–54), 4702–4706.

Calvez L., Ma H.L., Lucas, J. and Zhang, X.H. (2007b), 'Selenium-based glasses and glass-ceramics transmitting light from the visible to the far IR', *Advanced Materials*, 19, 129–132.

Cernoskova E., Ivanova Z.G. and Pamukchieva V. (1998), 'Crystallization kinetics of $Ge_{10}Sb_{30}Se_{60}$ glass', *Thermo-chimica Acta*, 316, 97–100.

Chen H.S. (1965), 'A new method of analysing thermogravimetric data', *Bulletin of the Chemical Society of Japan*, 38 (11), 1881–1886.

Cheng, J. (1982), 'Properties and structure of the infrared-transmitting arsenic-germanium-selenium-tin glass-ceramic system', *Huadong Huagong Xueyuan Xuebao*, 3, 337–351.

Cheng J. (1986), 'Phase separation and controlled crystallization of non-oxide', *Journal of Non-Crystalline Solids*, 80 (1–3), 52–68.

Dai S., Peng B., Zhang P., Xu T., Xunsi W., Nie Q. and Zhang X.H. (2010), 'The near- and mid-infrared emission properties of Tm^{3+}-doped GeGaS–CsI chalcogenide glasses', *Journal of Non-Crystalline Solids*, 356 (44–49), 2424–2428.

Dai S., Lin C., Chen F., Zhang X.H., Liu C., Xu K., Heo J., Shen X. and Wang X. (2011), 'Enhanced mid-IR luminescence of Tm^{3+} ions in Ga^2S^3 nanocrystals embedded chalcohalide glass ceramics', *Journal of Non-Crystalline Solids*, 357 (11–13), 2302–2305.

Delaizir G., Lucas P. and Zhang X.H. (2007), 'Infrared glass-ceramics with fine porous surfaces for optical sensor applications', *Journal of the American Ceramic Society*, 90, 2073–2077.

Delaizir G., Dollé M., Rozier P. and Zhang, X.H. (2010). 'Spark plasma sintering: an easy way to make infrared transparent glass–ceramics', *Journal of the American Ceramic Society*, 93, 2495–2498.

Dong G., Tao H., Xiao X., Lin C., Guo H. and Zhao X. (2007), 'Second harmonic generation in transparent microcrystalline α-$CdGa_2S_4$-containing chalcogenide glass ceramics', *Optics Communications*, 274 (2), 466–470.

Fayek S.A., Fouad S.S., Balboul M.R. and El-Bana M.S. (2007), 'Crystallization kinetics in As–Se–Sn glassy system', *Physica B: Condensed Matter*, 388 (1–2), 230–236.

Frumar M., Firth A.P. and Owen A.E. (1995), 'Optically induced crystal-to-amorphous-state transition in As_2S_3', *Journal of Non-Crystalline Solids*, 192–193, 447–450.

Griffiths J.E., Espinosa G.P., Remeika J.P. and Phillips J.C. (1981), 'Reversible reconstruction and crystallization of $GeSe_2$ glass', *Solid State Commun.*, 40, 1077–1080.

Griffiths J.E., Espinosa G.P., Phillips J.C. and Remeika J.P. (1983), 'Raman spectra and athermal laser annealing of $Ge(S_xSe_{1-x})_2$ glasses', *Phys. Rev. B*, 28 (8), 4444–4453.

Gu S.X., Hu H., Guo H. and Tao H. (2008), 'Second-harmonic generation in transparent surface crystallized GeS_2–Ga_2S_3–CdS chalcogenide glasses', *Optics Communications*, 281 (9), 2651–2655.

Guignard M., Nazabal V., Troles J., Smektala F., Zeghlache H., Quiquempois Y.,

Kudlinski A. and Martinelli G. (2005), 'Second-harmonic generation of thermally poled chalcogenide glass', *Optics Express*, 13 (3), 789–795.

Guignard M., Nazabal V., Zhang X.H., Smektala F., Moréac A., Pechev S., Zeghlache H., Kudlinski A., Martinelli G. and Quiquempois Y. (2007), 'Crystalline phase responsible for the permanent second-harmonic generation in chalcogenide glass-ceramics', *Optical Materials*, 30 (2), 338–345.

Guillevic E., Allix M., Zhang X.H., Adam J.L., Matzen G. and Fan X. (2010), 'Synthesis and characterization of chloro-sulphide glass-ceramics containing neodymium(III) ions', *Materials Research Bulletin*, 45, 448–455.

Guimond Y., Adam J.L., Jurdyc A.M., Mugnier J., Jacquier B. and Zhang X.H. (1999), 'Dy^{3+}-doped stabilized GeGaS glasses for 1.3 μm optical fiber amplifiers', *Optical Materials*, 12 (4), 467–471.

Guin J.P., Rouxel T., Sangleboeuf J.C., Melscoet I. and Lucas J. (2002a), Hardness, toughness, and scratchability of germanium–selenium chalcogenide glasses,' *Journal of the American Ceramic Society*, 85 (6), 1545–1552.

Guin J.P., Rouxel T., Keryvin V., Sangleboeuf J.C., Serre I. and Lucas J. (2002b), 'Indentation creep of Ge–Se chalcogenide glasses below T_g: elastic recovery and non-Newtonian flow', *Journal of Non-Crystalline Solids*, 298, 260–269.

Hammel J.J. (1974), US patent 3,843,341.

Haro E., Xu Z.S., Morhange J.F., Balkanski M., Espinosa G.P. and Phillips J.C. (1985), 'Laser-induced glass-crystallization phenomena of $GeSe_2$ investigated by light scattering', *Phys. Rev. B*, 32 (2), 969–979.

Hendy S. (2002), 'Light scattering in transparent glass-ceramics', *Appl. Physics Letters*, 81(7), 1171–1173.

Heo J. (2003), ''Rare-earth doped chalcogenide glasses for fiber-optic amplifiers', *Journal of Non-Crystalline Solids*, 326–327, 410–415.

Hewak D. (1998), *Properties, Processing and Applications of Glass and Rare Earth Doped Glasses for Optical Fibers*, INSPEC, London.

Holand W. and Beall G.H. (2002), *Glass-ceramic technology*, Westerville, OH, American Ceramic Society.

Hosono H., Zhang Z. and Abe Y. (1989), 'Porous glass-ceramic in the $CaO\text{-}TiO_2\text{-}P_2O_5$ system', *Journal of the American Ceramic Society*, 72, 1587–1590.

Hubert M., Delaizir G., Monnier J., Godart C., Ma H.L., Zhang X.H. and Calvez L. (2011), 'An innovative approach to develop highly performant chalcogenide glasses and glass-ceramics transparent in the infrared range', *Optic Express*, 19 (23), 23513–23522.

Hubert M., Calvez L., Lucas P. and Zhang X.H. (2013), 'Enhanced luminescence in Er^{3+}-doped chalcogenide glass-ceramics based on selenium', *Journal of Luminescence*, submitted.

Ivanova Z.G. and Cernoskova E. (2004), 'Study on glass transition and crystallization kinetics of $Ge_xSb_{40-x}Se_{60}$ glasses by differential thermal analysis', *Thermochimica Acta*, 411 (2), 177–180.

Johnson W.A. and Mehl R.F. (1939), 'Reaction kinetics in processes of nucleation and growth', *Transactions of the American Institute of mining and metallurgical Engineers*, 135, 416–462.

Khan S.A., Al-Hazmi F.S, Faidah A.S. and Al-Ghamdi A.A. (2000), 'Calorimetric studies of the crystallization process in a-Se75S25–xAgx chalcogenide glasses', *Current Applied Physics*, 9(3), 567–572.

Kissinger H.E. (1957), 'Reaction kinetics in differential thermal analysis', *Analytical Chemistry*, 29, 1702–1706.

Ledemi Y., Bureau B., Calvez L., Roze M., Guillevic E., Audebrand N., Poulain M., Messaddeq Y. and Zhang X.H. (2009), 'Innovating transparent glass ceramics based on Ga_2S_3–GeS_2–CsCl', *Optoelectronics and Advanced Materials – Rapid Communications*, 3 (9), 899–903.

Li Z., Xu T., Shen X., Dai S., Wang X., Nie Q. and Zhang X.H. (2011), 'Investigation of 2.9 μm luminescence properties and energy transfer in Tm^{3+}/Dy^{3+} co-doped chalcohalide glasses', *Journal of Rare Earths*, 29 (2), 105–108.

Lin C., Tao H., Pan R., Zheng X., Dong G., Zang H. and Zhao X. (2008), 'Permanent second-harmonic generation in $AgGaGeS_4$ bulk-crystallized chalcogenide glasses', *Chemical Physics Letters*, 460 (1–3), 125–128.

Lin C., Calvez L., Rozé M., Tao H., Zhang X.H., Adam J.L. and Zhao X. (2009), 'Crystallization behavior of $80GeS_2 \cdot 20Ga_2S_3$ chalcogenide glass', *Applied Physics A*, 97 (3), 713–720.

Lin C., Calvez L., Ying L., Chen F.F., Song B.A., Shen X., Dai S.X. and Zhang X.H. (2011a), 'External influence on third-order optical nonlinearity of transparent chalcogenide glass-ceramics', *Applied Physics A*, 104 (2), 615–620.

Lin C., Calvez L., Tao H., Allix M., Moréac A., Zhang X.H. and Zhao X. (2011b), 'Evidence of network demixing in GeS_2–Ga_2S_3 chalcogenide glasses: A phase transformation study', *Journal of Solid State Chemistry*, 184 (3), 584–588.

Liu C., Tang G., Luo L., and Chen W. (2009), 'Phase separation inducing controlled crystallization of $GeSe_2$–Ga_2Se_3–CsI glasses for fabricating infrared transmitting glass–ceramics', *Journal of the American Ceramic Society*, 92 (1), 245–248.

Ma H.L., Zhang X.H. and Lucas J. (2003), 'Infrared transmitting chalcogenide glass ceramics', *Journal of Non-Crystalline Solids*, 317, 270–274.

Mecholsky J.J. (1973), *Microstructural Investigations of a Chalcogenide Glass Ceramic*, Washington, DC, Catholic University of America.

Mecholsky J.J., Moynihan C.T., Macedo P.B. and Srinivasan G.R. (1976), 'Microstructure and properties of an infra-red transmitting chalcogenide glass-ceramic', *Journal of Materials Science*, 11, 1952–1960.

Mikla V. (1996), 'Photoinduced structural changes and related phenomena in amorphous arsenic chalcogenides', *J. Phys-Condens Matter*, 8, 429–448.

Mironov R.A., Karaksina E.V., Zabezhailov A.O., Shaposhnikov R.M., Churbanov M.F. and Dianov E.M. (2010), 'Mid-IR luminescence of Cr^{2+}: II-VI crystals in chalcogenide glass fibres', *Quantum Electronics*, 40 (9), 828–829.

Miyata N. and Jinno H. (1982), 'Discussion of the indentation hardness of a glass-ceramic with particulate microstructure', *J. of Mat. Science*, 17, 2693–2699.

Ozawa T. (1965), 'A new method of analyzing thermogravimetric data', *Bulletin Chemical Society of Japan*, 38, 1881–1886.

Park S.H., Lee D.C. and Heo J. (2000), 'Pr^{3+}/Er^{3+} codoped Ge-As-Ga-S glasses as dual-wavelength fiber-optic amplifiers for 1.31 and 1.55 μm windows', *Journal of the American Ceramic Society*, 83 (5), 1284–1286.

Phillips J.C. (1979), 'Topology of covalent non-crystalline solids I: Short-range order in chalcogenide alloys', *Journal of Non-Crystalline Solids*, 34(2), 153–181.

Qiu J., Jiang X., Zhu C., Shirai M., Si J., Jiang N. and Hirao K. (2004) 'Manipulation of gold nanoparticles inside transparent materials', *Angewandte Chem. Int. Edition*, 43 (17), 2230–2234.

Ray C.S. and Day D.E. (1990), 'Determining the nucleation rate curve for lithium disilicute glass by differential thermal analysis', *Journal of the American Ceramic Society*, 73 (2), 439–442.

Ray C.S., Day D.E., Huang W., Narayan K.L., Cull T.S., Kelton K.F. (1996), 'Non-isothermal calorimetric studies of the crystallization of lithium disilicate glass', *Journal of Non-Crystalline Solids*, 204 (1), 1–12.

Ray C.S., Fang X. and Day D.E. (2000), 'New method for determining the nucleation and crystal-growth rates in glasses', *Journal of the American Ceramic Society*, 83 (4), 865–872.

Rozé M., Calvez L., Ledemi Y., Allix M., Matzen G. and Zhang X.H. (2008), 'Optical and mechanical properties of glasses and glass-ceramics based on the Ge-Ga-Se system', *Journal of the American Ceramic Society*, 91 (11), 3566–3570.

Rozé M., Calvez L., Hubert M., Toupin P., Bureau B., Boussard-Pledel C. and Zhang, X.H. (2011), 'Molded glass–ceramics for infrared applications', *International Journal of Applied Glass Science*, 2, 129–136.

Sakai K., Uemoto T., Yokoyama H., Fukuyama A., Yoshino K., Ikari T. and Maeda K. (2000), 'Annealing time and temperature dependence for photo-induced crystallization in amorphous $GeSe_2$', *Journal of Non-Crystalline Solids*, 266–269, 933–937.

Scheffler M., Kirchhof J., Kobelke J., Schuster K. and Schwuchow A. (1999), 'Increased rare earth solubility in As–S glasses', *Journal of Non-Crystalline Solids*, 256–257, 59–62.

Seznec V., Ma H.L., Zhang X.H., Nazabal V., Adam J.L., Qiao X.S. and Fan X.P. (2006a), 'Preparation and luminescence of new Nd^{3+} doped chloro-sulphide glass–ceramics, *Optical Materials*, 29 (4), 371–376.

Seznec V., Ma H.L., Zhang X.H., Nazabal V., Adam J.L., Qiao X.S. and Fan X.P. (2006b), 'Spectroscopic properties of Er doped chalco-halide glass-ceramics', *Optical Components and Materials III*, SPIE, San Jose, CA, 61160B.

Shen X., Chen F., Lv X., Dai S., Wang X., Zhang W., Song B., Xu T., Nie Q., Liu C., Xu K. and Heo J. (2011), 'Preparation and third-order optical nonlinearity of glass ceramics based on GeS_2–Ga_2S_3–CsCl pseudo-ternary system', *Journal of Non-Crystalline Solids*, 357, 2316–2319.

Song S.M., Choi S.Y. and Lee Y.K. (1997), 'Crystallization property effects in $Ge_{30}Se_{60}Te_{10}$ glass', *Journal of Non-Crystalline Solids*, 217 (1–2), 79–82.

Stookey S.D. (1957), *Nucleation*. New York, John Wiley & Sons.

Stookey S.D. (1959), 'Catalyzed crystallization of glass in Theory and practice', *Industrial & Engineering Chemistry*, 51 (7), 805–808.

Stookey, S.D. (1964), 'Low expansion glass-ceramic and method of making it', US Patent 3157522, November 17.

Strnad Z. (1986), *Glass-Ceramic Materials*. Prague, Publishers of Technical Literature.

Sugai S. (1986), 'Two-directional photoinduced crystallization in $GeSe_2$ and $SiSe_2$ glasses', *Phys. Rev. Lett.*, 57, 456–459.

Tanaka K. and Shimakawa K. (2011), *Amorphous Chalcogenide Semiconductors and Related Materials*, New York, Springer.

Turnbull D. (1949), 'Rate of nucleation in condensed systems', *J. Chem. Phys.*, 17, 71–73.

Turnbull D. and Cohen M.H. (1960), 'Crystallisation kinetics and glass formation', in *Modern Aspects of the Vitreous State* (ed. J.D. Mackenzie), London, Butterworths.

Tverjanovich A, Grigoriev Ya.G., Degtyarev S.V., Kurochkin A.V., Man'shina A.A. and Tver'yanovich Yu.S. (2001), 'Up-conversion fluorescence in Er-doped chalcogenide glasses based on GeS_2–Ga_2S_3system', *Journal of Non-Crystalline Solids*, 286, 1–2, 89–92.

Tver'yanovich Y.S., Kim D.S., Rusnak A.N., Turkina E.Y. and Korzinin A.A. (2005), 'Formation of $CuCr_2Se_4$ ferromagnetic spinel microcrystals in a chalcogenide glass matrix', *Glass Physics and Chemistry*, 31 (2), 168–172.

Vázquez J., López-Alemany P.L., Villares P. and Jiménez-Garay R. (1998), 'A study on non-isothermal transformation kinetics: application to the crystallization of $Sb_{0.20}As_{0.32}Se_{0.48}$ alloy', *J. Alloys Compd.*, 270, 179–185.

Volmer M. and Marder M. (1931), 'Zur Theorie der linearen Kristallisationsgeschwindigkeit', *Z. Phys. Chem.*, 154, 97–112.

Weinberg M.C. and Zanotto E.D. (1991), 'Re-examination of the temperature dependence of the classical nucleation rate: homogeneous nucleation in glass', *Journal of Non-Crystalline Solids*, 74, 1905–1909.

Xia F., Zhang X., Ren J., Chen G., Ma H. and Adam J.L. (2006), 'Glass formation and crystallization behavior of a novel GeS_2-Sb_2S_3-PbS chalcogenide glass system', *Journal of the American Ceramic Society*, 89 (7), 2154–2157.

Xu Y., Guang Y., Wei W., Zeng H., Zhang X.H. and Chen G. (2008), 'Formation and properties of the novel $GeSe_2$-In_2Se_3-Csl chalcohalide glasses', *Journal of the American Ceramic Society*, 91(3), 902–905.

Yang A., Chen D., Yu Y., Lina H. and Wang Y. (2011), 'Crystallization behaviours of $In_{2.67}S_4$ nanophase in chalcohalide glasses', *CrystEngComm*, 13, 3008–3012.

Yang Z., Chen W., Ma H.L. and Zhang X.H. (2008), 'Controlled crystallization of $GeSe_2$–Ga_2Se_3–CsI chalcohalide glasses during molding', *Journal of the American Ceramic Society*, 91 (8), 2720–2723.

Zhang, X.H., Guimond, Y. and Bellec Y. (2003), 'Production of complex chalcogenide glass optics by molding for thermal imaging', *Journal of Non-Crystalline Solids*, 326–327, 519–523.

Zhang X.H., Ma H.-L., Lucas L., Guimond Y. and Kodjikian S. (2004a), 'Optical fibers and molded optics in infrared transparent glass-ceramic', *Journal of Non-Crystalline Solids*, 336, 49–52.

Zhang X.H., Ma H.L. and Lucas J. (2004b), 'A new class of infrared transmitting glass-ceramic based on controlled nucleation and growth of alkali halide in a sulphide based glass matrix', *Journal of Non-Crystalline Solids*, 337, 130–135.

Zhao D., Zhang X.H., Xia F., Wang H., Ma H.L., Adam J.L. and Chen G. (2005), 'Non isothermal study on crystallization kinetics of $GeSe_2$–As_2Se_3–CdSe chalcogenide glasses by differential scanning calorimeter', *Journal of Crystal Growth*, 285 (1–2), 228–235.

Zhu S., Ma H.L., Calvez L., Zhang X.H., Lucas J., Adam J.L., Shang H. and Rouxel T. (2007), 'Optical and mechanical properties of far infrared transmitting glass ceramics', *Journal of Non-Crystalline Solids*, 353(13–15), 1298–1301.

Part II

Applications of chalcogenide glasses

11
Rare-earth-doped chalcogenide glass for lasers and amplifiers

J. HEO, Pohang University of Science and Technology, Republic of Korea and W. J. CHUNG, Kongju National University, Republic of Korea

DOI: 10.1533/9780857093561.2.347

Abstract: Recent results regarding chalcogenide glasses doped with rare-earth ions for optical gain media such as lasers and amplifiers are reviewed in this chapter. Among various chalcogenide glasses, Ge-Ga-S glasses modified with alkali halides such as KBr, KI, CsBr and CsI show significant improvement in their emission properties in the optical communication range and their spectroscopic properties as an amplifier gain medium are introduced. A large increase of intensity and lifetimes of 1.31 μm and 1.48 μm emissions from Dy^{3+}- and Tm^{3+}-doped Ge-Ga-S glasses, respectively, were observed. Investigations are conducted to understand the structural origin of the emission property improvement. Dy^{3+} and Pr^{3+} doped Ge-Ga-Sb-Se glasses and optical fibers showed potential for high power mid-IR gain medium and their spectroscopic properties are also introduced.

Key words: rare-earth ions, emission, chalcogenide glasses, molecular structure, fiber-optic amplifiers.

11.1 Introduction

Chalcogenide glasses doped with rare-earth ions have been considered as potential materials for mid-infrared lasers and fiber-optic amplifiers due to their low multiphonon relaxation (MPR) compared to oxide glasses. The MPR process competes against the desired emission and reduces its efficiency. On the other hand, chalcogenide glasses have relatively low phonon energy (Savage, 1987) and therefore, can provide enhanced emission properties of certain fluorescence that are usually quenched in oxide hosts.

As_2S_3 glasses were initially used for hosts of several rare-earth ions such as Nd^{3+}, Ho^{3+}, Er^{3+} and Tm^{3+} (Heo, 1995, Shin *et al.*, 1995). Several fluorescences in the mid-infrared region were reported such as 2.98 μm and 4.40 μm from the cascaded transitions of ${}^6H_{13/2}\rightarrow{}^6H_{15/2}$ and ${}^6H_{11/2}\rightarrow{}^6H_{13/2}$, in Dy^{3+} respectively. Wei *et al.* demonstrated fluorescence from Dy^{3+} or Pr^{3+} in $Ge_{25}Ga_5S_{70}$ (at.%) glasses suitable for the amplification of the optical communication signals at 1.3 μm that was normally quenched in the oxide glasses (Wei *et al.*, 1994, 1995).

However, chalcogenide glasses, especially Ge-Ga-S glasses, still have high multiphonon relaxation for certain transitions of practical interest. For example, Dy^{3+} has been considered as an alternative active ion for the 1.31 μm window (Wei *et al*., 1994, Schaafsma *et al*., 1998). A stimulated emission cross section of this transition is 4.35×10^{-20} cm^2, that is approximately four times larger than that of the $^1G_4 \rightarrow ^3H_5$ transition in Pr^{3+} (Wei *et al*., 1995). The branching ratio of the transition exceeds 90%. However, the measured lifetime of the emission level ($^6F_{11/2} \cdot ^6H_{9/2}$) for the 1.31 μm fluorescence is only about 38 μs and it is approximately one tenth of the lifetime measured from the $^1G_4$ level in Pr^{3+}. As a result, the quantum efficiency of the emission from Dy^{3+} is low (–17%). Therefore, Dy^{3+} has not been considered as a practical choice for the 1.31 μm band amplifier and new host materials with considerably reduced phonon vibration need to be identified.

This work reviews the effect of alkali halide addition into sulfide glasses on the emission properties of rare-earth ions as well as potential of rare earth doped selenide glasses as a mid-IR (MIR) gain medium. First, enhancement on the emission properties of various rare-earth ions upon the CsBr addition will be reviewed. Then, rates of multiphonon relaxation in many different glasses will be analyzed together with the effect of CsBr on the multiphonon relaxation of Ge-Ga-S glasses. Results from the various spectroscopic techniques will be analyzed to understand the structural origin of the improvement in the radiative properties. A possible structural variation on the addition of CsBr in sulfide glasses will be proposed based on Raman and extended x-ray absorption fine structure (EXAFS) spectroscopies. Spectroscopic results on Dy^{3+} and Pr^{3+} doped Ge-Ga-Sb-Se glasses and optical fibers at MIR range will also be analyzed.

11.2 Rare-earth (RE)-doped chalcogenide glasses for optical fiber amplifiers

11.2.1 Experimental procedures: an overview

The Ge-Ga-S system was selected since it has high rare-earth solubility and good glass-forming ability (Wei *et al*., 1995). Three compositions, $Ge_{0.29}Ga_{0.05}S_{0.66}$, $Ge_{0.25}Ga_{0.10}S_{0.65}$ and $Ge_{0.18}Ga_{0.18}S_{0.64}$ (in atomic fraction) were selected and will be referred to as GGS1, GGS2, and GGS3, respectively. Alkali halides of KBr, KI, CsBr, and CsI were added with typical compositions of (1-*y*)GGS1(or GGS2, GGS3)-*y*MX(M=K, Cs and X=Br, I), where *y* is 0.05, 0.1, 0.16 and 0.20 in mole fraction. The concentrations of rare-earth ions in glasses were 0.01–0.2 at.% depending on the solubility of each rare-earth ion. Glass samples were prepared following the conventional melting-quenching method (Shin *et al*., 2001) using silica ampoules as crucibles. High purity (>4N) starting materials were weighed into silica ampoules inside a globe box under an Ar gas environment. Ampoules containing starting powders were sealed and were heated at –1,000°C for 12 h in a rocking furnace. Melts inside the ampoules were subsequently quenched and annealed at their respective glass transition temperatures for several hours.

Absorption spectra of rare-earth ions were recorded using an UV/VIS/NIR spectrophotometer. Fluorescence was recorded using a computer-controlled 1/4m monochromator and an InSb detector cooled by liquid nitrogen. The excitation wavelengths were within 800–920 nm, provided by a CW tunable Ti-Sapphire laser pumped by an Ar^+ laser. The excitation beam was modulated by a variable frequency chopper controlled by a lock-in amplifier. Fluorescence decay curves were recorded using a computerized digital oscilloscope and lifetimes were determined from the first e-folding time of emission intensity.

11.2.2 Pr^{3+}- and Dy^{3+}-doped chalcogenide glasses for the 1.3 μm (O-band) fiber amplifiers

Fluorescence spectra in Fig. 11.1 recorded from the glass doped with 0.1% Dy^{3+} show the interesting intersity changes of two fluorescence changes at 1.31 μm ($^6F_{11/2}$·$^6H_{9/2}$→$^6H_{15/2}$) and 1.75 μm ($^6H_{11/2}$→$^6H_{15/2}$) (Shin *et al.*, 2001). In Ge-Ga-S glass, the intensity of the 1.31 μm emission is considerably smaller than that centered at 1.75 μm; similar to the spectrum reported previously

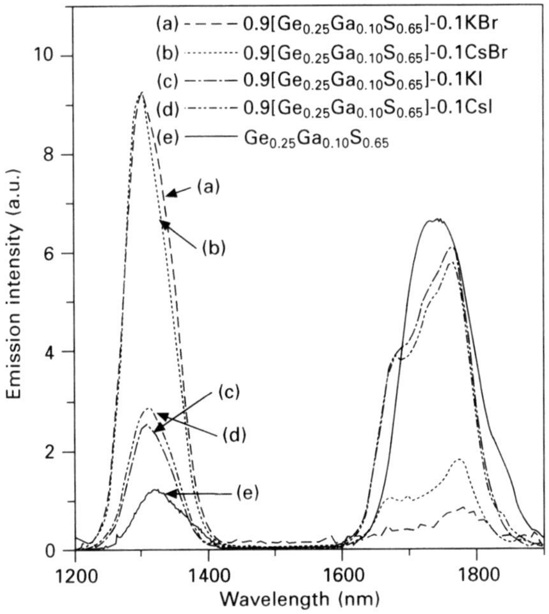

11.1 Emission spectra of the 1.31 μm ($^6F_{11/2}$, $^6H_{9/2}$→$^6H_{15/2}$) and 1.75 μm ($^6H_{11/2}$→$^6H_{15/2}$) transitions in Dy^{3+}-doped glasses (Shin *et al.*, 2001).

(Wei *et al.*, 1994). On the other hand, the 1.31 μm emission intensity clearly increased at the expense of the 1.75 μm fluorescence intensity in glasses containing bromides (KBr and CsBr). The similar phenomenon on the relative intensities of these two emissions was also observed in glasses containing iodides.

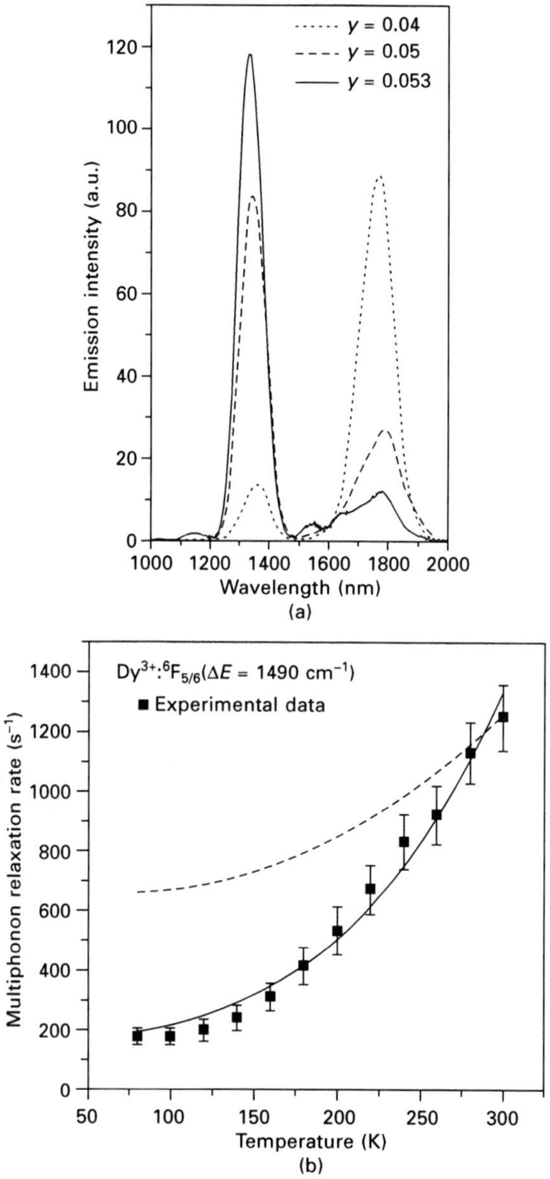

11.2 (a) Emission spectra and (b) decay curves of Dy^{3+}:$^6F_{11/2}$·$^6H_{9/2}$→$^6H_{15/2}$ (1.31μm) transition in (0.95-*y*)$Ge_{0.25}As_{0.1}S_{0.65}$-$0.05GaS_{3/2}$-*y*CsBr glasses.

Addition of CsBr into Ge-Ga-S glasses did not always result in the enhancement of the lifetimes and quantum efficiencies of the excited energy levels. For instance, Fig. 11.2(a) shows emission spectra for three Dy^{3+}-doped glasses in Ge-As-Ga-S with different amounts of CsBr. Emission spectra from glasses containing 4% and 5% CsBr were similar to those of Ge-Ga-S glasses (Wei *et al.*, 1994) with a weak 1.31 μm emission and a relatively strong 1.75 μm emission. On the other hand, the glass with >5% of CsBr showed a widely different fluorescence spectrum and became similar to that of the 10% CsBr-containing glass shown in Fig. 11.1.

One of the major drawbacks for Dy^{3+}-doped sulfide glasses has been the short lifetime (~ 38 μs, shown in Fig. 11.2(b)) of the upper emission level ($^6F_{11/2}{\cdot}^6H_{9/2}$) of the 1.31 μm fluorescence. This is due to the high multiphonon relaxation rate (~ 25,000 sec^{-1}) of the $^6F_{11/2}{\cdot}^6H_{9/2}{\rightarrow}^6H_{11/2}$ transition in sulfide glasses. When the appropriate amount of CsBr was added, the lifetime of the $^6F_{11/2}{\cdot}^6H_{9/2}$ level with 0.1 at.% Dy^{3+} doping was 1320 μs, and it is approximately 35 times longer than that measured from normal sulfide glasses (Shin *et al.*, 2002). Quantum efficiency (η) was less than 20% but glasses containing CsBr provided the quantum efficiency approaching 100%. Decay curves also showed notable changes with CsBr addition (Shin *et al.*, 2002).

11.2.3 Tm^{3+}-doped chalcogenide glasses for 1.45 μm (S-band) fiber amplifiers

Tm^{3+}-doped glasses have been investigated for the S-band (1460–1530 nm) amplifiers utilizing the 1.48 μm emission from the Tm^{3+}:$^3H_4{\rightarrow}^3H_6$ transition (Allain *et al.*, 1989). However, it has been difficult to realize efficient amplifiers because of the small energy gap between the emitting $^3H_4$ level and the one located at immediately below ($^3H_5$). Glasses with low phonon energy had to be used and they include sulfide (Shin *et al.*, 2000a), fluoride (Komukai *et al.*, 1995; Sakamoto *et al.*, 1995) and tellurite glasses (Naftaly *et al.*, 2000). Furthermore, the lifetime and population density of the Tm^{3+}:$^3H_4$ level also quenched via cross relaxation ($^3H_4$,$^3H_6{\rightarrow}^3F_4$,$^3F_4$) when Tm^{3+} concentration was high. Thus, the amount of Tm^{3+} in glasses must be kept low.

Three emission bands at 1.22, 1.48, and 1.82 μm in Fig. 11.3 are due to the $^3H_5{\rightarrow}^3H_6$, $^3H_4{\rightarrow}^3F_4$, and $^3F_4{\rightarrow}^3H_6$ transitions in Tm^{3+}, respectively (Song *et al.*, 2005). As the amount of CsBr exceeded 10 mol%, 1.48 μm emission intensity increased sharply similar to the case of Dy^{3+} (Shin *et al.*, 2001). Lifetimes of the $^3H_4$ level also increased rapidly from 0.23 to 1.23 ms when CsBr concentration increased from 5 to 10 mol% and the $^3F_4$ level lifetimes showed a similar change. It was considerably longer than the values observed in sulfide and tellurite glasses and similar to those from fluoride glass hosts. All glasses investigated showed 70–100% quantum efficiency.

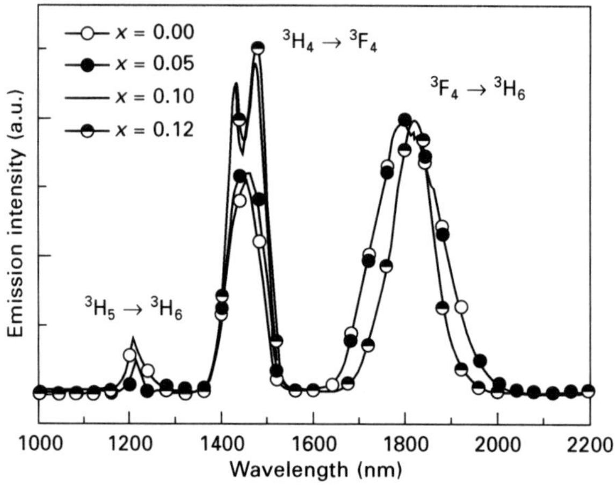

11.3 Emission spectra of Tm^{3+} in (1-*x*)($Ge_{0.25}Ga_{0.10}S_{0.65}$)-*x*CsBr glasses. Tm^{3+} concentration was 0.05 mol% for all glasses (Song *et al.*, 2005).

In particular, the non-radiative transition (or relaxation) from the $^3H_4$ level of Tm^{3+} doped into 0.90 ($Ge_{0.25}Ga_{0.10}S_{0.65}$)-0.10CsBr glass was negligible.

The measured lifetimes of the $^3H_4$ level decreased with increasing Tm^{3+} concentration in all glasses. For example, in $Ge_{0.25}Ga_{0.10}S_{0.65}$ glasses, $^3H_4$ level lifetimes decreased from 233 to 93 μs as the amount of Tm^{3+} increased from 0.05 to 0.50 mol%. These changes are due to the increased rate of cross relaxation (Tm^{3+}:$^3H_4$,$^3H_6$→Tm^{3+}:$^3F_4$,$^3F_4$) as the distance between Tm^{3+} ions decreased. The magnitude of increase in the cross relaxation rate seems to be larger in $Ge_{0.25}Ga_{0.10}S_{0.65}$ glasses than in 0.90 ($Ge_{0.25}Ga_{0.10}S_{0.65}$)-0.10CsBr glasses.

The peak wavelength of the $^3H_4 \rightarrow {}^3F_4$ transition in $Ge_{0.25}Ga_{0.10}S_{0.65}$ glass was observed at 1.46 μm with a FWHM (full width at half maximum) of ~ 87 nm. It shifted to 1.48 μm in 0.90 ($Ge_{0.25}Ga_{0.10}S_{0.65}$)-0.10CsBr with a FWHM of 91 nm. This is due to the splitting of this emission in 0.90 ($Ge_{0.25}Ga_{0.10}S_{0.65}$)-0.10CsBr glass. The peak splitting originated from the changes in Stark level splitting of $^3H_4$ and $^3F_4$ levels due to local structural modification. The Stark distributions of rare-earth ions are affected by the crystal field around the rare-earth ions. Therefore, changes in the local environment of rare-earth ions also result in changes in Stark distribution.

11.2.4 Ho^{3+}-doped chalcogenide glasses for 1.6 μm (U-band) fiber amplifiers

Ho^{3+} ions have several mid-infrared and visible emissions and their intensities are strongly dependent on host materials (Schweizer *et al.*, 1999). The 1.66 μm band emission from the $^5I_5 \rightarrow {}^5I_7$ transition was normally quenched in oxide hosts due to strong multiphonon interaction since the energy gap between the $^5I_5$ and $^5I_6$ levels is only 2,600 cm^{-1}. However, strong 1.66 μm band emission was realized in sulfide glasses (Lee *et al.*, 2004a,b). Estimation of gain showed that a large gain can be realized within the U-band (1,625–1,675 nm) window from sulfide glasses doped with Ho^{3+} and Tb^{3+} ions (Lee *et al.*, 2004a). Quantum efficiency, calculated from the ratio between the measured and radiative lifetimes of the $^5I_5$ level, was 57% for sulfide glasses. Since gain coefficients are directly proportional to the quantum efficiency of the specific transition, improvement in gain properties can be expected if host glasses with low phonon energies can be used. In this respect, selenide and chalcohalide glasses are attractive host materials due to their phonon vibration energies which are as low as ~ 250 cm^{-1} (Choi *et al.*, 2001, Shin *et al.*, 2000b).

Figure 11.4 shows strong emission at the U-band (1,625–1,675 nm) window from Ho^{3+} doped into chalcohalide glasses. The decay curves of the $^5I_5 \rightarrow {}^5I_7$ transition for specimen with 0.02 mol% Ho^{3+} showed a near-exponential

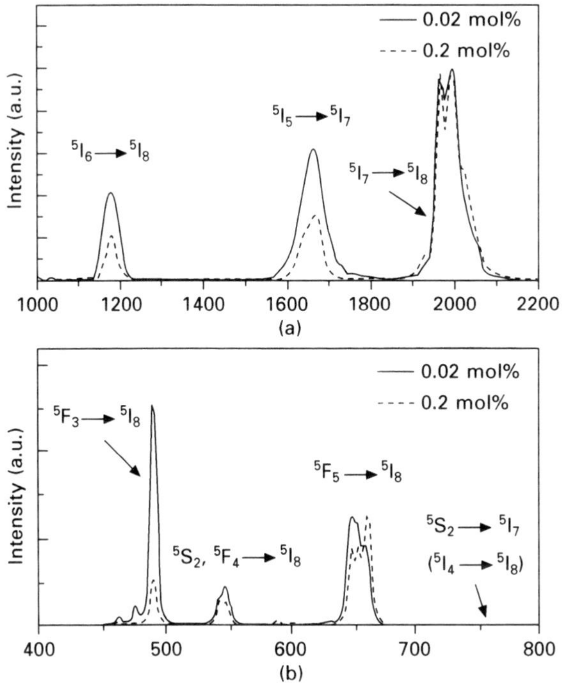

11.4 Emission spectra at (a) near infrared (NIR) and (b) visible (VIS) wavelengths from 0.95GGS1-0.05CsBr glasses doped with 0.02 and 0.2 mol% of Ho^{3+}. The excitation wavelength was 910 nm. Emission intensities were normalized to the intensities of the 2.00 and 0.66 μm emission bands for NIR and visible spectra, respectively (Lee and Heo, 2006).

decay with a measured lifetime of 9.38 ms. This value is comparable to the radiative lifetime (9.44 ms) calculated from the Judd–Ofelt analysis. This led to the high quantum efficiency of ~ 99%, clear evidence of the small phonon energy of chalcohalide glasses (~ 245 cm^{-1}) (Lee and Heo, 2005, 2006). As Ho^{3+} concentration increased, lifetime of the $^5I_5$ level started to decrease and decay curves deviated from the exponential behavior (Reisfeld and Jørgensen, 1987).

11.3 Local structure of RE ions

11.3.1 Multiphonon relaxation rates

Multiphonon relaxation (MPR) is one of the important properties that controls the lifetime and quantum efficiency. MPR rates (W_{mp}) were calculated from the measured (τ_m) and calculated (τ_R) lifetimes of several energy levels in rare-earth ions using Eq. [11.1] (Reisfeld and Jørgensen, 1987):

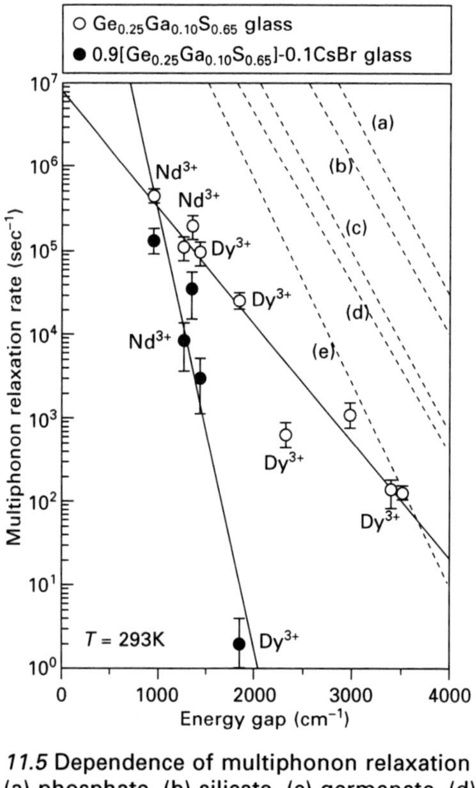

11.5 Dependence of multiphonon relaxation rates on energy gap for (a) phosphate, (b) silicate, (c) germanate, (d) tellurite and (e) fluoride glasses. Two solid lines are the result of the fitting for Ge-Ga-S and Ge-Ga-S-CsBr glasses (Shin *et al.*, 2001).

$$W_{mp} = \frac{1}{\tau_m} - \frac{1}{\tau_R} \quad [11.1]$$

Two solid lines in Fig.11.5 are the results for GGS2 and 0.9GGS2-0.1CsBr glasses from the least-squares fitting of the calculated W_{mp} and ΔE using the relationship $W_{mp} = W_0 \exp(-\alpha \Delta E)$ (Miyakawa and Dexter, 1970). In this equation, W_0 is the MPR rate extrapolated to zero energy gap and α is a constant depending on host materials. ΔE is the energy gap between the energy level of interest and the one immediately below. MPR rates from the excited levels decreased significantly with the addition of CsBr to the Ge-Ga-S glass (Shin *et al.*, 2001). For the Dy^{3+}:$^6F_{11/2}$·$^6H_{9/2}$ level which is the upper emission level of the 1.31 μm fluorescence, it decreased approximately four orders of magnitude. This difference in the MPR rates between two glasses explains, at least partially, such large changes in the intensity and lifetime of the 1.31 μm emission with CsBr addition.

11.3.2 Phonon modes controlling the multiphonon relaxation

To understand the changes in the local phonon mode quantitatively, the temperature dependence of the MPR rates from the excited states of Dy^{3+} was investigated (Fig. 11.6) (Shin *et al.*, 2000a). The following equation fits the measured MPR rates most precisely when the five-phonon relaxation process of the single vibration mode at 375 cm^{-1} was assumed:

$$W_{mp}(T) = W_0 \prod_i (n_i + 1)^{p_i} \quad [11.2]$$

Here, n_i is the Bose–Einstein occupation number, and p_i is the number of phonons needed to bridge the gap between the excited states and the energy level located immediately below. Vibration of 375 cm^{-1} is due to the asymmetric stretching vibration mode (ν_3) of the GeS_4 tetrahedra (Lucovsky *et al.*, 1974). However, it was not possible to obtain a good fit using the 375 cm^{-1} vibration for the glasses containing CsBr, as shown with a dotted line in Fig. 11.6. Instead, a solid line using the six-phonon process of the 245 cm^{-1} vibration was suitable. These results strongly indicate that the stretching vibration of Ga-Br bonds is responsible for the MPR process in glasses containing CsBr.

11.3.3 Phonon side band (PSB) spectroscopy

Phonon side band (PSB) modes represent Stokes or anti-Stokes fluorescence induced by phonon modes that are effectively coupled to the fluorescing energy level of rare-earth ions. A PSB spectrum of Eu^{3+}-doped $0.85(Ge_{0.18}Ga_{0.18}S_{0.64})$-0.15CsCl glass (all in atomic or mole fraction) was recorded by measuring

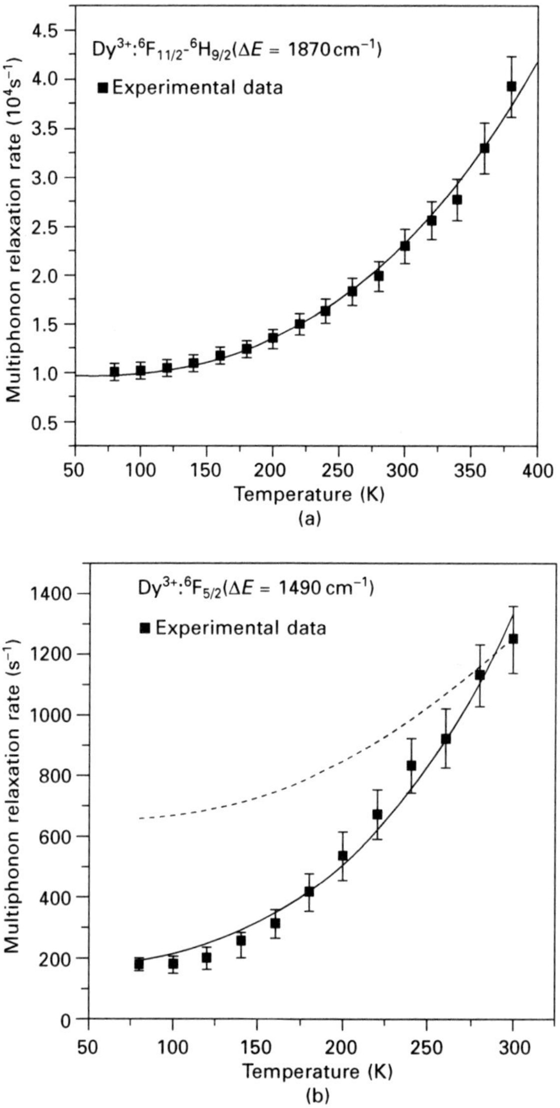

11.6 Temperature dependence of multiphonon relaxation rates in Dy^{3+} from the (a) $^6F_{11/2}$-$^6H_{9/2}$ level in GGS2 and (b) $^6F_{5/2}$ level in GGS2-CsBr glasses. See text for a detailed fitting procedure (Shin *et al.*, 2000a).

the fluorescence of the $^5D_0 \rightarrow {}^7F_2$ (~ 615 nm) at 10 K while continuously changing the pumping light energy in the vicinity of the Eu^{3+}:$^5D_0$ level (Chung and Heo, 2003). Five distinct phonon modes located at ~120, 149, 199, 226, and 259 cm^{-1} in Fig. 11.7 matched the vibration modes of Ga_2Cl_6 dimer and $CsGaS_{1.5}Cl$. This strongly supports the proposition that Eu^{3+} is

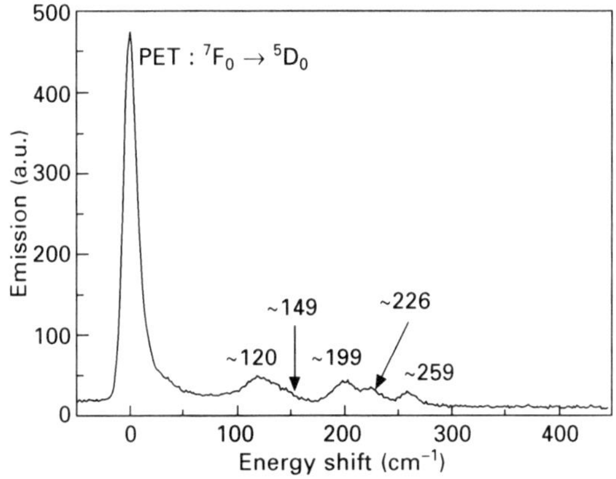

11.7 Phonon side band spectrum of 0.1 mol% Eu-doped $0.85(Ge_{0.18}Ga_{0.18}S_{0.64})$-0.15CsCl glass recorded by monitoring fluorescence change of $^5D_0 \rightarrow {}^7F_2$ transition (~ 615 nm) at 10K. PET = pure electronic transition (Chung and Heo, 2003).

located near to Ga_2Cl_6 or $[GaS_{3/2}Cl]$ structural units. It appears that Eu^{3+} ions in glasses have 6–8 Cl^- ions as their nearest neighbors as in $EuCl_3$ or they are bonded to the tetrahedral $[GaS_{3/2}Cl]^-$ subunit and/or Ga_2Cl_6. In all cases, Eu^{3+} is located next to Cl^- ions which probably form Eu-Cl bonds. These new Eu-Cl bonds reduced the bond strength compared to Eu-S bonds in normal sulfide glasses without CsCl.

11.3.4 Extended X-ray absorption fine structure (EXAFS) spectra

The local structure of Ge, Ga and Tm^{3+} ions in $(1-x)(Ge_{0.25}Ga_{0.10}S_{0.65})$-$x$CsBr glasses ($x$ = 0.00, 0.05, 0.10, and 0.12) were investigated using extended X-ray absorption fine structure (EXAFS) spectroscopy (Heo *et al.*, 2006; Song *et al.*, 2007a,b). Tm_2S_3 and $TmBr_3$ were measured with the transmission mode to prevent distortion of the EXAFS spectra. Glasses containing Tm^{3+} ions were recorded by the fluorescence method using the Lytle detector with a Co filter. The EXAFS data were analyzed following the standard method and detailed procedures can be found elsewhere (Heo *et al.*, 2006).

Figure 11.8 shows the Fourier transformed spectra of the Ga ions in glasses. The spectra in the figure show large changes in shape with the addition of CsBr. The main peak at an interatomic distance of ~ 2.28 ± 0.01 Å (after phase correction) is due to the Ga-S bonds in glasses (Loireau-Lozac'h *et al.*, 1996). Upon the addition of CsBr, intensity of this main peak increased and at the same time, a new shoulder appeared at a bond distance of ~ 2.36 ± 0.03 Å (after phase correction) due to the formation of Ga-Br bonds (Smirnov *et al.*, 2004). Results of the curve-fitting showed that Ga

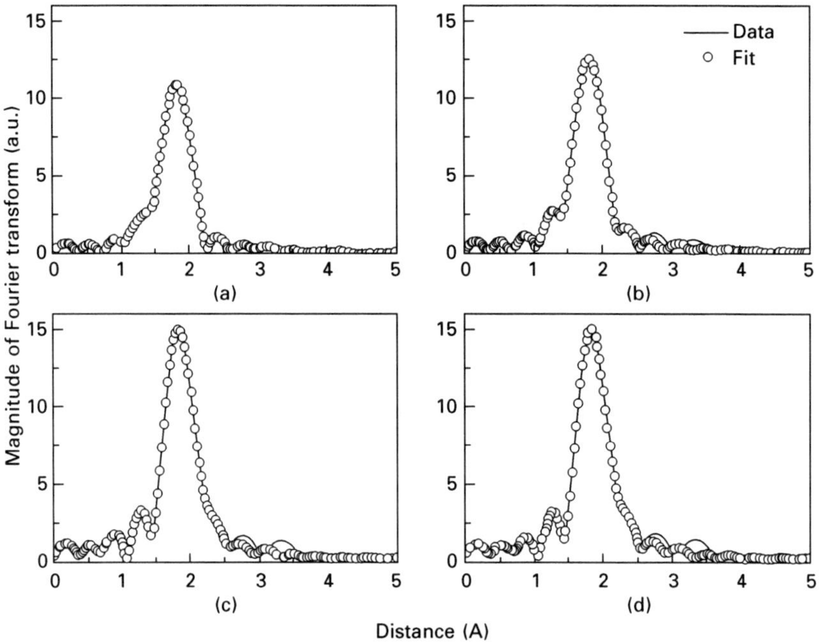

11.8 Radial distribution function (RDF) curves of Ga ions in (1-*x*)($Ge_{0.25}Ga_{0.10}S_{0.65}$)-*x*CsBr glasses where (a) $x = 0.00$, (b) $x = 0.05$, (c) $x = 0.10$ and (d) $x = 0.12$. Note that the phase shifts were not corrected (Song *et al.*, 2007b).

ions were surrounded by four sulfurs in $Ge_{0.25}Ga_{0.10}S_{0.65}$ glass. On the other hand, as the concentration of CsBr increased, average sulfur coordination surrounding Ga decreased while the coordination number of Br increased. In particular, in 0.90($Ge_{0.25}Ga_{0.10}S_{0.65}$)-0.10CsBr glass, the number of S ions in the nearest-neighbor shell of Ga decreased to three and the fourth ion bonded to Ga changed to Br. The peak position in the RDF curves of glasses changed with CsBr addition and became similar to that of $TmBr_3$ (White *et al.*, 1967; Brown *et al.*, 1968). These results indicate that Tm^{3+} ions are mainly surrounded by Br ions in 0.90($Ge_{0.25}Ga_{0.10}S_{0.65}$)-0.1CsBr glass. Results of the least-square fitting in Table 11.1 showed that Tm^{3+} ions were surrounded by ~ 6.77 (± 0.85) numbers of S ions in $Ge_{0.25}Ga_{0.10}S_{0.65}$ glass. Upon the addition of CsBr, the first coordination shell around Tm^{3+} ions were consisted with ~ 5.86 (± 1.58) Br ions. This result again supports that Tm^{3+} ions are predominantly surrounded by Br ions in 0.90($Ge_{0.25}Ga_{0.10}S_{0.65}$)-0.1CsBr glass (Song *et al.*, 2007b).

11.3.5 Proposed structural model

It was interesting to see that the addition of CsBr can result in such a large change in the emission properties. It suggested that the local environment

Table 11.1 Coordination numbers (*N*), bond distances (*R*) and Debye–Waller factors (σ^2) of Tm-S and Tm-Br bonds in crystals and glasses, along with *R*-factors showing the validity of fitting (reproduced from Song *et al.*, 2007b)

Bond	Composition	*R* (Å)	*N*	σ^2 (Å^2)	*R*-factor
Tm-S	Tm_2S_3 crystal	2.74 (0.01)	6.50 (fixed)	0.0115 (0.0015)	0.013
	$Ge_{0.25}Ga_{0.10}S_{0.65}$ glass	2.77 (0.01)	6.77 (0.85)	0.0107 (0.0014)	0.015
Tm-Br	$TmBr_3$ crystal	2.79 (0.01)	6.00 (fixed)	0.0066 (0.0003)	0.001
	0.90 ($Ge_{0.25}Ga_{0.10}S_{0.65}$)-0.10CsBr glass	2.79 (0.01)	5.86 (1.58)	0.0083 (0.0018)	0.020

*Values in parentheses are the estimated uncertainties.

surrounding rare-earth ions has experienced considerable modification with alkali halide addition. The backbone of Ge-Ga-S glass is composed of $GeS_{4/2}$ and $GaS_{4/2}$ tetrahedral units connected through the edge- or corner-sharing configuration (Loireau-Lozac'h *et al.*, 1996; Julien *et al.*, 1994). EXAFS analyses of $Ge_{0.25}Ga_{0.10}S_{0.65}$ glass also indicate that Ge and Ga ions are coordinated with four sulfur ions, in accordance with previous reports (Heo *et al.*, 2006).

EXAFS spectra on the Ga K-edge clearly show the number of Br ions in Ga tetrahedra increased with increasing CsBr content. In this case, the possible forms of Ga tetrahedral are $[GaS_{(4-x)/2}Br_x]^-$ (x = 1, 2, 3 or 4). In 0.90($Ge_{0.25}Ga_{0.10}S_{0.65}$)-0.10CsBr glass, where the ratio between Ga and CsBr is close to unity, most Ga ions form $[GaS_{3/2}Br]^-$tetrahedra. Each Ga tetrahedra is made of three S and one Br ion bonded to Ga. Shin *et al.* (2000a, 2001) and Tverjanovich *et al.* (1996) also proposed the formation of $[GaS_{3/2}Br$(or Cl)$]^-$ tetrahedral units in GeS_2-Ga_2S_3-CsBr (or Cl) glasses from analysis of their Raman spectra.

One can observe improvement on radiative properties of rare-earth ions only when the CsBr/Ga ratio in glass is equal to or greater than unity (Shin *et al.*, 2001). In these glasses, new structural units of $[GaS_{3/2}Br]^-$ were formed by substitution of sulfur by bromine. We believe that rare-earth ions are located near to these $[GaS_{3/2}Br]^-$ units to compensate for the charge balance. For example, in 0.90($Ge_{25}Ga_{0.10}S_{0.65}$)-0.10CsBr glass, most Ga ions form $[GaS_{3/2}Br]^-$ tetrahedral units as indicated from the EXAFS analyses. In this case, rare-earth ions are surrounded by non-bridging Br ions in the $[GaS_{3/2}Br]^-$ tetrahedra. The major phonon mode controlling the multiphonon relaxation then changes to the Ga-Br bond vibration from that of Ga-S, which leads to a large enhancement in radiative properties (Shin *et al.*, 2000a). Cs^+ ions in the glasses also are located near bromine ions with rare-earth ions. However, the exact role of Cs^+ has not yet been investigated.

The Tm L_3-edge EXAFS results further support that the first-shell around Tm^{3+} ions is composed of S ions in $Ge_{0.25}Ga_{0.10}S_{0.65}$ glass (Heo *et al.*, 2006; Song *et al.*, 2007a,b). Upon CsBr addition, the coordination environment of

Tm^{3+} showed pronounced changes and Tm^{3+} ions were surrounded mainly by Br ions in $0.90(Ge_{0.25}Ga_{0.10}S_{0.65})$-0.10CsBr glass. It should be noted that the number of S ions is approximately six times larger than Br ions in $0.90(Ge_{0.25}Ga_{0.10}S_{0.65})$-0.10CsBr glass. The preferred coordination of Tm^{3+} ions with Br instead of S are most probably related to the formation of $[GaS_{3/2}Br]^-$ structural units in the glass matrix (Heo *et al.*, 2006; Choi *et al.*, 2005). In $0.90(Ge_{0.25}Ga_{0.10}S_{0.65})$-0.10CsBr glass where the ratio between Ga and CsBr is close to unity, most Ga ions form $[GaS_{3/2}Br]^-$tetrahedra. In this case, $[GaS_{3/2}Br]^-$ tetrahedral units need charge compensator, and rare-earths ions can be dissolved by assuming the role of charge compensators for these tetrahedral units. Therefore, rare-earth ions experience the reduced multiphonon relaxation and local refractive index compared to the glasses without CsBr. As a result, the emission properties showed significant improvement as reported (Shin *et al.*, 2000a, 2001).

11.4 RE-doped chalcogenide glasses for mid-infrared lasers

The mid-infrared (MIR) wavelength range mostly refers to the wavelength band of 3–25 μm depending on the definition and corresponds to the characteristic molecular absorption or vibrational energy bands of most gases, liquids, and solid phases. Atmospheric windows at 3–5 μm and 8–12 μm also fall in this range which further emphasizes the importance of the MIR spectral range. Characteristic absorption of a certain molecule enables us to monitor specific gases and is very useful for monitoring hazardous gases in the air or chemical reactions during manufacturing processes. Biological tissues also have typical fingerprints within this spectral range and can be examined or modified with MIR light sources. High transmission at atmosphere, on the other hand, is a major requirement for military purposes such as secure communication, missile guidance, object detection, and so on. However, proper high powered light sources, e.g. lasers, have not been provided which has hindered the large-scale development of MIR applications. Based on those strong demands on MIR range, especially the 3–5 μm bands, numerous MIR laser sources have been widely studied for various applications including gas sensing, spectroscopy, chemical process monitoring and military uses (Werle *et al.*, 2002; Werle, 1998; Werle and Popov, 1999; Shaw *et al.*, 2001; Basiev *et al.*, 2002; Seddon *et al.*, 2010; Tittel *et al.*, 2003).

Semiconductor laser diodes (LDs) are dominant in visible (VIS) and near infrared (NIR) spectral ranges, including optical communication windows (1.3–1.6 μm) and also have been developed for gas sensing or chemical monitoring in the wavelength range longer than 3 μm (Werle *et al.*, 2002; Werle, 1998; Werle and Popov, 1999). Antimonide based III–V diodes lasers such as AlGaAsSb, InGaAsSb, and InAsSbP or lead salt based IV–VI diode lasers such as PbSSe or PbSnTe have been suggested but resulted in very low output power of less than 1 mW and low quantum efficiency. Quantum cascade (QC) structure can also give lasing at MIR range but requires low temperature and suffers from low output power, reaching only a few mW (Werle *et al.*, 2002; Tittel *et al.*, 2003). Frequency conversion of near-IR (NIR) sources using optical parametric oscillator (OPO) and difference frequency generation (DFG) can generate tunable mid-IR lasers, but requires complicated optics to give low output power (Tittel *et al.*, 2003). Realistic high-power fiber lasers (> 1 W) were reported with Ho^{3+} or Er^{3+}-doped ZBLAN optical fibers (Sumiyoshi *et al.*, 1999; Jackson, 2003; Linden, 2004) but operated at shorter wavelengths than 3 μm where few applications are found. ZBLAN glasses doped with Ho^{3+}, Er^{3+} and Dy^{3+} ions could show lasing action above 3 μm (Schneider *et al.*, 1997; Többen, 1992; Carbonnier *et al.*, 1998; Jackson, 2004). However, they could be achieved using high-powered pumping sources, but the wavelength is difficult to obtain and resulted in considerably low output power. Construction of a compact, high-powered and economic all-fiber MIR laser system is thus difficult with conventional fluoride glass fibers.

Unlike fluoride glasses, chalcogenide glasses composed of S, Se, or Te elements as network formers possess lower phonon energy (< 350 cm^{-1}) than fluoride glasses (> 500 cm^{-1}) and show enhanced transmission window above ~ 10 μm. As the quantum efficiency and radiative lifetimes of MIR transitions within rare-earth ions depend heavily on the energy gap to the next lower lying state and phonon energy of the host matrix (Layne *et al.*, 1977), various MIR emissions which are scarcely observed in oxide and fluoride glasses have been reported using chalcogenide glasses (Shaw *et al.*, 2001; Basiev *et al.*, 2002; Seddon *et al.*, 2010; Aggarwal and Sanghera, 2002; Sanghera and Aggarwal, 1999; Cole *et al.*, 1999). Increasing demands and interest in environmental, biomedical and military applications in recent years has further spurred research on chalcogenide glasses as an MIR gain medium (Shaw *et al.*, 2001; Basiev *et al.*, 2002; Seddon *et al.*, 2010; Tittel *et al.*, 2003; Lucas *et al.*, 2006; Zakery and Elliott, 2003; Seddon 1995; Kanamori *et al.*, 1984).

Various rare earth ions such as Pr^{3+}, Tb^{3+}, Dy^{3+}, Ho^{3+}, Er^{3+} and Tm^{3+}, have 4f-4f transitions of which wavelength corresponds to MIR range of 3–5 μm as shown in Fig. 11.9. However, it is also essential for them to have proper energy state distribution for effective lasing action such as a three- or four-level system, which is good for population inversion. In this regard, Dy^{3+} and Pr^{3+} have advantages in their electronic structure as well as their MIR emission properties. Dysprosium ions have strong emission bands near 3 μm with a three-level lasing system when pumped to the $^6H_{11/2}$ state. They can also give ~4.3 μm emission which has scarcely been reported due to high multiphonon relaxation rate in other glass hosts. Praseodymium covers

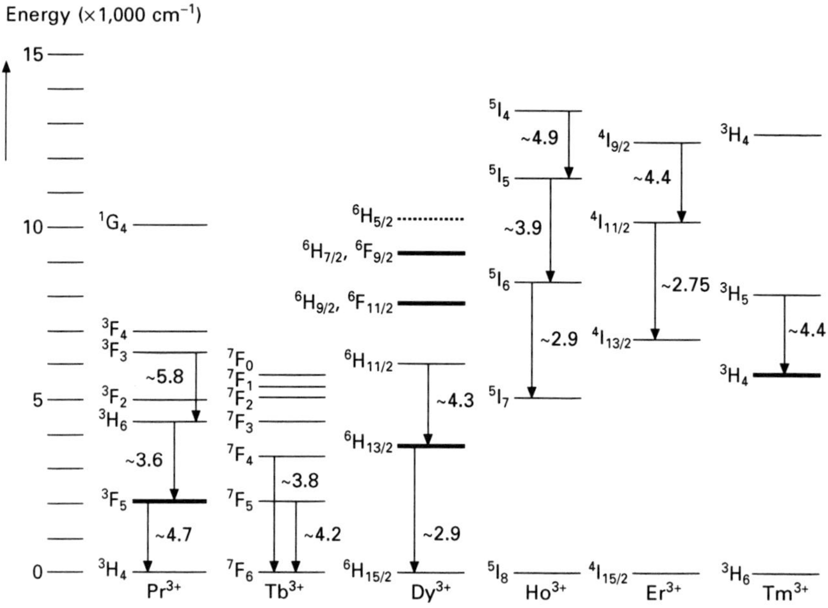

11.9 Schematic energy level diagrams of candidate rare-earth (RE) ions for mid-infrared (MIR) emissions and their transitions corresponding to 3–5 μm range.

a wide MIR band of 3.5–5.5 μm, but requires a suitable host matrix due to its narrow energy gaps between <4f> states (< 2,500 cm^{-1}). Sulfide glasses hosted those ions but mostly reported emissions at optical communication windows (1.3–1.7 μm) (Heo and Shin, 1996; Heo 2003; Wei *et al.*, 1994; Simons *et al.*, 1995; Park *et al.*, 2000). Practical MIR emissions from those ions were mostly reported from selenide glasses (Shaw *et al.*, 2001; Basiev *et al.*, 2002; Seddon *et al.*, 2010; Lucas *et al.*, 2006; Zakery and Elliott, 2003; Seddon 1995; Kanamori *et al.*, 1984; Park *et al.*, 2008a, 2008b) with lower phonon energy (< 250 cm^{-1}) than sulfides. Recent fabrication of low-loss single-mode optical fiber with Ge-Ga-Sb-Se glass (Chung *et al.*, 2005) along with the development of high-power laser diodes or fiber lasers operating at 1.48 μm and 2.0 μm both of which can directly excite the RE ions further improved the potential of MIR fiber lasers. In the following sections, thus, spectroscopic properties of Dy^{3+} and Pr^{3+} ions within the Ge-Ga-Sb-Se glasses and optical fibers for MIR gain media are introduced along with the study on co-dopants to enhance MIR emissions.

11.4.1 Dy^{3+}-doped chalcogenide glasses

Figure 11.10 shows the typical absorption bands due to the 4f-4f transition within Dy^{3+} ions in Ge-Ga-Sb-Se glass as well as the possible IR transitions

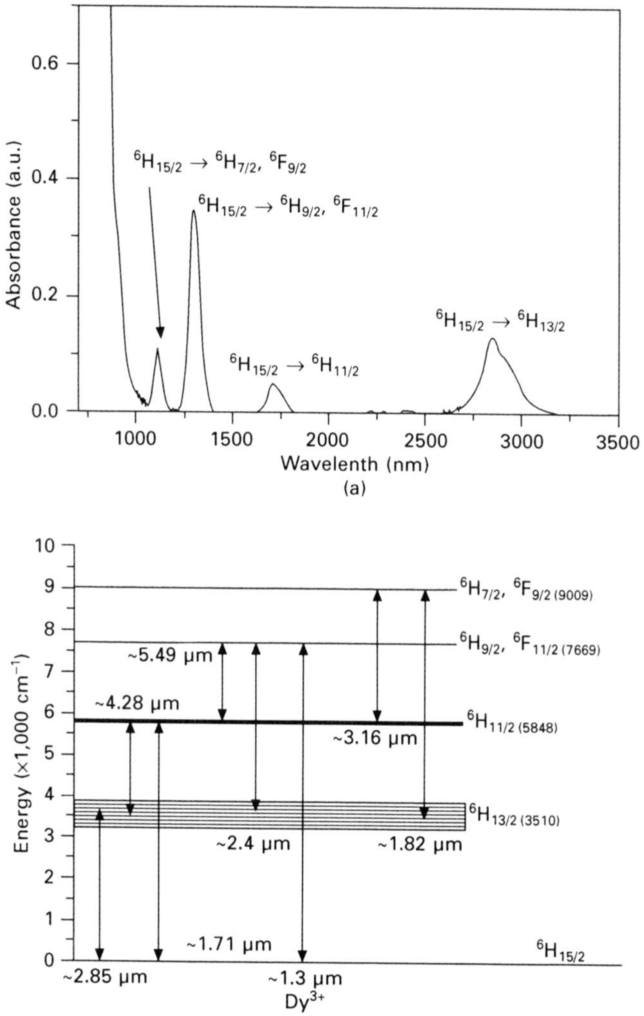

11.10 (a) Absorption spectrum of Dy^{3+} singly doped Ge-Ga-Sb-Se glass and (b) possible transitions within Dy^{3+}-ion whose wavelengths are estimated from the energy differences between the energy states (Park *et al.*, 2008a).

between them. As shown in the figure, various MIR emissions can be achieved with Dy^{3+}-ion and transitions from $^6H_{11/2}$ and $^6H_{13/2}$ to the ground state ($^6H_{15/2}$) corresponding to ~2.9 and ~4.3 μm, respectively are found to be promising for the lasing action due to their energy state distribution. Based on the measured absorption cross sections ($P(aJ;bJ')$) and the reduced matrix elements (Tanabe *et al.*, 1995; Carnall *et al.*, 1968, 1978), it is possible to study the host matrix dependence of those transitions using Judd–Ofelt (J-O) analysis which can be calculated as described previously (Tanabe *et al.*, 1995; Adam *et al.*, 1988) using the following relationship:

$$P(aJ;bJ') = \frac{8\pi^2 m v_{ab}}{3hg_a}\left[\frac{(n^2+2)^2}{9n}S_{ed} + nS_{md}\right] \quad [11.3]$$

where S_{ed} and S_{md} imply line strengths due to electric and magnetic dipole–dipole transition, respectively which are both considered in this calculation, m, v_{ab} and h represent electron mass, central frequency of the transition and the Plank constant, respectively. g_a is the degeneracy of the initial state while n is the refractive index of the glass. Table 11.2 summarizes the calculated J-O parameters of Dy^{3+}-ion in the Ge-Ga-Sb-Se glass along with parameters obtained from other chalcogenide glasses for comparison. It is known that $\Omega_{2,4,6}$ reflects the local environment of rare-earth ions such as asymmetry or polarization (Ω_2) and the covalency of the RE-ion (Ω_4 and Ω_6) (Tanabe *et al.*, 1992; Li *et al.*, 2002). Similar J-O parameters from various sulfide or selenide glasses suggest that Dy^{3+} ions tend to have similar bond characteristics and local symmetry in the chalcogenide glass system. The small discrepancies of Ge-As-Ga-Se glass is possibly due to the exclusion of magnetic dipole–dipole transition in its calculation (Shaw *et al.*, 2001).

Characterisitic MIR emissions from the Dy^{3+}-ion in selenide glasses ($Ge_{30}Ga_2Sb_8Se_{60}$) can be found in Fig. 11.11, which was excited using a 1.8 μm pumping source. Broad and strong emission bands of 2.9 μm and 4.3 μm can be attributed to the transition of Dy^{3+}:$^6H_{13/2}\rightarrow{}^6H_{15/2}$ and Dy^{3+}:$^6H_{11/2}\rightarrow{}^6H_{13/2}$, respectively, proving its potential as MIR activator. Absorption due to CO_2 during the measurement can be also found at a wavelength of ~4.24 μm (Shaw *et al.*, 2001). When singly doped, the maximum intensity was reached at 0.1 mol% of Dy^{3+}-ion concentration. It should be noted that the intensity ratio of 2.9 μm emission to the total emission $[I_{2.9}/(I_{2.9}+I_{4.3})]$ decreases with Dy^{3+} concentration when pumped with 1.8 μm ($^6H_{11/2}$ level). Cross relaxation (CR) between the $^6H_{13/2}\rightarrow{}^6H_{15/2}$ and $^6H_{11/2}\rightarrow(^6H_{7/2}, {}^6F_{9/2})$ transitions along with the excited state absorption (ESA) of $^6H_{13/2}\rightarrow(^6H_{7/2}, {}^6F_{9/2})$ manifolds can reduce the population of the $^6H_{13/2}$ energy state as Dy^{3+} concentration increases. The optimum concentration of Dy^{3+}-ion for MIR gain thus should be determined by taking into consideration those interactions between Dy^{3+}-ions.

Table 11.2 Judd–Ofelt parameters of Dy^{3+}-ion in various chalcogenide glasses (reproduced from Park *et al.*, 2008a)

(10^{-20}) cm^2	Ge-As-S	Ge-Ga-S	Ge-As-Ga-Se	Ge-Ga-Se	Ge-Ga-Sb-Se
Ω_2	10.53	11.86	9.4	12.64	9.67
Ω_4	3.17	4.00	2.7	2.94	3.16
Ω_6	1.17	1.47	2.6	1.48	1.03
Reference	Heo and Shin (1996)	Heo and Shin (1996)	Shaw *et al.* (2001)	Yang *et al.* (2005)	Park *et al.* (2008a)

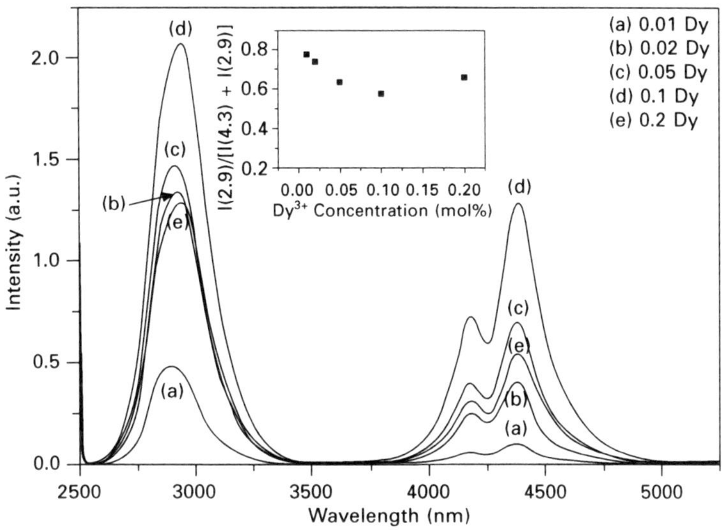

11.11 Emission spectra of Dy^{3+} in Ge-Ga-Sb-Se glasses with varying Dy^{3+} concentrations: (a) 0.01, (b) 0.02, (c) 0.05, (d) 0.1 and (e) 0.2 mol%, respectively. Pump wavelength was 1.8 μm. Inset figure shows the ratio of the 2.9 μm emission intensity ($I_{2.9}$) to the overall emission intensities ($I_{2.9}$+$I_{4.3}$) (Park *et al.*, 2008a).

Based on the measured emission results and calculated values, it is possible to estimate and compare stimulated emission cross section (σ_{se}) of the Dy^{3+}-doped selenide glass via the Füchtbauer–Ladenberg equation (Weber *et al.*, 1982) as follows:

$$\sigma_{se}(aJ;bJ') = \frac{\lambda_p^4}{8\pi c n^2 \Delta\lambda_{eff}} A(aJ;bJ') \qquad [11.4]$$

where λ_p and $\Delta\lambda_{eff}$ are the peak wavelength and effective linewidth of the transition, respectively (Weber *et al.*, 1982). *c* and *n* represent the speed of light and refractive index of the host matrix, respectively. *A(aJ;bJ')* is the transition probability between the energy levels. The obtained σ_{se} of the $^6H_{13/2} \rightarrow {}^6H_{15/2}$ transition from Ge-Ga-Sb-Se glass doped with 0.02 mol% Dy^{3+} was 1.195×10^{-20} cm^2 (Park *et al.*, 2008a). Considering the σ_{se} value of other excellent lasing transitions such as Nd^{3+}:$^4F_{3/2} \rightarrow {}^4I_{11/2}$ (~1.06 μm) (Weber *et al.*, 1982) or Er^{3+}:$^4I_{13/2} \rightarrow {}^4I_{15/2}$ (~1.54 μm) (Kumar *et al.*, 2004) in various host matrices showing 0.9–6 × 10^{-20} cm^2, the estimated σ_{se} of Dy^{3+}:$^6H_{13/2} \rightarrow {}^6H_{15/2}$ transition in selenide glass clearly indicates the high potential as a MIR gain source near 3 μm.

Further improvement of the emission can be made through introduction of a sensitizer, which can effectively transfer energy to the Dy^{3+}:$^6H_{13/2}$ state absorbing the pumping source. Similar energy levels to the Dy^{3+}:$^6H_{13/2}$ state and a large absorption cross section at pumping wavelength are required for the sensitizer to populate the Dy^{3+}:$^6H_{13/2}$ state. As shown in Fig. 11.9, Pr^{3+}, Tb^{3+}, Ho^{3+}, and Tm^{3+} are possible candidates satisfying those conditions and their contribution to Dy^{3+}-ion can be found in Figs 11.12 and 11.13. When

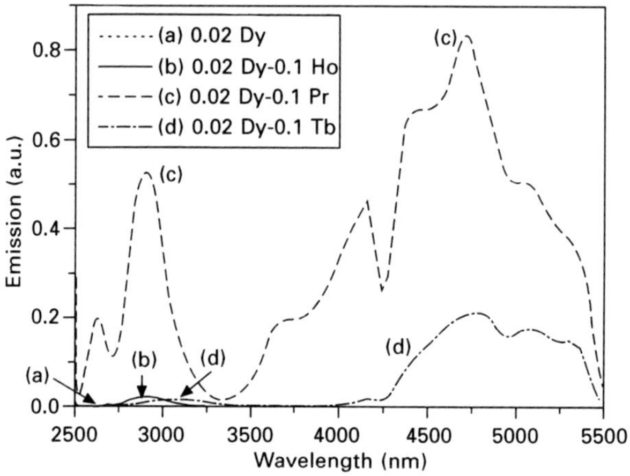

11.12 Emission spectra of Ge-Ga-Sb-Se glasses doped with (a) Dy^{3+} only and co-doped with (b) Ho^{3+}, (c) Pr^{3+} and (d) Tb^{3+}. Co-dopants were added with 0.1 mol% in concentration, while Dy^{3+} was fixed at 0.02 mol%. All glasses were pumped with 2.05 μm excitation source (Park *et al.*, 2008a).

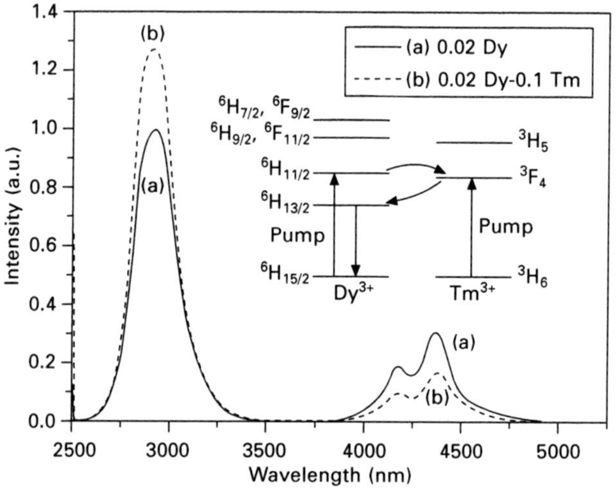

11.13 Emission spectra of Ge-Ga-Sb-Se glasses doped with (a) Dy^{3+} only (0.02 mol%) and (b) Dy^{3+} (0.02 mol%)-Tm^{3+}(0.1 mol%) ions. Excitation wavelength was 1.8 μm for both glasses. Inset represents the schematic energy transfer mechanism between Dy^{3+} and Tm^{3+} ions (reproduced from Park *et al.*, 2008a).

Pr^{3+}, Ho^{3+} and Tb^{3+} ions were excited with 2.05 μm excitation source, due to their absorption peak wavelength, Dy^{3+} showed the highest intensity of 2.9 μm emission with Pr^{3+}-ion (Fig. 11.12). No emission can be expected from Dy^{3+} single-doped glass due to less absorption band at the pumping wavelength. Effective energy transfer between the Pr^{3+}:($^3F_2$,$^3H_6$) and Dy^{3+}:$^6H_{13/2}$ levels may be responsible for the enhancement. However, there could be back energy transfer from Dy^{3+}:$^6H_{13/2}$ to Pr^{3+}:$^3H_5$, effectively depopulating the Dy^{3+}:$^6H_{13/2}$ levels when the concentration of Pr^{3+} increases. On the other hand, Tm^{3+}-ion can be pumped with 1.8 μm at which wavelength Dy^{3+} has absorption band and energy states near to Dy^{3+}:$^6H_{13/2}$ levels, suggesting its serious potential as sensitizer. As can be seen in Fig. 11.13, the emission intensity of ~ 2.9 μm was increased with the addition of Tm^{3+}-ion and can be mostly attributed to the effective energy transfer from the Tm^{3+}:$^3F_4$ to Dy^{3+}:$^6H_{13/2}$ level. Another energy transfer from Dy^{3+}:$^6H_{11/2} \rightarrow Tm^{3+}$:$^3F_4$ after simultaneous excitation by the pumping source can also exist. This can be observed from the decrease of ~ 4.3 μm emission band and allowing improved intensity ratio of $I_{2.9}/(I_{2.9}+I_{4.3})$. Dy^{3+}/Tm^{3+} co-doped selenide glass thus seems promising for MIR gain media at close to ~3.0 μm.

Such potential can be witnessed from the emission spectra measured with selenide optical fibers doped with Dy^{3+} (0.02 mol%) only and co-doped with Dy^{3+}(0.02 mol%) and Pr^{3+} (0.05 mol%) ions, respectively as shown in Fig. 11.14. Core-only selenide fibers with 25 cm in length can be prepared via modified double crucible method (Chung *et al.*, 2005) and examined. As schematically shown in Fig. 11.14(a), the fiber was spliced mechanically to silica fiber pig-tailed pumping sources and the MIR emission out of the fiber was monitored via condensing the light to the monochromator with band-pass filter and mechanical chopper. Strong emission centered around 2.95 μm out of Dy^{3+}-ion was clearly observed even from the fiber when excited with 1.8 μm, proving its potential for fiber laser operating at a wavelength of ~ 3 μm. Unlike bulk glasses, the MIR band at ~ 4.3 μm was reduced because of the long interaction length (25 cm) along with CO_2 absorption. The fiber co-doped with Pr^{3+} also showed different spectral behavior from the bulk glasses when pumped with 2.05 μm. The ~ 2.9 μm emission has been suppressed due to the energy relaxation from the Dy^{3+}:$^6H_{13/2}$ level back to the Pr^{3+}:$^3H_5$ state as described previously while traveling the fiber. The results from the fiber confirmed the detrimental effect of Pr^{3+}-ion as a sensitizer.

11.4.2 Pr^{3+}-doped chalcogenide glasses

Characteristic absorption bands of Pr^{3+} in selenide glass with a composition of $Ge_{30}Ga_2Sb_8Se_{60}$ can be found in Fig. 11.15(a) which was obtained via UV/VIS/NIR and FT-IR spectrometers. Possible MIR transitions between the |4f> states are also schematically represented in Fig. 11.15(b). Based on

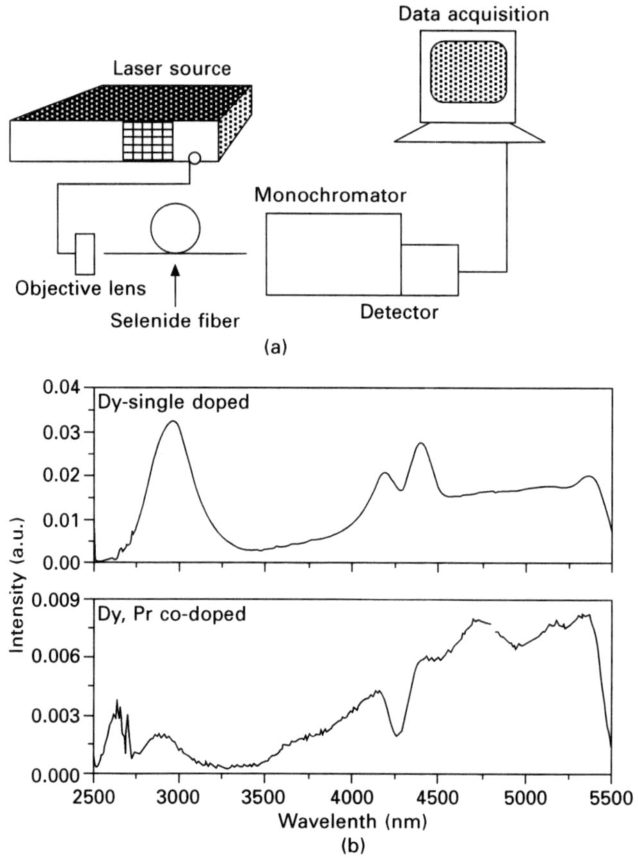

11.14 (a) Schematic diagram showing measurement set-up for emission spectra of selenide fibers and (b) emission spectra of the Ge-Ga-Sb-Se fiber doped with 0.02 mol% Dy^{3+} only (l = 25 cm, λ_{ex} = 1.8 μm; upper diagram) and co-doped with 0.02 mol% Dy^{3+} and 0.05 mol% Pr^{3+} (l = 15.5 cm, λ_{ex} = 2.05 μm; lower diagram) (reproduced from Park *et al.*, 2008a).

the energy level diagram, MIR transitions centered around 3.57 μm and 4.74 μm are expected and promising for the lasing actions when the Pr^{3+}-ion is excited up to the ($^3F_2$, $^3H_6$) state with a ~2.05 μm pump source. The strong and broad emission due to radiative transitions of ($^3F_2$, $^3H_6$)→$^3H_5$ (~3.7 μm) and $^3H_5$→$^3H_4$ (~4.7 μm) can be seen in Fig. 11.16. The absorption due to CO_2 at around 4.24 μm can also be found in the emission spectra and is corrected to give complete emission spectra (West *et al.*, 2000) as shown in Fig. 11.16(b). The emission covers a wide wavelength range of 3.5–5.5 μm, which is very useful for the broadband MIR sources and proves its serious potential as a MIR gain media. The broad emission of $^3H_5$→$^3H_4$ transition is mostly due to the many Stark split manifolds (Shaw *et al.*, 1997, 2001) along with the

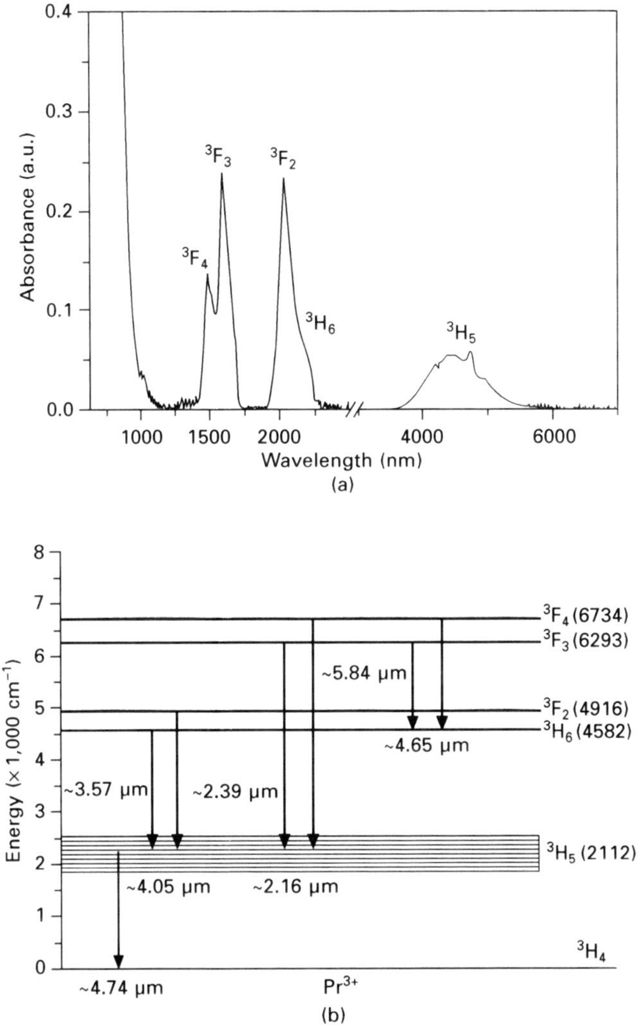

11.15 (a) Absorption spectra of Pr^{3+} singly-doped Ge-Ga-Sb-Se glass obtained by UV/VIS/NIR spectrophotometer and FT-IR spectrometer along with (b) energy levels of Pr^{3+} with possible emission transitions. The emission wavelengths were simply calculated from the energy gap between 4f energy levels (Park *et al.*, 2008b).

site inhomogeneity within the host matrix. It is interesting to note that the emission bandwidth increases as the concentration of Pr^{3+}-ion increases, while the emission band centered around 3.7 μm effectively reduces. Increased interaction between the excited states of Pr^{3+}-ions, such as energy transfer and spectral broadening of their emission, were responsible for the behavior (Park *et al.*, 2008b). As indicated in Fig. 11.16(b) after careful deconvolution

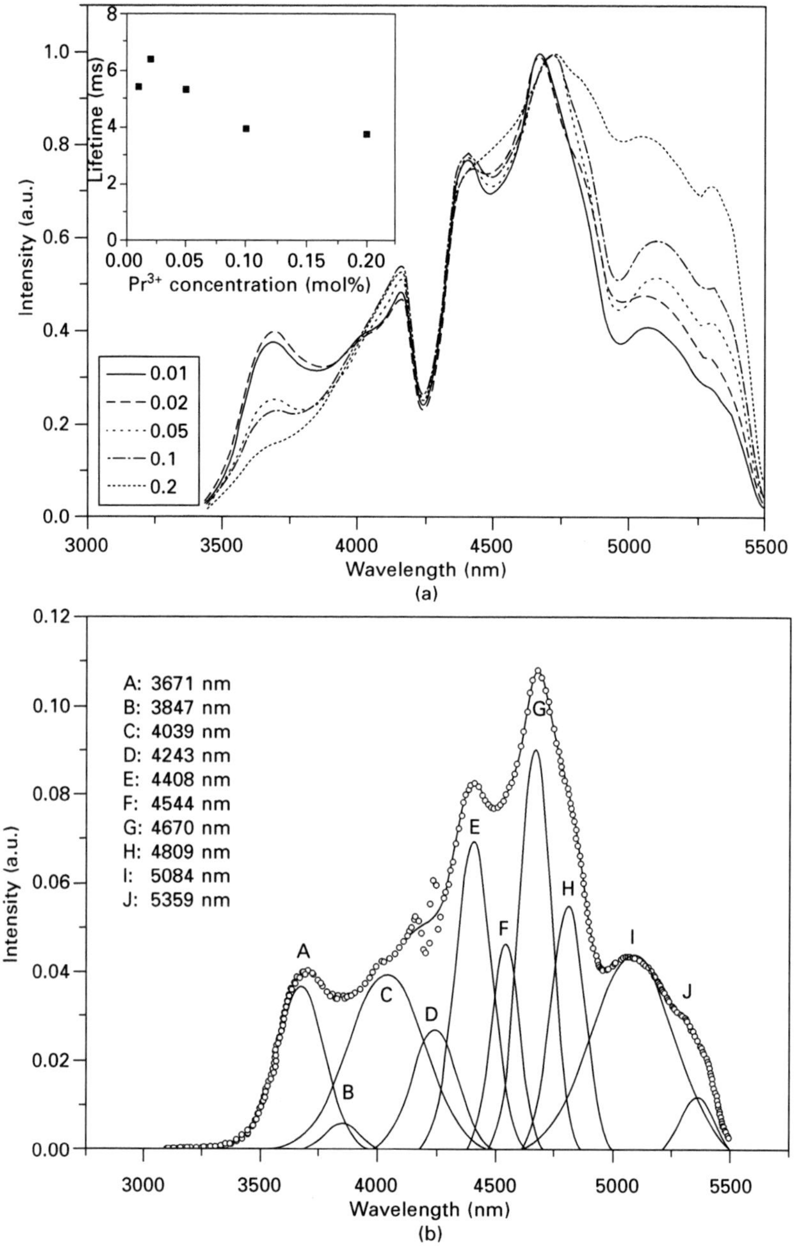

11.16 (a) Emission spectra of Pr^{3+} singly-doped Ge-Ga-Sb-Se glasses with varying Pr^{3+} concentration (0.01, 0.02, 0.05, 0.1, 0.2 mol%, respectively) when pumped at 2.05 μm. Intensities are normalized to their maximum intensity. The inset figure shows the lifetime change of the Pr^{3+}:$^3H_5$ state with varying Pr^{3+} concentration. (b) Peak deconvolution of MIR radiative transition of Pr^{3+}-ion after correction for CO_2 absorption (reproduced from Park *et al.*, 2008b).

of the emission spectrum (Park *et al.*, 2008b), emission bands of A and B due to the ($^3F_2$, $^3H_6$)→$^3H_5$ transition are effectively absorbed to populate the lower-lying energy level ($^3H_5$) increasing emission intensity of D to J corresponding to various Stark transitions between $^3H_5$ and $^3H_4$ manifolds. This process develops as the inter-ionic distance between Pr^{3+} decreased (CR1 process in Fig. 11.17(a)) and can be supported by the large spectral overlap between two transitions as shown in Fig. 11.17(b). Another cross relaxation between ($^3F_2$, $^3H_6$)→$^3H_5$ and ($^3F_2$, $^3H_6$)→($^3F_3$, $^3F_4$) transitions (Basiev *et al.*, 2002) can also consume A, B and C, bands supplying upconversion emission

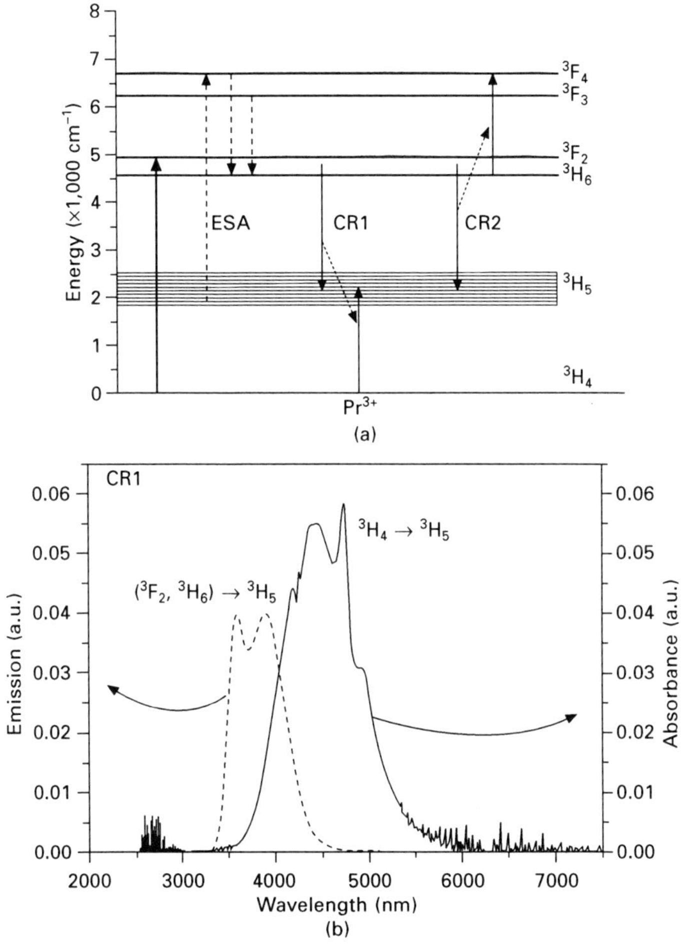

11.17 (a) Schematic diagram representing energy transfers between the excited states of Pr^{3+}-ions in selenide glasses. (b) Spectral overlap between Pr^{3+}:($^3F_2$, $^3H_6$)→$^3H_5$ emission obtained from the peak deconvolution and Pr^{3+}:$^3H_4$→$^3H_5$ absorption spectrum (Park *et al.*, 2008b).

bands out of $^3F_3$ and $^3F_4$ states as represented in Fig. 11.17 (CR2 process). Excited state absorption (ESA) of the pumping energy at $^3H_5$ level can also contribute to the emission bands of H, I and J and has been indicated by a dotted line in the figure. Thanks to those interactions between the excited states of Pr^{3+}-ions, it can be expected that the MIR emission at ~3.7 μm can hardly be used for the lasing action, while easy population inversion can be anticipated for the broad emissions out of $^3H_5$ state when the gain medium has long interaction length such as in optical fibers.

The measured fluorescence lifetime of the Pr^{3+}:$^3H_5$ state in selenide glasses was longer than 4 ms which was determined by the first e-folding time of the decay as found in the inset of Fig. 11.16(a). With the incorporation of Pr^{3+}- ions, it shows the maximum at 0.02 mol% and decreases afterwards. The CR1 process increases the electronic population of the $^3H_5$ states and is believed to be responsible for the initial increase with the Pr^{3+} concentration. It can be also witnessed by the increase of emission bands of A and B in Fig. 11.16(a). The following lifetime decrease can be elucidated by the typical energy transfer between Pr^{3+} ions due to the reduced inter-ionic distance with increasing concentration. Development of the emission bands at long wavelength (H, I, and J) with Pr^{3+} content supports the effects. Spatial radiation trapping (Chung *et al.*, 2004; Noginov, 1997) can also enhance the lower energy bands (H, I, and J) and cannot be completely ruled out, although the emission signals were recorded at the normal angle to the sample surface bypassing the pumping source. Considering the practical use of the Pr^{3+}-ion as a MIR activator, compromise between the lifetime and bandwidth should be reached. According to the results shown in Fig. 11.16, for example, suitable doping concentration of Pr^{3+} in Ge-Ga-Sb-Se glass could be 0.02–0.05 mol% to avoid interactions between Pr^{3+}-ions as described earlier.

As previously observed in Dy^{3+}, sensitizers can further improve MIR emissions out of the Pr^{3+}-ion. Proper sensitizers for this purpose are Tm^{3+} and Ho^{3+} ions which have similar absorption bands around 2 μm. They do not have any lower-lying energy levels than the Pr^{3+}:$^3H_5$ state which can eliminate possible back energy transfer from the Pr^{3+}-ion enhancing quantum efficiency of Pr^{3+}:($^3F_2$, $^3H_6$) and $^3H_5$ states. Emission spectra obtained from the co-doped selenide glasses is shown in Fig. 11.18. Noticeable improvement was achieved with Tm^{3+}-ion. Highly efficient energy transfer from the sensitizer ion due to the strong thermal coupling between the Tm^{3+}:$^3F_4$ and Pr^{3+}:($^3F_2$, $^3H_6$) states (Armagan *et al.*, 1991; Özen and DiBartolo, 2000) along with the high absorption cross section of the Tm^{3+}:$^3F_4$ state is mostly responsible for the improvement. Conversely, when pumped with 2.05 μm, Ho^{3+} decreased the MIR emissions out of the Pr^{3+}:($^3F_2$, $^3H_6$) level in spite of the close location of the Ho^{3+}:$^5I_7$ state in energy. This can be attributed to the back energy transfer from the Pr^{3+}:($^3F_2$, $^3H_6$) to Ho^{3+}:$^5I_7$ level suppressing the

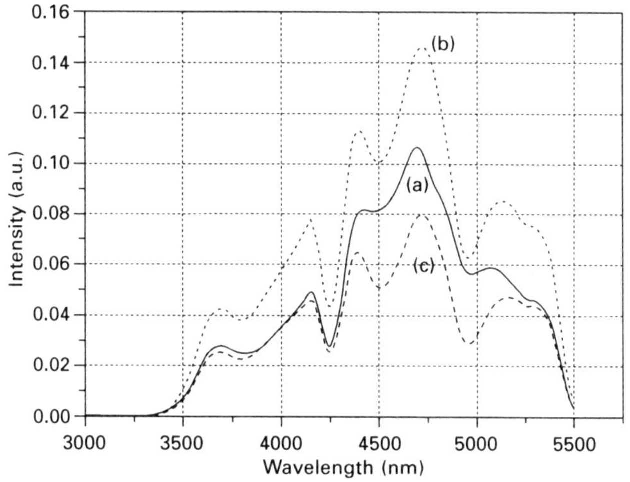

11.18 Emission spectra of selenide glasses doped with (a) 0.05 mol% Pr^{3+} only and co-doped with (b) 0.2 mol% Tm^{3+} and (c) 0.2 mol% Ho^{3+}. All glasses were pumped at 2.05 μm (Park *et al.*, 2008b).

radiative emission intensity. The lifetime of 4.7 μm emission was decreased with Ho^{3+} while no considerable change was observed with Tm^{3+} co-doped samples supporting the sensitizer effects. Those results suggest the high potential of Pr^{3+}/Tm^{3+} co-doped selenide glass as a MIR gain medium in the 3.5–5.5 μm band.

The feasibility for the fiber gain medium has also been examined with optical fibers doped with 0.02 mol% Pr^{3+} which was fabricated and tested as described earlier (Park *et al.*, 2008b, Chung *et al.*, 2005). 1.48 μm laser diode and 2.05 μm fiber laser excited the fiber to decide the proper pumping scheme for the fiber. As shown in Fig. 11.19, emission spectra similar to the bulk glass were obtained from the fiber proving its potential as a MIR gain medium when pumped at 1.48 and 2.05 μm. Emission peak at around 5.25 μm when pumped with 1.48 μm is the second-order emission of ($^3F_3$, $^3F_4$) $\rightarrow ^3H_5$ transition. However, the 2.05 μm pumping scheme is more favorable for the fiber laser with its distinct emission at ~4.7 μm and detrimental effects of 1.48 μm excitation. When the fiber is excited up to the ($^3F_3$, $^3F_4$) manifolds, excited state absorption (ESA) to the higher-lying energy levels such as $^1G_4$ level is likely to happen, depopulating the $^3H_6$ and $^3H_5$ levels and hampering gain development in the MIR range. Moreover, the maximum deliverable power of the 1.48 μm pump source can be limited since it can be absorbed by the selenide glass via two photon absorption or thermally coupled interaction between the lowest un-occupied molecular orbital (LUMO) and the excited energy states of Pr^{3+} through ESA. The 2.05 μm source eliminates those problems and could be a suitable choice for the fiber laser

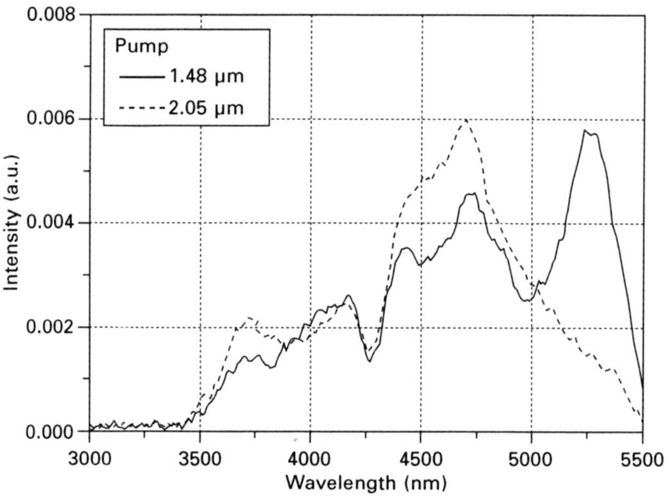

11.19 Fluorescence emission spectra of selenide optical fiber doped with 0.02 mol% Pr^{3+} varying the excitation wavelengths. Solid line shows the spectra pumped with 1.48 μm while the dashed line shows them with 2.05 μm (Park *et al.*, 2008b).

excitation. However, it should be mentioned that absorption due to impurities such as Ge-H (~4.95 μm) and Se-H (~4.5 and ~4.75 μm) (Shaw *et al.*, 1997; Kokorima, 1996) can seriously impair the MIR lasing actions when longer interaction length is required to improve accumulated stimulated emission (ASE). Such impurities should be removed with the controlled refinement processes during the glass and fiber fabrication. Atmospheric contamination of the selenide fiber should also be avoided through passivation of the fiber with proper jacket materials.

11.5 Conclusion and future trends

Chalcogenide glasses are highly promising materials for optical amplifiers and MIR laser gain media when doped with RE ions due to their low phonon energy and high transparency. Intensity, lifetime, and other radiative properties of the emission from several rare-earth ions experienced a large improvement upon the addition of alkali halides to Ge-Ga-S glasses and showed high potential for optical amplifier. For example, the lifetime of the 1.31 μm emission level ($^6F_{11/2}$-$^6H_{9/2}$) in Dy^{3+} increased as much as 35 times with 10 mol% addition of CsBr. Quantum efficiency also approached 100%. The effect of alkali halides was maximized when the ratio of MX(M=K,Cs;X=Br,I)/Ga was equal to or larger than unity. Several spectroscopic analyses on the local environment of rare-earth ions in Ge-Ga-S-CsBr glasses indicated that Br ions preferentially made bonds with Ga ions to form $[GaS_{3/2}Br]^-$tetrahedral. Rare-earth ions are located near these $[GaS_{3/2}Br]^-$ structural units. Rare-earth ions in the sulfide glass were mainly surrounded by approximately seven S ions while they were coordinated by approximately six Br ions in $0.90(Ge_{25}Ga_{0.10}S_{0.65})$-0.10CsBr glass. Consequently, they experienced the reduced electron-phonon coupling that led to the enhancement of the emission properties.

Selenide glasses provide lower phonon energy allowing MIR transitions to the doped RE ions than sulfide glasses which hardly generate MIR emissions. Characteristic ~3 and ~4.3 μm emission bands and broad emission bands of 3.5–5.5 μm were observed from Dy^{3+} and Pr^{3+} doped Ge-Ga-Sb-Se glasses, respectively. Optimum doping concentration of each ion should be determined by considering the emission intensity and lifetime of the excited state. Interactions between the excited states such as excited state absorption (ESA) and cross relaxation (CR) depopulate the lasing state and should also be considered, especially when the glasses are used in the form of optical fiber possessing long interaction length. The estimated stimulated emission cross section of Dy^{3+} and Pr^{3+} proved their potential as a high power gain medium in the MIR range. Various sensitizers such as Pr^{3+}, Ho^{3+}, Tb^{3+}, and Tm^{3+} ions can co-dope the selenide glass for Dy^{3+} but Tm^{3+} showed the most prominent effect due to its high absorption cross section at the pumping wavelength of 1.8 μm along with the efficient energy transfer to Dy^{3+}-ion. Tm^{3+} was also beneficial to Pr^{3+}-ion when pumped with 2.05 μm, enhancing the MIR emission due to $^3H_5 \rightarrow ^3H_4$ transition. Dy^{3+} and Pr^{3+} doped selenide optical fibers were fabricated and also showed MIR emissions supporting their serious potential as MIR laser materials.

Practical application of those chalcogenide optical amplifiers and MIR lasers can be achieved in the form of optical fibers. Although low-loss chalcogenide fibers have been produced recently, further refinement of the impurities is required due to significant absorption of impurities at the optical communication window and MIR range. Proper splicing between the gain medium and conventional single-mode silica fiber, in addition, should be provided to reduce the insertion loss induced by the large difference in refractive index between two fibers. For the realization of high powered all fiber MIR laser, it is also crucial to accommodate the high power pumping source at ~1.8 and ~2.0 μm, which can be supplied by the Tm^{3+}-doped silica fiber laser. Mirrors such as reflection coatings on the fibers or fiber gratings to compose the resonant cavity should also be offered.

Regarding optical fiber amplifier, although Pr^{3+}-doped fluoride fiber for O-band and Tm^{3+}-doped silica fiber for S-band are currently strong competitors, chalcogenide glass can also be a promising candidate with its significantly enhanced quantum efficiencies when modified with halides. The advantage of chalcogenide glasses can be further highlighted in the U-band (~1.6 μm), where other glasses cannot give proper emissions. As for MIR lasers, it is clear that the RE-doped selenide glass, having versatility in its fiber form, is a strong and competitive candidate for the high power MIR laser compared to other options such as semiconductors or lasers based on non-linear optics. Moreover, due to high potential for a gain medium of the Dy^{3+} and Pr^{3+}-doped selenide glasses, high powered MIR lasers can also be expected in the form of bulk or disc-type system by adapting conventional bulk optics.

11.6 References

Adam J L, Docq A D and Lucas J (1988), 'Optical transitions of Dy^{3+} ions in fluorozirconate glass', *J. Solid State Chem.*, 75, 403–412.

Aggarwal I D and Sanghera J S (2002), 'Development and applications of chalcogenide glass optical fibers at NRL', *J. Optoelectron. Advanced Mater.* 4, 665–678.

Allain J Y, Monerie M and Poignant H (1989), 'Tunable CW lasing around 0.82, 1.48, 1.88 and 2.35 μm in thulium-doped fluorozirconate fibre', *Electron. Lett.*, 25, 1660–1662.

Armagan G, Buoncristiani A M and DiBartolo B (1991), 'Energy transfer and thermalization in YAG:Tm, Ho', *J. Lumin.*, 48–49, 171–174.

Basiev T T, Orlovskii Yu V, Galagan B I, Doroshenko M E, Vorob'ev I N, Dmitruk L N, Papashivili A G, Skvortosov V N, Konyushkin V A, Pukhov K K, Ermakov G A, Osiko V V, Prokhorov A M and Smith S (2002), 'Evaluation of rare-earth doped crystals and glasses for 4–5-μm lasing', *Laser Phys.*, 12, 859–877.

Brown D, Fletcher S and Holah D G (1968), 'The preparation and crystallographic properties of certain lanthanide and actinide tribromides and tribromidehexahydrates', *J. Chem. Soc. A*, 8, 1889–1894.

Carbonnier C, Tobben H and Unrau U B (1998), 'Room temperature CW fibre laser at 3.22 μm', *Electron. Lett.*, 34, 893–894.

Carnall W T, Fields P R and Rajnak K (1968), 'Electronic energy levels in the trivalent lanthanide aquo ions I. Pr^{3+}, Nd^{3+}, Pm^{3+}, Sm^{3+}, Dy^{3+}, Ho^{3+}, Er^{3+}, and Tm^{3+}', *J. Chem. Phys.*, 49, 4424–4442.

Carnall W T, Crosswhite H and Crosswhite H M (1978), 'Energy level structure and transition probabilities in the spectra of the trivalent lanthanides in LaF_3', Argonne National Laboratory Report, ANL-78-XX–95.

Choi Y G, Kim K H, Park B J and Heo J (2001), '1.6 μm emission from Pr^{3+}:$(^3F_3,^3F_4) \rightarrow {}^3H_4$ transition in Pr^{3+}- and Pr^{3+}/Er^{3+}-doped selenide glasses', *Appl. Phys. Lett.*, 78, 1249–1251.

Choi Y G, Curry R J, Park B J, Kim K H, Heo J and Hewak D W (2005), 'Extreme hypersensitivity observed from $^6H_{15/2} \rightarrow {}^6F_{11/2}$ transition of Dy^{3+} in inorganic noncrystalline solids', *Chem. Phys. Lett.*, 403, 29–34.

Chung W J and Heo J (2003), 'Spectroscopic properties and local structure of Eu^{3+} in Ge-Ga-S-CsBr (or CsCl) glasses', *J. Am. Ceram. Soc.*, 86, 286–290.

Chung W J, Jha A, Shen S and Joshi P (2004), 'The effect of Er^{3+} ion concentration on the Er^{3+}:$^4I_{13/2}$->$^4I_{15/2}$ transition in tellurite glasses', *Phil. Mag.*, 84, 1197–1207.

Chung W J, Seo H S, Park B J, Ahn J. T and Choi Y G (2005), 'Selenide glass optical fiber doped with Pr^{3+} for U-band optical amplifier', *ETRI J.*, 27, 411–417 .

Cole B, Shaw L B, Pureza P C, Mossadegh R, Sanghera J S and Aggarwal I D (1999), 'Rare-earth doped selenide glasses and fibers for active applications in the near and mid-IR', *J. Non-Cryst. Solids*, 256–257, 253–259.

Heo J (1995), 'Optical characteristics of rare-earth-doped sulphide glasses', *J. Mat. Sci. Lett.*, 14, 1014–1016.

Heo J (2003), 'Rare-earth doped chalcogenide glasses for fiber-optic amplifiers', *J. Non-Cryst. Solids.*, 326–327, 410–415.

Heo J and Shin Y B (1996), 'Absorption and mid-infrared emission spectroscopy of Dy^{3+} in Ge-As(or Ga)-S glasses', *J. Non-Cryst. Solids*, 196, 162–167.

Heo J, Song J H and Choi Y G (2006), 'EXAFS investigation on the local environment of rare earth ions in Ge-Ga-S-CsBr glasses', *Phys. Chem. Glasses: Eur. J. Glass Sci. Technol. B*, 47, 101–104.

Jackson S D (2003), 'Continuous wave 2.9 μm dysprosium-doped fluoride fiber laser', *Appl. Phys. Lett.*, 83, 1316–1318.

Jackson S D (2004), 'Single-transverse-mode 2.5-W holmium-doped fluoride fiber laser operating at 2.86 μm', *Opt. Lett.*, 29, 334–336.

Julien C, Barnier S, Massot M, Chbani N, Cai X, Loireau-Lozac'h A M and Guittard M (1994), 'Raman and infrared spectroscopic studies of Ge-Ga-Ag sulphide glasses', *Mater. Sci. Eng. B*, 22, 191–200.

Kanamori T, Terunuma Y, Takahashi S and Miyashita T (1984), 'Chalcogenide glass fibers for mid-infrared transmission', *J. Lightwave Tech.*, LT-2, 607–613.

Kokorima V F (1996), *Glasses for Infrared Optics*, CRC Press, Boca Raton, FL.

Komukai T, Yamamoto T, Sugawa T and MiyajimaY (1995), 'Upconversion pumped thulium-doped fluoride fiber amplifier and laser operating at 1.47 μm', *IEEE. J. Quan. Elec.*, 31,1880–1889.

Kumar G A, Riman R, Chae S C, Jang Y N, Bae I K and Moon H S (2004), 'Synthesis and spectroscopic characterization of $CaF_2:Er^{3+}$ single crystal for highly efficient 1.53 μm amplification', *J. Appl. Phys.*, 95, 3243–3249.

Layne C B, Lowdermilk W H and Weber M J (1977), 'Multiphonon relaxation of rare-earth ions in oxide glasses', *Phys. Rev. B*, 16, 10–21.

Lee T H and Heo J (2005), '1.6 μm emission and gain properties of Ho^{3+} in selenide and chalcohalide glasses', *J. Appl. Phys.*, 98, 113510.

Lee T H and Heo J (2006), 'Energy transfer processes and Ho^{3+}:$^5I_5$ level population dynamics in chalcohalide glasses', *Phys. Rev. B*, 73, 144201.

Lee T H, Heo J, Choi Y G, Park B J and Chung W J (2004a), 'Emission properties of Ho^{3+}/Tb^{3+} co-doped in $Ge_{30}Ga_2As_8S_{60}$ glass', *J. Appl. Phys.*, 96, 4827–4832.

Lee T H, Heo J, Choi Y G, Park B J and Kim K H (2004b), 'Optimized combination of Ho^{3+} and sulfide glass for U-band fiber-optic amplifiers', *Chem. Phys. Lett.*, 384, 16–19.

Li H, Sundaram S K, Blanc-Pattison P A and Li L (2002), 'Spectroscopic study of neodymium(III) in sodium tellurite glass', *J. Am. Ceram. Soc.*, 85, 1377–1382.

Linden K J (2004), 'Fiber laser with 1.2-W CW-output power at 2712 nm', *IEEE Photon. Tech. Lett.*, 16, 401–403.

Loireau-Lozac'h A M, Keller-Besrest F and Bénazeth S (1996), 'Short and medium range order in Ga–Ge–S glasses: an X-ray absorption spectroscopy study at room and low temperatures', *J. Solid State Chem.*, 123, 60–67.

Lucas P, Riley M R, Boussard-Pledel C and Bureau B (2006), 'Advances in chalcogenide fibers evanescent-wave biochemical sensing', *Anal. Biochem.* 351, 1–10.

Lucovsky G, Galeener F L, Keezer R C, Geils R H and Six H A (1974), 'Structural interpretation of the infrared and Raman spectra of glasses in the alloy system $Ge_{1-x}S_x$', *Phys. Rev. B*, 10, 5134–5146.

Miyakawa T and Dexter D L (1970), 'Phonon sidebands, multiphonon relaxation of excited states, and phonon-assisted energy transfer between ions in solids', *Phys. Rev. B*, 1, 2961–2969.

Naftaly M, Shen S, and Jha A (2000), 'Tm^{3+}-doped tellurite glass for a broadband amplifier at 1.47 μm', *Appl. Optics*, 39, 4979–4984.

Noginov M A (1997), 'Reabsorption trapping of luminescence in laser crystals: enhancement of energy storage and upconversion', *Appl. Opt.*, 36, 4153–4158.

Özen G and DiBartolo B (2000), 'Energy transfer and thermalization in $LiYF_4$: Tm, Ho', *Appl Phys. B*, 70, 189–193.

Park B J, Seo H S, Ahn J T, Choi Y G, Heo J and Chung W J (2008a), 'Dy^{3+} doped Ge-Ga-Sb-Se glases and optical fibers for the mid-IR gain media', *J. Ceram. Soc. Jpn.*, 116, 1087–1091.

Park B J, Seo H S, Ahn J T, Choi Y G, Jeon D Y and Chung W J (2008b), 'Mid-infrared (3.5–5.5 μm) spectroscopic properties of Pr^{3+}- doped Ge-Ga-Sb-Se glasses and optical fibers', *J. Lumin.*, 128, 1617–1622.

Park S H, Lee D C, Heo J and Kim H S (2000), 'Pr^{3+}/Er^{3+} co-doped Ge-As-Ga-S glasses as dual-wavelength fiber optic amplifiers at 1.3 and 1.55 μm window', *J. Am. Ceram. Soc.*, 83, 1284–1286.

Reisfeld R and Jørgensen CK (1987), in *Handbook on the physics and chemistry of rare earths*, vol. 9, edited by Gschneider K A Jr., and Eyring L, Elsevier Science Publishers, Amsterdam.

Sakamoto T, Shimizu M, Kanamori T, Terunuma Y, Ohishi Y, Yamada M and Sudo S (1995), '1.4- μm-band gain characteristics of a Tm-Ho-doped ZBLYAN fiber amplifier pumped in the 0.8- μm band', *IEEE. Photo. Tech. Lett.*, 7, 983–985.

Sanghera J S and Aggarwal I D (1999), 'Active and passive chalcogenide glass optical fibers for IR applications: a review', *J. Non-Cryst. Solids*, 256–257, 6–16.

Savage J A (1987), 'Materials for infrared fiber optics', *Mater. Sci. Reports*, 2, 99–138.

Schaafsma D T, Shaw L B, Cole B, Sanghera J S and Aggarwal I D (1998), 'Modeling of Dy^{3+}-doped GeAsSe glass 1.3-μm optical fiber amplifiers', *IEEE Photon. Technol. Lett.*, 10, 1548–1550.

Schneider J, Carbonnier C and Unrau U B (1997), 'Characterization of a Ho^{3+}-doped fluoride fiber laser with a 3.9- μm emission wavelength', *Appl. Opt.*, 36, 8595–8600.

Schweizer T, Samson B N, Hector J R, Brocklesby W S, Hewak D W and Payne D N (1999), 'Infrared emission from holmium doped gallium lanthanum sulphide glasses', *Infrared Phys. & Tech.*, 40, 329–335.

Seddon A B (1995), 'Chalcogenide glasses: a review of their preparation, properties and applications', *J. Non-Cryst., Soldis*, 184, 44–50.

Seddon A B, Tang Z, Furniss D, Sujecki S and Benson T M (2010), 'Progress in rare-earth-doped mid-infrared fiber lasers', *Opt. Exp.*, 18, 26704–26719.

Shaw L B, Harbison B B, Cole B, Sanghera J S and Aggarwal I D (1997), 'Spectroscopy of the IR transitions in Pr^{3+} doped heavy metal selenide glasses', *Opt. Exp.*, 1, 87–96.

Shaw L B, Cole B, Thielen P A, Sanghera J S and Aggarwal I D (2001), 'Mid-wave IR and long-wave IR laser potential of rare-earth doped chalcogenide glass fiber', *IEEE J. Quantum Electron.*, 48, 1127–1137.

Shin Y B, Jang J N and Heo J (1995), 'Mid-infrared light emission characteristics of Ho^{3+}-doped chalcogenide glass and heavy-metal oxide glasses', *Opt. Quan. Electro.*, 27, 379–386.

Shin Y B, Heo J and Kim H S (2000a), 'Modification of the local phonon mode and electron-phonon coupling strengths in Dy^{3+} doped sulfide glasses for efficient 1.3 μm amplification', *Chem. Phys. Lett.*, 317, 637–641.

Shin Y B, Kim J H, Kim Y S and Heo J (2000b), 'Effect of Tb^{3+}co-doping on the electron population densities of Tm^{3+} in Ge–As–Ga–S glasses', *J. Appl. Phys.*, 88, 2515–2518.

Shin Y B, Heo J and Kim H S (2001), 'Enhancement of the 1.3- μm emission properties of Dy^{3+} doped Ge-Ga-S glasses with the addition of alkali halides', *J. Mat. Res.*, 16, 1318–1324.

Shin Y B, Yang C K and Heo J (2002), 'Optimization of Dy^{3+}-doped Ge-Ga-As-S-CsBr glass composition and its 1.31- μm emission properties', *J. Non-Cryst. Solids*, 298, 153–159.

Simons D R, Faber A J and de Waal H (1995), 'Pr^{3+}-doped GeS_x-based glasses for fiber amplifier at 1.3μm', *Opt. Let.*, 20, 468–470.

Smirnov P, Wakita H, Nomura M and Yamaguchi T (2004), 'Structure of aqueous gallium(III) bromide solutions over a temperature range 80–333 K by Raman spectroscopy, X-ray absorption fine structure, and X-ray diffraction', *J. Solution Chem.*, 33, 903–922.

Song J H, Heo J and Park S H (2005), '1.48-μm emission properties and energy transfer between Tm^{3+} and Ho^{3+}/Tb^{3+} in Ge-Ga-As-S-CsBr glasses', *J. Appl. Phys.*, 97, 083542.

Song J H, Choi Y G, Kadono K, Fukumi K, Kageyama H and Heo J (2007a), 'Emission properties and local structure of Tm^{3+} in Ge-Ga-S-Br glass', *J. Non-Cryst. Solids*, 353, 1676–1680.

Song J H, Choi Y G, Kadono K, Fukumi K, Kageyama H and Heo J (2007b), 'EXAFS investigation on the structural environment of Tm^{3+} in Ge-Ga-S-CsBr glasses', *J. Non-Crys. Solids*, 353, 1251–1254.

Sumiyoshi T, Sekita H, Arai T, Sato S, Ishihara M and Kikuchi M (1999), 'High-power continuous-wave 3- and 2-μm cascade ho^{3+}:zblan fiber laser and its medical applications', *IEEE J. Select. Topics in Quantum Electron.*, 5, 936–943.

Tanabe S, Ohyage T, Soga N and Hanada T (1992), 'Compositional dependence of Judd-Ofelt parameters of Er^{3+} ions in alkali-metal borate glasses', *Phys. Rev. B*, 46, 3305–3310.

Tanabe S, Hanada T, Watanabe M, Hayashi T and Soga N (1995), 'Optical properties of dysprosium-doped low-phonon-energy glasses for a potential 1.3 μm optical amplifier', *J. Am. Ceram. Soc.*, 78, 2917–2922.

Tittel F K, Richter D and Fried A (2003), 'Mid-infrared laser applications in spectroscopy', *Topics Appl. Phys.*, 89, 445–516.

Többen H (1992), 'Room temperature CW fibre laser at 3.5 μm in Er^{3+}-doped ZBLAN glass', *Electro. Lett.*, 28, 1361–1362.

Tverjanovich A, Tverjanovich Yu S and Loheider S (1996), 'Raman spectra of Gallium sulfide based glasses', *J. Non-Cryst. Solids*, 208, 49–55.

Weber M J, Ziegler D C and Angell C A (1982), 'Tailoring stimulated emission cross sections of Nd^{3+} laser glass: observation of large cross sections for $BiCl_3$ glasses', *J. Appl. Phys.*, 53, 4344–4350.

Wei K, Machewirth D P, Wenzel J, Snitzer E and Sigel G H Jr. (1994), 'Spectroscopy of Dy^{3+} in Ge-Ga-S glass and its suitability for 1.3- μm fiber-optical amplifier applications', *Opt. Lett.*, 19, 904–906.

Wei K, Machewirth D P, Wenzel J, Snitzer E and Sigel G H Jr. (1995), 'Pr^{3+}-doped Ge-Ga-S glasses for 1.3 μm optical fiber amplifiers', *J. Non-Cryst. Solids*, 182, 257–261.

Werle P (1998), 'A review of recent advances in semiconductor laser based gas monitors', *Spectrochimica Acta A*, 54, 197–236.

Werle P and Popov A (1999), 'Application of antimonide lasers for gas sensing in the 3–4- μm Range', *Appl. Opt.*, 38, 1494–1501.

Werle P, Slemr F, Maurer K, Kormann R, Mucke R and Janker B (2002), 'Near- and mid-infrared laser-optical sensors for gas analysis', *Opt. Lasers Eng.*, 37, 101–114.

West Y D, Schweizer T, Brady D J and Hewak D W (2000), 'Gallium lanthanum sulphide fibers for infrared transmission', *Fiber and Integrated Opt.*, 19, 229–250.

White J G, Yocom P N and Lerner S (1967), 'Structure determination and crystal preparation of monoclinic rare earth sesquisulfides', *Inorg. Chem.*, 6, 1872–1875.

Yang Z, Chen W and Luo L (2005), 'Dy^{3+}-doped selenide glasses for 1.3-mum optical fiber amplifiers', *J. Mater. Res.*, 20, 2597–2602.

Zakery A and Elliott S R (2003), 'Optical properties and applications of chalcogenide glasses: a review', *J. Non-Cryst. Solids*, 330, 1–12.

12
Chalcogenide waveguides for infrared sensing

C. BOUSSARD-PLÉDEL, University of Rennes, France

DOI: 10.1533/9780857093561.2.381

Abstract: This chapter deals with two types of chalcogenide glass fibers developed for infrared sensing. First, infrared chalcogenide single-index fibers, which are developed for fiber evanescent wave spectroscopy (FEWS). Many studies demonstrate the advantage of this technique for the detection, in real time, of chemical and biological molecules in the 3–12 μm region. Applications extend from the *in situ* follow-up of chemical reactions to the detection of pollutants for environmental purposes or of metabolism alterations in the medical domain. Second, single-mode chalcogenide fibers, which are developed for the detection of molecules such as water, ozone and carbon dioxide, attesting the presence of biological life on telluric earth-like exoplanets.

Key words: chalcogenide glasses, infrared fibers, evanescent wave spectroscopy, detection.

12.1 Introduction

Chalcogenide glasses have been revealed as good candidates for the elaboration of specialty fibers to be implemented in optical sensors. In particular, they are suitable for sensors based on an original spectroscopic method called fiber evanescent wave spectrocopy (FEWS). The infrared window of chalcogenide glasses extends from 2 to 25 μm, according to their composition, which corresponds to the vibration domain of most chemical and biological molecules; in addition, their thermo-mechanical properties permit the elaboration of fibers by drawing at low temperatures, as compared to silica. Moreover, a large range of glass formulations is available, which allows the optimization of fiber properties as a function of experimental conditions for FEWS measurement.

Chemical detection using chalcogenide glass fibers was initially reported in the late 1980s with the characterization of butanone (Compton *et al.*, 1988). Chemical analyses were then performed on acetone, ethanol, and sulfuric acid using Ge-Te-Se fibers (Heo *et al.*, 1991; Rodriguez and Sigel, 1991). It was shown that the detection-signal versus concentration obeys a linear Beer–Lambert law, allowing quantitative analysis of soluble species. A wider range of organic species, including carcinogens such as benzene, toluene, and trichloroethylene, was later detected (Sanghera *et al.*, 1994, 1995). In parallel, AgCl/AgBr polycrystalline fibers have also been developed as sensors (Eytan *et al.*, 1999; Spielvogel *et al.*, 1998, 1999, Katz *et al.*, 1994). They possess the required optical quality and transmit light up to 20 μm in the infrared range. Furthermore, from a mechanical point of view, they are flexible and easily handled. However, polycrystalline fibers are very sensitive to air contamination losing their properties of transparency. Moreover, they are obtained by extrusion methods, which are costly and difficult to implement. Finally, their sensitivity to detection is affected by their large diameter, around 1 mm.

More recently, chalcogenide fiber sensors have been successfully used and validated, mainly for environmental and biomedical purposes, in the framework of multidisciplinary research programs. Single index fibers have been developed, for spatial research programs, with the aim of detecting signs of life on earth-like planets.

12.2 Fiber evanescent wave spectroscopy

The elaboration of low-loss infrared optical fibers has permitted the development of a special spectroscopy called fiber evanescent wave spectroscopy (FEWS). In an optical fiber, the beam propagates by total internal reflection at the waveguide/sample interface. The interference between the incident and the reflected waves gives rise to standing waves inside the waveguide, perpendicular to the interface, and to the generation of an evanescent field inside the sample. The evanescent wave decreases exponentially as it propagates into the sample, away from the fiber. If the evanescent wave is absorbed partially by the sample at specific wavelengths, the reflectivity is lower than 1. This principle is called attenuated total reflection (ATR), and it is the basis for evanescent wave (EW) spectroscopy.

The advantage of this technique is to perform remote, real-time and *in situ* analysis. The system is composed of a Fourier transform infrared spectrometer. The beam, produced by a black body source, is focused at the input of the fiber by two parabolic mirrors. At the output of the fiber, the signal is again focused by two parabolic mirrors on the sensitive part of an MCT detector cooled by liquid nitrogen (Fig. 12.1). To avoid achromatism, lenses could be used to focus the infrared light. First, a reference is recorded and then the sample is placed in contact with the fiber and the signal is collected on the detector. The transmission spectrum is obtained by dividing the two signals as in classical transmission spectroscopy. In the same way, the absorbance spectrum is obtained by using Eq. [12.1]:

$$A = \log\left(\frac{I_{\text{ref}}}{I_{\text{s}}}\right) \qquad [12.1]$$

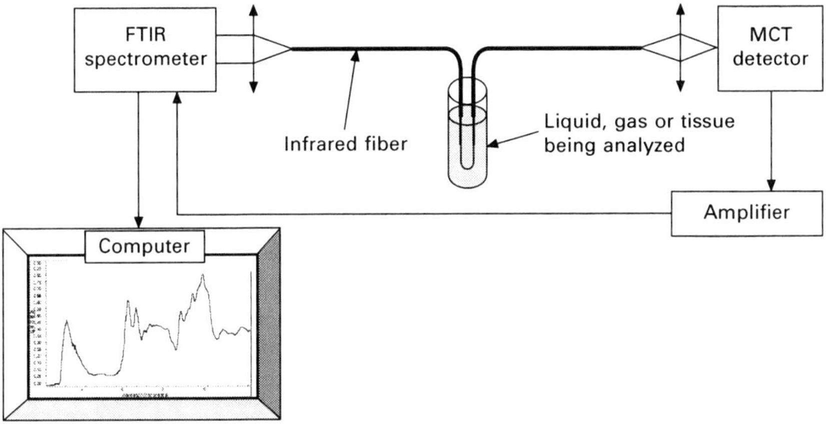

12.1 Principle of fiber evanescent wave specroscopy (FEWS). The light from the source is focused at the input of the chalcogenide optical fiber and collected on a MCT detector at the fiber output. The signal of the sample being analyzed is collected by putting it in contact with the fiber.

where I_{ref} is the intensity when the fiber is in air and I_s when the fiber is in contact with the sample being analyzed.

The FEWS method is quite simple to implement since the measurement necessitates only a standard spectrometer equipped with special kits for focusing the light and a MCT detector. The know-how consists in the control of glass compositions and of optical fiber engineering to obtain suitable properties for FEWS applications, that is transparency in a broad infrared wavelength range and low optical losses.

12.3 Fabrication of the fiber sensor

12.3.1 Selection of glass composition

The glass science and industry are widely dominated by oxides-based materials such as SiO_2, P_2O_5 or B_2O_3. These oxide glasses show an excellent chemical durability and, thanks to their high glass transition temperature, often above 500°C, good thermomechanical properties. On the other hand, they present a major inconvenience due to their strong chemical bonds and the weak atomic weight of their atoms, exhibiting high phonon energies and a transmission limited to the near-infrared domain. Beyond this spectral region, these glasses are opaque and cannot be used for searching and processing data. However, the mid-infrared spectral range contains the signature of most chemical and biological molecules through the fundamental vibration mode of their functional groups. For applications in mid-infrared spectroscopy, the requirements, concerning the choice of the composition, are: a large transmission window in the infrared, thermomechanical properties allowing the elaboration of optical

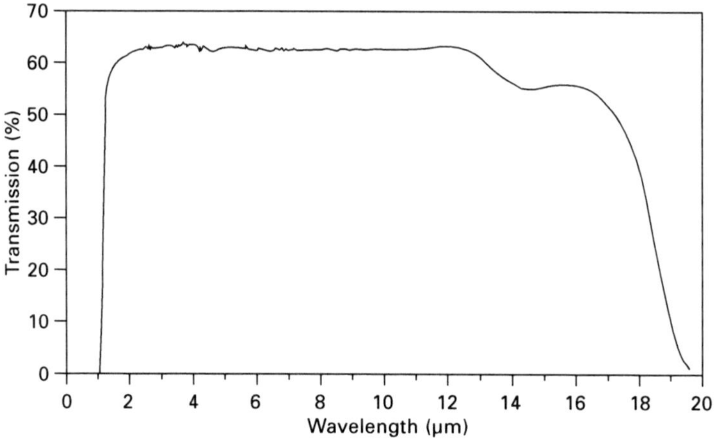

12.2 Optical transmission of the TAS glass (thickness 2 mm).

fibers, a glass transition temperature high enough to permit manipulations at room temperature, and a good chemical durability.

In the infrared, the optical transmission is limited by the chemical bond vibrations of the glassy network. Thus, the heavier the chemical elements are, the lower are the vibration frequencies of the chemical bonds. Glasses containing heavy elements possess a broad transparency window in the infrared. Tellurium-based glasses are the glasses that present the lowest absorption in the infrared, followed by sulphur and selenium-based glasses. This results in a red-shift of optical transmission in the S–Se–Te series. However, glasses containing high amounts of tellurium, more than 70 mol%, show a high tendency to crystallize and are difficult to shape into optical fibers. So, it is essential to add large amounts of selenium in Te-based glass compositions to achieve stability of the glassy state. The best compromise is the $Te_2As_3Se_5$ glass composition, referred to as TAS glass. The structure of this glass is constituted of Se and Te chains rigidified by the introduction of trivalent arsenic atoms that reticulate the chains. Thus, this glass is very stable thanks to its strong covalent bonds, and its thermomechanical properties allow an easy shaping into fiber (T_g = 137°C). Then, the transparency of glass disks, a few mm thick, extends from 2 to 16 μm (Fig. 12.2). Important steps of chemical purification are necessary to improve the transparency of this glass by reducing the presence of impurities such as oxygen, hydrogen, carbon, and silicon. This broad and optimized transparency is definitely needed to achieve the highest sensitivity with the sensor.

12.3.2 Elaboration of the fiber

Chalcogenide glass fibers are elaborated on special drawing towers, as shown in Fig. 12.3. Silica optical fibers used for telecommunication applications

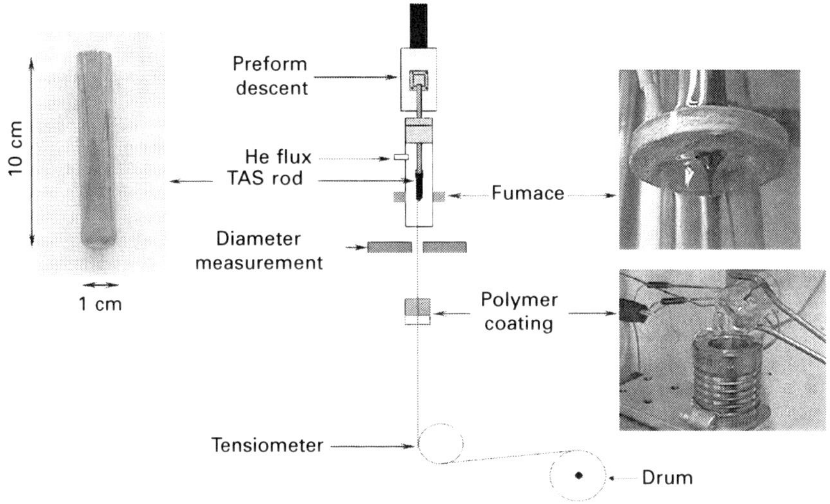

12.3 Fibering tower of the Glass and Ceramics Laboratory. From a TAS rod 10 cm long and 1 cm diameter, 20 meters of fiber 400 μm are obtained.

have a core/clad structure, with light propagating by total reflection in the core of the fiber whose refractive index is higher than that of the cladding. Infrared TAS fibers used for FEWS experiments are single-index fibers. In that case, light is guided by internal reflections at the interface between the glass, having a high refractive index (n = 2.8), and the air. Single-index fibers are obtained from single-glass rods. The rods are introduced in the drawing chamber equipped with a furnace having a small heating zone, which permits a local heating of the glass. Then, the rod is heated at nearly 300°C, at which temperature the viscosity is low enough to allow drawing. The fiber diameter is controlled during the entire fibering process and can be modified by adjusting the rotation of the drum or the preform descent speed. Typically, the fiber diameter is around 400 μm.

12.4 Characterization and optimization of the sensor

12.4.1 Fiber guiding properties

As shown in Fig. 12.1, the sensor consists in a unique single-index optical fiber, which has two functions: transporting the infrared signal from the source to the detector and detecting a sample by simple immersion or contact. Figure 12.4 shows the typical attenuation curve of TAS optical fiber. The attenuation is below 1 dB/m from 6 to 9 μm. This value, far from the attenuation of silica fibers (10^{-2} dB/Km at 1.55 μm), is low enough to permit short distance applications in a wavelength range extending from 2

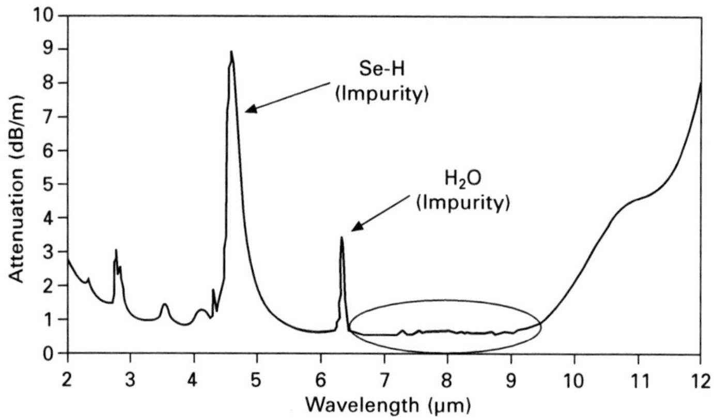

12.4 Attenuation curve of a TAS glass optical fiber between 2 and 12 μm; the minimum of attenuation is lower than 1 dB/m between 6.5 and 9 μm.

to 12 μm. With a 2-m long fiber, it is possible to record infrared spectra from 850 cm^{-1} to 5,000 cm^{-1}.

The number of guided modes M is calculated by the following formula:

$$M(\lambda) = \frac{2\pi^2 r^2 (n_2^2 - n_1^2)}{\lambda^2} \qquad [12.2]$$

where r is the fiber radius, n_2 is the glass refractive index, and n_1 the air refractive index. With a fiber diameter of about 400 μm and a difference between the refractive index of the glass ($n_2 = 2.8$) and the refractive index of the air ($n_1 = 1$), the fiber is multimode: from 37,500 modes at 12 μm (833 cm^{-1}) to 1,350,000 modes at 2 μm (5,000 cm^{-1}). In this case, and contrary to double-index or single-mode silica fibers, it is possible to model light propagation in the fiber by a classical model based on optical rays. The reflection coefficient R at the glass/sample interface is then a function of the incident angle θ, the wavelength λ, and the refractive indices n_1 and n_2. For non-absorbing substances like air, n_1 is a real number, and the reflection is total for angles superior to the critical angle. On the other hand, when an absorbing sample is in contact with the fiber, its refractive index is a complex number modifying the conditions of each reflection, which are no longer total. This is the attenuated total reflection, ATR, a well-known acronym in infrared spectroscopy applied to planar sensors, as demonstrated earlier with a ZnSe plate directly positioned in the sample compartment of a spectrometer (Harrick, 1979). It has been shown that for these highly multimode fibers, the experimental FEWS spectra can be recalculated using the ATR formalism without explicitly referring to the notion of evanescent wave or penetration depth (MacDonald and Bureau, 2003; MacDonald *et*

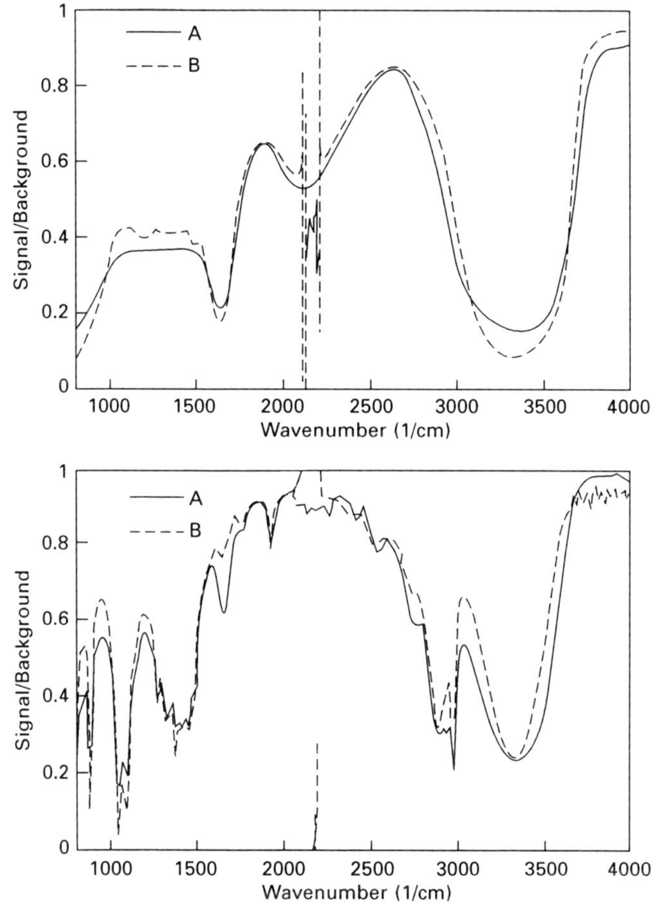

12.5 FEWS calculated (A) and experimental (B) spectra of (a) water, (b) ethanol, and (c) a 50% water – 50% ethanol mixture.

al., 2004). As an example, Fig. 12.5 shows the experimental and calculated FEWS spectra for water-alcohol mixture.

12.4.2 Penetration depth

The evanescent wave intensity decays exponentially with distance, z, from the surface of the fiber according to Eq. [12.3], where γ is the attenuation coefficient:

$$E = E_0 e^{-z\gamma} \tag{12.3}$$

The wave intensity is indeed mostly localized within 1 μm from the fiber surface. The penetration depth, d_p, is a function of the glass index as well as

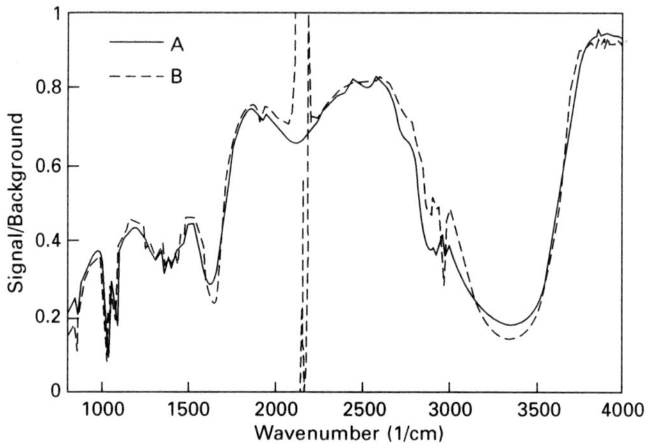

12.5 Continued

of the wavelength of the propagating light, according to Eq. [12.4], where λ is the wavelength, n_2 and n_1 are the refractive indices of the glass and surrounding, respectively, and θ_i is the angle of incidence of the wave in the fiber (Harrick, 1967; De Fornel, 1998):

$$d_p = \frac{\lambda}{2\pi\sqrt{n_2^2 \sin^2\theta_i - n_1^2}} \qquad [12.4]$$

The d_p parameter does not strictly represent the depth probed during the FEWS experiment but is rather defined as the depth at which the electric field drops to $1/e$ of its initial value at the surface. The wavelength dependence of d_p can be estimated, as shown in Fig. 12.6. The penetration of the evanescent wave increases linearly with the wavelength. This wavelength dependence must be considered when comparing transmission and FEWS spectra. The spectra collected in evanescent mode show typically lower intensities at shorter wavelengths in comparison with those of transmission spectra.

The high refractive index of chalcogenide glasses is an advantage because it increases both the penetration depth and the number of bound propagation modes (light ray trajectories) in the fiber, thereby increasing sensitivity.

12.4.3 Influence of fiber diameter on sensitivity

It has also been shown, Fig. 12.7(a), that the number of reflections at the glass/air interface, depends on the fiber diameter according to the formula:

$$N(\theta, d, L) = L * \frac{\tan(90 - \theta)}{d} \qquad [12.5]$$

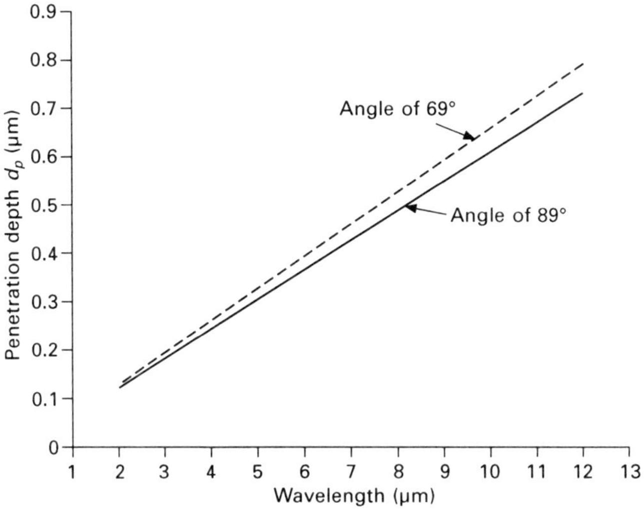

12.6 Variation in penetration depth (d_p) calculated for the two critical angles of propagation in a chalcogenide glass of index n_2 = 2.8 surrounded by air of index n_1 = 1.

where L is the contact length between the fiber and the sample, θ is the incident angle, and d is the fiber diameter in the sensing zone.

Clearly, the sensitivity of the sensor is increased when the fiber diameter is reduced (Le Foulgoc *et al.*, 1994). This diameter reduction can be operated by two routes (Hocdé *et al.*, 2000). The first one consists in increasing the drum speed for a fraction of a second during the fiber process (Bureau *et al.*, 2004) and the second one is an etching process with an acidic chemical solution, H_2O_2/H_2SO_4 (Le Coq *et al.*, 2001). With these two methods, the diameter could locally be reduced from 400 to 100 μm in the sensing zone in contact with the sample. The effect of diameter reduction on sensing efficiency has been studied by recording FEWS spectra of chloroform with different fiber diameters extending from 200 to 50 μm. In that case, the diameter was controlled by immersion of the fiber in the oxidizing solution as a function of time. These results, displayed in Fig. 12.7(b), show clearly that the sensitivity increases when the diameter decreases. Finally, in many studies, the fibers used for FEWS experiments show the following profile: 400 μm diameter in the transportation zone between the source of the spectrometer and the sample, 100 μm in the sensing zone where the sample is in contact with the fiber, and 400 μm, again, in the transportation zone between the sample and the detector. This possibility to reduce the fiber diameter of the TAS fibers is an asset, making these fibers a unique tool for FEWS spectroscopy.

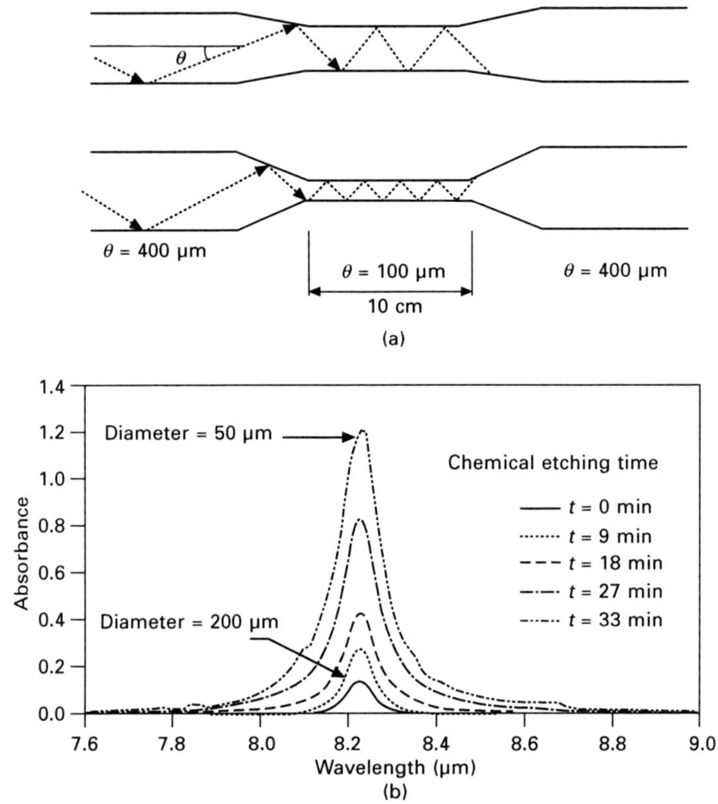

12.7 (a) Increasing the number of reflections in the fiber when the fiber diameter decreases. (b) Increasing the intensity for the chloroform absorption band at 1050 cm^{-1} versus the immersion time of the fiber in the oxidizing solution. The diameter decreases gradually from 200 to 50 μm.

12.4.4 Influence of concentration

In the transmission mode, the Beer–Lambert law is the linear relationship between absorbance and concentration of an absorbing substance. The general Beer–Lambert law is usually written as:

$$A(\lambda) = \varepsilon(\lambda).l.c \qquad [12.6]$$

where A is the measured absorbance, ε is the wavelength-dependent molar absorption coefficient, l is the path length, and c is the analyte concentration. For FEWS, a pseudo-Beer–Lambert law can be applied to qualify the sensitivity of the fiber. In this case, l is the length of the fiber in contact with the sample being analyzed. In 1991, Heo *et al.*, used Ge-Se-Te-based fibers to show that a linear relationship exists between the absorption by

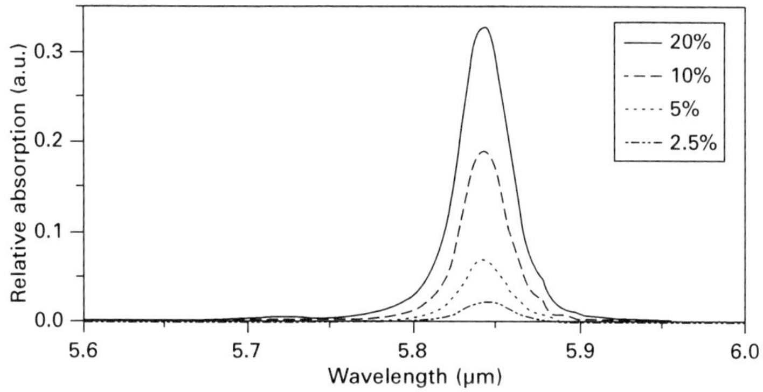

12.8 Detection of C=O band of acetone at different volumic percentages in methylene chloride using fiberoptic evanescent wave spectroscopy (FEWS).

the evanescent field and the concentrations of various analytes in acetone/water, sulphuric/water, ethanol/water mixtures. Later, Boussard-Plédel *et al.* (1999) and Hocdé *et al.* (2000) tested acetone/methylene chloride mixtures with acetone concentration varying from 20% to 10%, 5% and 2.5%. It was shown that the C=O absorption band in the 5.8–5.9 μm wavelength region increases with the concentration of acetone in the solvent (Fig. 12.8).

Other works on alcohol/water mixtures have been carried out (Sanghera *et al.*, 1994; Le Foulgoc *et al.*, 1994, Le Coq *et al.*, 2003), focusing on increasing the C-OH absorption band at 9.55 μm with alcohol concentration in water. The lowest detection limit is estimated around 0.5% of alcohol in water with a sensing zone diameter equal to 50 μm and a length of contact equal to 3 cm.

12.4.5 Influence of the fiber length in contact with the analyte

The influence of the fiber contact-length on absorbance has been studied by Le Foulgoc *et al.* (1994) and Hocdé *et al.* (2001a). For example, the absorbance spectra of absolute ethanol in the 8–12 μm region have been recorded for various immersed fiber lengths (Fig. 12.9). The intensity of 9.20, 9.53 and 11.37 μm absorption bands, respectively attributed to C-C-O stretching mode, C-O vibration mode, and O-H angular distortion outside the C-C-O plan, increased as the contact length increased, showing that the sensitivity of the probe is directly proportional to the immersed fiber length, as expected.

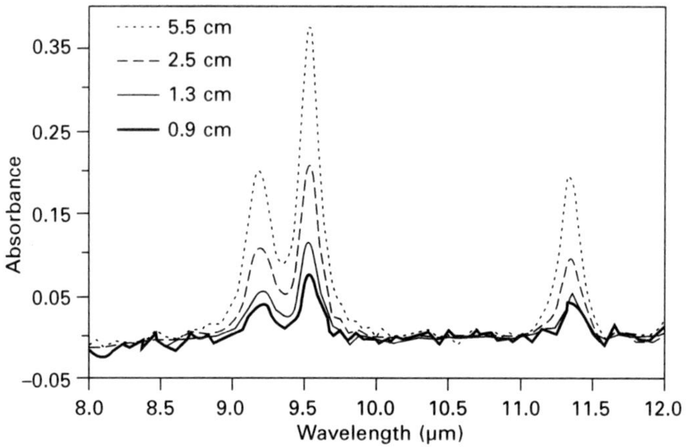

12.9 Absolute ethanol absorbance spectra for several lengths of detection cell.

12.4.6 Hydrophobicity

Deviation from the Beer–Lambert law has been observed by Michel *et al.* (2004) with ethanol/water mixtures, showing that the increase of the absorption band intensity at 1050 cm^{-1}, characteristic of alcohol, is not linear in the whole range of concentrations (Fig. 12.10). This result is attributed to the hydrophobic character of chalcogenide glasses, linked to the essentially covalent nature of chemical bonds in the glass fiber. Thus, surface bonds are softly polarized, which favours contact with non-polar targeted molecules, to the detriment of water molecules. This specific property of optical glass chalcogenide fibers is particularly important for applications in infrared spectroscopy, where spectra are often saturated with water fingerprints, which hide the vibration bands of the samples being studied. Nevertheless, the use of the Beer–Lambert law remains a good approximation for low concentrations in aqueous solution.

Another work (Lucas *et al.*, 2006) compared the water (1600 cm^{-1}) to ethanol (1050 cm^{-1}) peak ratio for ethanol-solution spectra collected by both chalcogenide FEWS and transmission spectroscopy. Again, the results showed an enhancement of the ethanol absorption peak when measured by the FEWS technique, as compared to transmission measurements.

But these results could also be explained by the nature of evanescent waves. According to Eq. [12.4], the penetration depth of evanescent waves is wavelength-dependent and, consequently, the absorbed intensity obeys the same dependence. Thus, with ATR technique, Harrick (1967) has demonstrated that absorption bands are enhanced at longer wavelengths (shorter wavenumbers).

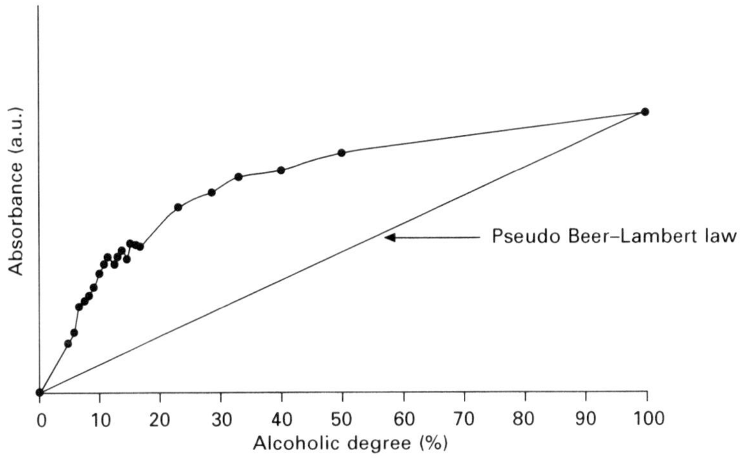

12.10 Evolution of the intensity of the alcohol absorption band at 1,050 cm^{-1} versus the percentage of alcohol in water. The straight line corresponds to an evolution following the pseudo-Beer–Lambert law. The hydrophobic character of the TAS glass shows the discrepancy to be in favor of alcohol and in detriment to water.

So, the effect of hydrophobicity could only be demonstrated for solvents showing an absorbing peak at short wavelength. This is the result obtained by Lucas *et al.* (2006) for acetonitirile, for which the peak at 2,260 cm^{-1} is enhanced.

12.5 Applications of the sensor

12.5.1 Monitoring of a chemical reaction

Remote analysis of fermentation processes

Zhang *et al.* (1994) and Le Coq *et al.* (2002) have tested fiber evanescent wave spectroscopy in the mid-infrared to follow up manufacturing processes. This technique offers a real assistance to improve the quality of the products by a better knowledge of their compositions. For example, during the alcoholic fermentation of grape juice, it has been shown that the absorption peaks of fructose at 1,063 cm^{-1} (9.40 μm) and glucose at 1,033 cm^{-1} (9.68 μm) disappear progressively, while the absorption bands at 1,045 cm^{-1} (9.56 μm) and 1,085 cm^{-1} (9.21 μm), characteristics of ethanol, increase (Fig. 12.11). A similar investigation was conducted with lactic fermentation, showing a decrease of lactose absorption peaks between 1,000 and 1,100 cm^{-1} during the progressive transformation of lactose into lactic acid.

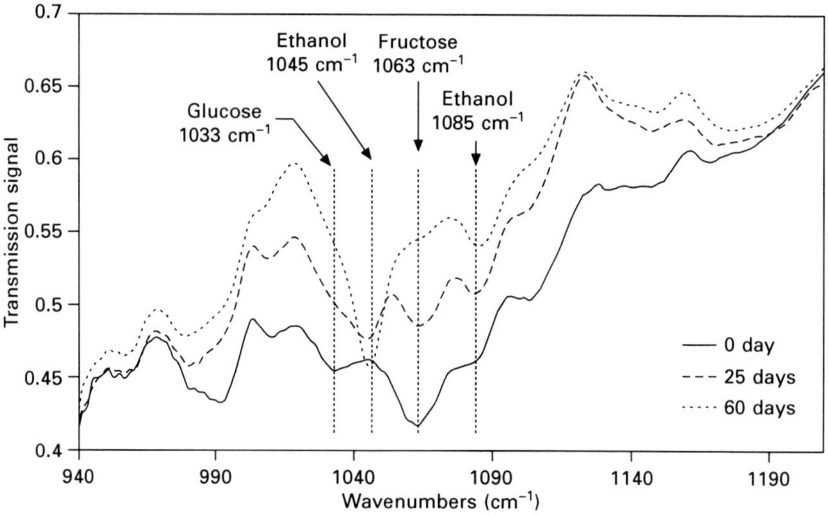

12.11 Evolution of the absorption band intensities during fermentation.

Monitoring of a chemical reaction in a microwave oven

Boussard-Plédel *et al.* (1999) has reported the monitoring of a chemical reaction inside a microwave oven. The same experimental set-up as described in Fig. 12.1 was used for this study, except that the FEWS probe was introduced in a microwave oven. A glass tube perforated by holes which permits liquid and gas circulation was used to support the fiber. A test experiment performed without reactants has verified that the fiber sensor was not altered by the microwave irradiations, the transmission of the fiber remaining unchanged. After this positive test, during an organic reaction, the evolution of C=O bands in the 1700–1750 cm^{-1} region was investigated as a function of time, under irradiation with 150 W microwave radiation. Figure 12.12 indicates a change in the infrared spectra of the different organic molecules when the time changes from 0 to 30 min for a constant temperature T = 70°C. So, it can be concluded that FEWS allows us to follow *in situ* the kinetics of a chemical reaction.

Monitoring of an industrial resin polymerization

In 2009, Anne *et al.* reported a study on the monitoring of polymerization of an industrial resin, by using a Sb-Se-Ga-Ge (2S2G) fiber (Anne *et al.*, 2009a). With a high T_g of 200°C, this fiber was suitable for the study of this curing process, where infrared spectra were recorded every 5 mn. A direct measurement of the integrated intensities of the relevant absorption bands, in

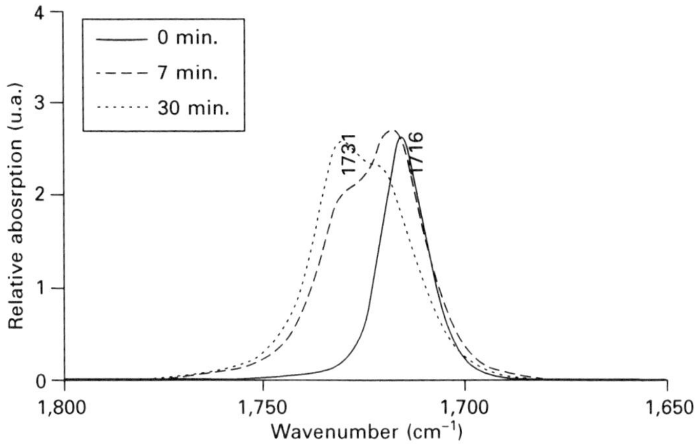

12.12 C=O band shift at different times during microwave assisted organic reaction. The kinetics is followed *in situ* by FEWS.

the 970–1030 cm^{-1} range, enabled the times when the curing process starts and when it is finished to be determined.

12.5.2 Environmental applications

Detection of pollutants in water

In the framework of a European project IMSIS (*In Situ* Monitoring of Landfill related Contaminants in Soil and Water by Infrared Sensing), the Glass and Ceramics Laboratory at the University of Rennes (Michel *et al.*, 2003, 2004) has studied the potential of infrared optical fiber sensors to detect pollutants in contaminated waters and in soils near dumps. For this application, the advantage of the fiber is to permit *in situ* and real-time analysis without sampling and/or addition of reagent. Furthermore, as every species have their own optical signatures, the technique permits a selective detection of the different polluting molecules. Several measurement campaigns have been carried out in different localities: in an artificial aquifer at the Technical University of Vienna, in a natural aquifer at the University of Munich, and on industrial wasteland near Paris. As an example, the first pilot tests were carried out in Vienna in an aquifer containing 1 m^3 water reserve filled with sand and gravel (Fig. 12.13). A water circulation allowed the pollutants to be introduced and the mixture to be homogenized. During the migration of the pollutants, infrared spectra were collected every 5 min and simultaneously samples were taken from the wells to perform later analytical measurements. The targeted molecule was the tetrachloroethylene C_2Cl_4. As the concentrations are very low, in the order of few ppm, the signal-to-noise ratio on the infrared spectra is very low too, but the absorption bands of the pollutant are clearly

12.13 Experimental set-up of the pilot scale tests exhibiting the TAS glass fiber connected to the FTIR spectrometer. The fiber dips in the tank thanks to a well.

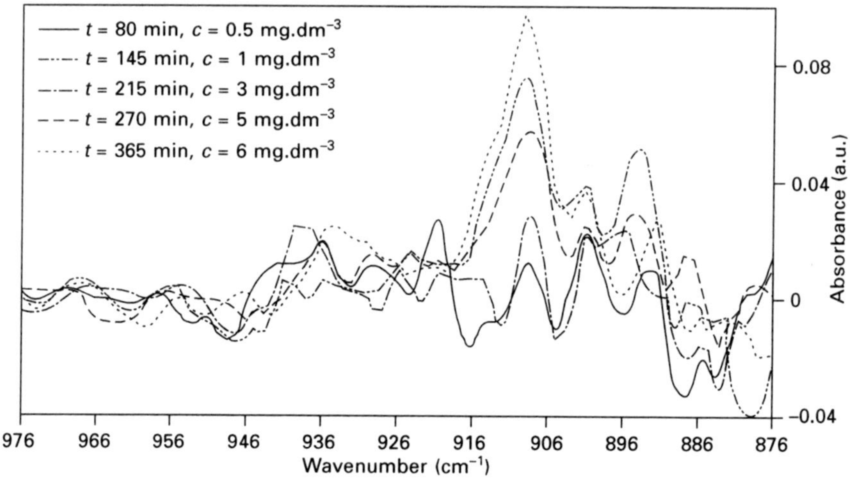

12.14 Evolution of the C-Cl absorption line of C_2Cl_4 at 910 cm^{-1} versus time for a length of immersion of the fiber equal to 20 cm.

present, zooming on the relevant spectral range (Fig. 12.14) and particularly the absorption band at 910 cm^{-1} due to C-Cl vibration. In parallel, the C_2Cl_4 concentrations were measured by chromatography. Figure 12.15 displays

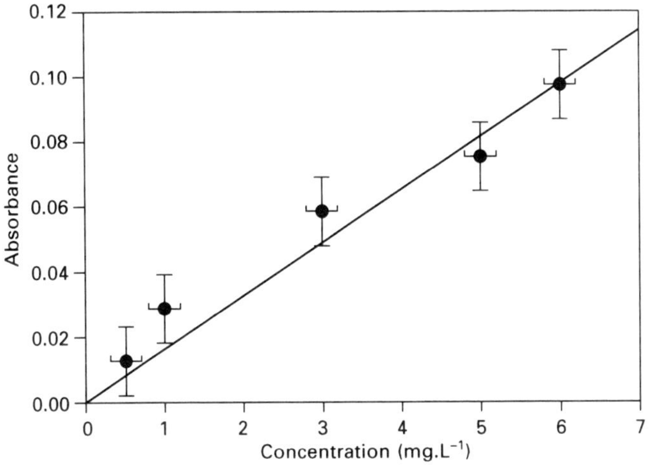

12.15 Evolution of the intensity of the line at 910 cm^{-1} versus the concentration of C_2Cl_4 in the tank.

the evolution of absorbance versus concentration, demonstrating that the pseudo-Beer–Lambert law could be applied.

Monitoring of carbon dioxide

The ability to detect and analyze CO_2 has become increasingly important in environmental processes. CO_2 is responsible for global warming, with increasing emissions every year. According to this evolution, a concentration as high as 1,000 ppm could be reached by the end of the twenty-first century. For comparison, today CO_2 concentration is around 390 ppm and some dramatic consequences for health could occur if the 1,000 ppm concentration is attained. So, some solutions must be found to reduce or control CO_2 emissions. One of them is the capture and the storage of CO_2 in natural underground geological formations, but this requires specific monitoring of the storage wells. The main infrared signature of carbon dioxide consists in a double absorption peak located at 4.26 μm. Charpentier (2009a) performed FEWS experiments using a tapered single-index As_2Se_3 fiber. The fiber was placed through a cylindrical gas cell and the reduced part of the fiber was in contact with the gas being analyzed (Fig. 12.16). Argon was used to purge the cell and then was mixed with the gas being analyzed for concentration-dependent experiments. Analyzed gases flow through the cell before being collected by a CO_2 analyzer that measures the exact concentration of carbon dioxide in the gas flux. Spectra were recorded for the following CO_2 concentrations: 0, 35, 57, 78, 91, and 100 vol.%. Intensity of the double absorption peak increases with the CO_2 concentration, as shown in Fig. 12.17. With this

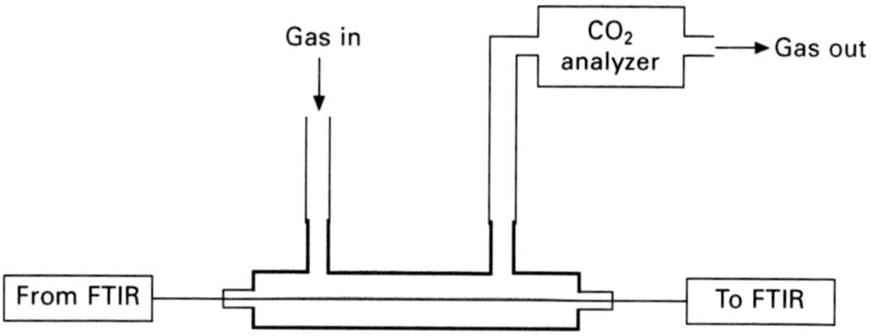

12.16 Schematic experimental set-up of CO_2 sensor using chalcogenide fiber.

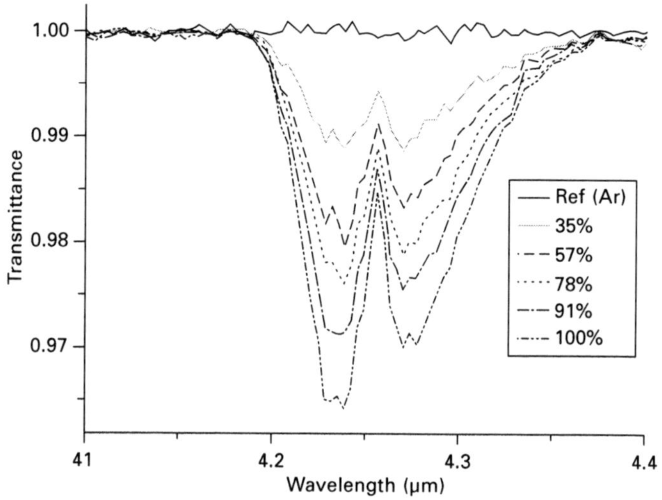

12.17 Transmittance spectra for tapered fiber as a function of CO_2 concentration (vol.%): 0, 35, 57, 78, 91, and 100.

technique, the detection limit is too high, around 20%, and the reversibility is not complete probably because of gas adsorption at the surface of the fiber. Better results were obtained with a set-up using two $GeSe_4$ chalcogenide fibers separated by a 4 mm spacing. The same experiments as above lead to a sensitivity threshold of 0.5 vol.% (Charpentier, 2009b), and the system was found totally reversible with response times and recovery times less than one minute. Furthermore, this system was used to perform CO_2 detection experiments on natural sites.

12.5.3 Biomedical applications

Advantage of FEWS for early medical diagnosis

The prognosis of numerous pathologies could be improved when the diagnosis is proactively realized. This requires non-invasive analysis methods which could be used repeatedly. The infrared spectroscopy is a well-adapted technique permitting the characterization of complex substances like proteins, nucleic acids, and lipids which are the main constituents of biological systems. So, it possesses aptitudes for characterizing precocious metabolic anomalies in different pathologic environments.

It is well established that the infrared fingerprint of biological molecules is located in the spectral region lying from 4,000 cm^{-1} to 800 cm^{-1} (2.5 μm–12.5 μm) (Bellamy, 1975; Cross and Jones, 1969; Fringeli and Günthard, 1981; Mendelsohn and Mantsch, 1986; Jackson and Mantsch, 1996; Naumann *et al.*, 1991), which corresponds exactly to the transparency range of TAS glass fibers. For example, C–H vibrations are located around 2,900 cm^{-1}, the amide I and II groups, characterized by the C=O and N–H bonds respectively, absorb in the 1,700 to 1,500 cm^{-1} region, while the P–O bonds have their IR signature in the 1,300 to 1,000 cm^{-1} spectral domain.

The advantage of optical fibers is, thanks to their length and flexibility, to permit the remote collection of IR signals; it could be implemented directly on patients by guiding the probe light onto the area of interest, rather than performing biopsies. Similarly, living organisms, cell cultures and tissue explants could be probed directly with an experimental setup designed to provide appropriate environmental conditions.

In the past, evanescent wave spectroscopy has been successfully applied to IR analysis of biological molecules by means of polycrystalline silver halide fibers (Kosower and Marom-Albeck, 1994; Kosower *et al.*, 1994). In putting into contact a small portion of the tapered fiber with living tissues, the detection of some IR characteristic bands was possible. In 1998, Brooks compared normal human skin tissue and acupuncture points of human skin tissues using polycrystalline optical fiber FTIR spectroscopy and pointed out differences between the relative intensities of the absorption bands of the two types of tissues (Brooks *et al.*, 1998). Spielvogel *et al.* (1998) used the FEWS based on polycrystalline fibers for studying infrared absorption in tumors and in neighboring healthy tissue. A significant and reproducible difference was observed between the ratio of the amide I at 1,642 cm^{-1} and amide II at 1,548 cm^{-1} between malignant and healthy tissues. This preliminary result allows us to envisage the opportunity for *in vitro* or *in vivo* analysis for cancer diagnosis.

In order to perform biomedical analysis using chalcogenide fibers, the biocompatibility of the fiber versus biological molecules must be proved. So, several tests have been conducted to demonstrate the inertia of the glass towards biological substances (Keirsse *et al.*, 2003a). It has been shown that no denaturation of proteins was induced after contact of a concentrated serum albumin solution with the fiber during 4 h. In the same way, no effect on the growth of living cells was observed by placing small fiber fragments in contact with a *P. mirabilis* bacteria biofilm.

Analysis of tissues

The infrared spectra of tissues taken from mouse liver have been investigated (Hocdé *et al.*, 2001b). The samples used for the studies are cryogenically sectioned slides of tissues having a thickness of about 10 μm. These slides of tissues were then deposited on the sensing zone of the IR fiber in order to provide a perfect contact between the liver tissues and the evanescent wave propagating at the surface of the fiber. Figure 12.18 shows the IR spectrum obtained using this procedure. The following absorption peaks are identified, from high to low frequencies: from 3,500 cm^{-1} to 3,000 cm^{-1} are the stretching bands of water, from 2,960 to 2,850 cm^{-1} are the bands due to CH_2 and CH_3 groups contained in phospholipids and protein components. The amides I and II absorptions dominate the spectrum respectively around 1,650 and 1,550 cm^{-1}. The amide I arises from C=O stretching vibrations weakly coupled to C–N stretching vibrations. The amide II, on the other hand, arises from N–H bending vibrations coupled to C–N stretching vibrations. The 1,500 to 1,000 cm^{-1} region is very rich and is known to correspond to absorption due to phospholipids and nucleic acids. These bands are attributed to vibrational modes of COO^- and PO_2^- groups. Finally, the 1,200–900 cm^{-1} domain is dominated by the C–O–C and C–O–P elongation bands of oligo- and polysaccharides. Despite the complexity of biological samples, this type of spectrum can be considered as representative of a living tissue in a regular metabolic mechanism.

For several years, studies have been conducted in order to identify spectral fluctuations in IR spectra related to the deregulation of metabolism,

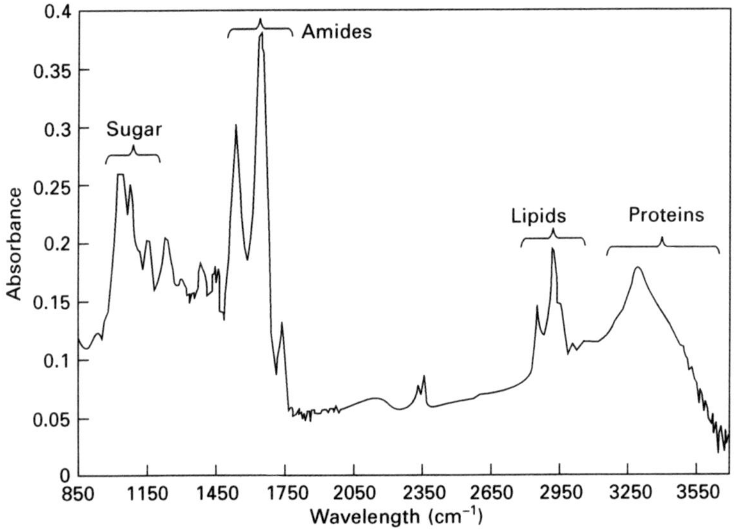

12.18 Typical FEWS spectrum of a liver cell.

especially deregulations due to the development of tumors or cancers. In 2003, a preliminary study by Keirsse *et al.* has shown the effect of starvation on mouse liver cellular constituents (Keirsse *et al.*, 2003b). Infrared spectra exhibited significant differences between fed and starved mice. In particular, a decrease of absorption bands at 1,080 and 1,030 cm^{-1}, assigned to C–O bond vibration of sugar rings, was observed. These data were confirmed by histological studies, which evidenced a decrease of glycogen. Thus, these results proved that FEWS can give information about a physiologic state, showing evidence of modifications in the hepatic metabolism.

Hocdé and coworkers (2001b, 2004) have shown the ability of the fiber to distinguish tumorous from healthy tissues by probing mouse liver biopsies deposited on the fiber. For this study, three groups of male mice received an injection of diethylnitrosamide, a liver tumor inductor, and iron-dextran which induces liver iron overload. After a few months, they developed macroscopic liver tumors. Tumoral and non-tumoral areas were examined by FEWS. As shown in the MIR spectra (Fig. 12.19), the most important effects observed in the tumoral cells are: an increase of the fatty acid, CH_3 and CH_2, vibrational bands and a 10 cm^{-1} upshift with a large increase in the C=O ester band at 1,746 cm^{-1}, which corresponds to the C=O vibration of triglyceride. This suggests that accumulation of triglyceride occurs during the tumoral process.

Another experimental work involved the monitoring of living human lung cells coating the surface of a chalcogenide fibre and the study of their behavior when exposed to a chemical agent such as a surfactant (Lucas *et*

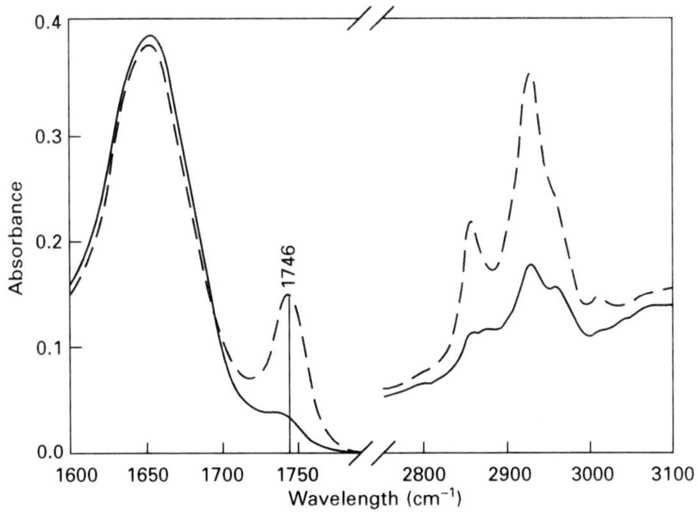

12.19 FEWS IR spectra of tumoral (dashed line) and non-tumoral (solid line) mice liver tissues.

al., 2005). To coat a layer of living cells on the fibre surface, the sensing zone was inserted into a pellet of centrifuged cell culture, which showed strong adherence after about one hour (Plate XI between pages 330 and 331). The vibrational spectrum of the cells was then monitored continuously. Upon introduction of micromolar quantities of Triton X-100 surfactant within the cells environment, a strong and rapid decay in the phospholipids signal occurred. This is consistent with the known mechanism of attack of surfactants on cell membranes. Metabolic variations after exposure to various toxic agents were also detected in other spectral regions (Riley, 2006a, 2006b).

Analysis of serums

Metabolic abnormalities due to diseases can also be detected in plasma or sera through the quantification of biochemical components such as glucose, cholesterol, and triglyceride. Most of the methods developed until now to quantify or characterize serum components give information for only one component and are not sufficient for establishing a diagnosis. Therefore, methods providing an extensive view of the serum species levels would be helpful for giving an early diagnosis, in addition to classical clinical, biological, and imaging findings. So, FEWS using glass fibers combined with adequate statistical analyses such as PCA (principal component analysis) or PLS (partial least squares) could provide spectral fingerprints containing the information necessary to discriminate between ill and healthy patients. PCA is a non-supervised analysis method that permits information to be obtained from spectra without any specific attribution of the bands. On PCA maps, each spectrum is represented by a point in two- or three-dimension space (Rouessac and Rouessac, 1998; Bertrand and Dufour, 2000). The PLS is an extension of the multiple linear regression model, allowing the derivation of multiclass decision rules and thus classification of data (Fort and Lambert-Lacroix, 2004).

In 2004, Keirsse *et al.* reported a first study on sera from mice developing obesity related to a homozygous mutation in the leptin gene, leading to hyperphagia and type II diabetes. The fiber evanescent wave spectra of these sera were compared with the spectra of mice that do not develop obesity, used as control spectra. All spectra, recorded by placing 10 μl of serum in contact with the fiber, were processed by PCA in the 1,100–1,000 cm^{-1} range, which corresponds to the sugar ring vibration bands. On the PCA map, the points corresponding to control spectra are well gathered and the spectra for obese mice are localized in a totally different area, spread over a larger zone. So, this experiment proved that the FEWS technique coupled with an analysis method like PCA is an efficient tool for distinguishing pathological sera from normal ones.

More recently, Anne *et al.* (2009b) published FEWS results on sera from patients exhibiting chronic liver diseases, such as hemochromatosis, alcoholic cirrhosis, and dysmetabolic hepatosiderosis (DYSH). These diseases present important metabolic disturbances and PLS, applied to the recorded spectra, permitted the discrimination of patients with cirrhosis or DSYH from a control group.

Monitoring of bacterial biofilms

FEWS was used to study bacterial biofilms (Le Coq *et al.*, 2002; Keirsse *et al.*, 2003a, 2006) and, in particular, the uropathogenic *Proteus mirabilis*, a Gram-negative bacterium considered as a model since it shows strong swarming abilities. These bacteria are able to spread over the entire surface of a Petri plate within a few hours by alternating periods of active mass migration and of cell division without colony expansion (consolidation). The TAS fibre was used to detect and monitor the biofilm's development on an agar gelose substratum *in situ* and in real time. An IR spectrum was collected every 15 min during 4 h. The experimental set-up consists in placing the U-turn of the fiber sensing zone in contact with the gelose and in recording before, during and after the moment when the periphery of the biofilm has come into contact with the fiber. Plate XII (between pages 330 and 331) shows the series of spectra recorded in the 1,800–850 cm^{-1} range. This domain is dominated by the water bending mode (1,636 cm^{-1}), the amide I (1,650 cm^{-1}) and amide II (1,550 cm^{-1}) bands arising from protein peptide bonds. Indeed, as the biomass increases, the 1,636 cm^{-1} component vanishes at the expense of the amide I mode. Drastic alterations of the IR spectra were observed during the swarming process in the 1,200–1,000 cm^{-1} frequency domain. A large frequency upshift from 1,068 to 1,082 cm^{-1} was indicated. This frequency domain is mainly assigned to carbohydrate backbones present in polysaccharides. This time-dependent frequency upshift is more obvious when one considers second derivative spectra (Fig. 12.20). It is shown that the frequency shift occurs at ca. 2 mm from the periphery indicating that swarmer cells are characterized by a 1,068 cm^{-1} band which is progressively replaced by a 1,082 cm^{-1} band as bacteria enter in the consolidation phase. Indeed, Gram-negative bacteria possess an outer membrane, in addition to the cytoplasmic one. The external leaflet of the asymmetric outer membrane is generally composed of only one type of amphipatic molecules, the lipopolysaccharides (LPS). Exopolysaccharides (EPS, also designated CPS for capsular polysaccharides) are the major components of the slime and present a high chemical similarity to the polysaccharide part of LPS. Even though it is not possible to distinguish EPS from LPS from their respective IR spectra, the observed upshift in the region assigned to the sugar backbone vibrations likely arises from an alteration in the EPS synthesis in relation to the alternation of migration and consolidation phases.

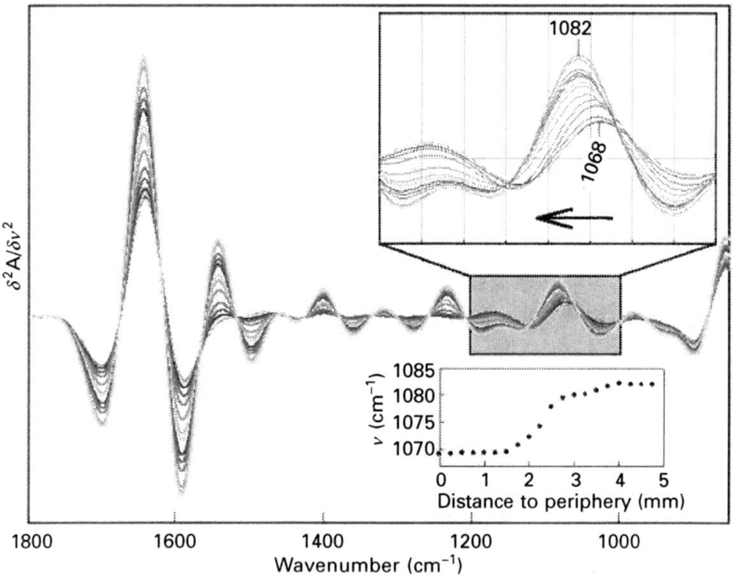

12.20 The corresponding inverted second derivative spectra are shown along with a zoom (upper inset) of the 1,200–1,000 cm^{-1} domain. This clearly shows the large frequency upshift which is observed as a function of time and hence also of distance from the periphery. The lower inset displays the evolution of the frequency as a function of distance to periphery.

This study clearly demonstrates that FEWS is sensitive enough, not only to detect, in real time, the colonization of a substratum by a bacterial biofilm, but also to detect minute biochemical alterations, i.e. in their polysaccharide membrane content, which reflect the switch from a vegetative phenotype to a swarming, pathogenic one. This study also demonstrates that distinct IR spectral signatures may contribute to map, in space and time, the behavior of complex biological systems. The need for integrated (i.e. living cells) and local information on complex biological systems may hence be partially fulfilled by the use of FEWS.

12.5.4 Food safety

Others applications concerning monitoring of tissues and fluids can be found in the food industry. For example, spectra collected for various chicken organs show significant variations in spectral features, allowing differentiation of organ/meat types based on a fast optical analysis. Similarly, analyses of different egg fractions showed clearly discernible absorption bands between egg white and egg yolk (Sanghera *et al*., 1999).

More recently, FEWS was used to classify various foodborne pathogens using chemometric methods such as principal component analysis (PCA) and logistic-partial least squares (logistic-PLS) (Brandily *et al.*, 2011). The results showed that PCA allows visualizing differentiation between pathogens and endogenous flora. In addition, logistic-PLS goes farther with the discrimination of pathogen strains with a classification error lower than 3.5%. This allows us focus on the potential of FEWS for early detection of pathogens in food matrices which could be extended to various applications in the health field.

12.6 Spatial area

In recent years, much effort has been paid to the development of infrared glasses that are transparent far into the infrared in order to detect signs of life on earth-like planets. Indeed, the DARWIN project from the European Space Agency (ESA) and the 'Terrestrial Planet Finder' from the NASA agency are two research programs whose goals are to find exoplanets that have the same conditions as Earth and are able to sustain life. The presence of life is materialized by the presence of water, oxygen, and ozone in the planet's atmosphere. These three molecules absorb in the infrared region at 6.2, 9 and 15 μm respectively. To demonstrate the presence of these IR signatures, a flotilla of telescopes will be launched in space and positioned with high precision to observe the targeted earth-like planets. To prevent blinding from strongly emitting nearby stars, efficient filtering is required to collect radiation information specifically from the exoplanet. This technique is called nulling interferometry. To fulfill these severe requirements, IR single-mode fibres (SMF) are needed. Their role consists in permitting only the propagation and the selection of light existing in the strategic wavelength window, which extends from 6 to 20 μm. As a consequence, the goal is to manufacture a fibre or several fibres that are transparent in the required optical domain typically from 6 to 20 μm. ESA, for example, has defined a two-wavelength band system to cover the operational wavelengths of the DARWIN mission. The short wavelength fiber will be designed for 6–12 μm and the long wavelength fiber will be used for 12–20 μm. Thus, a step-index fiber exhibiting single-mode propagation at 10.6 μm with an adequate coupling efficiency, as shown in Plate XIII (between pages 330 and 331), was elaborated (Houizot *et al.*, 2007). This fiber, with a core diameter of 22 μm and a cladding thickness of 250 μm for a cut-off wavelength of 3.7 μm, was prepared by the classical rod-in-tube method, and is suitable for the detection of water and ozone.

In recent years, to detect CO_2 at 15 μm, new materials containing large amounts of tellurium have been developed in the Te-Ge-Ga and Te-Ge-I systems (Danto *et al.*, 2006; Wilhelm *et al.*, 2007a,b). These compositions show a high tendency to crystallize, so high cooling rates are necessary to obtain vitreous materials. However, the advantage is related to the high atomic weight of tellurium, leading to low phonon energy and permitting transmission farther in the infrared than other chalcogenide glasses. More recently, the Te-Ge-Se system containing only a few percent of selenium, in order to keep a large transmission in the far infrared range, has been explored (Maurugeon *et al.*, 2009, 2010, 2011). These studies demonstrated the ability of a single-index fiber based on $Te_{76}Ge_{21}Se_3$ glass and of a multimode double-index fiber to detect the broad CO_2 absorbing band between 13 and 16 μm. The next step will be the elaboration of single-mode fiber fulfilling the requirements of ESA.

12.7 Conclusion

This chapter gives an overview of the use of chalcogenide glass optical fibers as sensors in many fields of application. The current state of the art of FEWS technique using chalcogenide fibers has been established. FEWS is performed with an infrared spectrometer equipped with fiber connectors and an infrared fiber. This method permits the collection, *in situ* and in real time, of an infrared vibrational spectrum of the sample being analyzed, without sampling or adding other substances as in fluorescence spectroscopy. This spectrum is a fingerprint of the molecule, obtained in a few minutes with very low quantities of material. As an example, 10 μl are sufficient for the analysis of a human serum. Many domains of application have been developed in environmental and biomedical domains. Supervised and unsupervised methods such as PCA and PLS have been successfully applied to studies in medical areas, permitting a classification between healthy and diseased sera. Improvements in both instrumentation and spectral analysis techniques are in progress, especially for biomedical applications. To increase sensitivity, quantum cascade laser sources, now available in many ranges of the infrared domain, could be used, permitting a miniaturization of the equipment. Other results, obtained in the framework of spatial programs, showed that infrared chalcogenide fibers could be the key element for nulling interferometry, allowing the detection of the three main molecules which attest the presence of life on earth-like planets.

12.8 References

Anne, M. L., Keirsse, J., Nazabal, V., Hyodo, K., Inoue, S., Boussard-Plédel, C., Lhermite, H., Charrier, J., Yanakata, K., Loreal, O., Le Person, J., Colas, F., Compere, C. and Bureau, B. (2009a) Chalcogenide glass optical waveguides for infrared biosensing. *Sensors*, 9, 7398–7411.

Anne, M. L., Le Lan, C., Monbet, V., Boussard-Plédel, C., Ropert, M., Sire, O., Pouchard, M., Jard, C., Lucas, J., Adam, J. L., Brissot, P., Bureau, B. and Loreal, O. (2009b) Fiber evanescent wave spectroscopy using the mid-infrared provides useful fingerprints for metabolic profiling in humans. *Journal of Biomedical Optics*, 14, 054033.

Bellamy, L. J. (1975) *The infrared spectra of complex molecules*, London, Meuthen.

Bertrand D. and Dufour, E. (eds) (2000) *La spectroscopie infrarouge et ses applications analytiques*, Paris, Tech.&Doc.

Boussard-Plédel, C., Hocdé S., Fonteneau G., Ma H.L., Zhang, X.H., Le Foulgoc K. and Lucas J. (1999) Infrared glass fibers for evanescent wave spectroscopy. In Katzir, A. and Harrington, J. A. (eds) *Specialty fiber optics for medical applications*, San José, CA, SPIE.

Brandily, M. L., Monbet, V., Bureau, B., Boussard-Pledel, C., Loreal, O., Adam, J. L., Sire and O. (2011) Identification of foodborne pathogens within food matrices by IR spectroscopy. *Sensors and Actuators B*, 160, 202-206.

Brooks, A. L., Bruch, R. F., Afanasyeva, N. I., Kolyakov, S. F., Butvina, L. N. and Ma, L. (1998) Investigations of normal human skin tissue and acupuncture points of human skin tissue using fiberoptical FTIR spectroscopy. In Katzir, A. (ed) *Surgical-Assist Systems*, San José, CA, SPIE.

Bureau, B., Zhang, X. H., Smektala, F., Adam, J. L., Troles, J., Ma, H. L., Boussard-Plédel, C., Lucas, J., Lucas, P., Le Coq, D., Riley, M. R. and Simmons, J. H. (2004) Recent advances in chalcogenide glasses. *Journal of Non-Crystalline Solids*, 345, 276–283.

Charpentier, F., Bureau, B., Troles, J., Boussard-Plédel, C., Michel-Le Pierres, K., Smektala, F. and Adam, J. L. (2009a) Infrared monitoring of underground CO_2 storage using chalcogenide glass fibers. *Optical Materials*, 31, 496–500.

Charpentier, F., Troles, J., Coulombier, Q., Brilland, L., Houizot, P., Smektala, F., Boussard-Plédel, C., Nazabal, V., Thibaud, N., Le Pierres, K., Renversez and Bureau, B. (2009b) CO_2 detection using microstructured chalcogenide fibers. *Sensor Letters*, 7, 745–749.

Compton, D. A. C., Hill, S. L., Wright, N. A., Druy, M. A., Piche, J., Stevenson, W. A. and Vidrine, D. W. (1988) In-situ FT-IR Analysis of A Composite Curing Reaction Using A Mid-Infrared Transmitting Optical Fiber. *Applied Spectroscopy*, 42, 972–979.

Cross, A. D. and Jones, R. A (1969) *An introduction to practical Infra-red spectroscopy*, London, Butterworths.

Danto, S., Houizot, P., Boussard-Plédel, C., Zhang, X. H., Smektala, F. and Lucas, J. (2006) A family of far-infrared-transmitting glasses in the Ga-Ge-Te system for space applications. *Advanced Functional Materials*, 16, 1847–1852.

De Fornel, F. (1998) *Les ondes évanescentes en optique et en optoélectronique*. Paris, Editions Eyrolles.

Eytan O., Bormashenko, E., Pogreb R., Sutovski S., Vaserman I., Seal B.A. and Katzir A. (1999) Fiberoptic evanescent wave spectroscopy (FEWS) for blood diagnosis: the use of polymer coated AgClBr fibers and neural network analysis. *Specialty fiber optics for medical applications*, San José, CA, SPIE.

Fort G. and Lambert-Lacroix S. (2004) Classification using partial least squares with penalized logistic regression. *Bioinformatics*, 21, 1104–1111.

Fringeli, U. P. and Günthard, Hs. H. (1981) In Grell, E. (ed.) *Membrane spectroscopy*. New York, Springer.

Harrick, N. J. (1967) *Internal refection spectroscopy*, New York, John Wiley & Sons.

Harrick, N. J. (1979) *Internal reflection spectroscopy*, New York, Harrick Scientific Corporation.

Heo, J., Rodrigues, M., Saggese, S. J. and Sigel, G. H. (1991) Remote Fiberoptic Chemical

Sensing Using Evanescent-Wave Interactions In Chalcogenide Glass-Fibers. *Applied Optics*, 30, 3944–3951.

Hocdé, S., Boussard-Plédel, C., Fonteneau, G., Lecoq, D., Ma, H. L. and Lucas, J. (2000) Recent developments in chemical sensing using infrared glass fibers. *Journal of Non-Crystalline Solids*, 274, 17–22.

Hocdé, S., Boussard-Plédel, C., Fonteneau, G. and Lucas, J. (2001a) Chalcogens based glasses for IR fiber chemical sensors. *Solid State Sciences*, 3, 279–284.

Hocdé, S., Loreal, O., Sire, O., Turlin, B., Boussard-Plédel, C., Le Coq, D., Bureau, B., Fonteneau, G., Pigeon, C., Leroyer, P. and Lucas, J. (2001b) Biological tissues infrared analysis by chalcogenide glass optical fiber spectroscopy. In Gannot, I., Gulyaev, Y. V., Papazoglou, T. G. and Vanswol, C. F. P. (eds) *Biomonitoring and endoscopy technologies*, San José, CA, SPIE.

Hocdé, S., Loreal, O., Sire, O., Boussard-Plédel, C., Bureau, B., Turlin, B., Keirsse, J., Leroyer, P. and Lucas, J. (2004) Metabolic imaging of tissues by infrared fiber-optic spectroscopy: an efficient tool for medical diagnosis. *Journal of Biomedical Optics*, 9, 404–407.

Houizot, P., Boussard-Plédel, C., Faber, A. J., Cheng, L. K., Bureau, B., Van Nijnatten, P. A., Gielesen, W. L. M., Pereira Do Carmo, J. and Lucas, J. (2007) Infrared single mode chalcogenide glass fiber for space. *Optics Express*, 15, 17, 12529–12538.

Jackson, M. and Mantsch, H. H. (1996) Biomedical infrared spectroscopy. In Mantsch H. H. and Chapman, D. (eds) *Infrared spectroscopy of biomolecules*, New York, Wiley-Liss.

Katz M., Katzir, A., Schnitzer I. and Bornstein A. (1994) Quantitative evaluation of chalcogenide glass fiber evanescent wave spectroscopy. *Applied Optics*, 33, 5888–5894.

Keirsse, J., Boussard-Plédel, C., Loreal, O., Sire, O., Bureau, B., Leroyer, P., Turlin, B. and Lucas, J. (2003a) IR optical fiber sensor for biomedical applications. *Vibrational Spectroscopy*, 32, 23–32.

Keirsse, J., Boussard-Plédel, C., Loreal, O., Sire, O., Bureau, B., Turlin, B., Leroyer, P. and Lucas, J. (2003b) Chalcogenide glass fibers used as biosensors. *Journal of Non-Crystalline Solids*, 326, 430–433.

Keirsse, J., Bureau, B., Boussard-Plédel, C., Leroyer, P., Ropert, M., Dupont, V., Anne, M. L., Ribault, C., Sire, O., Loreal, O. and Adam, J. L. (2004) Chalcogenide glass fibers for *in-situ* infrared spectroscopy in biology and medicine. In Culshaw, B., Mignani, A. G. and Riesenberg, R. (eds) *Optical Sensing*, San José, CA, SPIE.

Keirsse, J., Lahaye, E., Bouter, A., Dupont, V., Boussard-Plédel, C., Bureau, B., Adam, J. L., Monbet, V. and Sire, O. (2006) Mapping bacterial surface population physiology in real-time: infrared spectroscopy of *Proteus mirabilis* swarm colonies. *Applied Spectroscopy*, 60, 584–591.

Kosower, E. M. and Marom-Albeck, O. (1994) Protein FTIR-ATR On Silver-halide Fibers Under Reagent-Accessible Conditions. *Abstracts of Papers of the American Chemical Society*, 208, 156.

Kosower, E. M., Maron-Albeck, O., Paiss, I., Katzir, A., Kosower, N.S. (1994) Protein and cell FTIR-ATR on silver halide fibers under reagent-accessible conditions. *Biomedical fiber optic instrumentation*. Los Angeles, CA, SPIE.

Le Coq, D., Michel, K., Fonteneau, G., Hocde, S., Boussard-Plédel, C. and Lucas, J. (2001) Infrared chalcogen glasses: chemical polishing and fibre remote spectroscopy. *International Journal of Inorganic Materials*, 3, 233–239.

Le Coq, D., Michel, K., Keirsse, J., Boussard-Plédel, C., Fonteneau, G., Bureau, B., Le

Quere, J. M., Sire, O. and Lucas, J. (2002) Infrared glass fibers for *in-situ* sensing, chemical and biochemical reactions. *Comptes Rendus Chimie*, 5, 907–913.

Le Coq, D., Boussard-Plédel, C., Fonteneau, G., Pain, T., Bureau, B. and Adam, J. L. (2003) Chalcogenide double index fibers: fabrication, design, and application as a chemical sensor. *Materials Research Bulletin*, 38, 1745–1754.

Le Foulgoc K., Le Neindre, L., Zhang X.H. and Lucas J. (1996) Tapered TeX glass otical fibers for remote IR spectroscopic analysis. In Lieberman, R. A. (ed.) *Chemical, biochemical and envrironmental fiber sensors VIII*. Denver, CO, SPIE.

Lucas, P., Le Coq, D., Juncker, C., Collier, J., Boesewetter, D. E., Boussard-Plédel, C., Bureau, B. and Riley, M. R. (2005) Evaluation of toxic agent effects on lung cells by fiber evanescent wave spectroscopy. *Applied Spectroscopy*, 59, 1–9.

Lucas, P., Solis, M. A., Le Coq, D., Juncker, C., Riley, M. R., Collier, J., Boesewetter, D. E., Boussard-Plédel, C. and Bureau, B. (2006c) Infrared biosensors using hydrophobic chalcogenide fibers sensitized with live cells. *Sensors and Actuators B – Chemical*, 119, 355–362.

Macdonald, S. A. and Bureau, B. (2003) Fourier transform infrared attenuated total reflection and transmission spectra studied by dispersion analysis. *Applied Spectroscopy*, 57, 282–287.

Macdonald, S., Michel, K., Lecoq, D., Boussard-Plédel, C. and Bureau, B. (2004) Optical analysis of infrared spectra recorded with tapered chalcogenide glass fibers. *Optical Materials*, 25, 171–178.

Maurugeon, S., Bureau, B., Boussard-Plédel, C., Faber, A. J., Zhang, X. H., Geliesen, W. and Lucas, J. (2009) Te-rich Ge-Te-Se glass for the CO_2 infrared detection at 15 μm. *Journal of Non-Crystalline Solids*, 355, 2074–2078.

Maurugeon, S., Boussard-Plédel, C., Troles, J., Faber, A. J., Lucas, P., Zhang, X. H., Lucas, J. and Bureau, B. (2010) Telluride glass step index fiber for the far Infrared. *Journal of Lightwave Technology*, 28, 3358–3363.

Maurugeon, S., Bureau, B., Boussard-Plédel, C., Faber, A. J., Lucas, P., Zhang, X. H. and Lucas, J. (2011) Selenium modified $GeTe_4$ based glasses optical fibers for far-infrared sensing. *Optical Materials*, 33, 660–663.

Mendelsohn, R. and Mantsch, H. H. (1986) Fourier transform infrared studies of liquid-protein interactions. In Watts, A. and De Pont, J. J. H. H. M. (eds) *Progress in lipid protein interactions*, New York, Elsevier.

Michel, K., Bureau, B., Pouvreau, C., Sangleboeuf, J. C., Boussard-Plédel, C., Jouan, T., Rouxel, T., Adam, J. L., Staubmann, K., Steinner, H., Baumann, T., Katzir, A., Bayona, J. and Konz, W. (2003) Development of a chalcogenide glass fiber device for *in situ* pollutant detection. *Journal of Non-Crystalline Solids*, 326, 434–438.

Michel, K., Bureau, B., Boussard-Plédel, C., Jouan, T., Adam, J. L., Staubmann, K. and Baumann, T. (2004) Monitoring of pollutant in waste water by infrared spectroscopy using chalcogenide glass optical fibers. *Sensors and Actuators B – Chemical*, 101, 252–259.

Naumann, D., Helm, D., Labischinski, H. and Giesbrecht, P. (1991) The characterization of microorganisms by Fourier-transform infrared spectroscopy (FT-IR). In Nelson, W. H. (ed.) *Modern techniques for rapid microbiological analysis*, New York, Wiley-VCH.

Riley, M. R., Derosa, D., Blaine, J., Potter, B. G., Lucas, P., Le Coq, D., Juncker, C., Boesewetter, D. E., Collier, J. M., Boussard-Plédel, C. and Bureau, B. (2006a) Biologically inspired sensing: infrared spectroscopic analysis of cell responses to an inhalation health hazard. *Biotechnology Progress*, 22, 24–31.

Riley, M. R., Lucas, P., Le Coq, D., Juncker, C., Boesewetter, D. E., Collier, J. L., Derosa, D. M., Katterman, M. E., Boussard-Plédel, C. and Bureau, B. (2006b) Lung cell fiber evanescent wave spectroscopic biosensing of inhalation health hazards. *Biotechnology and Bioengineering*, 95, 599–612.

Rodrigues, M. and Sigel, G. H. (1991) Chalcogenide glass fibers for remote spectroscopic chemical sensing. *Infrared Fiber Optics III*, Boston, MA, SPIE.

Rouessac F. and Rouessac, A. (1998) *Analyse chimique: méthodes et techniques instrumentales modernes*, Paris, Dunod.

Sanghera, J. S., Kung, F. H., Pureza, P. C., Nguyen, V. Q., Miklos, R. E. and Aggarwal, I. D. (1994) Infrared Evanescent-Absorption Spectroscopy With Chalcogenide Glass-Fibers. *Applied Optics*, 33, 6315–6322.

Sanghera, J. S., Kung, F. H., Busse, L. E., Pureza, P. C. and Aggarwal, I. D. (1995) Infrared Evanescent Absorption-Spectroscopy Of Toxic-Chemicals Using Chalcogenide Class Fibers. *Journal of the American Ceramic Society*, 78, 2198–2202.

Sanghera, J. S., Shaw L. B., Busse, L. E., Talley, D. and Aggarwal, I. D. (1999) Infrared transmitting fiber optics for biomedical applications. In Katzir, A. and Harrington J. A. (eds *Specialty fiber optics for medical applications*, San José, CA, SPIE.

Spielvogel, J., Lobik, L., Nissenkorn, I., Hibst, R., Gotshal, Y. and Katzir, A. (1998) Cancer diagnosis using Fourier transform fiberoptic infrared evanescent wave spectroscopy (FTIR-FEWS). In Katzir, A. (ed.) *Surgical-Assist System*, San José, CA, SPIE.

Spielvogel, J., Reuter, S., Hibst, R. and Katzir, A. (1999) Monitoring the diffusion of topically applied drugs through human and pig skin using fiber evanescent wave spectrosocpy (FEWS). In Katzir, A. and Harrington, J. A. (eds) *Specialty fiber optics for medical applications*, San José, CA, SPIE.

Wilhelm, A., Boussard-Plédel, C., Lucas, P., Riley, M. R., Bureau, B. and Lucas, J. (2007a) New tellurium based glasses for use in bio-sensing applications. IN Gannot, I. (ed.) *Optical fibers and sensors for medical diagnostics and treatment applications VII*. San José, CA, SPIE.

Wilhelm, A. A., Boussard-Plédel, C., Coulombier, Q., Lucas, J., Bureau, B. and Lucas, P. (2007b) Development of far-infrared-transmitting Te based glasses suitable for carbon dioxide detection and space optics. *Advanced Materials*, 19, 3796–3800.

Zhang, X. H., Ma, H. L., Blanchetière, C., Le Foulgoc, K., Lucas, J., Heuzé, J., Collardelle, P., Froissard, P., Picque, D. and Corrieu, G. (1994) Tellurium halide IR fibers for remote spectroscopy. *Biomedical Fiber Optic Instrumentation*, Los Angeles, CA, SPIE.

13
Chalcogenide microstructured optical fibers for infrared applications

J. TROLES, University of Rennes, France and
L. BRILLAND, PERFOS, France

DOI: 10.1533/9780857093561.2.411

Abstract: This chapter describes firstly the originality and significance of chalcogenide microstructured optical fibers. The interest in combining the properties of microstructured optical fibers with those of chalcogenide glasses will be presented. Different methods for preparing preforms and fibers will be described, such as stack and draw, extrusion, drilling and molding techniques. Several examples of fiber realizations will be given. Finally, the applications of such fibers for nonlinear optics will be discussed through different modeling and nonlinear characterizations.

Key words: chalcogenide glasses, infrared, microstructured optical fibers, nonlinear optical properties.

13.1 Introduction

In the mid-1970s, the concept of holey fiber was first published [1]. This study was motivated by the search for fibers with very low attenuation for transmitting information for telecommunication. This technology was abandoned after the development of the modified chemical vapor deposition technique (MCVD), which, even now, is still the best method for the elaboration of low-loss silica fibers. It was not until the 1990s, based on the concept of photonic crystals and following the studies of Russel, Birks and Knight, that silica microstructured fibers experienced their glory days [2–4]. Since then, this new fiber design has spread over other glassy matrices [5], and since 2000 to chalcogenide glasses [6].

This chapter describes first the originality and significance of this new type of fiber. Then, the interest for combining the properties of microstructured optical fibers (MOFs) with those of chalcogenide glasses will be presented. Indeed, microstructured fibers possess original guiding properties associated with infrared transparency and the nonlinear properties of chalcogenide glasses. Different methods for preparing preforms and fibers are described in Section 13.3. These include stack and draw, extrusion, drilling and molding techniques. Several examples of fiber realizations are presented, including first attempts, large-core, small-core, single-mode, and hollow-core fibers. The realization of structured hybrid fibers made of silica-chalcogenide and polymer chalcogenide will also be presented. Several characterizations of chalcogenide microstructured fibers are discussed. Finally, the applications of such fibers for nonlinear optics are discussed through different modeling and nonlinear characterizations. Indeed, chalcogenide MOFs have shown their high potential for wavelength conversion for telecommunication, and for the realization of new coherent sources such as supercontinuum generation in the mid-infrared.

13.2 General principles of microstructured optical fibers

13.2.1 Origins

This concept of fiber is based on an analogy with the physical principle of photonic crystals, which is, as an example, at the origin of the color of butterfly wings. The first fibers arising from this concept are the Bragg fibers [7]. These fibers are composed of successive circular layers presenting two different refractive indices (Fig. 13.1a). Then, Birks and Russell worked on silica fibers, presenting a periodic array of cylindrical air holes running along their entire length that corresponds to a 2D photonic crystal (Figs 13.1b, c). In 1995 they demonstrated that such microstructured optical fibers can in theory present photonic band gap propagation [2]. Then, in 1997, they realized the first air-silica MOF [3] and in 1999, the first air-guiding MOF [8].The exceptional properties of these fiber guides quickly attracted the interest of the scientific community.

Then, two different types of fibers have been studied: hollow-core fibers (Fig. 13.1b) and solid-core fibers (Fig. 13.1c). In the first case, the periodic structure causes a real photonic band-gap effect. The light is trapped in the low-index core of the fiber. These fibers are called photonic band gap microstructured optical fibers (PBG-MOFs).

In the case of solid-core fibers (Fig. 13.1c), one inclusion or more are missing at the center, which will form the core of the fiber. The refractive

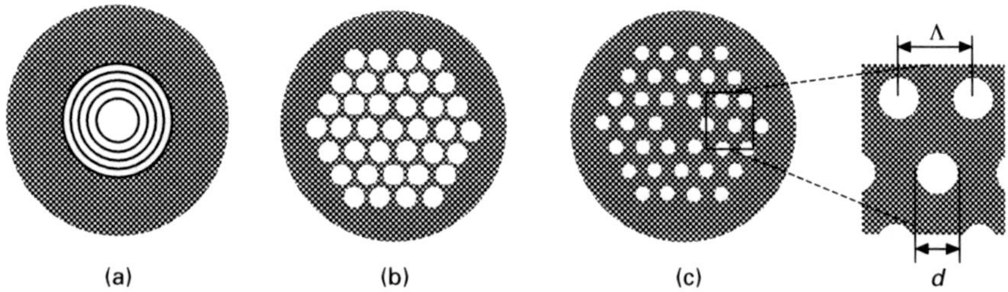

13.1 Transverse section: (a) Bragg fibers, (b) photonic band gap fiber, (c) solid core MOF.

index of the center is therefore higher than the effective index of the surrounding array composed of inclusions of lower refractive index. In this case, the guidance is quite similar to the guidance in a classical step-index fiber. This type of fibers is based on modified total internal reflection, and is thus called MTIR-MOF. In any case, for PBG-MOF or MTIR-MOF, the optical properties of the fibers are closely related to geometry: core size, number of rings, hole diameter, *d*, hole spacing Λ, and also the ratio *d*/Λ (Fig. 13.1). Numbers of guided modes, mode effective area, broadband single-mode guidance, nonlinearity, band-gap width, and dispersion are examples of properties that can be managed by the geometry of MOFs.

In comparison to double-step index fibers, MTIR-MOFs offer different properties:

- In most cases, only one material is required to ensure propagation of light in the core. The inclusions surrounding the core are generally filled with air. Contrary to the fabrication of standard fiber, there is no need to control precisely the index of refraction difference between core and cladding. This point is important because it offers the possibility to make MOF based on glasses for which the control of refractive index is difficult to achieve.
- The guiding regime depends only on the geometrical parameters. When the ratio *d*/Λ is lower than 0.42, the fiber is single-mode whatever the wavelength. This guiding regime is often called endlessly single-mode regime [3, 9].
- A smaller-core diameter can be achieved, which is an advantage for enhancing nonlinear effects. Furthermore, the zero-dispersion wavelength (ZDW) can be managed over a wider range of wavelengths.
- Conversely, a large-core diameter can be obtained to reduce thermal and nonlinear effects. Consequently, the transmitted power can be significantly increased.

The first MOFs were made of pure SiO_2. Quickly this new design of fibers was extended to other glasses, often called 'soft glasses' due to a lower transition temperature (T_g). If SiO_2 is well known for its high transparency, it is also known for its relative poor nonlinearity and transmission window limited by intrinsic material absorption around 2.5 μm. In most cases, soft glasses have a larger transmission window and a greater nonlinearity. So, it might be interesting to study soft glass MOFs in order to increase the fields of application. The fabrication of soft glasses such as lead and bismuth silicates [10–12], TeO_2-based glasses [13–15], fluoride glasses [16, 17] and of course what is interesting in this chapter, chalcogenide glasses [6, 18, 19] has been investigated.

The first chalcogenide MOF, based on Ga-La-S, was made in 2000, but no light propagation was demonstrated [6] (Fig. 13.2(a)). Several years later,

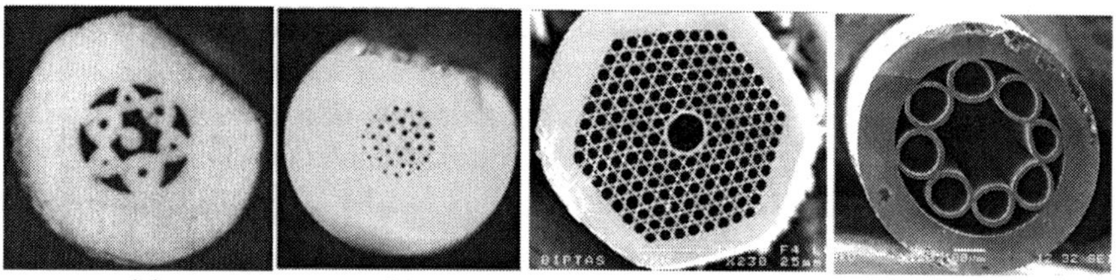

(a) (b) (c) (d)

13.2 First chalcogenide microstructured optical fibers: (a) first MTIR-MOF, no light propagation observed (2000 [6]), (b) first single-mode (1.55 μm) MTIR-MOF (2006 [18]), (c) first hollow-core fiber, no band gap effect demonstrated (2000 [24]), first hollow-core MOF with light guidance observed at 10.6 μm (2011 [25]).

light guidance was obtained in sulfur and selenide chalcogenide MOFs [19, 20]. In 2006, the demonstration of single-mode propagation was published, with a fiber based on Ge-Ga-Sb-S presenting three rings of holes [18] (Fig. 13.2(b)). Optical losses for this fiber were as high as 15–20 dB/m at 1.55 μm, due to the fabrication method [21]. So, the main challenge in the 2000s was to obtain low-loss chalcogenide MOF. Ten years after the making of the first fiber, the improvement in the elaboration of chalcogenide MOF (Section 13.4.1) means it is now possible to obtain fibers presenting losses less than 1dB/m [22], quite similar to the glass ones. Except for hybrid Bragg fibers [23], all chalcogenide MOFs were MTIR-MOFs until 2009. The first realization of a six-ring, hollow-core, fiber did not show any bang-gap propagation [24] (Fig. 13.2(c)). It was only in 2011 that Kosolapov *et al.* demonstrated guidance of 10.6 μm light from a CO_2 laser through a Te-As-Se glass hollow-core fiber [25] (Fig. 13.2(d)).

13.2.2 Interest in chalcogenide MOFs

Why is it of interest to prepare and study chalogenide MOFs? Compared to silica, chalcogenide glasses present a wider infrared transmission window and higher nonlinear optical properties.

Chalcogenide fibers based on sulfur, selenium and/or tellurium are transparent in the infrared windows from 0.5–1 to 8–12 μm depending on the compositions (Table 13.1). They are well known for their highly nonlinear properties. For example, the nonlinear refractive index of As_2Se_3 is more than 600 times that of silica and more than 100 times the refractive index of heavy-metal oxide glasses like TeO_2-based glasses (see Table 13.1).

The main interest is to combine the optical properties inherent in MOFs and in chalcogenide glasses, in order to extend the fields of application of chalcogenide fibers to new exciting domains. Thus, for nonlinear applications, small-core diameters enable non-linear effects to be enhanced with very low

Table 13.1 Optical properties of different glasses

Family glass	Glass	Fiber transparency window* (μm)	n_0	n_2 10^{-20} m^2/W	n_2/n_{2SiO2}	ZDW** (μm)
Oxide	SiO_2	0.2–2.4	1.44	2.2 [26]	1	1.26 [27]
Oxide	TeO_2	0.5–4.2	2.10	25–59 [15, 28]	11–27	2.24 [27]
Fluoride	ZBLAN	0.3–4.2	1.45	2.1 [29]	0.9	1.6 [30]
Chalcogenide	As_2S_3	0.7–8	2.44	480	220 [31]	4.8 [32]
Chalcogenide	As_2Se_3	1–11	2.81	1450	660 [31]	7.2
Chalcogenide	$Ge_{11}As_{22}Se_{67}$	1–11	2.6	1160	530 [33]	> 6

* Approximate wavelength when losses reach more than 10 dB/m, n_0 and n_2 are given at 1.55 μm.
**ZDW-zero dispersion wavelength.

power threshold, such as Brillouin, Raman or Kerr effects (e.g. four wave mixing and self-phase modulation). For example, wavelength conversion based on four wave mixing is of great interest for data processing in future high-bit-rate telecommunication networks. Chalcogenide MOFs could contribute to improve flexibility at very low power consumption.

A high nonlinearity combined with the management of ZDW will lead to the generation of supercontinuum in the mid-infrared region for spectroscopic applications. Also, Raman effect could result in wavelength conversion in the mid-infrared window.

Conversely, to increase transportation of power through the fiber, one would like to limit the nonlinear effects by significantly increasing the core diameter. A few watts of signal could be transmitted in the 3–5 or 8–12 μm windows, corresponding to an atmospheric window with relative transparency. In this case, chalcogenide MOFs could help to create remote laser sources.

In the area of chemical detection, chalcogenide MOFs could be utilized as sensors, since most molecules possess fundamental absorptions in the mid-IR. In that configuration, the holes are filled with the liquid or gas to be analyzed, to achieve overlap of the mode field with the medium. Furthermore, MOFs can be specially designed to increase the evanescent field around the core and hence the sensitivity.

13.2.3 Design and modeling of high-index MOFs

In this section, the main linear properties of MTIR-MOFs obtained from numerical simulations are described: guiding losses, single-mode regime, and chromatic dispersion properties. In conventional step-index fibers, the propagation constant of several modes is real (more particularly for the fundamental one). In this case, the imaginary part of the propagation constant is equal to zero. So the fundamental mode presents any guiding losses due to structure of the fiber. However, in a MTIR-MOF, all the modes (fundamental included) present guiding losses depending on the geometry (number of rings, d, Λ, and d/Λ). Indeed, the guiding losses can be equal to zero if the fiber presents an infinite number of rings (which is never the case in the real world!). In fact, in MTIR-MOFs, the question is: how many rings are necessary to obtain reasonable guiding losses compared to the intrinsic materials ones. It has been shown (Fig. 13.3) that, in high refractive index materials, three or four rings are enough to obtain losses significantly smaller than the intrinsic losses of chalcogenide glasses [9, 34]. As a comparison, more than six rings are needed in silica fibers to obtain negligible guiding losses, as compared to silica attenuation [35].

Another key property of MTIR-MOFs is the constant single-mode behavior [3]. Figure 13.4 describes an approximate phase diagram of the second mode for a high index matrix (2.5) with seven and an infinite number of rings [9]. Theoretically, when d/Λ is smaller than 0.42, the second mode is never confined in the core of the fiber (Fig. 13.4, domain I). The fiber is single mode whatever the wavelength: the fiber is constantly single mode. This limit of 0.42 does not depend on the refractive index of the glass matrix. When $d/\Lambda > 0.42$, the single-mode regime depends on the wavelength. In domain II, the second mode is not confined in the core (the guiding losses of the second mode are too high). In domain III, the second mode is confined in the core, the fiber is then multimode. One can note that the single-mode regime depends strongly on the geometry such as the number of hole rings and the ratio d/Λ. Consequently, in order to determine theoretically the single-mode character of a MOF, one must compute numerically the modal properties

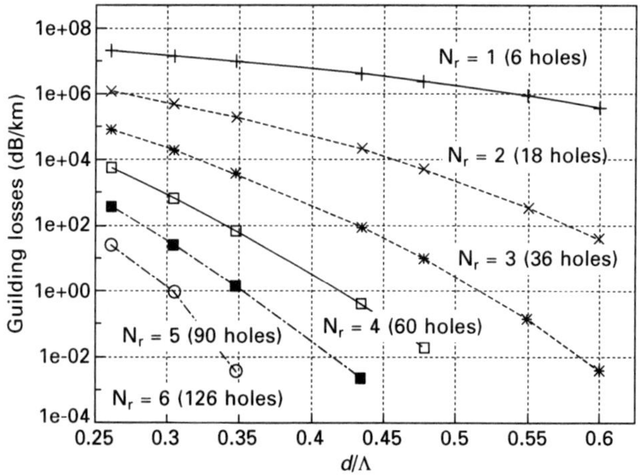

13.3 Fundamental mode losses as function of number of air hole ring N_r and the ratio d/Λ for MTIR-MOF with $n = 2.5$, $\Lambda = 2.3$ μm and $\lambda = 1.55$ μm [34].

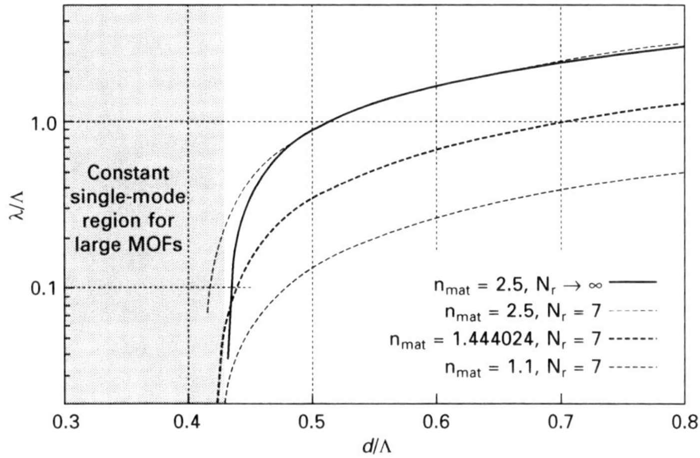

13.4 Schematic phase diagram for the second mode in a MTIR-MOF (adapted from [9]).

for a given opto-geometrical structure, especially if the structure presents a low number of hole rings.

Another original property of MOFs is the tunability of dispersion. Indeed, on first approximation, the total dispersion of a fiber is the sum of the intrinsic dispersion of the material (D_{mat}) and of the dispersion of the waveguide (D_{wave}). Thanks to the large number of possible geometries, various dispersion profiles can be obtained in MOFs [36]. As an example, specific geometries permit the highly negative dispersion of chalcogenide glasses in the 1.5–3 μm spectral range [32, 37, 38] to be compensated. In addition, these properties permit the ZDW (zero dispersion wavelength) of chalcogenide fibers to be shifted. Indeed, the intrinsic ZDW of chalcogenide glasses is above 4 μm.

13.3 Elaboration of chalcogenide microstructured optical fibers

The efficiency of optical functions depends strongly on the optical transmission of the glasses to be used for the preparation of fibers. So, before elaborating MOFs, highly-purified glasses must be fabricated. Chapter 1 is dedicated to the different processes that can be used for the realization of high-purity chalcogenide glasses. Fiber preparation is presented specifically in this chapter.

13.3.1 The 'stack and draw' technique

The 'stack and draw' technique is the most common method for the manufacturing of silica MOFs [39], and it was naturally the first one used for the elaboration of chalcogenide MOFs [6, 18]. For that purpose, it is necessary at first to shape chalcogenide tubes (Fig. 13.5). To do so, a tube of 10–12 mm outer diameter and around 5 mm inner diameter is initially fabricated. The tube is then elongated into capillaries of around 600–650 μm outer diameter. The capillaries are then stacked in a hexagonal lattice around a glass rod of an identical diameter, placed in the center (Fig. 13.5). The stack is then introduced in a jacket tube and this ensemble is drawn into a fiber. The presence of variable gas pressure in the holes and a negative pressure around the stack enables a better control of the geometry of the fiber during drawing [18]. It has been shown that the 'stack and draw' technique was not well suited for the preparation of chalcogenide glass MOFs. Indeed, strong excess losses, above 20 dB/m, were observed whatever the composition of the glass fiber (Ge-Sb-S, As-Se, Te-As-Se) [21]. During drawing, when the interstitial holes are collapsed, defects appear at the glass interfaces and induce strong optical losses. It has been shown that the transmission of the fiber can be improved if the interstitial holes stay partially uncollapsed [21] by decreasing the number of defects at the interfaces. In this case, the attenuation can be reduced below 4–5 dB/m. However, using this technique, an excess of losses was always observed, compared to the attenuation of a classic single-index chalcogenide fiber. Consequently, in order to reduce optical losses, alternative methods had to be investigated: molding process, drilling, extrusion.

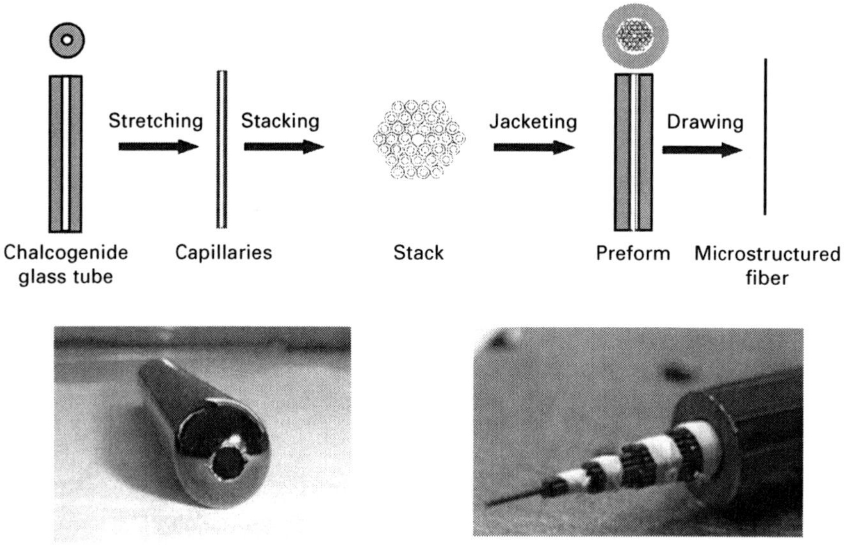

13.5 Stack and draw technique.

13.3.2 Molding method

The principle of the casting process is illustrated in Fig. 13.6. A purified glass rod is heated to a temperature at which it is almost liquid. The glass must be soft enough to flow in a structured silica mould [40]. Once the glass has filled the structured silica mould, the ampoule is quenched in air and annealed. The silica structure is then removed by hydrofluoric acid treatment. At this stage, a structured chalcogenide-glass preform is obtained (Fig. 13.6).

The mould is made entirely of silica capillaries thread in silica hexagonal guides. The silica guides are slices of a silica microstructured preform (Fig. 13.6(a)). Then, from these bulk chalcogenide preforms, fibers are drawn in a specific drawing tower under controlled dry atmosphere (He). Typically, the fiber diameter can vary from 300 μm to 100 μm. During the fibering step, the diameters of the holes are adjusted by applying pressure in the preform holes.

This method produces a variety of geometries (Fig. 13.7). Indeed, the fibers can have three rings of holes or a suspended core. The mode area can be controlled, and can be from 3 to 300 μm^2. With this method, MOF

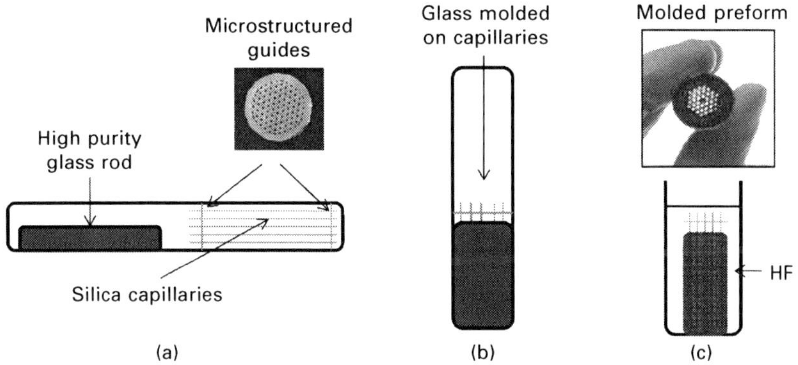

13.6 Molding method: (a) mold fabrication, (b) heating and flowing, (c) silica capillary removal.

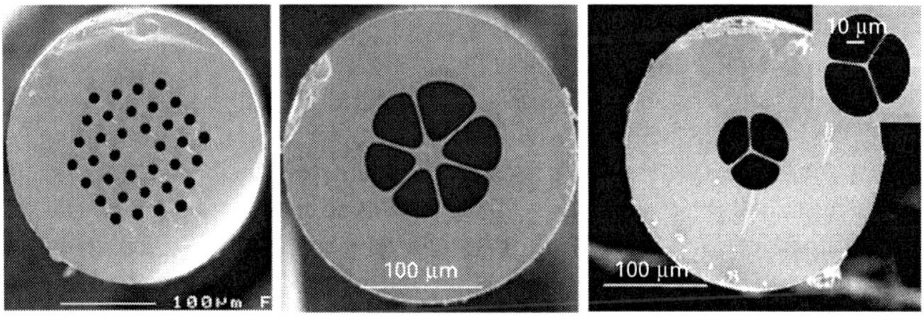

13.7 Geometries obtained by the molding technique.

fibers can present attenuation curves similar to those of classical single-index chalcogenide fibers. This is a real achievement compared to the results obtained with the 'stack and draw' technique. The combination of high-purity synthesis with the molding method reduces optical losses to smaller than 50 dB/km at 3.7 μm in a multimode MTIR-MOF [22].

13.3.3 Alternative methods including mechanical drilling and extrusion

The extrusion technique has shown its success with soft glasses like lead silicate [10], telluride [14, 41] and fluoride glasses [16]. For chalcogenide, no MOF has been published using this technique. However, a GLSO preform (Ga_2S_3-La_2S_3,-La_2O_3) has been extruded in 2005 [5], but the paper gave no results about the drawing of such extruded preform. For the elaboration of MOF preforms, drilling methods can be utilized. The first technique is the ultrasonic drilling used for the elaboration of lead silicate MOF preforms [5]. Nevertheless, any article about ultrasonic drilling concerns the preparation of chalcogenide MOF. However, mechanical drilling has been achieved in As_2S_3 to obtain complex preforms [42]. The diameter of the preform is about 12–16 mm and the diameter of the holes close to 0.8–1 mm and 40 mm in length. Various geometries can be obtained by mechanical drilling. Indeed, preforms can present 3, 6, 18, 36 or 64 holes (four rings of holes) [42]. The optical quality of fibers obtained by this technique is close to that of materials. As an example, the optical losses in those fibers can be as low as 0.35 dB/m at the telecom wavelength 1.55 μm [42].

13.3.4 Hybrid chalcogenide-silica fibers and all-solid fibers

Two types of all-solid chalcogenide MOFs have been published. The first family concerns hybrid fibers composed of silica and chalcogenide glasses. The second is a solid-core chalcogenide fiber where low-index inclusions of another chalcogenide glass substitute for the air inclusions.

In the first case, silica and chalcogenide glasses are drawn together. For silica, the drawing temperature is more than 1,800°C. At this high temperature, chalcogenide glasses would decompose. Consequently, the realization of such hybrid fiber necessitates the use of alternative methods.

Hybrid all-solid MOF can be elaborated by pressure-assisted filling of chalcogenide glasses into silica solid-core fibers [43, 44]. This process can produce several-centimeter long silica MOF filled with chalcogenide glasses (or telluride glasses [44]). The fibers can be filled with hole diameters of 0.2–15 μm. Thanks to this process, all-solid PBG-MOF can be obtained. The replacement of air inclusions by chalcogenide inclusions in a solid-core silica MOF, permits silica-core transmission by band-gap effect at several wavelengths in the visible spectrum (43). Supercontinuum generation, magneto-optics [45], and wavelength selective devices are example of applications that can be achieved by this original fabrication.

The second example of an all-solid chalcogenide MOF concerns the elaboration of an As_2Se_3 MTIR-MOF where air holes are replaced by As_2S_3 inclusions. In this case the, index contrast is largely smaller than in a conventional MTIR-MOF with air holes. However, core guiding can be obtained, as shown in Fig. 13.8.

13.4 Optical properties

13.4.1 Optical losses and transmission

Table 13.2 presents the optical losses of several chalcogenide MOFs. By using the 'stack and draw' technique, the minimum losses are 3 dB/m [21] and 10 dB/m [46] at 1.55 μm, for a multimode fiber and a single-mode fiber, respectively. Attenuations lower than 1 dB/m have been obtained by using mechanical drilling [32] or molding [22, 40].

The lowest attenuation, less than 0.05 dB/m, was observed at 3.7 μm in a six-hole suspended-core As-Se fiber, which presents multimode propagation [22]. This attenuation is close to the record of 0.012 dB/m measured in a classical step-index As_2S_3 fiber [54]. If we consider single-mode fibers at the telecom wavelength, then the best attenuation is 0.35 dB/m in a As_2S_3 suspended-core MOF and 1 dB/m in a Ge-As-Se MOF. Those attenuations are far from that for silica. However, for optical devices where short lengths

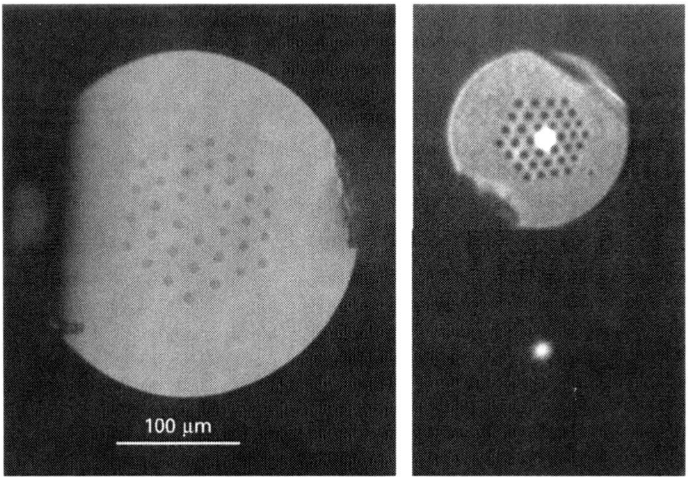

13.8 All-solid chalcogenide microstructured optical fiber: cross section and light guiding at 1.55 μm.

Table 13.2 Optical attenuation of different chalcogenide MOFs

Composition of the MOF	Process/core	Attenuation (dB/m)	Wavelength (μm)	Single mode[a]	Core ϕ (μm)	Year/[ref]
PES/$As_{40}Se_{60}$	Bragg fiber/Hollow-core	0.95	10.6	–	700–750	2002/[23]
$As_{40}Se_{60}$	S & D[b]/Solid-core	4.8	1.5	No	10	2006/[19]
$As_{40}Se_{60}$	S & D/Solid-core	5	1.55	No	13	2008/[21]
$As_{40}Se_{60}$	S & D/Solid-core	10	1.55	Yes	11	2009/[46]
$As_{38}Se_{62}$	Molding/Solid core	2	1.55	No	13	2010/[40]
$As_{38}Se_{62}$	Molding/Suspended core	4.5	1.55	Yes	2	2010/[47]
$As_{38}Se_{62}$	Molding/suspended core	0.8	1.55	No	3.5	2011/[48]
$As_{38}Se_{62}$	Molding/suspended core	0.4	1.55	No	22	2010/[22]
		< 0.05	3.7			
$As_{38}Se_{62}$	Molding/solid-core	1	1.55	Yes	12	2010/[22]
		0.55	1.995	Yes		
		0.2	3.7	Yes		
$As_{38}Se_{62}$	Molding/suspended core	0.9	1.55	–	1.13	2011/[49]
Tellurite/$As_{40}S_{60}$	S & D/Solid-core	18.3	1.55	Yes	18	2009/[28]
$As_{40}s_{60}$	Drilling/Suspended core	0.35	1.55	Yes	2.3	2010/[32]
$Ge_{15}Sb_{20}S_{65}$	S & D/Solid-core	15	1.55	No	22	2008/[50]
$Ge_{15}Sb_{20}S_{65}$	S & D/Solid-core	35	1.55	Yes	4	2008/[50]
$Ge_{15}Sb_{20}S_{65}$	S & D/Solid-core	3	1.5	No	13	2008/[21]
		4.5	3.39			
$Ge_{15}Sb_{20}S_{65}$	S & D/Solid-core	5	1.55	Yes	15	2009/[46]
$Te_{20}AS_{30}Se_{50}$	S & D/Solid-core	9	3.39	No	16	2009/[51]
		6	9.3			
$Te_{20}AS_{30}Se_{50}$	S & D/Solid-core	⊕ 20	3.39	Yes	28	2009/[51]
		⊕ 20	9.3			

$Te_{20}AS_{30}Se_{50}$	Molding/Solid core	0.4	3.39	Yes	12	2011/[52]
		0.4	7			
		4.5	4.5			
$Te_{20}AS_{30}Se_{50}$	S & D/Hollow core	11	10.6	–	380	2011/[25]
$Ge_{10}AS_{22}Se_{68}$	Molding/solid core	1	1.55	Yes	4	2012/[53]
		0.7	3.0			

[a]Relative optical losses of higher modes are (at least) one order more than the fundamental mode.
[b]Stack and draw.

of nonlinear fibers are needed, chalcogenide MOFs possess a high potential (see Section 13.5 on nonlinear properties).

For longer wavelengths in the IR, Fig. 13.9 compares the optical losses obtained by the 'stack and draw' and the molding techniques. Figure 13.9(a) presents the attenuation curve before and after a 'stack and draw' process. In Fig. 13.9(b), the optical losses of chalcogenide MOFs drawn from a molded preform are compared to the materials losses measured before the shaping process. One can note that the implementation of the molding technique avoids the occurrence of strong additional optical losses. The high optical losses observed in fibers prepared by the 'stack and draw' technique are explained by the poor quality of the interfaces when capillaries collapse upon drawing [21].

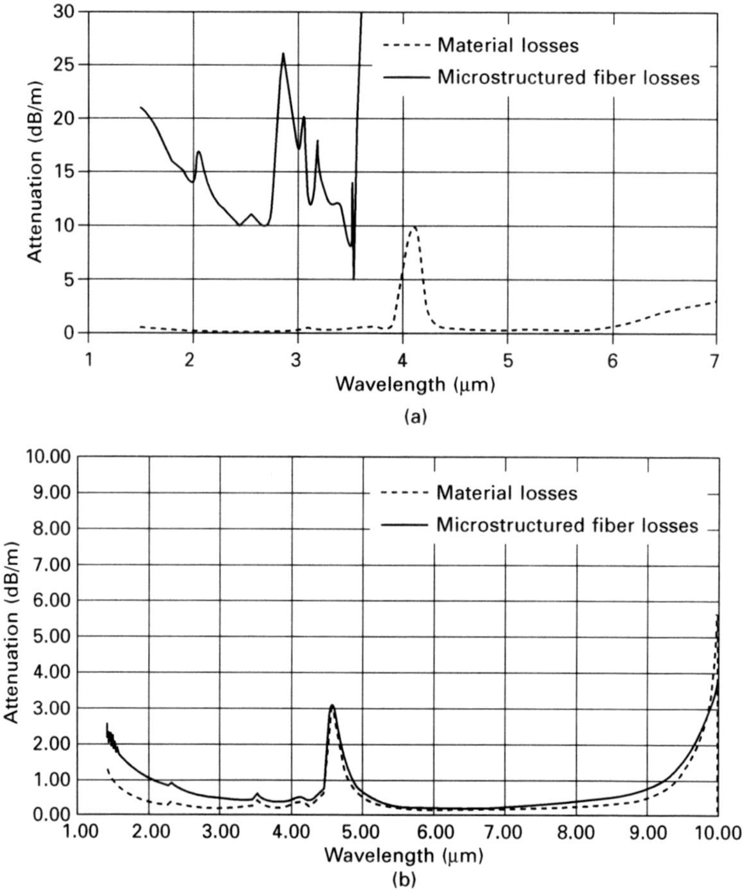

13.9 Optical losses of chalcogenide MOFs (straight lines) compared to the material ones (dotted lines): (a) using the 'stack and draw' technique, (b) using the molding technique.

13.4.2 Hollow-core fibers

The most famous hollow-core chalcogenide fibers are hybrid structured Bragg fibers (Fig. 13.1(a)). These hollow-core fibers are waveguides composed of successive layers of low-index material (PES: polyethersulphone) and high-index layers (As_2Se_3) [23]. In this case, fibers are photonic band-gap fibers where the light propagates in air. Depending on the geometry, the position of the band gap permits transmission at different wavelengths, in the visible or in the infrared, the optical attenuation at 10.6 μm (CO_2 laser) has been found to be close to 1 dB/m [23]. Typically, these Bragg fibers present large core (few hundreds μm) that is suitable for high energy transmission. However, the light propagation does not present a single-mode regime.

Only two publications concern the realization of all-chalcogenide hollow fibers [24, 25]. The first article presents the drawing of six-ring hollow-core fibers (Fig. 13.2(c)), which corresponds to theoretical geometries suitable for bang-gap guiding at 9.3 μm [24]. Unfortunately, due to numerous defects in the geometry of the fibers, no light propagation could be observed in the hollow core of the fibers. Another realization with a chalcogenide glass ($Te_{20}As_{30}Se_{50}$) fiber with negative curvature hollow core has been demonstrated [25] (Fig. 13.2(d)). In this case, the light propagation of a CO_2 laser at 10.6 μm was obtained with an attenuation of around 11 dB/m. The outer diameter of the obtained fiber was 750 μm, the core diameter was 380 μm, and the capillary wall thickness was 13 μm (Fig. 13.2(d)).

13.4.3 Single-mode fibers

For solid-core MOFs, the single-mode regime strongly depends on the geometry of the fibers. The single-mode regime depends on the ratio d/Λ, the number of rings, the core size and the wavelength. For structures with a large number of air-hole rings, the critical d/Λ ratio, which defines the constantly single-mode regime, is in the interval [0.42–0.425] both for silica and chalcogenide matrices [9, 36]. However, the number of rings for commonly fabricated chalcogenide MTIR-MOFs is limited to three or four. This parameter implies that the finite size effects can modify the d/Λ ratio needed for an constant single-mode behavior. Consequently, the single-mode behavior of a MOF must be numerically computed with the exact opto-geometrical structure. However, numerous chalcogenide MOFs, with a low number of rings of holes have shown a single-mode-behavior. Indeed, the single-mode regime can be observed because guiding losses for the second mode are much higher than that of the fundamental mode. As an example, single-mode behavior has been observed in a $Te_{20}As_{30}Se_{50}$ MTIR-MOF presenting three rings of holes [51]. In this fiber, the fundamental guiding losses, at 9.3 μm, were calculated to be 1.61 dB/m, while the second-mode guiding losses were computed at 750 dB/m. In this example, after 10 cm of fiber length, only $3 \times 10^{-6}\%$ of the second mode energy propagates in the core of the fiber.

Of course, for specific applications, this single-mode behavior can not be sufficient. For example, for interferometry or telecom optical functions, a better single-mode guiding may be required. So, the geometry of the fibers must be optimized according to the application.

13.4.4 Chalcogenide MOFs for sensing

Chalcogenide fibers have been successfully implemented in fiber evanescent spectroscopy experiments, for detection of biochemical molecules in various fields of application, including water pollution [55, 56], microbiology, and medicine [57, 58]. It has been shown that the infrared signature of gases embedded in the holes of chalcogenide MTIR-MOF can be detected. Indeed, the signature of CO_2 gas has been observed at 4.2 μm, through a 36-hole As_2S_3 MOF filled by pure CO_2 [59]. However, in this case, the design does not permit sufficient optical interactions between the core of the fiber and the chemical species. Indeed, to be detected, the chemical species must fill the fiber holes. Besides, a new configuration of microstructured fibers was developed: microstructured exposed-core fibers [60]. This design, which was proposed for the first time by Hoo *et al.* in 2003, consists of an optical fiber with a suspended micron-scale core that is partially exposed to the external environment [61]. In 2012, an exposed-core chalcogenide MOF was demonstrated for the first time [62]. This configuration has been chosen to elaborate a chalcogenide fiber suitable for chemical species detection. Figure 13.10 shows the infrared signatures of propan-2-ol molecules recorded with a 110 μm diameter single-index fiber and with a chalcogenide MTIR-MOF presenting a core diameter of 15 μm (with an outer diameter of 220 μm). Although evanescent wave absorption is inversely proportional to the fiber diameter, the result shows that an exposed-core fiber is much more sensitive than a single-index fiber with a twice-smaller external diameter.

13.5 Nonlinear optical properties

13.5.1 Telecommunication applications

The use of nonlinear effects has proved to be a promising technique to achieve all-optical fast processing such as data regeneration, time-domain demultiplexing, or wavelength conversion. Due to their high nonlinear refractive index (Table 13.1), and fast response time, chalcogenide glasses are one of the favorite choices for fabricating highly nonlinear components. These nonlinear properties can be further enhanced by drawing chalcogenide

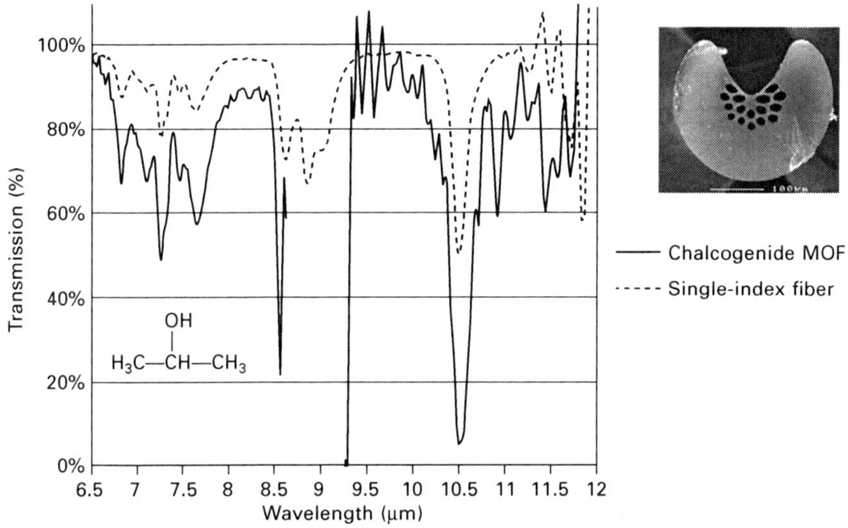

13.10 Infrared signature of propan-2-ol on the surface of a single-index fiber (outer diameter = 110 μm) and a chalcogenide exposed core MOF (outer diameter = 220 μm, core diameter = 15 μm). The inset shows the cross section of the chalcogenide exposed core MOF.

glasses into MOFs, due to the possibility of designing such fibers with a very small core diameter. The nonlinear parameter $\gamma = 2\pi n_2/\gamma\lambda A_{eff}$ (n_2 is the nonlinear refractive index, A_{eff} is the effective area of the mode and λ is the wavelength) allows the non-linear properties in different fibers to be compared.

As shown in Table 13.3, highly nonlinear coefficients are obtained in chalcogenide fibers. As a comparison, the nonlinear parameter is equal to 70 W^{-1} km^{-1} in silica MOF [63], while it can reach 46,000 W^{-1} km^{-1} in chalocogenide MOF. One can note that high nonlinear coefficient of 133,000 $W^{-1}.km^{-1}$ and 94,300 $W^{-1}.km^{-1}$ have been achieved with short-length chalcogenide microtapers [69] or nanowires, respectively [72]. Self-phase modulation (SPM) broadening has been observed in high nonlinear chalcogenides [42, 46, 47, 73]. A Mamyshef regenerator [74] has been investigated with a As_2S_3 suspended core MOF [42]. A transfer function was obtained, but the average power needed to observe all-optical regeneration was much higher than the damage threshold of the fiber. Concerning wavelength conversion, several studies have been devoted to four-wave-mixing (FWM) effects [47, 49, 64]. In these cases, both pulsed pump (λ_1) and CW source (λ_2) are injected in a chalcogenide MOF. At the output of fibers, new wavelengths (λ_3 and λ_4) appear in symmetrical pairs. The generated wavelengths depend on the wavelength detuning between the pulsed signal and the CW source, with $1/\lambda_3 + 1/\lambda_4 = 1/\lambda_1 + 1/\lambda_2$. Figure 13.11 shows an

Table 13.3 Nonlinear properties for different fibers at 1.55 μm

Fiber	γ ($W^{-1}.km^{-1}$)	α (dB/m)	Dc (ps/nm/km)	Raman gain (m/W)	Brillouin gain (m/W)	Ref
MTIR-MOF						
SiO_2	70	0.19	–30	–	–	[63]
As_2S_3	1500	2	–260	–	–	[64]
As_2S_3	2750	0.35	–240	–	–	[65]
$Ge_{15}Sb_{20}S_{65}$	227	5.5	–406	$1.8.10^{-11}$	8.10^{-10}	[46]
TeO_2/As_2S_3	9300	18.3	–250	–	–	[28]
$As_{38}Se_{62}$	31300	4.6	–820	–	–	[47]
$As_{38}Se_{62}$	46000	0.9	–300	–	–	[49]
$As_{38}Se_{62}$	7600	1	–	–	$5.5.10^{-9}$	[66]
$As_{38}Se_{62}$ (at 2 μm)	5500	0.8	–350	$1.6.10^{-11}$	–	[48]
$Ge_{10}As_{22}Se_{68}$	11000	2.5	–780	–	–	[67]
Others						
SiO_2 standard	1	0.0002		$6.7.10^{-14}$	$4.5.10^{-11}$	[68]
As_2Se_3microtaper	133000	1.8	–160	–	–	[69]
As_2Se_3 SIF	1600	0.84	–	–	$6.0.10^{-9}$	[70]
As_2Se_3 SIF	1600	0.7	–	$2.3.10^{-11}$	–	[71]

Dc = Chromatic dispersion at 1.55 μm; MTIR-MOF = Modified total internal reflexion-microstuctured optical fiber; SIF = Step-Index Fiber.

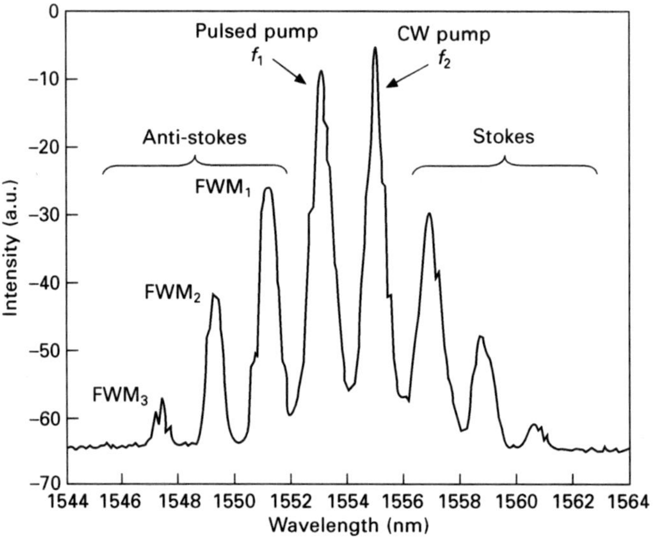

13.11 Cascaded four-wave-mixing in an As-Se suspended core fiber at 10 GHz [49].

example of FWM in a chalcogenide MOF presenting a nonlinear coefficient of around 46,000 $W^{-1}.km^{-1}$. In this fiber, successive orders of FWM can be obtained [49].

The conversion bandwidth that can be obtained strongly depends on the chromatic dispersion multiplied by the length of the device [64]. So, for large wavelength conversion by FWM, compromise between the nonlinear coefficient, the chromatic dispersion and the fiber length has to be considered. However, this approach permits wavelength conversion of a 42.7 Gbit/s return to zero (RZ) 33% signal in a 1 m long GeAsSe MOF with a total average power of 56 mW only [67]. Wavelength conversion has also been demonstrated in a chalcogenide chip, but with a total power of 400 mW [75].

13.5.2 Raman and Brillouin scattering

Stimulated Raman scattering (SRS) and stimulated Brillouin scattering (SBS) in optical fibers are important nonlinear interactions that can be exploited to obtain wavelength conversion, lasers and amplifiers.

The values of the Brillouin gain, gB, measured in chalcogenide MOFs are close to the gain measured in step-index fibers [66, 70] (Table 13.3). However, in both cases, the Brillouin gain can attain more than 120 times the gain measured in a silica fiber. For As-Se glasses, the two fiber designs have been investigated. And only Brillouin thresholds are significantly different. Indeed, the Brillouin threshold is 85 mW in a 4.9 m-long SIF [70], while a threshold of only 22 mW was observed with only 3-m long MOF [66]. The single frequency Brillouin laser made of this fiber has also shown a promising frequency noise reduction of around 6 dB compared to the pump laser.

One can note that the Brillouin shift is around 8.2 GHz in a sulfur fiber [46], whereas it is measured at 7.95 GHz in As-Se fibers (SIF or MOF) [66, 70].

Stimulated Raman scattering has been investigated as well, in chalcogenide fibers (SIFs and MOFs) [48, 71, 76, 77]. The observed Raman gains are around 200–300 times greater than that of silica at 1.55 μm. However, the Raman shift for chalcogenides (200–250 cm^{-1}) is around two times smaller than the Raman shift of silica (≈440 cm^{-1}) due to the presence of heavier atoms in chalcogenide glasses, as compared to silica. Raman gain is of the same order of magnitude in both chalcogenide MOF and SIF (Table 13.3). However, by pumping at 2 μm, third and fourth cascaded Raman shifts have been observed in an As-Se suspended core fiber [48]. The Stokes order appears at wavelength of 2092, 2205, 2330 and 2450 nm.

13.5.3 New IR sources and generation of supercontinuum

This application is one of the most promising applications for chalcogenide MOF due to their high nonlinearity and their broadband transmission in the infrared. Numerous simulations and computing studies in such glassy matrix have confirmed this high potential [37, 38, 78]. Indeed, simulations have shown that it is possible to reach wavelength greater than 5 μm in infrared by pumping with a smaller wavelength (2.5 μm) [37]. Several experimental studies have demonstrated the generation of supercontinuum in chalcogenide MTIR-MOF by pumping with either a fiber laser emitting at 1.55 μm (400 fs pulses) [42], or a thulium-doped silica fiber at 2 μm (4 ps pulses) [79] or a Ti:sapphire laser at 2.5 μm [19] (Fig. 13.12). The largest spectral broadening was observed with a As_2Se_3 MOF, from 2.1 μm up to 3.2 μm by pumping with a Ti:sapphire laser at a wavelength of 2.5 μm using 100 fs pulses.

To observe an efficient and broad supercontinuum, the pump wavelength has to be close to the ZDW of the fiber. However, the intrinsic ZDWs of chalcogenide bulk glasses are quite far in the mid-IR, 4.8 μm and 7.2 μm for As_2S_3 and As_2Se_3, respectively. Thanks to the realization of adequate MOF designs, the total chromatic dispersion can be tuned toward shorter wavelengths [48, 65, 71, 76, 77]. For example, in a As_2S_3 suspended-core fiber, the ZDW is equal to 2 μm while the core diameter is around 2.6 μm [65]. To obtain the same ZDW in a As_2Se_3 fiber, the core size needs to be close to 1 μm. To maintain reasonable conditions of injection at the fiber input, a small core is achieved in the middle part of the fiber only, by tapering techniques [79]. However, even with this special design, spectral broadening remains in the 1.3–2.6 μm range by pumping at a wavelength of 2 μm with pulse duration of 2 μm, as shown in Fig. 13.12(c) [79].

In conclusion, chalcogenide MOFs are candidates for generation of supercontinuum in the IR. However, the best compromise between the nonlinear coefficient, the core size, and the dispersion has to be found to obtain optimal conditions for the realization of new fibered IR sources.

13.6 Conclusion

Since the first realization of a chalcogenide microstructured optical fiber, strong improvements have been achieved: regular geometries, low-loss fibers, single-mode fibers and highly nonlinear fibers. These improvements open new domains of investigation in numerous fields of application such as telecom functions, interferometry, generation of new infrared sources and sensors. So, as shown in Fig. 13.13, we can consider the future of chalcogenide MOF with a smile.

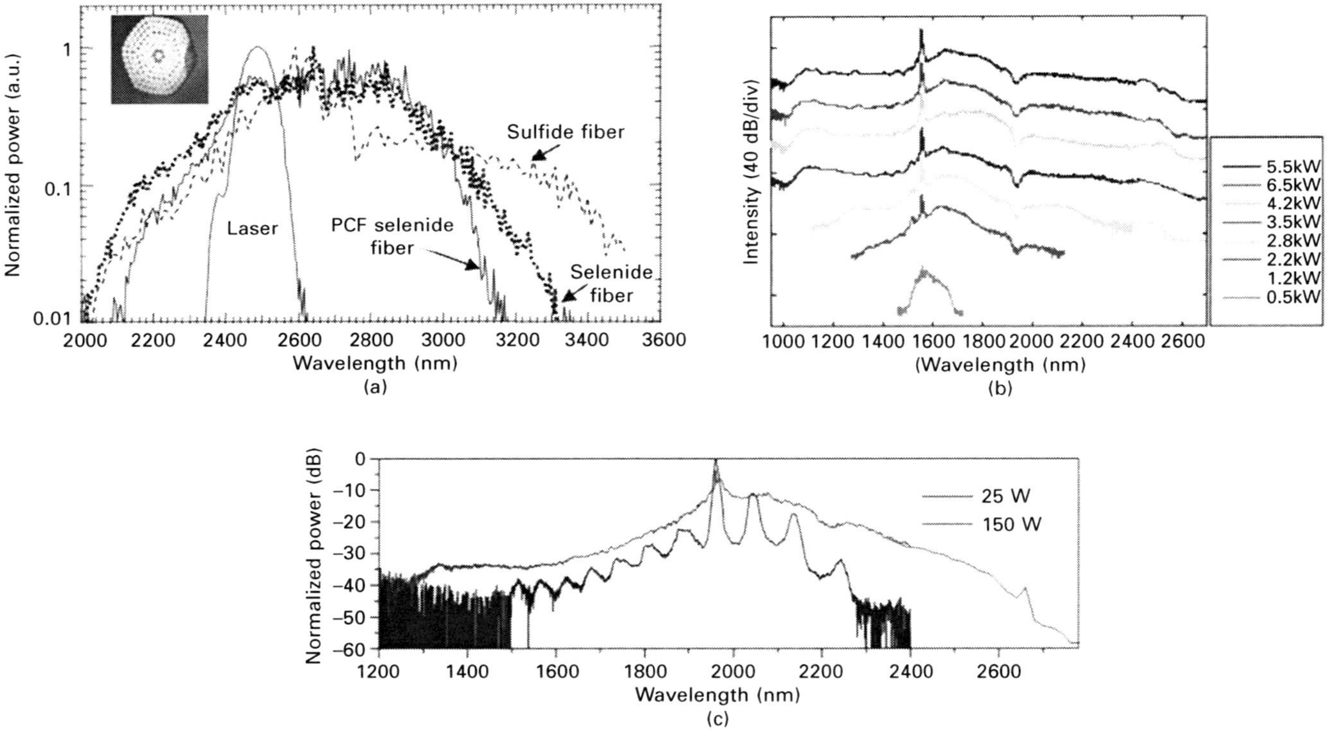

13.12 Supercontinuum obtained in chalcogenide MTIR-MOF by pumping: (a) with a Ti:sapphire at a wavelength of 2.5 μm [19], (b) with a fiber laser at 1.55 μm (400 fs pulses) [42], (c) with a thulium-doped silica fiber at 2 μm (4 ps pulses) [79].

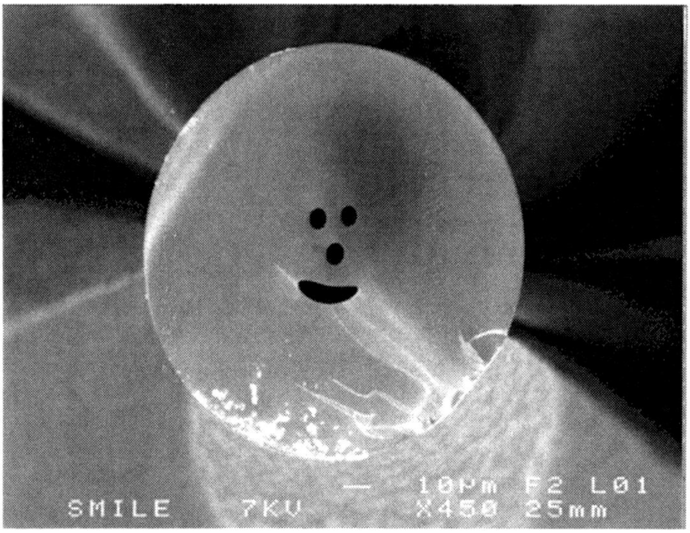

13.13 Original chalcogenide MOF!

13.7 References

1. P. V. Kaiser and H. W. Astle, 'Low-loss single-material fibers made from pure fused silica,' *Bell Syst. Tech. J* **53**, 1021–1039 (1974).
2. T. A. Birks, P. J. Roberts, P. S. J. Russell, D. M. Atkin and T. J. Shepherd, 'Full 2-D photonic bandgaps in silica/air structures,' *Electronics Letters* **31**, 1941–1943 (1995).
3. T. A. Birks, J. C. Knight and P. S. Russell, 'Endlessly single-mode photonic crystal fiber,' *Opt. Lett.* **22**, 961–963 (1997).
4. J. C. Knight, T. A. Birks, B. J. Manjan and P. S. J. Russell, 'Microstructured silica as an optical-fiber material,' *MRS Bulletin* **26**, 614–617 (2001).
5. X. Feng, A. K. Mairaj, D. W. Hewak and T. M. Monro, 'Nonsilica glasses for holey fibers,' *J. Lightwave Technol* **23**, 2046–2054 (2005).
6. T. M. Monro, Y. D. West, D. W. Hewak, N. G. R. Broderick and D. J. Richardson, 'Chalcogenide holey fibres,' *Electronics Letters* **36**, 1998–2000 (2000).
7. P. Yeh, A. Yariv and E. Marom, 'Theory of Bragg fiber,' *J. Opt. Soc. Am.* **68**, 1196 (1978).
8. R. F. Cregan, B. J. Mangan, J. C. Knight, T. A. Birks, P. S. Russell, P. J. Roberts and D. C. Allan, 'Single-mode photonic band gap guidance of light in air,' *Science* **285**, 1537–1539 (1999).
9. G. Renversez, F. Bordas and B. T. Kuhlmey, 'Second mode transition in microstructured optical fibers: determination of the critical geometrical parameter and study of the matrix refractive index and effects of cladding size,' *Opt. Lett.* **30**, 1264–1266 (2005).
10. V. V. R. Kumar, A. George, W. Reeves, J. Knight, P. Russell, F. Omenetto and A. Taylor, 'Extruded soft glass photonic crystal fiber for ultrabroad supercontinuum generation,' *Opt. Express* **10**, 1520–1525 (2002).
11. H. Ebendorff-Heidepriem, P. Petropoulos, S. Asimakis, V. Finazzi, R. Moore, K.

Frampton, F. Koizumi, D. Richardson and T. Monro, 'Bismuth glass holey fibers with high nonlinearity,' *Opt. Express* **12**, 5082–5087 (2004).

12. H. Ebendorff-Heidepriem and T. M. Monro, 'Extrusion of complex preforms for microstructured optical fibers,' *Opt. Express* **15**, 15086–15092 (2007).
13. X. Feng, W. H. Loh, J. C. Flanagan, A. Camerlingo, S. Dasgupta, P. Petropoulos, P. Horak, K. E. Frampton, N. M. White, J. H. Price, H. N. Rutt and D. J. Richardson, 'Single-mode tellurite glass holey fiber with extremely large mode area for infrared nonlinear applications,' *Opt. Express* **16**, 13651–13656 (2008).
14. V. Kumar, A. George, J. Knight, and P. Russell, 'Tellurite photonic crystal fiber,' *Opt. Express* **11**, 2641–2645 (2003).
15. P. Domachuk, N. A. Wolchover, M. Cronin-Golomb, A. Wang, A. K. George, C. M. B. Cordeiro, J. C. Knight and F. G. Omenetto, 'Over 4000 nm bandwidth of mid-IR supercontinuum generation in sub-centimeter segments of highly nonlinear tellurite PCFs,' *Opt. Express* **16**, 7161–7168 (2008).
16. H. Ebendorff-Heidepriem, T. C. Foo, R. C. Moore, W. Q. Zhang, Y. H. Li, T. M. Monro, A. Hemming and D. G. Lancaster, 'Fluoride glass microstructured optical fiber with large mode area and mid-infrared transmission,' *Opt. Lett.* **33**, 2861–2863 (2008).
17. P. McNamara, D. G. Lancaster, R. Bailey, A. Hemming, P. Henry and R. H. Mair, 'A large core microstructured fluoride glass optical fibre for mid-infrared single-mode transmission,' *J. Non-Cryst. Solids* **355**, 1461–1467 (2009).
18. L. Brilland, F. Smektala, G. Renversez, T. Chartier, J. Troles, T. N. Nguyen, N. Traynor and A. Monteville, 'Fabrication of complex structures of Holey Fibers in chalcogenide glass,' *Opt. Express* **14**, 1280–1285 (2006).
19. J. S. Sanghera, I. D. Aggarwal, L. B. Shaw, C. M. Florea, P. Pureza, V. Q. Nguyen and F. Kung, 'Nonlinear properties of chalcogenide glass fibers,' *J. Optoelectron. Adv. Mater.* **8**, 2148–2155 (2006).
20. J. Le Person, F. Smektala, T. Chartier, L. Brilland, T. Jouan, J. Troles and D. Bosc, 'Light guidance in new chalcogenide holey fibres from GeGaSbS glass,' *Mater. Res. Bulletin* **41**, 1303–1309 (2006).
21. L. Brilland, J. Troles, P. Houizot, F. Desevedavy, Q. Coulombier, G. Renversez, T. Chartier, T. N. Nguyen, J. L. Adam and N. Traynor, 'Interfaces impact on the transmission of chalcogenide photonic crystal fibres,' *J. Ceram. Soc. Jpn* **116**, 1024–1027 (2008).
22. J. Troles, Q. Coulombier, G. Canat, M. Duhant, W. Renard, P. Toupin, L. Calvez, G. Renversez, F. Smektala, M. El Amraoui, J. L. Adam, T. Chartier, D. Mechin and L. Brilland, 'Low loss microstructured chalcogenide fibers for large non linear effects at 1995 nm,' *Opt. Express* **18**, 26647–26654 (2010).
23. B. Temelkuran, S. D. Hart, G. Benoit, J. D. Joannopoulos and Y. Fink, 'Wavelength-scalable hollow optical fibres with large photonic bandgaps for CO_2 laser transmission,' *Nature* **420**, 650–653 (2002).
24. F. Desevedavy, G. Renversez, J. Troles, P. Houizot, L. Brilland, I. Vasilief, Q. Coulombier, N. Traynor, F. Smektala and J.-L. Adam, 'Chalcogenide glass hollow core photonic crystal fibers,' *Opt. Mater.* **32**, 1532–1539 (2010).
25. A. F. Kosolapov, A. D. Pryamikov, A. S. Biriukov, V. S. Shiryaev, M. S. Astapovich, G. E. Snopatin, V. G. Plotnichenko, M. F. Churbanov and E. M. Dianov, 'Demonstration of CO_2-laser power delivery through chalcogenide-glass fiber with negative-curvature hollow core,' *Opt. Express* **19**, 25723–25728 (2011).
26. A. Boskovic, S. V. Chernikov, J. R. Taylor, L. Gruner-Nielsen and O. A. Levring,

'Direct continuous-wave measurement of n_2 in various types of telecommunication fiber at 1.55 μm,' *Opt. Lett.* **21**, 1966–1968 (1996).

27. S. Fujino and K. Morinaga, 'Material dispersion and its compositional parameter of oxide glasses,' *J. Non-Crystal. Solids* **222**, 316–320 (1997).
28. M. Liao, C. Chaudhari, G. Qin, X. Yan, C. Kito, T. Suzuki, Y. Ohishi, M. Matsumoto and T. Misumi, 'Fabrication and characterization of a chalcogenide-tellurite composite microstructure fiber with high nonlinearity,' *Opt. Express* **17**, 21608–21614 (2009).
29. C. Xia, M. Kumar, O. P. Kulkarni, M. N. Islam, J. F. L. Terry, M. J. Freeman, M. Poulain and G. Mazé, 'Mid-infrared supercontinuum generation to 4.5 μm in ZBLAN fluoride fibers by nanosecond diode pumping,' *Opt. Lett.* **31**, 2553–2555 (2006).
30. C. Day, P. France, S. Carter, M. Moore and J. Williams, 'Fluoride fibres for optical transmission,' *Opt. Quant. Electron.* **22**, 259–277 (1990).
31. J. M. Harbold, F. Ilday, F. W. Wise, J. S. Sanghera, V. Q. Nguyen, L. B. Shaw and I. D. Aggarwal, 'Highly nonlinear As-S-Se glasses for all-optical switching,' *Opt. Lett.* **27**, 119–121 (2002).
32. M. El-Amraoui, J. Fatome, J. C. Jules, B. Kibler, G. Gadret, C. Fortier, F. Smektala, I. Skripatchev, C. F. Polacchini, Y. Messaddeq, J. Troles, L. Brilland, M. Szpulak and G. Renversez, 'Strong infrared spectral broadening in low-loss As-S chalcogenide suspended core microstructured optical fibers,' *Opt. Express* **18**, 4547–4556 (2010).
33. J. M. Harbold, F. O. Ilday, F. W. Wise and B. G. Aitken, 'Highly nonlinear Ge-As-Se and Ge-As-S-Se glasses for all-optical switching,' *IEEE Photonics Technology Letters* **14**, 822–824 (2002).
34. J. Troles, L. Brilland, F. Smektala, P. Houizot, F. Desevedavy, Q. Coulombier, N. Traynor, T. Chartier, T. N. Nguyen, J. L. Adam and G. Renversez, 'Chalcogenide microstructured fibers for infrared systems, elaboration modelization, and characterization,' *Fiber and Integrated Optics* **28**, 11–26 (2009).
35. F. Zolla, G. Renversez, A. Nicolet, B. Kuhlmey, S. Guenneau, D. Felbacq, A. Argyros and S. Leon-Saval *Foundations of Photonic Crystal Fibres*, 2nd edn *Imperial College Press*, London (2012).
36. G. Renversez, B. Kuhlmey and R. McPhedran, 'Dispersion management with microstructured optical fibers: ultraflattened chromatic dispersion with low losses,' *Opt. Lett.* **28**, 989–991 (2003).
37. J. Hu, C. R. Menyuk, L. B. Shaw, J. S. Sanghera and I. D. Aggarwal, 'Maximizing the bandwidth of supercontinuum generation in As_2Se_3 chalcogenide fibers,' *Opt. Express* **18**, 6722–6739 (2010).
38. B. Dabas and R. K. Sinha, 'Dispersion characteristic of hexagonal and square lattice chalcogenide As_2Se_3 glass photonic crystal fiber,' *Opt. Commun.* **283**, 1331–1337 (2010).
39. J. C. Knight, T. A. Birks, P. S. J. Russell and D. M. Atkin, 'All-silica single-mode optical fiber with photonic crystal cladding,' *Opt. Lett.* **21**, 1547 (1996).
40. Q. Coulombier, L. Brilland, P. Houizot, T. Chartier, T. N. Nguyen, F. Smektala, G. Renversez, A. Monteville, D. Méchin, T. Pain, H. Orain, J.-C. Sangleboeuf and J. Trolès, 'Casting method for producing low-loss chalcogenide microstructured optical fibers,' *Opt. Express* **18**, 9107–9112 (2010).
41. X. Feng, T. M. Monro, V. Finazzi, R. C. Moore, K. Frampton, P. Petropoulos and D. J. Richardson, 'Extruded single-mode, high-nonlinearity, tellurite glass holey fibre,' *Electronics Lett.* **41**, 835–837 (2005).

42. M. El-Amraoui, G. Gadret, J. C. Jules, J. Fatome, C. Fortier, F. Désévédavy, I. Skripatchev, Y. Messaddeq, J. Troles, L. Brilland, W. Gao, T. Suzuki, Y. Ohishi and F. Smektala, 'Microstructured chalcogenide optical fibers from As_2S_3 glass: towards new IR broadband sources,' *Opt. Express* **18**, 26655–26665 (2010).
43. N. Granzow, P. Uebel, M. A. Schmidt, A. S. Tverjanovich, L. Wondraczek and P. S. J. Russell, 'Bandgap guidance in hybrid chalcogenide–silica photonic crystal fibers,' *Opt. Lett.* **36**, 2432–2434 (2011).
44. N. Da, L. Wondraczek, M. A. Schmidt, N. Granzow and P. S. J. Russell, 'High index-contrast all-solid photonic crystal fibers by pressure-assisted melt infiltration of silica matrices,' *J. Non-Crystal. Solids* **356**, 1829–1836 (2010).
45. M. A. Schmidt, L. Wondraczek, H. W. Lee, N. Granzow, N. Da and P. S. Russell, 'Complex Faraday rotation in microstructured magneto-optical fiber waveguides,' *Adv. Mater.* **23**, 2681–2688 (2011).
46. J. Fatome, C. Fortier, T. N. Nguyen, T. Chartier, F. Smektala, K. Messaad, B. Kibler, S. Pitois, G. Gadret, C. Finot, J. Troles, F. Desevedavy, P. Houizot, G. Renversez, L. Brilland, and N. Traynor, 'Linear and nonlinear characterizations of chalcogenide photonic crystal fibers,' *J. Lightwave Technol.* **27**, 1707–1715 (2009).
47. D. M. Nguyen, S. D. Le, K. Lengle, D. Méchin, M. Thual, T. Chartier, Q. Coulombier, J. Troles, L. Bramerie and L. Brilland, 'Demonstration of nonlinear effects in an ultra-highly nonlinear AsSe suspended-core chalcogenide fiber,' *IEEE Photonics Technology Letters* **22**, 1844–1846 (2010).
48. M. Duhant, W. Renard, G. Canat, T. N. Nguyen, F. Smektala, J. Troles, Q. Coulombier, P. Toupin, L. Brilland, P. Bourdon, and G. Renversez, 'Fourth-order cascaded Raman shift in AsSe chalcogenide suspended-core fiber pumped at 2 μm,' *Opt. Lett.* **36**, 2859–2861 (2011).
49. S. D. Le, D. M. Nguyen, M. Thual, L. Bramerie, M. Costa e Silva, K. Lenglé, M. Gay, T. Chartier, L. Brilland, D. Méchin, P. Toupin and J. Troles, 'Efficient four-wave mixing in an ultra-highly nonlinear suspended-core chalcogenide $As_{38}Se_{62}$ fiber,' *Opt. Express* **19**, B653–B660 (2011).
50. F. Desevedavy, G. Renversez, L. Brilland, P. Houizot, J. Troles, Q. Coulombier, F. Smektala, N. Traynor and J. L. Adam, 'Small-core chalcogenide microstructured fibers for the infrared,' *Applied Optics* **47**, 6014–6021 (2008).
51. F. Desevedavy, G. Renversez, J. Troles, L. Brilland, P. Houizot, Q. Coulombier, F. Smektala, N. Traynor and J. L. Adam, 'Te-As-Se glass microstructured optical fiber for the middle infrared,' *Applied Optics* **48**, 3860–3865 (2009).
52. C. Conseil, Q. Coulombier, C. Boussard-Pledel, J. Troles, L. Brilland, G. Renversez, D. Mechin, B. Bureau, J. L. Adam and J. Lucas, 'Chalcogenide step index and microstructured single mode fibers,' *J. Non-Crystal. Solids* **357**, 2480–2483 (2011).
53. P. Toupin, L. Brilland, J. Trolès and J.-L. Adam, 'Small core Ge-As-Se microstructured optical fiber with single-mode propagation and low optical losses,' *Opt. Mater. Express* **2**, 1359–1366 (2012).
54. G. Snopatin, V. Shiryaev, V. Plotnichenko, E. Dianov and M. Churbanov, 'High-purity chalcogenide glasses for fiber optics,' *Inorganic Materials* **45**, 1439–1460 (2009).
55. K. Michel, B. Bureau, C. Pouvreau, J. C. Sangleboeuf, C. Boussard-Plédel, T. Jouan, T. Rouxel, J. L. Adam, K. Staubmann, H. Steinner, T. Baumann, A. Katzir, J. Bayona and W. Konz, 'Development of a chalcogenide glass fiber device for *in situ* pollutant detection,' *J. Non-Crystal. Solids* **326–327**, 434–438 (2003).

56. K. Michel, B. Bureau, C. Boussard-Pledel, T. Jouan, J. L. Adam, K. Staubmann and T. Baumann, 'Monitoring of pollutant in waste water by infrared spectroscopy using chalcogenide glass optical fibers,' *Sensors and Actuators B: Chemical* **101**, 252–259 (2004).
57. J. Keirsse, C. Boussard-Pledel, O. Loreal, O. Sire, B. Bureau, B. Turlin, P. Leroyer and J. Lucas, 'Chalcogenide glass fibers used as biosensors,' *J. Non-Crystal. Solids* **326–327**, 430–433 (2003).
58. M. L. Brandily, V. Monbet, B. Bureau, C. Boussard-Pledel, O. Loreal, J. L. Adam and O. Sire, 'Identification of foodborne pathogens within food matrices by IR spectroscopy,' *Sensors and Actuators B: Chemical* **160**, 202–206 (2011).
59. F. Charpentier, J. Troles, Q. Coulombier, L. Brilland, P. Houizot, F. Smektala, C. Boussard-Pledel, V. Nazabal, N. Thibaud, K. Le Pierres, G. Renversez and B. Bureau, 'CO_2 detection using microstructured chalcogenide fibers,' *Sens. Lett.* **7**, 745–749 (2009).
60. S. C. Warren-Smith, S. Afshar V and T. M. Monro, 'Theoretical study of liquid-immersed exposed-core microstructured optical fibers for sensing,' *Opt. Express* **16**, 9034–9045 (2008).
61. Y. L. Hoo, W. Jin, C. Shi, H. L. Ho, D. N. Wang and S. C. Ruan, 'Design and modeling of a photonic crystal fiber gas sensor,' *Appl. Opt.* **42**, 3509–3515 (2003).
62. P. Toupin, L. Brilland, C. Boussard-Pledel, B. Bureau and J. Troles, 'Comparison between single index and microstructured exposed-core chalcogenide glass fibers for chemical sensing,' in *ISNOG*, St. Malo, France. (2012).
63. J. H. Lee, W. Belardi, K. Furusawa, P. Petropoulos, Z. Yusoff, T. M. Monro and D. J. Richardson, 'Four-wave mixing based 10-Gb/s tunable wavelength conversion using a holey fiber with a high SBS threshold,' *IEEE Photonics Technology Letters* **15**, 440–442 (2003).
64. C.-S. Brès, S. Zlatanovic, A. O. J. Wiberg and S. Radic, 'Continuous-wave four-wave mixing in cm-long chalcogenide microstructured fiber,' *Opt. Express* **19**, B621–B627 (2011).
65. M. El-Amraoui, J. Fatome, J. C. Jules, B. Kibler, G. Gadret, I. Skripatchev, Y. Messaddeq, G. Renversez, M. Szpulak, J. Troles, L. Brilland and F. Smektal, 'Experimental observation of infrared spectral enlargement in As_2S_3 suspended core microstructured fiber,' in *Photonic Crystal Fibers IV*, K. Kalli and W. Urbanczyk (eds), SPIE Bellingham, WA (2010).
66. K. H. Tow, Y. Leguillon, P. Besnard, L. Brilland, J. Troles, P. Toupin, D. Mechin, D. Tregoat and S. Molin, 'Relative intensity noise and frequency noise of a compact Brillouin laser made of $As_{38}Se_{62}$ suspended-core chalcogenide fiber,' *Opt. Lett.* **37**, 1157–1159 (2012).
67. S. D. Le, D. M. Nguyen, M. Thual, L. Bramerie, M. Costa e Silva, K. Lenglé, M. Gay, T. Chartier, J.-C. Simon, L. Brilland, D. Méchin, P. Toupin and J. Troles, '42.7 Gbit/s RZ-33% wavelength conversion in a chalcogenide microstructured fiber', in *OFC*, Los Angeles, CA (2012).
68. G. P. Agrawal (ed.), *Nonlinear Fiber Optics*, 3rd edn, Academic Press, New York (2001).
69. C. Baker and M. Rochette, 'Highly nonlinear hybrid AsSe-PMMA microtapers,' *Opt. Express* **18**, 12391–12398 (2010).
70. K. S. Abedin, 'Observation of strong stimulated Brillouin scattering in single-mode As_2Se_3 chalcogenide fiber,' *Opt. Express* **13**, 10266–10271 (2005).

71. P. A. Thielen, L. B. Shaw, P. C. Pureza, V. Q. Nguyen, J. S. Sanghera and I. D. Aggarwal, 'Small-core As-Se fiber for Raman amplification,' *Opt. Lett.* **28**, 1406–1408 (2003).
72. D. I. Yeom, E. C. Mägi, M. R. E. Lamont, M. A. F. Roelens, L. Fu and B. J. Eggleton, 'Low-threshold supercontinuum generation in highly nonlinear chalcogenide nanowires,' *Opt. Lett.* **33**, 660–662 (2008).
73. D. Mechin, L. Brilland, J. Troles, Q. Coulombier, D. M. Nguyen, M. Thual and T. Chartier, 'Next-generation fibers: chalcogenide photonic-crystal fibers expand nonlinear applications,' *Laser Focus World* **46**, 41–47 (2010).
74. P. V. Mamyshev, 'All-optical data regeneration based on self-phase modulation effect,' *24th European Conference on Optical Communication*, **1** (1998).
75. M. D. Pelusi, F. Luan, S. Madden, D. Y. Choi, D. A. Bulla, B. Luther-Davies and B. J. Eggleton, 'Wavelength conversion of high-speed phase and intensity modulated signals using a highly nonlinear chalcogenide glass chip,' *IEEE Photonics Technology Letters*, **22**, 3–5 (2010).
76. O. P. Kulkarni, C. Xia, D. J. Lee, M. Kumar, A. Kuditcher, M. N. Islam, F. L. Terry, M. J. Freeman, B. G. Aitken, S. C. Currie, J. E. McCarthy, M. L. Powley and D. A. Nolan, 'Third order cascaded Raman wavelength shifting in chalcogenide fibers and determination of Raman gain coefficient,' *Opt. Express* **14**, 7924–7930 (2006).
77. R. T. White, and T. M. Monro, 'Cascaded Raman shifting of high-peak-power nanosecond pulses in As_2S_3 and As_2Se_3 optical fibers,' *Opt. Lett.* **36**, 2351–2353 (2011).
78. J. Hu, C. R. Menyuk, L. B. Shaw, J. S. Sanghera and I. D. Aggarwal, 'Computational study of 3–5 μm source created by using supercontinuum generation in As_2S_3 chalcogenide fibers with a pump at 2 μm,' *Opt. Lett.* **35**, 2907–2909 (2010).
79. M. Duhant, W. Renard, G. Canat, J. Troles, P. Toupin, L. Brilland, F. Smektala, A. Betourne, P. Bourdon, and G. Renversez, 'Mid-infrared strong spectral broadening in microstructured tapered chalcogenide AsSe fiber,' in *Fiber Lasers IX: Technology, Systems, and Applications*, (San Francisco, CA (2012).

14
Chalcogenide glass waveguide devices for all-optical signal processing

R. PANT and B. J. EGGLETON, The University of Sydney, Australia

DOI: 10.1533/9780857093561.2.438

Abstract: High bandwidth and high-speed communication networks are critical to meet the enormous increase in data demand. Electrical networks, because of their inherent limitations such as electromagnetic interference (EMI) and losses, are not suitable for realizing ultrahigh bandwidth networks. All-optical networks have been proposed to alleviate these problems. Optical networks not only provide large bandwidth but they also do not suffer from problems such as EMI. However, in order to exploit the full potential of all-optical networks it is important to perform signal processing at intermediate routing nodes in an optical domain. All-optical signal processing is, therefore, critical to realization of all-optical networks. This chapter discusses on-chip all-optical signal processing for a number of applications, e.g. tunable delay via slow light, radio-frequency spectrum analyser, microwave photonic filter, using two key nonlinear processes namely stimulated Brillouin scattering (SBS) and Kerr effect.

Key words: chalcogenide, on-chip, stimulated Brillouin scattering, all-optical signal processing, slow light, microwave photonics, automatic dispersion compensation, radio-frequency spectrum analyser, all-optical XOR logic.

14.1 Introduction

Increasing data demand has necessitated an increase in the capacity and speed of communication networks. Electrical networks, however, do not have the required bandwidth to meet this demand and at high-speed, electrical networks suffer from high loss and electromagnetic interference (EMI). Optical networks, on the other hand, provide inherently large data bandwidth (~ 10's THz) and they do not suffer from EMI and frequency-dependent losses. All-optical networks have therefore been developed to solve the network capacity and speed crunch. However, before the data reaches its destination, it has to go through a number of network nodes where it is converted from the optical-to-electrical (O/E) domain to carry out certain signal processing tasks, which are currently performed in the electrical domain. The processed data is then converted back to the optical domain via electrical-to-optical (E/O) conversion and transmitted to the destination using an optical network. This O/E and E/O conversion introduces latency and reduces the network speed. One way to overcome this problem is to perform signal processing in the optical domain, which avoids O/E and E/O conversion and the related latency.

Over the last few years, a number of techniques and platforms have been developed for realizing all-optical signal processing. Of all these approaches, exploiting the $\chi^{(3)}$ nonlinear processes in silica, chalcogenide and silicon platforms have been at the forefront of research [1–18]. Contributions to $\chi^{(3)}$ come from two sources: (i) the electronic part of susceptibility and (ii) electrostriction. The electronic part of $\chi^{(3)}$ has an instantaneous response (~fs) and gives rise to an intensity-dependent refractive index $n(I) = n_0 + n_2 I$ [19], where n_0 is the background index, n_2 is the nonlinear index of refraction and I is the intensity. The intensity-dependent index change is known as the Kerr effect, which manifests itself in the form of nonlinear processes such as self-phase modulation (SPM), cross-phase modulation (XPM) and four-wave mixing (FWM). All of these processes rely on phase shift ($\phi = \gamma PL$) induced by a pulse either on itself or other pulse, where the nonlinear coefficient γ depends on the nonlinear index of refraction n_2 according to ($\gamma = \omega n_2 / cA_{eff}$) with A_{eff} being the effective mode area, P is the peak pulse power and L is the interaction length. The nonlinear index n_2 depends on the material index according to Miller's rule and therefore optical materials such as silicon and chalcogenide, which have larger n compared to silica, exhibit larger n_2.

On the other hand, electrostriction-induced $\chi^{(3)}$ manifests itself in the form of stimulated Brillouin scattering (SBS) [19], which occurs at a slower time scale (~ ns) and depends on Brillouin gain $G = g_B I_P L_{eff}$, where g_B is the Brillouin gain coefficient and has a strong dependence on refractive index n ($g_B \propto n^7$), $I_P = P_p / A_{eff}$ is the pump intensity and $L_{eff} = (1 - \exp(-\alpha L))/\alpha$ is the effective length with α being the propagation loss and L the physical length of the device.

With the recent focus on all-optical integration [3–5, 6, 8–10, 15, 20], it is critical to harness both the Kerr effect and SBS in a chip-scale device to realize on-chip all-optical signal processing. The fundamental requirement to realize on-chip optical signal processing is to generate large $\phi = \gamma PL_{eff}$ and $G = g_B PL_{eff} / A_{eff}$ in a short length at moderate power levels. At a practical power level, large ϕ and G can be achieved in a chip-scale device by reducing A_{eff} and using a material with large n_2 and g_B. Silica glass, which is commonly used in fibre optic communications, has low n_2 and g_B on account of low index of refraction. Further, due to small index contrast between core and cladding, it is difficult to reduce effective mode area significantly, although with the advent of photonic crystal fibres this restriction is slightly relaxed. Therefore it is hard to reduce both the power requirement and the length at the same time.

Silicon exhibits large n_2 and large index contrast between the core and the waveguide cladding [21], which is typically silica. Silicon therefore allows the fabrication of a device with smaller A_{eff} and thus large ϕ can be achieved in a short device at relatively low power levels. However, silicon has large two-photon absorption (TPA) and free-carrier absorption (FCA), both of which reduce the efficiency of silicon devices [21–23]. While there has been no demonstration of SBS in silicon waveguides, a recent analysis by Rakich *et al.* [24] has predicted giant enhancement of SBS in silicon nanostructures due to the combined effect of electrostriction and radiation pressure induced forces.

Chalcogenide glasses, offer both large n_2 and g_B due to their large n [2, 17], have negligible TPA and FCA, and their large index of refraction provides large index contrast with cladding which allows fabrication of a device with smaller A_{eff}. Therefore, chalcogenide glass is a potential candidate for harnessing on-chip signal processing using SBS and the Kerr effect.

In this chapter, we discuss on-chip all-optical signal processing based on chalcogenide rib waveguides. In Section 14.2, we focus on all-optical processing exploiting on-chip stimulated Brillouin scattering (SBS). In Section 14.3, we review on-chip all-optical signal processing based on the Kerr effect followed by a summary of the chapter.

14.2 Stimulated Brillouin scattering (SBS) based on-chip processing

Stimulated Brillouin scattering (SBS) is a $\chi^{(3)}$ nonlinear process which results from the interaction between light and sound [19]. Figure 14.1 illustrates the light–sound interaction that occurs in the SBS process. A pump (ω_p) and a weak counter propagating signal (ω_s), called the Stokes signal, generate a travelling intensity modulation at the beat frequency ($\Omega_B = \omega_p - \omega_s$) in the pump direction as shown in Fig. 14.1(a). This beat signal then generates density fluctuations via the process of electrostriction: an effect which causes material compression in the region of strong light intensity [19]. These density fluctuations act as a sound wave of frequency (Ω_B) and create a travelling index grating via the elasto-optic effect. The pump gets scattered from the travelling index grating and generates a counter propagating signal, downshifted from the pump frequency by Ω_B due to the Doppler effect. The backscattered signal in turn beats with the pump and reinforces the index grating which results in an increase in the strength of backscattered Stokes signal and the process continues leading to the amplification of the Stokes signal [19, 25]. For a signal centred at the anti-Stokes frequency $\omega_{as} = \omega_p + \Omega_B$, absorption occurs. This process is inherently phase matched and satisfies both energy and momentum conservation as shown in Fig. 14.1(a).

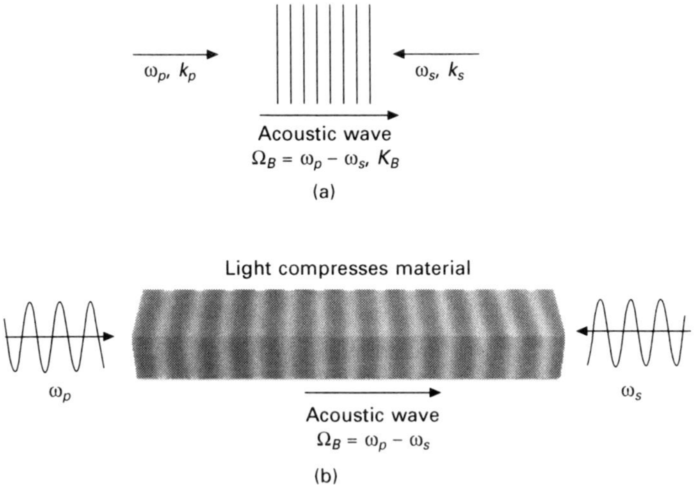

14.1 (a) Principle of stimulated Brillouin scattering where a pump (ω_p) and a counter-propagating Stokes (ω_s) wave interact via an acoustic wave (Ω_B). The three-wave interaction satisfies both energy and momentum conservation [19]. (b) Schematic of light–sound interaction via electrostriction and elasto-optic effect.

The phase matching condition can be written in terms of the frequency ($\nu_B = \Omega_B/2\pi$) of the sound wave as:

$$\nu_B = \frac{2nV_a}{\lambda_p} \qquad [14.1]$$

where V_a is the speed of sound, n is the effective mode index and λ_p is the pump wavelength.

The scattering cross-section (g_B) of the SBS process is given by the expression:

$$g_B = \eta \frac{2\pi n^7 P_{12}^2}{c\rho V_a \Delta\nu_B \lambda_p^2} \qquad [14.2]$$

where η is the overlap integral between the acoustic and optical mode, P_{12} is the elasto-optic coefficient, c is the speed of light in vacuum, ρ is the material density, $\Delta\nu_B$ is the natural Brillouin linewidth and λ_p is the pump wavelength. From Eq. [14.2] it is evident that the SBS gain coefficient strongly depends on the index of refraction and therefore a material with large n is more likely to have large g_B provided other parameters are comparable for different materials. The measured Brillouin gain coefficient in chalcogenide glasses has been shown to be ~100 times that of silica. Silicon has larger refractive index n than chalcogenide; however, exciting SBS in silicon has been a challenge. Recent analysis by Rakich *et al.* has predicted that great enhancement of SBS can be achieved in a silicon nanostructure due to the combined effect of electrostriction and radiation pressure induced forces [24].

SBS has been well characterized in silica-based single-mode fibres [26–29] and photonic crystal fibres [30, 31]. However, long length (tens of meters) is required to realize SBS at practical power levels due to the small g_B of silica [30, 31]. SBS lasers have been demonstrated using distributed feedback (DFB) in a 12.4 cm long fibre Bragg grating (FBG) at 1.55 μm at a power level of ~ 30 mW [32] and using a 6 mm diameter high-Q cavity disc [33]. SBS has also been demonstrated in calcium fluoride glass at 1,064 nm using a high-Q whispering gallery mode resonator where a low lasing threshold of ~3 μW was achieved [34]. Although it is possible to excite SBS in small length of silica and other materials at low power using high-Q cavity, realizing a compact SBS device using single-pass configuration requires very large power. It is therefore difficult to fabricate a compact SBS device without a cavity because of the small g_B of many optical materials.

Recently, Pant *et al.* demonstrated on-chip SBS, for the first time, using a travelling wave geometry [17]. The device used in these experiments was a 7 cm long chalcogenide rib waveguide (see Plate XIV between pages 330 and 331) and had a $g_B \sim 0.74 \times 10^{-9}$, cross-sectional area 4 μm × 850 nm and an effective mode area $A_{eff} \sim 2.3\ \mu m^2$. The large refractive index n of the chalcogenide glass and strong confinement of the optical and acoustic modes (see Plate XIV) in the chalcogenide photonic chip resulted in large g_B. The large index contrast between the chalcogenide rib and the cladding also leads to small A_{eff}. In earlier demonstrations of SBS in chalcogenide glasses, long (10's of meter) fibre lengths were used to achieve large $G = g_B P_p L_{eff} / A_{eff}$ because the large mode area (A_{eff}) in these fibres does not allow large intensity at moderate pump power [35, 36]. This on-chip demonstration of SBS using travelling wave geometry enables photonic integration of a number of applications, e.g. slow- and fast-light [37], microwave photonic filters [38, 39] and, Brillouin dynamic grating [40].

The measured backscattered spectra for the on-chip SBS are shown in Plate XIVa (between pages 330 and 331) for different input pump powers. The Stokes signal power starts to increase with increase in pump power. Residual backscattered pump was also observed in these experiments due to reflections at the interface between the chip and coupling fibre. Pulsed pump was used in the backscattering experiment and the average Brillouin threshold pump power was ~ 50mW which corresponds to a coupled peak power of ~ 2W. Although the Brillouin threshold is ~2W, it is ~ 1,000 times smaller than that required for single-pass SBS in a single-mode fibre of the same length. However, using cavity effects, the Brillouin threshold can be significantly reduced [32–34, 41, 42]. The output pump power in Plate

XIV (between pages 330 and 331) shows a change in its rate of increase at input pump power (~50 mW), at which the Stokes signal starts, to appear, demonstrating the scattering of pump to generate the Stokes wave. The Brillouin shift in on-chip SBS experiments was measured to be ~ 7.7 GHz, which was consistent with that in chalcogenide fibres.

The other important characteristic of SBS is its linewidth ($\Delta\nu_B$), which provides an estimate of the phonon lifetime (T_{phonon}). Linewidth can be measured using a pump-probe experiment where the SBS process is seeded with a weak CW signal whose frequency is tuned around the stokes frequency. Under the un-depleted pump condition, the propagation of a weak probe signal is governed by [43]:

$$E_s(\omega, L) = E_s(\omega, 0)\exp^{jk(\omega)L} \qquad [14.3]$$

where $E_s(\omega, 0)$ and $E_s(\omega, L)$ are the probe field at the input and output of the device and the complex propagation function $k(\omega)$ is given as [43]:

$$k(\omega) = \frac{g_B I_P \Gamma/2}{2((\omega - \omega_s) + j\Gamma/2)} \qquad [14.4]$$

where $\Gamma/2 = 2\pi\Delta\nu_B$.

From Eq. [14.4], it is evident that the imaginary part of the $k(\omega)$ results in gain (absorption) for a signal centred at $\omega_s\,(\omega_{as})$, while the real part gives rise to a frequency dependent phase response.

Figure 14.2 (a) shows the measured SBS gain profile for the optical chip along with a Lorentzian fit, which gives an estimate of the Brillouin linewidth ($\Delta\nu_B$). In the experiment carried out by Pant *et al.*, $\Delta\nu_B$ estimated from a Lorentzian fit to the measured data was 34 MHz [17]. The measured linewidth and Brillouin shift combined with Eqs [14.1] and [14.2] give an estimate of the SBS gain coefficient (g_B). For the rib structure used in [17] the g_B calculated using Eq. [14.2] was 0.74×10^{-9} m/W. At the Stokes frequency, the intensity of output Stokes signal is given by the expression [19]

$$I_s(\omega, L) = I_s(\omega, 0)\exp^{g_B I_P L} \qquad [14.5]$$

The Brillouin gain coefficient can therefore be estimated by fitting an exponential to the measured output Stokes intensity for different values of the pump intensity as shown in Fig. 14.2(b). The value of $\sim 0.715 \times 10^{-9}$ m/W obtained from the fit was consistent with the theoretical estimate of 0.74×10^{-9} m/W.

Harnessing SBS in a chip-scale device is critical to a number of signal processing tasks in microwave photonics such as true-time delay (TTD) and microwave photonic filter (MWPF), where multiple delay lines with tuneable delay of ~ 10 ns are desirable [44]. While tuneable delay is realized using slow

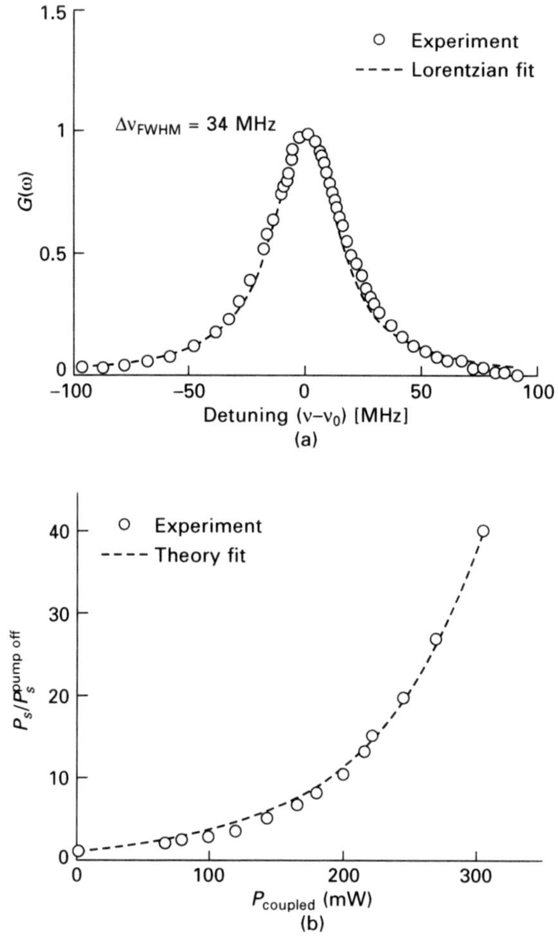

14.2 Measurement results for a (a) gain spectrum of rib waveguide and (b) Stokes power vs. input pump power showing an exponential probe gain as the pump power is increased [17]. An exponential fit to the measured data gives an estimate of the Brillouin gain coefficient.

and fast light [13], which relies on the phase response of the SBS process, microwave filter and TTD exploits both the gain and phase response of SBS [44]. The following subsections are devoted to on-chip tuneable delay using SBS-based slow and fast light and microwave photonic filters.

14.2.1 Slow and fast light

Slow (fast) light refers to a substantial reduction (increase) in the speed of a light pulse, called group velocity (v_g), compared to its value (c) in vacuum.

The group velocity is related to the vacuum speed of light according to relation [45]:

$$v_g = \frac{c}{n(\omega_o) + \omega \left.\frac{dn}{d\omega}\right|_{\omega=\omega_0}} = \frac{c}{n_g} \qquad [14.6]$$

where $n(\omega_0)$ is the index of refraction at the pulse centre frequency and n_g is the group index. From Eq. [14.6] one can note that slow and fast light can be realized using a device which offers large index slope. For the positive slope we get slow light, while for the negative slope fast light is achieved. This large index slope can be realized using an induced or structural resonance. Early research in this field achieved striking results by propagating pulses of light through atomic vapours and exploiting the coherent interaction of light with the 'frozen' atoms. Specifically, Hau *et al.* [46] demonstrated slow light by creating a hole in the absorption profile of sodium atoms, and slowed light down to bicycle speed. Since then a number of techniques have been developed to realize control over the speed of light [7–9, 12, 13, 20, 36, 43, 47–66]. Of all these techniques, the stimulated Brillouin scattering (SBS) based approach is the most flexible and practical for slow light-based devices due to its wavelength independent nature, large signal bandwidth range (MHz–GHz), room temperature operation and delay tuneability [7, 9, 12, 36, 43, 50–56, 62–66].

SBS-based slow and fast light exploits the spectral phase response induced by the real part of $k(\omega)$, which is related to the imaginary part via the Kramers–Kronig relation [45]. Figure 14.3 plots the imaginary and real parts of $k(\omega)$ for the Stokes and anti-Stokes frequency. The positive (negative) slope of dispersion at the Stoke (anti-Stokes) frequency gives rise to the slow (fast) light. The group delay (ΔT_d) associated with the slow and fast light can be expressed in terms of Brillouin gain coefficient, pump intensity and device length as [43]:

$$\Delta T_d(\omega_{s,as}) = \frac{L\Delta n_g}{c} = \frac{g_B I_p L}{\Gamma_B} \qquad [14.7]$$

where Δn_g is the change in the group index at gain (absorption) resonance. From Eq. [14.7], it is evident that change in group index depends on the Brillouin gain coefficient, pump intensity and Brillouin linewidth. Because the Brillouin linewidth for most of the optical materials is nearly the same, large Δn_g can be achieved in a chip scale device using large $g_B I_P$ product. As discussed earlier in this chapter, chalcogenide waveguide offers large $g_B I_p$ product because of its large g_B and small A_{eff}, and therefore a large reduction in the speed of a light pulse, centred at ω_s, can be achieved, which leads to large pulse delays in short length. Further, because $\Delta n_g \propto I_p$, group velocity and thus delay can be tuned by varying the pump power.

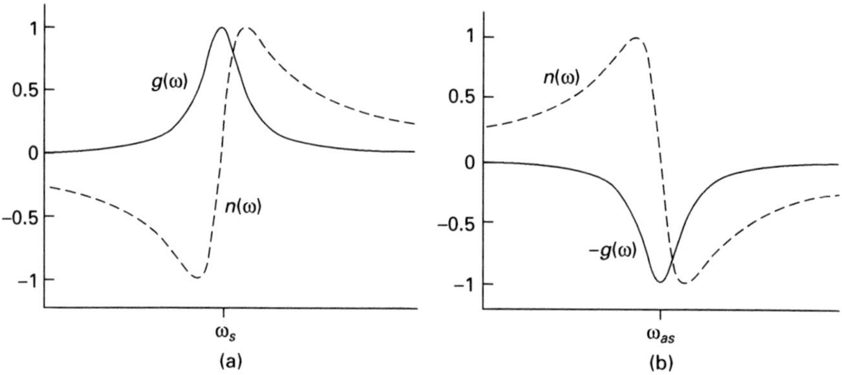

14.3 Real and imaginary parts of the complex wave propagation function in Eq. [14.4] showing (a) gain and associated dispersion at the Stokes frequency and (b) absorption and associated dispersion at the anti-Stokes frequency. The positive (negative) slope of dispersion at the Stokes (anti-Stokes) frequency gives rise to slow and fast light.

The narrow natural linewidth (~ MHz) of SBS resonance, however, restricts slow and fast light-based tuneable delay to narrow band signals [19]. However, SBS gain bandwidth can be broadened beyond its natural linewidth using a broadband pump. SBS slow and fast light has been demonstrated for signal bandwidths ranging from MHz to several GHz [13, 50, 64, 67–69]. Further, because the Stokes wavelength depends on the pump wavelength, the spectral position of the resonance can be moved to any desirable wavelength and, therefore, SBS allows wavelength independent tuneable delay device. Although large delay (~ 50ns) has been obtained using SBS in silica [13] and chalcogenide fibres, the actual group velocity in many of these studies was less than *c*/10 [36] and the major contribution to delay came from the long fibre length. Recently, however, SBS slow and fast light has been demonstrated in a 7 cm long chalcogenide rib waveguide where a group velocity of ~ *c*/130 was observed, which resulted in a delay of 23 ns for a 100 ns long Gaussian pulse [9]. Figure 14.4 shows the measurement of group index change in the rib waveguide device in the slow and fast light region using the pump-probe experiments [9].

The slow and fast light group indexes measured in these experiments were 68 and –44, respectively. From the measured pulse delay for a 100 ns long Gaussian pulse, a delay of 23 ns was obtained which resulted in a group index change of ~130. This large group index change results in the biggest ever reduction in the speed of light pulses using SBS. Slow, fast and negative group velocities were observed in these experiments.

Figure 14.5(a) shows the pulse delay for a Gaussian pulse with input full-width at half-maximum (FWHM) width of 25 ns. The pulses centred at the Stokes frequency undergo delay while the pulses centred at the anti-Stokes

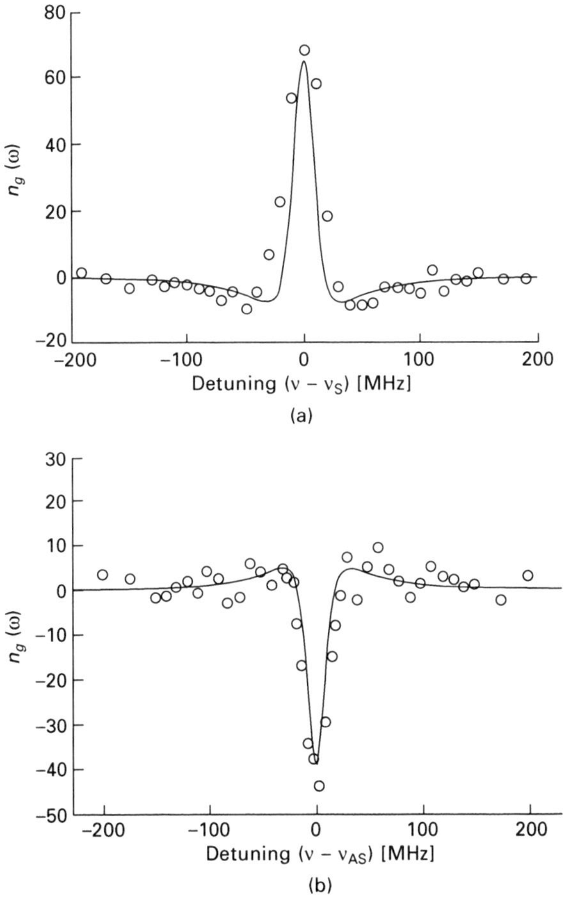

14.4 Group-index for the (a) slow-light (○: measured; solid line: theory) and (b) fast-light (○: measured; solid line: theory) regime, obtained by measuring the delay of low-frequency sinusoidal signals centered at the Stokes and anti-Stokes frequency [9].

frequency experience pulse advancement as predicted by Eq. [14.7]. However, an important point to note is that Eq. [14.7] gives the value of the delay at the resonance frequency and therefore it gives a good estimate of delay only for signal band widths narrower than the SBS gain band width. Figure 14.5(b) plots the measured delay for 25 ns and 100 ns pulses along with the theoretical delay estimate obtained from Eq. [14.7]. For a 100 ns pulse, the pulse spectral bandwidth is 10 MHz, which is small compared to the SBS bandwidth of 34 MHz, implying that the signal spectrum is concentrated near resonance. Therefore in Fig. 14.5(b) a better agreement between theoretical and measured delay is obtained for 100 ns pulse compared to that for a 25 ns

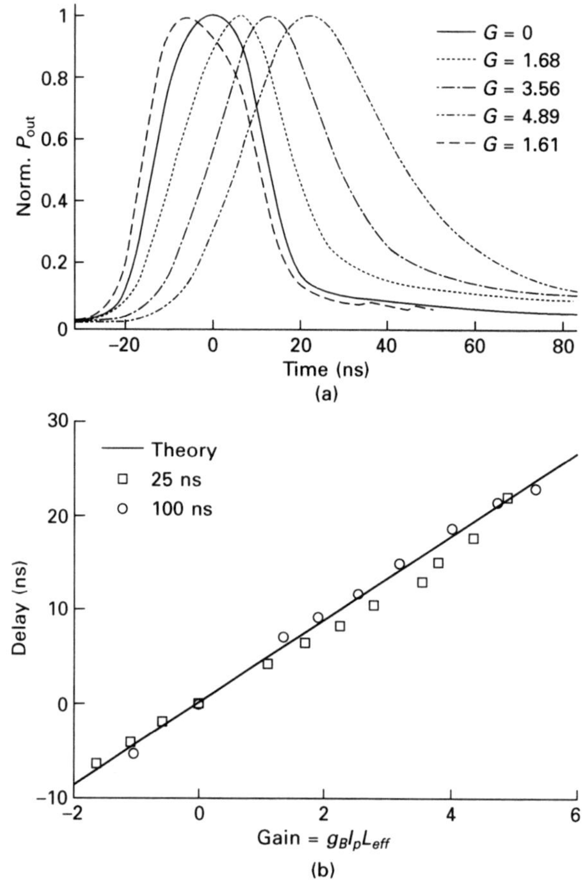

14.5 (a) Measured pulse delay and advancement for a 25 ns long pulse demonstrating delay tunability as the gain is varied and (b) comparison of the measured and theoretical delay, calculated using Eq. [14.7], for pulses with full-width at half-maximum width of 25 ns and 100 ns [9] demonstrating good agreement between theory and experiments.

pulse. On-chip tuneable delay of ~10 ns is critical for microwave applications such as phased array antenna where multiple tuneable delay lines are required to control the delay of radar elements [44, 70]. Demonstration of on-chip SBS slow (fast) light enables microwave photonic signal processing and tunable delay.

14.2.2 Microwave photonic filter

Microwave photonic (MWP) filter is a critical element in RADAR signal processing, mobile communications and radio-over-fibre (RoF) systems.

Implementation of these filters in the optical domain has a number of advantages such as reduced electro-magnetic interference (EMI), frequency and bandwidth tuneability, and reduced complexity [71]. There are two major criteria in designing a MWP filter (i) frequency tuneability and (ii) filter bandwidth. Recently, MWP filters with large frequency tuneability and bandwidth of ~ GHz have been demonstrated using a comb source with a dispersive medium, Mach-Zehnder Interferometer and ring resonators [72–76]. These techniques, however, do not provide narrow bandwidth filters (~10 MHz) with wide tuning range. Recently, the SBS process was used in long length optical fibres to demonstrate a single, narrow band pass (~ 20 MHz) MWP filter with wide tuning range (~ 20 GHz) and notch filters with periodic frequency response [44, 77–79, 80]. However, the long fibre length (~ 10 km) used in these experiments does not offer all-optical integration. Recently, Byrnes *et al.* demonstrated a photonic-chip based widely tuneable (2–12 GHz) single-pass band MWP filter with ~ 20 dB extinction and a narrow bandwidth of ~20 MHz [38, 39].

Figure 14.6 shows the principle of MWP filter where a microwave signal $A(\Omega)$ is encoded into an optical carrier (ω_o) using phase modulation. The phase encoding results in equal amplitude upper and lower sidebands that are out of phase by π. The beat signals between these out-of-phase sidebands and the carrier interfere destructively and, hence, no filtered signal appears in the microwave spectrum [78]. However, when one of the sidebands is amplified using SBS, the beat signals have different amplitudes and no longer cancel, resulting in the generation of a filtered signal in the microwave domain that has the spectrum and bandwidth of the SBS gain profile.

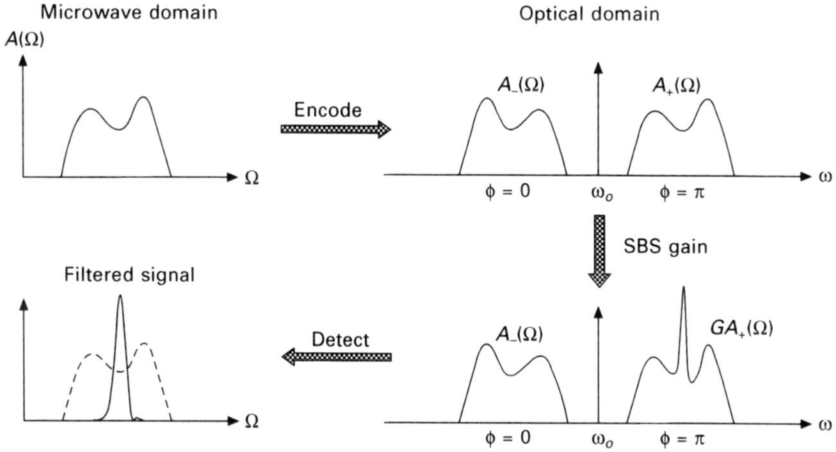

14.6 Principle of SBS-based microwave photonic filter demonstrating that amplifying one of the phase modulated sideband via SBS results in filtering of the microwave signal where the bandwidth and spectrum of the filtered microwave signal are determined by the SBS profile [38].

In the experiment of Byrnes *et al.*, the Brillouin pump frequency was kept fixed at the laser frequency (ω_L) while the frequency of the carrier, on which the RF signal (ω_{RF}) was phase encoded, was tuned using an intensity modulator (IM) [38]. Figure 14.7(a) shows the tuning response of the MWP filter which demonstrates a wide tuning range of –2–12 GHz. This approach, however, provides an unlimited tuning range because for a given microwave signal the frequency of the optical carrier can be moved such that one of the phase modulated sidebands is always at the Brillouin shift from the pump wavelength.

To tune the MPF centre frequency, both the optical carrier frequency and the phase modulator driver frequency were varied, in order to turn the center frequency, so that one PM sideband was always at the Brillouin shift from the pump. The amplitude and bandwidth response of the MWP filter over the tuning range were characterized and are shown in Fig. 14.7(b). A good amplitude (~20 ± 2 dB) and bandwidth (f_{3dB} ~23 ± 2 MHz) stability was demonstrated in these experiments. The inset in Fig. 14.7(b) shows the MWP filter profiles centred at RF frequencies of 4 and 11 GHz, obtained by scanning the phase modulator driver frequency over 1 GHz. The maximum Q-factor, defined as the MPF centre frequency divided by the f_{3dB}, was calculated to be ~520 for a RF centre frequency of 12 GHz.

It is well known that the bandwidth and shape of the SBS gain profile is determined by the pump profile and its bandwidth. Therefore by tailoring the pump profile, one can reconfigure the MWP filter bandwidth and shape factor (S), defined as the ratio of 20 dB to 3 dB bandwidth in this case. In the demonstration by Byrnes *et al.* [38], a dual-pump configuration was used to reconfigure the MWP filter profile. Figures 14.8 (a) and (c) show that when the pump spectrum was changed from a single- to dual-pump configuration, the resulting MWP filter profile (solid line in Fig. 14.8(c)), yielded a flat top and larger f_{3B}. Figures 14.8(b) and (d) show the measured MWP filter profiles for single and dual pumps (separated by 18 MHz), respectively. Using two pumps, a nearly flat-topped MPF profile was achieved with f_{3dB} increased from ~20 MHz to ~40 MHz, which in turn improved S from ~3.5 for a single pump to ~2 for the dual pump.

14.3 On-chip processing using the Kerr effect

The instantaneous response of Kerr nonlinearity is critical to realize all-optical signal processing, necessary for enabling ultrahigh-speed all-optical networks. As mentioned in the introduction, the Kerr effect arises from the intensity dependence of the refractive index [19]

$$n(I) = n_0 + n_2 I \qquad [14.8]$$

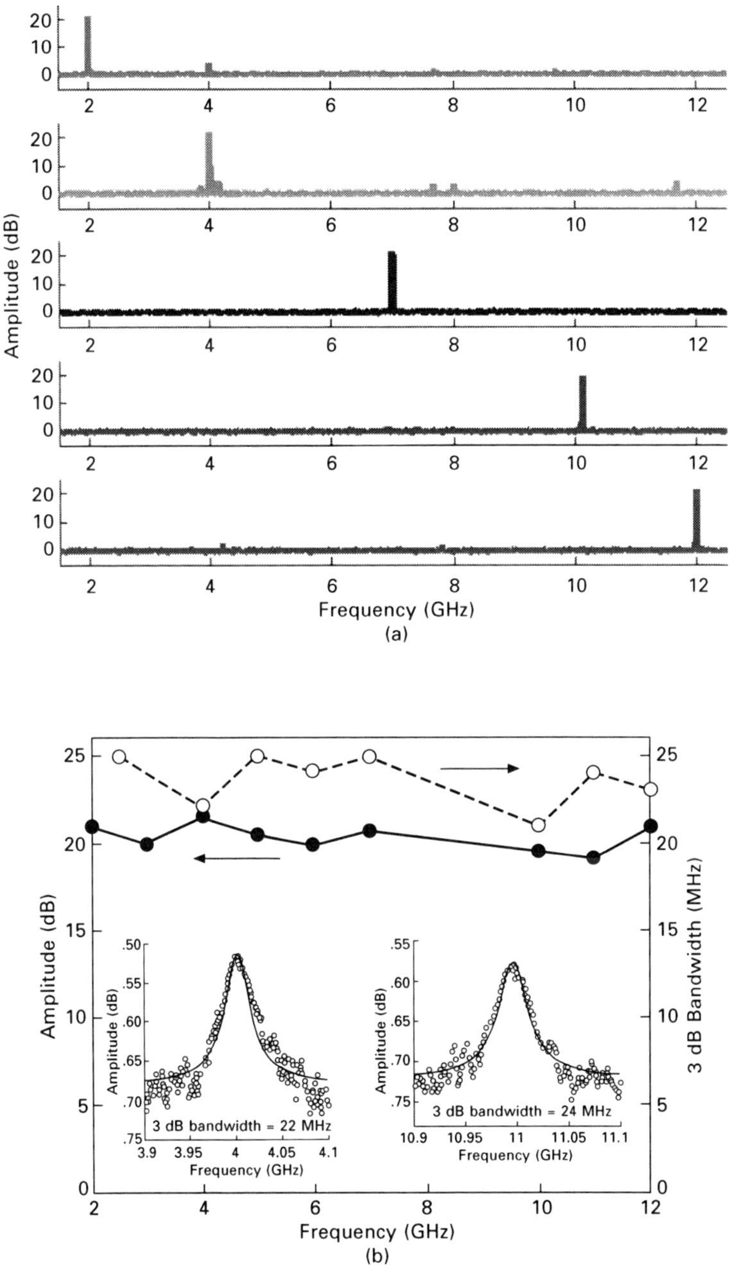

14.7 Measured response of the chip-based microwave photonic filter showing (a) wide tuning range of 2–12 GHz and (b) stable amplitude and bandwidth response over the tuning range [38].

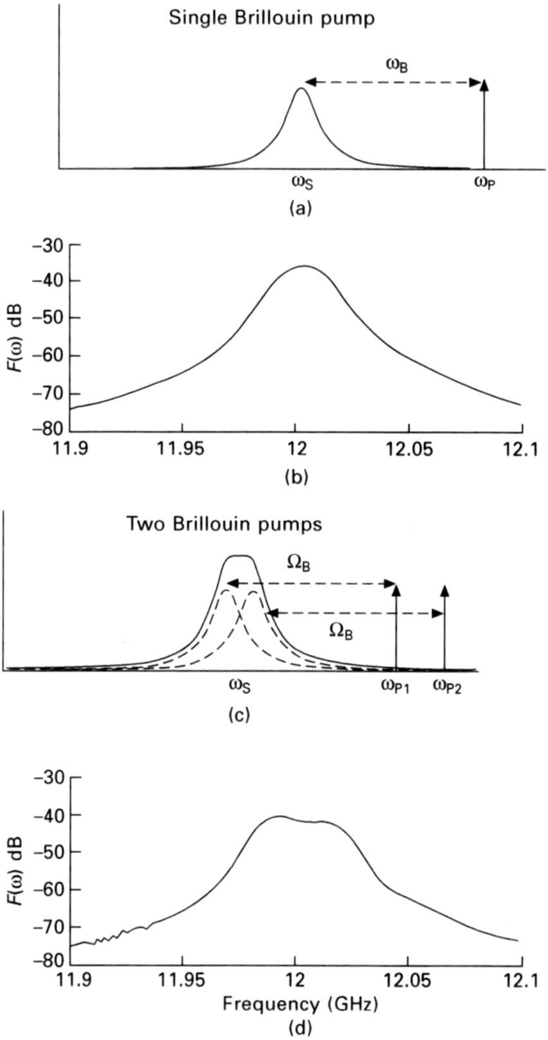

14.8 Re-configuration of microwave photonic filter bandwidth and shape using single-pump (a) theory; (b) measured) and dual-pump (c) theory; (d) measured), which results in the enhancement of 3 dB bandwidth from 20 MHz for the single-pump case to 40 MHz for the dual-pump. The corresponding shape factor was enhanced from S=2 for single-pump to S=3.5 for dual-pump [38].

This intensity dependence of the refractive index leads to an intensity-dependent change in the phase ($\phi(I) = \omega n(I)L/c$) of an optical signal. In the case of a pulse, the intensity-dependent index leads to a time-dependent change in the phase, which leads to a change in the instantaneous frequency of the pulse according to

$$\delta\omega(t) = \frac{d\phi}{dt} = \frac{\omega L}{c}\frac{dn(I)}{dt} = \frac{\omega L n_2}{c}\frac{dI}{dt} \qquad [14.9]$$

When the phase of a pulse is changed by the intensity of the pulse itself, the process is called self-phase modulation (SPM). However, when another pulse is used to change the phase and, thus, the frequency spectrum of a pulse, the process is called cross-phase modulation (XPM) [19]. A rather important manifestation of the Kerr effect takes place in the form of four-wave mixing (FWM), where three optical waves at different wavelengths interact to generate a fourth-wave (idler) and, hence, the name four-wave mixing. All these processes have an instantaneous response which makes them suitable for ultrafast all-optical signal processing. In the following subsections, three critical signal-processing applications are discussed in detail: (i) radio frequency spectrum analyser, (ii) automatic dispersion compensation of Tbaud/s signal, and (3) all-optical XOR logic.

14.3.1 RF spectrum analyser

A radio-frequency spectrum analyser (RFSA) is used to measure the spectral response of a RF signal. Commercial electrical RF spectrum analysers typically have bandwidth ~ 50 GHz and therefore are not suitable for measuring the RF spectrum of Tbit/s signals that exist in optical communications. On the other hand, optical systems inherently have ultrahigh bandwidth, which can be used to realize RF spectrum analysers for Tb/s signals. The RF spectrum of an optical signal is given according to

$$S_{RF}(\omega) = \left| \int_{-\infty}^{\infty} |E(t)|^2 e^{j\omega t} dt \right|^2 \qquad [14.10]$$

The Tbit/s signals are usually generated by multiplexing several low bit-rate data streams, typically 10–40 Gbit/s. The RF spectrum analysis of optical communication signals allows real-time tracking and correction of impairments, which are introduced by dispersion, noise, and pulse jitter in a data stream [81]. It is therefore important to have a RF spectrum analyser that can work for Tbit/s signals. Recently, an all-optical approach was proposed by Dorrer and Maywar [81] where an optical spectrum analyser (OSA) was used to measure the RF spectrum of a high-frequency RF signal using a probe signal. An OSA measures the power spectrum of an optical field, which is given as

$$S_o(\omega) = \left| \int_{-\infty}^{\infty} E(t) e^{j\omega t} dt \right|^2 \qquad [14.11]$$

From the definition of RF spectrum given by Eq. [14.10], it is evident that if we want to use an OSA to measure the RF spectrum of a signal, then the probe electric field must have a dependence on the intensity of signal under test (SUT), i.e. $E_P(t) \propto mI_S(t)$, where $I_S(t) \propto |E_S(t)|^2$. As discussed earlier, XPM can be used to induce an intensity-dependent phase shift in a probe field via an intense which then allows measurement of the signal RF spectrum.

Figure 14.9 shows the principle of an all-optical RF spectrum analyser where a highly nonlinear signal imparts an intensity-dependent phase shift to a CW probe via XPM. The probe electric field arising from the interaction between SUT and probe is given as [81]:

$$E(t) = e^{i\omega_o t} e^{imI(t)} \approx e^{i\omega_o t}(1 + mI(t)) \qquad [14.12]$$

Substituting Eq. [14.12] for the probe field ($E_P(t)$) into Eq. [14.11] for the spectrum obtained from an OSA, we note that the measurement of the probe spectrum on OSA contains information about the RF spectrum of the nonlinear signal.

This technique thus provides a simple and an easy-to-implement method for realizing a RF spectrum analyser for signals with THz bandwidth. An on-chip implementation of all-optical RF spectrum analyser is important to enable photonic integration of this rather critical component for all-optical signal processing and signal monitoring.

Recently, Pelusi *et al.* demonstrated a photonic chip-based RF spectrum analyser with bandwidth ~3 THz [10]. The key to this demonstration was the dispersion engineering of the chalcogenide rib waveguide which was achieved by reducing the height of the rib structure. The large waveguide dispersion induced by the reduction in rib height offsets the large material dispersion of –364 ps-nm^{-1} km^{-1} to give a net dispersion of ~ 40 ps-nm^{-1} km^{-1}. The dispersion engineered photonic chip offers large bandwidth by minimizing the phase and group velocity mismatch between the signal and the probe [10]

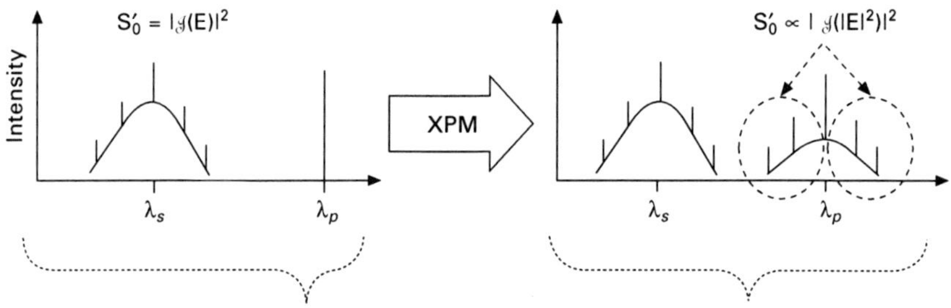

14.9 Schematic of a RF spectrum analyser exploiting cross-phase modulation in a chalcogenide photonic chip [10]. The signal under test (SUT) induces an intensity dependent phase shift to a continuous-wave probe signal via XPM. Measurement of probe spectrum on an OSA thus contains information about the RF spectrum of the SUT.

which is very difficult in long length optical fibres due to the non-uniformities induced by fibre dispersion along the length. The rib waveguide used in these experiments had a dimension of 3 μm × 870 nm and was 14 cm long. The nonlinear parameter γ for 3 μm wide rib waveguide was 7,250 W^{-1}-km^{-1} and the effective mode area was ~1.9 μm^2 [10].

Plate XV (between pages 330 and 331) shows the schematic of a photonic chip-based implementation of RF spectrum analyser where a signal, whose RF spectrum needs to be measured, is co-propagated with a continuous wave (CW) probe through an As_2S_3 rib waveguide. The probe spectrum is measured on an OSA, which gives information about the signal RF spectrum. The bandwidth of the photonic chip-based optical RF spectrum analyser was measured by measuring the strength of the XPM sidebands on a CW probe centred at a wavelength of 1,530 nm. A 100 mW signal centred at 1,550 nm was used to introduce XPM sidebands on the probe wave. For a 3 μm wide rib waveguide, the 3 dB bandwidth was ~2.5 THz in these experiments. Plate XV (b) shows the normalized power in the XPM sideband of the probe, demonstrating a 3 dB bandwidth of 3 THz and 3.2 THz for a 4 μm wide rib waveguide for probe wavelengths of 1,530 nm (squares) and 1,569 nm (circles), respectively.

Figure 14.10 (a) shows the eye diagrams for a 320 Gbit/s signal generated using (i) best and (ii) non-optimal multiplexing of 40 Gbit/s data streams. The 320 Gbit/s signal imparts a phase shift on a CW probe, centred at a wavelength of 1,546 nm, via XPM and generates sidebands as shown in the OSA traces (see Fig. 14.10(b)) for both the best and non-optimally multiplexed data streams. Converting these traces into frequency domain gives a measure of the RF spectrum. Figure 14.10(c) shows the measured OSA spectrum in the frequency scale (ν-ν_0), where ν_0 is the centre frequency of the CW probe, along with the simulated RF spectrum. The frequency domain spectrum shows a large tone at a frequency detuned from the probe frequency by 320 GHz. As mentioned earlier in this section, the RF spectrum is used to measure the signal impairment due to dispersion, jitter and noise. From Fig. 14.10(c)(ii), it is evident that for a non-optimally multiplexed data stream, the power in the 320 GHz tone is reduced and the power in other tones increases, showing a degradation in the RF spectrum compared to the optimum data stream. This degradation in RF spectrum can be used to ascertain the type and extent of signal degradation. Thus, exploiting the Kerr effect in a dispersion engineered chip allows implementation of a broadband all-optical RF spectrum analyser.

Recently, a novel application of RFSA was developed by Paquot *et al.*, where degradation in the data-fidelity of a Tbaud/s data stream due to higher-order dispersion was corrected by monitoring its RF spectrum. In the next section, automatic dispersion compensation of a 1.28 Tbaud/s optical time division multiplexed (OTDM) signal based on RFSA is discussed.

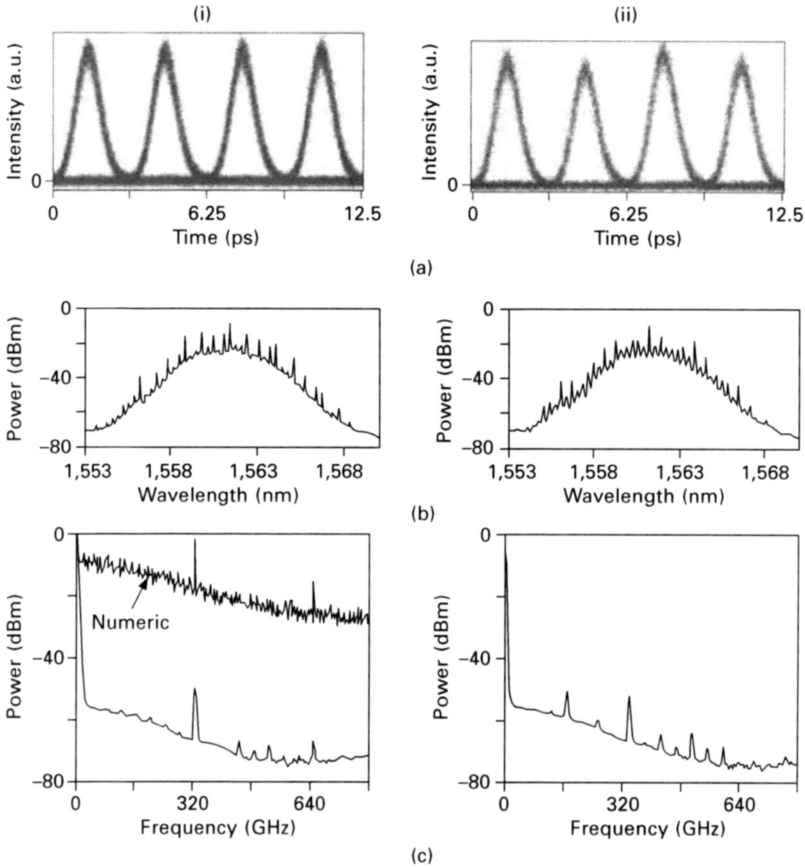

14.10 Characterization of the data fidelity of (i) optimally and (ii) non-optimally multiplexed 320 Gb/s data stream using (a) measured eye-diagram (b) measured optical spectrum after propagation through the photonic-chip and (c) photonic-chip based radio-frequency spectrum analyser, demonstrating that for the non-optimal multiplexing both the eye-diagram and RF tone power at 320 GHz show degradation [10].

14.3.2 Automatic dispersion compensation of 1.28 Tbaud/s signal

Dispersion-induced pulse distortion is one of the main impairments that is introduced as a result of signal propagation through an optical fibre. This impairment becomes severe as the pulse bandwidth increases or temporal pulse width reduces. In an optical time division multiplexed (OTDM) signal, several low baud rate data streams, typically ~10–40 Gbaud/s, are multiplexed to generate a Tbaud/s signal. This multiplexing scheme necessitates the use of temporally narrow pulses ~ 100's fs, which experience severe first- and higher-order dispersion resulting in signal impairment. Compensating dispersion in real time is critical to maintain the data fidelity in these ultra-high baud rate systems. A number of techniques have been developed to compensate first- and higher-order dispersion based on concatenating speciality fibres, using phase modulators, tuneable Bragg gratings, micro electro-mechanical switches (MEMS) [82], spatial light modulators (SPM) and parametric conversion [83]. However, many of these schemes are static in nature, operate over a narrow band, and fail to take into account the stochastic fluctuation in dispersion parameters due to temperature fluctuations. In order to compensate the fluctuations in dispersion parameters, real-time tracking and compensation of dispersion is necessary.

Recently Paquot *et al.* [18] developed a novel scheme for real-time tracking and compensation of dispersion-induced degradation of an ultra-high baud rate optical communication systems and used it to recover a 1.28 Tbaud/s signal from a dispersion impaired state to its original state.

Figure 14.11 shows the principle of this dispersion compensation scheme which is based on tracking the power of 1.28 THz tone in the RF spectrum of a 1.28 Tbaud/s signal. The power of the RF tone reduces as the dispersion-induced pulse distortion increases. A feedback loop is used to adjust the dispersion parameters of a spectral pulse shaper until the power in the RF tone is recovered to its original state.

This technique was exploited for automatic compensation of dispersion-induced pulse distortion in a 1.28 Tbaud/s data stream. The pulses in the data stream had a full-width at half-maximum (FWHM) pulse width of 275 fs. The RF spectrum was observed using the chip-based RF spectrum

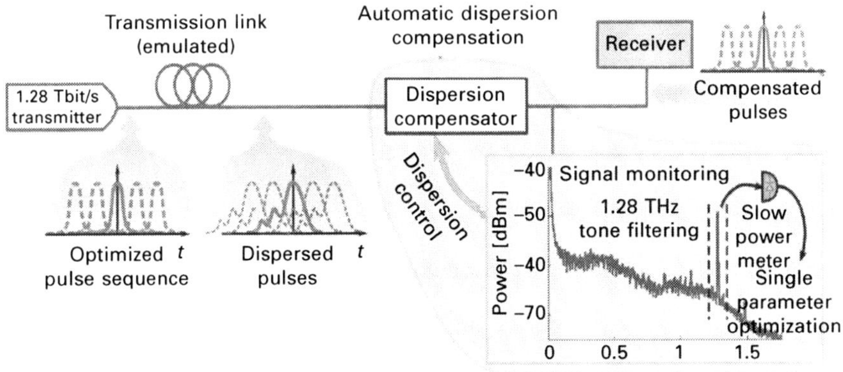

14.11 Schematic showing the principle of RF spectrum analyser based automatic dispersion compensator where dispersion induced reduction in the RF tone power is recovered by adjusting the dispersion parameters in a feedback loop [18].

analyser discussed in the previous section. As shown earlier, this RFSA has a 3 dB bandwidth of ~ 3 THz and therefore is suitable for tracking the RF spectrum of a Tbaud/s signal. This scheme simultaneously compensates for fluctuations in dispersion parameters β_2, β_3 and β_4.

Figure 14.12 shows the degradation in the RF spectrum of a 1.28 Tbaud/s signal as a result of fluctuations in β_2, β_3 and β_4 along with the optimal case (lowest row) where the tone power is the maximum. The tone power reduces as the dispersion parameter is detuned from its optimal value, and once the dispersion is adjusted back to its optimal value, the tone power recovers to its optimum state.

In the experiment of Paquot *et al.*, a Finisar waveshaper was used as a spectral pulse shaper (SPS) to emulate dispersion fluctuations and also to compensate for these fluctuations. Figure 14.13 shows the experimental set-up to realize RFSA-based automatic dispersion compensator. An As_2S_3 photonic chip was used as a RFSA in this experiment. A 1.28 Tbaud/s OTDM was used to impart XPM on a CW probe to generate sideband at a frequency detuned from the probe centre frequency by 1.28 THz. As the dispersion parameters are changed using the first SPS, the 1.28 THz tone power reduces. This reduction in tone power is used for tracking and providing feedback for automatic compensation using a second SPS which adjusts the dispersion parameters such that the tone power is recovered to its original value.

Figure 14.14 shows real-time recovery of 1.28 THz tone power after it is reduced due to a sudden change in a dispersion parameter. The time instant when the dispersion parameter is changed is shown by dashed lines. This time instant also corresponds to a sudden drop in tone power which

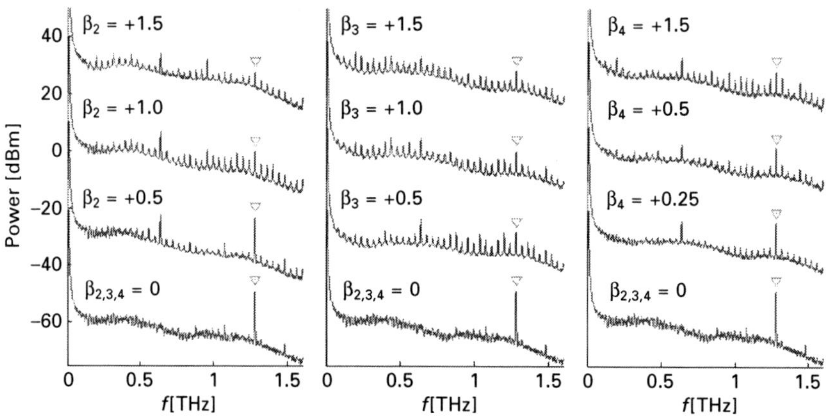

14.12 Degradation of the RF spectrum as a result of the fluctuation in the dispersion parameters $\beta 2$ (column 1), $\beta 3$ (column 2) and $\beta 4$ (column 3). The last row in each column corresponds to the optimized dispersion parameters showing that the tone power is the maximum for the optimal case [18].

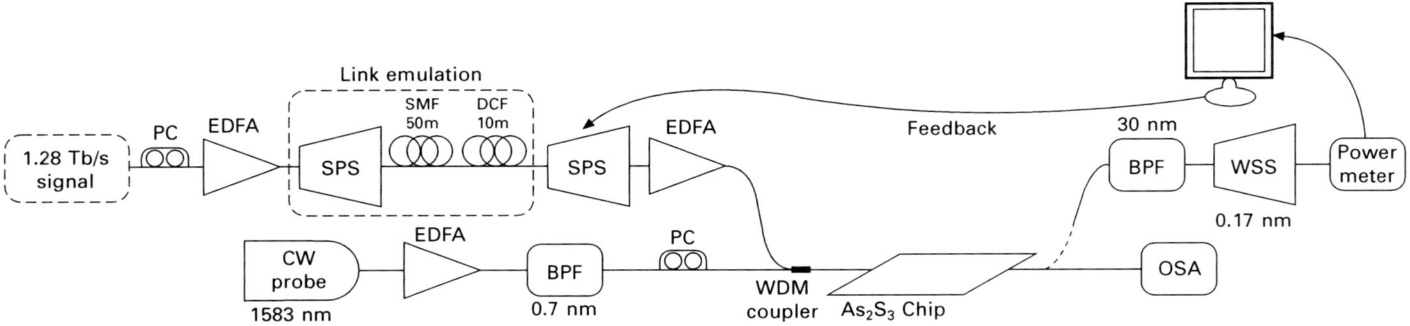

14.13 Experimental set-up to realize automatic dispersion compensation exploiting photonic chip-based RF spectrum analyser [18].

is then recovered to its original state by applying automatic compensation via second SPS. Photonic chip-based RFSA is, therefore, not only useful to measure the RF spectrum of Tbit/s signals, but it is also useful to perform signal monitoring and recovery. The dispersion engineering of the As_2S_3 chip used in these experiments is critical for ultrafast signal processing, as it enhances the bandwidth of operation as evident from Plate XV (between pages 330 and 331).

In applications where the pulses at two different wavelengths is interact via FWM or XPM, dispersion engineering is critical because any walk-off between the pulses due to group velocity dispersion will restrict the operating wavelength range. In the next section, we discuss one such signal processing task, an XOR gate, where temporal alignment between pulses centred at wavelengths λ_1 and λ_2 is crucial for efficient, wideband operation.

14.3.3 DPSK XOR logic

An XOR gate performs exclusive OR logic on the input bits (e.g., A and B), the output of which is summarized in Table 14.1. An XOR gate is the building block for a number of processing tasks such as pseudo-random bit sequence (PRBS) generation, switching, encoding, label swapping and data encryption. An all-optical XOR gate, besides offering ultrahigh speed and bandwidth, also enables encoder for spectrally sliced all-optical analog-to-digital converters (ADCs) [84]. The truth Table (14.1) is for an all-optical XOR gate-based on-off keying (OOK) data format. However, in the recent past, there has been increased interest to use advanced modulation formats such as differential phase shift keying (DPSK) and quadrature phase shift keying (QPSK) for data transmission. Recently, therefore, all-optical XOR gates using DPSK data format have been explored in different platforms

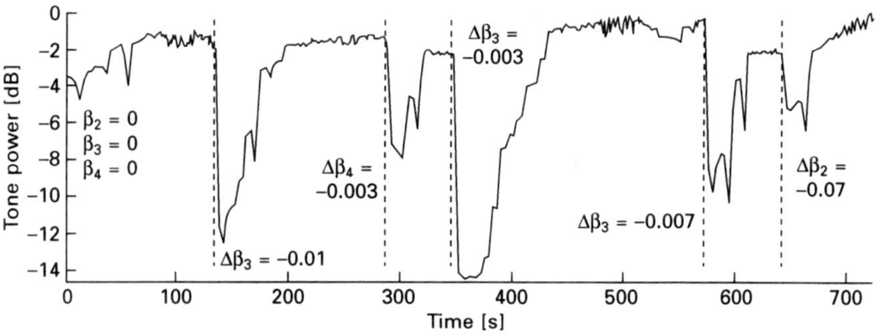

14.14 Degradation in RF tone power with time as a result of sudden fluctuation in dispersion parameters β_2, β_3 and β_4. The power recovers to its initial state after the initial drop due to automatic dispersion compensation [18].

Table 14.1 XOR logic truth table for OOK data format

A	B	$X = \bar{A}B + \bar{B}A$
1	1	0
1	0	1
0	1	1
0	0	0

Table 14.2 XOR logic truth table for DPSK data format

A	B	$X = \bar{A}B + \bar{B}A$
π	π	0
π	0	π
0	π	π
0	0	0

using FWM [15, 85, 86]. Here we discuss an all-optical 160 Gb/s DPSK XOR logic implemented on a chalcogenide photonic chip. In a DPSK signal, the data is encoded in the phase difference of the successive bits. The truth table for a DPSK XOR gate is shown in Table 14.2.

Figure 14.15 shows the principle of DPSK XOR logic, where phase-encoded data streams (Data1 and Data 2) centred at wavelengths λ_1 and λ_2 are launched into the chalcogenide photonic chip along with a continuous wave probe centred at wavelength λ_P. The phase of the two input data streams is denoted by ϕ_1 and ϕ_2 and the probe beam has a constant phase ϕ_p. The three waves undergo four-wave mixing in the chip and generate an idler signal centred at a wavelength λ_I with a phase $\phi_I = \phi_1 + \phi_2 - \phi_p$. For a CW probe with constant phase and $\phi_{1,2}$ taking values of 0 and π, the phase of FWM idler (Fig. 14.15(b)) follows the phase of the output in the DPSK XOR truth table. This happens because phase periodicity implies that a phase shift of 2π is equivalent to a 0 phase shift.

For realizing DPSK XOR logic at high bit rates, it is critical that efficient FWM occurs in the chip, which requires that the group velocity dispersion parameter of the chip is small so as to minimize the walk-off between the data pulses at two different wavelengths. In a recent experiment by Vo *et al.*, two DPSK data streams at 40 Gb/s are coupled to the As_2S_3 chip along with a CW probe and the FWM idler is demodulated to realize the XOR logic operation [15]. Figure 14.16 shows the experimental set-up along with the optical spectrum at the output of the chip for conditions when (a) probe frequency was fixed and (b) when probe frequency was scanned. The idlers result from different combinations in the FWM term and any of these can be used to realize the XOR logic. The wavelengths of the two DPSK signals

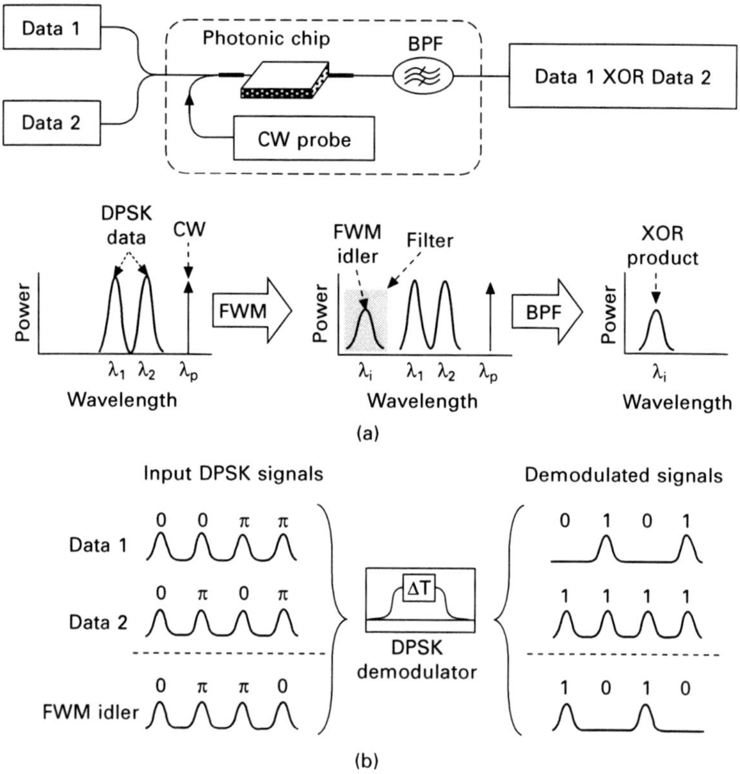

14.15 (a) Schematic of photonic chip-based DPSK XOR gate showing the phase-encoded data streams Data1 and Data2 launched into the chip along with the CW probe where they generated XOR idler via FWM. (b) Phase of the idler generated via FWM between the data inputs and CW probe. The demodulated data satisfies the truth table for XOR logic [15].

in this experiment were 1,557.36 nm and 1,558.90 nm and the wavelength of the CW probe was 1,565 nm. The average power of each DPSK signal inside the chip was P_{avg} ~ 10 mW and the CW probe power was ~ 30 mW. The rib waveguide dimensions were 2 μm × 850 nm and it was 7 cm long. The GVD and nonlinear parameter for the waveguide were 29 ps $(nm\text{-}km)^{-1}$ at 1550 nm and 10 $(W\text{-}m)^{-1}$, respectively.

The FWM bandwidth in this scheme determines the wavelength range of the XOR operation, which can be determined by scanning the probe wavelength. Uniform FWM conversion efficiency was achieved over a range of ~ 30 nm as shown in Fig. 14.16(c).

In optical communications, it is critical that the data fidelity is maintained in doing any operation. The data fidelity can be measured by performing the bit error rate (BER) analysis or using eye diagrams. Figure 14.17(a) shows

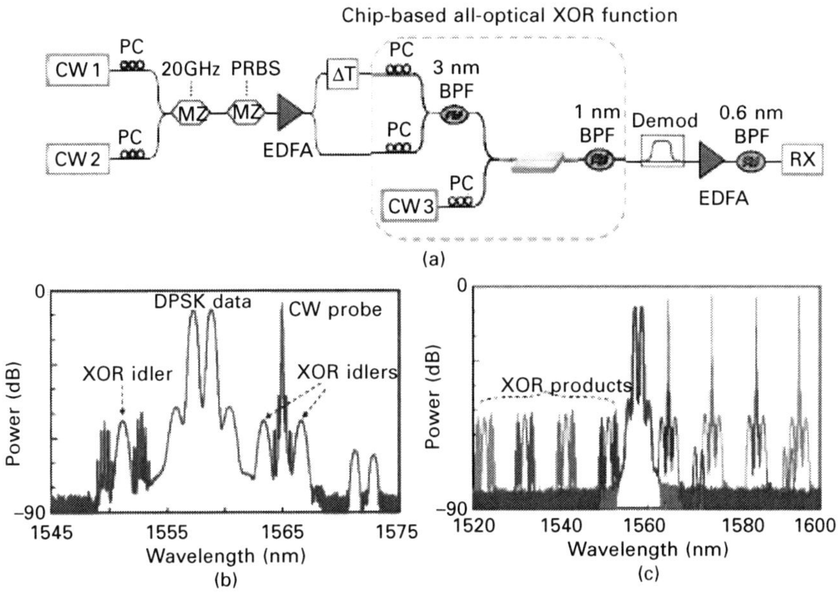

14.16 Experimental set-up for photonic chip-based 40 Gbit/s all-optical DPSK XOR gate where the first Mach–Zehnder modulator (MZM) was used to modulate two CW sources to generate 40 Gbit/s data streams and the second MZM was used to encode phase information on each of the input data streams. (b) Optical spectrum at the output of the photonic chip showing the XOR idler along with the spectrum for the input DPSK data and the CW probe signal. (c) XOR idler for different CW probe wavelengths showing uniform FWM efficiency over a broad wavelength region, demonstrating broadband operation of DPSK XOR gate [15].

the demodulated input data streams and their XOR product along with the eye diagrams and BER measurement. Large eye-opening and small BER ~10^{-9} in these experiments (see Fig 14.17(b)) demonstrate that the data fidelity was maintained on performing the XOR operation without incurring any power penalty. The potential for extending the XOR operation to higher bit rates was explored using 160 Gbit/s input DPSK data streams. Figure 14.18(a) shows the optical spectrum at the output of the chip showing the XOR idler. The eye diagrams for the demodulated 160 Gbit/s input data streams and the XOR product are shown in Fig. 14.18(b). The eye-diagrams show slightly reduced eye-opening for the XOR product which can be improved using low-noise amplifiers [15] and by reducing losses. The output pulses for the demodulated input and XOR product at 40Gbit/s, obtained by de-multiplexing the 160Gbit/s data, show much better eye-opening (Fig. 14.18(d)) and improved data fidelity, demonstrating the potential for high bit rate all-optical DPSK XOR logic using a chalcogenide platform.

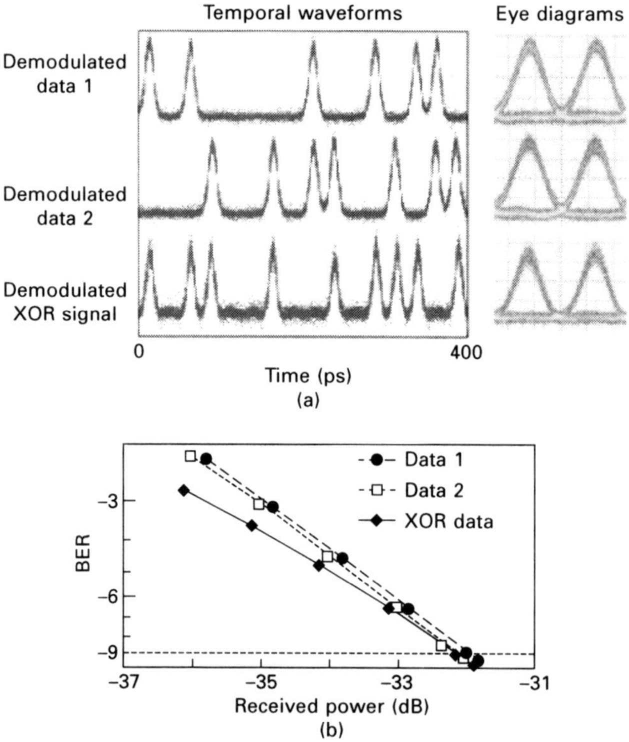

14.17 (a) Demodulated data and XOR pulses at the output of the photonic chip for a 40 Gbit/s all-optical DPSK XOR logic along with the measured eye diagrams. The clear eye-opening for the XOR product demonstrates high fidelity on-chip DPSK XOR logic at 40 Gbit/s and (b) measured BER for the demodulated data inputs and the XOR logic showing error-free operation [15].

All-optical XOR logic has been realized for OOK and DPSK signals in other platforms such as silicon slow-light photonic crystal waveguides (PhCWG) [85], silicon nanowires [86], semiconductor optical amplifier (SOA) [87] and periodically poled lithium niobate (PPLN) [88]. While PhCWG provides the compact device, the energy per bit is larger (~ 1000 fJ/bit) than the other schemes and the operating wavelength range is limited by the slow-light region bandwidth [85]. XOR gate at a data rate of 640 Gbit/s has been demonstrated in PPLN waveguides [88] using OOK data format. However, this technique requires control of waveguide temperature at ~92°C and the energy was 500 fJ/bit [88]. SOA-based XOR logic relies on cross-gain modulation (XGM), which limits the operating bit rate to ~ tens of Gbit/s due to the gain saturation and recovery [87]. Chalcogenide photonic chip enables broadband, room temperature XOR logic at high bit rates with energy per bit ~ 62.5 fJ/bit [15], which is facilitated by dispersion engineering of the waveguide.

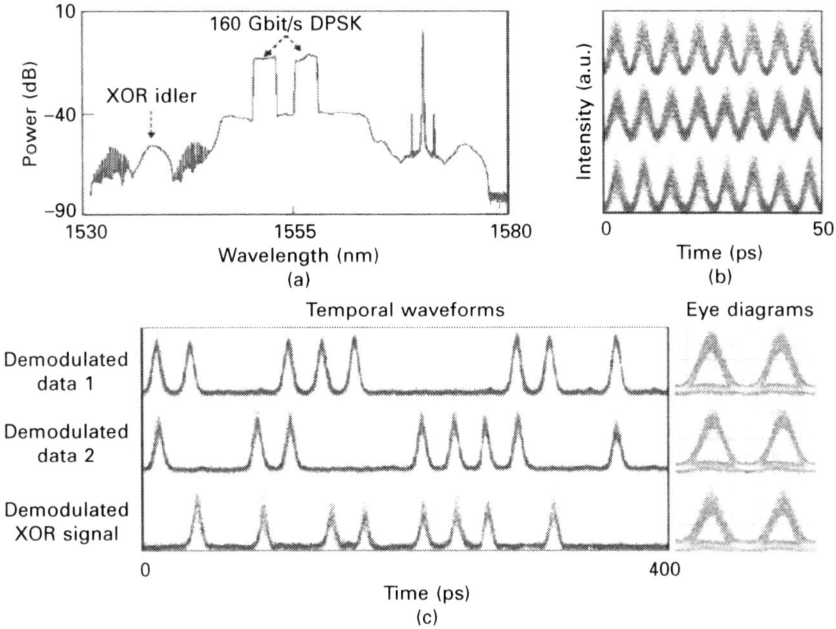

14.18 (a) Characterization of the XOR logic for two 160 Gb/s DPSK data streams showing (a) output optical spectrum (b) eye-diagram for the input data streams and the XOR product (c) de-multiplexed and demodulated 40 Gb/s data streams and their XOR product (d) eye-diagrams for the de-multiplexed data and XOR logic [15].

Although optics allows for ultrafast signal processing and ultrahigh bandwidth operation, these operations must be performed with energy consumption that is comparable to that of current electronic circuits. The energy per bit for the XOR operation shown in Fig. 14.18 was ~60 fJ/bit which is larger than the energy consumption for an electronic logic gate. However, energy consumption can be further reduced by minimizing the propagation loss and enhancing the intensity by reducing the effective mode area.

14.4 Conclusion

In conclusion, we have discussed on-chip all-optical signal processing using two third-order optical nonlinearities: Brillouin scattering and Kerr effect. While large Kerr effect has been observed and exploited in chip-scale devices fabricated from other optical materials, on-chip SBS using travelling wave geometry was first demonstrated in chalcogenide photonic chip. Besides exhibiting large Kerr effect and SBS nonlinearity, chalcogenide glass also has low two-photon absorption, which makes it a potential platform for all-optical signal processing. Demonstration of applications based on SBS (e.g.

slow and fast light, microwave photonic filter, Brillouin dynamic gratings) and Kerr effect (e.g. radio-frequency spectrum analyser and DPSK XOR gate) in a chalcogenide photonic chip makes it a potential platform for on-chip all-optical processing.

14.5 References

1. Corcoran, B., *et al.*, Green light emission in silicon through slow-light enhanced third-harmonic generation in photonic-crystal waveguides. *Nature Photonics*, 2009. **3**(4): 206–210.
2. Eggleton, B.J., B. Luther-Davies and K. Richardson, Chalcogenide photonics. *Nature Photonics*, 2011. **5**: 141–148.
3. Eggleton, B.J., *et al.*, Photonic chip based ultrafast optical processing based on high nonlinearity dispersion engineered chalcogenide waveguides. *Laser & Photonics Reviews*, 2012. **6**(1): 97–114.
4. Hillerkuss, D., *et al.*, 26 Tbit s(-1) line-rate super-channel transmission utilizing all-optical fast Fourier transform processing. *Nature Photonics*, 2011. **5**(6): 364–371.
5. Leuthold, J., C. Koos and W. Freude, Nonlinear silicon photonics. *Nature Photonics*, 2010. **4**(8): 535–544.
6. Levy, J.S., *et al.*, CMOS-compatible multiple-wavelength oscillator for on-chip optical interconnects. *Nature Photonics*, 2010. **4**(1): 37–40.
7. Okawachi, Y., *et al.*, Tunable all-optical delays via Brillouin slow light in an optical fiber. *Physical Review Letters*, 2005. **94**(15), 153902.
8. Okawachi, Y., *et al.*, All-optical slow-light on a photonic chip. *Optics Express*, 2006. **14**(6): 2317–2322.
9. Pant, R., *et al.*, Photonic chip based tunbale slow and fast light via stimulated Brillouin scattering. *Optics Letters*, 2012. **37**(5): 969–971.
10. Pelusi, M., *et al.*, Photonic-chip-based radio-frequency spectrum analyser with terahertz bandwidth. *Nature Photonics*, 2009. **3**(3): 139–143.
11. Pelusi, M.D., *et al.*, Optical phase conjugation by an As(2)S(3) glass planar waveguide for dispersion-free transmission of WDM-DPSK signals over fiber. *Optics Express*, 2010. **18**(25): 26686–26694.
12. Song, K.Y., M.G. Herraez and L. Thevenaz, Long optically controlled delays in optical fibers. *Optics Letters*, 2005. **30**(14): 1782–1784.
13. Thevenaz, L., Slow and fast light in optical fibres. *Nature Photonics*, 2008. **2**(8): 474–481.
14. Van Erps, J., *et al.*, Automatic dispersion compensation for 1.28Tb/s OTDM signal transmission using photonic-chip-based dispersion monitoring. *Optics Express*, 2010. **18**(24): 25415–25421.
15. Vo, T.D., *et al.*, Photonic chip-based all-optical XOR gate for 40 and 160 Gbit/s DPSK signals. *Optics Letters*, 2011. **36**(5): 710–712.
16. Huang, T.X.H., X.K. Yi and R.A. Minasian, Microwave photonic quadrature filter based on an all-optical programmable Hilbert transformer. *Optics Letters*, 2011. **36**(22): 4440–4442.
17. Pant, R., *et al.*, On-chip stimulated Brillouin scattering. *Optics Express*, 2011. **19**(9): 8285–8290.

18. Paquot, Y., *et al.*, Single parameter optimization for simultaneous automatic compensation of multiple orders of dispersion for a 1.28 Tbaud signal. *Optics Express*, 2011. **19**(25): 25512–25520.
19. Boyd, R.W., *Nonlinear Optics*. 2003, Academic Press, New York.
20. Wu, B., *et al.*, Slow light on a chip via atomic quantum state control. *Nature Photonics*, 2010. **4**(11): 776–779.
21. Suzuki, N., FDTD analysis of two-photon absorption and free-carrier absorption in Si high-index-contrast waveguides. *Journal of Lightwave Technology*, 2007. **25**(9): 2495–2501.
22. Abdollahi, S. and M.K. Moravvej-Farshi, Effects of heat induced by two-photon absorption and free-carrier absorption in silicon-on-insulator nanowaveguides operating as all-optical wavelength converters. *Applied Optics*, 2009. **48**(13): 2505–2514.
23. Liang, T.K. and H.K. Tsang, Role of free carriers from two-photon absorption in Raman amplification in silicon-on-insulator waveguides. *Applied Physics Letters*, 2004. **84**(15): 2745–2747.
24. Rakich, P.T., *et al.*, Giant enhancement of stimulated brillouin scattering in the subwavelength limit. *Physical Review X*, 2012. **2**(1): 011008.
25. Khurgin, J.B. and R.S. Tucker, *Slow Light: Science and Applications 2008*, CRC Press, Boca Raton, FL.
26. Azuma, Y., *et al.*, Wavelength dependence of Brillouin-gain spectra for single-mode optical fibers. *Electronics Letters*, 1988. **24**(5): 250–252.
27. Horiguchi, T., *et al.*, Brillouin gain variation due to a polarization-state change of the pump or Stokes fields in standard single-mode fibers. *Optics Letters*, 1989. **14**(6): 329–331.
28. Shibata, N., *et al.*, Identification of longitudinal acoustic modes guided in the core region of a single-mode optical fiber by brillouin gain spectra measurements. *Optics Letters*, 1988. **13**(7): 595–597.
29. Shibata, N., K. Okamoto, and Y. Azuma, Longitudinal acoustic modes and Brillouin-gain spectra for geo_2-doped-core single-mode fibers. *Journal of the Optical Society of America B – Optical Physics*, 1989. **6**(6): 1167–1174.
30. Dainese, P., *et al.*, Stimulated Brillouin scattering from multi-GHz-guided acoustic phonons in nanostructured photonic crystal fibres. *Nature Physics*, 2006. **2**(6): 388–392.
31. Beugnot, J.C., *et al.*, Complete experimental characterization of stimulated Brillouin scattering in photonic crystal fiber. *Optics Express*, 2007. **15**(23): 15517–15522.
32. Abedin, K.S., *et al.*, Single-frequency Brillouin distributed feedback fiber laser. *Optics Letters*, 2012. **37**(4): 605–607.
33. Lee, H., *et al.*, Chemically etched ultrahigh-Q wedge-resonator on a silicon chip. *Nature Photonics*, 2012. **6**(6): 369–373.
34. Grudinin, I.S., A.B. Matsko and L. Maleki, Brillouin lasing with a CaF_2 whispering gallery mode resonator. *Physical Review Letters*, 2009. **102**(4): 043902.
35. Abedin, K.S., Brillouin amplification and lasing in a single-mode AS(2)Se(3) chalcogenide fiber. *Optics Letters*, 2006. **31**(11): 1615–1617.
36. Abedin, K.S., G.W. Lu and T. Miyazaki, Slow light generation in singlemode Er-doped tellurite fibre. *Electronics Letters*, 2008. **44**(1): 16–21.
37. Pant, R., *et al.*, Photonic-chip-based tunable slow and fast light via stimulated Brillouin scattering. *Optics Letters*, 2012. **37**(5): 969–971.

38. Byrnes, A., *et al.*, Photonic chip based tunable and reconfigurable narrowband microwave photonic filter using stimulated Brillouin scattering. *Optics Express*, 2012. **20**(17): 18836–18845.
39. Pant, R., *et al.* Photonic chip based tunable and dynamically reconfigurable microwave photonic filter using stimulated Brillouin scattering. in *Postdeadline Advanced Photonics Congress*, Colorado Springs. 2012: Optical Society of America.
40. Pant, R., *et al.*, Observation of Brillouin dynamic grating in a photonic chip. *Optics Letters.*, 2013. **38**(3): 305–307.
41. Pant, R., *et al.*, Cavity enhanced stimulated Brillouin scattering in an optical chip for multiorder Stokes generation. *Optics Letters*, 2011. **36**(18): 3687–3689.
42. Tomes, M. and T. Carmon, Photonic micro-electromechanical systems vibrating at X-band (11-GHz) rates. *Physical Review Letters*, 2009. **102**(11): 113601.
43. Zhu, Z.M. and D.J. Gauthier, Numerical study of all-optical slow-light delays via stimulated Brillouin scattering in an optical fiber. *Journal of the Optical Society of America B – Optical Physics*, 2005. **22**(11): 2378–2384.
44. Chin, S., *et al.*, Broadband true time delay for microwave signal processing, using slow light based on stimulated Brillouin scattering in optical fibers. *Optics Express*, 2010. **18**(21): 22599–22613.
45. Jackson, J.D., *Classical Electrodynamics*. 2001, John Wiley & Sons, New York.
46. Hau, L.V., *et al.*, Light speed reduction to 17 metres per second in an ultracold atomic gas. *Nature*, 1999. **397**(6720): 594–598.
47. Baba, T., Slow light in photonic crystals. *Nature Photonics*, 2008. **2**(8): 465–473.
48. Bigelow, M.S., N.N. Lepeshkin and R.W. Boyd, Superluminal and slow light propagation in a room-temperature solid. *Science*, 2003. **301**(5630): 200–202.
49. Boyd, R.W. and D.J. Gauthier, Controlling the velocity of light pulses. *Science*, 2009. **326**(5956): 1074–1077.
50. Cabrera-Granado, E., *et al.*, Observation of large 10-Gb/s SBS slow light delay with low distortion using an optimized gain profile. *Optics Express*, 2008. **16**(20): 16032–16042.
51. Gonzalez-Herraez, M., K.Y. Song and L. Thevenaz, Optically controlled slow and fast light in optical fibers using stimulated Brillouin scattering. *Applied Physics Letters*, 2005. **87**(8): 081113.
52. Henker, R., *et al.*, Enhancement of maximum time delay in one fiber segment slow light systems based on stimulated Brillouin scattering. *2007 Conference on Lasers & Electro-Optics/Quantum Electronics and Laser Science Conference*. 2007. 1563–1564.
53. Herraez, M.G., K.Y. Song and L. Thevenaz, Arbitrary-bandwidth Brillouin slow light in optical fibers. *Optics Express*, 2006. **14**(4): 1395–1400.
54. Jauregui, C., P. Petropoulos and D.J. Richardson, Brillouin assisted slow-light enhancement via Fabry-Perot cavity effects. *Optics Express*, 2007. **15**(8): 5126–5135.
55. Lee, M.J., R. Pant and M.A. Neifeld, Improved slow-light delay performance of a broadband stimulated Brillouin scattering system using fiber Bragg gratings. *Applied Optics*, 2008. **47**(34): 6404–6415.
56. Lee, M.J., *et al.*, SBS gain-based slow-light system with a Fabry-Perot resonator. *Optics Communications*, 2008. **281**(10): 2975–2984.
57. Mok, J.T., C.M. de Sterke and B.J. Eggleton, Delay-tunable gap-soliton-based slow-light system. *Optics Express*, 2006. **14**(25): 11987–11996.

58. Mok, J.T., *et al.*, Dispersionless slow light using gap solitons. *Nature Physics*, 2006. **2**(11): 775–780.
59. Mok, J.T., *et al.*, Dispersionless slow light with 5-pulse-width delay in fibre Bragg grating. *Electronics Letters*, 2007. **43**(25): 1418–1419.
60. Poon, J.K.S., *et al.*, Slowing light with Fabry-Perot resonator arrays. *Journal of the Optical Society of America B – Optical Physics*, 2007. **24**(11): 2763–2769.
61. Schneider, T., K. Jamshidi and S. Preussler, Quasi-light storage: a method for the tunable storage of optical packets with a potential delay-bandwidth product of several thousand bits. *Journal of Lightwave Technology*, 2010. **28**(17): 2586–2592.
62. Shumakher, E., *et al.*, On the balance between delay, bandwidth and signal distortion in slow light systems based on stimulated Brillouin scattering in optical fibers. *Optics Express*, 2006. **14**(13): 5877–5884.
63. Song, K.Y., M.G. Herraez and L. Thevenaz, Observation of pulse delaying and advancement in optical fibers using stimulated Brillouin scattering. *Optics Express*, 2005. **13**(1): 82–88.
64. Song, K.Y. and K. Hotate, 25 GHz bandwidth Brillouin slow light in optical fibers. *Optics Letters*, 2007. **32**(3): 217–219.
65. Zhang, B., *et al.*, A single slow-light element for independent delay control and synchronization on multiple gb/s data channels. *IEEE Photonics Technology Letters*, 2007. **19**(13–16): 1081–1083.
66. Zhu, Y.H., *et al.*, High-fidelity, broadband stimulated-Brillouin-scattering-based slow light using fast noise modulation. *Optics Express*, 2011. **19**(2): 687–697.
67. Pant, R., *et al.*, Optimal pump profile designs for broadband SBS slow-light systems. *Opt. Express*, 2008. **16**(4): 2764–2777.
68. Pant, R., *et al.*, Maximizing the opening of eye diagrams for slow-light systems. *Applied Optics*, 2007. **46**(26): 6513–6519.
69. Shi, Z., *et al.*, Design of a tunable time-delay element using multiple gain lines for increased fractional delay with high data fidelity. *Optics Letters*, 2007. **32**(14): 1986–1988.
70. Capmany, J., I. Gasulla and S. Sales, Microwave photonics: harnessing slow light. *Nature Photonics*, 2011. **5**(12): 731–733.
71. Capmany, J., B. Ortega and D. Pastor, A tutorial on microwave photonic filters. *Journal of Lightwave Technology*, 2006. **24**(1): 201–229.
72. Lloret, J., *et al.*, Tunable complex-valued multi-tap microwave photonic filter based on single silicon-on-insulator microring resonator. *Optics Express*, 2011. **19**(13): 12402–12407.
73. Capmany, J., On the cascade of incoherent discrete-time microwave photonic filters. *Journal of Lightwave Technology*, 2006. **24**(7): 2564–2578.
74. Mora, J., L.R. Chen and J. Capmany, Single-bandpass microwave photonic filter with tuning and reconfiguration capabilities. *Journal of Lightwave Technology*, 2008. **26**(13–16): 2663–2670.
75. Mora, J., *et al.*, Photonic microwave tunable single-bandpass filter based on a Mach-Zehnder interferometer. *Journal of Lightwave Technology*, 2006. **24**(7): 2500–2509.
76. Supradeepa, V.R., *et al.*, Comb-based radiofrequency photonic filters with rapid tunability and high selectivity. *Nature Photonics*, 2012. **6**(3): 186–194.
77. Chang, Y.M. and J.H. Lee, High-Q tunable, photonic microwave single passband filter based on stimulated Brillouin scattering and fiber Bragg grating filtering. *Optics Communications*, 2008. **281**(20): 5146–5150.

78. Zhang, W.W. and R.A. Minasian, Widely tunable single-passband microwave photonic filter based on stimulated brillouin scattering. *IEEE Photonics Technology Letters*, 2011. **23**(23): 1775–1777.
79. Sancho, J., *et al.*, Dynamic microwave photonic filter using separate carrier tuning based on stimulated Brillouin scattering in fibers. *IEEE Photonics Technology Letters*, 2010. **22**(23): 1753–1755.
80. Vidal, B., M.A. Piqueras, and J. Marti, Tunable and reconfigurable photonic microwave filter based on stimulated Brillouin scattering. *Optics Letters*, 2007. **32**(1): 23–25.
81. Dorrer, C. and D.N. Maywar, RF spectrum analysis of optical signals using nonlinear optics. *Journal of Lightwave Technology*, 2004. **22**(1): 266–274.
82. Neilson, D.T., *et al.*, MEMS-based channelized dispersion compensator with flat passbands. *Journal of Lightwave Technology*, 2004. **22**(1): 101–105.
83. Kurosu, T., *et al.*, Parametric tunable dispersion compensation for the transmission of sub-picosecond pulses. *Optics Express*, 2011. **19**(16): 15549–15559.
84. Pant, R., *et al.*, Investigation of all-optical analog-to-digital quantization using a chalcogenide waveguide: a step towards on-chip analog-to-digital conversion. *Optics Communications*, 2010. **283**(10): 2258–2262.
85. Husko, C., *et al.*, Ultracompact all-optical XOR logic gate in a slow-light silicon photonic crystal waveguide. *Optics Express*, 2011. **19**(21): 20681–20690.
86. Li, F., *et al.*, All-optical XOR logic gate for 40Gb/s DPSK signals via FWM in a silicon nanowire. *Optics Express*, 2011. **19**(21): 20364–20371.
87. Kim, J.H., *et al.*, All-optical XOR gate using semiconductor optical amplifiers without additional input beam. *IEEE Photonics Technology Letters*, 2002. **14**(10): 1436–1438.
88. Bogoni, A., *et al.*, 640 Gbits/s photonic logic gates. *Optics Letters*, 2010. **35**(23): 3955–3957.

15
Controlling light on the nanoscale with chalcogenide thin films

D. W. HEWAK, N. I. ZHELUDEV and K. F. MACDONALD, University of Southampton, UK

DOI: 10.1533/9780857093561.2.471

Abstract: The phase change technology behind rewritable optical disks and the latest electronic memory storage devices has provided clear commercial and technological advances for the field of data storage. Furthermore, new generations of optoelectronic devices are being driven by the merging of optics and electronics, as photons and electrons begin to cooperate in a single material platform. As a part of this evolution, plasmonics and metamaterials bring with them the ability to focus and manipulate light on the nanoscale. Here, the use of chalcogenide glasses in active plasmonic switching devices is discussed with proof of principle demonstrations of nanophotonic switching in the infrared domain.

Key words: phase change, metamaterials, plasmonics, switching, electro-optic, all-optical.

15.1 Introduction

The phase change technology behind current rewritable optical disks and the latest generation of electronic memory storage devices has provided clear commercial and technological advances for the field of data storage, by virtue of the many key attributes chalcogenide materials offer (Strand, 2005). New generations of optoelectronic devices are being driven by the merging of optics and electronics, as photons and electrons begin to cooperate in a single material platform. As a part of this evolution, plasmonics and metamaterials bring with them the ability to concentrate and manipulate light on the nanoscale, far beyond the diffraction limit of conventional optics. With such strong credentials in the 'parent' fields of photonics and electronics, it is more than reasonable to assume that chalcogenides have much to offer in the plasmonic domain as well. With these considerations in mind, the use of chalcogenide glass media in active plasmonic switching devices are considered and proof of principle demonstrations of nanophotonic switching in the near and infrared domain based on reversible electronic and photo-induced changes in the optical properties of a chalcogenide thin film is presented (Sámson *et al.*, 2010; Gholipour *et al.*, 2013).

15.1.1 Fundamentals of photonic metamaterials

The selection of an appropriate metamaterial structure with strongly resonant optical properties is crucially important to the active metamaterials envisioned here. Nanostructured media with extraordinary electromagnetic properties not found in nature have become the subject of intense investigations (Ozbay, 2008; Zheludev, 2010, 2011), with the potential to impact largely across fields ranging from telecommunications and defence to renewable energy and healthcare. A remarkable array of new fundamental physical phenomena and functionalities have been demonstrated, e.g. asymmetric transmission (Fedotov *et al.*, 2007b), optical magnetism (Enkrich *et al.*, 2005), optical activity without chirality (Plum *et al.*, 2009b), negative refractive index (Smith *et al.*, 2004), induced transparency (Papasimakis and Zheludev, 2009), super-lensing (Zhang and Liu, 2008; Liu and Zhang, 2008), and cloaking (Pendry *et al.*, 2006; Pendry and Li, 2008).

The functionality of metamaterials, typically comprising an ensemble of identical metallic nanostructures embedded in or supported by a dielectric, is underpinned by a resonant response characteristic related to the structural dimensions and material composition. The studies described in this chapter employ a planar structure belonging to a class of metamaterials that supports trapped mode plasmonic excitations (Zouhdi *et al.*, 2009; Fedotov *et al.*, 2007a; Plum *et al.*, 2011). In such metamaterials, weak coupling of the excitation mode to free-space radiation modes creates exceptionally narrow reflection, transmission and absorption resonances with asymmetric, Fano-like dispersion (Luk'yanchuk *et al.*, 2010; Fano, 1961). Modelling illustrates how the presence of an asymmetry in a split ring resonator is crucial to the provision of narrow resonances, compared to the broad dipole resonance of a symmetrically split ring resonator (see Fig. 15.1).

One of the earliest examples of such a metamaterial displaying the so-called trapped mode resonance was a periodic array of metallic wire asymmetrically split ring (ASR) resonators that has found numerous applications where sharp spectral features are required (Plum *et al.*, 2009a, 2011; Lahiri *et al.*, 2009; Driscoll *et al.*, 2009; Dicken *et al.*, 2009). It has also very recently been shown both theoretically and experimentally that the strong interactions between the magnetic moments of the oscillating trapped mode currents in an array of ASR resonators are able to create a light emission with high spatial coherence fuelled by plasmonic current oscillations, in devices dubbed lasing spasers' (Plum *et al.*, 2009a; Zheludev *et al.*, 2008). Here, a complementary structure to the wire ASR metamaterial is used, a periodic array of ASR slits in a metal film.

As has been mentioned, the resonant characteristics of metamaterial structures are highly sensitive to the local environment. Because of this, a small change in the properties of an active medium adjacent to the metamaterial

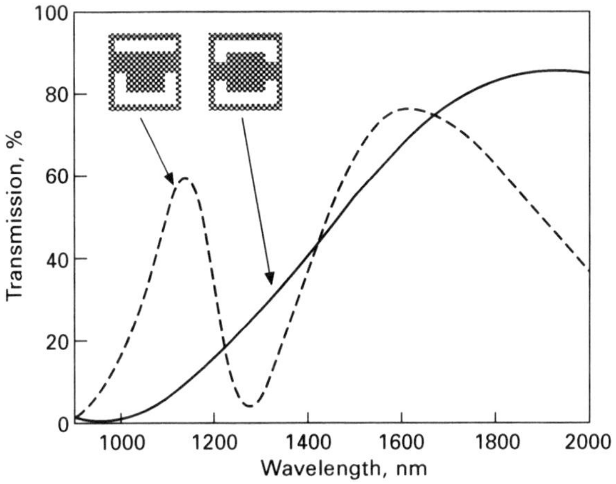

15.1 Calculated optical transmission spectra of arbitrary split ring metamaterials with asymmetric and symmetric splits. In the case of the symmetrically split ring resonator, one can observe a broad dipole resonance response (indicated by the solid curve), however if asymmetry is brought to the system, another current mode, the so-called trapped mode resonance appears along with the dipole mode, which creates the narrow band resonances between around 1200 and 1400 nm in the present case (dashed curve).

can translate into a dramatic change in the properties of the combined hybrid structure. This idea has already been implemented in some forms, such as using photo-excitation of semiconductors to tune the resonance characteristics of a metamaterial in the terahertz (THz) (Chen *et al.*, 2006, 2009) and near-infrared domains (Dani *et al.*, 2009), or by harnessing the metal-semiconductor transition in vanadium dioxide (VO_2), again in the terahertz (Driscoll *et al.*, 2009; Seo *et al.*, 2010) and infrared (Dicken *et al.*, 2009) domains, respectively, or by electrochemical modulation of metamaterials to tune the plasmonic resonances in the terahertz range. But when it comes to practical phase change functionality, the obvious switchable dielectric candidates for the realization of this concept are the chalcogenide glasses found in the heart of today's rewritable optical disc technologies (Yamada and Wuttig, 2007) and non-volatile electronic memories (Wuttig and Welnic, 2008). With threshold switching properties (an ability to be reversibly switched between amorphous and crystalline phases with markedly different electromagnetic properties) more robust than alternative phase change media (Driscoll *et al.*, 2009; Dicken *et al.*, 2009; Seo *et al.*, 2010), they have become the materials of choice for such commercial applications. It has now clearly been shown that their functionality can be brought to bear in the plasmonic domain to create switching devices (Sámson *et al.*, 2010; Gholipour *et al.*, 2013).

15.2 Chalcogenide-based active elements

15.2.1 Chalcogenides

The chalcogenides are a remarkable family of materials, displaying an extensive range of technologically relevant optical, electronic, thermal and mechanical effects (Popescu, 2005; Eggleton *et al.*, 2011). Chalcogenide glasses are sensitive to the absorption of electromagnetic radiation and therefore exhibit several types of photo-induced phenomena (Elliott and Zakery, 2003), including the ability to be reversibly switched on the nanosecond timescale between structural phases with markedly different electromagnetic properties ('Memory' or 'Ovshinsky effect' switching; Ovshinsky, 1968). They are also physically robust and their properties (e.g. refractive index, nonlinearity, resistivity, melting temperature, etc.) can be tuned substantially through compositional variation (see, for example, Fig. 15.2 and Simpson, 2008).

As such, chalcogenide glasses have been in widespread commercial use for many years in rewritable optical discs, where data is routinely encoded at more than 200 Mbit/s (Yamada and Wuttig, 2007) (here, the relative differences between low reflectivity amorphous state and high reflectivity crystalline state are exploited) and are the basis of next generation electronic memories (PRAM), where bit sizes may be as small as a few tens of nanometres and

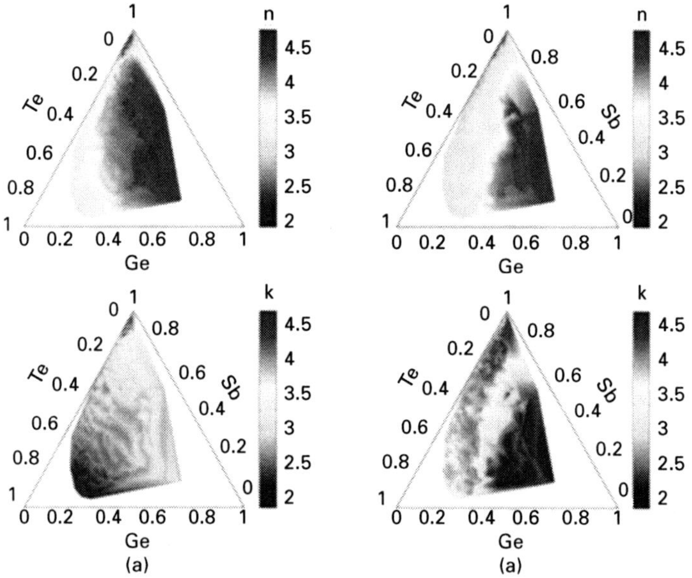

15.2 Tunability of the refractive index (top row) and extinction coefficient (bottom row) through compositional variation in (a) amorphous and (b) crystalline germanium antimony telluride (GST) samples. The images are courtesy of Dr Sam Guerin (Ilika Technologies Ltd, Southampton, UK.

switching energies are on the order of picojoules (Wuttig and Welnic, 2008) (here, the relative differences between high resistivity amorphous state and low resistivity crystalline state is exploited). Chalcogenide glasses also exhibit several other interesting phenomena (Eggleton *et al.*, 2011; Ta'eed *et al.*, 2007), including transparency into the mid-infrared wavelengths (Hilton, 1970) and third-order optical nonlinearities more than two orders of magnitude larger than silica (Vogel *et al.*, 1991; Harbold *et al.*, 2002). These characteristics can provide broad operational bandwidth and the possibility of useful plasmonic functionalities beyond the phase change modulation concept at the core of this chapter.

Apart from the great commercial significance in the field of data storage (Strand, 2005; Mathur and Greer, 2005), and the great promise they hold for plasmonics, chalcogenide glasses are also extensively studied for infrared devices (such as detectors, lenses, sensors and optical fibres operating in infrared wavelengths; Eggleton *et al.*, 2011; Seddon, 1995; Murugan, 2010), laser power delivery applications (such as industrial welding operations by, for example, high energy CO_2 laser and laser surgery in the wavelength coinciding with the peak absorption band of water; Seddon, 1995; Murugan, 2010; Temelkuran *et al.*, 2002), solar cell (Repsin *et al.*, 2008) applications and very recently introduced as a key element in future cognitive information processing and computing technology (Ovshinsky, 2004).

First, though, our attention turns to gallium lanthanum sulphide (GLS), a semiconducting, non-toxic chalcogenide glass with a transparency band extending from 500 nm to around 10 μm (see Fig. 15.3). Since its discovery in 1976 (Loireau-Lozac'h *et al.*, 1976), GLS glass has been recognized for its superior properties such as low toxicity (it does not contain the arsenic-based glass formerly found in the majority of commercially available chalcogenides) and exhibits great thermal stability. With a glass transition and melting temperature of around $T_g = 600°C$ and $T_m = 900°C$ respectively, it is well above other chalcogenide systems (e.g., $T_g = 100°C$ and $T_m = 600°C$ are the typical values for GST) that results in an enhanced environmental stability and is undoubtedly suited for practical devices (Simpson, 2008; Murugan, 2010; West *et al.*, 2000). Also, measurements indicate that among optical glasses, GLS offers one of the highest nonlinearies; with third-order nonlinear susceptibility of around $\chi^{(3)} = 1.15 \times 10^{-20}$ m^2/V^2 (Murugan, 2010; West *et al.*, 2000; Kang *et al.*, 1995), it is roughly two orders of magnitude more nonlinear than silica glass.

15.2.2 Phase change

In order to achieve reversible phase change between the amorphous and crystalline states of chalcogenide alloys, an energy barrier has to be overcome before the switch (phase transition) in the structure can be achieved. When

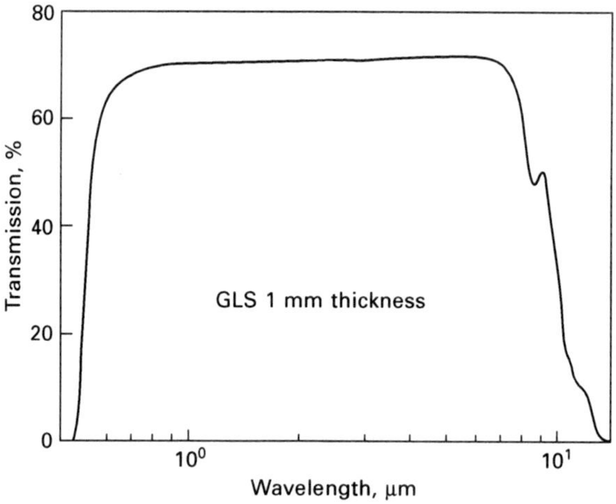

15.3 Optical transmission of a 1 mm thick amorphous GLS glass sample in the visible to near-IR range.

the energy applied to the material exceeds this threshold energy, the material is excited to a high mobility state, where it is possible to rearrange bond lengths and angles through the slight movement of the individual atoms within the lattice structure.

The energy provided to the material can be applied through a number of different mechanisms. A reversible transition between the two stable states of the material can be achieved by optical heating, through exposing the material to intense laser beams, thus achieving optical switching. This can also be replicated by the application of a current or voltage pulse or sweep to the material, achieving electrical switching of the material through joule heating. It should be noted that, in essence, both the above mechanisms excite the material by using thermal switching (Lankhorst *et al.*, 2005; Merton, 2001; Geppert, 2003; Mott and Davis, 1979). Both the optical and electrical power is converted to thermal energy within the material which allows the structural change to be achieved. Having said this, each type of switching mechanism has its own characteristics and intricacies which need to be optimized and are specific to optical and electronic properties of the material being reversibly switched from the glassy to crystalline state.

Electrical phase change

One of the most special properties of certain chalcogenide glasses is their ability to be rapidly and reversibly switched between amorphous and crystalline phases using electrical energy with markedly different electromagnetic properties. This was first investigated by Ovshinsky in 1968, who observed that the chalcogenide glasses could be switched from the high resistivity amorphous state to a conducting crystalline state at a critical point, termed 'the threshold voltage' (V_t). Ovshinsky noticed that increasing applied voltage leads to a point where abrupt threshold switching occurs and through localized joule heating, the amorphous phase is converted to the more conductive crystalline form. Such a dynamic response is highlighted in Fig. 15.4, showing that even after the switch, when the material is in the highly conductive state, the voltage can be increased or decreased without affecting the structural phase, that is, indicating a type of memory effect.

Optical phase change

Optical phase change of chalcogenide thin films has been an area of intense research for the past three decades bringing about great commercial success with the production of CDs, DVDs and Blu-ray discs. At the centre of these technologies is the inherent change in the optical properties of these thin films in the amorphous and crystalline states. This reversible non-volatile change appears well suited to the field of nanophotonics. Indeed, using the optical phase change of chalcogenide glasses as the switchable dielectric medium in contact with the metamaterial molecules can allow the drastic and controlled change of the resonant response of the metamaterial in a reversible manner.

The principle of phase-change optical storage is based on the concept that some physical property of a microscopic area of the recording layer on the disc surface is altered due to crystallographic structure changes when the films

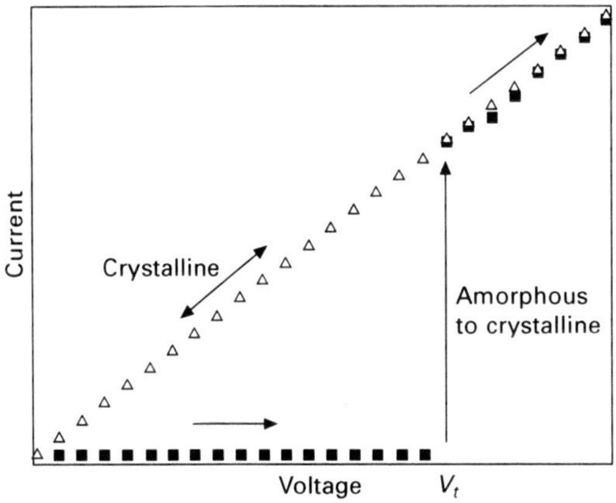

15.4 Generic electrical phase switching concept of chalcogenide glasses: the current–voltage diagram shows that the chalcogenide maintains a high (amorphous state) resistivity until the applied voltage reaches a critical level when abrupt threshold switching occurs.

are irradiated by laser pulses. The reproduction of the recorded information takes advantage of the difference in reflectivity due to the difference in refractive index and extinction coefficient between the two phases. Although there are two possible types of phase change (one is between amorphous and crystalline phases and the other is between two different crystalline phases), the one applied in rewritable phase-change optical discs is only the first type. Before recording data on the phase-change optical discs, the as-deposited writing process, the amorphous state is achieved by heating the phase-change thin films with sufficient laser power above its melting point and then rapidly quenching it to room temperature. Because the atoms of the phase-change material after melting are in a disordered state and the cooling rate of the area irradiated by laser pulses is very high, the time is not sufficient for the atoms to be arranged into the ordered crystalline structure. Thus amorphous marks are formed. The absolute minimum quenching rates required for amorphization are different for various materials, ranging from 10^6 to 10^{11} K/s.

In the erasing process as shown in Fig. 15.5, the crystalline phase is realized by annealing the phase-change films at the temperature between the crystallization temperature T_c and the melting point T_m with a medium power laser irradiation. During the irradiation period, the atoms of the phase-change media are rearranged into an ordered structure; thus the amorphous region can be changed to the crystalline state. The phase changes in the phase-change optical discs are accomplished by using the irradiation of laser light which typically has a diameter on the order of 1 μm. When a laser beam, having a 1 μm diameter, moves on the recording thin films at a linear velocity of 10 m/s, the irradiation time of a point on the film is only 100 ns. Hence, the available time for the energy deposition is also within this irradiation time duration.

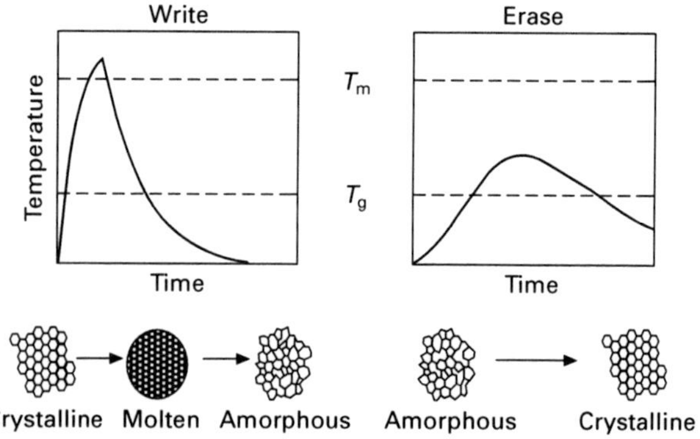

15.5 Schematic of the temperature–time profiles associated with formation of amorphous dots (left panel) and crystallization on (right panel) of amorphous marks in a crystalline layer.

For optical nanophotonic applications, $Ge_2Sb_2Te_5$ has been utilized as an optical phase-change dielectric. This material is an established commercial medium for optical data storage, having been utilized in CDs and DVDs in the marketplace. The crystallization of $Ge_2Sb_2Te_5$ is a nucleation-driven process as the optical power has been found to be constant for complete erase time (CET) of different diameter amorphous marks. This behaviour is indicative of a nucleation dominant process. As further proof, inspection of the crystallized marks with a TEM also revealed that as the crystallizing pulse power was increased, the number of crystallites present increased and, at the threshold power for complete erasure, the mark was found to be filled with non-crystalline grains with sizes ranging from 10 nm to 30 nm, indicating that the nucleation occurs before growth from the surrounding crystalline background (Almeida *et al.*, 2004).

The CET of $Ge_2Sb_2Te_5$ decreases with increasing film thickness; this is understood to be a result of competing interface and bulk effects. Below a threshold thickness, the film crystallization characteristics are dominated by interface effects. As the film is made thicker, the CET becomes less sensitive to these effects and bulk nucleation is predominantly the influencing factor. Consequently, below a critical thickness the interface material, surface energy and roughness all influence the nucleation rate of $Ge_2Sb_2Te_5$ (Almeida *et al.*, 2004).

For the purposes of electro-optic switching, the chalcogenide alloy gallium lanthanum sulphide (GLS) presents an ideal material for this purpose. GLS is a non-toxic, semiconducting chalcogenide glass with an optical transparency range extending from 500 nm to 11 μm. With glass transition, crystallization and melting temperatures of around 520, 730 and 830°C, respectively, it also offers exceptional thermal stability. It is receptive to high concentrations of noble metal and rare earth dopants, which opens up a range of application opportunities. Most importantly, as compared to other chalcogenide phase-change alloys, its ultra-high amorphous and crystalline electrical resistivity means that efficient heating of the active chalcogenide layer can be achieved through electrical joule heating for the purposes of crystallization and melt/quenching the layer in switching applications. This makes it the ideal material for electro-optic nanophotonic switches.

15.3 Nanoscale switches

Future nanophotonic applications, in particular photonic data processing circuits, will undoubtedly require active devices of sub-wavelength dimensions (Zheludev, 2006, 2007). Unfortunately, electro-optic modulation of light in a device of nanoscale thickness is not a trivial problem.

In conventional electro-optic modulators exploiting, for example, the Pockels effect, the polarization switching involved requires the interference of two propagating modes to develop over distances far in excess of the wavelength of light. The dimensions of such modulators in the propagation direction are often in the centimetre range (Yeh and Yariv, 2007).

Signal modulation via control of the waveguide absorption coefficient or refractive index is another possibility. However, this approach also requires substantial propagation lengths over which an amplitude or phase change accumulates, or it involves resonators and/or interferometric arrangements that are inherently longer than the wavelength of light. The typical dimensions of electro-absorptive quantum well (QW) structures, for example GaAs ring resonators, are only a few micrometres (Ravindran *et al.*, 2009). Similar compact (10–12 μm) structures based on refractive index modulation of silicon electro-optic modulators have been demonstrated (Almeida *et al.*, 2004; Xu *et al.*, 2005). It has recently been suggested that strong signal modulation may be achieved in truly nanoscale photonic devices, despite very short propagation lengths, through the use of materials that show a substantial change in absorption or refraction in response to a control excitation (Almeida *et al.*, 2004; Xu *et al.*, 2005). The relative change in the real and/or imaginary parts of the refractive index must be of the order of unity and this can only be achieved in metals, where phase changes can bring about significant changes in optical properties.

Such functionality has been extensively studied and demonstrated with elemental gallium, which can exist in several phases with radically different optical properties. In this case, phase changes led to the modification of the plasmon and interband absorption to provide a platform for various nanoscale active devices (Zheludev, 2002; Almeida *et al.*, 2004; Xu *et al.*, 2005; MacDonald *et al.*, 2006). Such demonstrated devices among others consist, for example, of a high-contrast plasmonic modulating device, where light control by light was achieved by changing the efficiency of light coupling into a plasmon polariton wave (Krasavin *et al.*, 2004; MacDonald *et al.*, 2007). Another device that has potential applications in phase-change memory and data storage technologies operating at extremely low power levels was demonstrated first by light-induced structural phase transitions in a single gallium nanoparticle (Soares *et al.*, 2005), followed by the demonstration of electron beam-induced structural phase transitions of individually addressable gallium nanoparticles (Denisyuk *et al.*, 2009).

In this chapter, another approach to nanoscale electro-optic and all optical modulation is considered that relies not on absorption modulation, but rather on a change in the refractive index of a material associated with a control-input-induced phase change. In a layer of nanoscale thickness, such a refractive index change would be insufficient to noticeably modulate the intensity or phase of a transmitted wave. However, it is now widely recognized that by combining a nanoscale layer of phase-change material with a planar plasmonic metamaterial (Fig. 15.6), one can exploit the fact that the position of narrow resonant absorption lines in certain metamaterials are strongly dependent on the dielectric environment (Luk'yanchuk *et al.*, 2010). Switching the dielectric layer in contact with such a metamaterial produces a massive change in its resonance frequency. Importantly, given that the planar metamaterial also has a sub-wavelength thickness, the whole electro-optic modulating structure can be much thinner than the light wavelength. Moreover, the resonance frequency, and therefore the spectral band of high contrast switching, can be engineered by design and located anywhere within the entire transparency range of the dielectric phase-change medium.

15.4 Modelled phase change functionality in metamaterials

When one considers that amorphous–crystalline phase transitions in chalcogenides may bring about a relative index change approaching 0.1 (Simpson, 2008) on nanosecond timescales, the great potential held by chalcogenides in the field of metamaterials and their role in nanoscale switching becomes clear. The first studies of the role of phase change chalcogenides in switching explored their functionality in the plasmonic domain both experimentally and through modelling. Using the transfer matrix method, an analysis was conducted with the focus on amorphous-crystalline phase change. The real and imaginary parts of the refractive index of amorphous

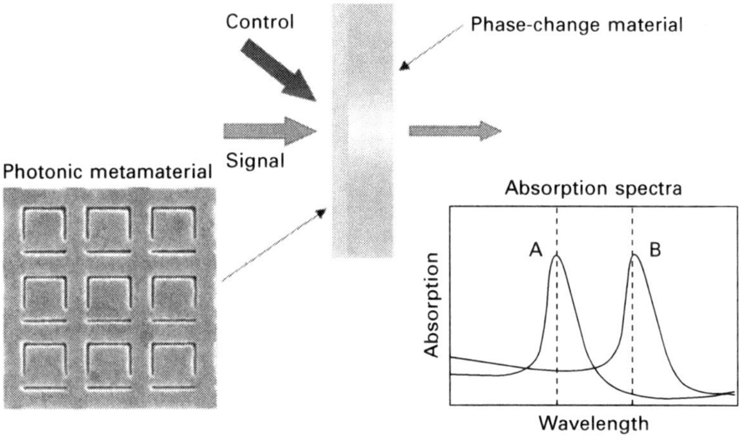

15.6 Nanoscale electro-optic modulator concept: changes in the properties of a phase-change medium adjacent to a plasmonic metamaterial, for example by electrical, optical or thermal means, leads to dramatic changes in the hybrid (metamaterial plus phase-change medium) structure resonance (changes, for example, in absorption, reflection, transmission). Also shown is the resulting absorption spectra of an arbitrary phase A and B.

GLS are shown in Fig. 15.7. For easy comparison it is repeated in Fig. 15.8 together with the data for the corresponding crystalline phase.

The biggest difference going from the as-deposited amorphous phase to crystalline phase is the reduction in the real part and increase in the imaginary part of the refractive index. In the crystalline GLS phase, the real part of the refractive index is n_c = 2.165 at wavelength 1550 nm, considerably smaller than in the amorphous phase (n_a = 2.281). The imaginary part of the refractive index is still very small, reaching only k_c = 0.023 at the wavelength of 1550 nm, as opposed to k_a = 0 at this wavelength (see Figs 15.8(a) and (b)).

For the two forms of GLS, reflectivity R as a function of incident angle θ and silver film thickness t for p-polarized incident light at a wavelength of 1550 nm can be seen in Fig. 15.9. The pronounced reflectivity dip associated with the excitation of SPPs on the metal/chalcogenide interface is present in both cases for silver film thicknesses of a few tens of nanometres. Between around 20 and 80 nm in the case of the amorphous phase, and around 10 and 60 nm in the case of the crystalline phase. Figure 15.9(c) shows cross-sections of these plots at t = 35* nm and illustrates the dramatic 8:1 shift in resonant coupling angle associated with the structural transition in GLS from amorphous (μ_{SPP} = 73.5° at R^A_{min}) to crystalline (θ_{SPP} = 65.4° at R^C_{min}) phase. It is immediately apparent that the amorphous-crystalline transition in GLS can provide high contrast switching functionality in this basic sample geometry: for example, at an incident angle of θ_{SPP} = 65.4, the phase change provides an optical reflectivity contrast ratio of almost 7:1. The corresponding

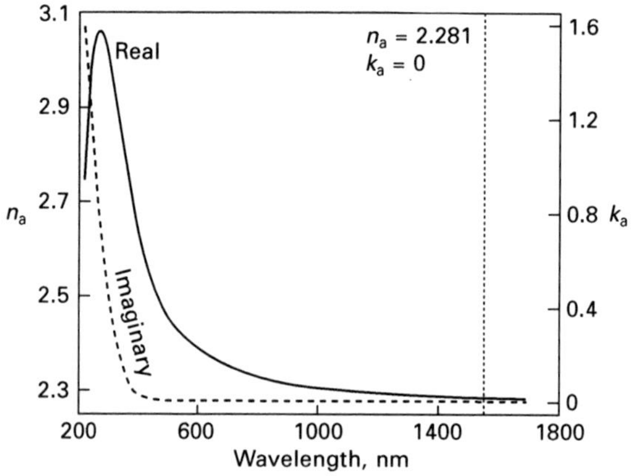

15.7 Real and imaginary parts of the refractive index of amorphous GLS in the visible to near-IR spectra. The dashed vertical line corresponds to 1550 nm used in the calculations, along with specific values of the refractive indices at this wavelength. The refractive index data were derived from ellipsometric measurements.

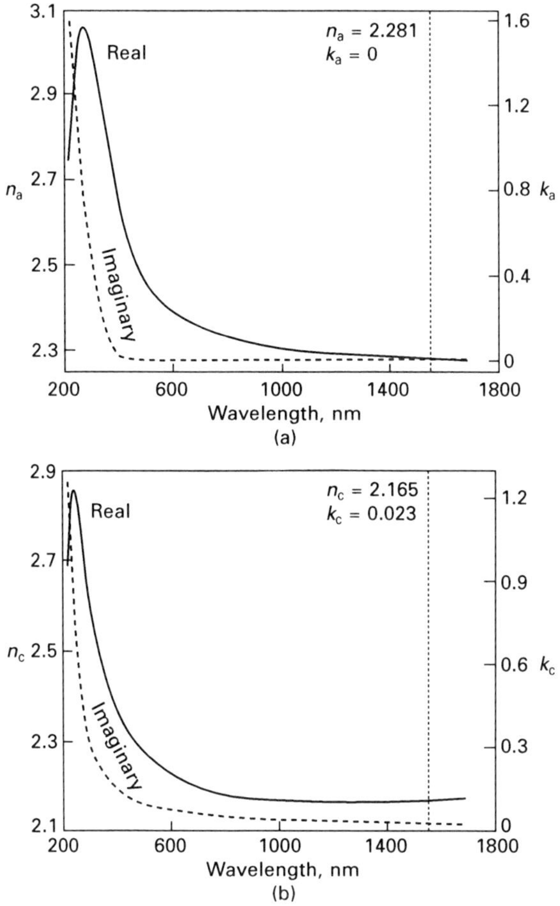

15.8 Real and imaginary parts of the refractive indices of (a) amorphous and (b) crystalline GLS in the visible to near-IR spectra. The dashed vertical lines correspond to 1550 nm used in the calculations, along with specific values of the refractive indices at this wavelength. The refractive index data were derived from ellipsometric measurements.

values of the reflectivity at this incident angle are $R_A = 0.733$ and R_C min $= 0.11$ for the amorphous and crystalline phase, respectively.

Because the resonant SPP coupling characteristics in the Kretschmann geometry depend strongly on the chalcogenide film thickness (t_2), it is beneficial to look at the changes during the full amorphous–crystalline phase transition as well. Figure 15.10(b) shows, for the two forms of GLS the resonant coupling angles. Both the amorphous and crystalline phases show similar trends and increases with increasing GLS thickness. However, the rate of increase is slower for the crystalline phase, leaving larger angle differences

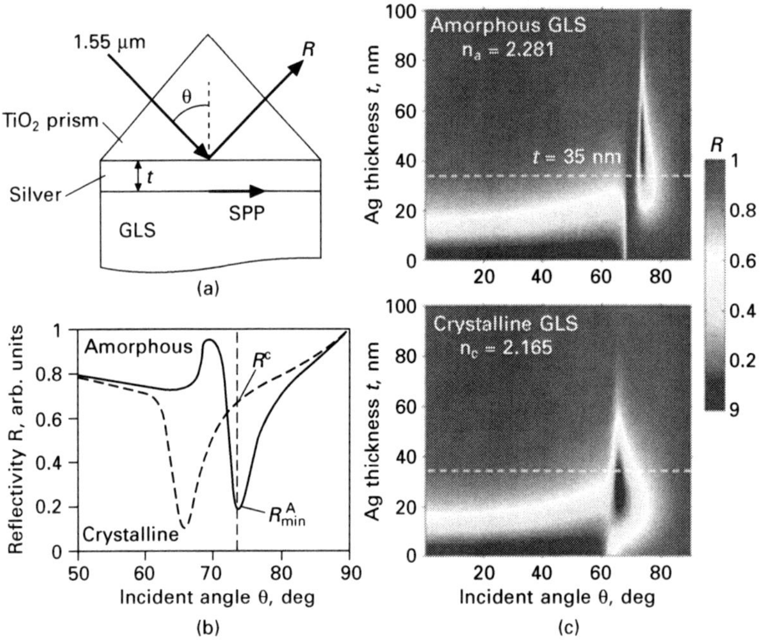

15.9 (a) Modelled Kretschmann geometry for SPP excitation on a silver/GLS interface via attenuated total reflection in a TiO_2 prism. (b) Reflectivity R of the structure shown in part (a), for p-polarized light at 1550 nm as a function of incident angle θ and silver film thickness t for the amorphous and crystalline forms of GLS. (c) Cross-sections of the plots in part (b) showing reflectivity as a function of incident angle at t = 35nm.

in the coupling resonance between the two phases. On the other hand, Fig. 15.10(c) shows the reflectivity for the two forms of GLS at specific thicknesses. As one can see from Fig. 15.10(c), by choosing larger chalcogenide film thicknesses, one can rely on larger plasmon property differences because of the large angle differences in the SPP excitation. However, this comes at the expense of the excess amount of energy needed to put into the system to switch larger areas. One may also define a resonance contrast ratio:

$$(R^{\uparrow}(\mathrm{C}, \mathrm{A}) - R_{\downarrow\min}{}^{\uparrow}(\mathrm{A}, \mathrm{C})/(\mathrm{R}_{\downarrow\min}{}^{\uparrow}(\mathrm{A}, \mathrm{C}))$$

where $(\mathrm{R}_{\downarrow\min}{}^{\uparrow}(\mathrm{A}, \mathrm{C})$ is the minimum reflectivity for a given phase (SPP coupling resonance) and R^{CA} is the level of the other phase at the same incident angle as illustrated for the amorphous phase minimum in Fig. 15.10(c). For a large range of GLS thicknesses, the contrast ratio is around 10 for both structural phases. However, in the case of an amorphous to crystalline transition, the contrast ratio can reach very high values at GLS thicknesses of around 150 nm due to the fact that the reflectivity minima

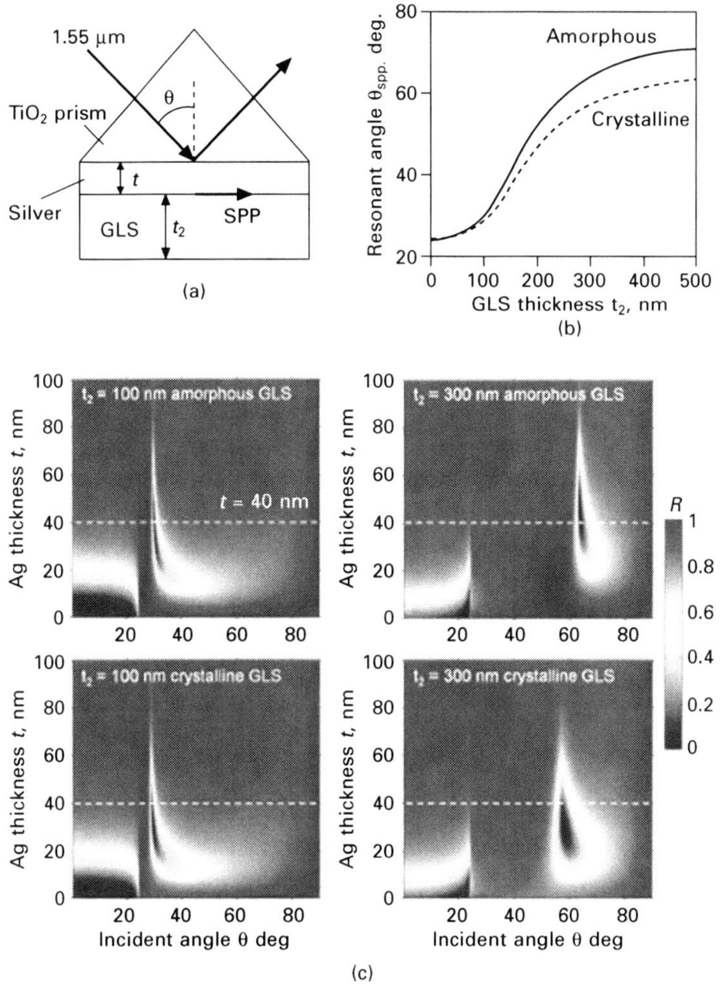

15.10 (a) Modeled SPP coupling geometry with finite GLS thickness t_2. (b) Angular position of SPP coupling resonance (min. R) as a function of t_2 at a fixed silver thickness $t = 40$ nm for the two forms of GLS. (c) Reflectivity R of the structure shown in part (a) for p-polarized light at 1550 nm as a function of incident angle θ and silver thickness t for different thicknesses t_2 and phases of GLS (as labelled).

here approaches zero. The resonance contrast ratios decline abruptly for GLS film thicknesses below about 70 nm (Fig. 15.11).

It has already been shown in Fig. 15.6 how modelling reproduces the characteristic resonance features of the basic metamaterial, but more importantly it allows us to explore the subtle effects of variations in the coverage of the active layer. For this reason, computational results using

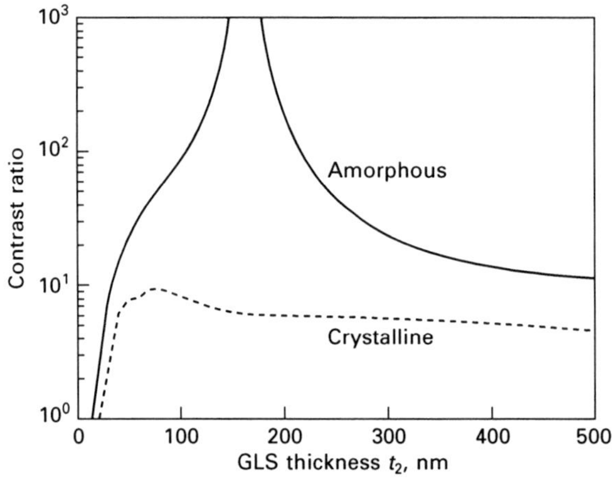

15.11 Switching contrast ratios at resonance as functions of GLS thickness for the two structural forms of GLS (at a fixed silver film thickness t = 40 nm).

a three-dimensional finite element method to solve Maxwell's equations (finite element multiphysics modelling) are presented here. The calculations employed experimentally determined values of dielectric parameters for GLS and data for gold and silicon nitride from Palik (1985). The modelled structures were such that they had a rectangular slit as a negative design penetrating all the way and through 50 nm of the silicon nitride substrate as shown in Fig. 15.12. The other characteristic design parameters were the same as in the real structure such as film thicknesses of 70 nm gold and 100 nm silicon nitride. The metamaterial unit cell of 375 nm was used with 300 × 300 nm^2 square ASR resonators having 115 nm wide asymmetric splits and 35 nm line widths, as highlighted in Fig. 15.13. For investigation of the GLS thickness dependence, two sets of analytical measurements were performed. In the first sets of calculations, the GLS thicknesses were tuned between 0 and 200 nm (see Fig. 15.14(a)), in order to study the exact spectral locations of the trapped mode resonances during transmission measurements. Several different scenarios were investigated: when no GLS was present or 50, 70, 100, 120 and 200 nm GLS fills the gap and the top gold layer. Figure 15.14(b) shows numerically calculated transmission resonance characteristics of the GLS hybridized metamaterial at the above-mentioned GLS thicknesses. In all cases, the curves resemble the unique Fano-shaped trapped mode resonances as expected, with increasing GLS thicknesses, the resonances red-shifted and broadened gradually. The GLS thickness dependence was also found to be more sensitive inside the slits of the negative ASR design and less sensitive outside of it on top of the gold film, as shown in Fig. 15.14(c). Between 50 and 120 nm covering the 70 nm Au nanofabricated layer, the resonance

15.12 Cross section of the modelled ASR unit cell used for the computational calculations, showing the rectangular slit penetrating through the substrate and the GLS filling distribution.

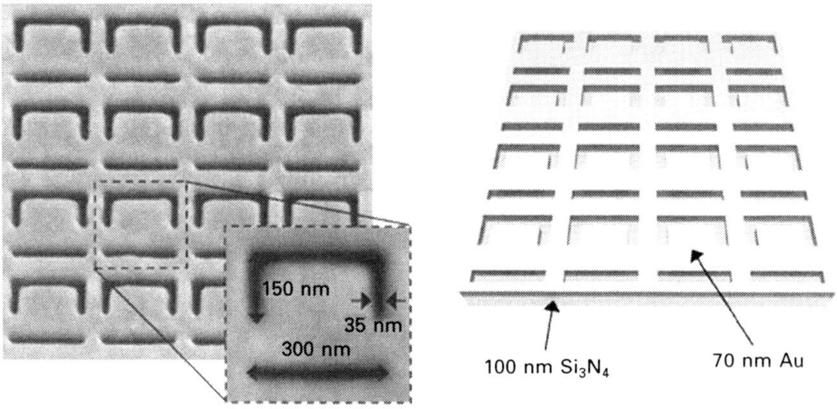

15.13 Section of an electron microscope image of the asymmetric split ring (ASR) slit array metamaterial. The inset shows the dimensions of an individual unit cell. Also shown is a computer generated image of the device structure with thicknesses of the individual layers.

shift is largest, indicated by the near linear increase, above this the rate of the shift is reduced.

The dependence of the spectral response of the device on the level which the metamaterial gaps are filled with the chalcogenide dielectric layer is further demonstrated by the second set of calculations, where two different GLS filling distributions were compared (see Fig. 15.15(a). Figure 15.15(b) shows

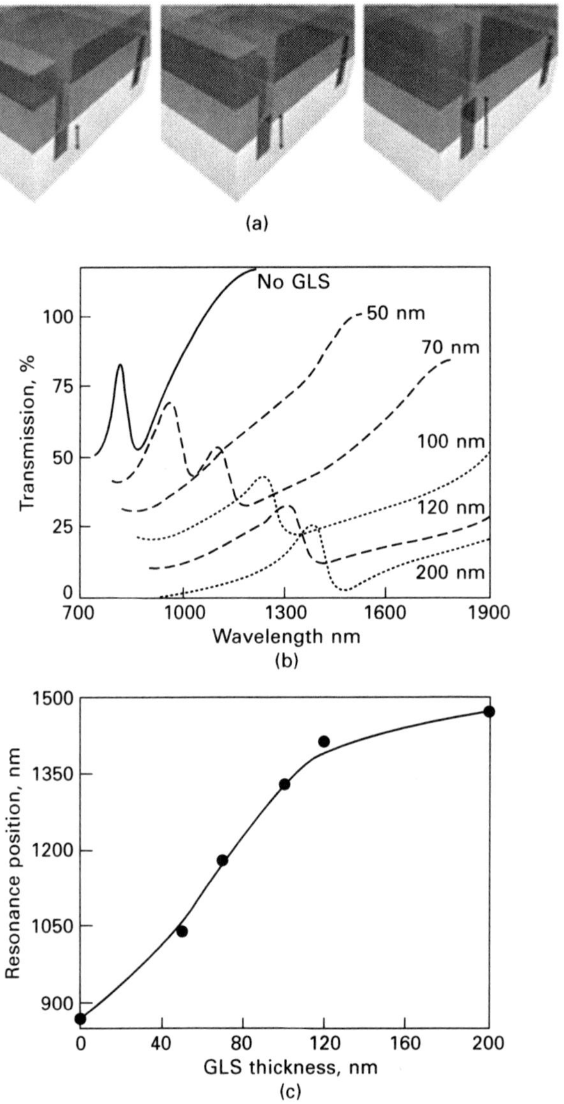

15.14 GLS thickness dependence: (a) various thicknesses of GLS layer covering the slit and the top layer of the hybrid device structure. (b) Transmission resonance spectra for the amorphous phase of the chalcogenide glass layer with varying thicknesses (as labelled). Graphs are vertically offset by 10% for clarity. (c) Approximate trapped mode resonance position as a function of GLS thickness.

the calculated trapped mode transmission lines of the metamaterial structure and the hybrid metamaterial/GLS device for amorphous and crystalline forms of GLS. This model compares a situation where the GLS completely fills

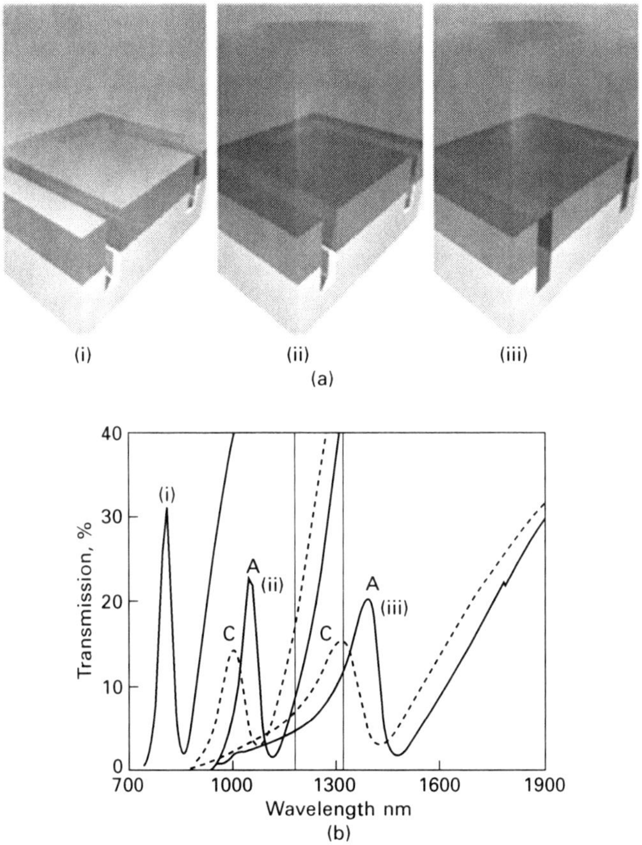

15.15 GLS thickness dependence: (a) various GLS distributions in the hybrid device structure. (b) Computed optical transmission spectra of the ASR structure (i) without the GLS film; (ii) when the GLS forms a layer over the metal nanostructure without filling the slits; (iii) when the GLS also fills the slits in the gold. The solid and dashed lines correspond to amorphous (A) and crystalline (C) phases of GLS respectively. The dashed vertical lines indicate the approximate spectral positions of the structure's trapped mode resonances from the experimental measurements.

the slits in the metal nanostructure and forms a 200 nm layer on top of the gold with one where the slits remain unfilled beneath the 200 nm top layer. Even though the chalcogenide layer thicknesses above the gold film are the same in both cases, just by having an incomplete fill, the structure's trapped mode resonance position considerably blue-shifts. A comparison with the experimentally measured positions of the transmission lines (indicated by the dashed vertical lines in Fig. 15.15(b); see Section 15.5) indicates that a partial filling of the nanoscale slits in the gold film has been achieved.

For comparison with the experiment, it is instructive to reproduce a family of transmission, reflection and absorption curves using the above-mentioned finite element numerical method. The chalcogenide thickness in this case is set to 200 nm (filling the slits of the metamaterial structure). Figure 15.16 shows the computed results of the hybrid metamaterial/GLS device for amorphous and crystalline forms of GLS. Similar to the experiments,

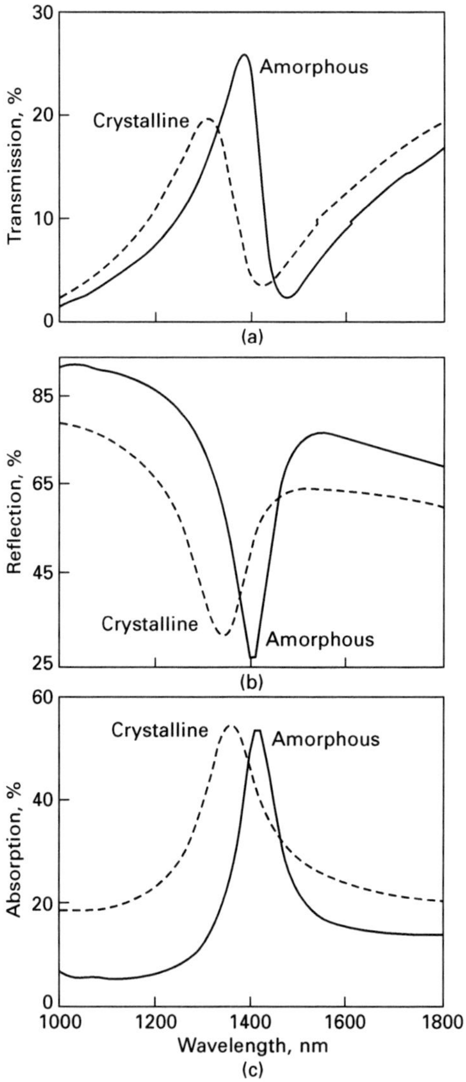

15.16 Numerically simulated (a) transmission, (b) reflection and (c) absorption resonance spectra for the amorphous and crystalline phases of the chalcogenide glass layer with thicknesses of 200 nm.

the polarization of the incident light was selected perpendicular to the split in the ring resonators to ensure excitation of the trapped mode resonance. The numerical results show a good qualitative agreement with experiments, reproducing the form of resonance features and the shifts associated with the GLS phase switching. Quantitative discrepancies between the two are attributed to manufacturing imperfections in ASR array fabrication that results in variations of line widths, depths, corner radii and imprecise knowledge of the ion beam profile resulting in non-rectangular slits.

Similar to the experimental investigations, the transmission and reflection modulation can be calculated resulting from the phase change in the computational domain as well. Figure 15.17 shows the changes during the amorphous to crystalline transition. In this particular computational situation, the changes are even more dramatic compared to the experimental case. Here, the transmission changes can be even higher than a factor of five at around 1400 nm, while reflectivity changes are also dominant and can reach up to a factor of two at 1330 nm. Again, it must be highlighted that the device thickness envisioned here is on the nanometre scale, demonstrating a truly nanoscale switching functionality.

15.4.1 Fabrication and characterization

The metamaterial structures are fabricated by focused ion beam (FIB) milling through a 50–70 nm thick gold film evaporated on a 100 nm thick Si_3N_4 membrane or a substrate which exhibits a high transparency in the part of the spectrum intended for operation. The gold film with surface roughness of 2–3 nm is obtained using low pressure (10^{-8} mbar) thermal

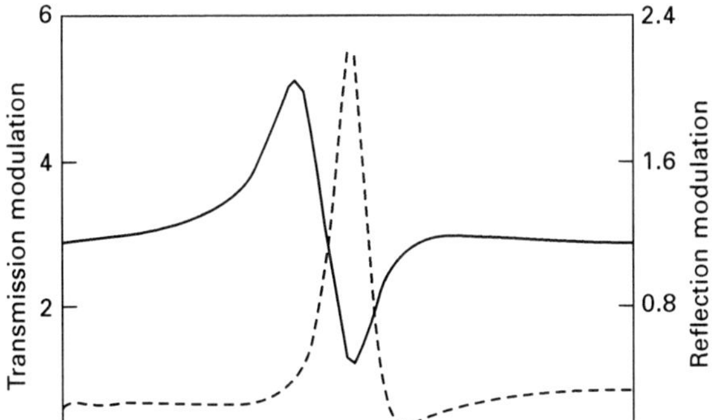

15.17 Spectral dispersion of transmission and reflection modulation contrast associated with chalcogenide phase switching in the hybrid structure during computational simulations.

evaporation. It is beneficial to use gold for this kind of application, as it does not oxidize at room temperature which could influence the metamaterial characteristics.

50 × 50 μm² negative asymmetric split ring metamaterial arrays with a square unit cell size relevant to the intended part of the spectrum were then manufactured in the gold by focused ion beam milling. The corresponding dielectric layers can be deposited through CVD, sputtering or PVD. Sputtering is the ideal choice due to the room temperature nature of the deposition process, ensuring that post-metamaterial fabrication, the sample does not endure harsh environments (Fig. 15.18).

15.4.2 Device performance

Optical characterizations of the metamaterials were performed using a microspectrophotometer, typically sampling a 5 μm² area within the metamaterial array. Near-infrared transmission spectra for normally incident polarizations parallel and perpendicular to the ring resonator split direction are recorded. Measurements on the uncoated gold ASR array (Fig. 15.19) reveal polarization-sensitive resonance features. For light polarized parallel to the resonator split direction (designated as the x-direction as defined in the inset to Fig. 15.19), there is a single broad transmission peak centered at 1170 nm. However, for the y-polarization, a transmission dip associated with the excitation of the metamaterial's trapped mode resonance is clearly seen at around 1030 nm. Hereinafter, a polarization direction of incident light perpendicular to the ring resonator split is always selected to ensure excitation of the structure's trapped mode resonance.

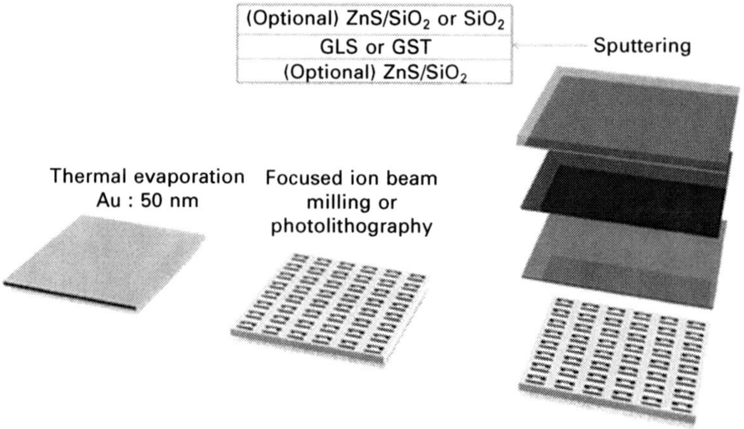

15.18 Process flow diagrams for the fabrication of phase change electro- and all-optical metamaterial switches.

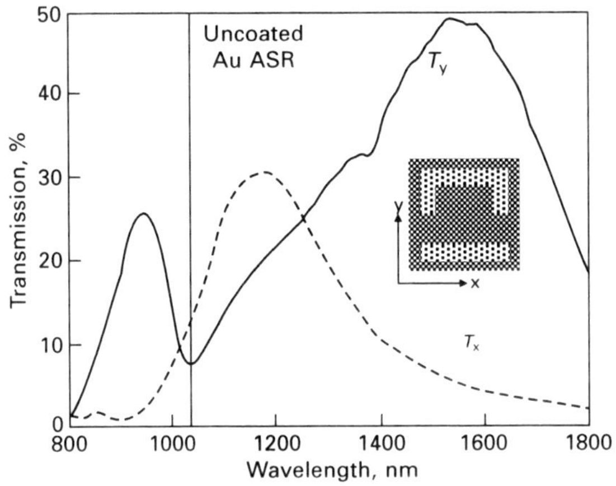

15.19 Optical transmission spectra for the uncoated gold asymmetric split ring metamaterial shown in Fig. 15.10. Absolute transmission data are presented for incident polarizations parallel (T_x) and perpendicular (T_y) to the resonator split, as illustrated by the inset. The dashed vertical line indicates the approximate spectral position of the structure's trapped mode resonance.

After the initial characterization of the device structure, an amorphous film of gallium lanthanum sulphide (GLS) was sputtered onto the nanostructured metal from a target prepared by melt quenching and annealing a mixture of high-purity gallium sulphide and lanthanum sulphide precursors, as noted earlier. Upon deposition with the chalcogenide layer, substantial changes in the metamaterial's optical properties are observed as a result of the chalcogenide functionalization. The presence of a 200 nm as-deposited GLS film substantially changed the near-field dielectric environment of the ASRs, and it has been found that the structure's trapped mode resonance red-shifted to a new centre wavelength as shown in Fig. 15.20.

Not only the transmission resonance, but all resonance features exhibited the anticipated red-shift resulting from the reduction of plasmon frequency due to the presence of the high refractive index GLS layer. The transmission dip at around 1320 nm as observed previously and in (Fig. 15.21(a) is the characteristic sign of the proximity location of the structure's trapped mode resonance. Such a transmission dip appears as a peak in reflection measurement as shown in Fig. 15.21(b) around the same wavelength. The absorption spectrum (see Fig. 15.21(c)), which shows a peak as well, is derived from the above measurements as:

$$A(\%) = 100\% - T(\%) - R(\%)$$

where T, R and A denotes transmission, reflection and absorption, respectively.

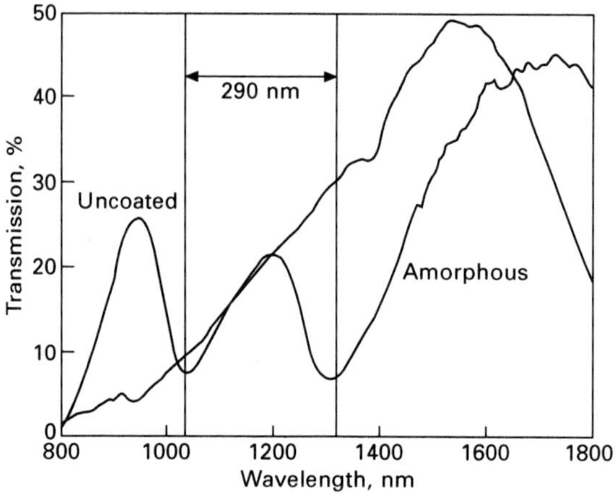

15.20 Transmission spectra comparison of the uncoated and the amorphous GLS functionalized Au ASR metamaterials. The presence of a 200 nm amorphous GLS film shifts the resonance by 290 nm. The dashed vertical lines indicate the approximate spectral positions of the structure's trapped mode resonances.

15.5 Electro-optic switches

15.5.1 Characterization

The electrical phase switching in a localized area of the GLS film in the hybrid (chalcogenide functionalized metamaterial) device was achieved by applying 10 ms pulses of incrementally increasing voltage between the nanofabricated gold metamaterial layer and a wire probe electrode brought into contact with the top surface of the GLS film. By using a Keithley 238 source measure unit (SMU) electrometer for this purpose, the electrical properties of the GLS film could be monitored in real time, providing easy identification of the transition point. The chalcogenide maintained a high resistivity amorphous state until the applied voltage reached approximately $V_t = 45$ V, at which point threshold switching occurred.

15.5.2 Performance

This transition to the more conductive crystalline state (occurring through localized joule heating) brings about a dramatic blue-shift of around 140 nm in the spectral position of the hybrid metamaterial's trapped mode resonance, which now is located at around 1180 nm. Figure 15.22 shows again the corresponding transmission, reflection and absorption characteristics of the GLS hybridized metamaterial, but in the new structural phase of crystalline GLS.

The resonance features were quantified again at normal incidence using

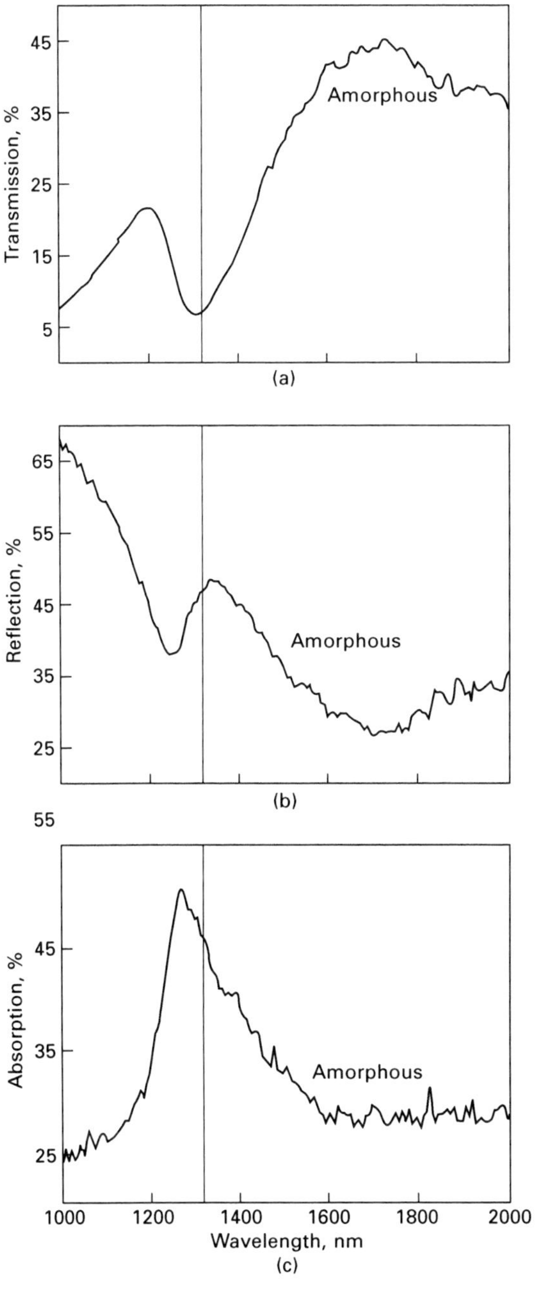

15.21 (a) Transmission, (b) reflection and (c) absorption resonance spectra of the amorphous GLS functionalized Au ASR metamaterial. The dashed vertical lines indicate the approximate spectral positions of the structure's trapped mode resonances.

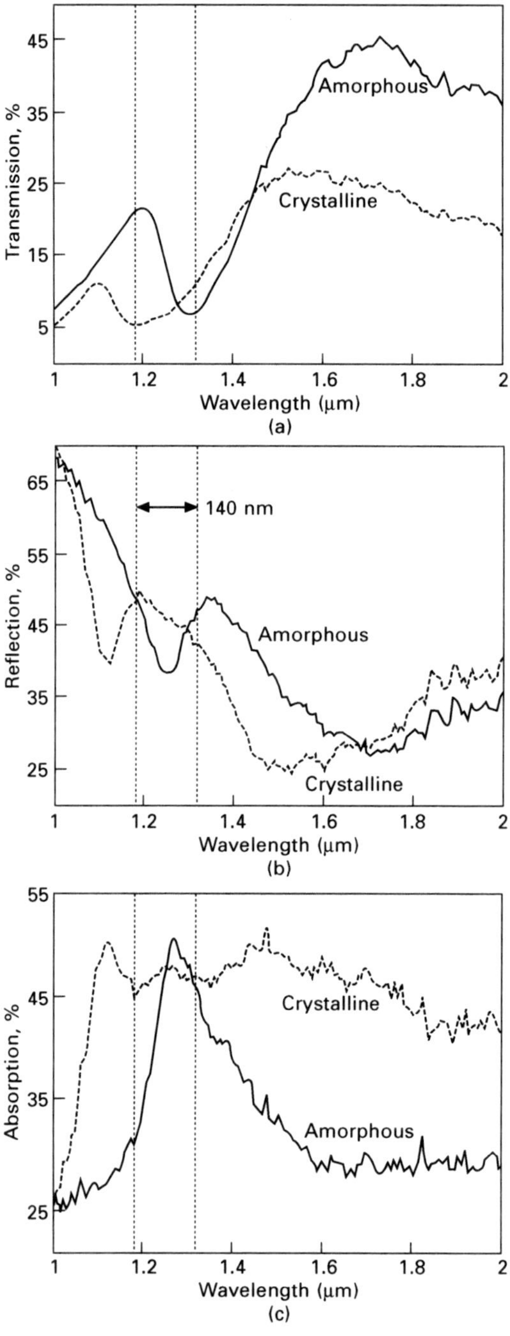

15.22 Metamaterial electro-optic modulator switching characteristics: (a) Transmission, (b) reflection and (c) absorption resonance spectra for the amorphous and crystalline phases of the chalcogenide glass layer. The dashed vertical lines indicate the approximate spectral positions of the structure's trapped mode resonances.

the above-mentioned microspectrophotometer. Also displayed in Fig. 15.22, the corresponding curves for the amorphous state for easy comparison and indication of the induced spectral shifts of around 140 nm. Following hybridization with GLS, the metamaterial's absorption spectrum, as indicated in Fig. 15.22(c), accrues a background associated with the additional absorption of the GLS layer and the broadening of the plasmon peak due to increased damping. This is more pronounced in the more conductive crystalline phase, which inherently has larger extinction coefficients as well (see Fig. 15.22).

The 140 nm resonance shift during structural phase transition from amorphous to crystalline (as highlighted in Fig. 15.23), brings about substantial changes in the transmission and reflection of the hybrid structure at certain wavelengths. The transmission and reflection modulation contrast associated with chalcogenide phase switching in the hybrid structure are given as:

$$T_{\mathrm{mod}} = \frac{T_{\mathrm{a}}}{T_{\mathrm{c}}},\ R_{\mathrm{mod}} = \frac{R_{\mathrm{a}}}{R_{\mathrm{c}}}$$

where a and c denote the amorphous and crystalline forms of GLS during transmission and reflection measurements. For example, the transmission changes by as much as a factor of four from around 20% to 5% within a wavelength band centred at around 1200 nm and near the structure's trapped mode resonances during crystalline phase – see Fig. 15.23). The reflectivity changes resulting from phase switch are also substantial, reaching almost 60% at around 1450 nm and 50% at around 1100 nm. Thus, the hybridized metamaterial acts as a transmission switch, providing a contrast ratio of 4:1 in a device structure less than one third of a wavelength thick (Sámson *et al.*)

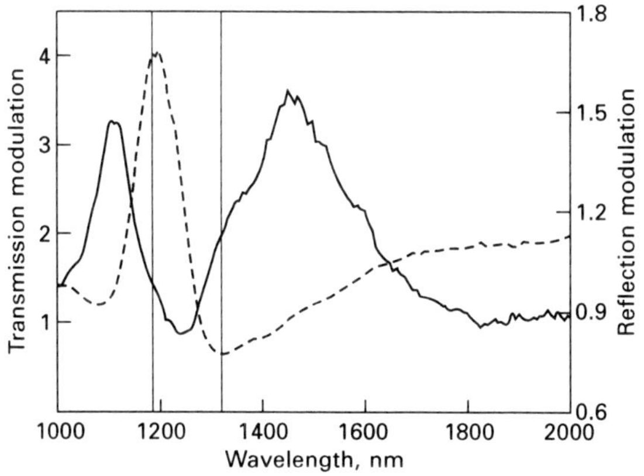

15.23 Spectral dispersion of transmission and reflection modulation contrast associated with chalcogenide phase switching in the hybrid structure. The dashed vertical lines indicate the approximate spectral positions of the structure's trapped mode resonances.

15.6 All-optical switches

When designing devices for nanophotonic phase change optical switches, one needs to take into account the incredible wealth of knowledge available on optical phase change of chalcogenides. By studying the structure of phase change optical discs, which can be seen in Fig. 15.24 a great deal of knowledge can be obtained. The phase change layer is typically sandwiched between two dielectric layers which protect the film and this is capped with a reflective layer for optical phase change recording and the whole structure is deposited on a polycarbonate disk. This is improved upon by also incorporating two interface layers which promote the nucleation rate as well as prevent the possible sulphur diffusion into the phase-change film from the sulphur-containing dielectric layers. It should be noted that the above-motioned methods to improve the nucleation rate are based on the growth-driven (GeSbTe) materials and may not be compatible for the nucleation-driven (doped SbTe) materials.

15.6.1 Structure

With the capability to optically switch large areas of chalcogenide thin films, hybrid plasmonic metamaterial devices for all-optical switching in the near- and mid-infrared ranges using germanium antimony telluride (GST) glasses have been demonstrated. These structures comprise:

- a substrate (CaF for mid-IR applications, fused quartz for near-IR);
- a metamaterial layer (an evaporated gold film with a typical thickness of 50 nm, patterned by photolithography with 50 × 50 μm square arrays of asymmetric split rings);

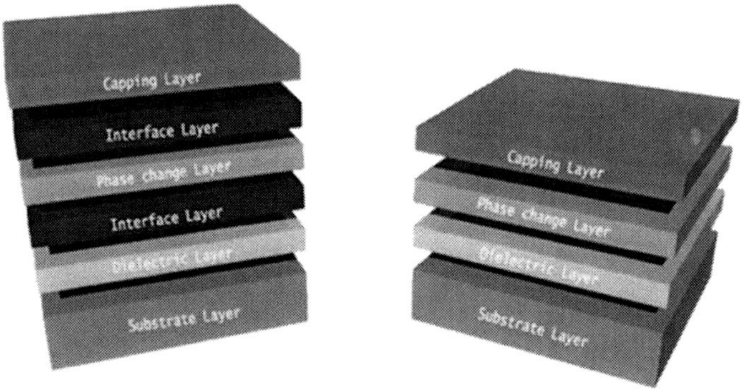

15.24 Schematic of the layer stack used in a conventional optical disk. The difference between the four- and six-layer stack is the introduction of the interface layers in the six-layer disks which aim to aid crystallization and prevent layer diffusion.

- a functional thin film of GST sputtered under an argon atmosphere to a thickness of 15–40 nm);
- an inert buffer layer of SiO_2 or ZnS/SiO_2 between the metamaterial and chalcogenide to prevent diffusion of metal into the chalcogenide;
- a capping layer, again of SiO_2 or ZnS/SiO_2, on top of the GST to ensure that the chalcogenide does not deteriorate or react with air (especially at elevated phase transition temperatures). (The samples remain under vacuum throughout the process of buffer, chalcogenide and capping layer deposition).

It should be noted that the choice of polished calcium fluoride as the substrates was made due to the very high transmission of CaF and as such its suitability in trying to minimize the effect of the substrate on the transmission performance of the final devices. The production of the metamaterial pattern on a gold film is followed by a triple layer structure (see Figs 15.24 and 15.25). The presence of the buffer and capping layers is essential to achieving reproducible, reversibly switching functionality, but they also affect the optical and thermal properties of the structure. For example, by spacing the metamaterial and phase-change layers, the buffer layer affects the spectral position and width of the structure's absorption resonance and is detrimental to contrast ratio. On the other hand, the presence of the buffer and capping layers allows more efficient optical joule heating of the samples, which enables switching with shorter optical pulses, thereby increasing the speed of the devices. Through practical testing and numerical simulation, one is able to optimize the layer structure of samples and associated metamaterial cell designs for mid- and near-IR applications or indeed any part of the transparency range of the dielectric in order to maximize switching contrast while retaining a physically robust device structure. However, the most important design consideration is to ensure that the chalcogenide phase change layer is deposited with two dielectric layers on either side to ensure that, during the phase transition from crystalline to amorphous, the phase-change layer does not interact with the atmosphere as well as not mixing with the gold layer underneath.

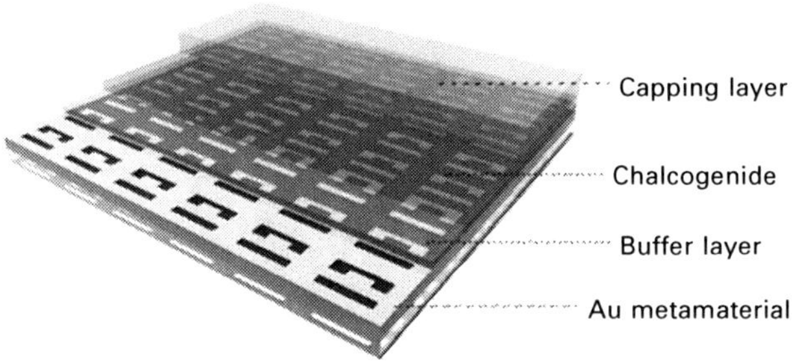

15.25 Layer structure of chalcogenide metamaterial optical switching device (artistic impression).

15.6.2 Characterization

The simplest way to achieve large area switching is restoring the phase transition excitation beam to completely phase switch a large area. However, as the raster method of large area switching was not very well suited to an integrated device, effort was made to create a setup that can be used for single-pulse switching. To this end, a system can be assembled for routine large-area optically-induced switching. This provides the capability for targeted excitation of structural transitions in thin film chalcogenide samples regardless of underlying metamaterial sample structure. This system (Fig. 15.26) is based on a 660 nm diode laser focused to a spot size at target of up to 50 μm diameter (giving intensities up to 1.25 mW/μm^2). Successful reversible switching of both germanium antimony telluride (GST) and gallium lanthanum sulphide (GLS) chalcogenide films has been demonstrated and switching of chalcogenide hybrid metamaterial devices has subsequently been achieved.

Laser pulses (durations down to 50 ns) were shaped using an Agilent 8110A pulse generator, while optical changes in the chalcogenide layer were monitored in real time to identify the phase transition points. In this way it was possible to optimize pulse energy and duration separately for the forward (amorphous-crystalline) and reverse (crystalline-amorphous) transition directions: the former requiring longer, lower power pulses; the latter shorter, higher power pulses.

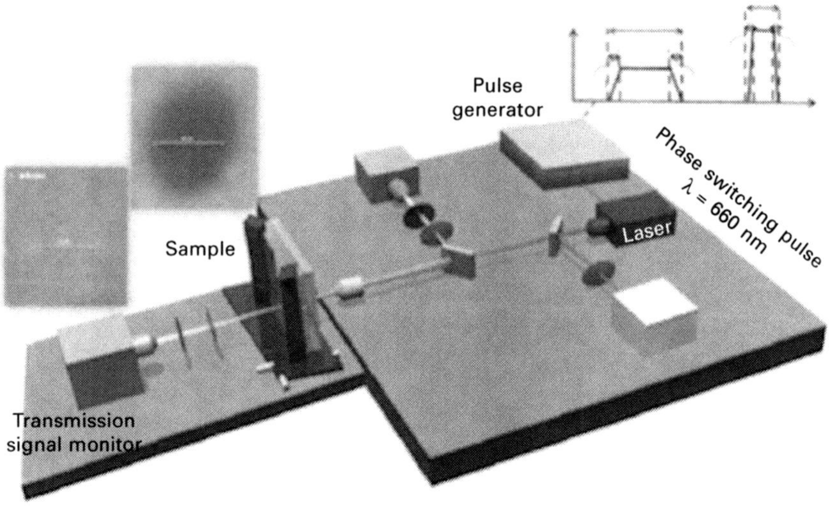

15.26 Experimental platform for single-pulse large-area optical phase switching of chalcogenide glasses.

15.6.3 Performance

The underlying concept used within these hybrid optical switches is the phase transition of GST between an amorphous and a crystalline state. It should be noted that in the first instance the phase switching behaviour of a GST thin film on a CaF is observed without the presence of a metamaterial array (Fig. 15.27).

When looking at the transmission contrast between the two phases of the material. One can observe that the modulation is between 1:1 to 2:1 across the range measured. The next step in the production of the device is the deposition of the confirmed phase change layer GST with the gold film used for metamaterial fabrication. Figure 15.28 shows the spectra for both transmission and reflection of a test device identical to the optical switching chips fabricated but without a metamaterial array. One can observe the

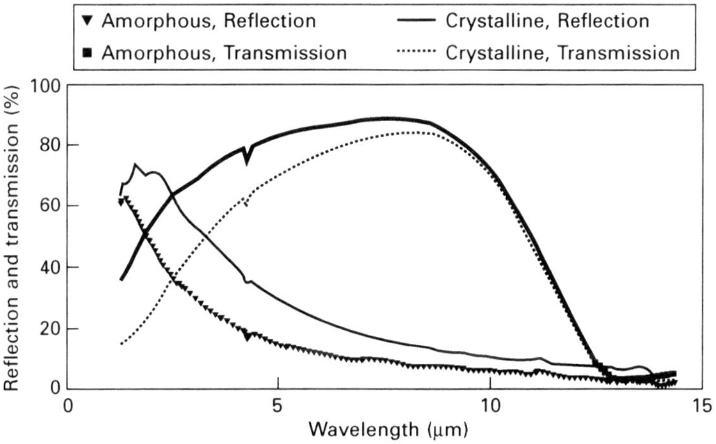

15.27 Transmission and reflection of GST thin film in both phases.

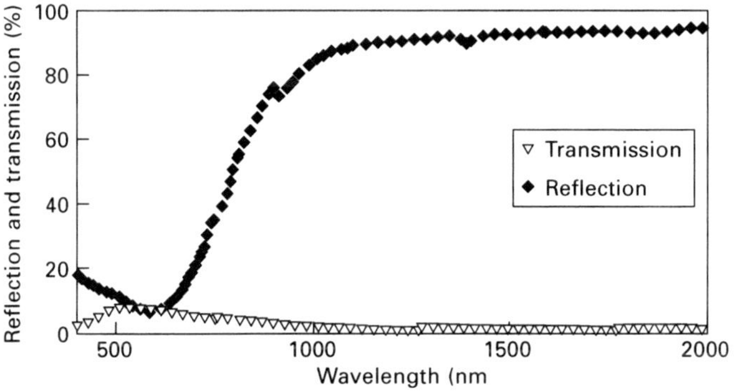

15.28 Spectra for GST/gold film without a metamaterial pattern.

enhancement in transmission and reflection at resonant wavelengths observed as a result of the introduction of the metamaterial arrays in both the near and mid-infrared (Fig. 15.29).

It had been anticipated that as a result of the optical phase change of the active GST layer, a bidirectional shift in the resonant peaks of the metamaterial should be observed. This is as a result of phase switching, the chalcogenide layer in the hybrid metamaterial produces a marked change in the resonance frequency of the structure. The GST transition from amorphous to crystalline increases its refractive index and red-shifts the resonance frequency of the metamaterial, bringing about a substantial change in optical transmission at wavelengths in the vicinity of the resonance.

In the case of a chalcogenide metamaterial hybrid device, this capability provided for single-pulse initiation of uniform phase transitions across entire nanostructured regions (Fig. 15.30), bringing about dramatic changes in the

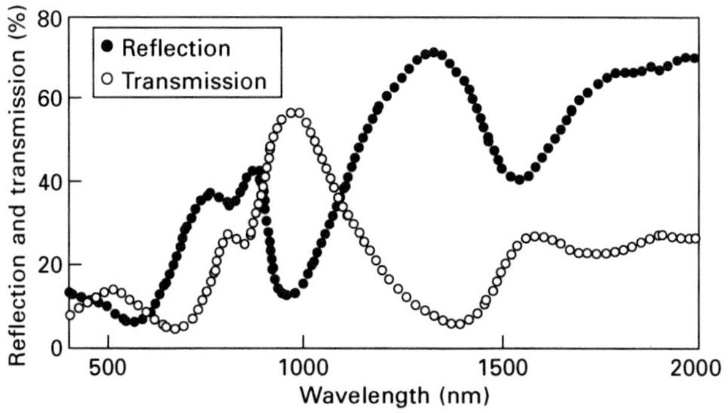

15.29 Typical as-deposited near-IR GST/gold metamaterial spectral response.

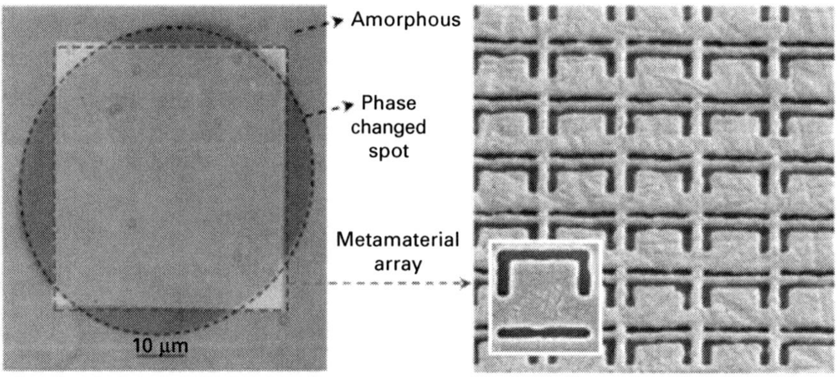

15.30 Large area optical switching in a chalcogenide thin film hybridized with a photonic metamaterial.

spectral position of the metamaterial, trapped mode absorption resonance and thereby the optical transmission of the structure. Figure 15.30 shows scanning electron microscopy (SEM) images of a metamaterial engineered for mid-infrared (2–10 μm) functionality.

The pattern was fabricated by focused ion beam milling through a 50 nm thick gold film deposited on a calcium fluoride (CaF_2) substrate by thermal evaporation. The metamaterial was then coated with buffer, GST and capping layers as described previously. Figure 15.30 shows an optical microscope image of the sample after single-pulse laser excitation to induce crystallization of the GST over an area (the darker circular region) almost entirely covering the patterned domain (the square region). Crystallization is confirmed by further SEM imaging (Fig. 15.31).

The metamaterial geometry can be scaled for functionality at any wavelength within the transparency range of the chalcogenide dielectric, and we have demonstrated high contrast modulation (up to a factor of six) in both the mid-infrared and near-infrared ranges (Fig. 15.32) using structures <150 nm thick (excluding substrate).

To summarize this section, for the first time, reversible, non-volatile, all-optical switching in a hybrid chalcogenide glass metamaterial structure is demonstrated. Using germanium antimony telluride (GST) glass and gold plasmonic nanostructures of sub-wavelength thickness, reversible switching in the near and mid-infrared spectral range with modulation contrast exceeding 4:1 has been achieved using nanosecond pulses at an optical excitation levels below 0.25 mW/μm^2 (Fig. 15.32) (Gholipour *et al.*, 2013).

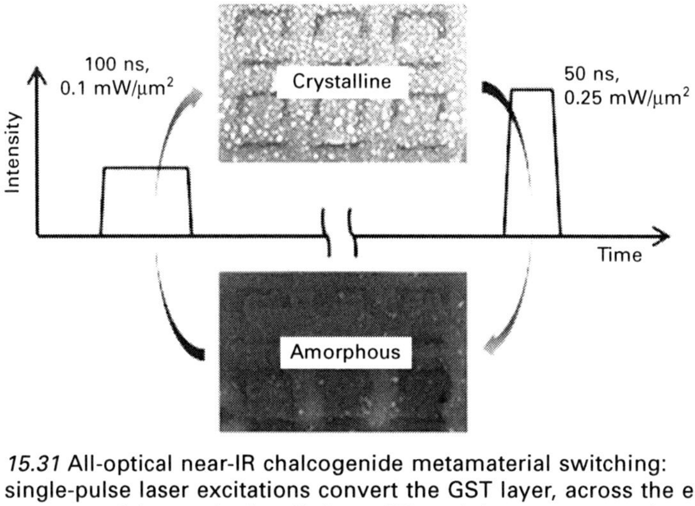

15.31 All-optical near-IR chalcogenide metamaterial switching: single-pulse laser excitations convert the GST layer, across the entire metamaterial array (unit cell size = 400 nm), between amorphous and crystalline state, thereby switching the reflectivity and transmission of the hybrid structure with high contrast at wavelengths close to the metamaterial resonance.

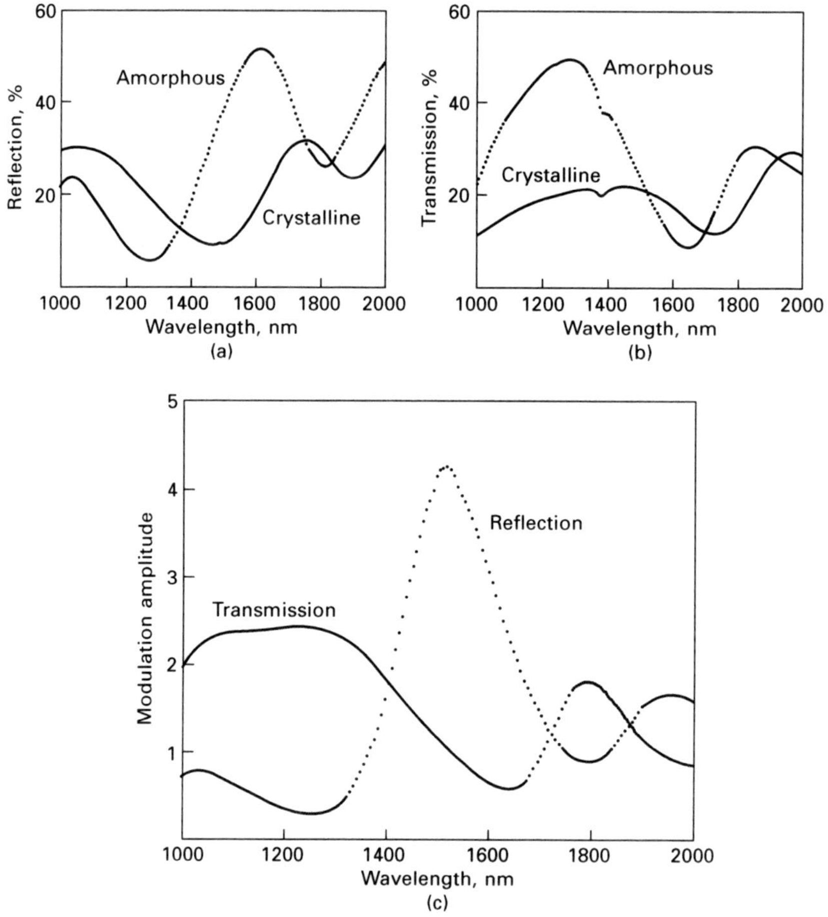

15.32 All-optical near-IR chalcogenide metamaterial switch: metamaterial transmission (a) and reflection (b) resonances are spectrally shifted when GST undergoes transitions between amorphous and crystalline states, giving strong modulation contrast (c) in wavelength bonds around these resonances (metamaterial unit cell seizes 400 nm).

15.7 Conclusion

Nanophotonic applications, in particular, photonic data processing circuits, require active devices of sub-wavelength dimensions. However, modulation of light in a device of nanoscale thickness is not a trivial problem. In conventional modulators exploiting the Pockels or Kerr effects, the polarization switching involved requires the interference of two propagating modes to develop over distances far in excess of the wavelength of light. The dimensions of such modulators in the propagation direction are often in the centimetre range.

Signal modulation via control of the waveguide absorption coefficient or refractive index is another possibility.

However, this approach also requires substantial propagation lengths over which an amplitude or phase change accumulates, or it involves interferometric arrangements that are inherently longer than the wavelength of light. It has been suggested that strong signal modulation may be achieved in nanophotonic devices, namely photonic metamaterial nanostructures, despite very short propagation lengths. Nanoscale electro-optic and all-optical modulation that relies not on absorption modulation, but rather on a change in the refraction of a material associated with a control input-induced phase change is pursued here. In a layer of nanoscale thickness, such a refractive index change would be insufficient to noticeably modulate the intensity or phase of a transmitted wave. However, it is widely recognized that by combining a nanoscale layer of phase change material with a planar plasmonic metamaterial, one can exploit the fact that the position of narrow resonant absorption lines in certain metamaterials are strongly dependent on the dielectric environment. Switching the dielectric layer in contact with such a metamaterial produces a massive change in its resonance frequency. Importantly, given that the planar metamaterial also has a sub-wavelength thickness, the whole electro-optic modulating structure can be much thinner than the light wavelength. Moreover, the resonance frequency, and therefore the spectral band of high contrast switching, can be engineered by design and located anywhere within the entire transparency range of the dielectric phase change medium.

15.8 References

Almeida, V. R., Barrios, C. A., Panepucci, R. R. and Lipson, M., 2004. All-optical control of light on a silicon chip. *Nature*, 431, 1081–1084.

Chen, H. T. *et al.*, 2006. Active terahertz metamaterial devices. *Nature*, 444, 597–600.

Chen, H. T. *et al.*, 2009. A metamaterial solid-state terahertz phase modulator. *Nature Photonics*, 3, 148–151.

Dani, K. M. *et al.*, 2009. Subpicosecond optical switching with a negative index metamaterial. *Nano Letters*, 9, 3565–3569.

Denisyuk, A. I., MacDonald, K. F., Garcfia, F. J. and Zheludev, N. I., 2009. Towards femtojoule nanoparticle phase-change memory. *Japanese Journal of Applied Physics*, 48, 03A065.

Dicken, M. J. *et al.*, 2009. Frequency tunable near-infrared metamaterials based on VO_2 phase transition. *Optics Express*, 17, 18330–18339.

Driscoll, T. *et al.*, 2009. Memory metamaterials. *Science*, 325, 1518–1521.

Eggleton, B. J., Luther-Davies, B. and Richardson, K., 2011. Chalcogenide photonics. *Nature Photonics*, 5, 141–148.

Elliott, A. and Zakery, S. R., 2003. Optical properties and applications of chalcogenide glasses: a review. *Journal of Non-Crystalline Solids*, 330, 1–12.

Enkrich, C. *et al.*, 2005. Magnetic metamaterials at telecommunication and visible frequencies. *Physical Review Letters* 95, 203901.

Fano, U., 1961. Effects of configuration interaction on intensities and phase shifts. *Physical Review*, 124, 1866–1878.

Fedotov, V. A. *et al.*, 2007a. Sharp trapped-mode resonances in planar metamaterials with a broken structural symmetry. *Physical Review Letters*, 99, 147401.

Fedotov, V. A. *et al.*, 2007b. Asymmetric transmission of light and enantiomerically sensitive plasmon resonance in planar chiral nanostructures. *Nano Letters* 7, 1996–1999.

Geppert, L., 2003. The new indelible memories. *IEEE Spectrum*, 40, 48–54.

Gholipour, B. *et al.* 2013. All-optical, Non-volatile, Bi-directional, Phase-change Meta-switch. *Advanced Materials*, 25, 3050–3054.

Harbold, J. M. *et al.*, 2002. Highly nonlinear As-S-Se glasses for all-optical switching. *Optics Letters*, 27, 119–121.

Hilton, A. R., 1970. Optical properties of chalcogenide glass. *Journal of Non-Crystalline Solids*, 2, 28–39.

Kang, I. *et al.*, 1995. Femtosecond measurement of enhanced optical nonlinearities of sulfide glasses and heavy-metal-doped oxide glasses. *Journal of the Optical Society of America B*, 12, 2053–2059.

Krasavin, A. V., MacDonald, K. F., Zheludev, N. I. and Zayats, A. V., 2004. High contrast modulation of light with light by control of surface plasmon polariton wave coupling. *Applied Physics Letters*, 85, 3369–3371.

Lahiri, B. *et al.*, 2009. Asymmetric split ring resonators for optical sensing of organic materials. *Optics Express*, 17, 1107–1115.

Lankhorst, M., Ketelaars, B. and Wolters, R., 2005. Low-cost and nanoscale non-volatile memory concept for future silicon chips. *Nature Materials*, 4, 347–352.

Liu, X. and Zhang, Z., 2008. Superlenses to overcome the diffraction limit. *Nature Materials*, 7, 436–411.

Loireau-Lozac'h, A. M., Guittard, M. and Flahaut, J., 1976. Glasses formed by rare earth sulphides La_2S_3 with gallium sulphide Ga_2S_3. *Materials Research Bulletin*, 11, 1489–1496.

Luk'yanchuk, B. *et al.*, 2010. The Fano resonance in plasmonic nanostructures and metamaterials. *Nature Materials*, 9, 707–715.

MacDonald, K. F., Soares, B. F., Bashevoy, M. and Zheludev, N. I., 2006. Controlling light with light via structural transformations in metallic nanoparticles. *IEEE Journal of Selected Topics in Quantum Electronics*, 12, 371–376.

MacDonald, K. F., Krasavin, A. V. and Zheludev, N. I., 2007. Optical modulation of surface plasmon-polariton coupling in a gallium/aluminium composite. *Optics Communications*, 278, 207–210.

Mathur, A. L. and Greer, N., 2005. Changing face of the chameleon. *Nature*, 437, 1246–1247.

Merton, C., 2001. *Encyclopedia of Materials: Science and Technology*. Amsterdam Elsevier Science.

Mott, E. A. and Davis, N. F., 1979. *Electronic Processes in Non-Crystalline Materials*. Oxford: Clarendon Press.

Murugan, G. S., 2010. *Photonic Glasses and Glass-Ceramics*. Kerala: Research Signpost.

Ovshinsky, S. R., 1968. Reversible electrical switching phenomena in disordered structures. *Physical Review Letters*, 21, 1450–1453.

Ovshinsky, S. R., 2004. Optical cognitive information processing: A new field. *Japanese Journal of Applied Physics*, 43, 4695–4699.

Ozbay, E., 2008. The magical world of photonic metamaterials. *Optics and Photonics News*, 19, 22–27.

Palik, E. D., 1985. *Handbook of Optical Constants of Solids*. London: Academic Press.

Papasimakis, N. and Zheludev, N. I., 2009. Metamaterial-induced transparency. *Optics and Photonics News*, 20, 22–27.

Pendry, J. and Li, J. B., 2008. Hiding under the carpet: a new strategy for cloaking. *Physical Review Letters*, 101, 203901.

Pendry, J. B., Schurig, D. and Smith, D. R., 2006. Controlling electromagnetic fields. *Science*, 312, 1780–1782.

Plum, E. *et al.*, 2009a. Towards the lasing spaser: controlling metamaterial optical response with semiconductor quantum dots. *Optics Express*, 17, 8548–8551.

Plum, E. *et al.*, 2009b. Metamaterials: optical activity without chirality. *Physical Review Letters*, 102, 113902.

Plum, E. *et al.*, 2011. A combinatorial approach to metamaterials discovery. *Journal of Optics*, 055102.

Popescu, M., 2005. Disordered chalcogenide optoelectronic materials: phenomena and applications. *Journal of Optoelectronics and Advanced Materials*, 7, 2189–2210.

Ravindran, S., Alameh, K. and Lee, Y. T., 2009. Design and analysis of electroabsorptive quantum well based double ring resonators for wavelength switching applications. *Optical and Quantum Electronics*, 41, 635–644.

Repsin, I. *et al.*, 2008. 19.9%-efficient ZnO/CdS/CuInGaSe_2 solar cell with 81.2% filll factor. *Progress in Photovoltaics: Research and Applications*, 16, 235–239.

Sámson, Z. L. *et al.*, 2010. Metamaterial electro-optic switch of nanoscale thickness. *Applied Physics Letters*, 96, 143105.

Seddon, A. B., 1995. Chalcogenide glasses: a review of their preparation, properties and applications. *Journal of Non-Crystalline Solids*, 184, 44–50.

Seo, M. *et al.*, 2010. Active terahertz nanoantennas based on VO_2 phase transition. *Nano Letters*, 10, 2064–2068.

Simpson, R. E., 2008. Chalcogenide thin film materials for next generation data storage. Ph.D thesis, University of Southampton.

Smith, D. R., Pendry, J. B. and Wiltshire, M. C., 2004. Metamaterials and negative refractive index. *Science*, 305, 788–792.

Soares, B. F., MacDonald, K. F., Fedotov, V. A. and Zheludev, N. I., 2005. Light-induced switching between structural forms with different optical properties in a single gallium nanoparticulate. *Nano Letters*, 5, 2104–2107.

Strand, D., 2005. Ovonics: from science to products. *Journal of Optoelectronics and Advanced Materials*, 7, 1679–1690.

Ta'eed, V. G. *et al.*, 2007. Ultrafast all-optical chalcogenide glass photonic circuits. *Optics Express*, 15, 9205–9221.

Temelkuran, B. *et al.*, 2002. Wavelength-scalable hollow optical fibres with large photonic bandgaps for CO_2 laser transmission. *Nature*, 420, 650–653.

Vogel, E. M., Weber, M. J. and Krol, D. M., 1991. Nonlinear optical phenomena in glass. *Physics and Chemistry of Glasses*, 32, 231–254.

West, Y. D., Schweizer, T., Brady, D. J. and Hewak, D. W., 2000. Gallium lanthanum sulphide fibers for infrared transmission. *Fiber and Integrated Optics*, 19, 229–250.

Wuttig, W. and Welnic, M., 2008. Reversible switching in phase-change materials. *Materials Today*, 11, 20–27.

Xu, Q., Schmidt, B., Pradhan, S. and Lipson, M., 2005. Micrometre-scale silicon electrooptic modulator. *Nature*, 435, 325–327.

Yamada, M. and Wuttig, N., 2007. Phase-change materials for rewriteable data storage. *Nature Materials*, 6, 824–832.

Yeh, A. and Yariv, P., 2007. *Photonics: Optical Electronics in Modern Communications*, 6th edn. New York: Oxford University Press.

Zhang, X. and Liu, Z., 2008. Superlenses to overcome the difraction limit. *Nature Materials* 7, 435–441.

Zheludev, N. I. 2002. Nonlinear Optics on the nanoscale. *Contemporary Physics* 43, 365–377.

Zheludev, N. I., 2006. Single nanoparticle as photonics switch and optical memory element. *Journal of Optics A: Pure and Applied Optics*, 8, S1–S8.

Zheludev, N. I., 2007. All change, please. *Nature Photonics*, 1, 551–553.

Zheludev, N. I., 2010. The road ahead for metamaterials. *Science*, 328, 528–583.

Zheludev, N. I., 2011. A roadmap for metamaterials. *Optics and Photonics News*, 22, 30–35.

Zheludev, N. I., Prosvirnin, S. L., Papasimakis, N and Fedotov, A. V., 2008. Lasing spaser. *Nature Photonics*, 2, 351–354.

Zouhdi, S., Sihvola, A. and Vinogradov, A. P., 2009. *Metamaterials and Plasmonics: Fundamentals, Modelling, Applications*. Dordrecht: Springer.

16

Second harmonic generation in chalcogenide glasses

V. NAZABAL, University of Rennes, France and I. KITYK, Częstochowa University of Technology, Poland

DOI: 10.1533/9780857093561.2.509

Abstract: This chapter is mainly devoted to the second-order nonlinear optical processes induced in optically isotropic chalcogenide glasses. The chapter is divided into two main parts related to the second-order nonlinear optical effects in near infrared and IR stimulated processes obtained in chalcogenide glasses. The different processes applied to break the centrosymmetry of glasses, which enables their use for the conversion of wavelength and for electro-optical purposes, are described: optical poling, poling assisted by e-beam or thermal poling. The origin and mechanisms of the phenomena induced by these processes, dedicated to the generation in glasses of a second-order nonlinear optical response, are discussed. In addition to these extrinsic methods, the interest in glass-ceramics containing non-centrosymmetric crystalline phases is considered. This chapter is also dedicated to some experiments describing IR-laser-stimulated changes of main linear and nonlinear optical constants in chalcogenide glasses. The origin of the observed phenomena is described within phenomenological and microscopic approaches, with special attention being paid to the contribution of phonons.

Key words: second-order nonlinear optical, second harmonic generation (SHG), poling, IR-laser-induced non-centrosymmetry, phonon contributions.

16.1 Introduction

The field of applications of nonlinear optics is as wide as the diversity of its phenomena, with equally strong interest in the fundamental studies of light–matter interactions, and in industrial developments. Therefore, nonlinear effects are at the basis of laser operation, extending the field of coherent sources (harmonic generation, parametric amplification, stimulated Raman or Brillouin scattering, etc.). They have also made considerable progress for spectroscopy, microscopy and biophotonics; there is currently great interest in the physics of ultrashort laser pulses possessing extremely high peak intensities to probe the properties of matter within extremely short timescales (see, e.g., Boyd, 1992).

The objective of this chapter is not to give a precise description of nonlinear effects and of their major fundamentals; we strongly encourage readers to immerse themselves in some well-known references in the field, such as the books by Butcher and Cotter (1990) or Boyd (1992). However, we will briefly describe all the necessary elements for the understanding of second-order nonlinear optical phenomena occurring in chalcogenide glasses.

16.2 General principles for the generation of second-order nonlinear optical effects in glasses

16.2.1 Basic notions of nonlinear polarization

The nonlinear optical processes express the property of dielectric materials to achieve, under an intense electromagnetic field, a polarization that depends nonlinearly on the electric field amplitude (Courtois, 1996). The polarization plays a key role in the description of nonlinear optical phenomena, considering a time-varying polarization acting as the source of new components to the electromagnetic field. Under the action of the electromagnetic field of a laser wave, oscillating at a frequency of about 10^{13}–10^{15} Hz, dielectric charges are subject to frequency oscillation forming a set of oscillating dipoles. The effect of magnetic field on charged particles is, in turn, much smaller and can be neglected. In addition, due to the large mass of positive ions as compared to that of electrons, one can consider with good approximation that electron movements only must be taken into account. Thus, in the presence of such electric field, the charged particles of the dielectric material, strongly related to each other although they are linked with some 'elasticity', show only transitory movements and deviate slightly from their original positions. This situation results in induced electric dipole moments, i.e. polarization. With pulsed laser, the applied electric field can reach an amplitude comparable to characteristic inter-atomic electric field strength that is typically greater than 100–10^4 MV/m. The usual notion of linear optical response characterized by a constant refractive index ceases to be valid. In most cases, the physical origin of nonlinear properties is strongly affected, at the microscopic level of dielectric materials, by the nature of atoms or molecules, of bond type and orientation, of crystalline structure, etc. In the perturbation limit, where the laser field magnitude is still small compared with that of the intra-atomic field, the resulting polarization including nonlinear phenomena can be described as a power of field strength, $\vec{E}(\omega)$ which is of the form:

$$\vec{P} = \varepsilon_0(\chi^{(1)}\vec{E}(\omega) + \chi^{(2)}\vec{E}(\omega)\vec{E}(\omega) + \chi^{(3)}\vec{E}(\omega)\vec{E}(\omega)\vec{E}(\omega) + \ldots) \qquad [16.1]$$

where the constant of proportionality, $\chi^{(1)}$, is the linear susceptibility, and $\chi^{(2)}$ and $\chi^{(3)}$ are the second- and third-order nonlinear optical susceptibilities, respectively, ($\chi^{(n)}$ is an $n + 1$ order tensor), ε_0 is the permittivity of free space and the electric field strength $E(\omega)$ is oscillating at the pulsation ω.

The quadratic polarization $\vec{P}^{(2)} = \varepsilon_0 \chi^{(2)} \vec{E}(\omega)\vec{E}(\omega)$ causes singular effects, essential for mixing wave phenomena, based on the generation of sum and difference frequencies (Butcher and Cotter, 1990). As a general rule, the amplitude of the nonlinear polarization component oscillating at pulsation ω_3 is defined by the relationship:

$$P_i^{(2)}(\omega_3) = \frac{\varepsilon_0}{2} \chi_{\mathrm{ijk}}^{(2)}(-\omega_3;\omega_1,\omega_2) E_{\mathrm{j}}(\omega_1) E_{\mathrm{k}}(\omega_2) \qquad [16.2]$$

These wave-mixing phenomena lead to many applications such as tunable sources, parametric amplifiers and oscillators. In the process of difference frequency generation (Fig. 16.1), the presence of radiations at frequency ω_2 or ω_3 can stimulate the emission of additional photons at these frequencies. Optical parametric oscillators based on this effect are mainly used as tunable infrared sources.

The process of second harmonic generation (SHG) can be described by considering a nonlinear material characterized by a second-order susceptibility assumed to be purely real, which is illuminated by a plane wave of pulsation ω and amplitude $E(\omega)$ propagating along the z axis. The electric field associated with the optical wave, without taking into account the polarization of the field, is of the form:

$$E(z,t) = E(\omega)e^{(-i\omega t + ik_\omega z)} + c.c \qquad [16.3]$$

In the special case of SHG (where $\omega_2 = \omega_1 = \omega$), and considering Eq. [16.1], the polarization is composed of two terms. The first term is given by Eq. [16.4], with a contribution at 2ω frequency and the second is given by Eq. [16.5] with a contribution at zero frequency:

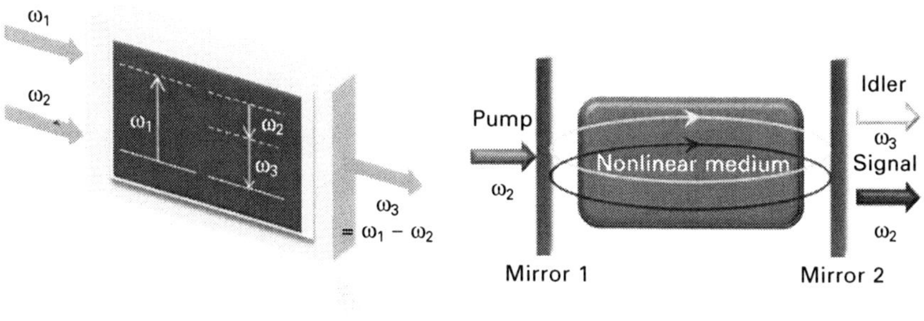

16.1 Scheme of difference frequency generation and optical parametric oscillator.

$$P_i^{(2)}(2\omega) = \frac{\varepsilon_0}{2}\chi_{\mathrm{ijk}}{}^{(2)}(-2\omega;\,\omega,\omega)\,E_{\mathrm{j}}(\omega)E_{\mathrm{k}}(\omega) \qquad [16.4]$$

This component of the polarization is at the origin of the frequency-doubling process or SHG. The generation of second harmonic at pulsation of 2ω was first observed by Franken *et al.* (Franken, Weinreich *et al.*, 1961); their work opened the era of experimental nonlinear optics.

$$P_i^{(2)}(0) = \frac{\varepsilon_0}{2}\chi_{\mathrm{ijk}}{}^{(2)}(0;\,\omega,-\omega)\,E_{\mathrm{j}}(\omega)E_{\mathrm{k}}(\omega) \qquad [16.5]$$

Equation [16.5] is associated with the phenomenon of optical rectification, resulting in the creation of a static electric field across the nonlinear material.

The components of the susceptibility tensor exhibit invariance in the permutation of *j* and *k* indices as both electric fields associated to the optical wave are indistinguishable. Since the medium is necessarily lossless when the applied field frequencies, ω_i, are very much smaller than the resonance frequency ω_0, Kleinman has shown that the $\chi^{(2)}$ tensor is symmetric with respect to full permutations of the three indices, *ijk* (Kleinman, 1962a, 1962b). This condition is valid when dispersion of the susceptibility can be neglected. The components of the polarization at 2ω in the Cartesian coordinate system (x, y, z) can be written in reduced matrix form:

$$\begin{bmatrix} p_x^{(2)}(2\omega) \\ p_y^{(2)}(2\omega) \\ p_z^{(2)}(2\omega) \end{bmatrix} = \frac{\varepsilon_0}{2}\begin{bmatrix} \chi_{111} & \chi_{122} & \chi_{133} & \chi_{123} & \chi_{113} & \chi_{112} \\ \chi_{211} & \chi_{222} & \chi_{233} & \chi_{223} & \chi_{213} & \chi_{212} \\ \chi_{311} & \chi_{322} & \chi_{333} & \chi_{323} & \chi_{313} & \chi_{312} \end{bmatrix}$$

$$\begin{bmatrix} E_x^2(\omega) \\ E_y^2(\omega) \\ E_z^2(\omega) \\ 2E_y(\omega)E_z(\omega) \\ 2E_x(\omega)E_z(\omega) \\ 2E_x(\omega)E_y(\omega) \end{bmatrix} \qquad [16.6]$$

16.2.2 Intensity of the second harmonic signal and phase matching

During the propagation of a light wave of pulsation ω in lossless medium exhibiting nonlinear optical properties, a second harmonic wave at 2ω will appear. Their propagation is submitted to Maxwell's equations of propagation. The approximation of the slowly varying amplitude takes into account low variations of $\vec{E}_\omega$, along the propagation axis, in amplitude and phase; assuming the light waves as monochromatic and plane leads to neglecting the phase term. Thus, the description only considers the amplitude of the electric fields in the direction of $\vec{z}$ propagation regarded as identical in a first approximation; the propagation equations at ω and 2ω can then be simplified accordingly:

$$\frac{\delta E_\omega}{\delta z} = \frac{i\omega^2\mu_0}{2k}\chi^{(2)}(-\omega;-2\omega,\omega)\cdot E_{2\omega}E_\omega e^{(i(k_2-2k)\cdot z)} \qquad [16.7]$$

$$\frac{\delta E_{2\omega}}{\delta z} - \frac{i(2\omega)^2\mu_0}{2k_2}\chi^{(2)}(-2\omega;\omega,\omega)\cdot E_\omega E_\omega e^{(i(2k-k_2)\cdot z)} \qquad [16.8]$$

If it is assumed that the transfer of intensity from the fundamental beam to its harmonic is weak (in the undepleted-pump approximation), the intensity along $\vec{z}$ can be regarded as constant (i.e., $\delta\vec{E}_\omega/\delta z = 0$). Consequently, the integration of the differential equation permits an expression to be obtained for the spatial dependence of the second harmonic field amplitude for a length of material equal to L:

$$I_{2\omega}(L) = \frac{(2\omega)^2}{8\varepsilon_0 c_3}\frac{|\chi^{(2)}_{\text{eff}}(-2\omega,\omega,\omega)|}{n_\omega^2 n_{2\omega}}\cdot I_\omega^2\cdot L_2\cdot \sin c^2\left((2k-k_2)\cdot\frac{L}{2}\right) \qquad [16.9]$$

where the quantity called the wave-vector (or momentum) mismatch $\Delta k = 2k - k_2 = 2\omega(n_{2\omega} - n_\omega)/c$ is introduced. It should be noted that the intensity of the second harmonic depends on two wave vectors $2k$ and k_2. The notion of forced wave or polarization wave (with wave-vector $2k$) can be introduced; it is locally created by the incident wave of frequency ω and, unlike the free wave (with wave-vector k_2), it does not propagate in the material (Butcher and Cotter, 1990). Forced and free waves will therefore interfere during the propagation in the nonlinear material. When the two waves are in phase agreement, the interferences are constructive, thus the medium is not dispersive ($n_\omega = n_{2\omega}$) and the energy transfer to the second harmonic is ultimate. In most cases, phase matching is not achieved ($\Delta k \neq 0$) and the interferences cause oscillation of the second harmonic amplitude during propagation proportional to $\sin c^2\left(\Delta k\cdot\frac{L}{2}\right)$. The notion of coherence length $\left(L_c = \frac{\pi}{\Delta k}\right)$ is related to the distance where the phase shift is equal to π. Thus, only one coherence length contributes efficiently to the generation of the second harmonic. If co-linearity between the two wave vectors and equality between the $n(2\omega)$ and $n(\omega)$ indices are possible, then the condition for phase matching is satisfied. This is actually achievable in some crystals. The condition of phase matching is a drastic one, as a very small deviation from the optimal direction of propagation seriously decreases the conversion efficiency. The conditions of quasi-phase matching (QPM) were investigated to allow conversion efficiency with more flexible conditions, implying a periodic variation of the sign of the second-order susceptibility (Armstrong, Bloembergen *et al.*, 1962). Here, it is necessary to optimize the spatial modulation of $\chi^{(2)}{}_{\text{eff}}$ along the waveguide to obtain efficient QPM. The periodic structure has to respect the coherence length, as is the case for waveguides in ferroelectrics NLO crystals, to improve the efficiency of wavelength conversion of optical telecommunication signals (Suhara and Fujimura, 2010). It should be pointed out that compared to nonlinear crystals ($LiNbO_3$, KTP, etc.), glasses have the advantage of high damage threshold, low propagation loss and small dispersion.

When the length (L) of the nonlinear medium is continuously changed by modifying the angle of incidence (θ) of the fundamental wave at frequency ω going through the material, the interaction between the free wave and the forced wave gives rise to interferences, propagating at a frequency 2ω. The transmitted intensity $I_{2\omega}$ (supposing $L >> L_c$) then presents an alternating sequence of maxima and minima forming so-called Maker fringes (Maker, Terhune *et al.*, 1962) (Fig. 16.2). The absence of Maker fringes indicates that the interference term in Eq. [16.9] does not show any oscillating character yet. The pump ω and second harmonic 2ω waves are usually linearly polarized in Maker fringe experiments, parallel or perpendicular to the plane of incidence of the pump wave (p and s, respectively) (Fig. 16.3). It is possible to get tensor coefficients of the quadratic polarizability by combining the results from different sets of polarization for the pump and second harmonic wave.

16.3 Second harmonic generation (SHG) in glasses: origin and mechanism

Ideally, optical devices need optical materials with tunable linear and nonlinear responses which can be transformed in optical waveguides with low optical losses. Silica-based glass optical fibres are extensively investigated because they are the backbone on which optical communications can develop. Over time, silicon photonics also became a topic of great interest with new solutions for optoelectronics. This trend is driven by both the computer industry and that of optical networks, which highlights an ongoing convergence between silicon technology and photonics. However, all optical functions cannot be performed by silicon alone, which presents, for example, a transparency only above 1.1 μm. Thus, a hybrid solution should be explored to compensate for some shortcomings of silicon. Opening new prospects for hybrid microphotonics, an interesting point lies in the choice

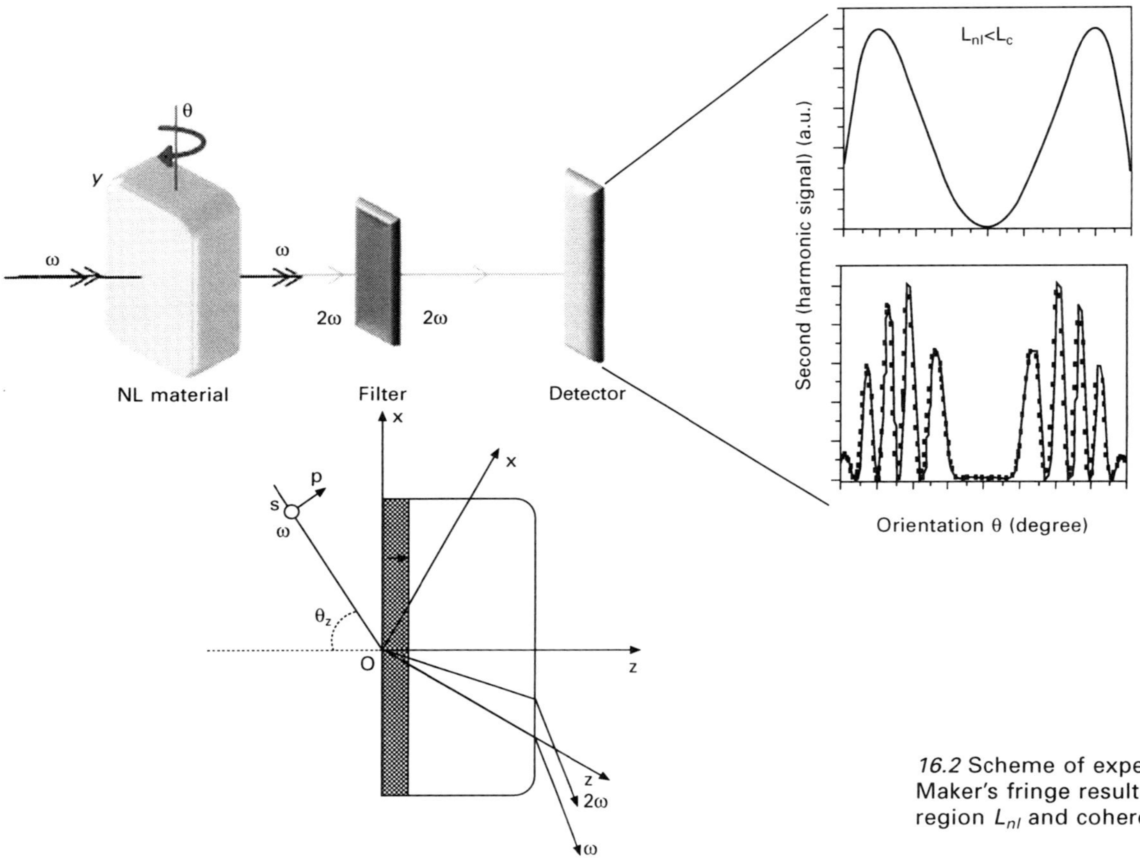

16.2 Scheme of experimental set-up and Maker's fringe results depending on nonlinear region L_{nl} and coherence length L_c.

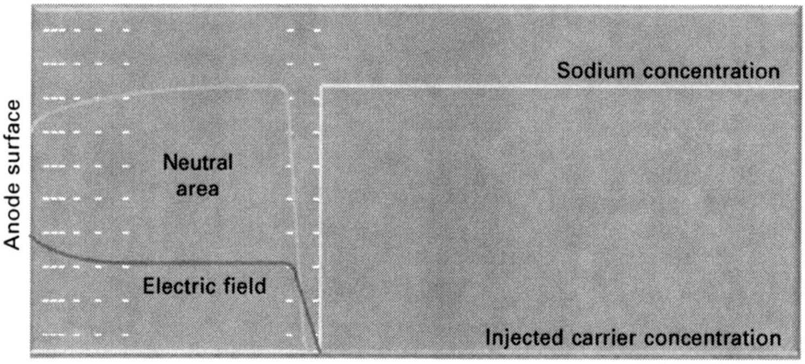

16.3 Simplified scheme of the positive charge carrier's concentration in the silica glass (grey and light grey lines), of the charge distribution (negatively charged and neutral area, white signs) and resulting electric field distribution for long poling times (black) (adapted from Kudlinski, Quiquempois *et al.*, 2005).

of glassy materials with efficient and tunable linear and nonlinear responses to take full advantage of glass integrated optical devices.

It is important to mention that the quadratic polarization does not exist in optical isotropic dielectric material as $\chi^{(2)}$ vanishes for such media with inversion symmetry. This symmetry prevents electro-optic effects and wavelength conversion in glass by second-order, nonlinear parametric processes. Nevertheless, several SHG experiments have been performed in oxide glass after specific treatments like optical poling (Ohmori and Sasaki, 1982; Osterberg and Margulis, 1987), thermal poling or electron beam irradiation (Myers, Mukherjee *et al.*, 1991; Kazansky, Kamal *et al.*, 1993); generation of three-dimensional (3D) optical second-order nonlinear microstructures by femtosecond laser has already been demonstrated (Choi, Bellec *et al.*, 2012), for example. Second-order nonlinear optical effects in glasses, from which optical fibres and channel waveguides can be produced, are of interest for the conception of optical devices based on frequency conversion. Indeed, the creation of a second-order nonlinearity in glasses offers the possibility of manufacturing cheap active or passive components, such as electro-optic modulators or switches.

In the case of optical poling, several models have been proposed to account for the rate of conversion between the pump wave and its second harmonic (Stolen and Tom, 1987; Dianov, Prokhorov *et al.*, 1989). For instance, when exposed to intense radiation for several hours, germanium-doped silica fibres present photosensitivity involving colour centres or defects. These photogenerated charges permanently change their optical properties by creating an internal electric field which is able to couple with the third-order nonlinear susceptibility inducing an effective second-order nonlinear susceptibility ($\chi^{(2)}_{\text{eff}} = \chi^{(3)}\vec{E}_{\text{dc}}$). Indeed, when an internal electric field, E_{dc}, exists in the material, the development of the macroscopic polarization, according to Eq. [16.1], reveals the creation of an electromagnetic wave at 2ω via the third-order nonlinear susceptibility $\chi^{(3)}(-2\omega; 0, \omega, \omega)$:

$$\vec{P}^{(2)}(2\omega) = \frac{\varepsilon_0}{2}(\chi^{(2)} + 3\chi^{(3)}\vec{E}_{dc}(0))\vec{E}(\omega)\vec{E}(\omega) \qquad [16.10]$$

At maximum, conversion efficiencies achieved by this method are of about 5% with a possible seeding process to improve the efficiency of the conversion.

Significant second-order nonlinear optical coefficients $\chi^{(2)}$ have been measured in oxide glassy materials (Fujiwara, Matsumoto *et al.*, 2000; Tanaka, Narazaki *et al.*, 2000a,b; Dussauze, Fargin *et al.*, 2005) opening up new horizons. Since the first demonstration of thermal poling in bulk silica glasses in 1991, this method has been extensively investigated for potential telecommunication applications; the generation of second harmonic has been reported in the literature for different types of commercial fused silica glasses, with $\chi^{(2)}$ estimated to be in the range of 0.1–1 pm/V. The thermal poling method remains the easiest technique to create a second-order nonlinearity in several kinds of isotropic materials from silicate to tellurite glasses, including borate and phosphate glasses, even if it is a challenge to induce reproducible and efficient signal (Myers, Mukherjee *et al.*, 1991; Tanaka, Narazaki *et al.*, 2000a,b; Nazabal, Fargin *et al.*, 2001; Narazaki, Tanaka *et al.*, 2002; Thamboon and Krol, 2003; Dussauze, Fargin *et al.*, 2005; Guimbretiere, Dussauze *et al.*, 2010). This technique, used to break the centrosymmetry of isotropic materials by generating an axial symmetry, consists in applying a high dc electric field while heating the sample at a temperature below its glass transition temperature (T_g). Mukherjee *et al.* (Mukherjee, Myers *et al.*, 1994) first proposed that the creation of second-order nonlinearity is connected with the existence of residual static electric fields frozen within the sample after cooling and removing the applied dc voltage. They proposed expressing the second-order susceptibility $\chi^{(2)}$ as follows:

$$\chi^{(2)} \propto 3\chi^{(3)}E_{dc} + \frac{Np\beta}{5kT}E_{loc} \qquad [16.11]$$

In Eq. [16.11], the first term represents the contribution resulting from the coupling between the internal static electric field, E_{dc}, and the third-order nonlinear susceptibility, $\chi^{(3)}$, of the material. In poled silica glasses, it has been proposed that E_{dc} was created under the surface of the anode by migration of mobile ions, mostly Na^+ cations, towards the cathode side of the sample (Myers, Mukherjee *et al.*, 1991).

The second term of Eq. [16.11] corresponds to the contribution of the orientation of bonds, associated with a permanent dipole moment, p. Further, in Eq. [16.11], N is the number of dipoles per volume unit, with a microscopic hyperpolarisability β, and E_{loc} corresponds to the local electromagnetic field of the oriented entities.

It has been assumed that dipolar reorientation was weak in oxide glasses, and thus second-order susceptibility $\chi^{(2)}$ can be expressed mainly as a function of the third-order susceptibility $\chi^{(3)}$ and of the internal static dc electric field:

$$\chi^{(2)} \propto 3\chi^{(3)}E_{dc} \qquad [16.12]$$

It is useful to point out that the macroscopic polarization is related to the microscopic polarization; any element of symmetry can be applied in the dielectric medium affecting the susceptibility tensor. According to the symmetry group of the material, the number of non-zero independent variables is reduced. Assuming Eq. [16.10] and the poled layer symmetry description by uniaxial group $C_{\infty v}$ oriented along the applied E_{dc} field, the polarization at 2ω presents only two components, $\chi^{(2)}_{ZZZ}$ and $\chi^{(2)}_{ZXX} = \chi^{(2)}_{XZZ}$. It should be noted that the $\frac{\chi^{(2)}_{ZZZ}}{\chi^{(2)}_{ZXX}}$ ratio depends on the origin of the SHG. If the signal is related to the reorientation of dipole, the $\frac{\chi^{(2)}_{ZZZ}}{\chi^{(2)}_{ZXX}}$ ratio increases as the orientation goes up. On the other hand, if the signal is due to the combination of an internal E_{dc} field and $\chi^{(3)}$, the $\frac{\chi^{(2)}_{ZZZ}}{\chi^{(2)}_{ZXX}}$ ratio is equal to 3.

According to the first proposed models (Myers, Mukherjee *et al.*, 1991; Kazansky, Kamal *et al.*, 1993; Le Calvez, Freysz *et al.*, 1998), the creation of second-order polarization in silica glass, under a very high electric field, is driven by thermally assisted charge movements leading to a depleted layer of singly charged mobile cations (Na^+) located a few micrometres below the surface of the anode (Fig. 16.3). Further, it is well established that the SHG signal originates mainly from the anodic region as the nonlinearity is induced within the depletion region by the presence of a high electric field E_{dc} with values close to the dielectric breakdown field. However, this simple model of charge carriers cannot explain the dynamic creation of second-order susceptibility in silica glass, as, for instance, the existence of an optimal polarization time for which the amplitude of the nonlinear susceptibility is maximum depending on the sample thickness (Faccio, Pruneri *et al.*, 2001). For longer polarization times, a decrease of the susceptibility and an expansion of the nonlinear region were observed (Triques, Carvalho *et al.*, 2003). In this case, the single carrier model no longer fits the experimental results, because other phenomena such as charge injection or ionization become predominant. To explain this distribution, Alley *et al.* (Alley, Brueck *et al.*, 1998) first proposed a model involving multiple charge carriers with hydrogenated species present in the atmosphere. The creation of a nonlinear second-order susceptibility based on the ion exchange between a fast charge carrier (Na^+) and a much slower charge carrier (H_3O^+/H^+) gives the ratio of the mobility, [H^+ ions]/[Na^+ ions], of the order of 10^{-4}. Under the influence of an external electric field, the Na^+ ions migrate rapidly towards the cathode. The (H_3O^+/H^+) ions adsorbed on the anode surface penetrate the glass with a constant injection rate. They can then slowly migrate to the cathode, thereby neutralizing some of the negative charges of the depleted zone.

As opposed to currently used models, a model – based on the work of Proctor and Sutton – taking into account charge dissociation and charge recombination occurring during the poling process has been developed (Proctor, Sutton, 1959; Quiquempois, Godbout *et al.*, 2002). As the main result, they suggested the existence of a voltage threshold below which no $\chi^{(2)}$ can be induced and a nonzero electric field is created within the entire sample. Following this work, a second model has been proposed in a thermally poled Infrasil™ silica glass (Kudlinski, Quiquempois *et al.*, 2005). The most mobile charge carriers in silica glasses are alkali ions (Li^+, Na^+, K^+), present as impurities in Infrasil™ glass (with a concentration of about few ppm). In silica glasses, the thickness of the depleted layer is typically in the range of 3–10 μm. The numerical simulations of this model are in good agreement with the experimental results. The dynamic creation of the nonlinear susceptibility was analysed and it was shown that it can be divided into two phases.

The first phase corresponds to the creation of the space charge; there is an increase of the amplitude of the susceptibility as well as the thickness of the nonlinear region associated with the migration of Na^+ ions. Once the optimal polarization time is reached, the induced electric field is maximum at the surface of the anode and the spatial profile is triangular. The increase in poling temperature leads to a higher mobility of Na^+ ions, therefore the space charge is created faster. During this first phase of space charge creation, the injection of positive charges is negligible. For longer periods of poling, one cannot overlook the injection of H^+ ions under the influence of the field located at the surface of the anode which corresponds to the second phase. These will replace the Na^+ ions in the depleted zone and tend to create a neutral zone. The electric field is no longer shielded in volume and there is a migration of Na^+ ions to the cathode. This is the cause of the spread of the nonlinear region for higher times and temperatures of poling. At the same time, the spatial distribution of the nonlinearity is modified and tends to a rectangular profile close to the surface of the anode. As a result, a decrease of the nonlinear susceptibility of two orders of magnitude is observed.

High-quality silica used for conventional or photonic crystal fibres (e.g., Suprasil silica) contains a very low concentration of Na^+ ions. For these high-quality pure silica glasses, the second-order coefficient was found to be 10^{-3}–10^{-2} pm/V for a NL thickness of about 30 μm. It has been proposed that the maximum E_{dc} field, proportional to $\sqrt{N_0}$ (N_0 = mobile positive charges) (Von Hippel, Gross *et al.*, 1953), is lower than in Infrasil silica and, consequently, the NL width which is inversely proportional to $\sqrt{N_0}$ is thicker. With a square alternating poling field using particular frequencies in the mHz regime (Kudlinski, Martinelli *et al.*, 2008), the magnitude of the second-order nonlinearity induced in Suprasil can reach that obtained for Infrasil.

Lately, it has been proposed that the intrinsic breakdown strength limits the internal E_{dc} fields supported by the glass from the poling process. Considering the rectification model of thermal poling (Smith, Lanagan *et al.*, 2012), larger second-order nonlinearity in glasses could be expected in glasses with intrinsically high dielectric breakdown strength. Two different commercial alkali-free alkaline-earth boroaluminosilicate glasses, utilized for displays, have been selected for their high intrinsic dielectric breakdown strengths of 11–14 MV/cm. However, only internal E_{dc} fields of the order of 2–4 MV/cm could be estimated after thermal poling treatment, leading to $\chi^{(2)}$ values less than 0.5 pm/V in these glasses. The internal field is probably affected by nonlinear electronic conduction effects taking place within the depletion region at the poling temperature. Consequently, the E_{dc} is limited to a fraction of the breakdown field. The space charge formation can be analysed by considering the poling treatment conditions (time, temperature, atmosphere, electrodes, etc.) as well as the glass compositions, the thickness of the glass, and, more specifically, the nature and density of the charge carriers. Due to the variety of parameters, as mentioned above, a large variety of results are obtained for systems more complex than silica. For instance, the analysis often proves the existence of much more complex mechanisms for glasses with different optical basicity or with various impurities acting on the mobility of the carriers, which may be electronic or ionic in nature. The mechanisms, interleaved and interacting with each other, cannot always be clearly distinguished and, consequently, it is difficult to outline the effects of each of them on the SHG. However, the general trends obtained for silica glasses can be found for other glasses, examined both experimentally and through modelling. The region generating the SHG signal is located mainly in a layer of a few μm to several tens of μm below the surface of the anode. In multicomponent oxide glasses, quantitative Maker fringe measurements, which provide an accurate evaluation of second-order nonlinear susceptibility, generally give maximum $\chi^{(2)}$ values of 0.1–1 pm/V. In a rich sodium niobium borophosphate glass, the second-order nonlinear optical efficiency was estimated from the simulation of Maker fringe patterns of both transmitted and reflected polarized waves with a stepwise decreasing profile from the surface of the anode. The nonlinear layer was found to be sodium-depleted up to 5 μm deep inside the anode side and presented a reproducible $\chi^{(2)}$ susceptibility value as high as 5.0 ± 0.3 pm/V, which is to date the highest value obtained in multicomponent oxide glasses (Dussauze, Fargin *et al.*, 2005). Apart from bulk glasses, investigations have also been engaged with films and multi-layered structures, with compositions that can be more complex than pure silica; the role of interfaces has been particularly identified in such structures (Li, An *et al.*, 2008; An and Fleming, 2012a).

16.4 Optical waveguide for electro-optic effects and quasi-phase matching second harmonic generation (QPM-SHG) in glass

Poling techniques have been used to break optical isotropy in bulk glasses as well as in silica fibres or planar and channel glass waveguides as several benefits are accessible by using waveguide technologies. Effectively, optical waveguides can be used potentially in a wide range of applications such as SHG, electro-optic modulation and switching and direct generation of polarization-entangled photon pairs by exploiting type-II phase-matched spontaneous parametric down-conversion (Kashyap, Veldhuis *et al.*, 1994; Liu, Digonnet *et al.*, 1994; Abe, Kitagawa *et al.*, 1996; Canagasabey, Corbari *et al.*, 2009; Helt, Zhu *et al.*, 2009; An and Fleming, 2012a; Zhu, Tang *et al.*, 2012). QPM grating structures can be fabricated in glass by periodic poling with patterned periodic electrodes or by periodic erasure of poled regions using focused electron beams (Kazansky, Kamal *et al.*, 1993), ultraviolet exposure (Carvalho, Margulis *et al.*, 1991) and two-photon absorption of near-IR laser radiation (Montant, Le Calvez *et al.*, 1999).

The first demonstration of electro-optic phase modulation in a fused-silica channel waveguide was measured at 633 nm with a phase shift of 32 mrad (Liu, Digonnet *et al.*, 1994). An electro-optic switching was also demonstrated in an integrated Mach–Zehnder interferometer based on GeO_2-doped silica channel waveguides on a silicon substrate using thermal poling. The switching voltage was 1700 V at 1.55 μm with an extinction ratio of 20 dB and switching response time less than 100 ns (Abe, Kitagawa *et al.*, 1996). Chen *et al.* reported second-order optical nonlinearity created in Ge ion implanted channel waveguides on fused silica substrate with thermal poling (Chao, Chen *et al.*, 2005). First order QPM-SHG was demonstrated in these channel waveguides by periodic UV erasure of the second-order optical nonlinearity. The SHG conversion efficiency of 1,064 nm to 532 nm was 6.1×10^{-4} %/W/cm^2 for a 7 mm channel waveguide, i.e. $\chi^{(2)}{}_{\mathrm{eff}}$ of about 0.01 pm/V. A device, presented in Fig. 16.4, consisted of a buried

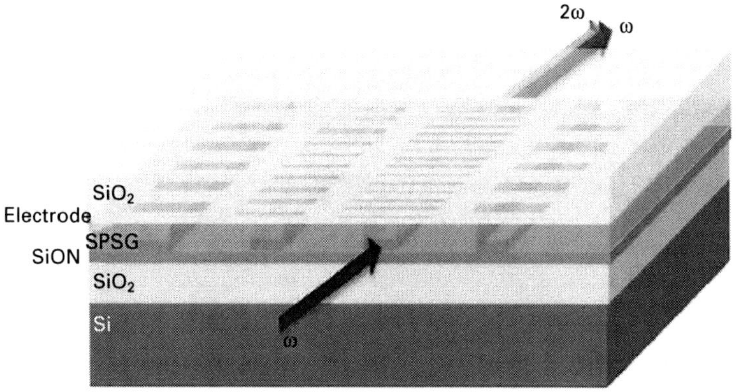

16.4 Scheme of a chip with QPM devices with different QPM periods (adapted from Fage-Pedersen, Jacobsen *et al.*, 2005).

silicon oxynitride (SiON) waveguide, and a 3 mm long QPM $\chi^{(2)}$ grating was fabricated showing a greater normalized conversion efficiency of 1.4×10^{-3} %/W/cm^2 (Fage-Pedersen, Jacobsen *et al.*, 2005). Later, a QPM-SHG at 532 nm was demonstrated in a channel photo-written by a femtosecond laser. The conversion efficiency of 2×10^{-5} %/W/cm^2 was obtained for a 1 cm waveguide for which SHG periodic erasure was carried out by electron beam (Li, Winick *et al.*, 2009).

In the same period, some researchers have reported the presence of the Pockels effect in poled silica or Ge:SiO_2 fibres induced by a temperature/electric field poling or UV-excitation poling (Long, Myers *et al.*, 1994; Fujiwara, Wong *et al.*, 1995; Kazansky, Russell *et al.*, 1995; Pruneri, Bonfrate *et al.*, 1999). To thermally pole an optical fibre, the normal practice is to use a twin-hole fibre in which two holes are present on either side of the core for holding metal wires as internal anode and cathode. Recently, Margulis *et al.*, reported that a strong and stable $\chi^{(2)}$ can be induced in twin-hole fibres with two internal electrodes, both as anodes (Margulis, Tarasenko *et al.*, 2009). Twin-hole fibres thermally poled with different internal electrode wire configurations in addition to the conventional anode–cathode configuration were investigated by An and Fleming (2012b). Experimental results show that both the single-anode and dual-anode configurations can achieve large $\chi^{(2)}$ values comparable to the conventional anode–cathode configuration, but the last configuration was less reproducible. Moreover, the position of the cathode hole in the fibre is probably not a severe requirement, which enables fibre fabrication with simplified design. Finally, an electro-optic silica fibre switch, based on a Sagnac interferometer, capable of selecting individual pulses from a mode-locked laser in the ~1 MHz range was reported (Malmström, Tarasenko *et al.*, 2012). The linear electro-optical effect was induced by thermal poling using two electrodes melted in fibre twin-holes of 78 cm length. The average nonlinear coefficient was found to be $\chi^{(2)}{}_{eff}$ = 0.26 pm/V. Laser pulses at wavelength of 1.5 μm were selected at a repetition rate up to 1 MHz from a laser running at 7 MHz with an extinction ratio as high as –30 dB, which represents encouraging results.

16.5 SHG in chalcogenide glasses: induced polarization by external stimulation

Chalcogenide amorphous materials are based on chalcogens (S, Se, and Te) covalently bound to other suitable elements like Ge, As, Sb used as network formers. These glasses are well known for their high third-order susceptibility, $\chi^{(3)}$, which can be a thousand times larger than in silica glass, as shown in Fig. 16.5 (Nasu, Kubodera *et al.*, 1990; Asobe, Kanamori *et al.*, 1992; Smektala, Quemard *et al.*, 1998; Cardinal, Richardson *et al.*, 1999; Lenz, Zimmermann *et al.*, 2000; Harbold, Ilday *et al.*, 2002a, 2002b; Ogusu, Yamasaki *et al.*, 2004; Yeom, Maegi *et al.*, 2008). That is the reason why chalcogenide glasses have been proposed for ultrafast switching in telecommunication systems.

Considering their high third-order nonlinearity Eq. [16.12], amorphous chalcogenides have great potential to generate large second order susceptibility to replace highly nonlinear single crystals in optical device. Second order nonlinear effects in amorphous semiconductors open up unique possibilities for the development of new devices for promising applications like parametric frequency converters or electro-optic modulators in the IR spectral region.

16.5.1 SHG induced by optical and optically assisted poling

The creation of a second-order nonlinearity in amorphous chalcogenides was first achieved using thermal, and also optically, assisted poling treatments of As_2S_3 chalcogenide thin film, converting the pump at 1.55 μm (Quiquempois, Villeneuve *et al.*, 2000). As_2S_3 amorphous films of about 4 μm thickness were deposited onto an indium tin oxide transparent thin electrode on BK7 substrate. A pressed-on silicon electrode was used to apply the high static electric field (~100–150 V). It is worth mentioning that after exposure to near band-gap illumination (488 and 514 nm, 5 J/mm^2, and 2.5 mW/mm^2), photosensitivity

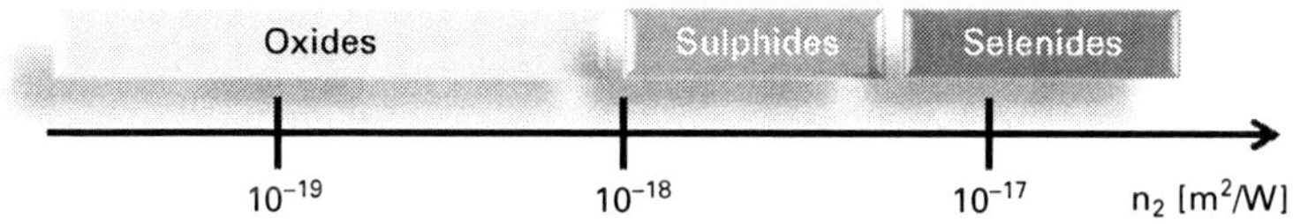

16.5 Scheme of nonlinear refractive index values of different glasses.

was observed. The $\chi^{(2)}_{zzz}$ coefficient was estimated to be superior to 0.6 pm/V, whatever the poling method used and the $\frac{\chi^{(2)}_{zzz}}{\chi^{(2)}_{zxx}}$ ratio is greater than 3. Therefore, the authors proposed that the main microscopic mechanism at the origin of second-order nonlinearity could be the hyperpolarizable dipole orientation Eq. [16.11].

In the case of optical poling with seeding technique, the induced nonlinearity originates from the generation of a permanent and micro-structured dc field resulting from the coherent photocurrent due to the beating between fundamental optical field and its harmonic. Photoinduced SHG was also observed in Ge-As-S glasses where fundamental and second-harmonic waves at 1064 nm and 532 nm, respectively, from a nanosecond laser were used to optically induce second-order nonlinearity (Qiu, Si *et al.*, 2001). Qiu *et al.* reported that the photoinduced SHG in $Ge_{20}As_{20}S_{60}$ glass was stable after optical poling at room temperature. They proposed that the origin of the SH effect is related to created defects in structures associated to the large $\chi^{(3)}$ of sulphide glasses. A second-order nonlinear susceptibility of 6×10^{-5} pm/V was measured using a ps source at 1064 nm (Fig. 16.6) in $50GeS_2$-$25Ga_2S_3$-25CsCl glasses (Lopez-Lago, Couderc *et al.*, 2001) for which $n_0 = 2.45$ and $n_2 = 0.7 \times 10^{-18}$ m^2/W.

However, in their case, the linear and nonlinear absorption of the second harmonic wave by the charge carriers causes a rapid decrease of the intensity of the second harmonic signal.

16.5.2 Polarization generated by implantation of charges

Kazansky *et al.* showed for the first time the possibility of second-order nonlinearity in lead silicate glass by implantation of electrons by means of exposure of low energy electron beam (0.3–10 nA with beam voltage between 5 and 40 kV and spot size of 0.5 μm) (Kazansky, Kamal *et al.*, 1993). This interesting technique, known for its high resolution, is used with polymers and also lithium niobate to create periodic structures. Nevertheless, conductivity and dielectric breakdown of glasses has been proposed as limitations of the second-order nonlinearity efficiency obtained by this method.

SHG has also been obtained in various silica thin films (Nakanishi, Sugihara *et al.*, 1999) and chalcogenide glasses from the Ge-As-S system ($Ge_{15}As_{25}S_{60}$, $Ge_{20}As_{25}S_{55}$, $Ge_{15}As_{30}S_{55}$) (Liu, Zhao *et al.*, 2001). The $\chi^{(2)}$ susceptibilities generated by induced polarization were found to be 0.5–1 pm/V in the case of silica glass. The poled Ge-As-S glass plates were characterized by using a pulsed Nd:YAG laser with a pulse width of about 10 ns. The value of effective second-order susceptibility in the case of sulphide glasses was estimated to be 0.8 pm/V by direct calculation using Eq. [16.12] Later,

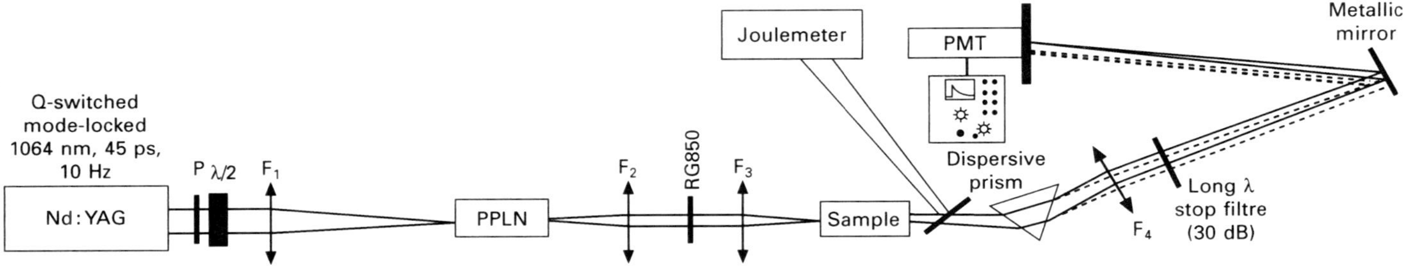

16.6 Experimental set-up of optical poling using seeding process.

SHG was also observed successfully in electron-beam-irradiated $0.95GeS_2$-$0.05In_2S_3$ and $0.56GeS_2$–$0.24Ga_2S_3$–$0.2KI$ chalcohalide glasses (Dong, Tao *et al.*, 2007c).

The well-known mechanism is based on the creation of a static electric field induced by the electron beam exposure concentrated near the surface of the electron-beam irradiated surface. The injection of electrons lead to the formation of a space-charge electrostatic field perpendicularly directed to the surface of the glass due to a macroscopic separation of charges. A negatively charged area is effectively formed by the implantation of electrons below the surface into the dielectric. The thickness of the nonlinear layer is limited by the implantation depth of electrons (proportional to the electron energy of the incident beam) (Kazansky, Kamal *et al.*, 1993). Secondary electrons are emitted from the surface of the sample which becomes positively charged and a static electric field is thus induced between the two regions. Due to the high spatial resolution of the electron beam, this technique allows the creation of complex optical patterns. As in the case of optical poling, the SHG is strongly related to the strong relaxation time of captive charges.

16.5.3 Thermal poling

Thermal poling is one of the most efficient processes used to break the centrosymmetry of amorphous material, enabling second-order susceptibility to be efficient. Thermal poling is carried out in a certain temperature range (depending on the composition) limited by the glass transition temperature (T_g) of the glasses under study. The high voltage (< 5 kV) is applied across the sample by means of electrodes, the sample is then cooled down to room temperature and the high voltage is finally removed. Various sets of poling temperatures, voltages and times were tested to optimize the SHG in sulphide glasses. Such a process was applied successfully to some sulphide glasses showing one of the most efficient second-order optical responses in glassy materials (Guignard, Nazabal *et al.*, 2005, 2006, 2007a; Jing, Guang *et al.*, 2006; Dussauze, Zheng *et al.*, 2012).

16.6 Thermal poling in chalcogenide glasses

16.6.1 Sulphide glasses

Although the third-order susceptibility, $\chi^{(3)}$, is higher in selenide or telluride glasses Eq. [16.12], sulphide glasses were specially selected for their transparency in the visible and near-IR, allowing the generation of second harmonic from near-IR and mid-IR laser sources. Sulphide glasses present lower electronic conductivity in order to minimize the occurrence of dc static electric relaxation required to break the centrosymmetry of the glass. The second-order NL optical properties induced by thermal poling were studied mainly in germanium- or arsenic-based sulphide glasses (Guignard, Nazabal *et al.*, 2007a; Dussauze, Zheng *et al.*, 2012). Experimental conditions of the poling treatment (temperature, voltage, time) were optimized for each glass composition (Fig. 16.6).

The second-order nonlinear signals were recorded by using Maker fringes experiment. Sulphide glasses are transparent from 0.5 to 12 μm, for a 1 mm thick sample. The coherence length calculated from n_ω and $n_{2\omega}$ for germanium-based sulphide glasses decreases from 17.0 μm for $Ge_{33}S_{67}$ glass to 11.1 μm for $Ga_5Ge_{20}Sb_{10}S_{65}$ glass. The optical band-gap range is about 460–580 nm. Therefore, the pump beam wavelength has been chosen in the near infrared region (~1.9 μm) in order to avoid any absorption of the SH signal, photoinduced effects or exaltation of SH signal by resonant interaction. Pump beam was focused on the sample, which can rotate from –85° to 85°. A second-order coefficient $\chi^{(2)}$ up to 8 pm/V was measured in $Ge_{25}Sb_{10}S_{65}$ glass. This value was obtained by using a simulation based on the accurate knowledge of the thickness of the nonlinear layer. If the analysis is performed only from SH data recorded in transmission configuration, it can be the cause of interpretation errors as a NL thickness, slightly greater than the coherence length, can lead to the same shape of Maker fringe pattern. It is therefore extremely important to have a confirmation of the thickness of the nonlinear region by another technique to extract the precise value of $\chi^{(2)}$. In order to determine the localization of the NL region, a chemical etching of the anode surface of several $Ge_{25}Sb_{10}S_{65}$ and $Ga_5Ge_{20}Sb_{10}S_{65}$ poled samples was performed with a sodium hydroxide solution (0.2 mol/L). It was found that the SH signal was generated within the first 15 micrometers under the anode side in both glasses for optimized poling parameters (Fig. 16.7).

Two mechanisms were proposed to explain the creation of a nonlinear layer under the anodic surface in germanium-based sulphide glasses: the formation and migration of charged defects towards the anode may occur mainly in $Ga_5Ge_{20}Sb_{10}S_{65}$ and $Ga_5Ge_{25}S_{70}$ glasses, whereas the migration of Na^+ ions towards the cathode may be responsible for the accumulation of negative charges under the anode in $Ge_{33}S_{67}$ and $Ge_{25}Sb_{10}S_{65}$ glasses. A higher electronic conductivity in gallium-containing glasses seems to be at the origin of the phenomenon. Even if SH signals were generated in all compositions, two distinct behaviours were clearly observed by measuring the total current between the two electrodes during a fixed thermal poling treatment (270°C, 5 kV, 30 min) (Fig. 16.8).

In the case of $Ga_5Ge_{20}Sb_{10}S_{65}$ and $Ga_5Ge_{25}S_{70}$ glasses, after high voltage application, the poling current increases slightly (30–40 μA). Considering the SHG efficiency, the SH signal increases continuously with increasing temperature, until damage (cracks and holes) appears on the cathode side

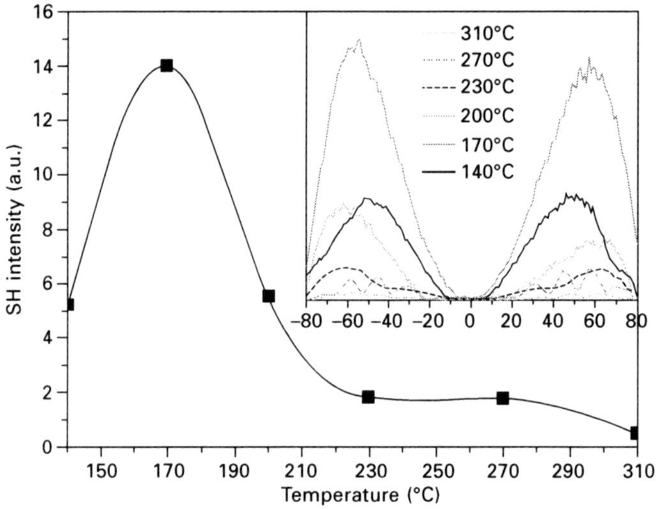

16.7 Maximum SH intensity versus the temperature used for the poling process of $Ge_{25}Sb_{10}S_{65}$ glass and MF patterns recorded for different temperatures from 170°C to 310°C (adapted from Guignard, Nazabal *et al.*, 2006).

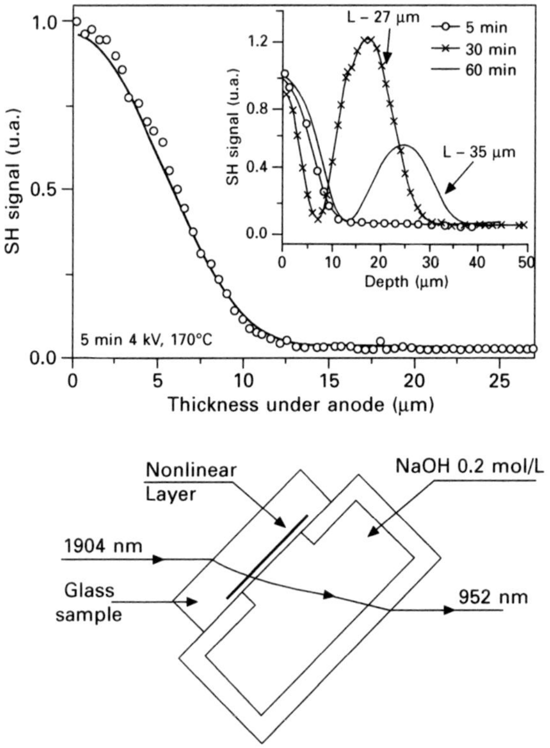

16.8 Determination of the thickness of NL layer of $Ge_{25}Sb_{10}S_{65}$ following the evolution of the SH intensity recorded at 952 nm versus the sodium hydroxide solution etching.

of $Ga_5Ge_{25}S_{70}$ and $Ga_5Ge_{20}Sb_{10}S_{65}$ glasses, that is around 45°C below the glass transition temperature for both glasses.

In the case of $Ge_{33}S_{67}$ and $Ge_{25}Sb_{10}S_{65}$ glasses, the current presents a severe decrease during the poling treatment from about ten to a few μA which can be related to the surface current and/or the residual current present within the materials (Fig. 16.9). The strong decay has earlier been related to a cation depleted layer at the anode side which acquires a much higher resistivity than the bulk. The maximal SH intensity seems to reach a maximum for a polarization temperature of about 100 and 170°C, respectively (Fig. 16.9). During poling treatment, a similar mechanism is assumed to occur in $Ge_{33}S_{67}$ and $Ge_{25}Sb_{10}S_{65}$ glasses based on the formation of local coordination defects allowing the migration of Na^+ ions (1–10 ppm in weight) towards the cathode. Indeed, a sodium depleted layer under the anode surface is clearly observable (dark part) by imaging the cross-section of the poled samples (Fig. 16.9) as has previously been reported in poled silica glasses (Alley, Brueck *et al.*, 1999).

Moreover, the progressive appearance of overmodulations, which were never observed for $Ga_5Ge_{20}Sb_{10}S_{65}$ and $Ga_5Ge_{25}S_{70}$ glasses, is highlighted on Maker fringes curves with increasing poling temperature for $Ge_{33}S_{67}$ and $Ge_{25}Sb_{10}S_{65}$ samples (Fig. 16.9). In such cases, overmodulations on Maker fringes pattern can be attributed to the nonlinear thickness spreading due to the Na^+ ions mobility increase with temperature. In parallel, the nonlinear

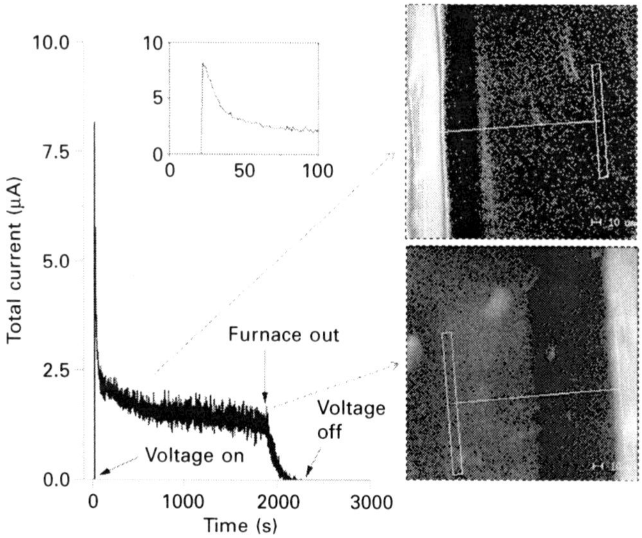

16.9 Total current registered during the poling treatment of the $Ge_{25}Sb_5S_{65}$ glass and SIMS images of sodium concentration analysis versus depth at different times of poling (5 or 30 min). (adapted from Guignard, Nazabal *et al.*, 2007a)

layer spreading observed when the poling time increases can be related to a rising number of removed charges (Fig. 16.9). Thus, the second-order nonlinear susceptibility is maximum for a poling time of a few minutes at about 140 and 170°C for glasses without Ga. As has already been mentioned, the decrease of $\chi^{(2)}$ coefficient with the increase of poling time in silica glasses is explained by the injection under the anode of a slower second charge carrier , H^+ ions, during the poling treatment, which progressively counteracts the nonlinear layer polarization. For sulphide glasses poled under Ar atmosphere, no clear proton migration has been evidenced by SIMS analyses and the contribution of other charge carriers as Li^+, K^+ or Ca^{2+} ions cannot be considered either. Thus, some ejection of electrons at the anode surface was proposed to explain the phenomenon.

SH intensities have been obtained in thermally poled (100-x)GeS_2-x(0.5Ga_2S_3-0.5CdS) chalcogenide glasses by Maker fringes using a 10-ns pulse pump beam at the wavelength of 1,064 nm. The SH intensity reached a maximum around an incident angle of ± 45° (Gu, Ma *et al.*, 2008); the authors concluded that the nonlinear layer is thinner than the coherence length L_c (ranging from 2.70 to 2.55 μm). Gu *et al.* proposed that the mechanism of SHG in GeS_2-Ga_2S_3-CdS glasses is related to the reorientation of dipoles originated from the structural defects within the glasses. Under given poling conditions (5 kV, 280°C, 30 min), a nonlinear relationship between $\chi^{(2)}$ and glass composition is proposed, based on the change of dipole reorientation ability with the evolution of glassy structure.

More recently, the second-order optical nonlinearity of poled $Ge_{25}Sb_{10}S_{65}$ glass was found to be strongly affected by femtosecond laser irradiation (Liu, Zhang *et al.*, 2009). The highest $\chi^{(2)}$ value was found to be 11.4 pm/V, 40–50% larger than before exposure. Raman spectral changes seem to indicate that defects were created by femtosecond light. These defects promoted the third-order optical nonlinearity of the glass. As a result, considering Eq. [16.12], its second-order nonlinearity was also greatly enhanced.

For arsenic-based glasses, three compositions of arsenic sulphide glasses were explored recently for SHG: pure As_2S_3, As_2S_3 doped with 10 ppm NaCl and As_2S_3 doped with 100 ppm of NaCl. The increasing concentration of sodium in very small quantities was carried out to observe a possible impact on the building of the static electric field E_{dc} at the origin of the SHG. Pure As_2S_3 glass certainly presents a deficit in sulphur due to several distillations carried out during glass synthesis, as compared to the two others; this is clearly observed in Raman spectra where the signatures of 'wrong' bonds – As-As homopolar bonds – appear at low wavenumber (Fig. 16.10, inset). Consequently, this sulphur deficit should also have an influence on the SHG response and its kinetic, compared to stoichiometric glasses with Na. Thermal poling was performed in air at a temperature of 100°C with a voltage of 5 kV applied for 30 min using n-doped silicon electrodes (soda lime glass

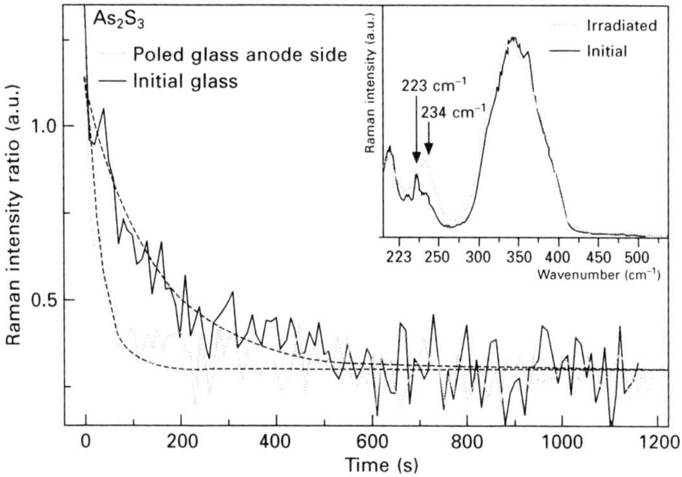

16.10 Raman intensity ($I_{223cm^{-1}}/I_{234cm^{-1}}$) ratio measured during the irradiation process for an unpoled and a poled As_2S_3 glass sample (inset: Raman spectra of As_2S_3 unpoled sample before and after irradiation). (adapted from Dussauze, Zheng *et al.*, 2012).

slides, 100 μm thick, introduced between the n-doped silicon electrodes). The efficiency and stability of the SHG response has been tested using laser irradiation at 1,550 nm with a maximum energy of 200 μJ at 10 ns pulse duration and 30 Hz repetition rate. It is noteworthy that only poling using soda lime glass slides between the electrodes and sulphide glass samples was effective leading to SH signals. In this case, the anode is not a blocking electrode and the mobile species, such as sodium ions, can migrate from the soda lime glass slide through the arsenic sulphide glass interface.

The SH signal of a poled pure As_2S_3 glass is clearly the least stable under 1,550 nm irradiation (and 775 nm generating by SHG to a lesser extent) compared to other poled samples. The irradiation procedure at 752 nm of the unpoled pure As_2S_3 glass has induced several spectral variations analysed by Raman scattering spectroscopy (Fig. 16.10, inset). The authors proposed that the poling treatment has a non-negligible influence on the kinetic of photoinduced structural rearrangements (Fig. 16.10). The variation of second harmonic signal intensity with time is strongly dependent on both the sulphur content in the glass and the light intensity during the SH measurement.

Indeed, Dussauze *et al.* proposed that the SHG response measured in arsenic sulphide glasses is affected by a multi-photon process induced by the charge defects created during poling treatment. Faster kinetics of the structural rearrangements induced by irradiation in the polarized glass is certainly related to the formation of charged defects in the nonlinear layer. The charged defects are induced by thermal-assisted electric field application in agreement with Shimakawa *et al.* on the acceleration of photodarkening under electric field (Shimakawa, Inami *et al.*, 1990, 1992). The kinetic origin of SHG could be related to the formation of local dipoles with charged defects and/or displacement, consequently compensated by bond-switching structural rearrangements leading to the formation of new neutral structural entities. The initial glass network, in the case of a slight sulphur deficit in As-S glass, is formed mainly by $AsS_{3/2}$ pyramids with a significant number of homopolar As-As bonds. These homopolar bonds, which are at the origin of the Raman modes at 190, 220 and 360 cm^{-1}, could be observed either in isolated/molecular As_4S_4 units or in $As_2S_{4/2}$ arrangements in which the homopolar bond is connected to the remaining glass network through sulphur atoms. The authors proposed that the irradiation at 752 nm should form As-As-As linkage similar to the one present in the pararealgar structure (with characteristic vibration mode at 230 cm^{-1}). To a lesser extent, it also occurs in the nonlinear layer during irradiation as suggested by monitoring the evolution of the Raman intensity ratio ($I_{223cm^{-1}}/I_{234cm^{-1}}$) under irradiation (Fig. 16.10). Thus, the differences observed in the SHG kinetic should also be explained by differences in sulphur content in glass, which directly influences the effectiveness of charged defect formation and structural rearrangements induced by illumination.

16.6.2 Chalcohalide glasses

With chalcohalide glasses, the objective is to select glasses that present low nonlinear absorption while preserving high nonlinear refractive indices. The expected band-gap edge shift of chalcohalide glasses to shorter wavelengths could reduce the two-photon absorption compared to that of chalcogenide glasses. The effects of halogen or halide addition on glass forming ability, structural organization or band-gap energy were mainly investigated in As-S, Ge-S, Sb-S and Ge-Ga-S-based chalcogenide glasses. The incorporation of caesium iodide in the Ge-In-S system, earlier explored by Boncheva-Mladenova and Ivanova (1978) was investigated. The decrease of the network connectivity of $(80-x)GeS_2-20In_2S_3-xCsI$ ($0 \leq x \leq 40$) glasses with the increase of CsI content is revealed by analysis of the Raman spectra and is consistent with the downward trend of the glass transition temperature and the evolution of the thermal expansion coefficient. The cut-off wavelengths were shifted towards the blue region of the spectrum as the electronegativity of the iodine atoms likely reduces electron delocalization in the glass network, and consequently, increases the band-gap energy. Although the introduction of halogenides generally decreases the nonlinear refractive index of chalcogenide glasses, iodine, among the halogens, presents the best compromise with an interesting hyperpolarizability (Guignard, Nazabal *et al.*, 2008). Therefore, nonlinear refractive index n_2 of the above-mentioned glasses is still 60 times higher than that of silica glasses without any nonlinear absorption. The different shape of Maker fringe patterns obtained for the two studied compositions can likely be attributed to distinct thickness of the nonlinear layer and might be related to a more perceptible ionic character in the case of $50GeS_2$-$20In_2S_3$-30CsI composition, as compared to $70GeS_2$-$20In_2S_3$-10CsI glass network (Fig. 16.11). The shape of the Maker fringes of $50GeS_2$-$20In_2S_3$-30CsI glass is the signature of a nonlinear layer thickness lower than the coherence length, which is about 15 μm in the materials discussed, with a pump beam at λ = 1,904 nm. On the other hand, the nonlinear layer thickness is certainly higher for the $70GeS_2$-$20In_2S_3$-10CsI glass causing the observation of several fringes or overmodulation in this case.

The origin of second-order nonlinearity in chalcohalide glasses may be related mainly to the migration of mobile cations, like Na^+, K^+, etc. In the extreme case of some alkali-rich oxide glasses, alkali ions can physically come out of the sample and a white oxide deposit is observed on the cathode side (Carlson, Hang *et al.*, 1972). An accumulation of negative charges leads to the creation of a high internal electric field E_{dc} near the anodic side resulting in the creation of SHG through contribution of the $\chi^{(3)}E_{dc}$ term. This mechanism could be followed by the reorientation of electrical dipoles or hyperpolarizable entities – such as the sulfur lone pairs – under the induced electric field, E_{dc}. The relatively high ionic and electronic conductivities could explain the spontaneous decay of the second harmonic signal likely caused by the relaxation of the space charge induced during poling.

SHG efficiency was then proposed in thermally poled $0.56GeS_2$-$0.24Ga_2S_3$-0.2KI chalcohalide glasses, containing iodine. The n_ω and $n_{2\omega}$ terms of the

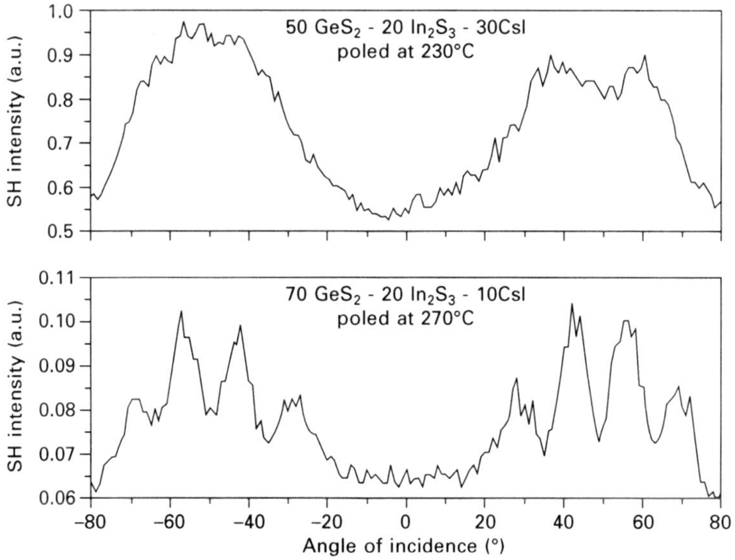

16.11 Maker fringes obtained for two GeS_2-In_2S_3-CsI sulphide glasses (10 and 30% of CsI). (adapted from Guignard, Nazabal *et al.*, 2008).

poled samples were estimated to be 2.10 and 2.20 μm, respectively, and a large SHG, recorded with a pump beam at 1,064 nm, was reported for an optimized thermal poling process (5.2 kV, 260°C for 120 min). A $\chi^{(2)}$ coefficient as large as 3.74 pm/V was calculated from the Maker fringe experimental results, although an unconventional shape of Maker fringe patterns was observed for this glass composition, suggesting that the value could be overestimated. For a similar system based on GeS_2-Ga_2S_3 glass matrix, addition of KBr was also explored and a thermally poled $60GeS_2$-$20Ga_2S_3$-20KBr glass showed SHG at 532 nm with non-symmetric Maker fringes (Jing, Guang *et al.*, 2006).

Another composition, $80GeS_2$-$10Ga_2S_3$-10AgCl, has also been examined because of its large n_2 value, suitable transparency and glass-forming ability in the (100-x)($80GeS_2$-$20Ga_2S_3$)-xAgCl glass system (Dong, Tao *et al.*, 2007a). Polarization of the glass was achieved by irradiation with an electron beam with moderate energy (25 kV, 25 nA, 15 min) (Dong, Tao *et al.*, 2007a). Raman scattering spectroscopy did not detect any structural change after irradiation, but according to the consideration described previously, a perpendicular built-in space-charge electrostatic field, E_{dc}, which results in generation of $\chi^{(2)}$, is certainly created in a few micron thin layers. From the Maker fringe pattern (Fig. 16.12), SH intensity of the irradiated glass is assumed to be about 1.5 times larger than that of Z-cut quartz as a reference, and $\chi^{(2)}$ is estimated to be greater than 6.1 pm/V.

16.7 Glass-ceramic samples

The nonlinearities originating from poling are frequently not in steady state. Another route for generating an optical second harmonic signal is the

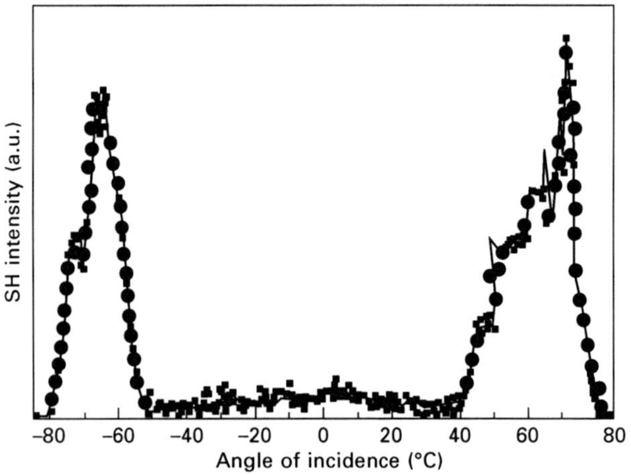

16.12 Maker fringe pattern of the $80GeS_2$-$10Ga_2S_3$-10AgCl glass poled by the electron beam (adapted from Dong, Tao *et al.*, 2007a).

presence of optically nonlinear crystals within a glass media, combining the glass processing ability with the optical properties of crystals. Controlled crystallization of a glass is known as an interesting method for the fabrication of transparent and dense materials with desired shapes and nano-structured optical functions. A macroscopic anisotropy can be introduced into glasses when the precipitating crystallites are non-symmetric. Instead of poling treatment, glass-ceramic materials might be promising candidates generating a stable SHG (Kao, Hu *et al.*, 1994; Shioya, Komatsu *et al.*, 1995; Benino, Takahashi *et al.*, 2004; Lee, Lee *et al.*, 2008; Iwafuchi, Fujiwara *et al.*, 2009; Lin, Le Calvez *et al.*, 2010; Sigaev, Ryzhenkov *et al.*, 2010; Vigouroux, Fargin *et al.*, 2012). Each crystal can generate independently a second harmonic signal under suitable light excitation. But, glass-ceramic materials must retain their transparency. So, ideally, crystals will be of small size relative to the fundamental and SH wavelengths, and show linear and nonlinear refractive indices close to that for the glass matrix, in order to limit light scattering. Various studies have already been performed for the development of intrinsic second-order nonlinear optical properties in transparent oxide glass-ceramics exhibiting SHG and/or ferroelectricity. Although high efficiencies could be obtained with surface-crystallized glasses, bulk-crystallized glasses have advantages worthy of interest for application: improved resistance to crack propagation and thermal shocks when compared to surface-crystallized materials. In the beginning, the electro-optical and ferroelectric activity of oxide glass-ceramic materials containing ferroelectric micro-crystallites were studied by Borrelli *et al.* (Borrelli, Herczog *et al.*, 1965). Crystallite sizes of 0.2 μm or more were required to obtain a measurable ferroelectric effect emphasizing the difficult combination of transparency and efficiency. However, it was demonstrated that an efficient phase matching in isotropic materials can be achieved avoiding the phase mismatch induced by optical dispersion of materials without orientation or specific polarization selection rules (Baudrier-Raybaut, Haidar *et al.*, 2004). The 'random quasi-phase-matching' approach is based on the random motion of the relative phases in highly-transparent polycrystalline materials. A linear dependence of the conversion yield is related to the sample thickness with a wavelength-dependent resonant size for the polycrystalline grains.

A SHG in transparent chalcogenide glass-ceramics was obtained in Cd-doped Ge-Sb-S system (Guignard, Nazabal *et al.*, 2007b). In Ge-Sb-S glass-forming systems, the $Ge_{25}Sb_{10}S_{65}$ glass composition was chosen for its stability against crystallization. Different amounts of metallic cadmium were then introduced in this glass in order to obtain non-centrosymmetric micro-crystals. Cd_4GeS_6 phase precipitated in $Ge_{23}Sb_{11}S_{65}Cd_1$ glass and the β-GeS_2 phase was found to be the most stable phase obtained during the crystallization process. The formation of germanium disulphide (GeS_2) ($\Delta_f H°_{298} = -125.4$ kJ/mol) is not thermodynamically favoured compared to diantimony trisulphide (Sb_2S_3) or cadmium sulphide (CdS), which have standard enthalpies of formation at 298 K of –138.4 kJ/mol and –161.9 kJ/mol. The large content of germanium in the glass composition, as compared to cadmium and antimony, may be responsible for the preferential crystallization of GeS_2 micro-particles with a size of a few micrometers. An increase of both the number and the size of such micro-particles is observed when duration of the heat treatment at 370°C increases, suggesting that both nucleation and growth of the crystals occur simultaneously. However, β-GeS_2 phase seems to crystallize into single-crystals with relatively homogeneous spatial distribution and size within the glass matrix. At this temperature, the viscosity may be low enough to allow particle growth by diffusion of germanium atoms towards the nuclei and the diffusion of antimony atoms away from the nuclei within a reasonable time, as was shown by NanoSIMS analyses (Fig. 16.13). The SHG feature recorded in glass-ceramics is not a classical fringe pattern. It is different from that observed for surface-crystallized glass or electrically poled glass. For the last case, we saw that the SH signal falls to zero in the region close to 0° (Fig. 16.11). The nonlinear optical behaviour of the obtained glass-ceramics is analogous to that of a polycrystalline powder with randomly oriented non-centrosymmetric particles dispersed within a glass media. The glass matrix might play the role of index-matching medium of the micro-crystallites, which would then permit the second harmonic not to be significantly scattered, due to reflection and refraction at particle interfaces. Consequently, the whole SH flux could be included in a narrow cone in the forward direction (Kurtz and Perry, 1968). Micro-crystallites precipitate in a high refractive index glass matrix (about 2.2). Thus, the decrease of the SH signal, observed in the Maker fringe pattern when the absolute incident angle of the laser beam increases, is mostly due to Fresnel losses at sample interfaces at large angles. According to Kurtz and Perry (1968), the SHG intensity in a polycrystalline powder can be analysed with regard to the r/l_c ratio, where r is the particle size and l_c is the coherent length of the particle,

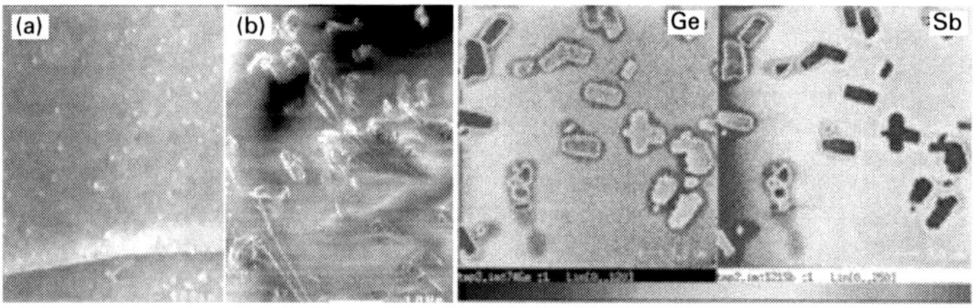

16.13 SEM images of $Ge_{23}Sb_{11}S_{65}Cd_1$ glass-ceramic annealed at 370°C (a) for 72 h and (b) for 144 h and mapping images of ^{74}Ge and ^{121}Sb elements obtained by nano-secondary ion mass spectrometry (Nano SIMS) from sample annealed for 144 h at 370°C (adapted from Guignard, Nazabal *et al.*, 2007b).

$l_c = \lambda_\omega/[4(n_{2\omega} - n_\omega)]$. Taking into account an approximation of the refractive index of both the crystallized phases in $Ge_{23}Sb_{11}S_{65}Cd_1$ glass-ceramics and the working wavelength, λ_ω, coherent length was estimated to be in the range of 10–40 μm for fundamental beam of $\lambda_\omega = 2$ μm. The SHG intensity of Cd_4GeS_6 powders (90 μm particles average size) measured with a reference of quartz powder has already been reported and presents an attractive $I^{2\omega}/I^{2\omega}_{quartz}$ value equal to 50 (at $\omega = 1.064$ μm). Glass-ceramics containing only nano-crystals of Cd_4GeS_6 (10 nm) with low crystallized volume fraction do not show any SHG. The existence of second-order nonlinear optical properties has clearly been related to the presence of β-GeS_2 in Cd-doped Ge-Sb-S glass-ceramics.

An increase of the SHG intensity versus the crystallized volume fraction clearly related to β-GeS_2 crystallites was qualitatively observed. SH intensities remain about 10 times lower than intensities recorded in thermally poled chalcogenide glasses and are about 0.01 to 0.06 times that of β-quartz, even if part of the generated SH signals is scattered out of the optical set-up. Indeed, the experimental set-up is not composed of any spherical mirror or integrating sphere to collect scattered photons in a backward direction, so reported SH intensities are thus underestimated. The effect of applying an electric field during the crystallization process was studied: an increase of the crystallization rate was demonstrated. Transparent GeS_2-Ga_2S_3-CdS and $80GeS_2$-$10Ga_2S_3$-$10CdI_2$ surface-crystallized glasses were also fabricated by heat treatment (Dong, Tao *et al.*, 2007b; Tao, Lin *et al.*, 2007). XRD and Raman spectroscopy have demonstrated the existence of α-$CdGa_2S_4$ nonlinear optical microcrystal at the origin of strong SHG in GeS_2-Ga_2S_3-CdS and $80GeS_2$-$10Ga_2S_3$-$10CdI_2$ samples. Nevertheless, for this family of glass-ceramics, the presence of Cd remains a weak point for future applications, due to its toxicity.

Further, an infrared transmitting surface-crystallized chalcohalide glass containing nontoxic and highly optically nonlinear $AgGaGeS_4$ crystallites was fabricated (Fig. 16.14). A strong green light (532 nm) of SHG was observed (Fig. 16.15) from $57GeS_2$-$43AgGaGeS_4$ glass-ceramics (Lin, Tao *et al.*, 2008). Only $AgGaGeS_4$ crystalline phase could be obtained when the Ge/(Ag+Ga+Ge) ratio is larger than 0.3 in the $AgGaS_2$-Ge_2S_4 system. A large second-order optical nonlinearity of 5.79 pm/V related to the presence of $AgGaGeS_4$ micro-crystallites was evaluated from the following relationship:

$$d_{eff}^{g-c} = d_{eff}^{g-c} \cdot \frac{P_{2\omega}^{g-c}}{P_{2\omega}^{quartz}} \frac{n^{g-c}}{n^{quartz}} \qquad [16.13]$$

where $$\eta = \frac{(n_\omega + 1)^3 (n_{2\omega} + 1)^3 (n_{2\omega} - n_\omega)^2 (n_{2\omega} + n_\omega)}{(n_{2\omega} R(0)}$$

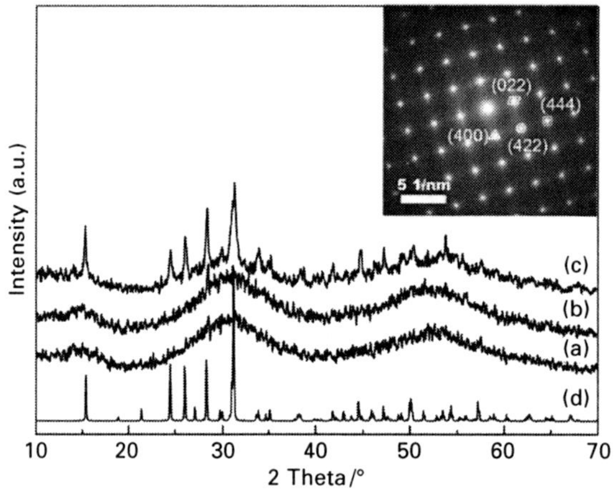

16.14 XRD patterns of samples after annealing at 350°C for (a) 12 h, (b) 24 h, (c) 48 h and (d) AgGaGeS4. The inset is the SAED pattern of the AgGaGeS4 crystal in sample heat treated at 350 °C for 24 h (adapted from Lin, Tao *et al.*, 2008).

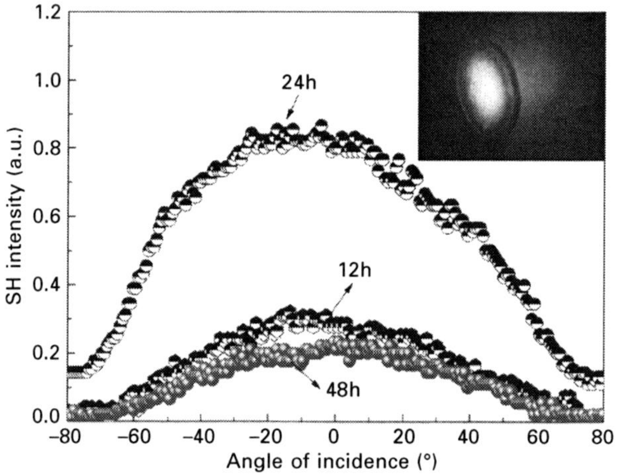

16.15 Maker fringe patterns of the crystallized samples fabricated at 350°C for 12h, 24 h and 48 h (adapted from Lin, Tao *et al.*, 2008).

Even though chalcogenide glass-ceramics cannot be used to fabricate waveguides because of scattering losses, they open a possible way to a cheap and efficient optical converter operating in the mid-IR spectral region.

The glass-ceramics are often almost exclusively crystallized at the surface, up to a few microns to several hundred microns. In the case of thin crystallized layers at both interfaces of the sample, narrow fringe patterns can be obtained due to interference of SH waves generated in the two surface layers separated by the optically linear volume of the glass. Further for thicker layers, broad-fringe pattern would be due to the lack of coherency of SH waves generated from surface crystalline layers ($L_{NL} > L_c$) and/or SH waves scattering by small crystallites possibly present also in the volume of the sample (Narazaki and Soga, 1998; Takahashi, Benino *et al.*, 2001). In order to explain the appearance of Maker fringe patterns presenting a minimum at normal incidence, the surface crystallized glasses were considered to be similar in terms of spontaneous polarization symmetry to that of electrically poled glasses with a perpendicular orientation to the surface. Moreover, the fringe patterns of SH signals obtained in pp- and sp-polarization modes showed a similar periodicity of the fine fringes and the SH intensity $I_{pp}/I_{sp} = 7$ ratio is close to that observed for thermally poled glasses ($I_{pp}/I_{sp} = 9$, according to ratio $\frac{\chi^{(2)}_{zzz}}{\chi^{(2)}_{zxx}} = 3$). Narazaki *et al.* proposed considering two possible sources of the second-order nonlinearity induced by surface crystallization: one is related to the polarization orientation of the non-centrosymmetric crystals and another is a stress caused by the crystallization process leading to a $C_{\infty v}$ symmetry (Narazaki, Tanaka *et al.*, 1999).

From the as-prepared $30GeS_2$-$35Ga_2S_3$-35AgCl chalcohalide glass at 350°C (T_g = 10°C) for 24 h (Zheng, Tao *et al.*, 2009), an intensive SHG was observed using the Maker fringe technique, and the origin of SHG was attributed mainly to the $AgGaGeS_4$ nonlinear optical crystallites contained in the surface crystallized layer. The thickness of the surface crystallized layer showing SHG activation is approximately 50 μm (Fig. 16.16). Taking into account that the composition of the glass and non-centrosymmetric crystal remains identical in GeS_2 glass, second-order nonlinear properties and better transparency could be achieved in β-GeS_2 glass-ceramics, insofar as the crystal size is small compared to the wavelength and as the refractive indices are slightly different from that of the glassy matrix. The surface

(a) (b)

16.16 SEM micrographs of the upper surface (a) and cross-section (b) of the $30GeS_2$-$35Ga_2S_3$-35AgCl glass-ceramic sample (adapted from Zheng, Tao *et al.*, 2009).

crystallization, presenting various Maker fringe patterns depending on the annealing process, leads to a second-order susceptibility estimated to be 7.3 pm/V.

Finally, the $APSe_6$ glass was proposed to generate SH intensities comparable to that of the infrared NLO material $AgGaSe_2$, without any poling treatments. Presenting crystal-glass phase change behaviour, the crystalline materials presents second-order coefficients, $\chi^{(2)}$, very high for phase-matching mid-IR nonlinear optical (NLO) materials, of about 151.3 and 149.4 pm/V for K^+ and Rb^+ salts, respectively. Starting from non-centrosymmetric phase change materials such as $APSe_6$ (A = K, Rb), it was claimed that optical glass fibres could have strong, intrinsic, and temporally stable second-order nonlinear optical (NLO) response (Chung, Jang *et al.*, 2010).

All these studies on glass and glass-ceramics are of major interest for the development of materials for frequency conversion devices.The great potential for optical converter operating in mid-IR of chalcogenide glass and glass-ceramics with nonlinear optical crystalline phase is presented in the next section.

16.8 Infrared (IR) stimulated processes in chalcogenide glasses

16.8.1 Physical mechanisms of IR-laser induced non-centrosymmetry of charge density

Study of the infrared photoinduced nonlinear optical effects in the mid-infrared (from 3 μm up to 12 μm) presents both fundamental and application interests. In the case of complex chalcogenide glasses, simultaneous excitation of phonon as well as of electron subsystems takes place (Kityk and Sahraoui, 1999). So, special attention must be paid to second-order nonlinear optical effects described by third-rank polar tensors, usually called second-order optical susceptibilities. This fact may have strong influence on photoinduced effects generated by popular CO_2 and CO lasers operating within the spectral range 3–12 μm. The glasses appropriate for this purpose should have enhanced optical polarizabilities to ensure larger IR-photoinduced changes. It was shown that ternary chalcogenide glasses possess such features (Kityk, 2004; Zakery and Elliott, 2007). In Contrast to traditional binary chalcogenide glasses, they show, for some specific properties and at given temperatures, enhanced electron–phonon interactions including anharmonic ones (Shpotyuk, Kasperczyk *et al.*, 1997). The main requirement for such glasses is to possess high ability to be polarized under influence of external IR-laser light (Qiu, Si *et al.*, 2001; Ogusu and Shinkawa, 2009). Most studies on infrared nonlinear optics are devoted to third-order optical effects, which do not require a non-centrosymmetric charge density in contrast to second-order optical effects described by second-order susceptibilities, which may be observed in crystals or partially crystallized glasses. This is crucial for the mid-IR spectral range where the main laser lines of CO_2 and CO lasers are situated (Kityk, Guignard *et al.*, 2007).

As seen previously, an additional factor for enhancing the SH conversion is the realization of phase matched conditions. It is well known that second- (or first-)order nonlinear optical effects described by third-rank polar tensors are basically forbidden in amorphous disordered materials, including glasses. To create a necessary non-centrosymmetric electronic charge density, an external dc electric field is usually applied (Chemla and Zyss, 1987). This approach often requires the application of high dc electric fields, which is probably not appropriate for narrow-band infrared chalcogenide materials, which possess relatively low resistance. However, applying external IR light with suitable parameters allows the required features to be achieved for the same glass (Kityk, Ozga *et al.*, 2008). In this case, the external dc electric field may be relatively low, so it does not cause any substantial occurrence of additional conductivity.

For crystals, second-order optical susceptibilities are fixed at given IR wavelengths and are determined by phase matching conditions (Petrov, Noack *et al.*, 2004; Miyata, Petrov *et al.*, 2007; Fedorchuk, Gorgut *et al.*, 2011). For disordered condensed matter, particularly for glasses, one can operate via second-order optical susceptibilities by using some IR laser treatment (Gautier, Lefumeux *et al.*, 2000; Tanaka, 2002). Such possibility opens many attractive applications like fabrication of IR-operated waveguides, triggers, memory devices, optically driven IR laser modulators, deflectors, etc. There are many studies devoted to optical poling performed in the near-IR (under 1.5 μm) spectral range (Zhao, Xu *et al.*, 1994; Antonyuk, 2000). A substantially smaller number of publications is devoted to mid-IR (3–12 μm) (Kityk, Kasperczyk *et al.*, 1999; Sahraoui and Kityk, 2003; Sanghera, Aggarwal *et al.*, 2006). Chalcohalide and chalcogenide glasses are particularly advantageous compared to other optical materials because they show relatively good and space-homogenous transparency as well as photo- and thermo-mechanical stability (Borisova, 1981).

The physical origin of the IR-induced second-order optical susceptibilities is different in the middle-IR spectral range, as compared to the visible and near-IR. This is due to close spectral positions of the corresponding laser wavelengths with respect to phonon resonances (Sanghera, Aggarwal *et al.*, 2005). Due to high IR laser power densities, a significant contribution of IR-stimulated anharmonic electron–phonon interactions (AEPI) is observed. These interactions are also described by third-rank polar tensors (Wasylak, Mefleh *et al.*, 2000). AEPI described by third-order space derivatives of the electrostatic potential (Geller, Dennis *et al.*, 2002) begin to be predominant for IR power densities larger than 0.3–0.4 GW/cm^2 (Napieralski, 1999).

For second-order optical effects in semiconductors (dielectrics) possessing an energy gap larger than ~ 0.2 eV, this large energy gap is crucial for partial nanocrystallization (Tao, Lin *et al.*, 2007). It must be emphasized that, due to the absence of crystalline periodic Brillouin zone, quasi-phonons are present in glasses; for convenience these vibrations may be described as phonons. However, the meaning of quasi-phonons is generally different when compared with the phonons in crystals that satisfy Bloch's translation rules. A contribution of the laser-excited phonons (due to non-resonant electrostriction-piezoelectric effects) may be described only within a framework of perturbation theory.

In this case, cascading virtual multi-photon processes are not caused prevailingly by electronic transitions, but possess dominant phonon contributions to electron–phonon broadened levels playing the role of additional relaxors. Sufficiently powerful IR-polarized laser beams favour the occurrence of IR-induced phonons in narrow-band glasses, with energies very close to electronic inter-band transitions. Multi-phonon excitations may also play an additional role in the enhanced electron–phonon anharmonicities (Kityk, Demianiuk *et al.*, 2004). Electron and harmonic electron–photon interactions can cause only re-scaling of the medium polarization. The formation of a non-centrosymmetric distribution of space charge density in chalcogenide or chalcohalide glasses, within a framework of (over)simplified approach, requires at least three phonons to participate. The contribution of one or two phonons leads only to the occurrence of the polarization of the medium, in which positive and negative polarization components are compensated, giving a zero value for the output polarization.

For clarification, a simplified scheme of charge density distribution in the IR-treated medium is depicted in Fig. 16.17 for different kinds of interactions (purely electronic, harmonic electron–phonon and anharmonic electron–phonon). One can see that pumping by IR electromagnetic waves should give a self-compensated contribution to the total macroscopic glass polarization (Fig. 16.17(a)). However, even in this case, the total polarization of the medium will also be compensated, which results in an output polarization equal to zero (Fig. 16.17(b)). Three phonons at least will permit the formation of non-zero polarization (Fig. 16.17(c)). Physically, it means that anharmonic electron–phonon interactions take place within a framework of third-order polar tensor. The efficiency of this process will be substantially dependent on pump power density. It should be highlighted that use of polarized IR light may cause more prominent displacements of particular ions from the equilibrium positions, with respect to unpolarized laser beam. For higher IR power densities, the IR laser field may cause a formation of anharmonic phonon modes described by third-rank tensors.

According to Fig. 16.17, AEPI should favour the occurrence of non-compensated total polarizability, responsible for local non-centrosymmetry,

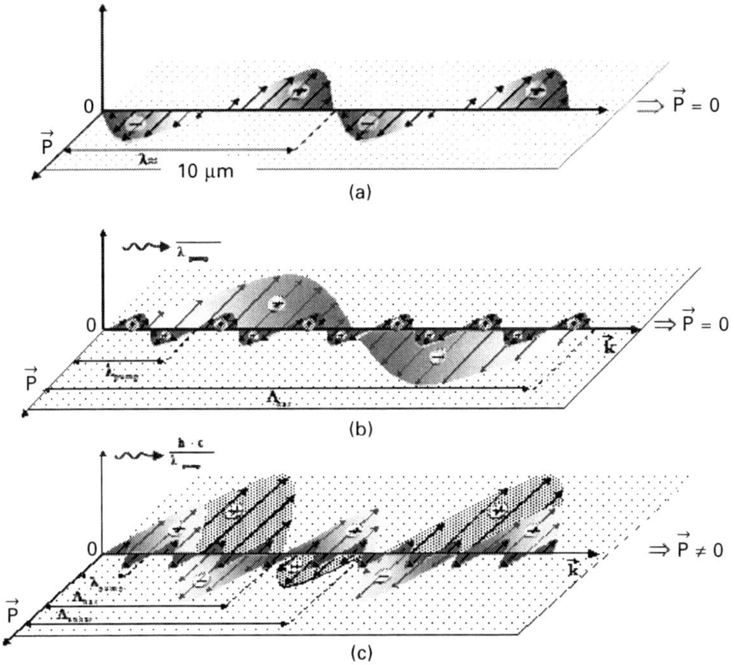

16.17 Scheme of charge density distribution in the IR-treated medium for different kinds of interactions: (a) pure electronic contribution, (b) electronic plus harmonic electron–phonon contribution, (c) three phonon contribution corresponding to electron–phonon anharmonic interactions.

which is necessary for the occurrence of second-order optical effects. These second-order effects could be more important for greater non-centrosymmetrical charge density distribution. Usually, all-optical poling is relatively weak and in order to enhance its manifestation, some additional dc electric field (up to 1 kV/cm) at increased material temperature is applied to achieve greater polarization of the glass. The required dc electric fields are strongly dependent on the of glass chemical composition.

Additionally, during the mentioned IR laser-induced phenomena, heating by infrared laser beam also plays some role. It may be somehow similar to thermal poling; however, in the case of laser heating, we are dealing with the local short-time heating of small surface areas by the laser beam (usually with a diameter from 1 up to 8 mm). Photo-thermal kinetics for the IR-induced SHG is completely different with respect to all the previous dependences. Under photo-stimulated heating or additional heating (like thermal poling), there is excitation of both electronic and multi-phonon subsystems. This phenomenon is dependent on the spectral positions of the IR-laser with respect to the phonon (electron) resonance frequencies. On the other hand, the high values of AEPI may be indicative of the presence of so-called low-phonon modes responsible for the phase transformations (soft modes) (Bussmann-Holder, Buttner *et al.*, 2007). In turn, phase transformations are responsible for the occurrence of long-range ordered spontaneous polarization (typical for ferroics) (Burns and Dacol, 1983), which are space-limited for glasses, contrary to crystalline materials.

To explain this effect, it is necessary to consider IR-stimulated effective temperature behaviour due to laser thermo-heating of the NLO susceptibilities, within the framework of simplified formalism. Following development of Maxwell's equations, the glass polarizability may be represented by two terms (linear and nonlinear):

$$\vec{P_i} = \vec{P_i^l} + \vec{P_i^{nl}} = \alpha_{ij} E_j^{(\omega)} + \beta_{ijk} E_j^{(\omega)} E_k^{(\omega)} + \gamma_{ijkl} E_j^{(\omega)} E_k^{(\omega)} E_l^{(\omega)} \vec{P_i^l} = \alpha_{ij} E_j^{(\omega)} \qquad [16.14]$$

$$\vec{P_i^{nl}} = \beta_{ijk} E_j^{(\omega)} E_k^{(\omega)} + \gamma_{ijkl} E_j^{(\omega)} E_k^{(\omega)} E_l^{(\omega)} \qquad [16.15]$$

where α_{ij}, β_{ijk}, γ_{ijkl} are microscopic susceptibilities (hyperpolarizabilities) related to macroscopic susceptibilities by the following equations:

$$\chi_{ij}^{(\omega)} = \alpha_{ij} L_i^{(\omega)} L_j^{(\omega)}$$

$$\chi_{ijk}^{(\omega)} = \beta_{ijk} L_i^{(\omega)} L_j^{(\omega)} L_k^{(\omega)} \chi_{ijkl}^{(\omega)} = \gamma_{ijkl} L_i^{(\omega)} L_j^{(\omega)} L_k^{(\omega)} L_l^{(\omega)}$$

$$L_z^{(\omega)} = \left(\frac{n_z^2 - 2}{n_z^2 + 3} \right) \frac{1}{\rho} \qquad [16.16]$$

where i, j, k indicate Cartesian indices for the given set-up.

$\chi_{ijk}^{(\omega)}$ is a macroscopic second-order NLO susceptibility responsible, for example, for photoinduced SHG (PISHG), and $\chi_{jkl}^{(\omega)}$ is the macroscopic third-order NLO susceptibility.

Using two-band and three-band (over)simplified models, microscopic hyperpolarizabilities may be represented as:

$$\alpha_{ij} \cong \frac{\vec{\mu}_i \vec{\mu}_j}{E_g^2}, \beta_{ijk} \cong \frac{\vec{\mu}_i \vec{\mu}_j \Delta\vec{\mu}_k}{E_g^3}, \gamma_{ijkl} \cong \frac{\vec{\mu}_i \vec{\mu}_j \Delta\vec{\mu}_k \Delta\vec{\mu}_l}{E_g^4} \qquad [16.17]$$

where $\vec{\mu}_{i,j}$ are transition dipole moments and $\Delta\vec{\mu}_{k,l} = \vec{\mu}_{k,l}^{(ex)} - \vec{\mu}_{k,l}^{(gr)}$ are differences between excited and ground state dipole moments.

The microscopic hyperpolarizabilities, β_{ijk}, define the efficiency of photo-stimulated second-order susceptibilities, which, in this case, is crucially dependent on the excited optically aligned dipole moments (Wasylak, Kucharski *et al.*, 1999). The latter are sensitive to long-range phonon critical temperature, T_c (analogously to Curie temperature), also called soft-phonon mode and determine the observed temperature dependences for the macroscopic susceptibility.

Simultaneously, the conductivity of free electrons may be described as:

$$\sigma = \frac{n \cdot e^2 \bar{l}}{\bar{V}} \qquad [16.18]$$

where $\bar{l} = \bar{\tau} \cdot \bar{V}$ stands for free length path, $\bar{V} = \sqrt{\frac{3k_\beta T}{m}}$ is the average velocity and n is the electron density.

Dipole moment may be expressed as $\vec{l}.e = \vec{\mu}$, and so:

$$\vec{\sigma} = \frac{n.e.e.\vec{\bar{l}}}{\bar{V}} = \frac{n.e}{\sqrt{3mk_\beta T}} \cdot \vec{\mu} \qquad [16.19]$$

Dipole moments have two contributions: $\vec{\mu} = \vec{\mu}_{el}, \vec{\mu}_{ph}$, where $\vec{\mu}_{ol}$ corresponds to the electronic part of the dipole momentum, which is less dependent on temperature, and $\vec{\mu}_{ph}$, the phonon contribution to dipole momentum, which is more sensitive to the temperature and presents bosons sensitive to the AEPI near the occurrence of some soft phonon modes (Delaire, Ma *et al.*, 2011). The latter term is described by third-order space derivatives of the anharmonic potential $\gamma_{ijk} = \frac{\partial^3 U}{\partial x_i \partial x_j \partial x_k}$,

or $$U = \frac{1}{2!}\alpha_{ij}x^2 - \frac{1}{3!}\beta_{ijk}x^3 - \frac{1}{4!}\gamma_{ijkl}x^4 \qquad [16.20]$$

or $\mu_{ph} \propto \gamma_{ijk}$; this anharmonic phonon interaction gives an additional contribution near the critical points of soft phonons, which is described by the following expression:

$$\gamma_{ijk} \propto \frac{2A}{\pi} \frac{\omega}{4(T - T_c)^\zeta + \omega^2} \qquad [16.21]$$

where A defines the half-width dispersion and T_C corresponds to the temperature of the occurrence of the soft phonon mode, causing effects similar to that of ferroic partially ordered dipoles and stimulating occurrence of non-zero polarizability.

Within a Maxwell phenomenological approach, the IR-induced SHG effect may be described by the following equation (Sahraoui and Kityk, 2003):

$$P_i = \chi_{ijklmn} E_j^\omega E_k^\omega E_l^{\Omega_1} E_m^{\Omega_2} E_n^{\Omega_3} \qquad [16.22]$$

where $E_{l,m,n}^{\Omega\omega 1,2,3}$ are the effective electric strengths of interacting electromagnetic IR laser beams created by SHG and IR-induced phonon dc polarized displace fields, particularly due to IR-induced electrostriction and piezoelectric effects.

The main requirement for observing such effects is to achieve the necessary degree of coherence between the interacting excited polaritons (quasi-phase matched conditions), which are determined by effective dispersion of the corresponding refractive indices, which, in turn, are determined by Kerr effects (Kityk and Sahraoui, 2005). Experimentally, it may be fulfilled upon splitting of the same laser pulses into two coherent pulses, which are automatically coherent, allowing the observation of the necessary nonlinear optical effects. For mid-IR lasers, this wavelength coherence is substantially higher than for near-IR laser beams, which is an essential factor to realize such kinds of interactions more effectively.

The observed IR laser-induced effect is different from that of all-optical poling in glasses (Balakirev, Kityk *et al*., 2003), where coherent bi-color interaction is important and the principal role is played by the $\chi^{(2)}$ grating, ensuring the phase matching conditions.

16.8.2 Experimental realization of IR laser-induced charge density acentricity in glasses

Following the reasons presented previously, the glasses satisfying the above-mentioned conditions, i.e. high AEPI and good local hyperpolarizabilities, belong to complex systems, usually ternary or quaternary. Most chalcogenide glasses show such complex features (Kityk and Sahraoui, 2001). The transparency spectrum for such kinds of glass, for example $As_{1.1}Bi_{0.9}Te_3$-$Ca_{0.5}Pb_{0.5}Br_2$ (called ABTC), is extended from 1 to 11 μm (Kityk and Sahraoui, 2005).

Specific feature of these glasses consists in a possibility to achieve relatively high AEPI corresponding to a local temperature increase by 40–80 K stimulated by incident IR laser light. For IR-stimulated nonlinear optical effects, illumination with pulsed 100 ps CO_2 laser with pulse energy up to 40 mJ and frequency about 50 Hz was used. The use of IR optical parametrical generator (OPG) based on chalcogenide crystals allows the tuning of IR laser wavelength within the 3–12 μm range (Fig. 16.18).

The IR semi-transparent mirror (M1) splits the input laser pulses into two different channels. The first one (fundamental laser channel) includes the shutter (Sh), polarizer (P_2) and ZnSe lens (L_2) with varying focus distances. The L_2 lens, together with the polarizer, was used to vary continuously the input fundamental power density up to 3 GW/cm^2 and laser wavelengths were changed within the 7–12.7 μm range. The second channel (writing beam channel) includes the set of IR reflecting mirrors M_2, M_3 and M_4. Mirrors were used to change angles between writing and fundamental laser beams. These two beams have used different parts of the lens L_2 to perform an optimization of laser beam geometries. The main part of the writing beam's channel is formed by doubled frequency Ag_3AsS_3 nonlinear single crystal.

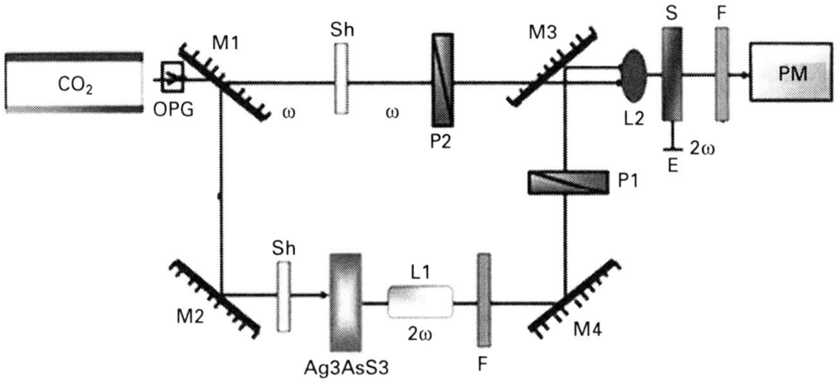

16.18 Principal experimental set-up for performance of IR-induced SHG measurements.

Filter F that cuts the fundamental laser beam and is transparent to the second beam. The angle of doubling-frequency crystal was varied to achieve maximal output of the doubled-frequency signal. The different filters were used to cut spectrally the fundamental laser beams with respect to the doubled frequency ones. The electric dc field (up to 1 kV/cm) was applied to the photo-treated sample together with laser treated temperature. Sometimes, this process may be presented as produced by stimulated-emission light through a two-step phonon-assisted process triggered by the optical near-field (Baxter, 2011).

The angles between the pump and probe laser beams have been varied within the 14–42° range. Polarizations and intensities of the pump, probe and SHG output beams were defined by polarizers P_1, P_2 and P_3, respectively.

During measurement of the ratio between the doubled-frequency signal and the fundamental beams, Fresnel losses, Gaussian-like sequence of the beam profile, optical attenuation and the photo-induced birefringence were taken into account. A parasitic fluorescence signal was observed below 1.12 μm, which corresponds to emission from the trapping intra-band levels. To study the laser photo-thermal heating, a laser profilometer was used. It was established that the local heating may give an additional enhancement of the second-order susceptibility in chalcogenide glasses (Guo, Zheng *et al.*, 2009). In the all-optical poling process, the temperature increase is varied in the range of the 2–10°C and stimulated phonons begin to play a major role in the process. Measurements were performed at two stages. During the first stage of 3–5 minutes, simultaneous IR laser treatment of the samples by two laser beams (fundamental and doubled frequency) was performed. Every 20–40 seconds, the writing channel was shut to control the SHG output with the simultaneously applied dc electric field. When the intensity of the output SHG varied less than 0.5%, it was considered that the process of IR-optical poling was almost finished. Such procedure was repeated for different wavelengths to obtain spectral dependencies of SHG output.

Typical spectral dependencies of IR-stimulated SHG output signal (proportional to $\chi^{(2\omega)}$), for different pumping IR power densities within the 3–8.9 μm spectral range, are presented in Fig. 16.19. For each pumping wavelength, the variation of power densities was carried out to achieve stabilization of the corresponding power density values. This behaviour confirms an effective contribution of IR-stimulated phonons.

16.8.3 Microscopic aspects of the IR-induced acentricity

Following the conclusion presented above, one can state that the observed effects are unambiguously dependent on the level of AEPI. To understand the origin of the effects, additional semi-empirical quantum chemical simulations were carried out. The semi-empirical quantum chemical simulations, performed by quantum chemical computer package HYPERCHEM 7.0, have shown that [Bi-As]–Te chemical bonds play a crucial role in the observed IR-induced nonlinear optical susceptibilities. For Pb–Br and Ca–Br chemical bonds, IR-induced changes are relatively small (less than 1.88%). For example, the matrix dipole moments for As–Bi bonds are equal to ~8.76 Debye (D) while for Pb–Br and Ca–Br bonds, total dipole moments are equal to 0.28 and 0.79 D, respectively. Our calculations, performed by norm-conserving pseudo-potential method, have shown that maximal level of AEPI is achieved with As–Bi(Te) clusters (up to 0.75 eV/A^3). From Fig. 16.20, one can clearly see that variations of As/Bi ratio from 1.1/0.9 up to 1.5/0.5 leads to substantial re-distribution of the IR-induced charge density. Space charge density distribution is very sensitive to substitution of Bi by As and should

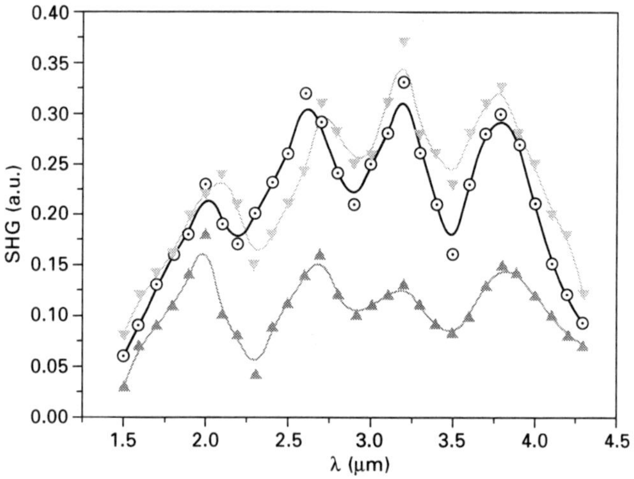

16.19 Spectral dependences of the ouput SHG for ABTC glasses at different power densities of fundamental CO_2 laser pulses: triangles, 0.4 GW/cm^2; inverted triangles, 0.6 GW/cm^2; rings, 0.8 GW/cm^2.

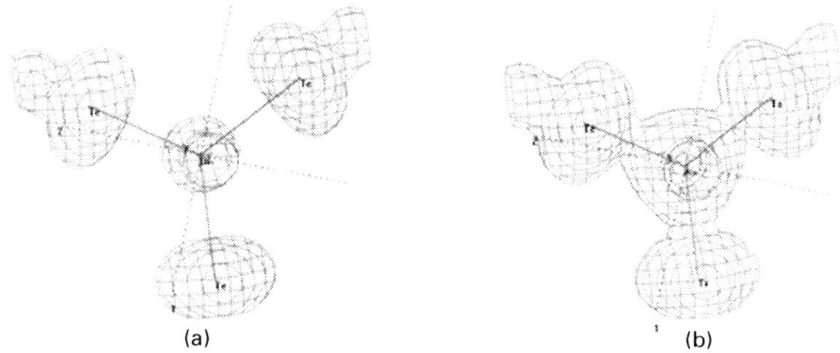

(a) (b)

16.20 Effective charge density distribution within the Bi(As)-Te cluster for the two limiting cases: (a) case of Bi atoms, (b) case of As atoms.

lead to substantial changes of third-order space derivatives defining the AEPI. Thus, by varying the As/Bi ratio, the AEPI can be changed and this factor is principal for the searching of glasses with improved secon-order susceptibilities.

Additionally, as results from Fig. 16.19 show, spectral SHG modulation has been observed corresponding to the inter-band quasi-resonances, which may originate from the enhanced oscillator strength due to anharmonic modes with several IR spectral maxima. The performed group-theory analysis together with quantum chemical simulations indicate that the observed phonon modes originate from LO + TO modes of the As-Te-Bi structural fragment formed by two 2A1 + B1 non-centrosymmetry bending and A1 + 2B2 stretching modes corresponding to the C_3 site point group of the As-Te clusters (Table 16.1). The principal anharmonic phonon modes are presented in the table.

The corresponding anharmonic IR phonon modes should cause the observed spectral shift of the SHG output compared to the actual phonon modes (Fig. 16.20) and may be responsible for additional spectral redistribution of the SHG output. Increase of the pump power density also causes an increase of IR SHG output signal. It must be noted that, at pump power densities below 0.11 GW/cm^2, the IR-induced SHG signal is comparable with the background noise.

16.8.4 Time kinetics and external field influence on the IR-induced effects

An additional important factor demonstrating the essential role of AEPI is the high sensitivity of SHG output versus pump-probe delay time (Fig. 16.21). According to Fig. 16.21, one can see that the maximum of the SHG signal is observed for pump-probe delay times equal to 55–60 ps, which are typical for relaxation times of AEPI relaxors. In this case, one of the specific features of the pump-probe delay dependence lies in the occurrence of a flat pump-

Table 16.1 Principal parameters of the IR-induced anharmonic phonon modes

Notation of peak	Wavelength (μm)	Anharmonic phonon mode 1 (cm^{-1})	Anharmonic phonon mode 2 (cm^{-1})	Anharmonic phonon mode 3 (cm^{-1})
1	2.52	1855	986	985
2	3.15	1251	1253	813
3	3.64	1247	1249	214

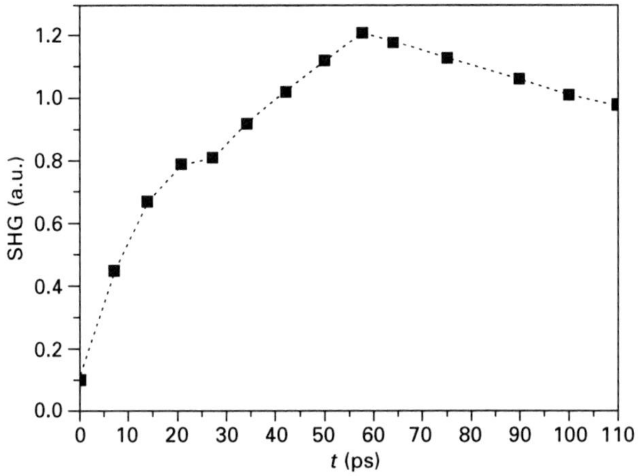

16.21 Typical pump-probe delay time dependence of the IR SHG for ABTC glasses.

probe delayed SHG maximum. This fact reflects a participation of a higher number of effective phonons to the observed effect. As a consequence, the quasi-phase-matched pump-probe conditions were realized at higher pump-probe delay times (Varga, Szingvari *et al.*, 2002).

The realization of partial SHG coherence is achieved due to close values of the IR-induced frequencies and overlapping effective energy values of IR energy gaps, which is proportional to the AEPI level. The second condition is achieved by the induction of acentric charge density anisotropy due to IR-induced anharmonic phonons (Ebothe, Imiolek *et al.*, 2008). Experimentally, the first factor may be observed through spectral SHG dependences versus anharmonic phonon frequencies. The second one may be shown through the monitoring of the pump-probe nonlinear optical kinetics occurring at delay times of ~55–60 ps, contrary to the ultrafast pure electronic relaxation time (below 1 ps) that is observed during the common multi-photon processes typical for chalcogenide glasses (Sanghera, Shaw *et al.*, 2000).

To show directly the role of anharmonic phonon modes, typical changes of IR-induced oscillator strengths are obtained from FTIR spectra. They correspond to the difference of IR mode maxima, evaluated from IR-absorption, at different IR-pump powers, as shown in Fig. 16.22. This behaviour is typical for the mid-IR nonlinear optical fibres (Sanghera, Shaw *et al.*, 2009). The main reason for the observation of such effects is the large enhancement of the phonon modes with respect to external mid-IR stimulations in chalcogenide glasses (Kohoutek, Yan *et al.*, 2011).

One can observe increasing anharmonic phonon mode strengths, which demonstrates a correlation with intensity dependences of the SHG presented in Fig. 16.19.

Within a framework of Maxwell phenomenology, the interaction of the fundamental IR laser beam $E_{pr}^{(\omega)}(r,t)$ with the investigated medium may be described as

$$E(r,t) = E_{pr}^{(2\omega)}(r,t) + E_{pr}^{(\omega)}(r,t) + E_{pump}^{(\omega_p)}(r,t) + E^{(\Omega_1 \pm \Omega_2 \pm \Omega_3)}(r,t-\tau) \quad [16.23]$$

where τ is the pump-probe delay time; and *pump* and *pr* indices correspond to pumping and probing IR beams, respectively. The interacting photon effective electric field strengths are presented within the plane wave approximation:

$$E_{pr}^{(\omega)}(r,t) = E^{(0)}\exp[i(\omega t + k\mathbf{r} + \phi_1)] \quad [16.24]$$

where $E^{(0)}$ and ϕ_1 are the amplitude and phase of the IR electromagnetic fundamental wave, respectively, possessing frequency (ω) at instant time (t) and the point of space ($\mathbf{r}$). Generally, there are many other slighter effects which may also give some contributions (Christodoulides, Khoo *et al.*, 2010). However, for this case, we will limit our consideration only to the prevailing contribution of the IR-stimulated subsystem.

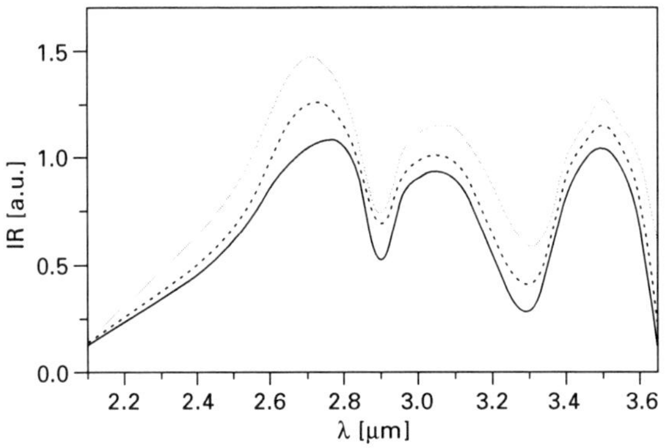

16.22 Typical IR-induced spectra of ABTC chalcogenide glasses at different laser power densities: grey, 0.6 GW/cm^2; dotted, 0.8 GW/cm^2; black, 0.4 GW/cm^2. All the IR-induced treatment was performed after applied dc electric field of about 1 kV/cm.

A term with frequency 2ω is caused by IR-induced non-centrosymmetry; it is similar to the process of interference for two coherent waves with fundamental and doubled frequencies (optical poling). The temporarily averaged output nonlinear polarization for the doubled frequency may be expressed as follows:

$$P^{(2\omega)}(r) \cong \varepsilon_0 |E_{pr}^{(\omega)}(r, t-\tau)|^2 \{< E_{pump}^{\omega p}(r,t) \cdot [E^{\Omega_1 \pm \Omega_2 \pm \Omega_3}(r,t)]^2 >$$

$$+ < E_{pr}^{\omega}(r,t)[E^{\Omega_1 \pm \Omega_2 \pm \Omega_3}(r,t)]^2 >< E_{pr}^{2\omega}(r, t-\tau)$$

$$\times [E^{\Omega_1 \pm \Omega_2 \pm \Omega_3}(r,t)]^2 >\} \quad [16.25]$$

where the brackets mean averaging over time.

The first term corresponds to the interaction between the pumping laser beam and the modulated IR-excited non-centrosymmetric charge density distribution due to IR-induced anharmonic displacement phonon modes possessing combined frequencies $n_1\Omega_1 \pm n_2\Omega_2 \pm n_3\Omega_3$. The electric field strength amplitudes of the photoinduced anharmonic displacement modes $(E^{(\Omega_1 \pm \Omega_2 \pm \Omega_3)}(r,t))$ are proportional to the IR-inducing power density and as a consequence can determine the photoinduced changes (Shin and Heo, 1999).

From general phenomenological consideration, one can conclude that the SHG output will be maximal for parallel directions of polarization for pumping and fundamental beams (Balakirev, Smirnov *et al.*, 2003). This conclusion reflects the fact that the space direction of the anharmonic charge density shift should be parallel to the polarization of the pumping (IR-induced) beam polarization.

Following the above-mentioned ideas, the output SHG should be dependent on the ratio between the fundamental and the writing beams; some optimal ratio should exist. To confirm this assumption, Fig. 16.23 presents the pump-power dependences of the SHG at different ratios, which unambiguously shows that the optimal SHG exists at the ratio equal to about 6:1. Such high ratio may be caused by specific features of chalcogenide glasses in the mid-IR (Sanghera, Aggarwal *et al.*, 2001).

The signal is prominent at dc electric field about 1 kV/cm and for further increase of the dc assisted electric field strength, the effect is limited by the occurrence of conductivity.

Another requirement is associated with the satisfaction of phase matching conditions. In the case of the cascading process discussed, phase matching conditions are fulfilled in a larger range of angles compared to traditional SHG. The second-order non-centrosymmetric effect may be observed because a range of phonon frequencies satisfying conditions $\omega\text{-}\omega_1 \pm n\Omega_1 \pm g\Omega_2 \pm h\Omega_3 = 0$ ($n, g, h = \pm 1, \pm 2, \pm 3, \ldots$) exists. They are essential for the appearance

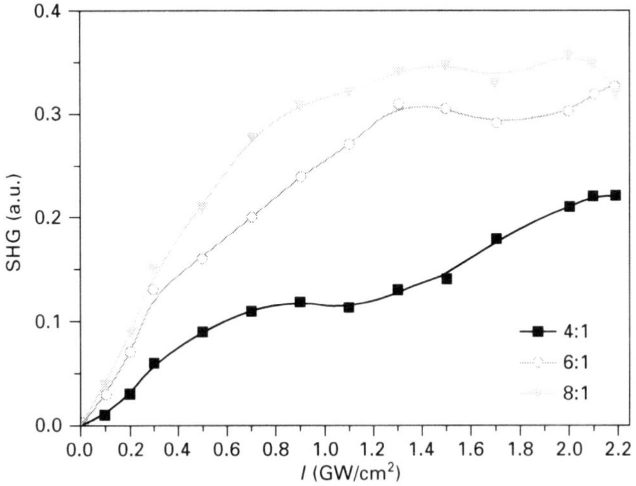

16.23 The output SHG versus the fundamental/writing beam ratio.

of non-centrosymmetric tensor components during the IR-picosecond photoexcitation satisfying the conditions of coherent interaction. It must be emphasized that the constant of electron–phonon nonlinear interactions is at least 20 times higher in chalcogenide glasses than in classical glasses (Ogusu, Li *et al*., 2004).

For the higher order cascading nonlinear effects, higher order interactions between the fundamental ($k_{pr}^{(\omega)}$), doubled frequency ($k^{(2\omega)}$) and pumping $k_{pump}^{(\omega_p)}$ beams and photoinduced combined IR-induced anharmonic frequencies of the medium with wave vectors ($K^{(\Omega_1 \pm \Omega_2 \pm \Omega_3)}$) should be taken into account:

$$p^{(2\omega)}(r,t) \cong \varepsilon_0 \chi^{(5;2\omega)}(2\omega;\omega,\omega,\Omega_1,\Omega_2,\Omega_3)|E_{pr}^{(\omega)}(r,t)|^2 \cdot \exp[-i\Delta kr] \quad [16.26]$$

where $\Delta k = k^{(2\omega)} - k_{pr}^{(\omega)} \pm (k_{pump}^{(\omega_p)} + K^{(\Omega_1 \pm \Omega_2 \pm \Omega_3)})$.

As a consequence, the range of angles satisfying phase matching conditions should be relatively wide because a large number of phonon modes fulfilling these conditions within the spectral range 2–9 µm exists.

Finally, the role of the dc electric field is shown in Fig. 16.24. Following the reasons presented above, one can state that the principal mechanisms are connected with the IR-induced medium polarization and the photoexcited anharmonic phonon subsystem, which contribute to the enhancement of the observed SHG. Here, the IR-induced phonons due to anharmonicities have a predominant effect (Forst, Manzoni *et al*., 2011).

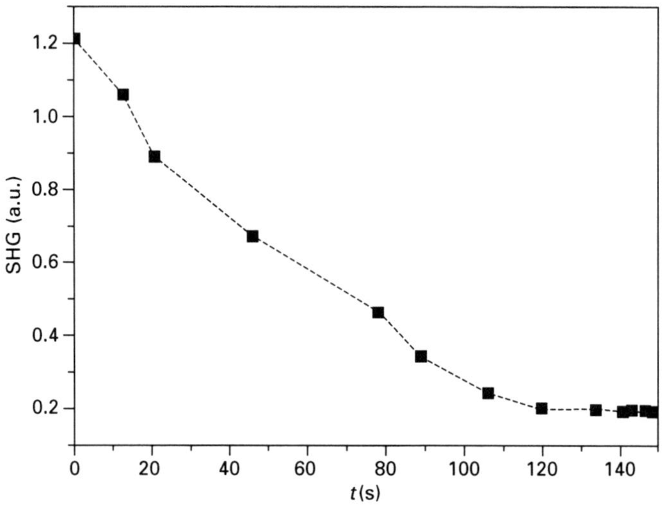

16.24 SHG evolution after switching off the external dc-electric field.

16.9 Conclusion

This chapter described the second order nonlinear optical processes induced in optically isotropic chalcogenide glasses. It presented both second order nonlinear optical effects in near infrared and IR stimulated processes obtained in chalcogenide glasses. The different processes applied with the aim of breaking the centrosymmetry of glasses, which enables their use for the conversion of wavelength and for electro-optical purposes were described: optical poling, poling assisted by e-beam or thermal poling. The origin and mechanisms of the phenomena induced by these processes, dedicated to the generation in glasses of a second order nonlinear optical response were discussed. In addition to these extrinsic methods, the interest of glass-ceramics containing non-centrosymmetric crystalline phases was considered. This chapter was also dedicated to some experiments describing IR-laser stimulated changes of main linear and nonlinear optical constants in chalcogenide glasses. The origin of the observed phenomena was described within phenomenological and microscopic approaches, with special attention being paid to the contribution of phonons.

16.10 References

Abe, M., T. Kitagawa, *et al.* (1996). 'Electro-optic switch constructed with a poled silica-based waveguide on a Si substrate.' *Electronics Letters* **32**(10): 893–894.

Alley, T. G., S. R. J. Brueck, *et al.* (1998). 'Space charge dynamics in thermally poled fused silica.' *Journal of Non-Crystalline Solids* **242**(2–3): 165–176.

Alley, T. G., S. R. J. Brueck, *et al.* (1999). 'Secondary ion mass spectrometry study of space-charge formation in thermally poled fused silica.' *Journal of Applied Physics* **86**(12): 6634–6640.
An, H. L. and S. Fleming (2012a). 'Controlling spatial distribution of thermal poling induced second-order optical nonlinearity with multilayered structures.' *Applied Physics Letters* **101**(10): 101101.
An, H. L. and S. Fleming (2012b). 'Investigating the effectiveness of thermally poling optical fibers with various internal electrode configurations.' *Optics Express* **20**(7): 7436–7444.
Antonyuk, B. P. (2000). 'All optical poling of glasses.' *Optics Communications* **181**(1–3): 191–195.
Armstrong, J. A., N. Bloembergen, *et al.* (1962). 'Interactions between light waves in a nonlinear dielectric.' *Physical Review* **127**(6): 1918–1939.
Asobe, M., T. Kanamori, *et al.* (1992). 'Ultrafast all-optical switching using highly nonlinear chalcogenide glass-fiber.' *IEEE Photonics Technology Letters* **4**(4): 362–365.
Balakirev, M. K., I. V. Kityk, *et al.* (2003). 'Anisotropy of the optical poling of glass.' *Physical Review A* **67**(2): 023806.
Balakirev, M. K., V. A. Smirnov, *et al.* (2003). 'Giant increase of the second harmonic radiation's absorption during optical poling of oxide glass.' *Journal of Modern Optics* **50**(8): 1237–1244.
Baudrier-Raybaut, M., R. Haidar, *et al.* (2004). 'Random quasi-phase-matching in bulk polycrystalline isotropic nonlinear materials.' *Nature* **432**(7015): 374–376.
Baxter, J. (2011). 'Light-emitting diodes: phonon-assisted emission.' *Nat Photon* **5**(9): 512.
Benino, Y., Y. Takahashi, *et al.* (2004). 'Second order optical non-linearity of transparent glass-ceramic materials induced by alternating field.' *Journal of Non-Crystalline Solids* **345**: 422–427.
Boncheva-Mladenova, Z. and Z. G. Ivanova (1978). 'Study of glass phases of germanium-sulfur-indium system.' *Journal of Non-Crystalline Solids* **30**(2): 147–153.
Borisova, Z. U. (1981). *Glassy Semiconductors*. New York, Plenum Press.
Borrelli, N. F., A. Herczog, *et al.* (1965). 'Electro-optic effect of ferroelectric microcrystals in a glass matrix.' *Applied Physics Letters* **7**(5): 117–118.
Boyd, R. W. (1992). '*Nonlinear Optics*.' San Diego, CA, Academic Press.
Burns, G. and F. H. Dacol (1983). 'Glassy polarization behavior in ferroelectric compounds $Pb(Mg_{1/3}Nb_{2/3})O_3$ and $Pb(Zn_{1/3}Nb_{2/3})O_3$.' *Solid State Communications* **48**(10): 853–856.
Bussmann-Holder, A., H. Buttner, *et al.* (2007). 'Polar-soft-mode-driven structural phase transition in $SrTiO_3$.' *Physical Review Letters* **99**(16): 167603.
Butcher, P. N. and D. Cotter (1990). *The Elements of Non-linear Optics*.' Cambridge, Cambridge University Press.
Canagasabey, A., C. Corbari, *et al.* (2009). 'High-average-power second-harmonic generation from periodically poled silica fibers.' *Optics Letters* **34**(16): 2483–2485.
Cardinal, T., K. A. Richardson, *et al.* (1999). 'Non-linear optical properties of chalcogenide glasses in the system As-S-Se.' *Journal of Non-Crystalline Solids* **257**: 353–360.
Carlson, D. E., K. W. Hang, *et al.* (1972). 'Electrode polarization in alkali-containing glasses.' *Journal of the American Ceramic Society* **55**(7): 337–341.
Carvalho, I. C. S., W. Margulis, *et al.* (1991). 'Erasure of frequency doubling gratings in optical fibers by ultraviolet-light excitation.' *Electronics Letters* **27**(17): 1497–1498.

Chao, S., H. Y. Chen, *et al.* (2005). 'Quasi-phase-matched second-harmonic generation in Ge-ion implanted fused silica channel waveguide.' *Optics Express* **13**(18): 7091–7096.

Chemla, D. S. and J. Zyss, eds (1987). *Nonlinear Optical Properties of Organic Molecules and Crystals*. New York, Academic Press.

Choi, J., M. Bellec, *et al.* (2012). 'Three-dimensional direct femtosecond laser writing of second-order nonlinearities in glass.' *Optics Letters* **37**(6): 1029–1031.

Christodoulides, D. N., I. C. Khoo, *et al.* (2010). 'Nonlinear refraction and absorption: mechanisms and magnitudes.' *Adv. Opt. Photon.* **2**(1): 60–200.

Chung, I., J. I. Jang, *et al.* (2010). 'Strongly nonlinear optical glass fibers from noncentrosymmetric phase-change chalcogenide materials.' *Journal of the American Chemical Society* **132**(1): 384–389.

Courtois, J.-Y. (1996). 'Optique non linéaire.' *EPJ Photovoltaics* **4**: 87–172.

Delaire, O., J. Ma, *et al.* (2011). 'Giant anharmonic phonon scattering in PbTe.' *Nature Materials* **10**(8): 614–619.

Dianov, E. M., A. M. Prokhorov, *et al.* (1989). 'Photoinduced 2nd-harmonic generation in optical fibers.' *JETP Letters* **50**(1): 13–15.

Dong, G., H. Tao, *et al.* (2007a). 'Study on the third and second-order nonlinear optical properties of GeS_2-Ga_2S_3-AgCl chalcohalide glasses.' *Optics Express* **15**(5): 2398–2408.

Dong, G., H. Tao, *et al.* (2007b). 'Second harmonic generation in transparent microcrystalline alpha-$CdGa_2S_4$-containing chalcogenide glass ceramics.' *Optics Communications* **274**(2): 466–470.

Dong, G., H. Tao, *et al.* (2007c). 'Mechanism of electron beam poled SHG in $0.95GeS_2 \cdot 0.05In_2S_3$ chalcogenide glasses.' *Journal of Physics and Chemistry of Solids* **68**(2): 158–161.

Dussauze, M., E. Fargin, *et al.* (2005). 'Large second-harmonic generation of thermally poled sodium borophosphate glasses.' *Optics Express* **13**(11): 4064–4069.

Dussauze, M., X. L. Zheng, *et al.* (2012). 'Photosensitivity and second harmonic generation in chalcogenide arsenic sulfide poled glasses.' *Optical Materials Express* **2**(1): 45–54.

Ebothe, J., W. Imiolek, *et al.* (2008). '$GeSe_2$-Ga_2S_3-PbI_2 as materials for IR-stimulated optical second harmonic generation.' *Journal of Materials Science – Materials in Electronics* **19**(3): 233–236.

Faccio, D., V. Pruneri, *et al.* (2001). 'Dynamics of the second-order nonlinearity in thermally poled silica glass.' *Applied Physics Letters* **79**(17): 2687–2689.

Fage-Pedersen, J., R. Jacobsen, *et al.* (2005). 'Planar glass devices for efficient periodic poling.' *Optics Express* **13**(21): 8514–8519.

Fedorchuk, A. O., G. P. Gorgut, *et al.* (2011). 'IR operated novel $Ag_{0.98}Cu_{0.02}GaGe_3Se_8$ single crystals.' *Journal of Physics and Chemistry of Solids* **72**(11): 1354–1357.

Forst, M., C. Manzoni, *et al.* (2011). 'Nonlinear phononics as an ultrafast route to lattice control.' *Nature Physics* **7**(11): 854–856.

Franken, P. A., G. Weinreich, *et al.* (1961). 'Generation of optical harmonics.' *Physical Review Letters* **7**(4): 118–119.

Fujiwara, T., S. Matsumoto, *et al.* (2000). 'Origin and properties of second-order optical non-linearity in ultraviolet-poled GeO_2-SiO_2 glass.' *Journal of Non-Crystalline Solids* **273**(1–3): 203–208.

Fujiwara, T., D. Wong, *et al.* (1995). 'Electrooptic modulation in germanosilicate fiber with uv-excited poling.' *Electronics Letters* **31**(7): 573–575.

Gautier, C. A., C. Lefumeux, *et al.* (2000). 'Spectral content of time resolved electronic and phonon induced non-linearities.' *Optics Communications* **178**(1–3): 217–224.

Geller, M. R., W. M. Dennis, *et al.* (2002). 'Theory of electron-phonon dynamics in insulating nanoparticles.' *Physica B – Condensed Matter* **316**: 430–433.

Gu, S., Z. Ma, *et al.* (2008). 'Second-harmonic generation in the thermal/electrical poling (100-x)GeS_2·x(0.5Ga_2S_3·0.5CdS) chalcogenide glasses.' *Journal of Physics and Chemistry of Solids* **69**(1): 97–100.

Guignard, M., V. Nazabal, *et al.* (2005). 'Second-harmonic generation of thermally poled chalcogenide glass.' *Optics Express* **13**(3): 789–795.

Guignard, M., V. Nazabal, *et al.* (2006). 'High second-order nonlinear susceptibility induced in chalcogenide glasses by thermal poling.' *Optics Express* **14**(4): 1524–1532.

Guignard, M., V. Nazabal, *et al.* (2007a). 'Chalcogenide glasses based on germanium disulfide for second harmonic generation.' *Advanced Functional Materials* **17**: 3284–3294.

Guignard, M., V. Nazabal, *et al.* (2007b). 'Crystalline phase responsible for the permanent second-harmonic generation in chalcogenide glass-ceramics.' *Optical Materials* **30**: 338–345.

Guignard, M., V. Nazabal, *et al.* (2008). 'Optical and structural properties of new chalcohalide glasses.' *Journal of Non-Crystalline Solids* **354**(12–13): 1322–1326.

Guimbretiere, G., M. Dussauze, *et al.* (2010). 'Correlation between second-order optical response and structure in thermally poled sodium niobium-germanate glass.' *Applied Physics Letters* **97**(17): 171103.

Guo, H. T., X. L. Zheng, *et al.* (2009). 'Large second-order nonlinearity in thermally poled Ge-Sb-Cd-S chalcogenide glass.' *Optical Materials* **31**(6): 865–869.

Harbold, J. M., F. O. Ilday, *et al.* (2002a). 'Highly nonlinear Ge-As-Se and Ge-As-S-Se glasses for all-optical switching.' *IEEE Photonics Technology Letters* **14**(6): 822–824.

Harbold, J. M., F. O. Ilday, *et al.* (2002b). 'Highly nonlinear As-S-Se glasses for all-optical switching.' *Optics Letters* **27**(2): 119–121.

Helt, L. G., E. Y. Zhu, *et al.* (2009). 'Proposal for in-fiber generation of telecom-band polarization-entangled photon pairs using a periodically poled fiber.' *Optics Letters* **34**(14): 2138–2140.

Iwafuchi, N., T. Fujiwara, *et al.* (2009). 'Optical characteristics of nanocrystallized glass fiber with second-order optical nonlinearity.' *Journal of the Ceramic Society of Japan* **117**(1362): 143–146.

Jing, R., Y. Guang, *et al.* (2006). 'Second-harmonic generation in thermally poled chalcohalide glass.' *Optics Letters* **31**(23): 3492–3494.

Kao, Y. H., Y. Hu, *et al.* (1994). '2nd-harmonic generation in transparent barium berate glass-ceramics.' *Journal of Non-Crystalline Solids* **167**(3): 247–254.

Kashyap, R., G. J. Veldhuis, *et al.* (1994). 'Phase-matched 2nd-harmonic generation by periodic poling of fused-silica.' *Applied Physics Letters* **64**(11): 1332–1334.

Kazansky, P. G., A. Kamal, *et al.* (1993). 'High 2nd-order nonlinearities induced in lead silicate glass by electron-beam irradiation.' *Optics Letters* **18**(9): 693–695.

Kazansky, P. G., P. S. Russell, *et al.* (1995). 'Pockels effect in thermally poled silica optical fibers.' *Electronics Letters* **31**(1): 62–63.

Kityk, I. V. (2004). 'IR-stimulated second harmonic generation in Sb_2Te_2Se-BaF_2-$PbCl_2$ glasses.' *Journal of Modern Optics* **51**(8): 1179–1189.

Kityk, I. V., M. Demianiuk, *et al.* (2004). 'IR-induced second-harmonic generation in PbSe microcrystallites.' *Journal of Physics – Condensed Matter* **16**(21): 3533–3544.

Kityk, I. V., M. Guignard, *et al.* (2007). 'Manifestation of electron-phonon interactions in IR-induced second harmonic generation in a sulphide glass-ceramic with beta-GeS_2 microcrystallites.' *Physica B – Condensed Matter* **391**(2): 222–227.

Kityk, I. V., J. Kasperczyk, *et al.* (1999). 'Two-photon absorption and photoinduced second-harmonic generation in Sb_2Te_3-$CaCl_2$-$PbCl_2$ glasses.' *Journal of the Optical Society of America B – Optical Physics* **16**(10): 1719–1724.

Kityk, I. V., K. Ozga, *et al.* (2008). 'Optical and DC-electric poling on AgX (X = Cl, I)-doped chalcohalide glasses.' *Laser Physics* **18**(6): 780–782.

Kityk, I. V. and B. Sahraoui (1999). 'Photoinduced two-photon absorption and second-harmonic generation in As_2Te_3-$CaCl_2$-$PbCl_2$ glasses.' *Physical Review B* **60**(2): 942–949.

Kityk, I. V. and B. Sahraoui (2001). 'Infrared-induced nonlinear optical effects in chalcogenide glasses.' *Journal of Chemical Physics* **114**(18): 8105–8112.

Kityk, I. V. and B. Sahraoui (2005). 'Phonon-assisted second harmonic generation in $As_{1-x}Bi_xTe_{3-x}CaBr_2$-$PbBr_2$ glasses.' *Journal of Physical Chemistry B* **109**(8): 3163–3168.

Kleinman, D. A. (1962a). 'Nonlinear dielectric polarization in optical media.' *Physical Review* **126**(6): 1977–1979.

Kleinman, D. A. (1962b). 'Theory of second harmonic generation of light.' *Physical Review* **128**(4): 1761–1775.

Kohoutek, T., X. Yan, *et al.* (2011). 'Enhanced Raman gain of Ge-Ga-Sb-S chalcogenide glass for highly nonlinear microstructured optical fibers.' *Journal of the Optical Society of America B – Optical Physics* **28**(9): 2284–2290.

Kudlinski, A., G. Martinelli, *et al.* (2008). 'Dynamics of the second-order nonlinearity induced in Suprasil glass thermally poled with continuous and alternating fields.' *Journal of Applied Physics* **103**(6): 063109.

Kudlinski, A., Y. Quiquempois, *et al.* (2005). 'Modeling of the phi((2)) susceptibility time-evolution in thermally poled fused silica.' *Optics Express* **13**(20): 8015–8024.

Kurtz, S. K. and T. T. Perry (1968). 'A powder technique for the evaluation of nonlinear optical materials.' *Journal of Applied Physics* **39**(8): 3798–3813.

Le Calvez, A., E. Freysz, *et al.* (1998). 'A model for second harmonic generation in poled glasses.' *European Physical Journal D* **1**(2): 223–226.

Lee, H. W., Y. S. Lee, *et al.* (2008). 'Preparation and second harmonic generation of $Ba_2TiSi_2O_8$ films.' *Materials Letters* **62**(14): 2053–2056.

Lenz, G., J. Zimmermann, *et al.* (2000). 'Large Kerr effect in bulk Se-based chalcogenide glasses.' *Optics Letters* **25**(4): 254–256.

Li, G. Y., K. A. Winick, *et al.* (2009). 'Quasi-phase matched second-harmonic generation through thermal poling in femtosecond laser-written glass waveguides.' *Optics Express* **17**(11): 9442–9453.

Lin, C., H. Tao, *et al.* (2008). 'Permanent second-harmonic generation in AgGaGeS4 bulk-crystallized chalcogenide glasses.' *Chemical Physics Letters* **460**: 125–128.

Li, W. T., H. An, *et al.* (2008). 'Second-order optical nonlinearity in thermally poled multilayer germanosilicate thin films.' *Electronics Letters* **44**(10): 639–640.

Lin, C., A. Le Calvez, *et al.* (2010). 'Second-order optical nonlinearity and ionic conductivity of nanocrystalline GeS_2-Ga_2S_3-LiI glass-ceramics with improved thermo-mechanical properties.' *Physical Chemistry Chemical Physics* **12**(15): 3780–3787.

Liu, A. C., M. J. F. Digonnet, *et al.* (1994). 'Electrooptic phase modulation in a silica channel waveguide.' *Optics Letters* **19**(7): 466–468.

Liu, Q. M., X. J. Zhao, *et al.* (2001). 'Second-harmonic generation in Ge-As-S glasses by electron beam irradiation and analysis of the poling mechanism.' *Optics Communications* **198**(1–3): 187–192.

Liu, W., Q. M. Zhang, *et al.* (2009). 'Enhancement of second-order optical nonlinearity in photo-darkened Ge25Sb10S65 chalcogenide glass by ferntosecond laser light.' *Optics Communications* **282**(10): 2081–2084.

Long, X. C., R. A. Myers, *et al.* (1994). 'Measurement of linear electrooptic effect in temperature electric-field poled optical fibers.' *Electronics Letters* **30**(25): 2162–2163.

Lopez-Lago, E., V. Couderc, *et al.* (2001). 'All-optical poling of a chalcohalogenide glass.' *Optical Materials* **16**(4): 413–416.

Maker, P. D., R. W. Terhune, *et al.* (1962). 'Effects of dispersion and focusing on production of optical harmonics.' *Physical Review Letters* **8**(1): 21–22.

Malmström, M., O. Tarasenko, *et al.* (2012). 'Pulse selection at 1 MHz with electrooptic fiber switch.' *Optics Express* **20**(9): 9465–9470.

Margulis, W., O. Tarasenko, *et al.* (2009). 'Who needs a cathode? Creating a second-order nonlinearity by charging glass fiber with two anodes.' *Optics Express* **17**(18): 15534–15540.

Miyata, K., V. Petrov, *et al.* (2007). 'Phase-matching properties for AgGaGeS4.' *Applied Optics* **46**(23): 5728–5731.

Montant, S., A. Le Calvez, *et al.* (1999). 'Light-controlled erasure of induced chi((2)) in thermally poled glasses.' *Applied Physics Letters* **74**(18): 2623–2625.

Mukherjee, N., R. A. Myers, *et al.* (1994). 'Dynamics of 2nd-harmonic generation in fused-silica.' *Journal of the Optical Society of America B – Optical Physics* **11**(4): 665–669.

Myers, R. A., N. Mukherjee, *et al.* (1991). 'Large 2nd-order nonlinearity in poled fused-silica.' *Optics Letters* **16**(22): 1732–1734.

Nakanishi, M., O. Sugihara, *et al.* (1999). 'Electron-beam irradiation-induced nonlinearity in silicate glass films and fabrication of nonlinear optical gratings.' *Journal of Applied Physics* **86**(5): 2393–2396.

Napieralski, J. (1999). 'Photoinduced phase transformation in Y-Ba-Cu-O ceramics.' *Ferroelectrics* **220**(1–2): 17–29.

Narazaki, A. and N. Soga (1998). 'Second harmonic generation in $BaTiO_3$ film crystallized on tellurite glass surface.' *Journal of Materials Science Letters* **17**(13): 1063–1065.

Narazaki, A., K. Tanaka, *et al.* (1999). 'Optical second-order nonlinearity of transparent glass-ceramics containing $BaTiO_3$ precipitated via surface crystallization.' *Journal of Materials Research* **14**(9): 3640–3646.

Narazaki, A., K. Tanaka, *et al.* (2002). 'Surface structure and second-order nonlinear optical properties of thermally poled WO_3-TeO_2 glasses doped with Na^+.' *Journal of the Optical Society of America B – Optical Physics* **19**(1): 54–62.

Nasu, H., K. Kubodera, *et al.* (1990). '3rd-harmonic generation from some chalcogenide glasses.' *Journal of the American Ceramic Society* **73**(6): 1794–1796.

Nazabal, V., E. Fargin, *et al.* (2001). 'Thermally poled new borate glasses for second harmonic generation.' *Journal of Non-Crystalline Solids* **290**(1): 73–85.

Ogusu, K., H. P. Li, *et al.* (2004). 'Brillouin-gain coefficients of chalcogenide glasses.' *Journal of the Optical Society of America B – Optical Physics* **21**(7): 1302–1304.

Ogusu, K. and K. Shinkawa (2009). 'Optical nonlinearities in As_2Se_3 chalcogenide glasses doped with Cu and Ag for pulse durations on the order of nanoseconds.' *Optics Express* **17**(10): 8165–8172.

Ogusu, K., J. Yamasaki, *et al.* (2004). 'Linear and nonlinear optical properties of Ag-As-Se chalcogenide glasses for all-optical switching.' *Optics Letters* **29**(3): 265–267.

Ohmori, Y. and Y. Sasaki (1982). '2-wave sum-frequency light generation in optical fiber.' *IEEE Journal of Quantum Electronics* **18**(4): 758–762.

Osterberg, U. and W. Margulis (1987). 'Experimental studies on efficient frequency doubling in glass optical fibers.' *Optics Letters* **12**(1): 57–59.

Petrov, V., F. Noack, *et al.* (2004). 'Phase-matching and femtosecond difference-frequency generation in the quaternary semiconductor $AgGaGe_5Se_{12}$.' *Applied Optics* **43**(23): 4590–4597.

Proctor, T. M., P. M. Sutton (1959). 'Static space-charge distributions with a single mobile charge carries.' *J. Chem. Phys.* **30**: 212–220.

Pruneri, V., G. Bonfrate, *et al.* (1999). 'Greater than 20%-efficient frequency doubling of 1532-nm nanosecond pulses in quasi-phase-matched germanosilicate optical fibers.' *Optics Letters* **24**(4): 208–210.

Qiu, J. R., J. H. Si, *et al.* (2001). 'Photoinduced stable second-harmonic generation in chalcogenide glasses.' *Optics Letters* **26**(12): 914–916.

Quiquempois, Y., N. Godbout, *et al.* (2002). 'Model of charge migration during thermal poling in silica glasses: evidence of a voltage threshold for the onset of a second-order nonlinearity.' *Physical Review A* **65**(4): 043816.

Quiquempois, Y., A. Villeneuve, *et al.* (2000). 'Second-order nonlinear susceptibility in As_2S_3 chalcogenide thin glass films.' *Electronics Letters* **36**(8): 733–734.

Sahraoui, B. and I. V. Kityk (2003). 'Mid-infrared light-induced second-harmonic generation in specific glasses.' *Journal of Optics A – Pure and Applied Optics* **5**(3): 174–179.

Sanghera, J. S., I. D. Aggarwal, *et al.* (2001). 'Applications of chalcogenide glass optical fibers at NRL.' *Journal of Optoelectronics and Advanced Materials* **3**(3): 627–640.

Sanghera, J., I. D. Aggarwal, *et al.* (2005). 'Chalcogenide optical fibers target mid-IR applications.' *Laser Focus World* (April): 83.

Sanghera, J. S., I. D. Aggarwal, *et al.* (2006). 'Nonlinear properties of chalcogenide glass fibers.' *Journal of Optoelectronics and Advanced Materials* **8**(6): 2148–2155.

Sanghera, J. S., L. B. Shaw, *et al.* (2000). 'Development and infrared applications of chalcogenide glass optical fibers.' *Fiber and Integrated Optics* **19**(3): 251–274.

Sanghera, J. S., L. B. Shaw, *et al.* (2009). 'Chalcogenide glass-fiber-based mid-IR sources and applications.' *IEEE Journal of Selected Topics in Quantum Electronics* **15**(1): 114–119.

Shimakawa, K., S. Inami, *et al.* (1990). 'Reversible photoinduced change of photoconductivity in amorphous chalcogenide films.' *Phys. Rev. B Condens. Matter* **42**(18): 11857–11861.

Shimakawa, K., S. Inami, *et al.* (1992). 'Origin of photoindued metastable defects in amorphous chalcogenides.' *Phys. Rev. B* **46**(16): 10062–10069.

Shin, Y. B. and J. Heo (1999). 'Mid-infrared emissions and multiphonon relaxation in Dy^{3+}-doped chalcohalide glasses.' *Journal of Non-Crystalline Solids* **253**: 23–29.

Shioya, K., T. Komatsu, *et al.* (1995). 'Optical-properties of transparent glass-ceramics in K_2O-Nb_2O_5-TeO_2 glasses.' *Journal of Non-Crystalline Solids* **189**(1–2): 16–24.

Shpotyuk, O. I., J. Kasperczyk, *et al.* (1997). 'Mechanism of reversible photoinduced optical effects in amorphous As_2S_3.' *Journal of Non-Crystalline Solids* **215**(2–3): 218–225.

Sigaev, V. N., V. S. Ryzhenkov, *et al.* (2010). 'Glasses and their crystallization in the $(1\text{-}x)KNbO_3{\cdot}xSiO_2$ system at low glass-forming oxide contents, $0 < x < 0.35$.' *Journal of Non-Crystalline Solids* **356**(20–22): 958–965.

Smektala, F., C. Quemard, *et al.* (1998). 'Chalcogenide glasses with large non-linear refractive indices.' *Journal of Non-Crystalline Solids* **239**(1–3): 139–142.

Smith, N. J., M. T. Lanagan, *et al.* (2012). 'Thermal poling of alkaline earth boroaluminosilicate glasses with intrinsically high dielectric breakdown strength.' *Journal of Applied Physics* **111**(8): 083519.

Stolen, R. H. and H. W. K. Tom (1987). 'Self-organized phase-matched harmonic-generation in optical fibers.' *Optics Letters* **12**(8): 585–587.

Suhara, T. and M. Fujimura, eds. (2010). *Waveguide Nonlinear-Optic Devices*. Berlin, Springer-Verlag.

Takahashi, Y., Y. Benino, *et al.* (2001). 'Second harmonic generation in transparent surface crystallized glasses with stillwellite-type $LaBGeO_5$.' *Journal of Applied Physics* **89**(10): 5282–5287.

Tanaka, K. (2002). 'Two-photon absorption spectroscopy of As_2S_3 glass.' *Applied Physics Letters* **80**(2): 177–179.

Tanaka, K., A. Narazaki, *et al.* (2000a). 'Large optical second-order nonlinearity of poled WO_3-TeO_2 glass.' *Optics Letters* **25**(4): 251–253.

Tanaka, K., A. Narazaki, *et al.* (2000b). 'Poling-induced structural change and second-order nonlinearity of N^{a+}-doped Nb_2O_5-TeO_2 glass.' *Journal of Physics – Condensed Matter* **12**(30): L513–L518.

Tao, H., C. Lin, *et al.* (2007). 'Optical second-order nonlinearity of the infrared transmitting $82GeS_2 \cdot 18CdGa_2S_4$ nanocrystallized chalcogenide glass.' *Applied Physics Letters* **91**(1): 011904.

Thamboon, P. and D. M. Krol (2003). 'Second-order optical nonlinearities in thermally poled phosphate glasses.' *Journal of Applied Physics* **93**(1): 32–37.

Triques, A. L. C., I. C. S. Carvalho, *et al.* (2003). 'Time evolution of depletion region in poled silica.' *Applied Physics Letters* **82**(18): 2948–2950.

Varga, J., Y. Szingvari, *et al.* (2002). 'IR-poled second-harmonic generation in glass.' *Optics and Laser Technology* **34**(6): 471–473.

Vigouroux, H., E. Fargin, *et al.* (2012). 'Synthesis and multiscale evaluation of $LiNbO_3$-containing silicate glass-ceramics with efficient isotropic SHG response.' *Advanced Functional Materials* **22**(19): 3985–3993.

Von Hippel, A., E. P. Gross, *et al.* (1953). 'Photocurrent, space-charge buildup, and field emission in alkali halide crystals.' *Phys. Rev.* **91**: 568–579.

Wasylak, J., J. Kucharski, *et al.* (1999). 'Photoinduced effects in the Sb_2Se_3-$BaCl_2$-$PbCl_2$ glasses.' *Journal of Applied Physics* **85**(1): 425-431.

Wasylak, J., A. Mefleh, *et al.* (2000). 'Photoinduced nonlinear optical phenomena in Sb_2Se_3-$BaCl_2$-$SnCl_2$ glasses.' *Journal of Materials Science* **35**(1): 215–219.

Yeom, D. I., E. C. Maegi, *et al.* (2008). 'Low-threshold supercontinuum generation in highly nonlinear chalcogenide nanowires.' *Optics Letters* **33**(7): 660–662.

Zakery, A. and S. R. Elliott (2007). *Optical Nonlinearities in Chalcogenide Glasses and their Applications*. Berlin, Springer-Verlag.

Zhao, X. J., L. Y. Xu, *et al.* (1994). 'Glass-formation in Sb_2Se_3-MX_n(metal halides).' *Journal of Non-Crystalline Solids* **167**(1–2): 70–73.

Zheng, X., H. Tao, *et al.* (2009). 'Second harmonic generation in surface crystallized $30GeS_2 \cdot 35Ga_2S_3 \cdot 35AgCl$ chalcohalide glasses.' *Optical Materials* **31**(10): 1434–1438.

Zhu, E. Y., Z. Tang, *et al.* (2012). 'Direct generation of polarization-entangled photon pairs in a poled fiber.' *Physical Review Letters* **108**(21): 213902.

17
Chalcogenide glass resists for lithography

H. JAIN, Lehigh University, USA, A. KOVALSKIY, Austin Peay State University, USA and M. VLCEK, University of Pardubice, Czech Republic

DOI: 10.1533/9780857093561.2.562

Abstract: The structure of chalcogenide glasses (ChG) is readily modified by exposure to near bandgap or UV light, X-rays, or electron/ion beams. Furthermore, their composition can be modified selectively by exploiting radiation-induced diffusion or dissolution of certain metals like silver – a unique property of ChG. These characteristics, together with wet or dry etching, make ChG superior resists for high resolution, grayscale lithography. The underlying mechanisms of radiation-induced modifications, optimum glass compositions, and processing conditions of ChG are critically reviewed with photo and e-beam resist applications in mind, and compared with those for widely used polymer resists. Along with several illustrative examples, an assessment is offered for the prospects of ChG resists in device fabrication.

Key words: resist, nano lithography, grayscale, chalcogenide glass, photo-doping.

17.1 Introduction

Lithography, in the context of modern technology, refers to the process which replicates a micropattern or an electronic circuit rapidly. The manufacture of numerous electronic, optoelectronic and photonic devices uses lithography as a necessary technique for fabricating micrometer size features in a substrate. For example, in microelectronics technology, transistors and other electrical circuit elements are produced in silicon. The desired modification of the electrical properties is accomplished by doping a predetermined region with appropriate dopants. By comparison, the fabrication of microelectromechanical system (MEMS) requires removal of pre-selected regions of the substrate. Optoelectronic and photonic devices may require compositional as well as topological modification of the substrate (Jain and Vlcek, 2008).

Overall, lithography involves an exposure (irradiation) source, a mask, a resist, and a series of fabrication steps that would accomplish pattern transfer from the mask to a resist and then to a substrate in which the device is fabricated. The creation of a mask or the use of a resist is based on materials that are sensitive to specific radiation. Complex patterns are transferred into a resist by irradiating its selected regions. Due to its radiation-sensitivity the chemical reactivity of exposed regions is modified with respect to unexposed regions. A developer then preferentially etches out either the exposed (positive resist) or unexposed (negative resist) region, thus faithfully transferring the original pattern into the resist. Over the years, the feature size of the elements to be fabricated by lithography has been decreasing continuously. With the push for nanofabrication of nano-electromechanical systems (NEMS) (van den Boogaart *et al.*, 2004), single electron devices (Vieu *et al.*, 2000), electrical connection of individual molecules (Tsutsumi *et al.*, 2000), etc., there is an emerging need for fabricating sub-10 nm size structures. The ability to alter the structure and properties of a resist at this length scale will naturally introduce new technology for ultra-high density storage. With this motivation, we examine the current status of chalcogenide glass (ChG) resists in this chapter, especially their resolution limitations and prospects for applications.

The transfer of patterns into silicon may not always be needed or even desirable, especially when the photoresist itself may work as an optical component (e.g., a refracting surface). A 3-D optical element patterned through a grayscale mask on the right resist material may itself possess the required optical properties (Rogers *et al.*, 2004), thus simplifying the device structure and fabrication considerably. The direct application of the patterned resists should significantly reduce the cost and time of fabrication as well. Different types of glass resists (sol-gel films, ChG) that are much harder than polymers, are often proposed for such applications. Moreover, polymers are not always suitable for ultrathin and nanosize structures, when one needs a resist material that is highly sensitive to irradiation, easily structured on the nanoscale, and hard enough to be used as the object/structural material or transferred into silicon substrate. ChG thin film photoresists satisfy most of these requirements. They are also expected to yield better control of the geometry than is possible with polymer resists.

Chalcogenide glasses have an additional remarkable property of radiation-induced diffusion of silver, which opens new opportunities for grayscale lithography. Light, x-ray, or e-beam irradiation causes complete, strongly transversal dissolution of thin silver layer deposited on top of a ChG film, gradually changing the chemical resistance of the latter depending on the time and intensity of irradiation and thickness of the silver layer. These characteristics form the basis of high-resolution, grayscale performance of ChG resists.

17.2 Resist materials for lithography

Most lithography processes employ light-sensitive resists. The resolution of these photoresists is limited by the diffraction limit of the laser beam. Therefore, attempts are made to develop lithography processes that employ shorter and shorter wavelength light, with a corresponding optimization of polymer resists (Gates *et al.*, 2005). Considering the difficulty with the focusing of X-rays, there is a practical limit to the resolution using short wavelength light. As a result, the most promising nano-resolution lithography process of the future is likely to be based on electron beam lithography (EBL). Commercial EBL equipment is available, which allows patterning with ~1 nm diameter electron beam. Note that EBL is a relatively slow process, which is more suitable for making master masks than large-scale manufacturing. However, there has been parallel progress in stamping technology, which in combination with such masks may indeed produce features of the size of several nm (Taniguchi *et al.*, 2004).

There is a variety of organic polymers and inorganic materials that are sensitive to radiation, and hence suitable as resists. On structural grounds, inorganic resists may be divided into two categories: (a) glassy resists and (b) polycrystalline resists. Silica glass, chalcogenide glasses, and amorphous peroxopolytungstic acid are examples of the former, whereas polycrystalline fluorides and Al–Al_2O_3 cermets are examples of the latter. In addition, there are resists that include both inorganic and polymer components, e.g. spin-on-glass (SOG), that is a –Si–O polymer with attached methyl groups, and the composite of polymer resist with silica nano-particles (Fedynyshyn *et al.*, 2000). At present, polymethyl methacrylate (PMMA)-based polymer resists containing small amounts of light-sensitive active agents are used in nearly all microelectronic processing. Structures <20 nm in width have been prepared using polymer EB resists, but 100 nm is a more typical dimension of the features manufactured by current technology. The polymer resists are generally applied on the substrate by a convenient, inexpensive spin coating process. The process has a limit of 3:1 for the aspect ratio (image height to width ratio), which can be obtained in a feature (Fedynyshyn *et al.*, 2000). Thus, for fabricating 25 nm features, the thickness of the resist should be <100 nm, which is too small for plasma etch transfer of the pattern into underlying layers. As a result, it appears that the finest features that can be fabricated using polymer resists are ~25 nm, unless new polymers are developed with a much higher selectivity to etching than possible at present. In view of these limitations of the polymer resists, inorganic resists are being considered, especially for high resolution nano-lithography.

The most meaningful and precise judgment on the performance of resists can be obtained from fine line lithography (Ito, 2005). Then the applicability of a resist is determined by a combination of several general and specific requirements. Traditionally, the former include: reasonable price, easy synthesis, good storage stability and low health and environmental hazards. The specific requirements may include: mainly high resolution capability, high sensitivity to specific beam source, selective solubility and high etching contrast in "safe" solvent or gas, and good adhesion to various substrates that are to be structured.

The main resist performance parameters viz. sensitivity (E_0), contrast (γ) and developer selectivity (n) are typically obtained from sensitivity (or contrast) plots, wherein the thickness of the exposed resist film (normalized to the initial thickness) remaining after development is plotted as a function of logarithmic exposure dose (mJ/cm^2 for UV and X-ray resists and mC/cm^2 for e-beam resists). The sensitivity of positive resists is defined as the dose at which the exposed areas are cleanly developed to the substrate with a minimal thickness loss in the unexposed regions. In the case of negative resist, the sensitivity is generally defined as the dose at which 50% of the thickness is retained in the exposed areas. The contrast (γ) is given by the slope of the linear portion of the curves (see Fig. 17.1).

In general, higher contrast leads to higher resolution, which is the Holy Grail of microlithography (Ito, 2005). It should be mentioned that in special situations like grayscale lithography, some deviations from this consideration may apply. For example, the resist with lower contrast is desired to provide the necessary level of smooth variation needed in grayscale structures.

As we will find below, several ChG compositions fulfill well the above-mentioned requirements of a good resist and therefore can be used as highly sensitive, cost-effective high and/or low contrast resists with extremely high resolution for dry and/or wet transfer of micro/nanostructures into desired substrates. Recently, there has been renewed interest in the development of ChG photoresists due to the fact that 3D micro/nano photo-structured chalcogenide glasses are materials of choice for fabricating infrared optical

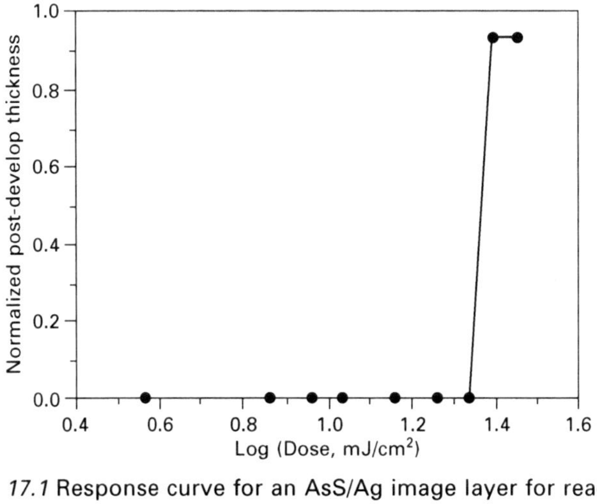

17.1 Response curve for an AsS/Ag image layer for reactive ion etching in CF_4 (Kozicki *et al.*, 1991).

components, optical recording in DVDs and CDs, photonic devices such as waveguides, Bragg gratings, etc.

ChG doped with some metals, such as silver, can also be used as effective resists. In 1971 Sakuma *et al.* published the article 'A new method for producing low-electrical-resistivity patterns in insulating chalcogenide glasses' (Sakuma *et al.*, 1971). It introduced the application of ChG in lithography, based on observation of the unusual phenomenon of photoinduced dissolution of silver discovered by Kostyshyn *et al.* (1966).

17.3 Basics of chalcogenide glass resists

For any material to be useful as a photoresist, it must be sensitive to suitable radiation. Important advantages of ChG for resist applications in micro- and nano-applications arise from the fact that they are sensitive to a broad spectrum of radiation that would produce one or more of the desired effects. They are sensitive to bandgap light, which is usually at visible (VIS) wavelengths. Such photosensitivity makes them suitable for inexpensive contact and proximity microlithography. The same glasses are sensitive also to higher energy radiation, including ultraviolet (UV) light, electron beam, flux of ions and/or to X rays, which allows their application in maskless lithography, nanopatterning (using e-beam lithography, for example) – fabrication of nanostructures well below 100 nm is not difficult. The sensitivity of ChG to various radiations is at least comparable to that of the traditional organic resists; it can be significantly higher when using short pulse laser exposure.

The high sensitivity of a ChG to various radiation sources is a consequence of its significantly weaker and less rigid network structure in comparison to common silicate glasses that are generally perceived as hard, brittle, and chemically resistant. Therefore, its internal structure can be easily modified by exposure to suitable radiation, which in turn changes the optical and physicochemical properties including chemical resistance. Besides the structural modification, the radiation may activate a ChG's interaction with the environment leading to, for example, local photo-oxidation, photo-dissolution or photo-diffusion of certain metals, which then result in even more significant changes in properties due to the fact that they are based on photoinduced compositional changes, and not just the reorganization of given atoms. These characteristics of ChG form the basis of their outstanding application in various kinds of high-resolution/high-sensitivity lithographic processes and other applications in visible (VIS) or IR optics, optoelectronics, high density re-writable data storage, etc.

The following radiation-induced processes may occur in ChG when subjected to suitable radiation (photons, electrons or other charged particles), which can be exploited readily in lithographic applications:

- Structural changes
  - changes in local atomic configuration
  - polymerization or depolymerization
  - phase changes, including crystallization
- Physico-chemical changes:
  - decomposition
  - vaporization
  - enhanced dissolution or diffusion of certain metals
  - thermoplastic changes.

Sensitivity to radiation alone is not sufficient for a solid to be useful as a resist. It should be resistant, at least for a reasonable time, to chemical agents that are to be used for substrate etching and at the same time be amenable for selective dry/wet etching. It means chemical reactivity of exposed parts should be sufficiently modified by radiation relative to the unexposed parts. Then such a glass can be useful as a positive (or negative) resist in which the exposed part etches at a faster (slower) rate than the unexposed part. ChGs are relatively easily etched by alkaline solutions. Most importantly, the etch rate is significantly affected by exposure to suitable radiation, thus providing the basis for ChGs, usefulness as resists. Some chalcogenide glasses are able to function both as a positive or a negative photoresist simply by changing the composition of the wet or dry developer (Kozicki *et al.*, 1991; Vlcek *et al.*, 1998). Considering that most frequently etched substrates in microelectronics (Si, SiO_2, Si_3N_4, Cr) require etching in acidic bath, ChGs with high resistance to acids can be excellent candidate resists for their processing. By tuning the compositions of ChG and etchant solution/gas, very high selectivity of positive or negative etching can be achieved (Jain and Vlcek, 2008). Finally, ChGs offer exceptionally high resolution of the etched pattern, which is perhaps one of their greatest advantages over the classical organic photoresists (Fig. 17.2).

17.3.1 Radiation-induced structural changes

So far, mainly sulfides and selenides of arsenic and germanium binary and/or more complex compositions have been evaluated for photoresist applications with As_2S_3 being the prototypic model composition (Jain and Vlcek, 2008). A few studies have also investigated the effect of doping of these glasses with Ga, In, I_2 or Tl. Arsenic-based glasses are preferred for lithography, where radiation-induced structural changes are exploited. Germanium-based chalcogenide glasses, in which the cations are tetrahedrally coordinated, have more rigid and compact structure than the corresponding As-chalcogenides, wherein As atoms are three-fold coordinated. So the etching selectivity of Ge-ChG is significantly lower than that of As-ChG.

17.2 Scanning electron micrograph of an $As_{35}S_{65}$ nano-grating at a stage tilt of 35° at 5 kV (Neilson *et al.*, 2007).

In principle, stoichiometric bulk As_2S_3 glass in metastable equilibrium should consist of As atoms that are bonded to three S atoms, and the S atoms that are bonded to two As atoms, all atoms being present in relatively stable heteropolar As–S bonds. The selective vaporization during the formation of films, such as the ones used in lithography, introduces structural units with bonding defects. In the case of thermally deposited As_2S_3 thin films, energetically less stable, 'wrong' bonds are formed within structural units, e.g. S–S bonds within S_n chains, S_8 rings, S_2 bridges, etc., or As–As bonds within As_4S_4 and As_4S_3 cages. (Note that the bulk samples of off-stoichiometric compositions will also have less stable homopolar bonds, but in thin film form the concentration of wrong bonds is significantly enhanced.) Depending on light intensity, light energy and sample thermal prehistory, different mechanisms of photostructural changes are possible. Usually, the homopolar bonds are transformed into As–S stable bonds by exposure to light (Kovalskiy *et al.*, 2009a):

$$\text{S–S} + \text{As–As} \rightarrow 2\text{As–S} \quad [17.1]$$

Different concentration of these bonding defects in as-evaporated (unexposed) and exposed films is the fundamental basis of selective etching of As-based chalcogenide glasses. The Raman spectra of as-evaporated thin film ChG (Fig. 17.3) show, beside the main band at 345 cm^{-1} (corresponding to the presence of expected AsS_3 pyramids), very intense bands between 135 cm^{-1} and 234 cm^{-1} and also at 363 cm^{-1}, which can be assigned to the

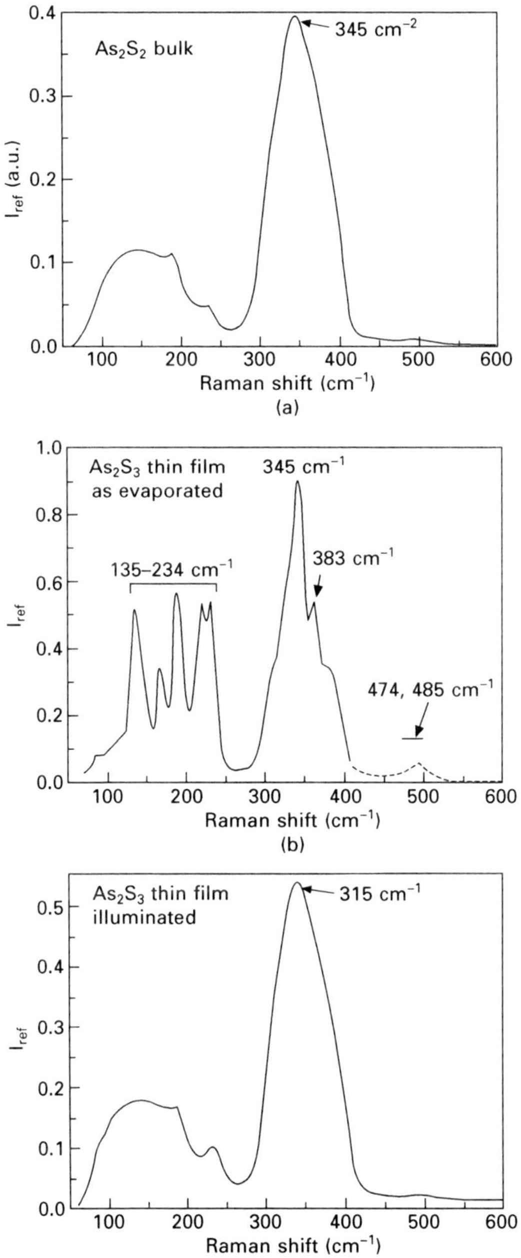

17.3 Raman spectra of As_2S_3 samples: (a) bulk, (b) as-evaporated thin film, and (c) thin film after 30 min exposure with halogen lamp (Kovalskiy *et al.*, 2009b).

vibration of As-As containing units, such as As_4S_4. Additional relatively strong double bands with maxima at 495 cm^{-1} and 474 cm^{-1} provide evidence of S-S containing structural units such as –S-S– chains (495 cm^{-1}) and S_8 rings (474 cm^{-1}) that are present as such or connect individual AsS_3 pyramids.

The concentration of atomic defects, and hence potential for radiation-induced structural changes, in as-prepared Ge-ChG is relatively small. Unlike arsenic-based ChG, where structural changes occur via defect reaction like in Eq. [17.1], the photosensitivity of Ge-based ChG films by this mechanism is inadequate for lithography applications. This is why for high-selectivity and high-resolution applications of Ge-based resists, one exploits radiation-induced physicochemical rather than simple structural changes. Specifically, radiation-induced dissolution of silver is exploited as it combines the advantages of both radiation-induced structural and compositional changes.

17.3.2 Radiation-induced compositional changes

The photodissolution (or photodoping) of some metals (Ag, Cu, Zn, In) into ChG thin films is caused by light-, electron-, X-ray- or ion-beam exposure. In the case of light, it takes place at band gap or higher photon energies. Among various metals, Ag dissolves the fastest and its mechanism has been investigated and debated in hundreds of publications in the past three decades. Basically, this phenomenon is understood as an intercalation chemical reaction, where silver diffuses as positively charged Ag^+ ions driven by internal electric field (or counterflow of holes; Tanaka, 1991), formed during the irradiation of Ag/ChG bi-layer. The process is characterized by a step-like front with Ag concentration at 25–40%, depending on the composition of ChG, and takes place with an Ag layer on top of ChG film or below it. There are several comprehensive review papers discussing the possible mechanisms and applications of this effect (Kolobov and Elliott, 1991; Wagner, 2002; Frumar and Wagner, 2003; Kluge, 1987). The following facts about silver photodoping are generally accepted:

- The thickness of ChG film is usually 100–1,000 nm and that of the metallic layer below 100 nm.
- The depth of Ag penetration can be quite significant and may even reach 20 μm (e.g., in $Ag/As_{16}S_{80}Te_4$; Kolobov, 2003).
- Energy sensitivity of the radiation is in the range of $1–10^2$ cm^2/J.
- The 'metal Ag–doped ChG' and 'doped ChG–undoped ChG' interfaces are the photoactive areas during photodoping.
- There is superposition of light-induced and temperature-induced effects.

The temperature dependence of sensitivity, S, is of the following form (Kostyukevych, 1998):

$$S(T) = S_0 \exp(-E_a/kT) \quad [17.2]$$

where E_a is the activation energy of the process (E_a for photodoping is much smaller than for thermal diffusion).

The rate of photodissolution depends on light intensity and time of irradiation. Effectiveness of the photodoping depends also on the chemical composition of ChG, parameters of film preparation, prehistory of ChG film (preliminary illumination, annealing or oxidation), light wavelength and intensity, etc. (Kolobov and Elliott, 1991). A summary of the effectiveness of photodoping for different ChG compositions and metals is given in Table 17.1 (Tanaka and Shimakawa, 2009).

Evidently, photoinduced dissolution is an effective mechanism to change local chemical composition, and hence the resistance of ChG thin films to etching with high resolution, which makes it very attractive for lithographic applications (Jain and Vlcek, 2008; Lyubin, 2009; Mizushima and Yoshikawa, 1982; Yoshikawa *et al.*, 1976; Janai, 1981), grayscale patterning (Kovalskiy *et al.*, 2009b), etc. Thus, for example, for 1 μm $Se_{80}Ge_{20}$ thin films a 10 nm thick Ag layer is sufficient to make the former insoluble in alkaline solution after photodissolution (Yoshikawa *et al.*, 1976). For lithography purposes, ChG is deposited initially on the substrate, followed by the Ag layer. The reverse order is adopted for holographic applications (Indutnyi *et al.*, 1995a).

The kinetics of photodissolution, which is important for lithography applications, depends on the conditions of sample preparation (e.g., Ag layer on top or underneath, contact with oxygen between depositions of separate layers, etc.), type and parameters of irradiation, and the metal being photodoped (Goldschmidt and Rudman, 1976; Kolobov and Tanaka, 2001; Popescu *et al.*, 2009). Three stages of photodissolution are usually reported when ChG film is exposed to air before Ag layer deposition. At the first stage (induction period) the photodissolution is very slow, if at all. The duration of this induction period is approximately inversely proportional to the intensity of irradiation (Fig. 17.4) and is often associated with the formation of an intermediate oxidation layer at the Ag/ChG interface. However, some data on Zn photodoping contradict this conclusion and report the existence of

Table 17.1 Efficiency of photo-reaction between chalcogenide films and metallic layers (Tanaka and Shimakawa, 2009)

Composition	Cu	Ag	Zn	Cd	Ga	In	Bi	Te	Cr	Ni
As_2S_3	Medium (M)	Good	–	M	M	–	M	M	–	M
As_2Se_3	M	M	–	M	–	M	M	–	–	M
As_2Te_3	–	–	–	–	–	–	–	–	–	–
Sb_2S_3	–	–	–	M	–	–	–	–	–	–
Sb_2Se_3	–	–	–	–	–	–	–	–	–	–

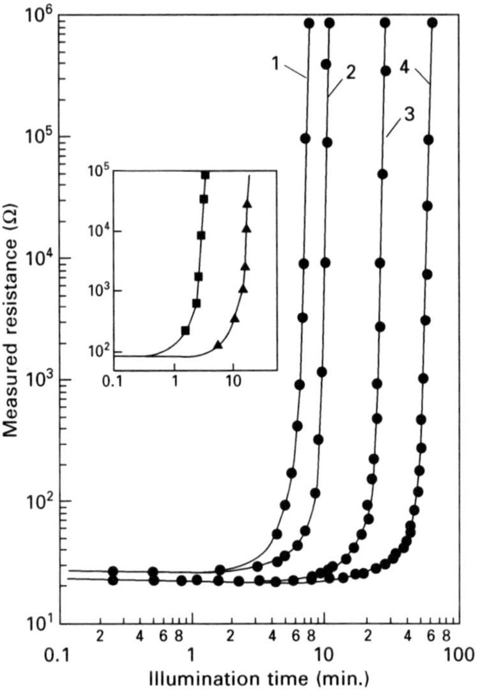

17.4 Kinetics of Ag dissolution into 150 nm thick As_2S_3 film (λ = 435 nm; light intensities: 1, 5.51 mW/cm^2; 2, 3.84 mW/cm^2; 3, 1.34 mW/cm^2; 4, 0.61 mW/cm^2). Inset: effect of ChG pre-irradiation on dissolution rate (▲, as-evaporated film; ■, exposed film) (Kolobov and Tanaka, 2001).

an induction period in samples that were not exposed to air (Arsh *et al.*, 2002). Our observations of kinetics on X-ray-irradiated Ag/As_2S_3 (deposited and characterized *in situ* within X-ray photoelectron spectroscopy (XPS) spectrometer without any contact with air) did not show any induction period. Furthermore, we noticed that the initial stage has a higher dissolution rate than the subsequent stage of the process (Fig. 17.5).

The second stage of silver photodoping is characterized by high dissolution rate with time, which strongly depends on the irradiation dose (light intensity and irradiation time) (see Fig. 17.4). Linear time dependence of the photodissolution rate is typical for photoinduced chemical reaction (Yaji and Kurita, 1983). At the same time, square root time dependence is a characteristic of diffusion controlled processes. Both forms have been reported (Kostyshin and Ushenin, 1982), leaving open the true nature of this stage.

Metallic Ag was thermally deposited onto ChG film in the earliest photodoping experiments. However, later it was revealed that thermal deposition itself contributes to photodoping of the deposited metal due to the

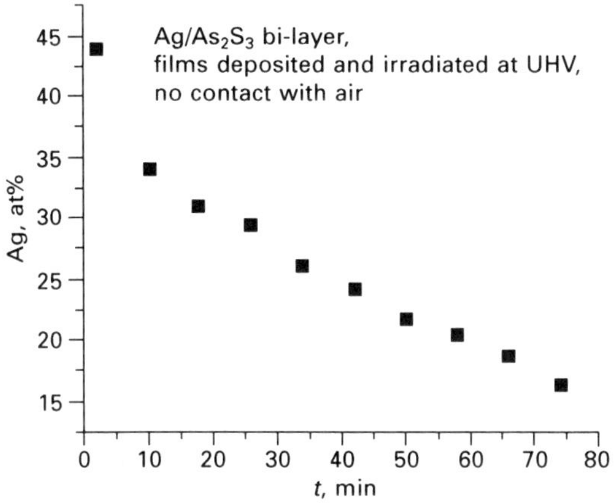

17.5 XPS data on kinetics of X-ray induced Ag dissolution into As_2S_3 thin film (ChG thickness, ~150 nm, Ag thickness, 3 nm) (Kovalskiy *et al.*, 2009b).

light emitted from evaporation source. This continuously doped layer becomes a significant problem for lithography, especially in the case of wet etching procedure, causing decreased selectivity (solubility ratio of the exposed and unexposed parts). To circumvent this problem, alternative approaches are adopted including the use of silver halide instead of metallic silver (Kolwicz and Chang, 1980) or dipping of ChG film in an aqueous solution of $AgNO_3$ (Masters and Goldberg, 1981).

For the As-S/Ag system, it was pointed out that Ag_2S and $AgAsS_2$ compounds form as a result of photodiffusion depending on ChG composition (Wagner *et al.*, 1993). The latter is a known ternary composition within the glass formation region (Fig. 17.6) and can be considered as the most probable product of the reaction. For the Ge-Se system, the chemical reaction of photodiffusion strongly depends on the chemical composition of the ChG.

17.3.3 Etching of chalcogenide glass resists

Both wet and dry etching techniques, as described next, are used for patterning ChG-Ag photoresists.

Wet etching

As mentioned above, mostly As-based ChG are used when exploiting radiation-induced structural changes for accomplishing high etching selectivity. For example, consider the thin film of As_2S_3 as a model, which is deposited on a silicon or glass substrate by vacuum evaporation. Experiments show that high selectivity, positive wet etching can be achieved in stoichiometric

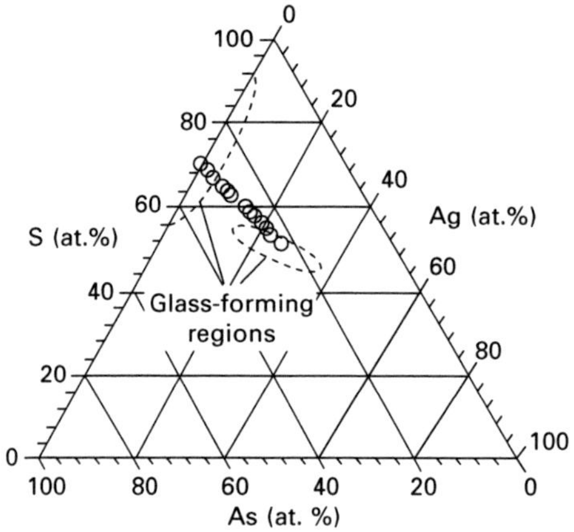

17.6 Ternary Ag–As–S diagram showing composition of the prepared Ag–As–S films together with glass-forming regions (Wagner *et al.*, 2001).

and As-rich compositions. By contrast, negative wet etching is observed in stoichiometric and chalcogen-rich compositions. Thus the etching mechanism of stoichiometric As_2S_3 is basic to understand the origin of positive and negative wet etching of As-S films.

The presence of As_4S_4 units in an as-prepared film lowers its dissolution in alkaline medium. The positive photo-etching results from the lower solubility of the As-rich clusters relative to As_2S_3, and the fact that an insoluble As thin film is formed on the surface of the layer during etching according to Russo *et al.* (2005):

$$3As_4S_4 + 24OH^- = 4AsS_3^{3-} + 4AsO_3^{3-} + 4As + 12H_2O \qquad [17.3]$$

The presence of atomic units or clusters containing hydrophobic S-S bonds further reduces the dissolution rate of as-evaporated films. By comparison, the exposed samples are chemically closer to the ideal, stoichiometric structure and hence dissolve easily according to:

$$As_2S_3 + 6OH^- = AsO_3^{3-} + AsS_3^{3-} + 3\ H_2O \qquad [17.4]$$

Thus positive-type selective etching takes place when previously exposed As_2S_3 film is etched in aqueous alkaline solvents. This mechanism is supported by the results of Raman spectroscopy. We note the disappearance of bands at 364 cm^{-1} and 167–235 cm^{-1} representing As-As bonds, and the 495 cm^{-1} band which arises from the presence of S-S bonds in the glass structure (see Fig. 17.3). Similar behavior is found also for non-stoichiometric films of the As-S system, where structural units with homopolar bonds are expected in both as-prepared and exposed films, but in different concentrations. Since primarily the presence of As_4S_4 units in the structure of as-prepared films is responsible for positive selective etching, it is no surprise that selectivity of etching is high mainly in As-rich films, with values as high as $\gamma \approx 15$–20. By comparison, for S-rich samples it is much less, for example for $As_{35}S_{65}$, $\gamma \approx 3$ (Russo *et al.*, 2005).

An increase in the selectivity of positive etching of As-based films in aqueous alkaline solvents is observed when surface active substances (SAS) are added in the etching bath. Due to its commercial importance, limited information is published on this process and precise composition of SAS is not described (Vlcek *et al.*, 1998). We suspect that for enhancing the selectivity of etching of As-based ChG in alkaline solvents, an SAS molecule must be a positively charged unit (or cation) and contain long hydrophobic chains as well. The positively charged part of the molecule would clamp on to 'active' negatively charged defects on the film surface and the hydrophobic chain would protect it from OH^- ions in the etching bath. The selective interaction of cationic SAS mainly protects the surface of as-evaporated, unexposed regions, and thus results in improved positive etching.

Furthermore, the interaction of SAS with As_4S_4 fragments on the surface of film would reduce its etching rate. Specifically, the As atoms in As_4S_4 structural units can be protective via chemical absorption and/or coulombic interaction. For the first possibility, note that As atoms in an As_4S_4 fragment have lower oxidation state (II) than the atoms in As_2S_3 structural units (III). They appear as stronger Lewis base when interacting with the cations of SAS and form a bond preferentially with these structural units. Thus, the cations from SAS adsorb practically only on the unexposed region of the ChG film and prevent it from etching. Alternatively, the As atoms in lower oxidation state in As_4S_4 with higher negative charge would have a stronger coulombic attraction to the positive charge centers of SAS than those in polymeric As_2S_3 fragments created during the exposure. From our experiments, we have found quaternary nitrogen (i.e., nitrogen atom connected by covalent bonds to four different functional groups) as a possible positively charged center within the SAS molecule, which effectively clamps on to As_4S_4. Increased selectivity of etching is then a result of the fact that hydrophobic long chains of SAS molecules span out to the etching bath and repel OH^- ions in the etchant solution.

When the stoichiometric As_2S_3 (or sulfur-rich As-S) films are etched in non-aqueous, mainly amine-based solvents, the consequence of photo-induced structural changes is just the opposite. In this case, a negative resist behavior is observed with the exposed part having lower solubility than the exposed regions. This behavior, opposite to that in aqueous alkaline medium, can be explained following Zenkin *et al.* (1997). Accordingly, etching of As-based chalcogenide films in non-aqueous solvents is a kinetically controlled process. The ultimate composition of the etching products is a function of the rate of all elementary processes. As discussed above, the structure of exposed stoichiometric films contains mostly heteropolar As-S bonds as in bulk As_2S_3. Such a structure dissolves in amines via slow ammonolysis of heteropolar As-S bonds by the following reaction (Zenkin *et al.*, 1997):

$$As_2S_3 + 6(C_2H_5)_2NH = [(C_2H_5)_2NH_2]_3AsS_3 + As[(C_2H_5)_2N]_3 \quad [17.5]$$

By comparison, the structure of an as-prepared film also contains a high concentration of fragments with homopolar bonds, such as S_n, As_4S_4, As_4S_3, etc. Its dissolution has other, more complex mechanisms, which can be described as a breakup of the polymeric network at homopolar bonds according to the following reactions (Zenkin *et al.*, 1997):

$$(C_2H_5)_2NH + S_n = (C_2H_5)_2NH^+S_n^-, \quad [17.6]$$

$$(C_2H_5)_2NH^+S_n^- + As_2S_{4/2} = (C_2H_5)_2NH_2^+S^-AsS_{2/2} + (C_2H_5)_2NAsS_{2/2}. \quad [17.7]$$

According to Zenkin, the processes described by Eqs [17.6] and [17.7] are much faster than ammonolysis of heteropolar As-S bonds (Eq. [17.5]) and thus the unexposed part of films is etched at a faster rate. The deconvolution of Raman spectra of as-prepared As-S films has shown that even up to $As_{33}S_{67}$ sulfur-rich composition, the as-prepared films contain As_4S_4 structural units that participate in etching by the steps described by reactions [17.6] and [17.7]. The concentration of As_4S_4 is significant even in these films that have even better selectivity of etching in amine-based solvents than stoichiometric $As_{40}S_{60}$.

Similarly selective etching can be achieved in Ge-based chalcogenide glasses, albeit their photosensitivity is generally lower in comparison with As-based glasses (requiring longer exposures/higher doses), and lower selectivity of etching is achieved. However, as in the case of As-based glasses, the addition of suitable SAS can improve selectivity and/or even change the character of etching. For example, for one of the most photosensitive and selective etching capable glasses in the Ge-Sb-S ternary system with a composition $Ge_{30}Sb_{10}S_{60}$, one observes negative etching in aqueous KOH solutions (Vlcek and Sklenar, 2002). By the addition of some specific SAS, such as rhodamine, the character of etching switches and becomes positive.

In the case of ternary systems, the mechanism of dissolution itself and especially the role of SAS are very complex, involving multiple parameters. The conditions of thin film preparation and thermal prehistory and/or pretreatment of thin films before exposure can also play a significant role. Kumar *et al.* (1987), for example, reported that the solubility difference of Ge-Sb-Se films (negative etching) can be enhanced by increasing the angle of film deposition. For the As_xS_{100-x} system it was shown that the etching of the layers in aqueous alkaline solvents was positive, but dipping the films with $x \geq 40$ into CS_2 resulted in negative etching in the same aqueous alkaline solvents (Vlcek *et al.*, 1994). Negative etching of Ge-Sb-S thin films in aqueous KOH solutions can be switched to positive etching by pre-annealing the films at a temperature close to their T_g (Kumar *et al.*, 1987).

Ag-chalcogenide materials are known to be resistant to alkaline aqueous or amine-based non-aqueous etchants, which normally dissolve pure chalcogenide thin films. The mechanism of wet etching for pure ChG at different etchants is described above. However, there are very few publications on the mechanism of interaction of Ag-doped phase with the etchants. In one study the kinetics of $AgAsS_2$ and $AgGeS_2$ glasses in alkaline solutions ($NaOH + NaNO_3$) is considered (Shkol'nikov, 2001). It is concluded that the etching of glass proceeds with the gradual formation of a dark deposit (Ag_2S precipitate) on the sample surface, which inhibits further dissolution. As a result, mechanical removal of the precipitate from the surface is needed every six hours to proceed with etching.

Evaluation of different etchants (NaOH, Na_2S, $(NH_4)_2S$, Na_2CO_3 and NaCN) showed that the only effective etchant solution for Ag-doped ChG thin film is NaCN (Orava *et al.*, 2006). The selective wet-etching characteristic of three $Ag_x(As_{0.33}S_{0.67})_{100-x}$, $Ag_x(As_{0.33}S_{0.335}Se_{0.335})_{100-x}$, $(As_{0.33}Se_{0.67})_{100-x}$ compositional series in alkaline aqueous solution of NaCN is depicted in Fig. 17.7. (Orava *et al.*, 2006). Interestingly, a gradual increase of the Ag content in $Ag_x(As_{33}S_{67})_{100-x}$ system changes from negative patterns to unique positive analog at Ag concentration of ~2.5 at.%. It was established that

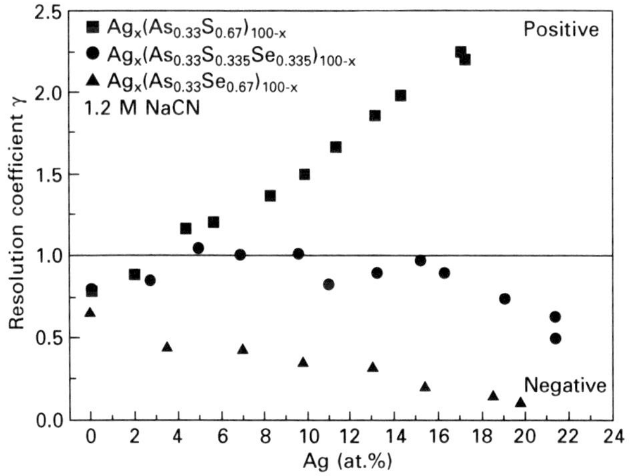

17.7 Contrast coefficient γ of $Ag_x(As_{0.33}S_{0.67})_{100-x}$, $Ag_x(As_{0.33}S_{0.335}Se_{0.335})_{100-x}$ and new possibility of selective etching of $Ag_x(As_{0.33}Se_{0.67})_{100-x}$ films in aqueous solution of 1.2 M NaCN (Orava *et al.*, 2006).

the films of $Ag_x(As_{0.33}S_{0.67})_{100-x}$ remain positive resist with increasing Ag content. However, $Ag_x(As_{0.33}S_{0.335}Se_{0.335})_{100-x}$ behaves as positive resist up to ~10–12 at.% of Ag and then turns to negative resist behavior. Similar 'switching' of resist behavior was revealed recently for dry etching of Ag/As_2S_3 thin films (Kovalskiy *et al.*, 2009a). This switching of resist response may be explained by the change of role of silver in the matrix (Yoshida and Tanaka, 1998). Up to ~10 at.% silver prefers to bind with the ends of sulfur or selenium chalcogenide chains, i.e. the Ag ion plays the role of glass modifier. At higher concentration, silver breaks the structure consisting of chains and/or AsS_3, $AsSe_3$ pyramids, and connects them with new linkages such as AsS_3 or $AsSe_3$ units connected by silver –S–Ag–S–, –Se–Ag–Se–. After that, silver plays the role of a 'glass forming element' which makes the structure stronger (Yoshida and Tanaka, 1998).

As a practical tip, note that the measurements of selective etching of $Ag_x(As_{0.33}Se_{0.67})_{100-x}$ thin film are made possible using hexamethyldisilazane as an adhesive layer. Without this layer the adhesion of Ag-doped $As_{33}Se_{67}$ to glass substrate is very poor and therefore any previous attempts at selective etching were unsuccessful. With this addition, the etching can be performed up to silver content of ~20 at.% in films. Higher contents of Ag lead to poor adhesion of the thin film to substrate during wet etching (Orava *et al.*, 2006).

Dry etching

In principle, dry reactive ion etching (RIE) can allow all stages of the lithographic process to be performed under high vacuum in ultra-clean conditions. Therefore, often an all-dry process is desired from the manufacturing point of view (Kovalskiy *et al.*, 2006). Generally, CF_4 and $SF_6 + O_2$ plasmas are used for ChG or ChG-Ag binary system patterning, which behave as negative resist (Kostyukevych, 1998; Tai *et al.*, 1980). Notwithstanding, a way to obtain positive resist with RIE is proposed, where the $As_{33}S_{67}$-Ag structure is dry etched in S plasma (Kozicki *et al.*, 1991).

High etching rates coupled with precise profile control and smooth surfaces are the main expectations from RIE. However, so far the RIE processes for ChG photoresists have not been optimized or studied sufficiently. Previous results of RIE of ChG without silver photodiffusion show that good wall inclination angles (70° for CF_4/O_2 and 60° for CF_4 atmospheres, respectively) can be obtained (Balan *et al.*, 2003). In this case, the etching rate was twice as fast as for the other etching techniques, and the etched part of the upper layer was rougher than the non-etched part. To transfer etch patterns into the Si substrate, the structures etched in ChG photoresists must have high selectivity of etching. In principle, this can be achieved by incorporating Ag photodiffusion into the lithography process (Kovalskiy *et al.*, 2006). The quality of structures obtained in our laboratory depended on the parameters of sample preparation, light exposure and dry etching (Kovalskiy *et al.*, 2006, 2007). The longer the time of exposure (or the higher light intensity), the deeper were the etched structures. However, the increased etching depth also enhanced the 'grass' roughness of the etched surface.

The smooth microstructures of varying forms in the ChG film can be transferred into Si by dry etching due to higher chemical resistance of the glass in comparison with polymers used in traditional lithography. The first example of such transfer of grayscale patterns was presented in 1988 at AT&T Bell Laboratories (Vadimsky, 1988). Subsequently, high-quality grayscale microlens arrays for IR application have been fabricated utilizing Ag photodoping into ChG thin films (Eisenberg *et al.*, 1996).

The exploratory experiments to transfer the grayscale patterns into Si substrate via RIE have given very encouraging results with an etch selectivity of 10:1 (etch rate in Si/etch rate in ChG film), which is a large increase over conventional polymer masking materials with an etch selectivity in parallel-plate RIE below 2:1 for anisotropic etch profiles using a similar etch recipe (Kovalskiy *et al.*, 2009b). The etch selectivity with ChG-Ag photoresist is likely to be further improved by the optimization of surface composition and RIE parameters, and decreased silicon loading by masking the carrier wafer. The higher etch selectivity of ChG-Ag films over traditional polymer films is believed to be due to the significant increase in chemical resistance from Ag photodiffusion, which reduces the rate of physical sputtering by the energized ions responsible for RIE.

17.4 Examples of chalcogenide resist applications

17.4.1 Photoresists

Optical gratings using photostructural changes

Amorphous As-S, As-Se and As-S-Se films have been used successively for photolithography in the visible spectral region with high resolution (about 5,000 lines/mm) and satisfactory light sensitivity (about 10 mJ/cm^2). As an example, gratings with a period down to 370 nm prepared in As-based films using interference lithography technique on 514 nm (Fig. 17.8) or micro-optical devices for IR such as cylindrical micro-lens array (Eisenberg *et al.*, 2006) have been demonstrated on these materials.

Fresnel lens array using photostructural changes

For some applications in Si technology such as focal plane arrays for energy concentration or remote sensing, ultrathin (~600 nm) Fresnel lenses are

17.8 SEM picture of sinusoidal grating with a period 2 μm (a) and crossed grating with a period 370 nm (b) formed in As-based chalcogenide thin layers. (b) Fresnel lens array using photostructural changes (Vlcek *et al.*, 2003).

desired (Kovalskiy *et al.*, 2009b). They should be capable of high focus (small focal length) with wide aperture (diameter of Fresnel lens) within a thin layer of material. The suitability of ChG photoresists was demonstrated with inexpensive and reliable fabrication of ultrathin Fresnel lenses that are transparent in the visible as well as in the IR region (Kovalskiy *et al.*, 2009b, Eisenberg *et al.*, 1996). The pictures of the individual Fresnel lens of 400 μm diameter focused on the surface (a) and at the focal point (b) are shown in Plate XVI (between pages 330 and 331) (Kovalskiy *et al.*, 2009b). The focal length of the lenses is 5.4 mm. The focal length for the Fresnel lenses within the 480 × 480 μm^2 arrays Plate XVII (between pages 330 and 331) is measured to be ~300 μm (Kovalskiy *et al.*, 2009b).

Holographic gratings using silver photodissolution

Significant change of the chemical reactivity of the ChG thin films after silver photodissolution was first reported by Shimizu *et al.* (1973). The procedure for the development of lithographic masks based on Ag-ChG photoresists was proposed by Yoshikawa *et al.* (1976). Its key steps are presented in Fig. 17.9 (Jain and Vlcek, 2008).

Holographic diffractive gratings are the most common features patterned on ChG-Ag photoresists (Wagner *et al.*, 2003; Indutnyi *et al.*, 1995b). The relief depths of gratings recorded on such structures and their groove profiles are determined by the thickness distribution of the photodoped layer. The Ag layer must be thick enough to provide the necessary relief depth but thin enough to avoid the loss of light intensity due to reflection. For grating fabrication, post-exposure treatment includes removal of undoped ChG by selective etching and, usually, refractive Al coating (Indutnyi *et al.*, 1995b). Typical gratings have 1,200 to 3,600 lines/cm and diffraction efficiencies up to 85% in polarized light and 60% in unpolarized light. The ChG-Ag structures are also suitable for fabricating diffraction gratings with asymmetrical (blazed) groove profiles. Grating writing on Ag-doped ChG can be realized also using electron beam irradiation (Nordman *et al.*, 2001; Nordman and Nordman, 2001), as described in Section 17.4.2.

Photonic crystals using silver photodissolution

Complex photoresist composition $As_{40}Ge_{10}Se_{15}S_{35}$-Ag was proposed for fabricating planar corrugated one-dimensional photonic crystal (1D PC) with a period of 570 nm by holographic method (Lee and Yao, 2002). Such 1D

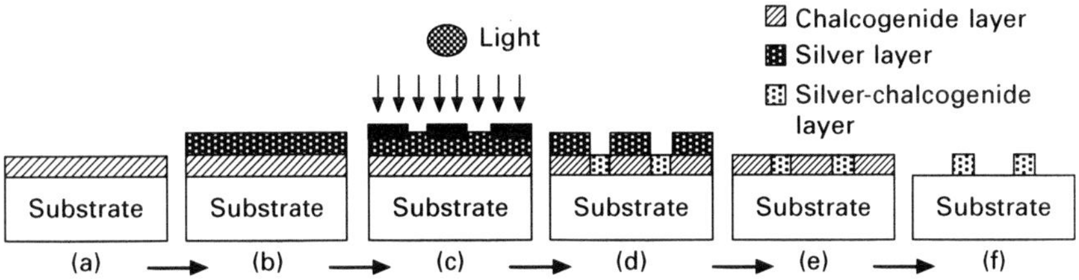

17.9 Major steps of Ag photodissolved ChG photoresist preparation: (a) deposition of chalcogenide layer; (b) deposition of silver layer; (c) exposure through mask; (d) silver diffusion; (e) removal of remaining silver by dissolving in acid or dry etching and (f) removal of ChG regions by wet or dry etching to create photoresist (Jain and Vlcek, 2008).

PCs can totally reflect arbitrary polarized light at all angles within a given frequency region. Since such 1D omni reflection can be realized with simple planar deposition techniques or conventional lithographies, it is directly applicable for conventional planar integration techniques.

It was shown that HeCd laser illumination at low temperatures (77 K) significantly improves the sensitivity and contrast of negative ChG resist, amplifying edge-sharpening effect (Indutnyi *et al.*, 1991). This feature can be especially useful in fabricating a fine planar 1D PC structure where more stepped shape of holographic gratings is desired. Such photodoping-holographic method has potential for the fabrication of 2D PCs too.

Strip waveguides using silver photodissolution

Lithographic fabrication of strip waveguides is a promising direction for the application of ChG-Ag structures (Balan *et al.*, 2003; DeCorby *et al.*, 2005). Highly anisotropic and vertical structures can be realized in this case. Metal doping of ChG produces large index change ($\Delta n > 0.3$ is possible). Another motivation to study photodoped waveguides is that Ag-doped glasses are a potential medium for third-order nonlinear optics having ultrafast Kerr nonlinearities – three to four orders of magnitude higher than for silica glass. The multimodal strip waveguides based on As_2Se_3-Ag structures have been patterned using a liftoff process (DeCorby *et al.*, 2005). These structures show strong guidance of 980 nm light, however, losses at this wavelength are high, e.g. close to 4 dB/cm.

Nanopatterns using electrical field-induced silver diffusion

With the application of scanning tunneling microscope (STM) lithography with two tip-scanning methods (vector-scan and raster-scan), it is possible to write complicated nanopatterns on Se-Ag resist (Utsugi, 1992). In the vector scan mode, the tip follows a zigzag path to fill in certain areas. However, for this mode there is a danger of a double scan when writing circular patterns and the effect of a strong field during tip jumping from pattern to pattern must be considered. In the raster scan mode, the feedback loop during periods of pulse application must be shut down to ensure that the gap between the tip and the surface remains nearly constant. The depth of grooves is changed by varying the input pitch of pulses (Utsugi, 1992).

Mask discs and tracking guides of optical discs using silver photodissolution

ChG-Ag thin film structures can be used also for the preparation of mask disks and tracking guides of optical disks for high-density information storage

(Indutnyi *et al.*, 1995a). By choosing the ChG with different optical band gap E_g^{opt}, it is possible to obtain registering media with sensitivity to UV, VIS and NIR parts of spectrum. The advantages of this method are (Kostyoukevych *et al.*, 1995): (i) increased light sensitivity of the media under the influence of sharp focused laser irradiation; and (ii) paths with substantially smaller width than the size of the laser light spot. These features are introduced presumably by the local heating of the resist under irradiation with laser beam of high power density.

Formation of tracking guides for optical disks includes deposition of 100 nm Cr layer on the disk substrate, 100–300 nm thick ChG film (As_2S_3 or $GeSe_2$) and 20–60 nm Ag thick coat on the top (Indutnyi *et al.*, 1995a). After treatment in selective etchant, two variants are possible:

1. etch the underlying Cr layer to obtain a Cr photomask of the tracking guides that can be used under lithographic copying;
2. use the pattern on the resist layer for copying, without etching the Cr layer.

Integrated circuits using RIE

A planarized arsenic/sulfur/silver lithographic scheme has been proposed for fabricating integrated circuits (Kozicki, *et al.*, 1991). Thus, for example, ChG is deposited onto a spun layer, which is not photoactive. CF_4 plasma is used to etch negative resist. Photo- or electron-beam irradiation is utilized for Ag photodoping. If electron-beam lithography provides a linewidth of ~35 nm, the maximum achievable contrast at photodoping is close to 13 in 100 nm $As_{33}S_{67}$ with 30 nm of Ag structures (see Section 17.4.2). It is concluded that the ChG layer should be 3–5 times thicker than the Ag layer. Use of shorter wavelength λ is suggested to avoid reflection from the surface. It is also noted that having Ag on the top has the additional benefit of reducing charging during e-beam exposure. A problem for the fabrication of integrated circuits arises from the fact that higher numerical aperture lenses provide better resolution but also result in decreasing depth of focus.

Grayscale MEMS structures

Silver photo-diffusion can be exploited for fabricating grayscale mask for micromachining of advanced micro-electromechanical systems (MEMS) and microelectronic technologies (Kovalskiy *et al.*, 2007). For such applications, lithographic masks with gradual change of transparency should be used. Grayscale lithography in ChG is based on the fact that not only exposure time, but light intensity too affects the etching rate. The typical change of etching depth with gradual variation of transparency is shown in Fig. 17.10.

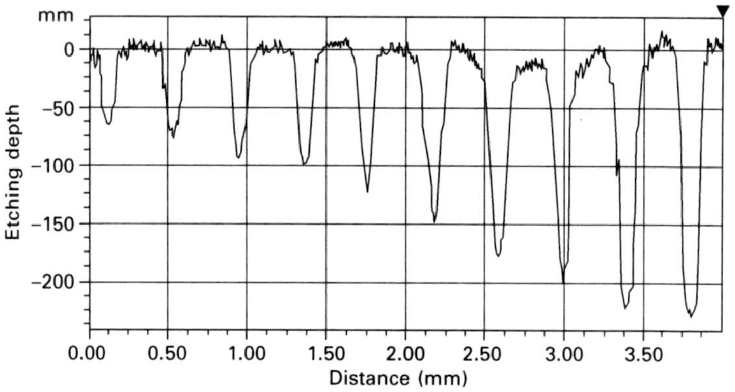

17.10 Profilogram demonstrating the change of etching depth with gradual variation of transparency within a mask (Kovalskiy *et al.*, 2007).

This dependence is nearly linear, thus demonstrating a uniform diffusion of silver to different depths that depend simply on the dose of irradiation. So a smooth shaping of more complex grayscale patterns in the optical mask can be accomplished.

A recent example of a complex grayscale structure is shown in Fig. 17.11 (Kovalskiy *et al.*, 2009b). Here, ChG thin film was irradiated with a halogen lamp through a HEBS5 calibration plate containing grayscale features with gradual smooth change of transparency as well as lens-like structures. After exposure, the residual undiffused silver was removed from unexposed parts using dilute HNO_3. Dry etching was carried out with a parallel-plate RIE system using CF_4 as the etchant gas. The figure shows a section of complex grayscale structures, which was obtained by negative dry etching.

17.4.2 Electron beam resists

Due to the fact that the chalcogenide glass network structure is made of small molecular units (rather than long chains), extremely high resolution can be expected. However, it cannot be exploited using visible or UV light due to diffraction limitations. On the other hand, it can be exploited using the sensitivity of ChG to electron beam (e-beam) exposure (Neilson *et al.*, 2007). Figure 17.2 shows an example of one of the finest structures written in a glass, consisting of 27 nm wide and 80–90 nm tall lines separated by just ~7 nm. It was wet etched in an $As_{35}S_{65}$ thin layer after electron beam exposure.

An additional advantage of electron beam sensitivity of ChGs in combination with their ability to be etched selectively is the linear relationship between exposure dose and increase of chemical resistance of exposed glass to wet etching. Such linear dependence is an important consideration for ultrafine

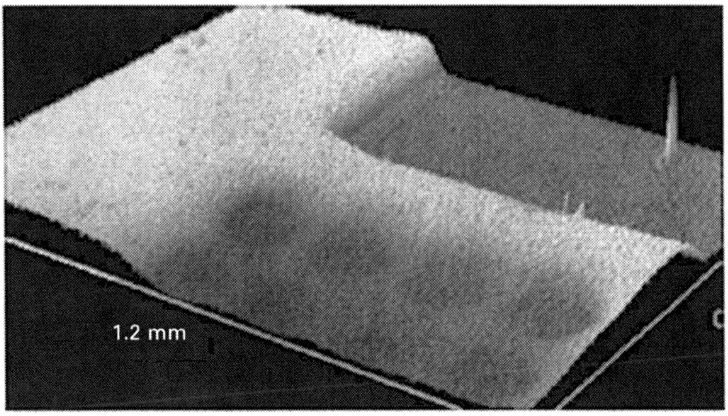

17.11 Complex grayscale structure, obtained from Ag/As_2S_3 layer by negative RIE in CF_4 plasma (Kovalskiy *et al.*, 2009b).

e-beam grayscale lithography. To establish the fidelity of patterning and the ability to form grayscale features, a wheel-shape structure, as shown in Fig. 17.12 was formed (Vlcek and Jain, 2006). The radial lines were written following the circle. Note the greater height of etched structure due to higher electron dose at the center, where the lines intersect each other. Thus the height of the etched structure depends linearly on the dose of electron irradiation (Fig. 17.12). It is also expected to yield better control of the geometry than is possible with traditional organic polymer resists.

The mechanism of e-beam induced structural transformations strongly depends on the dose of irradiation (Vlcek and Jain, 2006; Kovalskiy *et al.*, 2008). For example, study of S-rich $As_{35}S_{65}$ thin films shows that low dose electron beam irradiation (~10^{-3} mC/cm^2) activates the surface, which contributes to the oxidation of As atoms in 'wrong' As–As bonds, thereby decreasing the S/As ratio from 1.92 to 1.83 in the very top parts of thin ChG film. Electron irradiation with higher doses (~ 5 mC/cm^2) destroys the surface of the film. On subsequent exposure to air, the under-coordinated As atoms react with oxygen and decrease the S/As ratio dramatically from 1.92 to 1.54, forming a distinct As-enriched surface layer (Kovalskiy *et al.*, 2008).

The scattering trajectories of the incident electrons in As-S film obtained from Monte Carlo simulations are presented in Fig. 17.13 (Neilson *et al.*, 2007). We can see from Fig. 17.13(a) that for the trajectories solely in the thickness of the As–S film, the radius representing the laterally scattered beam trajectories is between 5 and 15 nm, which is comparable to the resolution seen in the grating in Fig. 17.2. The total interaction volume of the incident electrons at a much larger scale is shown in Fig. 17.13(b) – the electrons, which scatter back through the chalcogenide film, have a lateral radius of ~2 μm. So, for fabricating the highest resolution structures by e-beam, the critical

17.12 Three-dimensional AFM picture of 'nanowheel' and detail of its center, where 'nanospokes' intersect each other (Vlcek and Jain, 2006).

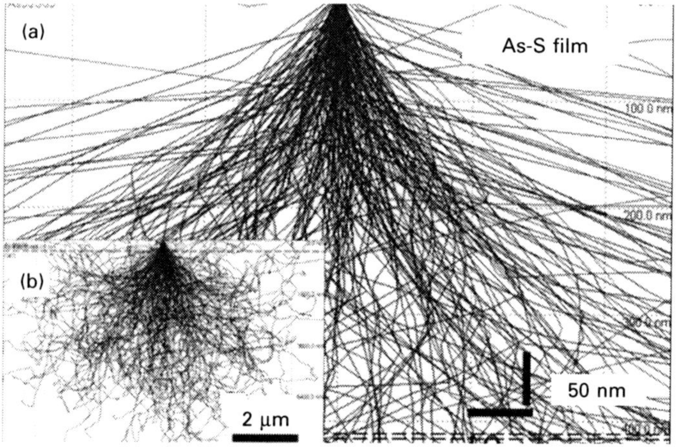

17.13 Incident electron trajectory obtained from Monte Carlo simulation at 30 kV of 400 nm $As_{35}S_{65}$ on 5 nm Cr on a SiO_2 substrate within the thin ChG (a) and the overall trajectory (b) (Neilson *et al.*, 2007).

objective is minimizing backscattered electrons. This can be achieved using an additional layer (e.g., Cr) between the ChG film and substrate. Decreasing the depth of electron penetration (taking into account conical character of electron trajectories) is the way to improve resolution; however, we should keep balance between the resolution and reasonable thickness needed for the transfer of pattern into the substrate.

The first reported ChG electron resist doped with Ag was based on 270 nm thick sputtered $Se_{85}Ge_{15}$ film (Yoshikawa *et al.*, 1977). A silver layer (10 nm) was formed by dipping in $AgNO_3$. Electron-beam exposure was carried out with a scanning electron microscope at an accelerating voltage of 15 kV with an electron dose 3×10^{-4} C/cm^2. This type of electron resist exhibits among the highest contrast, which favors fine pattern processing. The γ value may even reach as high as 8, effectively independent of accelerating voltage. However, the sensitivity of these electron resists shows strong dependence on accelerating voltage (E_0). For $E_0 < 20$ keV the electron dose, D_s, necessary to leave behind 70% of the initial film thickness after development, closely fit to the following empirical law (Yoshikawa *et al.*, 1977):

$$D_s \propto E_0^{1.5} \qquad [17.8]$$

In this definition of D_s, the sensitivity increases as the value of D_s decreases. Strong dependence of the width and height of the electron-induced lines on the accelerating voltage and deposition order has also been observed (Mitzsch and Fitzgerald, 2000).

One of the possible interpretations of electron-induced diffusion is that the indiffusing Ag^+ ion is accelerated by the so-produced free carriers just as in photodoping (Yoshikawa *et al.*, 1977). The excitation density near the interface is proportional to the incident electron dose per unit area/ penetration range. There are also some additional effects influencing the process of electron-beam induced Ag diffusion into ChG thin film. For example, e-irradiation causes accumulation of Ag atoms at distances near 50 μm from the irradiation spot and can be associated with Ag migration due to the electric field, which is generated by irradiated electrons and depends on the acceleration voltage (Kostyukevych, 1998; Mitzsch and Fitzgerald, 2000; Yoshida and Tanaka, 1997). We also observed similar effects from e-irradiation of As_2S_3-Ag structures (Kovalskiy *et al.*, 2007), but this process was time dependent. Accumulation of silver upon electron irradiation is followed by its dissolution, in agreement with data presented by Romero *et al.* (2004). It was confirmed also by Debnath *et al.* (2005) that accelerating voltage is a crucial variable in electron-induced diffusion.

Particular advantages of electron resists based on ChG-Ag structures are (Kostyoukevych *et al.*, 1995):

- no macromolecular structures;
- no swelling deformations;
- no distortions from charging of resist film under electron irradiation.

17.4.3 Other resists

Application of X-ray lithography and dry etching to silver halide–ChG resists can easily resolve 100 nm minimum linewidth (Kolwicz and Chang, 1980). Silver halide decomposes into metallic silver and halogen gas upon absorption of radiation. The mechanism of X-ray-induced photodoping is associated with the production by X-ray-induced shower of secondary electrons, which carry most of the energy of incident photon. However, the pattern-forming layer is relatively thin in the case of X-ray photodoping (50 nm or less, despite the very high sensitivity of ChG to X-rays, twice as sensitive as organic PMMA) and can be used only for very specific purposes.

Ag^+ ions can be directly implanted into ChG matrix. An example of such irradiation of $GeSe_2$ negative resist with Ag^+ doses from 10^{14} to 2×10^{16} cm^{-2} is given by Kluge *et al.* (1989). As was shown by Klebes *et al.* (1988), in the case of ion bombardment, not only negative but also positive resist can be obtained from the dry etching of $GeSe_2$-Ag structure.

Electron irradiation of the ChG-Ag (As_2Se_3-Ag and Sb_2S_3-Ag) structures using an SEM or STEM is especially convenient for fabricating X-ray masks (Debnath *et al.*, 2005). When these bilayers are supported on an X-ray transparent membrane, the exposed regions exhibit a different X-ray absorption behavior from regions unexposed to the electron beam, i.e. the silver accumulated regions have higher X-ray absorption due to higher atomic number of Ag. The parameters of the X-ray masks strongly depend on the accelerating voltage. Low accelerating voltage ($\leq$10 keV) produces silver deficient trough-like patterned structure, which can transmit more X-rays than the unexposed region. Conversely, higher accelerating voltages ($\geq$20–30 keV) cause small increase of Ag concentration and fabrication of protruding patterned structures. At higher accelerating voltages, fine structures can be generated due to smaller spot size and the maximum current achieved. Increasing the voltage improves the contrast of the X-ray mask, as well as the height and width of the patterns.

17.5 Advantages and disadvantages of chalcogenide resists

There are two fundamental reasons that, in principle, inorganic resists should give higher resolution compared to organic polymers. Firstly, organic polymers are made of much larger, long chain molecules than inorganic solids. Typical polymer resists have molecular weights of (5–20) $\times 10^3$ units, giving molecular pixel size of 4–9 nm. By comparison, homogeneous inorganic glasses have building blocks of size $\leq$1 nm, which will ultimately limit their resolution. Obviously, to achieve such high resolution, it will be necessary to produce films of homogeneity on the 1 nm scale. Secondly, in EBL using a finely focused beam, the resolution is primarily limited by the structural changes produced by the secondary electrons. Because the polymer resists are made of light elements and weak van der Waals bonding, the undesirable region affected by secondary electron scattering is expected to be significantly larger than that in the inorganic resists that are made of stronger covalent–ionic bond and often heavier elements. The operation of polymer resists is based on the fact that their degree of polymerization and hence subsequent reactivity to an etchant is altered by irradiation. Modification of atomic structure is also the basis of inorganic resists. However, some inorganic resists undergo compositional changes under irradiation, in addition to structural changes. Then the selectivity of such resists is greatly enhanced.

In regard to the application of resist on substrate, the inorganic ChG resists can be as good as or superior to currently used polymer resists in a number of applications. Like polymers, inorganic glasses can be fabricated as uniform film over large surfaces, both planar and non-planar. They have been prepared by vacuum evaporation, chemical vapor deposition (CVD), plasma enhanced CVD, radio frequency (RF) sputtering methods, etc. For over four decades, chalcogenide glasses have been known to be photosensitive. The effects of exposure to bandgap light on the physical and chemical properties of these materials have been investigated extensively.

The lack of fibrous features, an undesirable characteristic of some organic photoresists, is an advantage of silver-doped ChG resists. Initially it was assumed that the theoretical linewidth limit for photodoping lithography is near 10 nm, the lateral diffusion length of the photoexcited carriers (Jain and Vlcek, 2008; Das *et al.*, 1990). Experimentally, the ultimate resolution of the photodoping process was found to be much more impressive, ~1 nm, which is comparable to the medium range order in ChG (Yoshikawa *et al.*, 1981).

So, the main advantages of ChG-Ag photoresists are (Kostyukevych, 1998): uniformity of the layers, high values of contrast coefficient γ (as high as 10.5 was observed for $Ag_2Se/GeSe_2$ system; Polasko *et al.*, 1985), extremely high resolution capability (Yoshikawa *et al.*, 1981) and absence of shrinkage after treatment. ChG films show strong resistance to many acid solutions, such as HF, HF-NH_4F, H_3PO_4, HCl, H_2SO_4, which are conventionally employed in the etching of SiO_2 or Si_3N_4 in microfabrication technology. At the end of the process, ChG films containing Ag are easily removed in hot H_2SO_4 (Janai, 1981). Other important advantages of ChG-Ag photoresists are (Janai, 1981):

- pinholes are less of a problem due to the dry deposition process;
- dust sticking is less serious than for organic polymers;
- a high optical absorption coefficient (~$10^5\,cm^{-1}$ in the blue and ultraviolet) allows the use of thinner films, which results in dimensional accuracy in fine pattern delineation without any interference fringe effects;
- possibility of formation of grayscale hard photomasks.

There are also a few demerits and challenges related to the use of ChG resists (Janai, 1981):

- although generally acceptable, the sensitivity is slightly lower than that of conventional polymer resists;
- direct adhesion to aluminum is poor.

It is believed that during photodoping, silver diffuses perpendicular to the plane of film surface, and lateral photodiffusion can be neglected, especially in comparison with thermal diffusion (Utsugi, 1992). However, it is known that photodoping causes edge sharpening, which is associated with Ag accumulation on the border between non-irradiated and irradiated parts of the sample (see Fig. 17.14) (Kovalskiy *et al.*, 2007; Liang and Tada, 1988). Lateral diffusion is the most probable explanation of this effect. Edge sharpening is a negative factor for the formation of high-fidelity lithographic patterns, especially grayscale and 3D structures. Interestingly, edge sharpening has been suggested also as a way to obtain nanometer size features by photodiffusion, much finer than the wavelength employed for diffusion, due to the compensation of diffraction and interference effects (Tai *et al.*, 1980). We have also shown that the edge sharpening effect depends on the method of preparation (Kovalskiy *et al.*, 2006). Its magnitude can be decreased, for example, by preliminary illumination of ChG thin film before Ag deposition (Kolobov and Elliott, 1991).

Another unwanted feature of photodiffusion in ChG resist is the so-called 'grass' effect (Fig. 17.15), which is more clearly observed with dry etching. The size of the 'grass' is usually close to 100 nm. So it is a potential problem especially for submicron and nanoscale lithography. Possible causes of the grass effect include: (a) formation of As_2O_3 during exposure to the ambient or in the oxygen-containing atmosphere within the reactive ion etching (RIE) chamber (Bloking *et al.*, 2001); (b) inhomogeneity of the freshly prepared ChG films (Ston *et al.*, 2003).

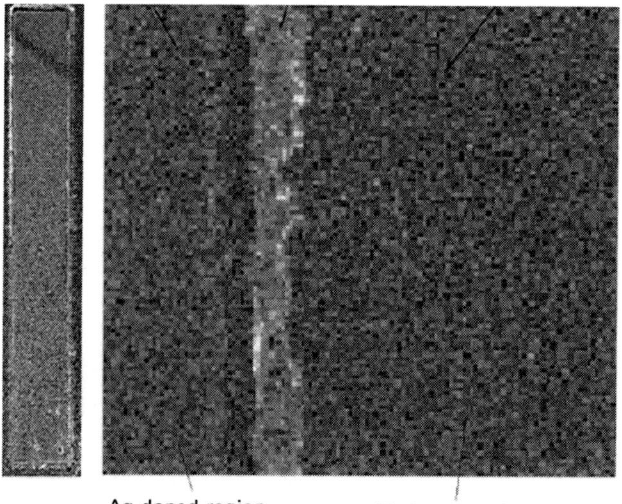

17.14 Ag concentration profile on the surface of dry etched As_2S_3-Ag sample. Note its segregation at the border of the non-irradiated and irradiated parts (edge effect) (Kovalskiy *et al.*, 2007).

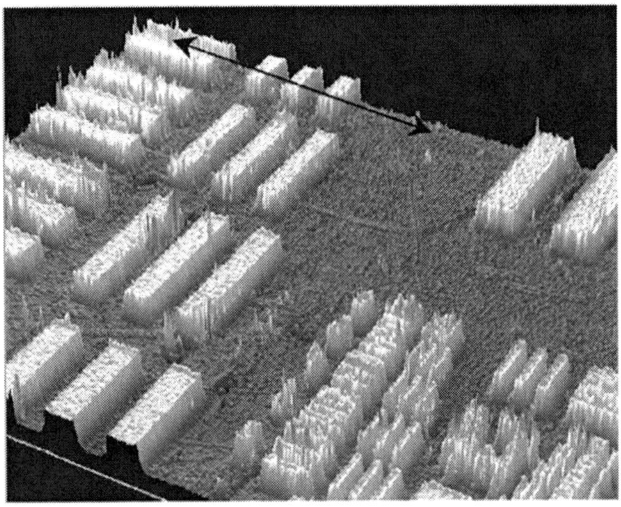

17.15 Demonstration of 'grass' effect, obtained on As_2S_3-Ag structure (Kovalskiy *et al.*, 2006).

Photodoping of ChG can help overcome some of the known lithographic problems encountered with polymer photoresists in the submicron region. These problems and the ways to resolve them are listed in Tai *et al.* (1980) and summarized below:

1. The modulation transfer function ($MTF = (I_{max} - I_{min})/(I_{max} + I_{min})$) in the submicron region does not provide enough margin over the minimum *MTF* required for image formation in polymer resists (here I_{max} and I_{min} are light intensity levels achievable for a given pattern with a particular optical projection system). As feature size is reduced, it becomes increasingly difficult to find a single exposure condition that is satisfactory for different feature sizes and geometries. Image development requires that the optically determined *MTF* exceeds some critical value related to the contrast characteristic of the resist material: $CMTF = (I_{100} - I_0)/(I_{100} + I_0)$, where I_{100} is minimum exposure energy for 100% exposure, and I_0 is maximum exposure energy for zero exposure. *CMTF* relates to the contrast of resist γ as (Tai *et al.*, 1980):

$$CMTF = (10^{1/\gamma} - 1)/(10^{1/\gamma} + 1) \qquad [17.9]$$

The minimum size printable feature using any combination of resist and lens is then determined by the condition MTF ≥ CMTF. The *CMTF* value for ChG-Ag bilevel structures is much lower than for polymers (Tai *et al.*, 1980). For example, for Ag_2Se/GeSe structures $CMTF = 0.2$ in comparison with 0.6 for polymers. It means that we can print much finer structures in the case of photodoped ChG. The effective *MTF* can be further increased by exploiting the effect of edge-sharpening from Ag diffusion.

2. A problem with polymer lithography is that the presence of topographic steps (variation of photoresist thickness near the step) on the wafer surface severely limits resolution and linewidth control. A way to overcome this difficulty with ChG-Ag structures is through the use of O_2 plasma during dry etching. In this case, we obtain high aspect ratio and vertical-walled features independently on stepped surfaces.
3. The standing wave effect degrades resolution of polymer resist-based lithography. However, since the ChG resists absorb UV light totally, the standing wave problem is eliminated.

17.6 Conclusion and future trends

With the observation of photosensitivity, the potential of chalcogenide glasses as a photoresist was recognized over 50 years ago. Subsequently, they were shown to be suitable resists also for use with UV, X-ray, electron and ion beams. During this period, numerous studies have been conducted to understand underlying mechanisms, and detailed processing methods established for practical applications. Both wet etching using alkaline and amine-based developers, and dry etching are available for fabricating high resolution (~10 nm) grayscale structures, which can be adapted to commonly used manufacturing conditions.

Radiation-induced changes in local structure (bonding defects) and photoinduced diffusion of silver are the two primary mechanisms that provide resist functionality to ChG. The latter mechanism is a unique aspect of ChG, which is shown to provide many advantages over currently used polymer resists. Arsenic sulfide compositions are considered most promising substitutes for polymer resists, but many other compositions are available that can provide additional/optimum functionality.

In spite of the numerous advantages offered by ChG resists, they have been rarely exploited in the manufacturing of actual devices. The primary reason for this apathy, especially from microelectronics industry, appears to be 'the fear of the unknown' – ChG resists are made of elements that are either unfamiliar to or are contaminants feared by the silicon device industry. Secondly, chalcogenide glasses often consist of toxic/heavy elements and the industry avoids their usage as long as any alternative is available (even if it is technically inferior). Notwithstanding the lack of interest in the past, we expect increasing usage of ChG in the future. At least for certain small to medium volume applications, ChG resists will be required due to their special characteristics that are not provided by conventional resists. Furthermore, the use of ChG has increased significantly in recent years in other applications such as phase-change memories, infrared optics, etc. This has helped develop protocols for the use of ChG in manufacturing, which should also contribute to the usage of ChG resists.

17.7 Acknowledgements

Separate parts of this work were supported by the Pennsylvania Department of Community and Economic Development through the Ben Franklin Technology Development program, and NSF's International Materials Institute for New Functionality in Glass (IMI-NFG) through Grant No. DMR 0844014. M.V. thanks the Czech Science Foundation for support under Grant No. P204/11/0832.

17.8 References

Arsh, A., Froumin, N., Klebanov, M. and Lyubin, V., 2002. Photodoping of glassy arsenic chalcogenide by zinc, *J. Optoelectron. Adv. Mater.*, 4, 27–32.

Balan, V. *et al.*, 2003. Chalcogenide glass-based rib ARROW waveguide. *J. Non-Cryst. Solids*, 326–327, 455–459.

Bloking, J.T. *et al.*, 2001. Photoinduced changes in the surface morphology of $As_{50}Se_{50}$ chalcogenide glass films. *Opt. Mater.*, 17, 453–458.

Das, A., Bello, A. and Jishi, R.A., 1990. Theoretical considerations relating to the characteristic curves of the silver chalcogenide glass inorganic resists. *J. Appl. Phys.*, 68, 3957–3963.

Debnath, R.K., Nusbar, N. and Fitzgerald, A.G., 2005. Electron beam induced chemical modification of amorphous chalcogenide-metal bilayer and its application. *Appl. Surf. Sci.*, 243, 228–231.

DeCorby, R.G. *et al.*, 2005. High index contrast waveguides in chalcogenide glass and polymer. *IEEE J. Sel. Top. Quant.*, 11, 539–546.

Eisenberg, N.P. *et al.*, 1996. Fabrication and testing of microlens arrays for the IR based on chalcogenide glassy resists. *J. Non-Cryst. Solids*, 198–200, 766–768.

Eisenberg, N.P. *et al.*, 2006. New micro-optical devices for the IR based on three-component amorphous chalcogenide photoresists. *J. Non-Cryst. Solids*, 352, 1632–1636.

Fedynyshyn, T.H. *et al.*, 2000. Prospects for using existing resists for evaluating 157-nm imaging systems. *Proc. SPIE*, 3999 (2), 627.

Frumar, M. and Wagner, T., 2003. Ag doped chalcogenide glasses and their applications. *Current Opinion Solid State Mater. Sci.*, 7, 117–126.

Gates, B.D. *et al.*, 2005. New approaches to nanofabrication: molding, printing, and other techniques. *Chem. Rev.*, 105, 1171–1196.

Goldschmidt, D. and Rudman, P.S., 1976. The kinetics of photodissolution of Ag in amorphous As_2S_3 films, *J. Non-Cryst. Solids*, 22, 229–243.

Indutnyi, I.Z., Kostishyn, M.T., Romanenko, P.F. and Stronski, A.V., 1991. Recording of holographic diffraction gratings on lightsensitive semiconductor-metal system. *J. Inform. Rec. Mater.*, 19, 239–248.

Indutnyi, I.Z. *et al.*, 1995a. Application of inorganic resists in high density information storage technology, *Proc. SPIE*, 2451, 456–465.

Indutnyi, I.Z. *et al.*, 1995b. Holographic optical element fabrication using chalcogenide layers. *Opt. Eng.*, 34, 1030–1039.

Ito, H., 2005. Chemical amplification resists in microlithography. *Adv Polym Sci.*, 172, 37–245.

Jain, H. and Vlcek, M., 2008. Glasses for lithography. *J. Non-Cryst. Solids*, 354, 1401–1406.

Janai, M., 1981. Photolithographic processes in amorphous semiconductors. *J. Physique*, 42, C4-1105–C4-1114.

Klebes, R., Thomas, A. and Kluge, G., 1988. Utilization of Ag:$GeSe_2$ films as inorganic positive resists. *Phys. Status Solidi A*, 110, K81–K84.

Kluge, G., 1987. A new interpretation of the photodoping effect in amorphous As- and Ge-chalcogenides. *Phys. Status Solidi A*, 101, 105–114.

Kluge, G., Thomas, A., Klabes, R. and Suptitz, P., 1989. Applicability of silver implanted amorphous $GeSe_2$ as inorganic negative resist. *Phys. Status Solidi A*, 113, K171–K175.

Kolobov, A.V., 2003. *Photo-induced metastability in amorphous semiconductors*. Weinheim: Wiley-VCH.

Kolobov, A.V. and Elliott, S.R., 1991. Photodoping of amorphous chalcogenides by metals, *Adv. Phys.* 40, 625–684.

Kolobov, A.V. and Tanaka, K., 2001. Photoinduced phenomena in amorphous chalcogenides: from phenomenology to nanoscale. In: H.S. Nalwa, ed. 2001. *Handbook of advanced electronic and photonic materials and devices*, San Diego, CA: Academic Press, 47–90.

Kolwicz, K.D. and Chang, M.S., 1980. Silver-halide chalcogenide glass inorganic resists for X-ray lithography. *J. Electrochem. Soc.*, 127, 135–138.

Kostyshin, M.T. and Ushenin, Y.V., 1982. Diffusion-controlled sensitivity of As_2S_3-Ag system to the irradiation through metal layer. *Fiz. Tekhn. Poluprovodn.*, 16, 19.

Kostyshin, M.T., Mikhaylovskaya, E.V. and Romanenko, P.F., 1966. Photographic sensitivity effect in thin semiconducting films on metal substrates. *Sov. Phys.–Sol. State*, 8, 451–452.

Kostyukevych, S.A., 1998. Investigations and modeling of physical processes in inorganic resists for use in UV and laser lithography. *Proc. SPIE*, 3424, 20–27.

Kostyoukevych, S.A., Shepeljavi, P.E., Stronski, A.V. and Indutnyi, I.Z., 1995. Investigation and modelling of physical processes in high density information recording with the help of organic resists. *Proc. SPIE*, 2647, 167–173.

Kovalskiy, A. *et al.*, 2006. Development of chalcogenide glass photoresists for gray scale lithography. *J. Non-Cryst. Solids*, 352 (6–7), 589–594.

Kovalskiy, A. *et al.*, 2007. On the mechanism of gray scale patterning of Ag-containing As_2S_3 thin films. *J. Phys. Chem. Solids*, 68, 920–925.

Kovalskiy, A. *et al.*, 2008. Comparative study of electron- and photo-induced structural transformations on the surface of $As_{35}S_{65}$ amorphous thin films. *Thin Solid Films*, 516, 7511–7518.

Kovalskiy, A. *et al.*, 2009a. Chalcogenide glass thin film resists for grayscale lithography. *Proc. SPIE*, 7273, 72734A-1–12.

Kovalskiy, A. *et al.*, 2009b. Chalcogenide glass e-beam and photoresists for ultrathin grayscale patterning. *J. Micro/Nanolith. MEMS&MOEMS*, 8, 043012.

Kozicki, M.N., Hsia, S.W., Owen, A.E. and Ewen, P.J.S., 1991. PASS – a chalcogenide-based lithography scheme for I.C. fabrication. *J. Non-Cryst. Solids*, 137–138, 1341–1344.

Kumar, A., Malhotra, L.K. and Chopra, K.L., 1987. Negative resist behavior of obliquely deposited Ge-Sb-Se thin films. *J. Vacuum Sci. Techn. A*, 5, 354–357.

Lee, H.-Y. and Yao, T., 2002. Wet etching selectivity of Ag-photodoped AsGeSeS thin films and the fabrication of a planar corrugated one-dimensional photonic crystal by a holographic method. *J. Vac. Sci. Technol. B*, 20, 2017–2023.

Liang, Y.-C. and Tada, K., 1988. Silver distribution in wet-sensitized $As_{10}Ge_{22.5}Se_{67.5}$ films after electron-beam exposure. *J. Appl. Phys.*, 64, 1378–1283.

Lyubin, V., 2009. Chalcogenide glassy photoresists: history of development, properties, and applications. *Phys. Status Solidi B*, 246, 1758–1767.

Masters, J. and Goldberg, G., 1981. *Process of doping silver image in chalcogenide layer.* U.S. Patent 4,269,935.

Mitzsch, K. and Fitzgerald, A.G., 2000. Electron-beam-induced patterning of thin film arsenic-based chalcogenides. *Appl. Surf. Sci.*, 162–163, 464–468.

Mizushima, Y. and Yoshikawa, A., 1982. Photoprocessing and lithographic applications. *Japan Annual Reviews in Electronics, Computers & Telecommunications*, 2, 277–295.

Neilson, J. R. *et al.*, 2007. Fabrication of nano-gratings in arsenic sulphide films. *J. Non-Cryst. Solids*, 353, 1427–1430.

Nordman, O. and Nordman, N., 2001. Electron and photo induced metal diffusion in amorphous As50Se50 thin films coated with two metal layers. *Opt. Commun.*, 193, 147–151.

Nordman, O., Nordman, N. and Pashkevich, V., 2001. Influence of ultraviolet-light exposure on electron-beam written gratings in amorphous As–Se thin films coated with different metals. *Appl. Phys. Lett.*, 79, 2004–2006.

Orava, J. *et al.*, 2006. Selective wet-etching of undoped and silver photodoped amorphous thin films of chalcogenide glasses in inorganic alkaline solutions. *J. Non-Cryst. Solids*, 352, 1637–1640.

Polasko, K.J., Pease, R.F.W., Marinero, E.E. and Cagan, M.R., 1985. Excimer laser exposure of $Ag_2Se/GeSe_2$: high contrast effect. *J. Vac. Sci. Technol. B*, 3, 319–322.

Popescu, M. *et al.*, 2009. Silver/amorphous As_2S_3 heterostructure. *J. Optoelectr. Adv. Mater.*, 11, 1586–1594.

Rogers, J.D. *et al.*, 2004. Realization of refractive microoptics through grayscale lithographic patterning of photosensitive hybrid glass. *Opt. Express*, 12, 1294–1303.

Romero, J.S., Fitzgerald, A.G. and Rose, M.J., 2004. A transmission electron microscope study of metal/chalcogenide amorphous thin films. *Appl. Surf. Sci.*, 234, 369–373.

Russo, L., Vlcek, M. and Jain, H., 2005. Arsenic sulphide glasses as novel photoresist materials. *Glass Technol., Part A*, 46, 94–98.

Sakuma, H., Shimizu, I., Kokado, H. and Inoue, E., 1971. A new method for producing low-electrical-resistivity patterns in insulating chalcogenide glasses. *Bull. Chem. Soc. Jpn.*, 44, 1723.

Shimizu, I., Kokado, H. and Inoue, E., 1973. The photo-erasable memory switching effect of Ag photo-doped chalcogenide glasses. *Bull. Chem. Soc. Japan*, 46, 3662–3665.

Shkol'nikov, E.V., 2001. Kinetics of etching of $AgAsS_2$ and $AgGeS_2$ glasses in alkaline solutions. *Glass Phys. Chem.*, 27, 279–281.

Ston, R., Vlcek, M. and Jain, H., 2003. Structure and photoinduced changes in. bulk and films of As-Ge-S system. *J. Non-Cryst. Solids*, 326–327, 220–225.

Tai, K.L. *et al.*, 1980. Submicron optical lithography using an inorganic resists/polymer bilevel scheme. *J. Vac. Sci. Technol.*, 17, 1169–1176.

Tanaka, K., 1991. Physics and applications of photodoping in chalcogenide glasses, *J. Non–Cryst. Solids*, 137–138, 1021–1026.

Tanaka, K. and Shimakawa, K., 2009. Chalcogenide glasses in Japan: A review on photoinduced phenomena. *Phys. Status Solidi B*, 246, 1744–1757.

Taniguchi, J. *et al.*, 2004. 3D imprint technology using substrate voltage change. *Appl. Surf. Sci.*, 238, 324–330.

Tsutsumi, T. *et al.*, 2000. Fabrication technology of a Si nanowire memory transistor using an inorganic electron beam resist process. *J. Vac. Sci. Technol. B*, 18, 2640–2644.

Utsugi, Y., 1992. Chemical modification for nanolithography using scanning tunneling microscopy. *Nanotechnology*, 3, 161–163.

Vadimsky, R.G., 1988. Three-dimensional photolithography with conformal GeSe resist. *J. Vac. Sci. Technol. B*, 6, 2221–2223.

van den Boogaart, M.A.F. *et al.*, 2004. Deep-ultraviolet–microelectro-mechanical systems stencils for high-throughput resistless patterning of mesoscopic structures. *J. Vac. Sci. Technol. B*, 22, 3174–3177.

Vieu, C. *et al.*, 2000. Electron beam lithography: resolution limits and applications. *Appl. Surf. Sci.*, 164, 111–117.

Vlcek, M. and Jain, H., 2006. Nanostructuring of chalcogenide glasses using electron beam lithography. *J. Optoelectr. Adv. Mater.*, 8, 2108–2111.

Vlcek, M. and Sklenar, A., 2002. *Transparent and semitransparent diffractive elements, particularly holograms and their making process*. U.S. Patent 6,452,698.

Vlcek, M., Prokop, J. and Frumar, M., 1994. Positive and negative etching of As-S thin layers. *Int. J. Electronics*, 77, 969–973.

Vlcek, M., Ewen, P.J.S. and Wagner, T., 1998. High efficiency diffraction gratings in As–S layers. *J. Non-Cryst. Solids*, 227–230, 743–747.

Vlcek M. *et al.*, 2003. Selective etching of chalcogenides and its application for fabrication of diffractive optical elements. *J. Non-Cryst. Solids*, 326–327, 515–518.

Wagner, T., 2002. Photo- and thermally-induced diffusion and dissolution of Ag in chalcogenide glasses thin films. *J. Optoelect. Adv. Mater.*, 4, 717–727.

Wagner, T. *et al.*, 1993. Kinetics and reaction products of the photo-induced solid state chemical reaction between silver and amorphous $As_{33}S_{67}$ layers. *J. Non-Cryst. Solids*, 164–166, 1255–1258.

Wagner, T. *et al.*, 2001. The tailoring of the composition of Ag–As–S amorphous films using photo-induced solid state reaction between Ag and $As_{30}S_{70}$ films. *Solid State Ionics*, 141–142, 387–395.

Wagner, T., Schroeter, S., Glaser, T. and Vlcek, M., 2003. Holographic grating preparation in Ag/$As_{30}S_{70}$ multilayer and bilayer structures. *J. Non-Cryst. Solids*, 326–327, 500–504.

Yaji, T. and Kurita, S., 1983. Photodoping sensitivity of Ag into amorphous As_2S_3 films. *J. Appl. Phys.*, 54, 647–651.

Yoshida, N. and Tanaka, K., 1997. Ag migration in Ag-As-S glasses induced by electron-beam irradiation. *J. Non-Cryst. Solids*, 210, 119–129.

Yoshida, N. and Tanaka, K., 1998. Electronic and chemical Ag-crystallization in Ag–As–S glasses. *J. Non-Cryst. Solids*, 227–230, 779–783.

Yoshikawa, A. *et al.*, 1976. A novel inorganic photoresist utilizing Ag photodoping in Se-Ge glass films. *Appl. Phys. Lett.*, 29, 677–679.

Yoshikawa, A., Ochi, O., Nagai, H. and Mizushima, Y., 1977. A new inorganic electron resist of high contrast. *Appl. Phys. Lett.*, 31, 161–163.

Yoshikawa, A. *et al.*, 1981. Angstroms resolution in Se-Ge inorganic photoresists. *Jpn. J. Appl. Phys.*, 20, L81–L83.

Zenkin, S.A. *et al.*, 1997. Mechanism for interaction of amine solutions with monolithic glasses and amorphous films in the As-S system. *Glass Phys. Chem.*, 23, 393–399.

18
Chalcogenide for phase change optical and electrical memories

B. HYOT, CEA, LETI, France

DOI: 10.1533/9780857093561.2.597

Abstract: Phase change (PC) materials are some of the most promising materials for data-storage applications. They are already used in the three generations of rewritable optical discs (CD, DVD and Blu-ray) and offer great potential as an emerging non-volatile electronic memory. The development of optimized PC alloys and the fine-tuning of their properties require a deep understanding of the physics of phase change. Study of the phenomena involved requires consideration of many different aspects. The first part of the chapter discusses thermodynamics of crystal nucleation and crystallization kinetics in $Ge_2Sb_2Te_5$ thin films, especially in the case of optical storage. The second part of this chapter shows that insights, at the atomic scale, in the PC material thanks to coupled fine structural analyses (X-ray diffraction, X-ray absorption fine structure) and *ab initio* simulations are essential to understand and improve materials for future advanced memory applications.

Key words: phase change (PC) materials, optical data storage, electronic non-volatile memory, thermodynamics of crystal nucleation and crystallization kinetics, X-ray absorption fine structure (XAFS); *ab initio* calculations.

18.1 Introduction: the basics of rewritable phase change data storage

Recent advances in multimedia would have been impossible without computer performance getting faster and the stored information denser. The ability of media to be used repeatedly, that is, to be rewritable, is also an important requirement. One of the most promising media for rewritable applications is phase change (PC) materials. The idea to use an amorphous-to-crystalline phase transition for information storage dates back to the 1960s when Ovshinsky (1968) suggested a memory switch based on changes in the electrical properties of amorphous and crystalline phases of multicomponent chalcogenides. The idea was subsequently extended to optical memories where differences in optical reflectivity serve as a basis for data storage. In 1987, the breakthrough discovery of the GeTe-Sb_2Te_3 pseudo-binary alloys by Yamada *et al.* (1987), removed the difficulties that had inhibited PC materials from being used in practical applications. A landmark was achieved by Ohta (2001), of Matsushita/Panasonic, who developed phase change optical disc technology that remained stable over a million use cycle. This technology became the mainstream in optical disc production and in the late 1990s resulted in the commercialization of 4.7 GB digital versatile disc random access memory (DVD-RAM). The relentless pursuit of material optimization driven by the market has led to the development of three generations of rewritable optical media; compact disc (CD) for music, digital versatile disc (DVD) for standard definition video and Blu-ray discs (BD) for high-definition TV (HDTV), respectively.

The advances in rewritable optical storage made possible by the identification of 'superior' materials have also led to a renaissance of non-volatile electronic memories, given up conceptually during the 1960s. Nowadays large semiconductors companies such as Samsung (Cho *et al.*, 2005), Intel (Hudgens and Johnson, 2004), IBM (Chen *et al.*, 2006), Macronix (Chen *et al.*, 2006), STMicroelectronics (Bez and Pirovano, 2004) and many others are actively investigating non-volatile memories based on phase change materials.

The concept of phase change recording is straightforward. An intense laser or current pulse heats and melts the crystalline material, which is subsequently quenched into the amorphous state. To switch the material back into the crystalline state, a longer but less intense pulse heats the material above the crystallization threshold, thus reversing the structural change. To read out the recorded information, light or electric pulses of much lower intensity are used.

The two properties that characterize phase change materials are their remarkable crystallization kinetics and their contrast between the amorphous and crystalline phases. Let us remind ourselves of a few orders of magnitude. For data integrity, it is crucial that the amorphous regions be stable for at least 10 years at room temperature. However, those amorphous regions should recrystallize within 10^{-8} s somewhere in the range between 250 and 550°C. In other words, the timescale needs to change by roughly 16–17 orders of magnitude on just increasing the temperature by a few hundred Kelvin.

In Section 18.2, we will focus on the crystallization process of the best known composition, $Ge_2Sb_2Te_5$, which combines all the requirements of fast crystallization, amorphous stability at room temperature and excellent contrast for optical data storage applications. The role of the chemical nature of the adjacent materials in contact with the PC layer on the crystalline nucleation process will be mainly discussed. In Section 18.3, the consequences of thinning PC layers up to the nanometer scale (for the development of multi-level rewritable optical media) on the stabilization of the amorphous state will be treated on the basis of a purely thermodynamical approach. Finally, in Section 18.4, we will focus on non-volatile electronic memories with the development of specific materials for embedded applications, e.g. in the automobile industry where chips are located in vicinity of a hot engine.

18.2 Crystal nucleation in chalcogenide $Ge_2Sb_2Te_5$ alloys: application to optical memories

Phase change materials with compositions on the tie-line GeTe-Sb_2Te_3 (Yamada *et al.*, 1987, 1991) are frequently used, for example in DVD-RAM. The discussions proposed below will concern mainly the best known composition, namely $Ge_2Sb_2Te_5$.

In the following, we focus on the crystallization process that takes place below the melting temperature but above the glass-transition temperature. We do not discuss any undesired recrystallization of amorphous marks below the glass transition temperature (relaxation phenomena).

Two processes are involved in the crystallization of an amorphous solid (hereafter: parent phase). First, crystallization is initiated by crystal nucleation. In a simple case, this occurs in the interior of the parent phase, i.e., without the involvement of a 'foreign' substance. This is called *homogeneous nucleation*. If the parent phase is in contact with a foreign substance that acts as a preferred nucleation site (like an impurity or a boundary surface), nucleation is called *heterogeneous*. All such interfaces have free energies and the elimination of part of the interface when an embryo is formed favors heterogeneous nucleation. Subsequently, a stable crystal nucleus can grow to macroscopic size (crystal growth).

The structure of an optical disc includes, in particular, a phase change layer, of the order of 10 nm, sandwiched between two dielectric layers (Meinders *et al.*, 2006). The dimension of the amorphous region being at least 10 times higher than the phase change layer thickness (Fig. 18.1), we can reasonably assume that interface effects can influence notably the heterogeneous nucleation process as discussed in Sections 18.2.1 and 18.2.2.

18.2.1 Thermodynamics of crystal nucleation and kinetics of crystallization

Heterogeneous crystal nucleation

High resolution transmission electron microscopy images (Fig. 18.2) show the presence of numerous very small nanocrystals of the order of 1 or 2 nm

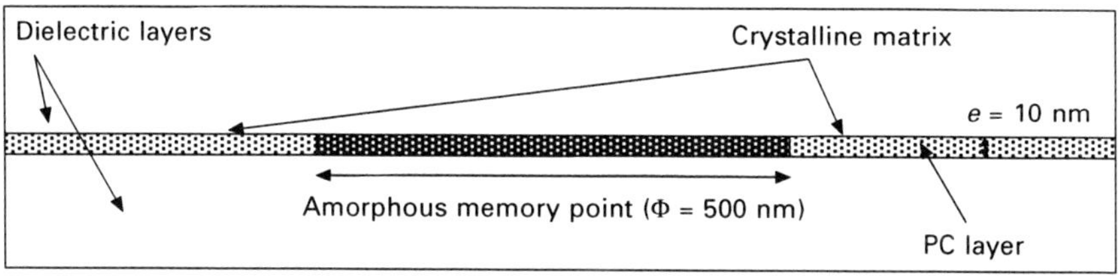

18.1 Schematic scale representation of a phase change recording stack.

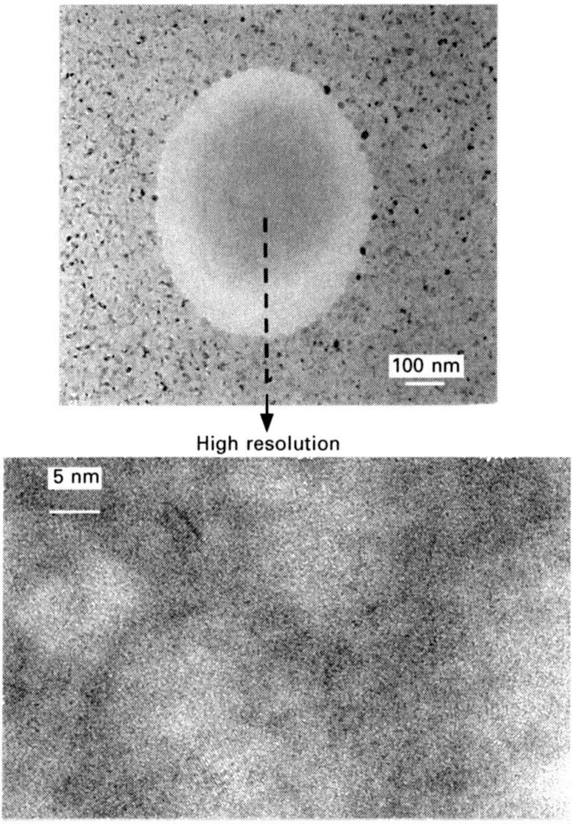

18.2 TEM observation of a 'melt-quenched' amorphous memory bit recorded in a thin layer (20 nm) of $Ge_2Sb_2Te_5$ PC material (upper image). High resolution-TEM observations of nanocrystals of 1–2 nm inside the amorphous phase (lower image).

in melt-quenched amorphous spots of a $Ge_2Sb_2Te_5$ thin layer of 20 nm. This observation allows the assumption of a very small size of the critical nuclei of the order of 1 nm.

Classically, the 'spherical-cap' model (Fig. 18.3) proposed by Volmer (1929) is used to treat the heterogeneous nucleation mechanism. For critical radii of nanometric size, this model is no longer relevant because the height of the cap (h in Fig. 18.3) is only one or two interatomic distances! It is clear that a macroscopic wetting angle (θ in Fig. 18.3) can no longer be defined rigorously. Nevertheless, it seems possible to analyze the heterogeneous nucleation mechanism from a simple but approximate way on the basis of 'equivalence'. We propose an approach based on a different geometric representation of the nucleus where the critical crystalline embryo is assumed to be cubic (Fig. 18.4).

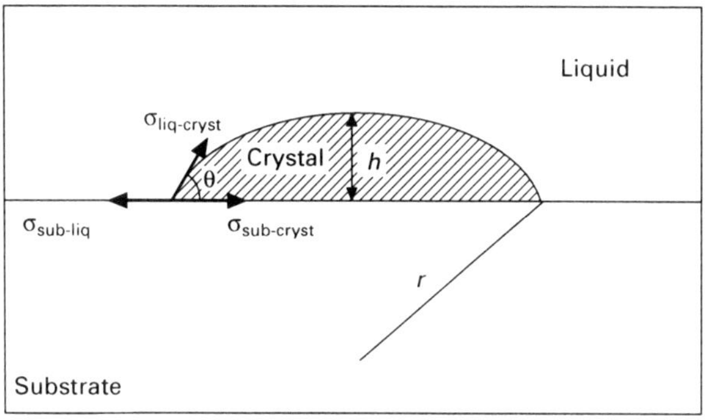

18.3 'Spherical-cap' model for heterogeneous crystalline nucleus formation. The quantities $\sigma_{liq\text{-}cryst}$, $\sigma_{liq\text{-}sub}$ and $\sigma_{cryst\text{-}sub}$ are the interfacial energies between the three substances involved.

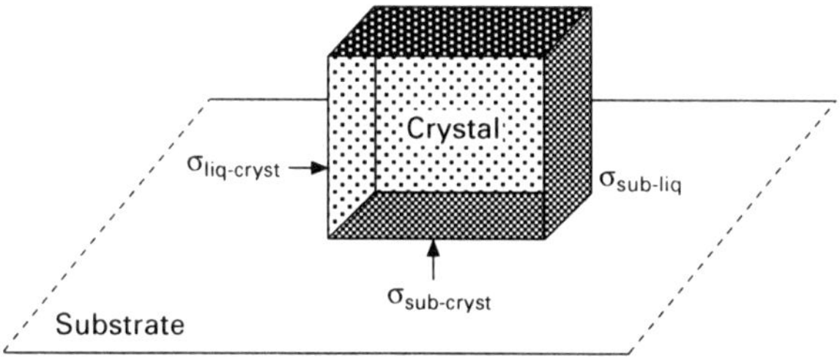

18.4 Cubic model for heterogeneous crystalline nucleus formation.

The change in Gibbs energy related to the formation of a cubic embryo is given by:

$$\Delta G = \Delta G_v a^3 + 5a^2\sigma_{liq\text{-}cryst} + a^2\chi \qquad [18.1]$$

with $$\chi = \sigma_{sub\text{-}cryst} - \sigma_{sub\text{-}liq} \qquad [18.2]$$

where a is the edge of the cube, ΔG_v the Gibbs energy difference per unit volume between the crystalline phase and the undercooled liquid (the driving force for crystallization, $\Delta G_v < 0$), and $\sigma_{liq\text{-}cryst}$, $\sigma_{sub\text{-}liq}$ and $\sigma_{sub\text{-}cryst}$ are, respectively, the liquid-crystal, substrate-liquid and substrate-crystal interfacial energies.

The parameter $\chi = \sigma_{sub\text{-}cryst} - \sigma_{sub\text{-}liq}$ measures the energy lowering due to the substrate–crystal interface formation. χ measures in a way the substrate effect on the nucleation process. The more negative χ is, the better is the 'wettability' of the crystal nuclei on the substrate and the easier is the nucleation process.

Setting $\partial\Delta G/\partial a$, the 'equivalent' critical size a^* of the cubic nucleus is:

$$a^* = \frac{-2(\chi + 5\sigma_{liq\text{-}cryst})}{3\Delta G_v} \qquad [18.3]$$

The barrier of nucleation related to the formation of the cubic crystal embryo is then:

$$\Delta G^* = \frac{4(\chi + 5\sigma_{\text{liq-cryst}})^3}{27\Delta G_v^2} \qquad [18.4]$$

The nucleation process implying $\Delta G^* > 0$, the model leads to $\chi > -5\sigma_{\text{liq-cryst}}$. Moreover, the 'substrate effect' being active only for $\chi < 0$, one may have:

$$-5\sigma_{\text{liq-cryst}} < \chi < 0 \qquad [18.5]$$

According to the Kelton and Greer model (Kelton *et al.*, 1983), the steady state nucleation rate I_s can be given by the following expression:

$$I_s = NO_{n^*}\gamma Z \exp\left(-\frac{\Delta G^*}{k_B T}\right) \qquad [18.6]$$

where N is the number of heterogeneous nucleation sites per unit volume, O_{n^*} is the number of surface atoms in the critical nucleus, and γ is the arrival rate of the amorphous phase atoms to the critical crystalline cluster (number of arrivals per atoms on the nucleus surface per unit time). Z is the Zeldovich factor according to Kelton *et al.* (1983):

$$Z = \left(\frac{-\Delta G_v V_{\text{mol}}}{6\pi RTn^*}\right) \qquad [18.7]$$

with n^*, the number of atoms in the critical nucleus and V_{mol} the molar volume.

The kinetic pre-factor γ is realistically chosen to describe a diffusion-limited crystallization process. Diffusion-limited refers to the nature of a local rearrangement (diffusive jump of an atom across the liquid–crystalline interface), not to the presence of a long-range diffusion field.

The frequency of diffusive jump γ across the interface atom is in general obtained from the diffusion coefficient D in the liquid ($\gamma \propto D/d^2$, where d is the average interatomic distance). The most relevant property to access the processes prevalent in undercooled liquid being the viscosity; the Stokes–Einstein equation which links the viscosity η, and the coefficient of diffusion D $\left(\eta D \propto \frac{k_B T}{d}\right)$, allows I_s to be expressed in terms of the liquid shear viscosity:

$$I_s \propto NO_{n^*}\frac{k_B T}{\eta d^3} Z \exp\left(-\frac{\Delta G^*}{k_B T}\right) \qquad [18.8]$$

The PC materials showing a fragile behaviour, the viscosity of the undercooled liquid can be described by the Vogel–Fulcher law (Vogel, 1921; Fulcher,

1925). This viscosity is considerable lower than the Arrhenius behavior would predict, because the undercooled liquid is able to adapt the equilibrium configuration suitable for a given temperature. The stronger the temperature dependence of the viscosity in the undercooled liquid, the more fragile the glass is (Spaepen, 1986).

In order to check the role of the χ parameter on the nucleation process, Fig. 18.5 shows the steady-state nucleation rate, according to the Kelton and Greer model, for two values of χ: $\chi = -0.3$ J/m^2 and $\chi = -0.2$ J/m^2. We see that an increase of χ (less negative value) leads to a decrease of the nucleation rate associated with a shift of the maximum position towards the low temperatures.

This approach is of particular interest because it allows changing the chemical nature of the substrate in order to modify the nucleation conditions. We will see in Section 18.2.2 that the values of the parameter χ ($\chi = -0.3$ J/m^2 and $\chi = -0.2$ J/m^2) allow interpreting at best the heterogeneous nucleation effect on the crystallization speed measured experimentally when a thin layer of $Ge_2Sb_2Te_5$ phase change material is in contact or not with a nitride interface.

Transient nucleation

The previous subsection dealt with the nucleation phenomena when the stationary regime is established. However, we have to note that treatments in the steady-state regime are applicable only if the statistic equilibrium corresponding to the distributions of the crystalline embryos is able to evolve sufficiently rapidly with regard to temperature variations. And this condition is no longer respected in particular during thermal annealing generated by a laser pulse of a few tens of nanoseconds. When the heating (or cooling) rates are very high, of the order of 10^{10}–10^{11} K/s, it becomes necessary to use the nucleation theory in the transient state. This theory was established in 1942 by Zeldovich who proposes the following expression of the nucleation rate:

$$I(t) = I_s(T)\exp\left(-\frac{\tau}{t}\right) \quad [18.9]$$

This expression introduces a corrective factor to the steady-state treatment. $I_s(T)$ is the steady-state nucleation rate, τ is the transient time, the characteristic time to reach the steady state (this time corresponds in fact to the process limiting the atomic transport from the parent phase to the embryo of the new phase) and t is the isothermal holding time (which is a function of the heating or cooling rate).

By using the expression proposed by Kashchiev (1969) for the calculation of τ:

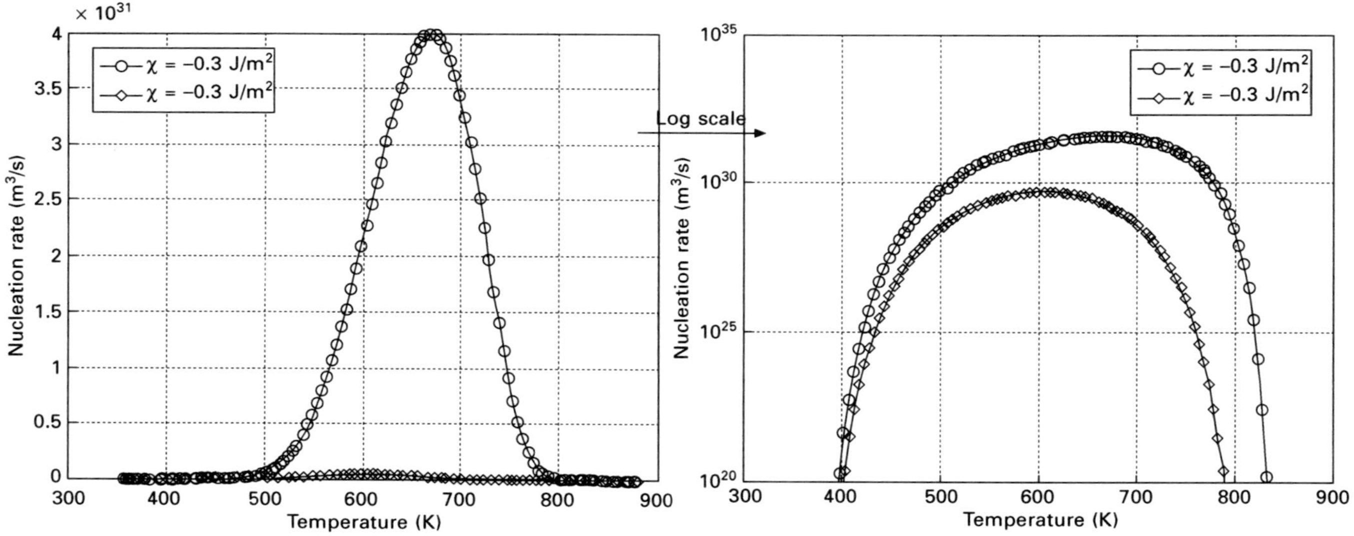

18.5 Steady-state heterogeneous nucleation rates (vs temperature) for $\chi = -0.3$ J/m^2 and $\chi = -0.2$ J/m^2.

$$\tau = \frac{4}{\pi^3 O_{n*} \gamma Z^2} \quad [18.10]$$

where the parameters O_{n*}, γ, and Z are as described above, Fig. 18.6 shows the correction which has to be brought to the nucleation rate when we take into account both the transient time and the thermal profiles calculated within the PC layer during a crystallization laser pulse. We see that the transient effects become significant between 473 and 573 K with inhibition of the nucleation process due to steep temperature rises of the order of 10 K/ns (10^{10} K/s) in this temperature range.

Figure 18.7 shows the theoretical evolution of the nucleation rate for different heating rates (20, 50, and 100 K/ns). It appears that for very steep heating rates (100 K/ns), the nucleation process may be completely by-passed, the atoms having no time to follow such a thermal excitation.

Crystal growth

The nucleation theory presented above describes the formation of stable crystal nuclei. Those can then grow to macroscopic size. In the case of the PC materials, there is no phase separation upon crystallization, the composition of the parent phase and crystal nuclei are the same at all times; it is a polymorphic transformation. Therefore, growth is controlled by a rearrangement process only at the liquid–crystalline interface and is called *interface-controlled*. The

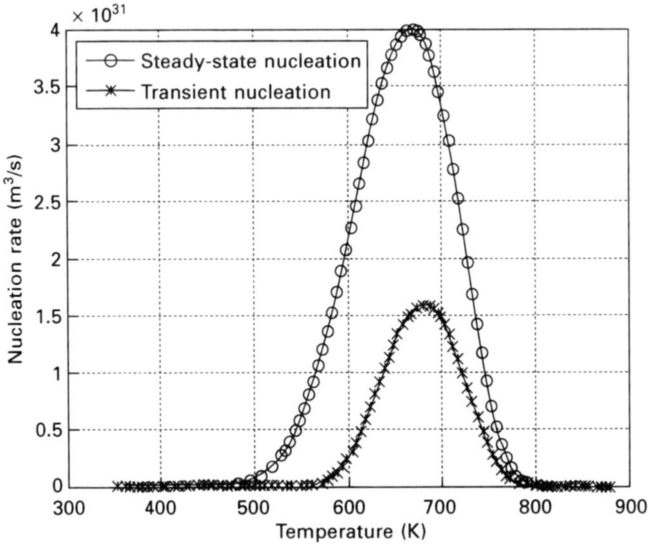

18.6 Comparison between steady-state and transient heterogeneous nucleation rates (vs temperature) for χ = –0.3 J/m^2.

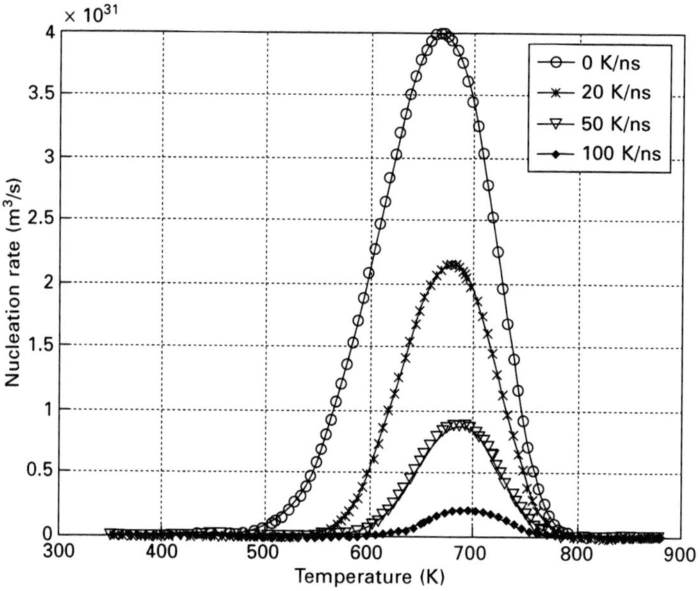

18.7 Theoretical evolution of the transient heterogeneous nucleation rates (vs temperature) for various heating rates from 0 to 100 K/ns for $\chi = -0.3$ J/m^2.

growth velocity is governed by the rate of advance of the crystal/undercooled liquid interface and is given by the following equation:

$$v = \gamma d\left(1 - \exp\left(\frac{\Delta G_{\mathrm{v}} V_{\mathrm{mol}}}{RT}\right)\right) \qquad [18.11]$$

The term in parentheses results from thermally activated transfer across the liquid–crystalline interface in both directions. The frequency of diffusive jump γ across the interface atom is the same as for nucleation.

18.2.2 Modelling of $Ge_2Sb_2Te_5$ thin films crystallization in optical media: relation between microstructure and crystallization speed

Role of interfacial layers

The strong interaction between the optical, thermal, and phase transformation aspects involved in optical storage led us to couple these three different aspects within a global multi-physics simulation. Our approach integrates the thermal modelling of the system (access to the temperature variations generated by the laser beam inside the PC layer) to the nucleation and growth laws (described in Section 18.2.1) driving the phase transformation.

Figure 18.8 shows the simulated crystalline microstructure (of an initially amorphous memory bit) obtained at the end of a crystallization pulse for two values of the parameter χ ($\chi = -0.3$ J/m^2 and $\chi = -0.2$ J/m^2). Note: all the others parameters are, on the other hand, kept the same between the two simulations. The first difference between these simulations is the crystallization time; only 52 ns with $\chi = -0.3$ J/m^2 and 88 ns with $\chi = -0.2$ J/m^2. Moreover, this result seems correlated to the microstructure of the material since we

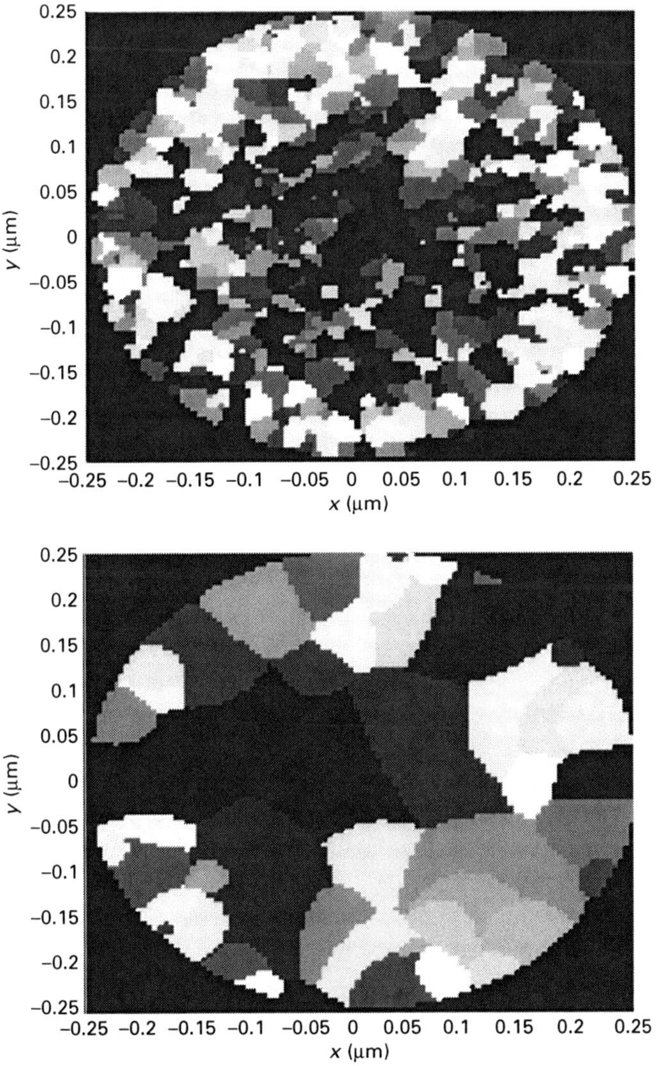

18.8 Simulated crystalline microstructure of an initially amorphous memory point at the end of a crystallization laser pulse for two values of the parameter χ: $\chi = -0.3$ J/m^2 and $\chi = -0.2$ J/m^2.

obtain a very fine microstructure with $\chi = -0.3$ J/m^2 and a crystal structure with bigger grains with $\chi = -0.2$ J/m^2. These simulated results illustrate reasonably the experimentally observed behaviors. When the GeSbTe layer is in direct contact with the nitride interface (GeN), the crystallization time is of the order of 40 ns and the observed microstructure by transmission electron microscopy (TEM) is finer than in the case where the GeSbTe layer is in contact with the classical ZnS-SiO_2 dielectric material used in optical storage. In the second case, the crystallization of the amorphous spot requires more than 100 ns (Fig. 18.9). It appears clearly that the chemical nature of the interfaces in contact with the PC layer influences significantly the crystallization speed of the GST material. A more negative value of χ ($\chi = -0.3$ J/m^2) in the simulation reveals, in fact, a better adhesion of the GST crystals on the nitride interfaces, which clearly increases the heterogeneous nucleation rate.

These results exhibit the crucial role of the 'substrate effect'; in the present case, the nitride interface layers on the crystallization time. The effect of the nitride layers can be interpreted as heterogeneous nucleation sites leading to a finer crystalline microstructure indicating a more efficient crystallization process (with a decrease of the crystallization time).

In order to test the pertinence of our approach to the nucleation process, we have observed, thanks to the cross-section TEM images in Fig. 18.10, the microstructure of a thin layer (10 nm) of $Ge_2Sb_2Te_5$ material at the early stages of crystallization. This thin layer is either in contact with nitride (GeN) interface layers or directly in contact with the classical ZnS-SiO_2 dielectric. It appears clearly that the crystallization starts on GeN/$Ge_2Sb_2Te_5$ interfaces. Several grains which do not cross the PC layer appear 'stuck' on both interfaces, showing a good wettability of the GeN 'substrates' by the crystals of $Ge_2Sb_2Te_5$. On the other hand, without those interface layers, the grains appear more localized within the PC layer (not shown here).

This study shows clearly that the chemical nature of the materials in contact with the thin PC layer plays a crucial role in the crystallization process of the GeSbTe layer. A low chemical affinity between GeSbTe and the adjacent materials has little influence on the crystallization process and the PC material must crystallize mainly by itself. On the other hand, when the chemical affinity between the two materials is high, the interface may act as a continuous region of potential heterogeneous nucleation sites leading to a more efficient nucleation process.

Modelling tool

The development of a multi-physics simulation tool allows a better understanding of the PC process and becomes of particular interest in forecasting the behavior of a material or in estimating its potentiality under

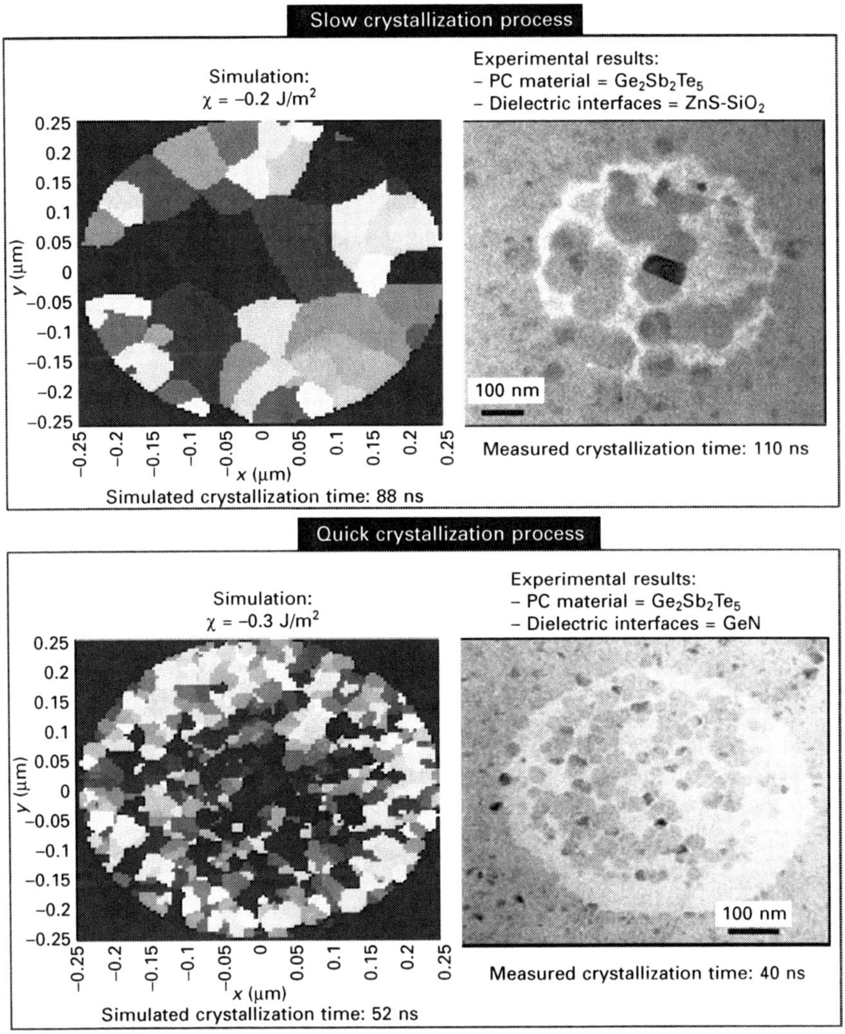

18.9 Simulation-experiment comparison of the crystallization behavior (crystallization time and microstructure) of a $Ge_2Sb_2Te_5$ PC layer either in contact with $ZnS\text{-}SiO_2$ dielectric layers or in contact with nitride (GeN) interfaces. In the case of GeN interfaces, the crystallization time is shorter than with $ZnS\text{-}SiO_2$ dielectric (40 ns vs 110 ns) and is coupled with a finer microstructure. The simulated microstructures and crystallization times with $\chi = -0.3$ J/m^2 and $\chi = -0.2$ J/m^2 illustrate quite well the experimental crystallization behaviors with GeN or $ZnS\text{-}SiO_2$ interfaces respectively.

extraordinary conditions (specific laser heating, for example). It is obvious that whatever the wavelength of the laser, the thermodynamic and kinetic behaviors of the phase transformation of the material remain unchanged at least, when the thickness of the thin layer is of the order of 10 nm. We will

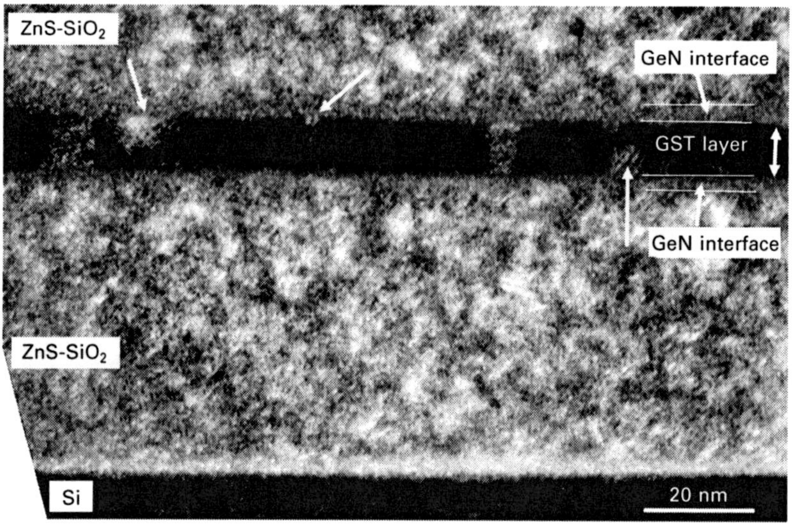

18.10 TEM image of a thin $Ge_2Sb_2Te_5$ layer (10 nm) sandwiched between two GeN interfaces at the early stages of crystallization.

see in Section 18.3 that a stabilization of the amorphous state of thin PC layers may occur when their thicknesses are only of a few nanometers.

For the moment, let us return to the simulation tool. The context of this work takes place between 2004 and 2005 when the R&D laboratories developed the dual-level Blu-ray RW (ReWritable) with two superimposed recording levels. The development of the semitransparent level is tricky; the optical constraints (increase of the stack transparency) require the thickness of the PC layer to be reduced to about 10 nm (more precisely 7–8 nm) and the transfer rate of the Blu-ray format requires in particular a rapid crystallization process of the order of 30–50 ns. The first technological trials with the standard $Ge_2Sb_2Te_5$ were unsuccessful; this material did not show a sufficiently rapid crystallization process.

It is in this context of Blu-ray RW development that we tested the robustness of our approach to the phase transition in PC materials. The simulation, in dynamical conditions, of the crystallization process of $Ge_2Sb_2Te_5$ in the semi-transparent structure of the Blu-ray, illustrates very realistically the experimentally observed behaviors. Figure 18.11 shows that whatever the erasing power, the crystallization of the previously amorphized domains is not complete. Even with the presence of nitride interfaces, introduced in the experimental structure as well as taken into account in the nucleation model, the $Ge_2Sb_2Te_5$ material does not succeed in satisfying the crystallization conditions imposed by the Blu-ray format. An increase by a factor of two of the heterogeneous nucleation rate in the simulation shows, this time, a complete crystallization of the amorphous marks. Under this condition, we see that the final microstructure is thinner than with the conventional

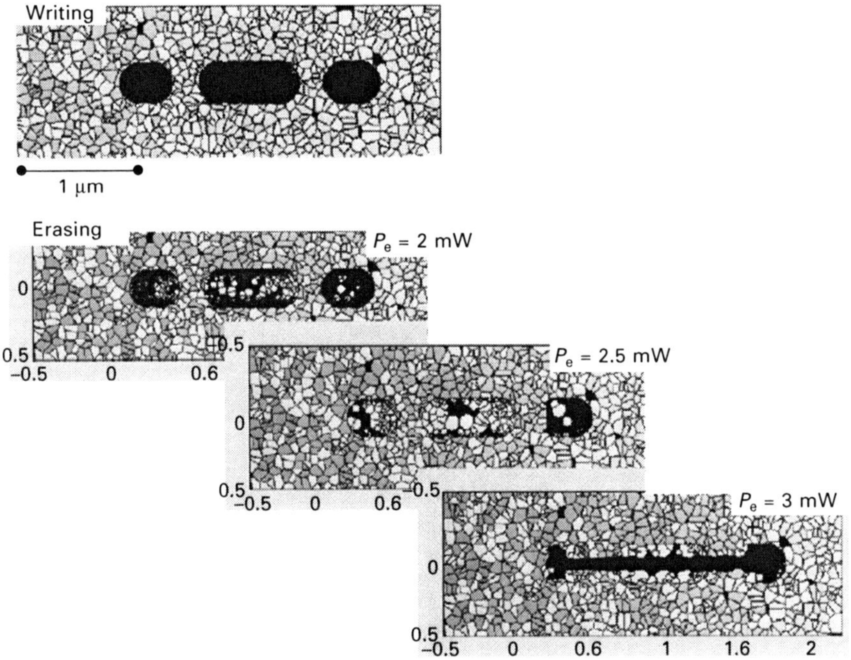

18.11 Simulation of the writing (amorphization) and erasing (crystallization) processes of a thin layer (8 nm) of $Ge_2Sb_2Te_5$ as active material in the case of the semi-transparent level of a dual-level Blu-ray RW disc.

$Ge_2Sb_2Te_5$ (Fig. 18.12). Consequently, a material with a more efficient nucleation process seems to be a conceivable solution. This is precisely the moment that Kojima and Yamada (2001) of Matsushita/Panasonic proposed substituting the classical $Ge_2Sb_2Te_5$ by a new tin-based chalcogenide alloy ($Ge_2Sb_2Sn_2Te_7$). The increase of the metallic nature of the GeSbTe alloy, via tin addition, appears to be a means to improve its crystallization ability. Its experimental behavior, in terms of crystallization speed, answers favourably to the Blu-ray specifications. This material allows ensuring an erasing process with a very fine crystalline microstructure, compatible with our proposition of 'reinforced' nucleation.

18.3 Stability of very thin amorphous chalcogenide layers

18.3.1 Crystallization speed as a function of film thickness

As discussed previously, the development of multi-level RW optical recording media requires using thinner and thinner PC layers (less than 10 nm). The

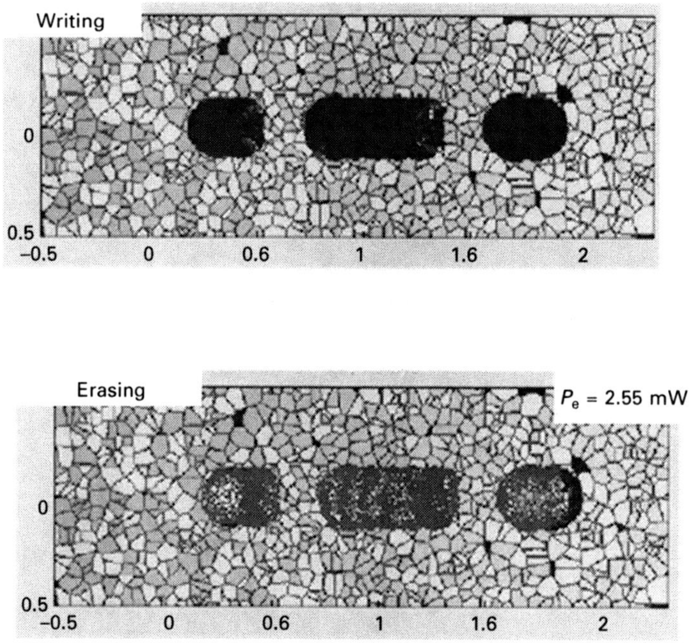

18.12 Simulation of the writing (amorphization) and erasing (crystallization) processes of a GeSbTe-based material with a nucleation rate twice higher than that of $Ge_2Sb_2Te_5$ shown in Fig. 18.11. The thin (8 nm) PC layer is included as in the case of Fig. 18.11 in the semi-transparent structure of a dual-level Blu-ray RW disc.

feasibility of a dual-level Blu-ray disc was demonstrated with a 7 nm-thick PC layer. The question is now to know if it is possible to develop media with more than two superimposed recording levels. For example, optical constraints for the development of a four-level disc imply using ultra-thin phase change layers of about 2–3 nm. However, it was reported by Kojima and Yamada (2001) that the crystallization ability of the GeSbTe material is strongly reduced with thinning of the film thickness below 5 nm (Fig. 18.13). It should be noticed that this phenomenon is not specific to the materials used in optical recording. Similar behaviors have been observed by Zacharias and Streitenberger (2000) in Si/SiO_2 or Ge/SiO_2 superlattices in the limit of ultrathin films. A German team led by Herr *et al.* (1999) has also demonstrated the stabilization of amorphous metallic (Fe-early transition metal) thin films when the thickness layer is 2 nm.

We describe in Section 18.3.2 a thermodynamic model explaining in an indirect way this phenomenon of thermal stabilization of very thin PC amorphous layers.

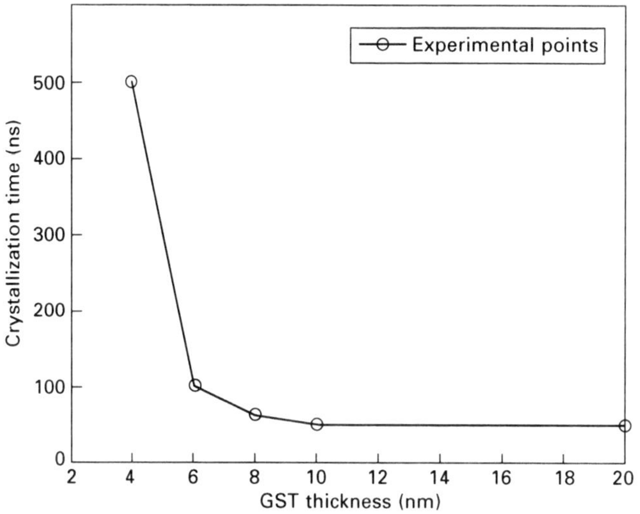

18.13 Measured crystallization time vs $Ge_2Sb_2Te_5$ layer thickness according to Kojima and Yamada (2001).

18.3.2 Thermodynamic considerations

Figure 18.14 is a schematic representation of the system used in the model where an amorphous thin layer is sandwiched between two adjacent crystalline layers (having the same composition as the glassy layer). Given the nanometric thickness of the films, the two amorphous–crystal interfaces interact via the glassy layer and the system, including these interfaces and the amorphous layer, must be treated globally. It is considered that the crystalline field acts on the atomic structure of the amorphous layer creating a gradient of the atomic density across the amorphous layer in the direction x perpendicular to the planar interfaces. The inhomogeneity (variation of the atomic density) of the amorphous layer is measured by an order parameter $\omega(x) = n(x)/n_c$ (where $n(x)$ is the atomic density in the disordered phase as a function of x and n_c is the atomic density in the crystal). Near the interfaces, due the crystalline field effect, the atomic density of the first glassy atomic layers is higher than in the central zone of the film. The order parameter $\omega(x)$ is consequently smaller in the centre of the film and shows a symmetric variation with regard to $x = 0$ ($\omega(x) = \omega_0$ in the center of the glassy layer and $\omega(x) = \omega_1$ in the glassy layer at the order–disorder interfaces near the crystals). The discontinuity of the topological order parameter at the interface ($\omega_1 \neq \omega_c$) is compatible with the 'first order' nature of the amorphous–crystal transition.

The expression of the Gibbs energy of this inhomogeneous system is treated on the basis of the Ginsburg–Landau potential. The minimization of this energy function leads to the definition of a layer thickness e^* corresponding

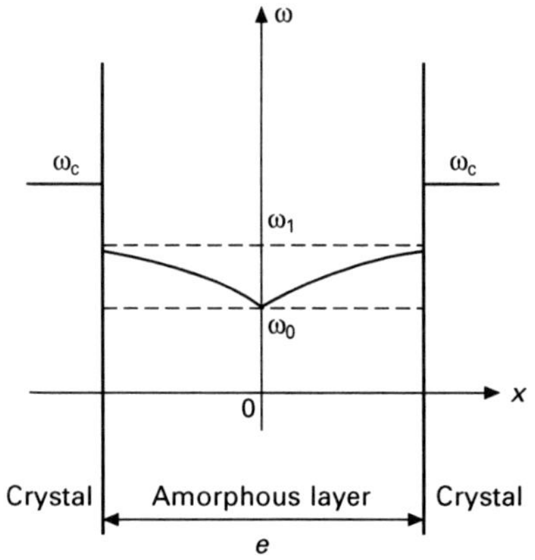

18.14 Schematic representation of the topological order parameter χ in the inter-crystals amorphous layer vs x in the direction perpendicular to the sharp plane interfaces.

to a minimum of the Gibbs energy, namely corresponding to a metastable state of the system:

$$e^* = 2\xi \arcsin(\varphi^{1/2}) \qquad [18.12]$$

with $$\varphi = \frac{G(\omega_0) - G(\omega_1)}{G(\omega_0) - G(\omega_c)}$$

where ξ is the correlation length measuring the interaction range, G is the Gibbs energy of the system.

In fact, the existence of such a minimum of the Gibbs energy and thus the existence of a characteristic thickness e^* is the consequence of two antagonist effects: on one hand, an attractive effect between the two interfaces tends to decrease the thickness of the amorphous layer because of its metastability (with regard to the crystal phase), and on the other hand, as thermodynamics requires, a repulsive effect tends to increase the thickness of the layer to decrease the gradient of the order parameter ω.

In the case of metallic and semi-metallic systems, where the interatomic interactions extend over relatively short distances, it seems reasonable to choose ξ of the order of 2 nm. Figure 18.15 shows the Gibbs energy of the system as a function of the layer thickness e (with $\varphi = 0.3$ and $\xi = 1, 2$ and 3 nm).

Such a minimum of the Gibbs energy (for $e = e^*$) describes a metastable state of the amorphous layer. This metastability contributes to a decrease in the driving force at the origin of the amorphous–crystal transition and to inhibit the crystallization process as long as the gradient of the order parameter is maintained. At higher temperature, where the atomic mobility

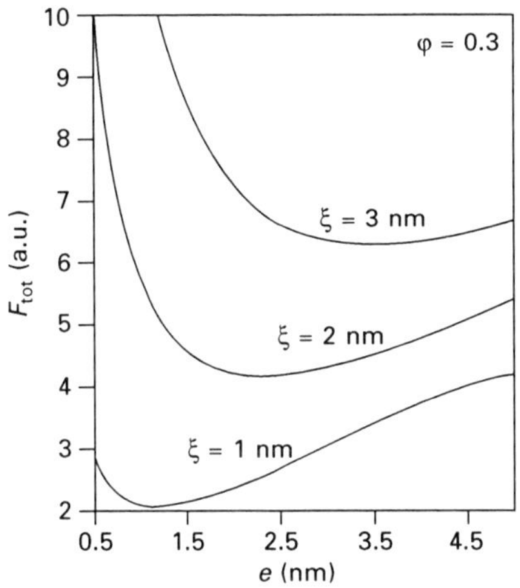

18.15 Gibbs energy of the system vs layer thickness *e* for different values of ξ when $\varphi = 0.3$.

becomes sufficient, this system can leave this metastability leading to the subsequent crystallization of the amorphous layer.

18.4 Influence of nitrogen on GeTe crystallization ability: application to embedded electrical memories

18.4.1 Context

Commercialization of electrical phase change memories (PCM) based on the well-known $Ge_2Sb_2Te_5$ compound has recently been started, tailored for consumer applications. Despite other excellent performance characteristics (e.g., low-power, scalability, etc.), data retention is assured up to 85°C, still limited for the automotive market segment where chips have to operate at elevated temperature in immediate vicinity of the hot engine. Alternative active material able to comply with the stringent requirements of automotive applications should exhibit possible higher crystallization temperature as well as higher activation energy with respect to GeSbTe. Recent literature showed that GeTe provides better retention (Fantini *et al.*, 2009; Raoux *et al.*, 2009), while several research works put in evidence how data retention is enhanced by inclusion in pure host materials (Horii *et al.*, 2003; Matsuzaki *et al.*, 2005; Morikawa *et al.*, 2007).

We propose to investigate in the next paragraphs, the impact of nitrogen doping in GeTe, as an alternative active material to boost the data retention performances of PCM.

18.4.2 Electrical characterization of phase change memory cell

The first criterion to select a potential material with improved data retention at high temperature (data retention = thermal stability of the melt-quenched amorphous state) is the crystallization temperature, T_χ. An increase of this temperature may, *a priori*, lead us to expect an increase of the thermal stability of the corresponding amorphous phase. Figure 18.16 shows that raising the nitrogen content (a few percent) in the GeTe binary compound, allows raising the crystallization temperature (with 10% of nitrogen, an increase of 75°C of T_χ is measured on the nitrogen-doped GeTe with regard to the pure GeTe).

To test the real potential of those materials, they have been integrated in simple test structures, named heater structure (Fig. 18.17), with a 300 nm

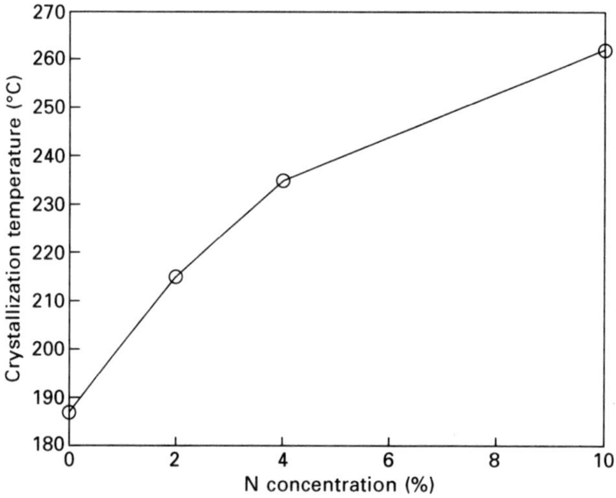

18.16 Crystallization temperature vs N concentration in the GeTe binary compound.

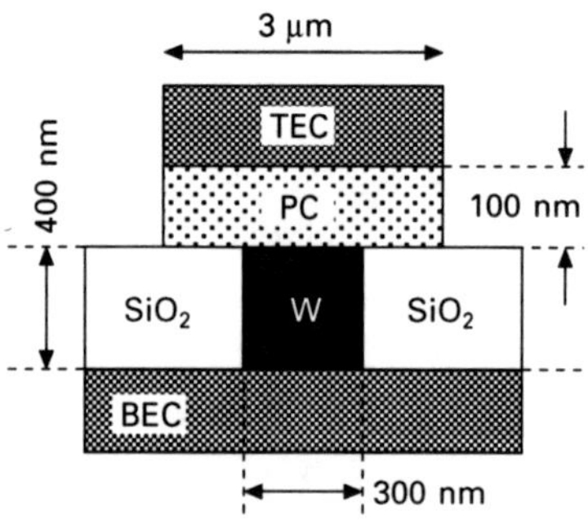

18.17 Electrical test structure named 'heater structure' with a 300 nm W plug.

plug. In Fig. 18.18, R-I SET-to-RESET characteristics are represented: all devices with different PC alloys show similar RESET current. The program curves are represented in Fig. 18.19: note that program speed is similar for

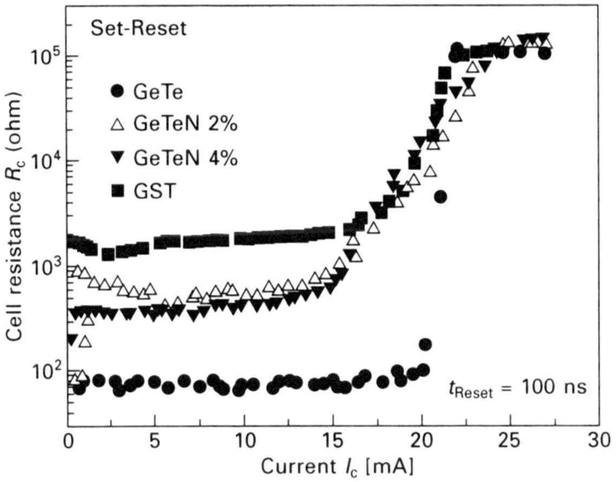

18.18 GST, GeTe and N-doped GeTe R-I SET-to-RESET characteristics (average values on 3 cells for each alloy). t_{RESET} = 100 ns.

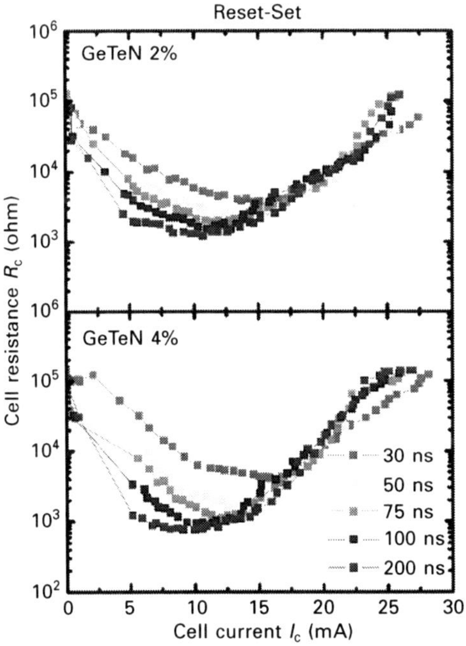

18.19 R-I program characteristics of N(2%)-doped GeTe and N(4%)-doped GeTe. 100 ns SET speed able to provide ~2 decades programming window. Almost no difference in write speed by different N doping.

both N doping (2 and 4%) and a one decade programming window can be achieved in 30 ns. In comparison to program performance of GeSbTe (Fig. 18.20), N-doped GeTe is faster. Eventually, the estimated data retention performance of different PC alloys are represented in Fig. 18.21. The extraction of the activation energy and an Arrhenius extrapolation at 10 years allow the fail temperature to be extrapolated. We see that the data retention of N-doped GeTe devices has been improved with respect to that of pure GeTe; in particular N (2%)-doped GeTe is demonstrated to perform the best with an extrapolated fail temperature, after 10 years, of 154°C, one of the

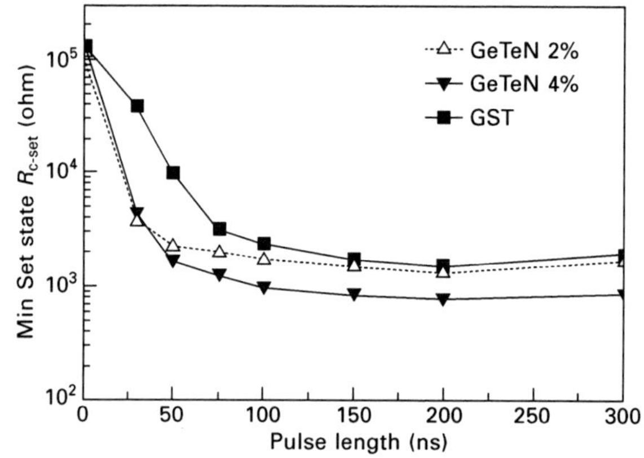

18.20 R-t_{SET} characteristics. GeTeN devices are faster than GST.

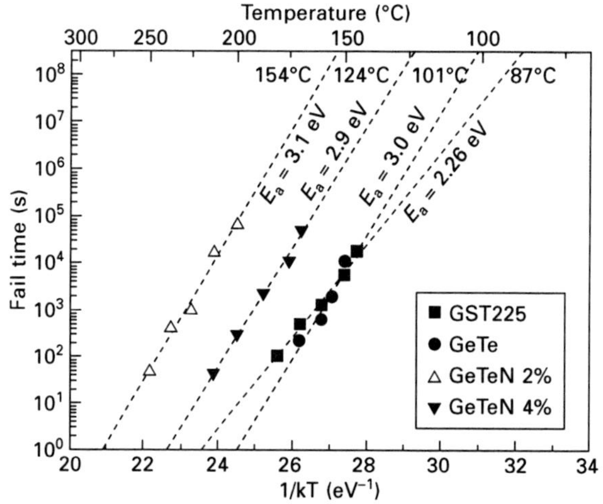

18.21 Data retention performances of the different PC alloys. Arrhenius extrapolation @ 10 years for GST, GeTe and N-doped GeTe devices: N(2%)-doped GeTe extrapolated fail temperature @ 10 years is 154°C.

highest ever shown in the literature. These results show that light nitrogen addition stabilizes the amorphous phase, without sacrificing the programming performance.

Note: Some interesting behavior deserves to be underlined; the extrapolated temperature of data retention (@10 years) is higher with N (2%)-doped GeTe than with N (4%)-doped GeTe. Contrary to what was expected according to the regular increase of T_χ with nitrogen concentration, the retention performance of the N-doped GeTe materials show a non-regular behavior with poorer performance when nitrogen concentration is higher than 2%. The results of electrical characterization have been published in Fantini *et al.* (2010).

In order to understand the role of nitrogen on the previous behaviors (crystallization temperature enhancement, increasing stability of the amorphous phase for light nitrogen doping), we have worked along two routes. This first route is an experimental way based on synchrotron experiments to investigate the structure of those materials (X-ray diffraction measurements to probe the global crystalline structure of the materials, XAFS experiments to probe the local structure of the materials around a chosen element, and also XPS experiments to get chemical information on the probed element). In parallel, we are doing *ab initio* simulations on different doped GeTe. Comparison of the total energies for structures with different location of the dopant (nitrogen) provides further grounds for the determination of the most stable structure containing N dopant.

18.4.3 Effect of nitrogen 'doping' on global and local order in crystalline GeTe

X-ray diffraction measurements on N-doped samples

XRD measurements were conducted on the BM32 (IF) beamline at the European Synchrotron Radiation Facility (Grenoble, France) using a GMT goniometer. 25 mm long samples were illuminated at 1.4° grazing incidence with a 300 × 300 micron 18 keV monochromatic beam. Diffraction (Bragg) peaks were collected in-plane with a Ge(111) analyzer crystal placed in front of a photomultiplier. Including X-ray beam convergence, the measured angular resolution is 9.7/1000°, a value that is negligible compared to the diffraction peak FWHM. Samples were heated under primary vacuum with 10°C/min heating or cooling rates over a temperature range from room temperature to 500°C.

A reversible behavior between a rhombohedral low-temperature phase and a rocksalt high-temperature phase in pure GeTe has been reconfirmed. Figure 18.22 shows a part of the XRD spectra measured at different temperatures that contains characteristic features of the two phases, namely, the [104] and [110] peaks of the rhombohedral phase (located at $q = 2.932$ Å^{-1} and q

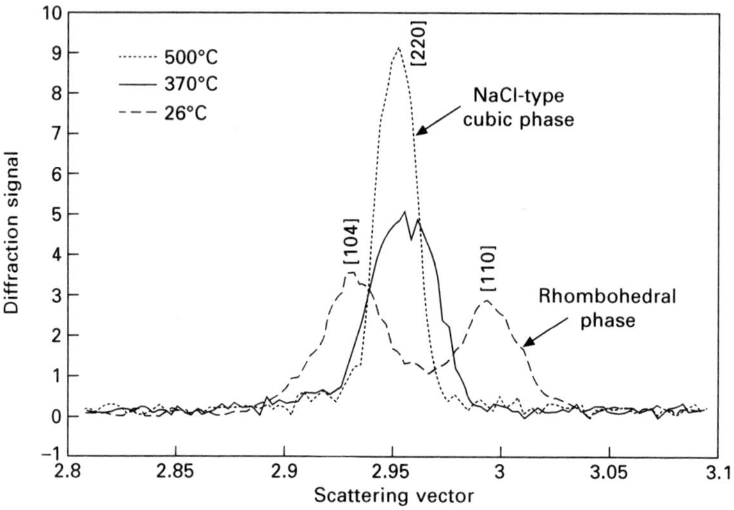

18.22 X-ray diffraction patterns of pure GeTe at different temperatures.

= 2.994 Å^{-1} at room temperature) that merge into the single [220] peak of the rock-salt phase (located at q = 2.952 Å^{-1} at 500°C).

The behavior drastically changes upon doping with nitrogen. As demonstrated in Fig. 18.23, just 4 at.% of dopants already leads to a significant smearing of the transition and in samples with ~ 10 at.% doping, only one peak is observed throughout the temperature range studied, indicating that globally the structure became cubic as a result of doping. It should also be noted that the Curie temperature decreases with the increased concentration of dopants. While the approximately 4% doped samples appear cubic at temperatures significantly lower than the Curie temperature of undoped GeTe, the sample that contains more than 10 at.% doping exhibits a single peak at room temperature and above which is characteristic of the cubic phase.

From these results it follows that the structure of N-doped GeTe becomes cubic. At the same time, Bragg diffraction (by definition) measures the average structure and is not sensitive to local distortions (Stern and Yacoby, 1996). For example, while for pure GeTe the rhombohedral-to-cubic transition has been reported to occur based on neutron diffraction studies, subsequent EXAFS measurements demonstrated that locally the structure remain distorted, i.e. the transition that takes place is not of displacive but rather of order–disorder nature according to Fons *et al.* (2010). In order to verify if the same was possible for the case of doping we have performed XAFS studies at Ge K-edge on N-doped samples. The results are described in the next section. Ge K-edge XAFS measurements on N-doped samples

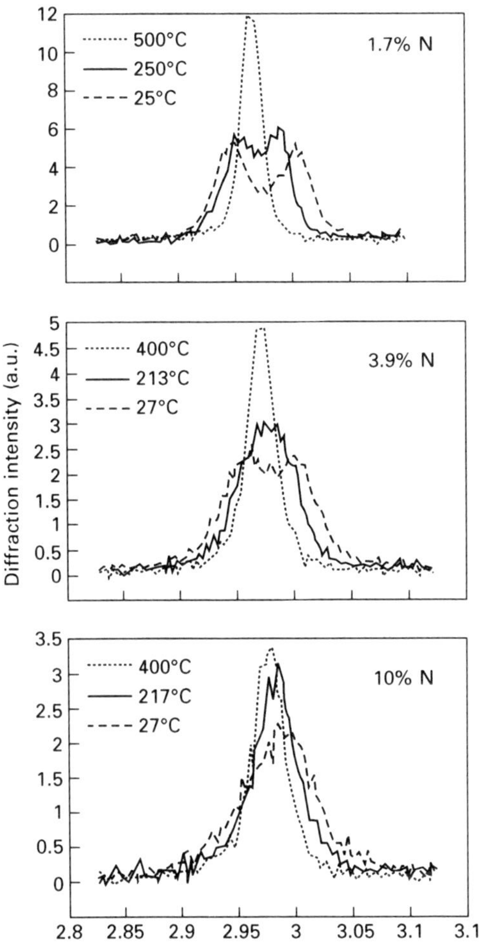

18.23 X-ray diffraction patterns of N-doped GeTe at different temperatures.

EXAFS measurements were performed at beamline BL01B1 at SPring-8 (Hyogo, Japan). The samples (GeTe-500 nm/Al foil-15 um/GeTe-500 nm) were stacked to obtain the optimum thickness for the edge jump of unity at the Ge K-edge. Measurements were done over a temperature range of 10–300 K. The obtained spectra were analyzed using ARTEMIS and ATHENA packages. The Fourier transformed EXAFS spectra are shown in Fig. 18.24. The shown data unambiguously demonstrate that the GeTe bond lengths remain unchanged after doping, i.e. the rhombohedral local distortion is preserved. Fitting of the experimental EXAFS data yields the values for the Ge-Te bond length and disorder parameter (mean-square relative displacement, MSRD) that are summarized in Table 18.1. The MSRD values are shown for spectra

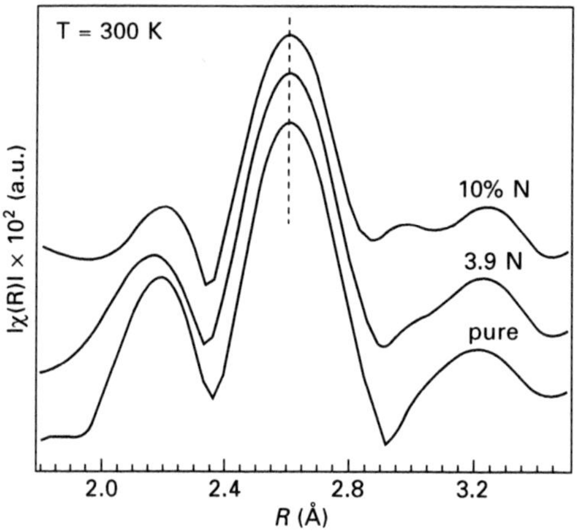

18.24 Fourier transformed EXAFS spectra of N-doped GeTe measured at room temperature. The dashed line (guide for the eyes) marks the shorter Ge-Te bond. The peak located at ~ 2.2Å originates predominantly from Ge-Ge bonds (Kolobov *et al.*, 2003).

Table 18.1 Ge-Te bond-length and 'static' MSRD at 10 K for undoped and N-doped GeTe

Composition	Ge-Te distance (Å)	MSRD ($Å^2 \times 10^3$)
GeTe (undoped)	2.83	2.7
GeTe (3.9% N)	2.84	3.2
GeTe (10% N)	2.85	4.1

taken at ~ 10 K in order to emphasize the static disorder due to dopants. One can see that the Ge-Te bond length remains essentially unchanged (2.83–2.85 Å) after doping. At the same time, the disorder becomes larger in the doped sample.

The reported results demonstrate that despite a significant change in the global structure of nitrogen-doped GeTe probed by XRD that manifests itself as a rhombohedral-to-rocksalt transition; the local structure remains unchanged, i.e. rhombohedrally distorted, across the ferroelectric-to-paraelectric transition temperature similar to the cases of undoped GeTe (Fons *et al.*, 2010) and the underlying mechanism can be viewed as follows. At low temperatures and/or low doping levels, the long-range forces ensure coherent displacements of all Ge atoms with respect to the Te sublattice which determines the rhombohedral structure of GeTe.

As the disorder is increased, either due to thermal effects or compositional disorder, the rather weak long-range forces are no longer sufficient to preserve the coherently distorted structure and the distortions become stochastic (Fig. 18.25). XRD which is 'looking at' the ensemble average structure interprets

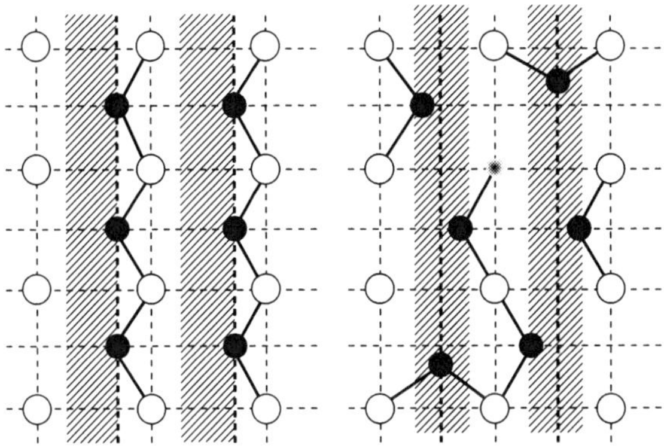

18.25 Schematic presentation of the undoped (a) and doped (b) GeTe structures. The subset of the shorter bonds is shown by solid lines. In the undoped GeTe, all Ge atoms are displaced coherently with the resulting shorter and longer interplane distances. Doping generates perturbation of the ordering with the resulting randomization of the distortions generating a structure with on average equal interplane spacings.

this change as a transition to a cubic phase when Ge atoms are displaced into an equidistant position with respect to the first-nearest Te neighbors. EXAFS, on the other hand, is a true local probe that 'sees' the local structure and can reliably detect local distortions.

The results of XRD and XAFS studies have been published in Biquard *et al.* (2011).

18.4.4 Atomic nitrogen or molecular nitrogen?

Soft X-ray absorption studies of N-doped GeTe

The location of the dopant inside GeTe and its role in stabilizing the amorphous phase is not known and this information is necessary in order to understand the mechanism of the allegedly improved device performance as well as to engineer novel materials. The aim of these absorption studies is to investigate the local structure around N species in doped GeTe. Measurements were performed at beamline BL27SU at SPring-8. Figure 18.26 shows the XANES spectra for N (5%)-doped GeTe samples in as-deposited amorphous and crystalline states. The spectra for N (10%)-doped GeTe are very similar and not shown here. Pronounced differences in the line shape demonstrate that local structure around nitrogen species is significantly different.

In parallel to the experimental part, first-principles calculations (using density function theory (DFT) for lattice relaxation and Green function

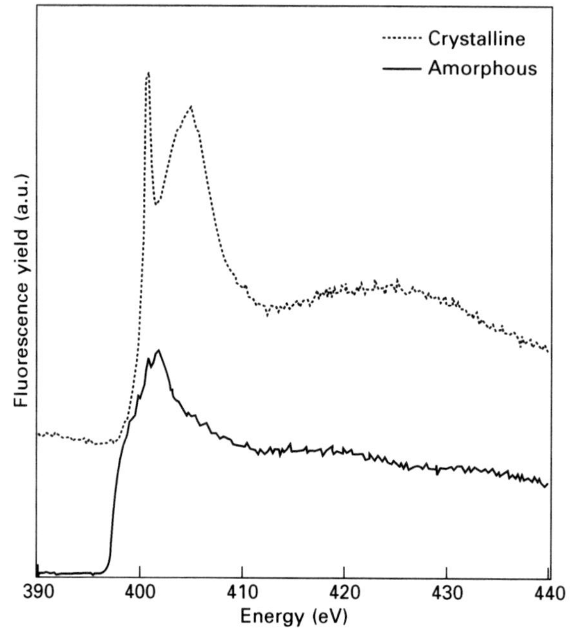

18.26 Nitrogen K-edge XANES spectra of as-deposited amorphous and crystallized N (5%)-doped GeTe.

based full multiple scattering) are done to simulate XANES spectra. The simulations show that XANES spectra at nitrogen K-edge depend strongly on the location of N in the lattice. Example calculations are included in Fig. 18.27 which shows N on Ge and Te sites (Figs 18.27a and b) with lattice relaxation (at 0 K) and Fig. 18.27(c) which shows the simulated XANES spectra. In Fig. 18.27(c), we use the notation of the following kind: N(Ge) stands for N-doped GeTe with the N atom located on a Ge site. Comparison of the experimentally measured spectra with simulated spectra assuming different location of N in crystalline GeTe allows us to suggest that N is predominantly bonded to Ge species.

Of special interest is an appearance of a sharp feature around 400 eV in the annealed spectra. While a somewhat similar feature exists in the simulated spectrum, it is highly possible that it corresponds to formation of molecular nitrogen. To address this issue, we have measured a high-resolution spectrum around the energy of interest. The results are shown in Fig. 18.28. One can clearly see the characteristic molecular nitrogen features in the high-resolution spectrum (Fons *et al.*, 2006) which unambiguously demonstrate that upon annealing a large fraction of nitrogen is transformed into molecular nitrogen. Our estimates suggest that doping above 1–2 at.% results in a material that contains molecular nitrogen.

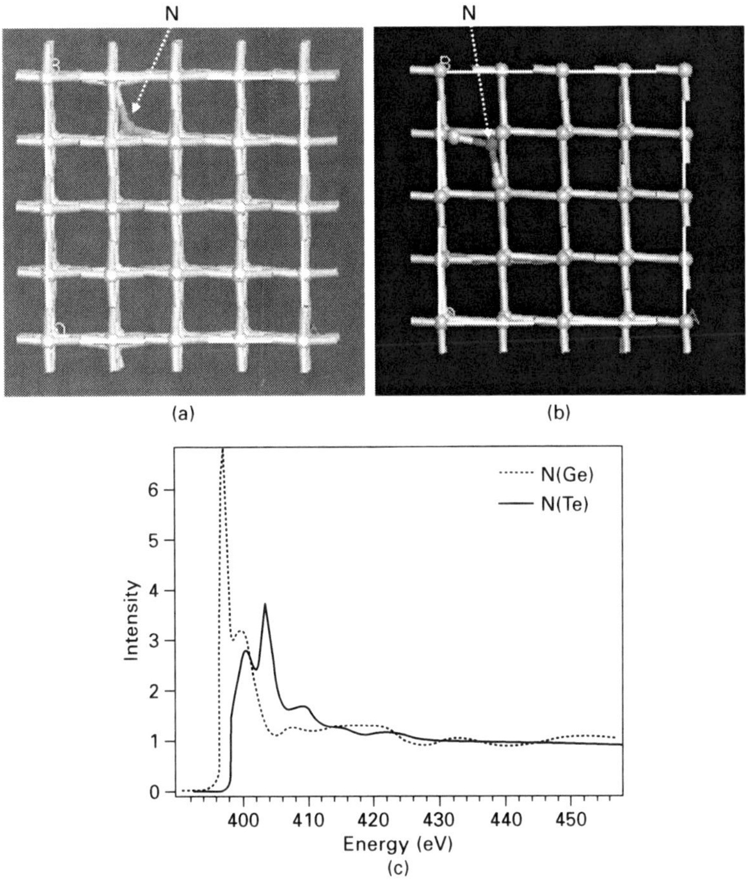

18.27 Relaxed structures for isolated N dopants on Ge (a) and Te (b) sites and corresponding simulated XANES spectra (c). In part (c), N(Ge) stands for N-doped GeTe with the N atom located on a Ge site. A strong difference between the spectra is clear.

High energy X-ray photoemission measurements of N-doped GeTe

In order to get further insights into the local structure around N dopants, we have additionally performed synchrotron-based X-ray photoemission experiments. The important advantage of using synchrotron radiation over conventional laboratory equipment is the escape depth. While excitation using laboratory sources probes a region that is very close to the surface (a couple of nanometers at best), use of high energy X-rays allows one to investigate regions 20–30 nm deep, i.e. to probe the bulk properties. Measurements were done at SPring-8 at beamline BL15XU.

Te 4d spectra (not shown) display very little change upon crystallization, only a slight narrowing of spectral features indicating a higher degree of

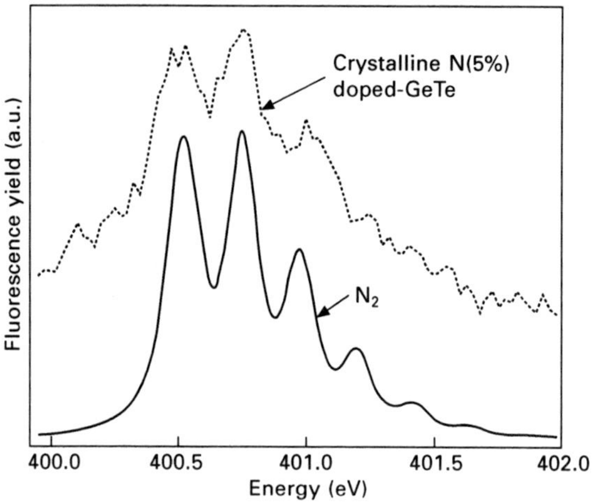

18.28 Comparison of high-resolution spectrum of the annealed N(5%)-doped GeTe with that of molecular nitrogen demonstrates the formation of N_2 molecules upon crystallization.

order in the structure is observed for the spectra of the crystalline samples. On the other hand, Ge spectra, especially those of the Ge 3d level (Fig. 18.29(a)), exhibit a pronounced difference between the amorphous and crystalline phases. The peak located at ca. 33 eV corresponds to Ge oxide (Yashina *et al.*, 2008). This peak gets more intense in the annealed samples and demonstrates that Ge species get easily oxidized despite annealing in inert atmosphere. Of higher interest are the features in a 29–32 eV range that correspond to Ge species in GeTe proper (Yashina *et al.*, 2008). The broad spectrum for the amorphous samples is likely to be composed of two (or more) major contributions. As demonstrated by previous XAFS measurements and simulations, Ge in the amorphous phase may be present in different bonding geometries, namely, either bonded to tellurium in Ge(4):Te(2) and Ge(3):Te(3) local configurations, or forming Ge-Ge bonds. In N-doped samples there is the additional possibility of Ge-N bonds. The broad amorphous phase spectrum is thus comprised of contributions from the various Ge species. Its shift to higher energies with respect to crystalline GeTe is in agreement with the location of Ge3d peak (30.2 eV according to Maeda *et al.*, 2006) in pure Ge_3N_4. In the spectrum of the crystalline phase, a single spin-orbit split Ge feature is clearly present. The obtained spectrum is in agreement with previous studies of crystalline GeTe (α-GeTe(111) surfaces according to Yashina *et al.*, 2008) and suggests that there are almost no Ge-N bonds in the crystalline phase, i.e. Ge-N bonds existing in the amorphous phase get broken upon crystallization.

This conclusion is further strengthened by the evolution of the N 1s spectra upon transformation from the amorphous phase to the crystalline phase (Fig. 18.29(b)). The spectra for the amorphous samples are dominated by the feature that agrees with the value characteristic of nitrogen bonded to Ge (397.5 eV according to Soukup *et al.*, 1996, and 398.1 eV according to Maeda *et al.*, 2006). The spectrum is very different for the crystalline samples. Upon crystallization, the low binding energy component essentially disappears and the N 1s spectra is dominated by the feature located at ca. 398.2 eV, i.e. close to the energy range that is usually associated with molecular nitrogen (398.9–400 eV according to Fukuda and Nagoshi, 1988), similar to the case reported by Fons *et al.* (2006) for ZnO. We believe that a shift to lower energies is due to the fact that molecular nitrogen is located not on the surface but inside the material and hence experiences a stronger interaction with the matrix.

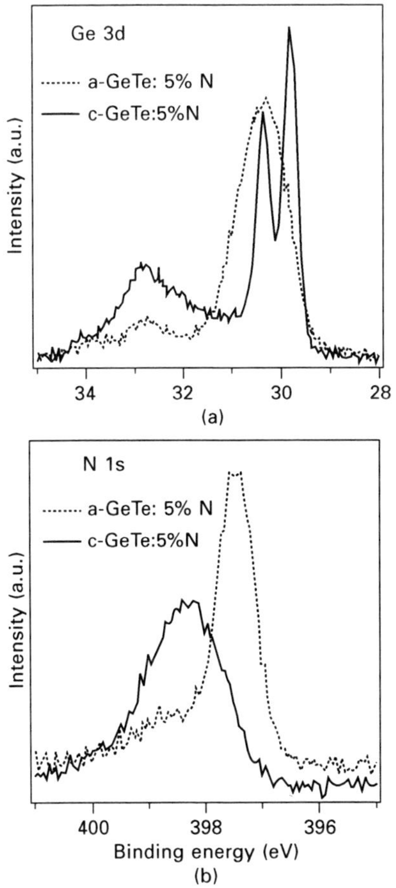

18.29 Ge 3d (a) and N 1s (b) XPS spectra of as-deposited amorphous and crystallized N (5%)-doped GeTe samples.

The results of soft X-ray absorption studies and bulk XPS measurements unambiguously demonstrate that while in as-deposited amorphous phase, N atoms are chemically bonded to Ge species, upon crystallization most N atoms are transformed into molecular nitrogen.

Ab-initio *studies of the structure of N-doped GeTe*

The previous results demonstrate that the concentration of nitrogen that can be stably incorporated into the matrix of GeTe is of the order of 1–2 at.%. Any nitrogen in excess of this concentration is transformed into molecular nitrogen during the first crystallization cycle. We now address an issue of the location of N in the crystalline phase. One possibility is that molecular nitrogen is incorporated into the crystalline phase proper, and the other one is that it is located at grain boundaries. Figure 18.30 shows the structures obtained by 0 K relaxation of the crystalline GeTe with two nitrogen atoms placed on adjacent Te sites so that N-Ge bonds exist in the starting model. Upon relaxation, the N atoms *move away from each other* suggesting that the formation of molecular nitrogen within *the crystalline phase proper* is unlikely. The N_2 molecules observed experimentally are most likely at the grain boundaries.

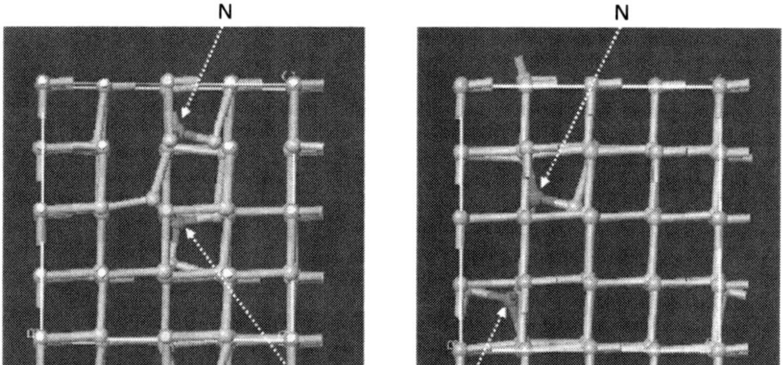

18.30 Relaxed crystalline GeTe:N structures with two nitrogen atoms located next to each other, one on a Ge site and one on a Te site (left) and both on Te sites (right) within the GeTe unit cell. In both cases, during the relaxation the Ge atoms moved away from each other suggesting that the formation of molecular nitrogen within the crystalline phase proper is very unlikely.

18.4.5 Discussion on the role of nitrogen: influence on amorphous phase stability or data retention performance

We have clearly identified that from 4%N included into the material, there is formation of N_2 molecules during the first crystallization cycle. As the energy necessary to dissociate molecular nitrogen is significantly larger that the energy needed to amorphize the PC material, it seems reasonable to assume that these very stable N_2 molecules are going to persist during the RESET (amorphization) cycles. These molecules can then act as heterogeneous nucleation sites and fail the stability of the amorphous phase. The decrease of the crystalline grain size with the nitrogen concentration seems to strengthen this hypothesis (Fig. 18.31).

It appears, consequently, that the optimal nitrogen concentration is the one allowing the formation of Ge-N bonds in order to increase the thermal stability of the amorphous phase but without molecular nitrogen formation at the origin of possible phenomena of hererogeneous nucleation unfavorable to data retention.

The N concentration should be, according to the previous results, less than 4%. And we see that this scenario allows interpreting the data retention behavior shown in Section 18.4.2, where the thermal stability of the melt-quenched amorphous phase (data retention) decreases when the N concentration is higher than 2%.

The results of XANES and XPS have been published in Kolobov *et al.* (2012).

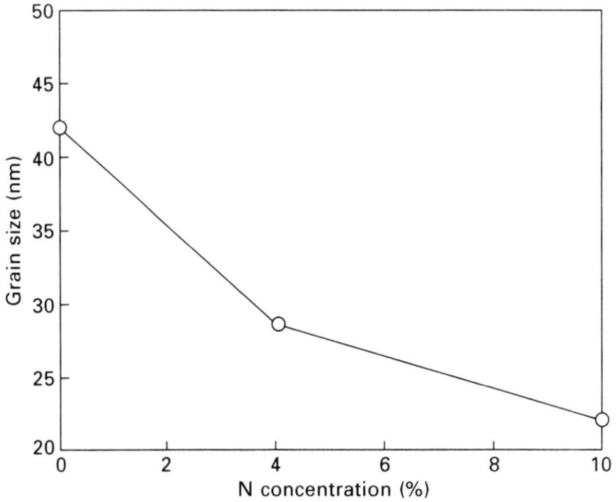

18.31 Crystalline grain size vs N concentration in GeTe.

18.5 Conclusion

Driven by commercial applications, materials research for optical recording has focused in the past 30 years on higher data rates and higher density formats. With the realization of the high-speed DVD+RW and Blu-ray discs, research on optical phase change materials seems to have slowed down. However, at the same time two interesting developments can be observed. On one hand, developments of non-volatile electronic memories are increasing. Here, a lot can be learned from the past research on optical memories. On the other hand, there is an increasing effort to improve understanding of the fundamental properties of phase change materials. This may lead to new insights that will help to explore these exciting materials even further.

18.6 References

Bez R and Pirovano A (2004), *Mater. Sci. Semicond. Proc.* 7, 349–355.

Biquard X, Krbal M, Kolobov A V, Fons P, Simpson R E, Hyot B, Andre B, Tominaga J and Tomaya U (2011), *Appl. Phys. Lett.*, 98, 231907.

Chen Y C, Rettner C T, Raoux S, Burr G W, Chen S-H, Shelby R M, Salinga M, Risk W P, Happ T D, McClelland G M, Breitwisch M, Schrott A, Philipp T B, Lee M H, Cheek R, Nirschl T, Lamorey M, Chen C F, Joseph E, Zaidi S, Yee B, Lung H L, Bergmann R and Lam C (2006), *IEDM Tech. Digest* 777–780.

Cho W Y, Cho B-H, Choi B-G, Oh H-R, Kang S, Kim K-S, Kim K-H, Kim D-E, Kwak C-K, Byun H-G, Hwang Y, Ahn S, Koh G-H, Jeong G, Jeong H and Kim K (2005), *IEEE J. Solid-State Circuits* 40, 293–300.

Fantini A, Perniola L, Armand M, Nodin J F, Sousa V, Lhostis S, Jahan C, Maitrejean S, Toffoli A, Kies R, Persico A, Reimbold G, De Salvo B, Mazoyer P, Benshael D and Boulanger F (2009), *IMW conference (International Memory Workshop)*, Monterey, CA.

Fantini A, Sousa V, Perniola L, Gourvest E, Bastien J C, Maitrejean S, Braga S, Pashkov N, Bastard A, Hyot B, Roule A, Persico A, Feldis H, Jahan C, Nodin J F, Toffoli A, Reimbold G, Fillot F, Roussel H, Pierre F, Annunziata R, Benshael D, Mazoyer P, Vallée C, Billon T, Hazart J, De Salvo B and Boulanger F (2010), *IEDM conference (International Electron Devices Meeting)*, San Francisco, CA.

Fons P, Tampo H, Kolobov A V, Ohkubo M, Niki S, Tominaga J, Carboni R, Boscherini F and Friedrich S (2006), *Phys. Rev. Lett.* 96, 045504.

Fons P, Kolobov A V, Krbal M, Tominaga J, Andrikopoulos K S, Yannopoulos S N, Voyiatzis G A and Uruga T (2010), *Phys. Rev. B*, 82, 155209.

Fukuda Y and Nagoshi M (1998), *Surf. Sci.* 203, L651–L655.

Fulcher G S (1925), *J. Am. Ceram. Soc.* 8, 339–355.

Herr U, Geisler H, Ippach H and Samwer K (1999), *Phys. Rev. B*, 59, 21, 13719.

Horii H, Yi J H, Park J H, Ha Y H, Baek I G, Park S O, Hwang Y N, Lee S H, Kim Y T, Lee K H, Chung U I and Moon J T (2003), *VLSI (Very Large Scale Integration) Technology Conference* (Technical Digest).

Hudgens S and Johnson B (2004), *Mater. Res. Soc. Bull.* 29, 829–832.

Kashchiev D (1969), *Surf. Sci.* 14, 209–220.

Kelton K F, Greer A L and Thompson C V (1983), *J. Chem. Phys.* 79 (12): 6261–6276.

Kojima R and Yamada N (2001), *Jpn. J. Appl. Phys.*, 40, 5930–5937.

Kolobov A V, Tominaga J, Fons P and Uruga T (2003), *Appl. Phys. Lett.*, 82, 382–384.

Kolobov A V, Fons P, Hyot B, André B, Tominaga J, Tamenori Y, Yoshikawa H and Kobayashi K, (2012), *Appl. Phys. Lett.*, 100, 061910.

Maeda T, Yasuda T, Nishizawa M, Miyata N, Morita Y and Takagi S (2006), *J. Appl. Phys.*, 100, 014101.

Matsuzaki N, Kurotsuchi K, Matsui Y, Tonomura O, Yamamoto N, Fujisaki Y, Kitai N, Takemura R, Osada K, Hanzawa S, Moriya H, Iwasaki T, Kawahara T, Takaura N, Terao M, Matsuoka M and Monowa M (2005), *IEDM (Electron Devices Meeting)* Technical Digest.

Meinders E R, Mijiritskii A V, Van Pieterson L and Wuttig M (2006), *Optical Data Storage. Phase-Change Media and Recording*. Dordrecht: Springer.

Morikawa T, Kurotsushi K, Kinoshita M, Matsuzaki N, Matsui Y, Fujisaki Y, Hanzawa S, Kotabe A, Terao M, Moriya H, Iwasaki T, Matsuoka M, Nitta F, Moniwa M, Koga T and Takaura N (2007), *IEDM (Electron Devices Meeting)* Technical Digest.

Ohta T (2001), *J. Optoelectron. Adv. Mater.* 3, 609–626.

Ovshinsky S R (1968), *Phys. Rev. Lett.* 21, 1450–1453.

Raoux S, Cheng H-Y, Munoz B and Jordan-Sweet J (2009), *EPCOS conference (European Phase Change and Ovonics Symposium)*, Aachen, Germany.

Soukup L, Perina V, Jastrabik L, Sicha M, Pokorny P, Soukup R, Novak M and Zemek J (1996), *Surf. Coat. Tech.*, 78, 280–283.

Spaepen F (1986), *Mat. Res. Soc. Symp. Proc.* 57, 161–184.

Stern E and Yacoby Y (1996), *J. Phys. Chem. Solids*, 57, 1449–1455.

Vogel H (1921), *Phys. Z.*, 22, 645–646.

Volmer M (1929), *Zeitschrift für Elektrochemie*, 35, 555–561.

Yamada N, Ohno E, Akahira N, Nishiuchi K, Nagata K and Takao M (1987), *Jpn J. Appl. Phys.*, Suppl. 26-4, 61–66.

Yamada N, Ohno E, Nishiuchi K, Akahira N and Takao M (1991), *J. Appl. Phys.*, 69, 2849–2856.

Yashina L V, Puettner R, Neudachina V S, Zyubina T S, Shtanov V I and Poygin M V (2008), *J. Appl. Phys.*, 103, 094909.

Zacharias M and Streitenberger P (2000), *Phys. Rev. B*, 62, 12, 8391–8396.

Zeldovich Y B (1942), *Zurnal Eksperimental'noj I Teoreticeskoj Fiziki*, 12, 525–538.

19
Chalcogenide glasses as electrolytes for batteries

M. TATSUMISAGO and A. HAYASHI,
Osaka Prefecture University, Japan

DOI: 10.1533/9780857093561.2.632

Abstract: Recent developments of sulfide glass electrolytes and solid-state batteries with these electrolytes are reviewed. Sulfide-based glass and glass-ceramic electrolytes in the systems Li_2S-SiS_2-Li_4SiO_4 and Li_2S-P_2S_5 have the advantages of high Li^+ ion conductivity, wide electrochemical window, and low grain-boundary resistance as a compressed powder pellet. Bulk-type all-solid-state batteries with sulfide electrolytes exhibited excellent cycle performance at room temperature. High temperature operation and surface coating of active materials such as $LiCoO_2$ with oxide thin films were useful for improving rate performance of solid-state batteries. Sulfur was used as an active material for solid-state batteries and they exhibited a large reversible capacity of more than 1000 mAh g^{-1} with a good cyclability. Sulfide thin films were formed on $LiCoO_2$ particles by pulsed laser deposition and a favorable electrode–electrolyte interface was effective in enhancing the energy density of solid-state batteries.

Key words: solid electrolyte, sulfide electrolyte, lithium ion conductor, glass electrolyte, glass-ceramic electrolyte, all-solid-state lithium battery.

19.1 Introduction

Chalcogenide glasses, in particular sulfide glasses, are well known to be a useful solid electrolyte because sulfide glasses with high Li^+ ion concentration have much higher Li^+ conductivity than oxide glasses (Julien and Nazri, 1994; Fergus, 2010). All-solid-state batteries with inorganic solid electrolytes have been studied in recent decades. Solid-state batteries with inorganic solid electrolytes and electrodes, which have advantages such as being safe (no flammability and no leakage of liquids) and having long cycle lives and versatile geometries, are recognized as next-generation power sources for a wide range of applications such as eco-cars and distributed home power storage (Julien and Nazri, 1994; Minami *et al.*, 2005). Bulk-type solid-state batteries composed of compressed powder layers of electrode and electrolyte have an advantage of achieving high energy density by using large amounts of active materials in an electrode layer, compared to thin-film solid-state batteries. Bulk-type solid-state batteries with the sulfide electrolytes exhibited excellent cycle performance at room temperature (Minami *et al.*, 2006; Tatsumisago and Hayashi, 2008).

In this chapter, after a brief introduction of the superiority of sulfides compared to oxides, recent development of sulfide glass and glass-ceramic electrolytes with high Li^+ ion conductivity is reported. Fabrication and characterization of bulk-type solid-state batteries with sulfide electrolytes are demonstrated. Several approaches to improve battery performance developed by our research group are discussed.

19.2 Advantages of sulfide glasses as solid electrolytes

Oxide and sulfide inorganic solid electrolytes have been widely studied. Inorganic solid electrolytes are classified into crystalline, glass, and glass-ceramic electrolytes. Crystalline oxide and sulfide materials with high Li^+ ion conductivity have been reported. Oxide crystalline materials such as $Li_{1+x}Al_xTi_{2-x}(PO_4)_3$ (NASICON) (Aono *et al.*, 1990), $Li_{0.5-3x}La_{0.5+x}TiO_3$ (perovskite) (Ito *et al.*, 1994) and $Li_7La_3Zr_2O_{12}$ (garnet) (Murugan *et al.*, 2007) have conductivities of 10^{-4}–10^{-3} S cm^{-1} at 25°C. Effective sintering at high temperatures (over 1,000°C) to decrease grain boundary is important to increase total conductivity of oxide crystalline electrolytes because grain-boundary resistance is larger in total resistance. Sulfide crystalline materials such as $Li_{3.25}Ge_{0.25}P_{0.75}S_4$ (Kanno and Murayama, 2001) and $Li_{10}GeP_2S_{12}$ (Kamaya *et al.*, 2011) have conductivities of 10^{-3}–10^{-2} S cm^{-1}. Sulfide electrolytes have a great merit of easy reduction of grain-boundary resistance by conventional cold-press of electrolyte powders.

On the other hand, glasses have several advantages as solid electrolytes for all-solid-state batteries. Glass electrolytes show a relatively high conductivity in a wide composition range, whereas crystalline electrolytes have a high conductivity at a limited composition giving favorable conduction paths for lithium ions. A principal strategy to develop the conductivity of glass electrolytes is to increase the number and mobility of lithium ions. Figure 19.1 shows the composition dependence of ionic conductivity at 25°C for the oxide and sulfide glasses in the systems Li_2O-SiO_2, Li_2O-P_2O_5, Li_2S-SiS_2, and Li_2S-P_2S_5. By increasing lithium ion concentration in the glasses, the conductivity of the glasses in all the systems monotonously increased. The ionic conductivity drastically increased by changing the glass matrix from oxides to sulfides. Lithium ions acting as a 'hard acid', which is classified from the viewpoint of the 'hard and soft acids and bases theory' of Pearson, would be more compatible to sulfide ions acting as a 'soft base' (Pearson, 1968). Further increase in conductivity of sulfide glasses was achieved by the addition of lithium halides such as LiI (Zhang and Kennedy, 1990) and lithium ortho-oxosalts such as Li_4SiO_4 (Hirai *et al.*, 1995). The latter oxysulfide glasses will be demonstrated in the next section.

Glasses have another benefit of easy softening at glass transition temperature (T_g). A dense sulfide glass pellet can be prepared by hot-press in the vicinity of T_g. For example, softening adhesion among $80Li_2S{\cdot}20P_2S_5$ glass particles occurred by pressing at around T_g and obvious grain boundaries and voids almost disappeared (Kitaura *et al.*, 2011). The hot-pressed pellet showed higher conductivity of 8.8×10^{-4} S cm^{-1} at 25°C compared to the conductivity of 3.7×10^{-4} S cm^{-1} in the cold-pressed pellet. The electrical conductivity increased by decreasing grain boundaries and voids. Moreover, this softening technique was applicable to form intimate contact at the electrode/electrolyte interface in all-solid-state batteries. Using viscous flow of supercooled liquid of electrolyte onto solid-state active materials was effective in forming liquid–solid interface, and the interface would give close solid–solid contact by cooling down to room temperature.

By heating of a supercooled liquid of electrolyte, a metastable or high-temperature phase with high conductivity tends to precipitate as a primary crystal. As a typical example, α-AgI (high-temperature phase) with extremely high Ag^+ ion conductivity was stabilized at room temperature by crystallization of AgI-based oxyhalide glasses (Tatsumisago *et al.*, 2001). For Li^+ ion conductors, high-temperature phase of $Li_7P_3S_{11}$ was prepared by a careful heat-treatment of Li_2S-P_2S_5 glass electrolytes (Mizuno *et al.*, 2005) and the obtained glass-ceramic electrolytes exhibited extremely high conductivity of 5.4×10^{-3} S cm^{-1} (Hayashi *et al.*, 2010a) as shown in Fig. 19.1. Therefore,

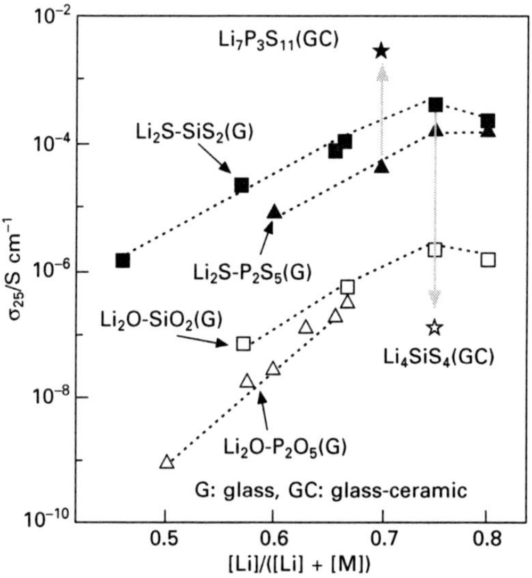

19.1 Composition dependence of conductivity at 25°C for oxide and sulfide glass-based electrolytes.

glass electrolytes are quite useful as a precursor for precipitating superionic crystals, which are difficult to be synthesized by a conventional solid-state reaction. The relationship between crystalline phase and conductivity will be reported in detail later.

19.3 Development of sulfide electrolytes for battery application

19.3.1 Sulfide and oxysulfide glass electrolytes

In general, it is difficult to prepare sulfide-based glass electrolytes with high Li^+ ion concentration because they are easily crystallized during the melt quenching process. To avoid crystallization, the glasses can be obtained by a rapid quenching technique using a twin-roller quenching apparatus (the cooling rate is larger than 10^5 K s^{-1}) (Tatsumisago and Minami, 1987). On the other hand, those glasses were also prepared via a mechanochemical process using a planetary ball mill apparatus. This process has several advantages as follows: the whole process is performed at room temperature, and fine electrolyte powders, which can be directly applied to solid-state batteries, are obtained without an additional pulverizing procedure. For most sulfide systems, the melting reaction has to be carried out in sealed quartz tubes because of the high vapor pressure of the sulfide starting materials at high temperatures. These glasses can be mechanochemically synthesized at room temperature and normal pressure (Morimoto *et al.*, 1999).

Figure 19.2 shows the differential thermal analysis (DTA) curves of the $70Li_2S{\cdot}30P_2S_5$ (mol%) glasses prepared by ball milling with two different experimental conditions (Hayashi *et al.*, 2010a). The milling process was carried out in an Ar atmosphere at room temperature. The DTA curve of the $70Li_2S{\cdot}30P_2S_5$ glass prepared by melt quenching (in a sealed quartz tube) is also shown for comparison. The glass prepared by milling using an Al_2O_3 vessel and 10 Al_2O_3 balls (10 mm in diameter) at a rotation speed of 370 rpm showed the glass transition temperature (T_g) of 200°C and the crystallization temperature (T_c) of 235°C. These temperatures were somewhat lower than those of the glass prepared by melt quenching. On the other hand, the glass prepared by milling with a ZrO_2 vessel and 500 ZrO_2 balls (4 mm in diameter) at a rotation speed of 510 rpm exhibited the T_g of 218°C and the T_c of 255°C, which were almost the same as those of the glass prepared by melt quenching. The glass prepared with ZrO_2 media showed clearer glass transition and crystallization than the glass prepared with Al_2O_3 media. It is noteworthy that a homogeneous glass with narrowly distributed T_g was obtained by using ZrO_2 media.

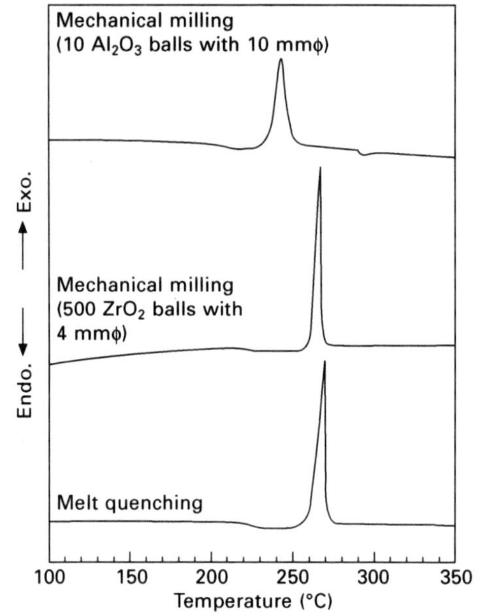

19.2 DTA curves of the $70Li_2S{\cdot}30P_2S_5$ glasses prepared by mechanical milling with ZrO_2 or Al_2O_3 media. The curve of the glass prepared by melt quenching is also shown for comparison.

The local structure around phosphorus atoms of the milled glass was almost the same as that of the melt quenched glass. Figure 19.3 shows the Raman spectra of the $70Li_2S{\cdot}30P_2S_5$ (mol%) glasses prepared by mechanical milling and melt quenching (Hayashi *et al.*, 2010b). The spectrum of the milled glass was quite similar to that of the quenched glass, and both the glasses were composed of two thiophosphate ions of PS_4^{3-} and $P_2S_7^{4-}$.

A wide glass-forming region to a high-alkali content direction is another merit of mechanochemistry (Hayashi *et al.*, 2001). For example, the $75Li_2S{\cdot}25P_2S_5$ (mol%) glass at the ortho-composition was prepared via mechanochemistry, while glass preparation for this composition is difficult by melt quenching. Sulfide-based glasses with high conductivity are a promising solid electrolyte for all-solid-state lithium batteries. However, they are inferior to oxide electrolytes in chemical stability in air, because hydrolysis of sulfides by water molecules generates harmful H_2S gas. Sulfide solid electrolytes with a moderate stability in air are desired from the viewpoint of safety in solid-state batteries. Chemical stability of sulfide electrolytes was thus investigated by examining structural changes that the electrolytes underwent in the atmosphere. The $75Li_2S{\cdot}25P_2S_5$ glass resulted in relatively high chemical stabilities and conductivities in the atmosphere. The amount of H_2S generated from the Li_2S-P_2S_5 glasses after being exposed to air was dependent on the glass composition, and it was minimized at the $75Li_2S{\cdot}25P_2S_5$

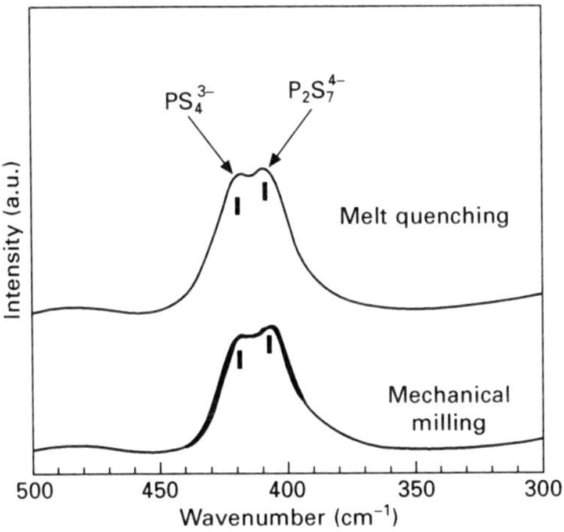

19.3 Raman spectra of the $70Li_2S{\cdot}30P_2S_5$ glasses prepared by mechanical milling and melt quenching.

composition. Figure 19.4 shows Raman spectra of the $75Li_2S{\cdot}25P_2S_5$ glass before and after exposure to air for 1 day (Muramatsu *et al.*, 2011). Three peaks at 270, 418 and 550 cm^{-1} attributed to the PS_4^{3-} ion were observed for the glass before and even after exposure to air, suggesting that the glass composed of ortho-thiophosphate anion (PS_4^{3-}) was tolerant of hydrolysis in the atmosphere. The glass-ceramic at the same composition also had a good chemical stability, suggesting that sulfide electrolytes based on the isolated thiophosphate anion (PS_4^{3-}) were an appropriate electrolyte from the viewpoint of safety in solid-state batteries.

Conductive behaviors of oxide, sulfide and oxysulfide glass electrolytes were compared in detail. Figure 19.5 shows the temperature dependence of conductivities of the glasses at the compositions $60Li_2O{\cdot}40SiO_2$, $60Li_2S{\cdot}40SiS_2$, and $95(0.6Li_2S{\cdot}0.4SiS_2){\cdot}5Li_4SiO_4$ (mol%). The conductivities were measured by ac impedance method for flake-like glasses with a thickness of 20 μm prepared by rapid melt quenching. The conductivities of the three glasses obeyed the Arrhenius law expressed by $\sigma = \sigma_0 \exp(-E_a/RT)$; where σ_0 is the pre-exponential factor, E_a the activation energy for conduction, R the gas constant, and T the absolute temperature. The conductivity of the $60Li_2S{\cdot}40SiS_2$ sulfide glass was in the order of 10^{-4} S cm^{-1} at room temperature, which was two orders of magnitude higher than that of the conductivity of the $60Li_2O{\cdot}40SiO_2$ glass with the same lithium ion concentration. The activation energy E_a of the sulfide glass (33 kJ mol^{-1}) was also lower than that of the oxide glass (48 kJ mol^{-1}), suggesting that the mobility of lithium ions was increased by replacing oxide matrix with a sulfide one.

The addition of oxide to sulfide was expected to monotonically decrease conductivity because oxide glasses have much lower conductivity than

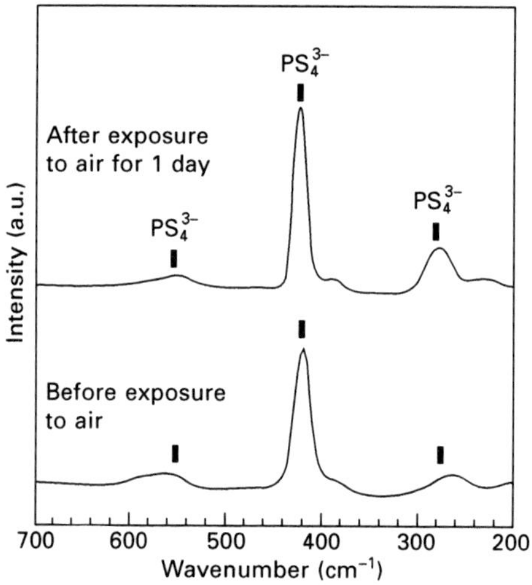

19.4 Raman spectra of the $75Li_2S{\cdot}25P_2S_5$ glass before and after exposure to air for 1 day.

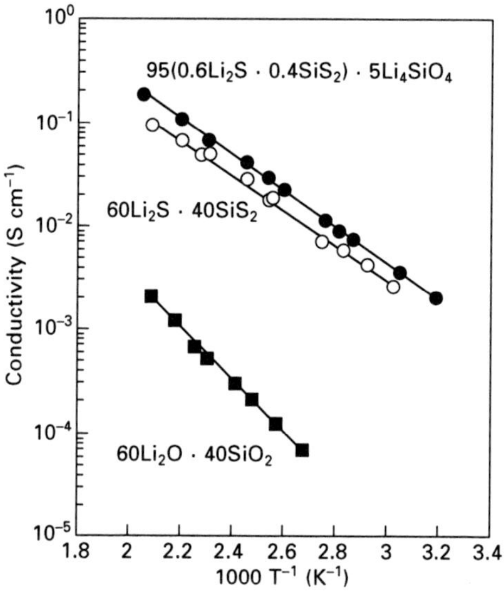

19.5 Temperature dependence of conductivities of the glasses at the compositions $60Li_2O{\cdot}40SiO_2$, $60Li_2S{\cdot}40SiS_2$ and $95(0.6Li_2S{\cdot}0.4SiS_2){\cdot}5Li_4SiO_4$ (mol%).

sulfide glasses. However, the addition of small amounts of a lithium ortho-oxosalt Li_3PO_4 to Li_2S-SiS_2 sulfide system was reported to enhance the conductivity (Kondo *et al.*, 1992). The addition of Li_4SiO_4 also enhanced the conductivity of the sulfide glass to the 10^{-3} S cm^{-1} level as shown in

Fig. 19.5. The addition of larger amounts of the Li_4SiO_4 to the Li_2S-SiS_2 sulfide system decreased the conductivity on the basis of low conductivity of oxide glasses (Minami *et al.*, 2000). It is noteworthy that the oxysulfide glass with 5 mol% of Li_4SiO_4 has a high conductivity of 10^{-3} S cm^{-1}. To elucidate the conductivity enhancement, the local structure of the oxysulfide glass with 5 mol% Li_4SiO_4 was analyzed by NMR and XPS (Minami *et al.*, 2000). Structural units of $Si_2OS_6^{6-}$ where silicon atoms were coordinated with six nonbridging sulfur atoms and a bridging oxygen atom were mainly observed in the glass, and thus the structural units would give a favorable conduction path for Li^+ ions in the oxysulfide glass. A slight increase in Li concentration by the addition of Li_4SiO_4 is also effective in increasing the conductivity of the oxysulfide glasses.

Conductivity enhancement was also reported in the Li_2S-GeS_2-GeO_2 oxysulfide system. The replacement of 5 mol% GeS_2 with GeO_2 increased conductivity and decreased activation energy for conduction of the $50Li_2S{\cdot}(50\text{-}x)GeS_2{\cdot}xGeO_2$ oxysulfide glasses. Conduction mechanism of Li^+ ion in the oxysulfide glass was proposed on the basis of the Anderson and Stuart energy barrier model. Figure 19.6 shows a simplified pictorial view of the ionic conduction energy for the $50Li_2S{\cdot}45GeS_2{\cdot}5GeO_2$ oxysulfide glass (Kim *et al.*, 2006). Ionic conduction requires two energy components of the electrostatic binding energy (ΔE_B) and the strain energy (ΔE_S). The ΔE_B is the energy for separating the Li^+ ion from its charge compensating anion site (nonbridging sulfur) to the next site. The ΔE_S is the energy for opening up the doorway (doorway radius: r_D) in the structure large enough for the ions to pass through. The introduction of bridging oxygen instead of bridging sulfur expands a conduction path, and a slightly increasing r_D in the strain energy term is proposed as cause of the decreasing activation energy, which resulted in increasing ionic conductivity. The same conductivity enhancement would be achieved in the Li_2S-SiS_2-Li_4SiO_4 oxysulfide system.

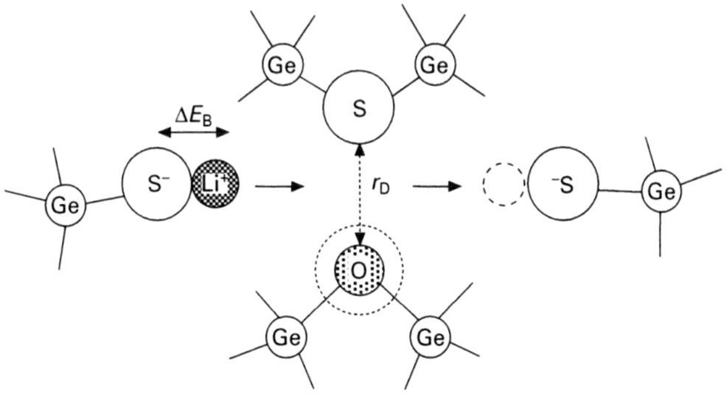

19.6 Simplified pictorial view of the ionic conduction energy for the $50Li_2S{\cdot}45GeS_2{\cdot}5GeO_2$ oxysulfide glass.

19.3.2 Sulfide glass-ceramic electrolytes

An alternative technique to increase conductivity of sulfide glass electrolytes is the precipitation of metastable crystals by crystallization of precursor glasses. The conductivity of glass-ceramics depends on the precipitated crystalline phases, which is determined by glass compositions as well as thermal annealing. As briefly mentioned in Fig. 19.1, crystallization of the $70Li_2S{\cdot}30P_2S_5$ (mol%) glass increased conductivity by the precipitation of the $Li_7P_3S_{11}$ phase, while the conductivity of the $60Li_2S{\cdot}40SiS_2$ (mol%) glass went down by the crystallization of the Li_4SiS_4 phase.

Figure 19.7 shows XRD patterns of (a) the $70Li_2S{\cdot}30P_2S_5$ glass, (b) the glass-ceramic, and (c) the crystal obtained by solid-state reaction (Mizuno *et al.*, 2005). XRD peaks in (c) are attributable to both thermodynamically stable crystals Li_3PS_4 and $Li_4P_2S_6$, and these crystals are known to show lithium ion conductivities smaller than 10^{-7} S cm^{-1} at room temperature. On the other hand, the XRD pattern of the glass-ceramic in (b) was different from that of thermodynamically stable crystals in the Li_2S-P_2S_5 system such as Li_3PS_4, $Li_4P_2S_6$ and Li_7PS_6. The XRD pattern shown in Fig. 19.7(b) is attributable to the $Li_7P_3S_{11}$ phase, which was not prepared by the solid-state reaction. The $Li_7P_3S_{11}$ crystal was prepared by crystallization from not only the glass but also the melt at the composition $70Li_2S{\cdot}30P_2S_5$ (Minami *et al.*,

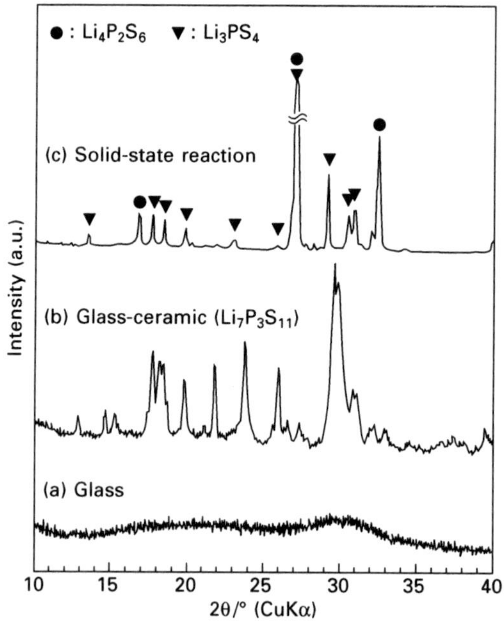

19.7 XRD patterns of (a) the $70Li_2S{\cdot}30P_2S_5$ glass, (b) the glass-ceramic and (c) the crystal obtained by solid-state reaction.

2010), suggesting that the $Li_7P_3S_{11}$ crystal is a high-temperature phase at this specific composition. The crystal structure of $Li_7P_3S_{11}$ was clarified by the Rietveld analysis based on synchrotron XRD measurements (Yamane *et al.*, 2007). The compound crystallized in a triclinic cell with space group of *P*-1, and contains P_2S_7 ditetrahedra and PS_4 tetrahedra similar to the case of Ag^+ ion conductor $Ag_7P_3S_{11}$ (monoclinic, space group C2/c).

Glass-ceramic electrolytes in the system Li_2S-P_2S_5 were prepared at several compositions around 70 mol% Li_2S. Crystalline phases precipitated from the glass state and electrical conductivities of the Li_2S-P_2S_5 glass-ceramics are listed in Table 19.1. The glass-ceramic with 67 mol% Li_2S showed a low conductivity of 7.6×10^{-6} S cm^{-1} because the $Li_4P_2S_6$ crystal with low conductivity of below 10^{-7} S cm^{-1} was precipitated (Mizuno *et al.*, 2006). On the other hand, the glass-ceramics with 70 mol% Li_2S or more had a high conductivity of over 10^{-4} S cm^{-1}. For the glass-ceramic with 70 mol% Li_2S, preparation conditions of glass-ceramics with $Li_7P_3S_{11}$ were investigated and high conductivities of 5.4×10^{-3} S cm^{-1} at room temperature was achieved (Hayashi *et al.*, 2010a). The crystal analogous to the Li_4GeS_4-Li_3PS_4 solid solution called 'thio-LISICON (LIthium Super Ionic CONductor)' was precipitated in glass-ceramics with 75 and 80 mol% Li_2S (Mizuno *et al.*, 2006). The thio-LISICON $Li_{4-x}Ge_{1-x}P_xS_4$ was reported to be divided into three composition regions with different types of cation ordering: region I ($0 < x < 0.6$), region II ($0.6 < x < 0.8$) and region III ($0.8 < x < 1.0$) (Kanno and Murayama, 2001). The XRD patterns of the glass-ceramics with 75 and 80 mol% Li_2S were respectively similar to those of thio-LISICON III ($Li_{3.2}Ge_{0.2}P_{0.8}S_4$) and thio-LISICON II ($Li_{3.25}Ge_{0.25}P_{0.75}S_4$). It was then presumed that the $Li_{3.2}P_{0.96}S_4$ and $Li_{3.25}P_{0.95}S_4$ with phosphorus deficiency, which would be analogous to thio-LISICON III and thio-LISICON II, were precipitated in the $75Li_2S{\cdot}25P_2S_5$ and $80Li_2S{\cdot}20P_2S_5$ glass-ceramics, respectively, without Ge atoms. These glass-ceramics respectively exhibited the ambient temperature conductivity of 3.2×10^{-4} and 1.3×10^{-3} S cm^{-1}, which were affected by the conductivity of the precipitated crystals of the thio-LISICON III (6.4×10^{-4} S cm^{-1}) and the thio-LISICON II (2.2×10^{-3}

Table 19.1 Crystalline phases and electrical conductivities of the Li_2S-P_2S_5 glass-ceramics

Li_2S content (mol%)	Crystalline phase	Conductivity at 25°C (S cm^{-1})	Reference
67	$Li_4P_2S_6$	7.6×10^{-6}	Mizuno *et al.* (2006)
70	$Li_7P_3S_{11}$	5.4×10^{-3}	Hayashi *et al.* (2010a)
75	thio-LISICON III analog ($Li_{3.2}P_{0.96}S_4$)	3.2×10^{-4}	Mizuno *et al.* (2006)
80	thio-LISICON II analog ($Li_{3.25}P_{0.95}S_4$)	1.3×10^{-3}	Mizuno *et al.* (2006)

S cm^{-1}). Precipitation of a high-temperature phase $Li_7P_3S_{11}$ or a metastable phase $Li_{3.25}P_{0.95}S_4$ as a primary crystal from super-cooled liquid state of mother glass is key to achieving a superior glass-ceramic electrolyte with conductivity of over 10^{-3} S cm^{-1} at room temperature.

An electrochemical window of sulfide glass-ceramic electrolytes was examined by cyclic voltammetry. A stainless-steel disk as a working electrode and a lithium foil as a counter electrode were attached on each face of a pelletized glass-ceramic electrolyte with $Li_7P_3S_{11}$. The potential sweep was performed using a potentiostat/galvanostat device with a scanning rate of 1 mV s^{-1}. A cyclic voltammogram of the $70Li_2S{\cdot}30P_2S_5$ glass-ceramic at the first cycle is shown in Fig. 19.8. A cathodic current flow due to lithium deposition and an anodic current peak due to lithium dissolution were observed reversibly at around 0 V (vs Li^+/Li). No large current peak except these peaks appeared over the whole range from – 0.1 V to 5.0 V. It is noted that the glass-ceramic electrolyte has a wide electrochemical window of over 5 V and a good compatibility with lithium metal.

19.4 All-solid-state lithium secondary batteries with sulfide electrolytes

19.4.1 Fabrication of bulk-type all-solid-state batteries

A bulk-type solid-state battery composed of compressed electrode and electrolyte powders has been studied. Compared to a thin film battery, a bulk-type battery attracts much attention because the battery is suitable for

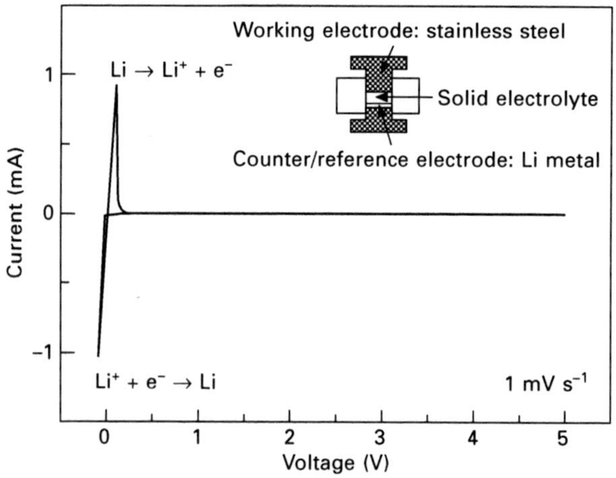

19.8 Cyclic voltammogram of the $70Li_2S{\cdot}30P_2S_5$ glass-ceramic at the first cycle.

large-sized energy-storage devices. Li_2S-based sulfide materials with high Li^+ ion conductivity are promising solid electrolytes for bulk-type solid-state batteries. The electrochemical performance of solid-state In/$LiCoO_2$ cells with the Li_2S-SiS_2-Li_3PO_4 oxysulfide glasses was reported in 1994 (Aotani *et al.*, 1994) and these cells with sulfide electrolytes have subsequently been developed.

In a cell with a liquid electrolyte, a favorable electrode–electrolyte interface is easily formed just by soaking electrodes in a liquid electrolyte, while in a cell with a solid electrolyte, electrode and electrolyte powders should be properly mixed to form intimate contact at the solid–solid interface. A composite electrode composed of an active material, a solid electrolyte, and a conductive additive is commonly used as a working electrode in bulk-type solid-state cells, in order to form continuous lithium ion and electron conducting paths to active material particles.

A schematic diagram of a typical all-solid-state electrochemical cell In/$LiCoO_2$ is shown in Fig. 19.9. The cell consists of a three-layered pellet prepared by uniaxial pressing at room temperature. The first layer is an indium foil as a negative electrode. The second layer is the $80Li_2S{\cdot}20P_2S_5$ glass-ceramic powder as a solid electrolyte (SE). The third layer is a composite powder as a positive electrode. In order to achieve smooth electrochemical reaction in the cell, the composite positive electrode composed of three kinds of powders was used: the active material ($LiCoO_2$), the SE powder providing lithium ion conduction path, and the conductive additive (acetylene black, AB) providing electron conduction paths. A cross-sectional SEM image of the three-layered pellet in the cell and schematic of the positive composite electrode are shown in Fig. 19.10. The composite positive electrode consisting of $LiCoO_2$, SE, and AB powders with a weight ratio of 20:30:3 was used for all-solid-state cells. Obvious grain-boundary in the composite electrode and solid electrolyte was not observed and intimate contacts between electrode

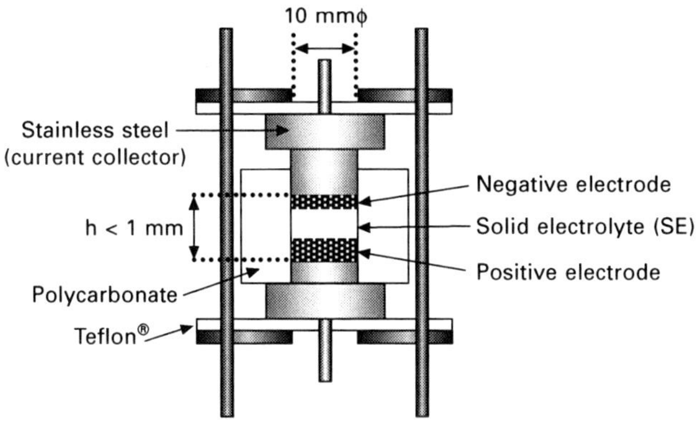

19.9 Schematic diagram of a typical all-solid-state electrochemical cell.

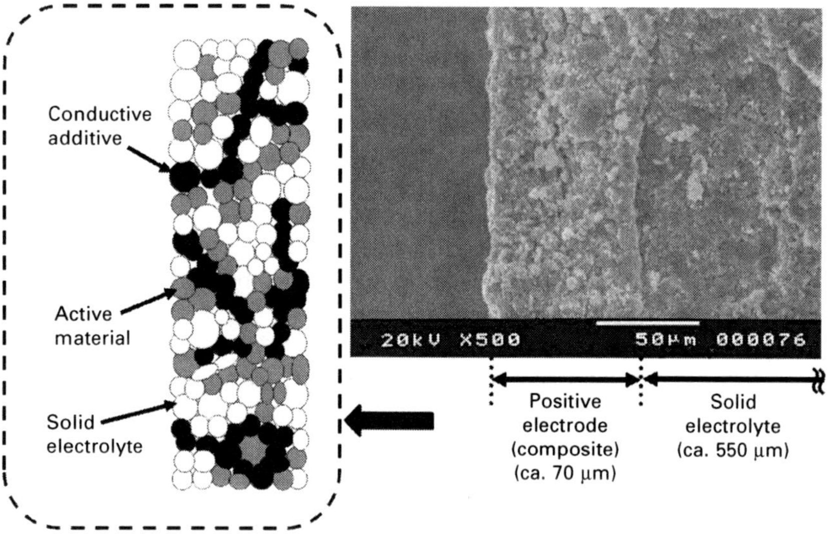

19.10 Cross-sectional SEM image of the three-layered pellet in the In/Li_2S-P_2S_5 glass-ceramic/$LiCoO_2$ cell and schematic of the positive composite electrode.

and electrolyte were achieved by uniaxial cold-press. A typical thickness of positive electrode and solid electrolyte layers were ~70 μm and ~550 μm, respectively.

A typical charge–discharge performance of bulk-type solid-state cells with the $80Li_2S{\cdot}20P_2S_5$ glass-ceramic electrolyte is shown in Fig. 19.11 (Tatsumisago and Hayashi, 2008). $LiCoO_2$ and $Li_{4/3}Ti_{5/3}O_4$ were used as active materials. Both the cells operated as a secondary battery at 25°C under a limited current density of 0.064 mA cm^{-2} and kept a charge–discharge efficiency of 100% for several hundred cycles. The bulk-type solid-state cell exhibited an excellent cycle performance without capacity loss at room temperature.

19.4.2 Li/S batteries with high capacity

A lithium–sulfur battery attracts much attention because of its high energy density due to the large theoretical capacity (1672 mAh g^{-1}) of sulfur active material (Marmorstein *et al.*, 2000; Ji and Nazar, 2010). However, the Li/S batteries with a conventional liquid electrolyte suffer from rapid capacity fading on cycling. This is mainly because polysulfides formed during a discharge process are dissolved into liquid electrolytes. The use of inorganic solid electrolytes resolves the key issue in Li/S batteries. The glass-based solid electrolytes in the systems Li_2S-SiS_2 (Machida *et al.*, 2004) and Li_2S-P_2S_5 (Hayashi *et al.*, 2003; Nagao *et al.*, 2011) were used for solid-state Li/S

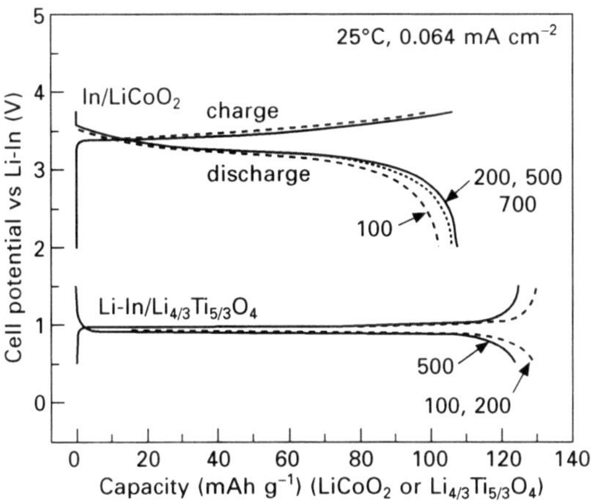

19.11 Charge–discharge performances of the all-solid-state cells of In/Li_2S-P_2S_5 glass-ceramic/$LiCoO_2$ and Li-In/Li_2S-P_2S_5 glass-ceramic/$Li_{4/3}Ti_{5/3}O_4$.

batteries. Sulfur is an insulator and thus electronic conductivity should be given to sulfur by combining it with metal or nanocarbon particles. Sulfur-based composite electrodes prepared by ball milling were applied to all-solid-state Li/S cells with the $80Li_2S \cdot 20P_2S_5$ glass-ceramic electrolyte. Figure 19.12 shows charge–discharge curves of all-solid-state Li-In/S cells with sulfur-copper (S-Cu) (Hayashi *et al.*, 2003) or sulfur-carbon (S-C) (Nagao *et al.*, 2011) composite electrode. The curve of all-solid-state In/$LiCoO_2$ cells is also shown for comparison. The x-axis denotes the cell capacity normalized by the weight of active material. The cells using S-Cu and S-C exhibited a much higher capacity of over 1,000 mAh g^{-1}, compared to the cell with $LiCoO_2$. The Li-In/S cells maintained a high capacity more than 1,000 mAh g^{-1} and a coulombic efficiency of 100% for 200 cycles (Nagao *et al.*, 2011). The use of inorganic solid electrolytes such as the Li_2S-P_2S_5 glass-ceramics is effective in achieving a good cycle performance of Li/S batteries. On the other hand, lithium sulfide Li_2S, which is a reaction product after discharge for sulfur active material, was also used as a positive electrode in all-solid-state batteries (Hayashi *et al.*, 2008; Takeuchi *et al.*, 2010); the merit of Li_2S instead of S is the compatibility with various negative electrodes without lithium sources such as graphite. Development of all-solid-state Li/S batteries will meet the demand of battery application with high energy density.

19.4.3 High-temperature operation

High-temperature operation at 100°C brings about high rate performance of solid-state cells because of decreasing cell resistance. Figure 19.13 shows cycle performance of the all-solid-state cell Li-In/Li_2S-P_2S_5 glass-ceramic/

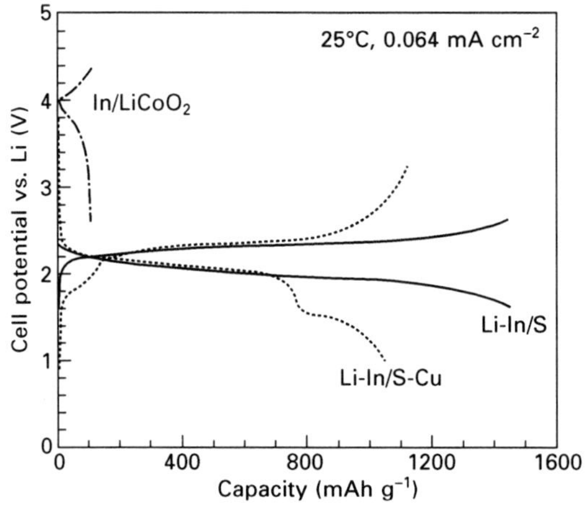

19.12 Charge–discharge curves of all-solid-state Li-In/Li_2S-P_2S_5 glass-ceramic/S cells with sulfur-copper (S-Cu) or sulfur-carbon (S-C) composite electrode.

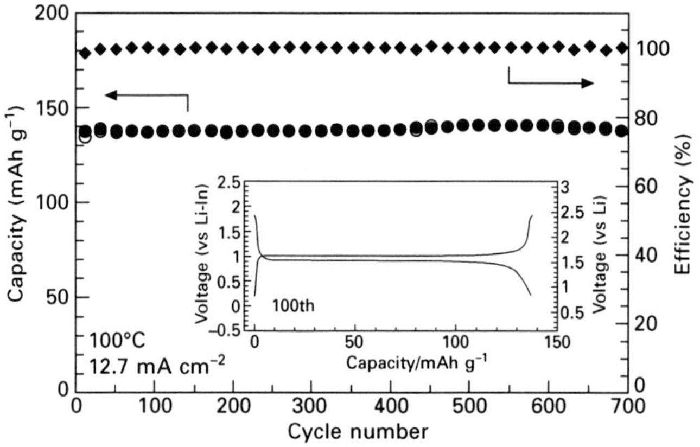

19.13 Charge–discharge curve at the 100th cycle (inset) and cycle performance of the all-solid-state Li-In/Li_2S-P_2S_5 glass-ceramic/ $Li_{4/3}Ti_{5/3}O_4$ cell at 100°C under the current density of 12.7 mA cm^{-2}.

$Li_{4/3}Ti_{5/3}O_4$ at 100°C (Minami *et al.*, 2011). The inset figure indicates the charge–discharge curve at the 100th cycle. The cell operated reversibly for 700 cycles with a charge–discharge efficiency of 100% even at a high current density of 12.7 mA cm^{-2}. The cell showed the discharge and charge capacity of about 140 mAh g^{-1} and exhibited an excellent cycle life without capacity fading for 700 cycles. It is beneficial to use all-solid-state batteries with glass-ceramic electrolyte for high-temperature application.

19.4.4 Interfacial modification between electrode and electrolyte

A bulk-type all-solid-state battery with a $LiCoO_2$ positive electrode has a superior cyclability in a limited current density as mentioned in the Section 19.4.1. To enhance rate performance of the battery, an interfacial problem between $LiCoO_2$ active material and Li_2S-P_2S_5 electrolyte should be resolved. Surface coatings of $LiCoO_2$ particles with oxide thin-films such as $LiNbO_3$ are effective in decreasing interfacial resistance caused by the charging process (Ohta *et al.*, 2006; Sakuda *et al.*, 2009).

Interfacial resistance of all-solid-state cells was examined by an AC impedance technique and the effects of oxide coatings on cell performance were investigated. Figure 19.14 shows Nyquist plots of all-solid-state In/Li_2S-P_2S_5 glass-ceramic/$LiCoO_2$ cells after the initial charge process (Sakuda *et al.*, 2009). The composite electrode consisting of $LiCoO_2$ and the $80Li_2S{\cdot}20P_2S_5$ electrolyte with the weight ratio of 70:30, and acetylene black as conductive additive was not added to simplify impedance analysis. The cell was charged at 25°C under the current density of 0.064 mA cm^{-2} up to 4.2 V vs Li. The cell with $LiCoO_2$, which was coated with a Li_2SiO_3 thin film (thickness of ~2 nm) by the sol-gel method in advance, showed a smaller semicircle at 500 Hz than the cell with noncoated $LiCoO_2$. The semicircle was not observed before the charge process, and it was attributable to the interfacial resistance between $LiCoO_2$ electrode and the sulfide electrolyte (Sakuda *et al.*, 2009). It is noteworthy that oxide coatings on $LiCoO_2$ were effective in decreasing the interfacial resistance in the solid-state cells.

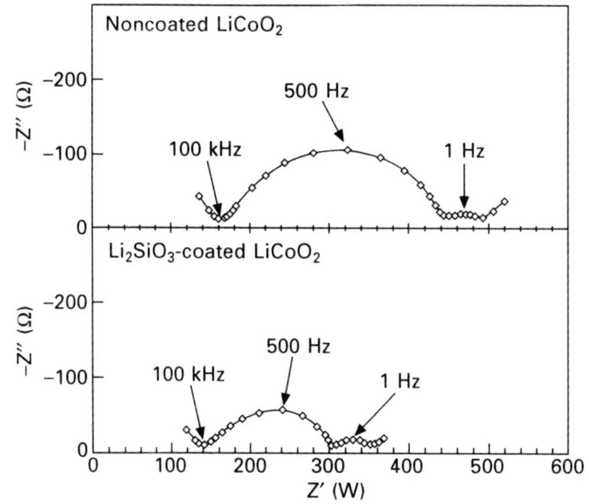

19.14 Nyquist plots of all-solid-state In/Li_2S-P_2S_5 glass-ceramic/$LiCoO_2$ cells after the initial charge process.

Figure 19.15 shows cross-sectional high-angle annular dark field (HAADF)-STEM images and EDX line profiles of the interface between the $LiCoO_2$ and Li_2S-P_2S_5 electrolytes in the composite electrodes layer after the initial charging (Sakuda *et al.*, 2010a). Interfacial layer with a different contrast was observed between noncoated $LiCoO_2$ and Li_2S-P_2S_5 electrolyte and a mutual diffusion of Co, P and S elements occurred at the interface. It is noted that Co was detected in the electrolyte even at a distance of 50 nm from the interface. On the other hand, an obvious interfacial layer was not observed in the TEM image, and the mutual diffusion was suppressed at the interface between Li_2SiO_3-coated $LiCoO_2$ and Li_2S-P_2S_5 electrolyte. The signal of Si for the Li_2SiO_3 thin-film was detected at the interface, and thus the oxide film would act as a buffer layer for inhibiting side-reaction at the electrode–electrolyte interface during a charging process. This coating effect on interfacial structure is closely related to the large decrease in interfacial resistance as shown in Fig. 19.14.

Figure 19.16 shows discharge curves of the all-solid-state In/Li_2S-P_2S_5 glass-ceramic/$LiCoO_2$ cells at 25°C under a relatively high current density of 6.4 mA cm^{-2}. The cell with Li_2SiO_3-coated $LiCoO_2$ exhibited higher discharge plateau and larger capacity than the cell with noncoated $LiCoO_2$. Moreover, the discharge curve of the cell with $Li_{3.5}Si_{0.5}P_{0.5}O_4$-coated $LiCoO_2$ is also shown in this figure. The $Li_{3.5}Si_{0.5}P_{0.5}O_4$ ($50Li_4SiO_4{\cdot}50Li_3PO_4$) film with a higher Li^+ ion conductivity of 10^{-6} S cm^{-1} was coated on $LiCoO_2$ particles by pulsed laser deposition (PLD) (Sakurai *et al.*, 2011). The cell

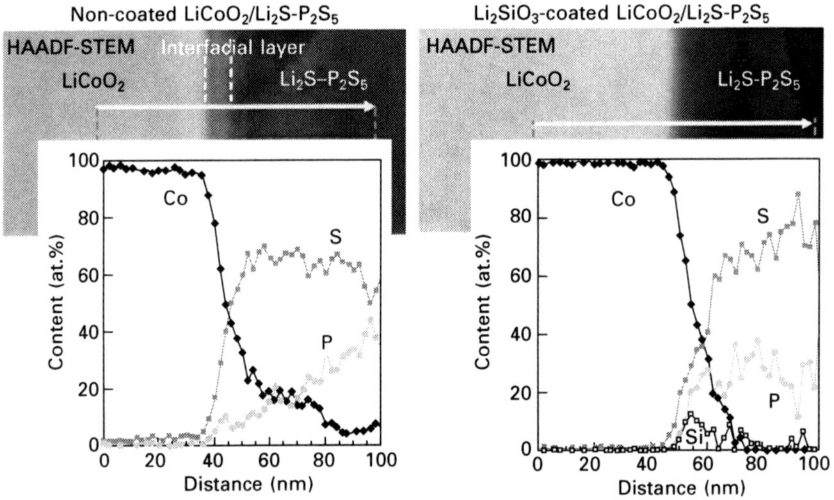

19.15 Cross-sectional high-angle annular dark field (HAADF)-STEM images and EDX line profiles of the interface between $LiCoO_2$ and Li_2S-P_2S_5 electrolyte in the composite electrode layer after the initial charge process.

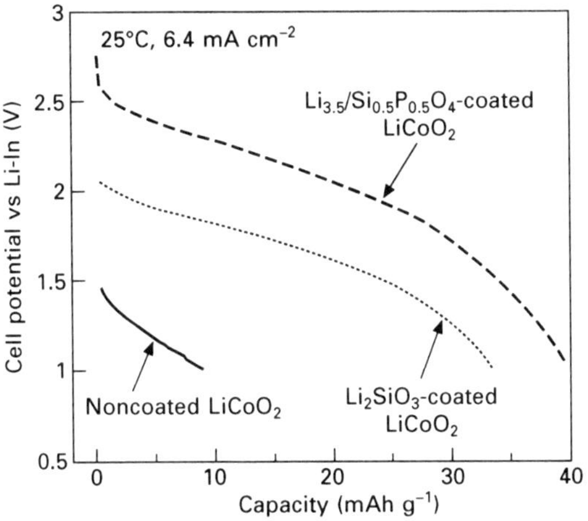

19.16 Discharge curves of the all-solid-state In/Li_2S-P_2S_5 glass-ceramic/$LiCoO_2$ cells at 25°C under a relatively high current density of 6.4 mA cm^{-2}.

with $Li_{3.5}Si_{0.5}P_{0.5}O_4$-coated $LiCoO_2$ exhibited superior discharge behavior than the other cells because of a larger decrease in interfacial resistance. It is noted that the use of coating films with higher Li^+ ion conductivity is effective in enhancing rate capability of solid-state cells with $LiCoO_2$ active material.

19.4.5 Improvement of energy density by PLD sulfide coating

Formation of a favorable contact between electrode and electrolyte is important to improve electrochemical performance of bulk-type solid-state batteries. To achieve a large contact area between the $LiCoO_2$ active material and the Li_2S-P_2S_5 solid electrolyte, SE coatings on $LiCoO_2$ particles were developed by pulsed laser deposition (PLD) (Sakuda *et al.*, 2011).

Figure 19.17 shows schematic diagrams of typical bulk-type all-solid-state cells (a) and all-solid-state cells using SE-coated $LiCoO_2$ (b). In typical cells, a composite electrode composed of $LiCoO_2$ electrode and Li_2S-P_2S_5 solid electrolyte was used as a working electrode to provide lithium ion conducting paths to the $LiCoO_2$. At least 20–30 wt.% of solid electrolyte particles were needed in the composite electrode. A decrease in solid electrolyte content by SE-coating on $LiCoO_2$ particles is useful for increasing the energy density of solid-state cells. In fact, a part of $LiCoO_2$ particles are directly connected by cold pressing for assembling cells, and electron conduction paths are formed even in the SE-coated $LiCoO_2$ electrode layer.

The coatings of the $80Li_2S{\cdot}20P_2S_5$ (mol%) SE on $LiCoO_2$ particles were carried out using PLD with a KrF excimer laser (λ = 248 nm). The

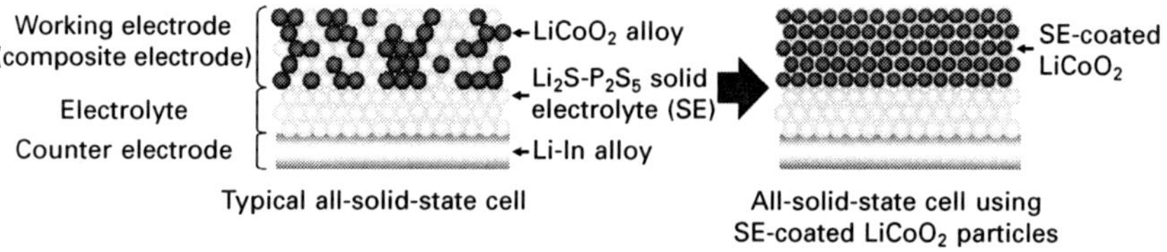

19.17 Schematic diagrams of typical bulk-type all-solid-state cells (a) and all-solid-state cells using solid electrolyte (SE)-coated $LiCoO_2$ (b).

schematic of PLD coatings on $LiCoO_2$ particles is shown in Fig. 19.18. A pelletized mixture of Li_2S and P_2S_5 crystalline powders with a molar ratio of 80:20 was used as a target. The $LiCoO_2$ particles used in this study were coated with $LiNbO_3$ film in advance to decrease the interfacial resistance after the charging process (Ohta *et al.*, 2007). During deposition of the solid electrolyte, $LiCoO_2$ particles were fluidized by a vibrator in order to form the SE layer uniformly on $LiCoO_2$ particles. Detailed PLD conditions were reported (Sakuda *et al.*, 2010b) and the SE films with the conductivity of $\sim 10^{-4}$ S cm^{-1} were prepared.

Plate XVIII (between pages 330 and 331) shows a cross-sectional high-angle annular dark field (HAADF)-STEM image and EDX mappings for the composite electrode consisting of only SE-coated $LiCoO_2$ after charge–discharge measurements for 450 cycles (Sakuda *et al.*, 2011). The $LiCoO_2$ particles coated with the Li_2S-P_2S_5 SE for 40 min by PLD were used for the working electrode and the SE particles were not added to the electrode. It is suggested from the EDX mappings of five elements of O, P, S, Co and Nb that the surface of $LiNbO_3$-coated $LiCoO_2$ particles was successfully covered with the Li_2S-P_2S_5 SE thin film, which would play a role as a lithium ion conduction path to $LiCoO_2$.

Bulk-type solid-state cells using a working electrode composed of SE-coated $LiCoO_2$ particles were fabricated and their performance was compared to that of the cell with noncoated $LiCoO_2$ particles. Figure 19.19 shows the charge–discharge curves of the solid-state In/$80Li_2S{\cdot}20P_2S_5$ glass-ceramic/$LiCoO_2$ cells under the current density of 0.13 mA cm^{-2} at 25°C. The Li_2S-P_2S_5 SE particles were not added to the $LiCoO_2$ working electrode. The cell with noncoated $LiCoO_2$ as a working electrode did not operate because lithium ion conduction paths to $LiCoO_2$ would be insufficient. On the other hand, the cells with SE-coated $LiCoO_2$ were charged and discharged and the charge–discharge capacity was increased with an increase in the SE coating time, which corresponds to the coating amounts of SE on $LiCoO_2$. The cell with SE-coated $LiCoO_2$ (coating time of 40 min) showed a discharge capacity of ~ 30 mAh g^{-1} (normalized by the weight of $LiCoO_2$), while the cell with SE-coated $LiCoO_2$ (coating time of 2 h) had the capacity of ~65 mAh g^{-1}. The latter cell retained the capacity for 10 cycles. The weight ratios of SE coatings to $LiCoO_2$ were only 1% and 3% in the working electrode

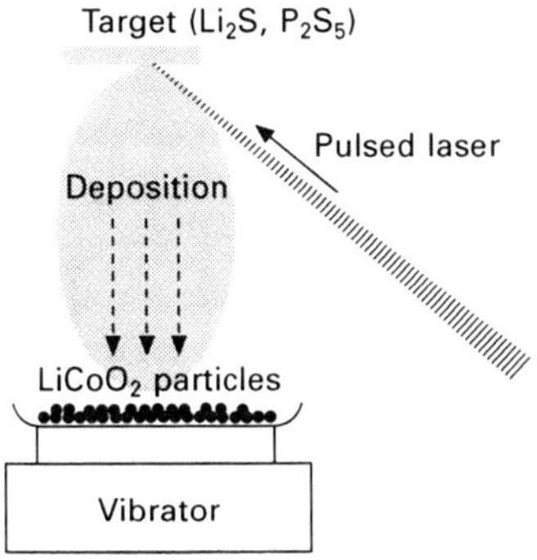

19.18 Schematic of PLD coatings on $LiCoO_2$ particles.

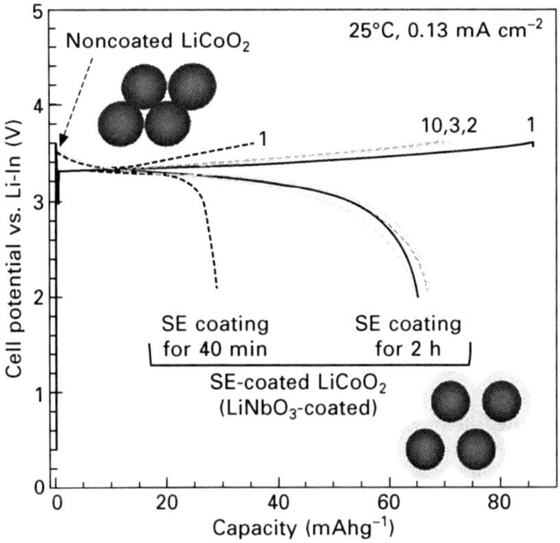

19.19 Charge–discharge curves of the solid-state cells In/Li_2S-P_2S_5 glass-ceramic/$LiCoO_2$ under the current density of 0.13 mA cm^{-2} at 25°C. Noncoated and SE-coated $LiCoO_2$ were used as an active material and the SE particles were not added to the $LiCoO_2$ working electrode.

for the former and the latter cells, respectively. It is noteworthy that the SE amount in the working electrode was drastically decreased by SE coatings on $LiCoO_2$, because the addition of almost 30 wt.% of SE particles to the working electrode was needed to operate conventional solid-state In/$LiCoO_2$ cells. Therefore, SE coatings by PLD are a useful technique to enhance the energy density of bulk-type solid-state batteries.

19.5 Conclusion

Recent developments to sulfide-based solid electrolytes and solid-state batteries with the electrolytes were reviewed. Sulfide glass electrolytes have several advantages like high conductivity, wide electrochemical window and low grain-boundary resistance.

Bulk-type all-solid-state batteries with Li_2S-P_2S_5 glass-ceramic electrolytes exhibited excellent cycle performance for several hundred cycles at room temperature. High-temperature operation at 100°C and the use of sulfur-active material with high capacity were large advantages of all-solid-state batteries, compared to conventional lithium-ion batteries with an organic liquid electrolyte. Surface coatings of $LiCoO_2$ with oxide thin-films were effective in enhancing rate capability of solid-state cells. Moreover, an increase in energy density of solid-state cells was achieved by sulfide electrolyte coating on $LiCoO_2$.

To realize bulk-type all-solid-state lithium rechargeable batteries with high energy density, high rate capability, and long cycle life, further increase of Li^+ ion conductivity of glass-based solid electrolytes and the formation of good electrode/electrolyte interface achieving rapid charge transfer by appropriate surface modification are needed. Control of size, morphology and dispersibility of both solid electrolyte and active material in an electrode layer is also important for establishing favorable solid–solid interfaces. Sulfide glass-ceramics are now a most suitable electrolyte for bulk-type batteries. Further development of solid-state batteries with sulfide electrolytes is desired in the near future.

19.6 References

Aono H, Sugimono E, Sadaoka Y, Imanaka N and Adachi G (1990), 'Ionic conductivity of solid electrolytes based on lithium titanium phosphate', *J. Electrochem. Soc.*, 137, 1023–1027.

Aotani N, Iwamoto K, Takada K and Kondo S (1994), 'Synthesis and electrochemical properties of lithium ion conductive glass, Li_3PO_4-Li_2S-SiS_2', *Solid State Ionics*, 68, 35–39.

Fergus J W (2010), 'Ceramic and polymeric solid electrolytes for lithium-ion batteries', *J. Power Sources*, 195, 4554–4569.

Hayashi A, Hama S, Morimoto H, Tatsumisago M and Minami T (2001), 'Preparation of Li_2S-P_2S_5 amorphous solid electrolytes by mechanical milling', *J. Am. Ceram. Soc.*, 84, 477–479.

Hayashi A, Ohtomo T, Mizuno F, Tadanaga K and Tatsumisago M (2003), 'All-solid-state Li/S batteries with highly conductive glass-ceramic electrolytes', *Electrochem. Commun.*, 5, 701–705.

Hayashi A, Ohtsubo R, Ohtomo T, Mizuno F and Tatsumisago M (2008), 'All-solid-state rechargeable lithium batteries with Li_2S as a positive electrode material', *J. Power Sources*, 183, 422–426.

Hayashi A, Minami K, Ujiie S and Tatsumisago M (2010a), 'Preparation and ionic conductivty of $Li_7P_3S_{11-z}$ glass-ceramic electrolytes', *J. Non-Cryst. Solids*, 356, 2670-2673.

Hayashi A, Minami K and Tatsumisago M (2010b), 'Development of sulfide glass-ceramic electrolytes for all-solid-state lithium rechargeable batteries', *J. Solid-State Electrochem.*, 14, 1761–1767.

Hirai K, Tatsumisago M and Minami T (1995), 'Thermal and electrical properties of rapidly quenched glasses in the system Li_2S-SiS_2-Li_xMO_y (Li_xMO_y=Li_4SiO_4, Li_2SO_4)', *Solid State Ionics*, 78, 269–273.

Ito M, Inaguma Y, Jung W H, Chen L and Nakamura T (1994), 'High lithium ion conductivity in the perovskite-type compounds $Ln_{1/2}Li_{1/2}TiO_3$ (Ln = La, Pr, Nd, Sm)', *Solid State Ionics*, 70–71, 203–207.

Ji X and Nazar L F (2010), 'Advances in Li-S batteries', *J. Mater. Chem.*, 20, 9821–9826.

Julien C and Nazri G A (1994), *Solid State Batteries: Materials Design and Optimization*. Kluwer Academic Publishers, Boston, MA.

Kamaya N, Homma K, Yamakawa Y, Hirayama M, Kanno R, Yonemura M, Kamiyama T, Kato Y, Hama S and Kawamoto K (2011), 'A lithium superionic conductor', *Nat. Mater.*, 10, 682–686.

Kanno R and Murayama M (2001), 'Lithium ionic conductor thio-LISICON; the Li_2S-GeS_2-P_2S_5 system', *J. Electrochem. Soc.*, 148, A742–A746.

Kim Y, Saienga J and Martin S W (2006), 'Anomalous ionic conductivity increase in $Li_2S + GeS_2 + GeO_2$ glasses', *J. Phys. Chem. B*, 110, 16318–16325.

Kitaura H, Hayashi A, Ohtomo T, Hama S and Tatsumisago M (2011), 'Fabrication of electrode-electrolyte interfaces in all-solid-state rechargeable lithium batteries by using a supercooled liquid state of the glassy electrolytes', *J. Mater. Chem.*, 21, 118–124.

Kondo S, Takada K and Yamamura Y (1992), 'New lithium conductors based on Li_2S-SiS_2 system', *Solid State Ionics*, 53, 1183–1186.

Machida N, Kobayashi K, Nishikawa Y and Shigematsu T (2004), 'Electrochemical properties of sulfur as cathode materials in a solid-state lithium battery with inorganic solid electrolytes', *Solid State Ionics*, 175, 247–250.

Marmorstein D, Yu T H, Striebel K A, McLarnon F R, Hou J and Cairns E J (2000), 'Electrochemical performance of lithium/sulfur cells with three different polymer electrolytes', *J. Power Sources*, 89, 219–226.

Minami T, Hayashi A and Tatsumisago M (2000), 'Preparation and characterization of lithium ion-conducting oxysulfide glasses', *Solid State Ionics*, 136–137, 1015–1023.

Minami T, Tatsumisago M, Wakihara M, Iwakura C, Kohjiya S and Tanaka I (2005), *Solid State Ionics for Batteries*. Springer-Verlag, Tokyo.

Minami T, Hayashi A and Tatsumisago M (2006), 'Recent progress of glass and glass-ceramics as solid electrolytes for lithium secondary batteries', *Solid State Ionics*, 177, 2715–2720.

Minami K, Hayashi A and Tatsumisago M (2010), 'Preparation and characterization of superionic conducting Li7P3S11 crystal from glass liquids', *J. Ceram. Soc. Jpn*, 118, 305–308.

Minami K, Hayashi A, Ujiie S and Tatsumisago M (2011), 'Electrical and electrochemical properties of glass-ceramic electrolytes in the systems Li_2S-P_2S_5-P_2S_3 and Li_2S-P_2S_5-P_2O_5', *Solid State Ionics*, 192, 122–125.

Mizuno F, Hayashi A, Tadanaga K and Tatsumisago M (2005), 'New, highly ion-conductive crystals precipitated from Li_2S-P_2S_5 glasses', *Adv. Mater.*, 17, 918–921.

Mizuno F, Hayashi A, Tadanaga K and Tatsumisago M (2006), 'High lithium ion conducting glass-ceramics in the system Li_2S-P_2S_5', *Solid State Ionics*, 177, 2721–2725.

Morimoto H, Yamashita H, Tatsumisago M and Minami T (1999), 'Mechanochemical synthesis of new amorphous materials of $60Li_2S{\cdot}40SiS_2$ with high lithium ion conductivity', *J. Am. Ceram. Soc.*, 82, 1352–1354.

Muramatsu H, Hayashi A, Ohtomo T, Hama S and Tatsumisago M (2011), 'Structural change of Li_2S-P_2S_5 sulfide solid electrolytes in the atmosphere', *Solid State Ionics*, 182, 116–119.

Murugan R, Thangadural V and Weppner W (2007), 'Fast lithium ion conduction in garnet-type $Li_7La_3Zr_2O_{12}$', *Angew. Chem. Int. Ed.*, 46, 7778–7781.

Nagao M, Hayashi A and Tatsumisago M (2011), 'Sulfur-carbon composite electrode for all-solid-state Li/S battery with Li_2S-P_2S_5 solid electrolyte', *Electrochim. Acta*, 56, 6055–6059.

Ohta N, Takada K, Zhang L, Ma R, Osada M and Sasaki T (2006), 'Enhancement of the high-rate capability of solid-state lithium batteries by nanoscale interfacial modification', *Adv. Mater.*, 18, 2226–2229.

Ohta N, Takada K, Sakaguchi I, Zhang L, Ma R, Fukuda K, Osada M and Sasaki T (2007), '$LiNbO_3$-coated $LiCoO_2$ as cathode material for all-solid-state lithium secondary batteries', *Electrochem. Commun.* 9, 1486–1490.

Pearson R G (1968), 'Hard and soft acids and bases, HSAB, Part 1: Fundamental principles', *J. Chem. Educ.*, 45, 581–586.

Sakuda A, Kitaura H, Hayashi A, Tadanaga K and Tatsumisago M (2009), 'Modification of interface between $LiCoO_2$ electrode and Li_2S-P_2S_5 solid electrolyte using Li_2O-SiO_2 glassy layers', *J. Electrochem. Soc.*, 156, A27–A32.

Sakuda A, Hayashi A and Tatsumisago M (2010a), 'Interfacial observation between $LiCoO_2$ electrode and Li_2S-P_2S_5 solid electrolytes of all-solid-state lithium secondary batteries using transmission electron microscopy', *Chem. Mater.*, 22, 949–956.

Sakuda A, Hayashi A, Hama S and Tatsumisago M (2010b), 'Preparation of highly lithium-ion conductive $80Li_2S \cdot 20P_2S_5$ thin-film electrolytes using pulsed laser deposition', *J. Am. Ceram. Soc.*, 93, 765–768.

Sakuda A, Hayashi A, Ohtomo T, Hama S and Tatsumisago M (2011), 'All-solid-state lithium secondary batteries using $LiCoO_2$ particles with pulsed laser deposition coatings of Li_2S-P_2S_5 solid electrolytes', *J. Power Sources*, 196, 6735–6741.

Sakurai Y, Sakuda A, Hayashi A and Tatsumisago M (2011), 'Preparation of amorphous Li_4SiO_4-Li_3PO_4 thin films by pulsed laser deposition for all-solid-state lithium secondary batteries', *Solid State Ionics*, 182, 59–63.

Takeuchi T, Kageyama H, Nakanishi K, Tabuchi M, Sakaebe H, Ohta T, Senoh H, Sakai T and Tatsumi K (2010), 'All-solid-state lithium secondary battery with Li_2S-C composite positive electrode prepared by spark-plasma-sintering process', *J. Electrochem. Soc.*, 157, A1196–A1201.

Tatsumisago M and Hayashi A (2008), 'All-solid-state lithium secondary batteries using sulfide-based glass-ceramic electrolytes', *Funct. Mater. Lett.*, 1, 31–36.

Tatsumisago M and Minami T (1987), 'Lithium ion conducting glasses prepared by rapid quenching', *Mater. Chem. Phys.*, 18, 1–17.

Tatsumisago M, Saito T and Minami T (2001), 'Stabilization of α-AgI at room temperature by heating of AgI-Ag_2O-MoO_3 glasses', *Chem. Lett.*, 790–791.

Yamane H, Shibata M, Shimane Y, Junke T, Seino Y, Adams S, Minami K, Hayashi A and Tatsumisago M (2007), 'Crystal structure of a superionic conductor, $Li_7P_3S_{11}$', *Solid State Ionics*, 178, 1163–1167.

Zhang Z and Kennedy J H (1990), 'Synthesis and characterization of the B_2S_3-Li_2S, the P_2S_5-Li_2S and the B_2S_3-P_2S_5-Li_2S glass systems', *Solid State Ionics*, 38, 217–224.

Index